D0773625

Reinforced
Masonry
Design

PRENTICE-HALL CIVIL ENGINEERING AND ENGINEERING MECHANICS SERIES

N. M. Newmark and W. J. Hall, Editors

Reinforced Masonry Design

ROBERT R. SCHNEIDER

Professor of Civil Engineering
California State Polytechnic University, Pomona

WALTER L. DICKEY

Consulting Structural Engineer
Higgins Brick Company

PRENTICE-HALL, INC., Englewood Cliffs, New Jersey 07632

Library of Congress Cataloging in Publication Data

Schneider, Robert R
Reinforced masonry design.

Bibliography: p.
Includes index.
1. Reinforced masonry. I. Dickey, Walter L.,
II. Title.
TA670.S37 624'.183 79–15528
ISBN 0–13–771733–4

Editorial/production supervision and interior design
by Barbara A. Cassel
Manufacturing buyer: Gordon Osbourne

Printed in the United States of America

10 9 8 7 6 5 4 3 2

PRENTICE-HALL INTERNATIONAL, INC., *London*
PRENTICE-HALL OF AUSTRALIA PTY. LIMITED, *Sydney*
PRENTICE-HALL OF CANADA, LTD., *Toronto*
PRENTICE-HALL OF INDIA PRIVATE LIMITED, *New Delhi*
PRENTICE-HALL OF JAPAN, INC., *Tokyo*
PRENTICE-HALL OF SOUTHEAST ASIA PTE. LTD., *Singapore*
WHITEHALL BOOKS LIMITED, *Wellington, New Zealand*

Contents

6 ENGINEERING DESIGN 135

7 REINFORCED WALLS 172

Preface

Masonry is one of man's oldest building materials and probably one of the most maligned and most certainly the least understood. Such misconceptions have led over the years to a serious misuse of the material through inadequate or even nonexistent design procedures and poor construction practices. However, perhaps because of the considerable amount of information and data available today, both as to its properties and structural performance, sound design techniques and vastly improved construction practices have evolved within recent years, all of which make for optimum use of the material's capabilities. This is in no small way due to the effort continually being exerted toward this evolution by such diverse agencies as the International Conference of Building Officials (ICBO) and the Masonry Institute of America (MIA). Thus, the authors felt that if this voluminous amount of information were to be used properly, a text was needed not only to develop and demonstrate the basic principles involved, but also to identify and clarify the many code requirements stipulated for the analysis and design of reinforced masonry structures. Also, we wished to show that masonry is a totally different and distinct type of construction material, not one that is "sort of like reinforced concrete." It is not, and should not, be treated as such. Furthermore, the wind, seismic, and structural performance research carried on during the recent past has resulted in building codes of increasing complexity. This, in turn, has led to more sophisticated and comprehensive methods of design. So a need evolved for an authoritative source which could interpret these results and provide an understanding of the different design methods, both for the practitioner and for the senior or graduate civil engineering student. Hopefully, this will lead to the adoption of courses in reinforced masonry in many colleges and universities where it is not now taught.

Essentially, the material lends itself to five major divisions as follows:

1. Masonry materials: properties and performance.
   Chapters 1, 2, 3, 4, and 18
2. Design criteria and methods in reinforced masonry.
   Chapters 5, 6, 7, 8, 9, 10, and 11
3. Design applications: low- and high-rise masonry buildings and retaining walls.
   Chapters 12, 13, 14, and 15
4. Environmental features of masonry.
   Chapter 16
5. Masonry research: past and present.
   Chapter 17

Actually, this breakdown follows a natural pattern in the development of the total design concept and use of reinforced masonry with all its ramifications.

Masonry is primarily a hand-placed material whose performance is highly influenced by factors of placement. Hence, knowledge of the basic ingredients (i.e., mortar, grout, masonry unit, and reinforcement) is essential if a practical and efficient design conception is to be achieved. In addition, if the design is to be brought to a successful fruition, as its designer conceived it, proper inspection procedures must be followed to ensure that its delivery will be more certain (Chapter 18). Furthermore, before anyone can hope to turn out an adequate design of any sort, he or she must possess a rudimentary knowledge of the properties and performance of the materials being employed. All of this is described in Chapters 1 through 4.

Next in the process comes the need for a description of the various load sources and intensities, a presentation of the fundamental precepts, and the development of the very basic design and analysis expressions as they evolve from the basic structural mechanics without reference to code limitations or empirical rules (Chapter 6). The many code requirements must then be incorporated into these basic expressions and relations to produce an integrated design procedure, one that will result in very practical solutions to the engineering problems normally encountered by structural engineers in everyday practice (Chapters 7, 8, 9, 10, and 11).

Observe that, although methods for sizing individual structural components (i.e., a wall, a column, a beam) are discussed, the emphasis in this book is placed on the *total* design of a masonry building or structure (Chapters 12, 13, 14, and 15). Such an approach begins with a consideration of the preliminary and nonstructural aspects of masonry bearing on the case study, such as its fire-resistive or environmental features (Chapter 16). Following this examination comes the determination of the live, dead, seismic, and wind loads (Chapter 5)—their magnitudes and stress paths from point of application to the ground. Finally, the member sizes and reinforcing requirements are selected, adequate connections are devised, and the system is detailed such that it can be readily constructed. The latter is an extremely important consideration, but one too often slighted or ignored. It was the authors' intention to show as much of this complete design detail as was practical in a book of this type. In the past, this total concept has not been given the emphasis it deserves.

The authors wish to make one more point. Many textbooks seem to ignore the aspect of stability of the total framing system as it resists lateral loads, focusing instead on the behavior of the individual beams, columns, walls, and other elements comprising the system. Certainly, modern buildings almost everywhere are subjected to significant lateral loads of one type or another to varying degrees of magnitude. The placement of the entire country into seismic zones of various degrees of probability and intensity has only served to accentuate this critical factor. To ignore it is folly, as some have found to their chagrin. It really is not overly difficult to design a building to withstand gravity loads. But developing a lateral-force-resisting system (frame, shear walls, or combination thereof) requires skill and imagination—a process that taxes the ingenuity of structural engineers to come up with solutions that are in all ways safe, practical, and yet economical. Insofar as masonry structures are concerned, this concept is thoroughly examined in Chapters 11, 12, and 13. Also, some highly condensed and very useful design aids are to be found in the Appendix.

If, in the opinion of the instructor, too much material appears in the text to be readily consumed in a quarter or even a semester at the senior or graduate level, the authors recommend that at least Chapters 1, 2, 3, 4, 5 (selected parts), 6, 7, 8, 9, 10, 11, and 12 (selected portions) be included in a first course in masonry design. Perhaps the remainder of the material (remaining portions of Chapters 5 and 12, plus Chapters 13, 14, 15, 16, 17, and 18) could be presented in a subsequent graduate course.

Regardless of precisely what time frame the material is covered in, the discussion should offer a comprehensive guide for both those uninitiated in the complexities of reinforced masonry design as well as those experienced in engineering practice. Properly used, the material contained within these various chapters will provide the basis for a sound professional practice in the field of applied reinforced masonry design and construction.

ACKNOWLEDGMENTS

The number of persons to whom the authors are indebted for the growth of this textbook are too numberous to mention. However, some of the major contributors warrant special comment.

The tolerance and cooperation of the staff of Cal Poly and of Mrs. Walter Dickey, who made available to the authors the hours of time involved in typing and compiling the text, is expressly appreciated and was absolutely essential for its development.

Appreciation is also expressed to the Masonry Institute of America, which provided much of the voluminous information that they have compiled, developed, and established over the years. This organization is undoubtedly the most exhaustive source of information on masonry performance and design anywhere today.

Also, we wish to thank the International Conference of Building Officials, which, through its Uniform Building Code, has provided the most up-to-date compilation of sound practice and theory in reinforced masonry available today, for it has been engaged in the development of this material since 1933.

Richard Kato, senior engineering student at Cal Poly, is to be commended for the preparation of the numerous design and analysis examples that appear throughout the text.

Specific thanks are also due Donald Strand of the Consulting Structural Engineering firm of Brandow and Johnson, who advised on numerous items and who also provided the design for the multistory reinforced brick building described in Chapter 13.

ROBERT R. SCHNEIDER
WALTER L. DICKEY
Palm Springs, California

Reinforced Masonry Design

1

Modern Masonry Construction

Although masonry is one of man's oldest building materials, it has probably remained the least understood of his major construction materials, at least as far as its structural behavior is concerned. One of the major purposes of this text will be to describe and analyze the behavior of masonry in its modern context and to present a rational method of masonry design which originates from the following logic:

1. A utilization of the basic structural mechanics of the assumed force system.
2. An incorporation into the analysis of those pertinent material properties of strength and stiffness.
3. A modification of these theoretical results by the appropriate building code empirical limitations.

Hopefully, a feasible masonry design approach will evolve from this fundamental knowledge. It will also be the intent to dispell numerous misconceptions surrounding this type of construction, many of which have come about through misuse as a result of improper design or poorly controlled construction practices. A paramount consideration in any such discussion lies in the distinction between plain and engineered reinforced masonry. *Plain masonry* refers to that form of construction whose strength depends solely upon the mortared masonry unit (concrete block or clay brick), with its high degree of compressive resistance. These, in turn, must act as an integral combination in order to resist both vertical compressive forces and shear forces. Essentially, these are bearing-wall structures, capable of carrying massive vertical loads, since their very considerable weight makes for an extremely stable structure, with considerable resistance to overturning. In contrast, lighter-weight structures offer very little gravity resistance to any type of lateral load, such as wind, blast, or the inertial forces induced by seismic motion. This serious limitation precludes the use of plain masonry except in the most stable geographical locations.

But the achievement of lateral stability by gravity places a practical economic limit on the size of the structure. This has led designers and builders to seek ways in which these massive bearing walls could be decreased in thickness without losing their stability in the process. Perhaps the dilemma faced by the designer of another era is most dramatically portrayed by the 16-story Monadnock Building, a brick bearing-wall structure built in Chicago in 1889–1891, shown in Figure 1-1. Its 6-ft-thick unreinforced masonry walls at the base of the building provided the re-

FIGURE 1-1. Monadnock Building, Chicago, 1891. (Courtesy Hedrick-Blessing, Chicago.)

FIGURE 1-2. Hanalei Hotel, San Diego.

quired stability against wind loads. Such structures made it rather apparent at the turn of the century that a size limit had been reached on masonry structures using methods then currently employed. If modern design techniques and construction methods had been available at that time, the designers of the Monadnock might have used instead masonry walls 1 ft or less in thickness. Compare this massive structure with the modern Hanalei Hotel in San Diego, California, with its 8-in.-thick walls at the bottom story (see Figure 1-2).

As buildings increased in height, another factor began to be recognized as a formidable limitation on height. This was the lateral force imposed upon the face of the structure by the high wind velocities that occur in many parts of the country. In other sections, particularly on the West Coast, the inertial forces induced on the bearing-wall system by seismic movements, frequent in those areas, provided an even greater challenge to the structural capability of the larger buildings. As a matter of fact, this challenge turned out to be so severe, as demonstrated by the 1933 Long Beach, California, earthquake, that the use of plain masonry was prohibited from use in the Pacific Coast region. As a consequence, the demands for the continued use of masonry led to the evolvement of plain masonry into the composite system that we now term *reinforced masonry*. The present concept utilizes floors and roofs as diaphragms acting as horizontal flanged girders, to distribute the lateral forces to walls, which in turn provide the horizontal shear resistances needed, in addition to carrying the normal vertical live and dead loads. This type of structure may be defined as a *box system* or a *shear-wall system*. (Refer to Figure 1-3 and Table 23-I in the

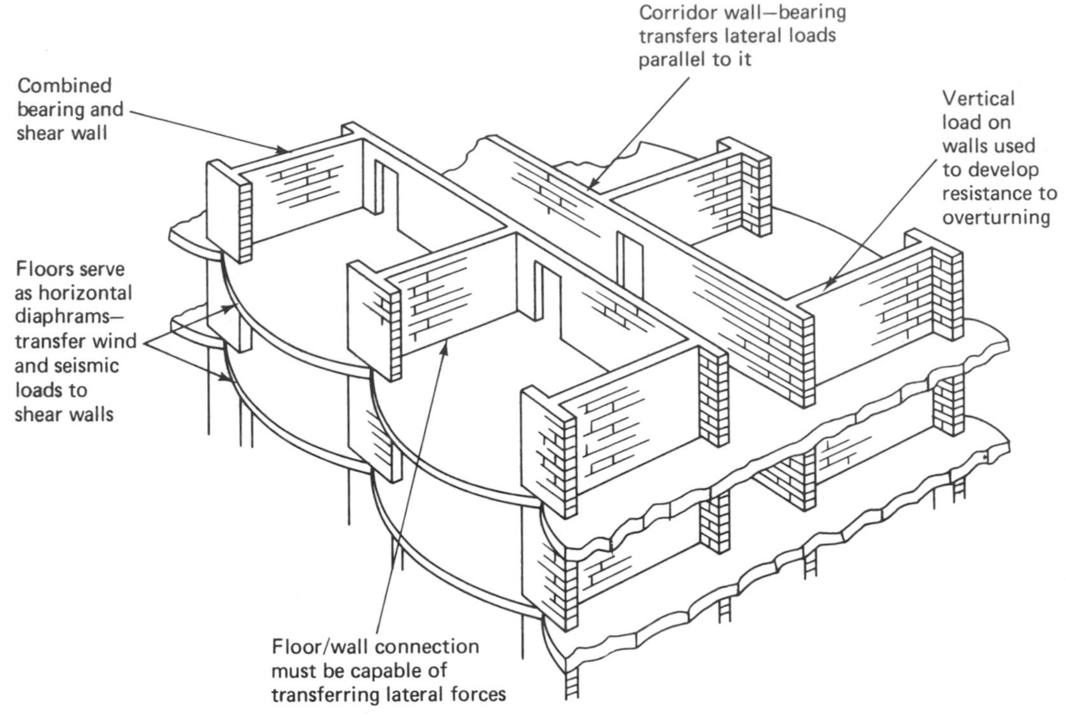

FIGURE 1-3. Concept of shear-wall system.

Uniform Building Code.) These walls, if constructed of plain masonry, would be incapable of resisting the magnitude of the horizontal shear and bending forces imposed upon them. For this reason, modern reinforced masonry contains reinforcing steel to resist the shear and tensile stresses so developed. When these walls are subjected to lateral forces acting normal to them, they behave as flexural members spanning vertically between floors or horizontally between pilasters. Therefore, reinforcing must also be provided to develop the resisting forces on the tension side of the element.

The combination of masonry and reinforcing is a very compatible one. The masonry brings to the system a high degree of compressive resistance, weathering durability, fire protection, and stability. Its stiffness and mass distribution minimizes flexural and shear deflections, and its composite heterogeneous nature tends to maximize the damping response to dynamic vibratory forces. Although, principally, reinforcing develops the flexural tensile resistance so necessary to resist lateral loads, it also imparts to this composite material the resilience and ductility needed in high-rise construction. This modern concept of engineered reinforced masonry thus makes it possible for owners to continue to derive the benefits of economical construction combined with all the advantages of sound control, fire resistance, and low maintenance costs. These features have provided a real impetus for the recent evolvement of high-rise masonry construction as a highly desirable building form.

The transition from plain to reinforced masonry construction is most vividly demonstrated in modern building codes. Since no national building code exists as such, it has been customary for municipalities to either adopt an applicable regional code or to develop one of their own, which the larger cities do. These codes have evolved over the years primarily in response to the demand that building standards be developed for the protection of the public health and safety. Such standards, when intelligently drafted and followed in both office and field, cannot help but result in the development of design procedures wherein economy, esthetics, and function can all be achieved without compromising structural capability.

HISTORICAL DEVELOPMENT

The use of plain masonry can be traced back to before the early Egyptians, Romans, and Greeks. Examples of masonry structures from this ancient era abound. The pyramid of Cheops in Egypt, for example, rising 475 ft in the air, is still one of the largest single structures ever built by man. A more spectacular and utilitarian example of the ancient wonders of the Pharoahs was the Pharos of Alexandria, the light house that stood watch over the navigators in the Mediterranean, guiding them to safe harbor. The structure was as high as the Washington Monument (550 ft), and the fire that burned atop could be seen for 35 miles. That landmark withstood the elements for over 1500 years before it was destroyed by an earthquake in the thirteenth century. The temple of Artemis at Ephesus in Lydia, 350 ft by 150 ft in size, was built about 350 B.C. and housed the alter and statue of Artemis, Goddess of Fertility and Harvest. Another early masonry structure, the tomb of Mausolus, King of Caria, was completed at about the same time. The fragmentary remains indicate that the structure, which had a high basement, was crowned by a pyramid. As a point of interest, the word "mausoleum," the modern term for an aboveground burial vault, stems from that ancient wonder of the world. We should recognize, too, that the small brick, with its relative ease of handling, probably enhanced the development of the arch and the vault. Syria seems to have been the first home of the arch, since the Syrians were accomplished brick layers.

Further along in time, the Dark Ages of Western civilization produced many castles and cathedrals which were actually true masonry structures, in that they were completely devoid of metal or wood for structural support. Also, consider the architectural style that has been achieved in these monuments through the use of brick. The Midi (southern France), centering around Toulouse and Albi, where possibly brick has been used with greater skill and more taste than anywhere else in the world, truly exemplifies the esthetics of brick masonry in its highest state. Brick in the Midi indeed transcends the merely utilitarian role and achieves, as in the cathedral of Albi, an architectural quality that could not have been attained in marble or stone. To cite another example, the dome of St. Sophia in Istanbul is built of bricks that are 24 to 27 in. square and 2 in. thick. The mortar joints are almost the thickness of the brick.

In this instance, the brick serves a purely utilitarian purpose, for it is sheathed on the exterior with lead and on the inside with glass mosaic.

These applications indeed offer some evidence that even during those times builders and designers began to understand, in an intuitive way at least, how forces were transferred from the structural components to the ground. Remember that it was Aristotle who correctly explained how an arch resisted gravity loads. Leonardo da Vinci broadened the theory by explaining the internal mechanics of the action between the elements of the arch. The present assumption of a triangular loading over masonry lintels had its origin in this classic explanation. Arches and columns were the mainstays of ancient construction. No doubt this stems from the fact that the load-carrying capacity of these structural elements were the first to be defined mathematically. For example, it was Euler, a Swiss mathematician, who in the middle of the eighteenth century defined the load-carrying capacity of a column in terms of its ultimate buckling load. In spite of many subsequent efforts to make it more complex, this simple mathematical definition remains today the fundamental expression for the elastic load-carrying capacity of a column of any type of material. The early users of stone and clay masonry also undoubtedly applied their engineering knowledge, perhaps still more in an "intuitive" way than with any theoretical soundness, in cathedral and fortress construction. The lintel and beam, the arches with flying buttresses, and the vaulted roofs are examples of the structural elements used in such structures. The walled cities of the twelfth century, still standing, certainly demonstrate the ancient's appreciation of the excellent compressive action of masonry. In building the walled city of Rothenburg, these early stone masons utilized another form of structural element, which we now refer to as deep-wall beams. Concentrated within the wall was a form of a flat arch, whereby the entire wall would, in effect, span between the isolated spread footing pads. The modern concept of this action is described in detail in Chapter 7. In addition, they made use of another currently used element, the pilaster, which is simply a column within a wall, as described in Chapter 8. They also devised arches laid up in such a way that the material was subjected only to axial compressive forces (i.e., virtually no tensile stresses existed)—a tribute certainly to their "intuitive" approach to design. Some of their structural elements were actually the forerunners of modern prestressed masonry.

Brick is actually the oldest manufactured building material remaining in use today. In the premodern era, the development of brick masonry reached its fruition in the United States and Europe. The successful use of this ancient material is certainly demonstrated in many early American brick structures, such as the Monadnock Building in Chicago, discussed earlier. But, as mentioned, its very massiveness discouraged further use of unreinforced masonry bearing walls for high-rise buildings. This condition remained unchanged for nearly 50 years, awaiting the advent of modern reinforced masonry. The Modadnock represented the watershed, in America at least, of the use of plain masonry bearing walls.

Further developments in this field, then, awaited the introduction of reinforced masonry. To Marc Brunel, once chief engineer for the City of New York, goes the credit for its initial discovery and use. He apparently conceived of the idea in about

1813, but he did not put the conception into practice until 1825. At that time, as a part of the Thames River Tunnel project, two 50-ft-diameter brick shafts, 70 ft high with 30-in.-thick walls, were built. These were reinforced vertically with 1-in.-diameter wrought iron bolts built integrally with the brick work. Iron hoops were also placed within the mortar joints as the work progressed—a rather crude forerunner of modern reinforced masonry construction, but certainly a beginning. Brunel fathered the initial research on this material form by conducting a test program to investigate the influence of steel reinforcement upon the structural behavior of the monolithic masonry assemblage. His tests were shortly thereafter supplemented by those of other engineers, among them, a Colonel Pasley of the Royal Corps of Engineers in 1837. Later, a series of such tests received wide publicity, perhaps because they were conducted at the Great Exposition in London in 1851. If for no other reason, these tests are notable because a "new" type of cement, known as portland cement, was the cementitious agent. From these and earlier data, including Brunel's, N. B. Corson computed the tensile stresses of unreinforced masonry and, on the basis of his results, brought forth a recommendation for an allowable tensile stress for use in designing masonry lintels. During the 1880s, other builders made occasional use of reinforced brick masonry in various types of buildings, not the least significant of which was the Church of Jean de Montmartre in Paris, designed by M. A. Boudot. The exterior brick walls, $4\frac{1}{2}$ in. thick, were reinforced with both vertical wires extending through holes in the brick and horizontal wires in the mortar joints—another early forerunner of modern reinforced masonry construction.

The initial United States venture into the investigation of the behavior of reinforced beams of any consequence seems to be the work of Hugo Filippi in 1913. He later formalized his discoveries in a sort of brick engineering handbook. In this, Filippi describes various types of structures that were constructed in the United States during the early 1930s. During this period, the development and actual use of reinforced brick masonry in the United States was in its very early stages. Filippi's work was followed by the important reinforced brick beam tests of L. J. Mensch, wherein the reinforcement was placed in a bed of mortar. During this same period, a reinforced grout core, formed between two wythes of brick, made its appearance in the United States, marking the initial use of grout in masonry construction. Prior to this time, all reinforcement had been located in the mortar bed joints.

In 1923, Under-Secretary A. Brebner, Public Works Department of the Government of India, published his report describing a series of extensive tests that had taken over 2 years to complete. It is said that this report marked the true beginning of the modern development of reinforced brick masonry. It is highly significant to note that his report, which also included a rational design theory, brought about a surge in the use of this material form in buildings throughout such earthquake-plagued countries as India and Japan. Engineers in these countries found that reinforced masonry, if properly constructed, offered excellent resistance to seismic forces, and they turned to its use in many instances. Brebner indicated that nearly 3 million square feet of reinforced brick masonry were placed in the 3 years prior to 1922. S. Kanamori, civil engineer, Department of Home Affairs, Imperial Japanese Government, reported in July 1930 that, "There is no question that reinforced brick work should be used instead of [unreinforced] brick work when any tensile stress would be incurred in the structure. We can make them more safe and stronger, saving much cost. Further, I have found that reinforced brick work is more convenient and economical in building than reinforced concrete, and what is still more important, there is always a very appreciable saving in time." In those early years, applications included a variety of structures: public and private buildings, retaining and sea walls, chimneys, bridges and culverts, and storage bins.

The mid-1930s brought forth a deluge of reinforced masonry construction in the United States. This was due in no small way to various technical agencies devoted primarily to promoting the development and use of reinforced masonry, such as the Associated Brick Manufacturers of Southern California and the Concrete Masonry Association, followed in later years by the Masonry Institute of America. The most numerous examples of reinforced masonry construction probably are to be seen in California. These would include such significant applications as warehouses, low-rise commercial buildings and banks, churches, public and parochial schools, state buildings, and, in particular, high-rise hotel and apartment structures. Other significant structures include such buildings as the Pacific Gas and Electric Company power plant constructed in 1951 at Antioch, California (designed by Walter Dickey); the Veterans Administration Sepulveda Hospital, a 26-building complex, built in 1952 in Los Angeles, California; and many others. It is interesting to note that the 26 buildings in the Veterans Hospital complex survived the 1971 San Fernando earthquake, incurring no appreciable structural damage. But five other hospitals in the area (not of masonry construction) were either badly damaged or demolished. In addition to buildings, use was made of reinforced brick masonry in such structures as retaining walls, storage bins, and highway bridge structures.

MODERN TECHNIQUES OF DESIGN AND CONSTRUCTION

Probably the most advanced state of the art of masonry construction in its present form is also to be found in California. With its long history of earthquake activity, this is not surprising. The 1933 Long Beach earthquake proved conclusively that unreinforced masonry, with its lime-mortar joints, cannot adequately withstand seismic shocks because of the lack of tensile and shear resistance. This fact provided the impetus for further development of design techniques for reinforced masonry as well as for improved high-rise construction methods by using a structurally integrated masonry element (masonry unit + mortar + grout + reinforcement), thereby producing much greater lateral load resistance. This was an absolutely imperative step if brick masonry were to remain a major construction material in California under the revised building codes. Otherwise, it would have been "codified" out of existence.

These advanced techniques of design and construction are embodied in modern high-rise buildings being constructed throughout the United States. Prime examples of these are to be found in such diverse structures as the 165-ft-high 17-story and 9-story Park Mayfair East apartment buildings in Denver, Colorado (Figure 1-4); a dormitory building at the U.S. Naval Base, San Diego, California (Figure 1-5); and the 5-story Sportsmans Lodge Hotel in North Hollywood, California (Figure 1-6), among many others. This trend back to the bearing-wall structure has revitalized the entire concept of load-bearing masonry walls for multistory buildings.

As described earlier, in this new concept of high-rise reinforced masonry, the walls function to carry both the gravity and the lateral forces, and the floors and roofs serve as horizontal diaphragms to transfer wind or seismic loads to the bearing-shear walls. To achieve this behavior, however, the floor-to-wall connections must be capable of transferring all lateral forces to or from those walls. In addition, the gravity loads on the walls are counted upon to develop resistance to overturning. This concept results in reduced construction time while providing a final building form that is esthetically pleasing. This type of construction has the advantages of sound control, thermal inertia, fire resistance, and low maintenance—features long associated with masonry structures, and ones that carry a high priority in our present energy-deficient era. The sound-resisting qualities of masonry walls are rated well above those required for comfort levels. The heat-transmission resistance of masonry (thermal inertia) has

FIGURE 1-4. Park Mayfair East apartment building. Park Mayfair East is a 17-story apartment building, one of a three-building complex, whose 11 in. thick brick bearing walls rise 165 ft into the Denver skyline. The building contains a total of 128 units with a parking garage for 110 automobiles. Bids received on a steel frame, with the necessary additional fireproofing required, proved that it would be more costly than a brick bearing-wall structural system. In addition, the owner's previous experience with brick bearing walls had proved the significant advantage of masonry bearing walls acting as sound barriers, with the advantages of low maintenance and speed of construction. (Courtesy SCPI.)

FIGURE 1-5. Dormitory, U.S. Naval Base, San Diego. This is a bearing-wall structure containing interior bearing walls of concrete brick, and reinforced brick exterior walls, consisting of blue face brick exterior wythes, with economical backup wythes.

FIGURE 1-6. Sportsman's Lodge, Los Angeles, California. This was one of the first of the precast concrete slab, concrete block bearing-wall apartment buildings. It was designed by Albyn Mackintosh for fast economical construction in a major seismic area. It suffered no distress in the San Fernando Valley earthquake, although it was near some structures that were severely damaged.

been recognized for years. This heat storage capacity results in drastic reductions in heating or cooling requirements in areas subject to large diurnal fluctuations. In essence, masonry construction offers many environmental advantages, combined with a freedom of architectural expression that is unsurpassed by any other building material.

Several types of floor systems can be utilized with reinforced masonry load-bearing walls. One of the more dramatic applications utilizes concrete slabs precast at ground level and lifted into place. The slab functions as the ceiling for the story below and as the floor for the story above (see Figure 1-7). In this system, each precast floor also provides a work platform for the next story. Fewer building trade crafts are needed, shoring and scaffolding layouts are reduced, and finishing operations can be completed quickly and easily in the enclosed building. In addition, while work is being carried on in the upper stories, the lower stories can be rented, providing an additional cost saving to the owner. Examples of this form abound, such as the hotel in San Diego, which was completed and opened within 4 months after the project was begun. Other types of floors include cast-in-place slabs; plant-produced precast elements, such as prestressed T's; hollow-core or voided slabs; or combination systems using precast units with a cast-in-place topping. For buildings with long spans, prestressed precast concrete slabs are suitable.

FIGURE 1-7. Precast floor systems. (a) Precast prestressed plank, factory-produced, are very efficient floor slabs for erection on sites which have limiting conditions, such as lack of space or lack of access. They also provide an economical floor system. As on precast tees, a topping slab is poured, providing the finish and the lateral force tie and diaphragm action. (b) The lift-on slab has proved most economical and efficient for sites where there is access for pouring and crane handling, and where the slabs are not too large to be lifted into position by cranes.

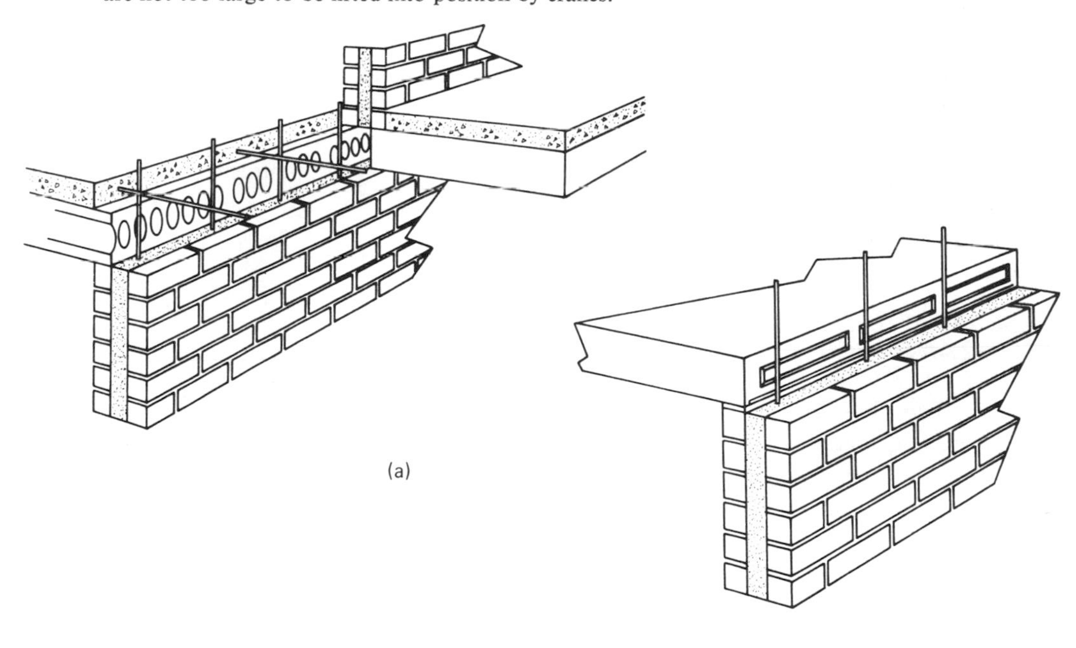

The masonry bearing-wall system is particularly suited to structures that require a fixed pattern of separation of areas repeated from floor to floor, such as those in motels, hotels, apartments, dormitories, barracks, or convalescent homes. This repetition permits increased speed of construction, with resulting economy for multistory construction, and provides for direct transfer of loads to the foundation. Excellent job efficiency is obtained, since floors can be poured simultaneously with excavation and foundation work. So when walls are built, floors can be placed immediately. There is a sequence of one activity following another in a repetitive fashion, which helps reduce errors. A more complete description of the details of high-rise masonry structures may be found in Chapter 13.

CODES AND STANDARDS

There have been many codes and standards developed throughout the world for the design and control of the quality of masonry materials and methods. In the United States, these have been developed by such agencies as the American Society for Testing and Materials (ASTM), American National Standards Institute (ANSI), American Society of Civil Engineers (ASCE), American Concrete Institute (ACI), Structural Engineers Association of California (SEAOC), International Conference of Building Officials (ICBO), and the California Office of the State Architect (OSA), among others. Most prominent among these, at least in the western part of the United States, has been ICBO. They have streamlined many of the old ASTM standards, added new ones that specifically relate to many recent developments such as reinforced masonry, and have incorporated these regulations and requirements into what is known as the Uniform Building Code (UBC) Standards. As a matter of fact, in this text the Uniform Building Code will be taken as the governing specification since, for reinforced masonry at least, it is by far the most advanced and widely used of its kind. For materials sampling, testing, and property evaluation, both the UBC Standards and ASTM Specifications are followed. The OSA requirements were developed to apply strictly to public buildings constructed in California. They are delineated in the California Administrative Code, and they take precedence over any local building codes where hospitals or state buildings are concerned.

CLASSIFICATIONS OF MASONRY CONSTRUCTION (UBC STANDARDS)

The classifications of masonry construction and the types of masonry walls appear in Chapter 24 of the UBC. The distinction between these various categories must be thorougly understood by anyone who intends to design masonry under UBC jurisdiction. For this reason, they are thoroughly delineated in the following sections.

Masonry construction is classified as follows: (1) "reinforced masonry," which must be engineered on the basis of sound theoretical principles combined with a set of empirical rules and limitations set forth by the Building Code, plus sound engineering judgment stemming from long experience; (2) "partially reinforced masonry," which was introduced into the Uniform Building Code primarily for those areas in which all the requirements of reinforced masonry were not needed, since the seismicity of the locale did not so dictate; (3) "unreinforced engineered masonry," which was developed in the East as an attempt to improve on past practices, many of which were unsound; and (4) "traditional masonry," which encompasses the use of masonry as it evolved over the years from certain arbitrary limitations and past practices without any real consideration for theoretical design characteristics; although it did provide for a generally conservative and safe type of construction for the majority of conditions.

TYPES OF MASONRY WALLS

Unburned clay masonry (UBC Section 2405)

Unburned clay masonry consists of unburned clay units, commonly referred to as "adobe" in the southwestern part of the United States. In earlier and less sophisticated days, it did perform quite satisfactorily where no seismic activity of any magnitude occurred. The early adobe was actually reinforced with straw and often also contained an emulsion that provided for greater compressive strength and durability. No particular energy problem was posed here, since it was sun-dried. Structural connections were a problem and, of course, the adobe possessed practically no tension value. It can still be used in restricted areas with Type M or S mortar. At any rate, it served a very important function in the early days, as the numerous missions in California and throughout the Southwest will attest. It also was utilized as a important housing material. The church shown in Figure 1-8, which is located in the Los Angeles Plaza, was built of sun-dried clay and protected by plaster.

FIGURE 1-8. Church of Nuestra Senora La Reina de Los Angeles, an adobe church of La Puebla at the plaza, the center where the present city of Los Angeles began.

Gypsum masonry (UBC Section 2406)

Gypsum masonry consists of gypsum block or gypsum tile units laid up with gypsum mortar. It has been used in the past, with considerable success, for interior partitions, primarily because of the ease with which it can be formed around ducts, window openings, and other discontinuities. It is also permitted in some "partially reinforced" walls.

Gypsum tile is laid up in gypsum mortar, similar to that used in plaster. The proportions consist of approximately one part gypsum to three parts sand, mixed with a sufficient amount of water to provide a good workable mortar. Since gypsum is fast-setting, the mortar sets up so rapidly that it has a limited "board" life. Thus, it is generally necessary to add a retarder of some sort, composed usually of certain organic materials.

Gypsum tile, like unreinforced brick masonry, received a bad press after the 1933 Long Beach earthquake, again because of material misuse, not because of any property deficiencies. Without adequate reinforcing and connections, these materials cannot stand up against earthquake load intensities. One method of bolstering the capability of the gypsum wall would be achieved by attaching a "chicken-wire" reinforcing mesh directly to both sides of the partition. Plaster is then applied over these surfaces. The resulting wire plaster facing has proved to be effective, at least in resisting the perpendicular horizontal loads on a nonbearing wall. A similar approach was advanced back in the 1950s, by the Los Angeles Board of Education, ostensibly as a means of rehabilitating pre-1933 unreinforced masonry school buildings.

Glass masonry (UBC Section 2408)

Glass masonry units are used in the openings of non-load-bearing exterior or interior walls. These filler panels must be at least 3 in. thick and the mortared surfaces of the blocks have to be treated to provide an adequate mortar-bonding effect. This is usually achieved by applying a roughened surface adhesive to the glass edges.

The panels themselves must be restrained laterally to resist the lateral-force effects of winds or earthquakes. Also, the sizes of the exterior panels are arbitrarily limited to a maximum vertical or horizontal dimension of 15 ft and an area of 144 ft^2. For interior glass block panels, these limits are 25 ft and 250 ft^2. Exceptions are permissible if calculations can substantiate the deviations.

The glass blocks must be laid in Type S mortar with both vertical and horizontal joints being between $\frac{1}{4}$ and $\frac{3}{8}$ in. thick. Reinforcement, as required by calculations, is provided. Exterior glass block panels have to be provided with $\frac{1}{2}$-in. expansion joints at the sides and at the top, and they must be entirely free of mortar so that the space can be filled with a resilient material to provide for needed movement. The expansion joint, of course, must also provide for lateral support while permitting expansion and contraction of the glass panel.

Stone masonry (UBC Section 2409)

Stone masonry is that form of construction made with natural or cast stone as the basic masonry unit, set in mortar with the joints thoroughly filled. See Figure 1-9 for an example of its use in residential construction.

In ashlar masonry, the bond stones are uniformly distributed and have to cover at least 10% of the area of the exposed facets. Rubble stone masonry, 24 in. or less in thickness, will have bond stones spaced a maximum of 3 ft both vertically and horizontally. Should the thickness exceed 24 in., the bond stone spacing is increased to 6 ft on both sides.

There are other limits, arbitrarily established. The maximum height/thickness ratio is 14, and the minimum wall thickness is 16 in. If regularly cut or shaped stones are used, they may be laid as solid or grouted brick masonry.

FIGURE 1-9. Residence of field stone. The arches provide an effective span over the openings. The abutments provide stability and relief of the straight lines or corners.

Cavity wall masonry (UBC Section 2410)

Cavity wall masonry is construction using brick, structural clay tile, concrete masonry, or any combination thereof, in which the facing and the backing wythes are completely separated except for metal ties that serve as cross ties or bonding elements, as shown in Figure 1-10. This is the type now permitted by the UBC in lieu of an earlier type of cavity wall masonry, in which the two faces of the walls were separate but bonded together with transverse solid masonry units. The

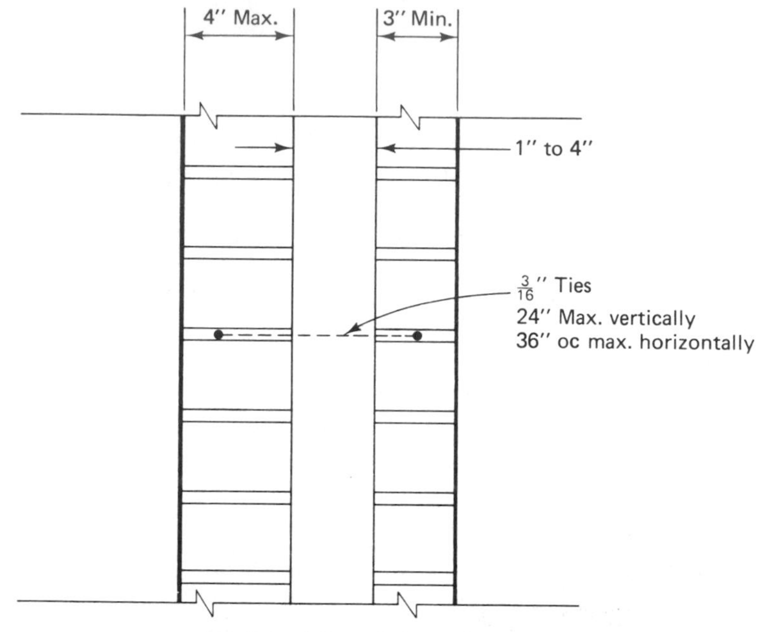

FIGURE 1-10. Cavity wall masonry.

maximum height/thickness ratio is limited to 18, with the minimum thickness being 8 in.

The cavity wall facing and backing wythes cannot be less than 4 in. in thickness, except that when both are constructed with clay or shale brick the limit decreases to 3 in. nominal thickness. The separating cavity must be between 1 and 4 in. in width; however, special calculations on variations in tie size or spacing may permit the use of greater or lesser cavities.

The two wythes have to be bonded together with $\frac{3}{16}$-in. metal ties embedded in the horizontal mortar joints. Tie spacing is limited such that they support no more than $4\frac{1}{2}$ ft^2 of wall area for cavity widths up to $3\frac{1}{2}$ in. Where the cavity width exceeds $3\frac{1}{2}$ in., this limit becomes 3 ft^2 of wall area. The tie spacing is always staggered in alternate courses, with the maximum vertical distance between ties being 24 in. and the maximum horizontal spacing 36 in. For hollow masonry units laid with the cells vertical, the ties have to be rectangular in shape. Where other types of units are used, a 90° bend provides the special anchorage. Additional bonding ties must be placed at all openings, spaced at 3 ft maximum around the perimeter of the openings, within 12 in. of the openings.

Hollow unit masonry (UBC Section 2411)

Hollow unit masonry describes a type of wall construction that consists of hollow masonry units set in mortar as they are laid in the wall (Figure 1-11). All units have to be laid with full-face shell mortar beds, with the head or end joints filled solidly with mortar for a distance in from the face of the unit not less than the thickness of the longitudinal face shells. This type of construction usually refers to an unreinforced

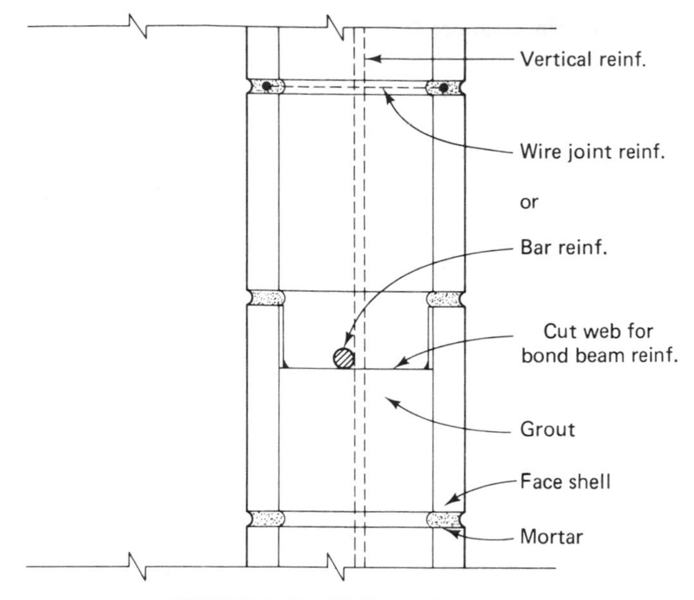

FIGURE 1-11. Hollow unit masonry.

state, although it actually can be reinforced as described in UBC Section 2415. The original intent of this masonry section was to provide a standard for hollow concrete block per ASTM standard C90 or C129 except that hollow clay units are now included (e.g., this can refer to hollow clay tile as per ASTM C34, C126, C212, C652, or the Western States Clay Products Association Standard).

Where the wall thickness consists of two or more hollow units placed side by side, the stretcher unit (Figure 2-13) must be bonded at vertical intervals not to exceed 34 in. This bonding is accomplished by lapping a block at least 4 in. over the unit below, or by lapping them at vertical intervals not to exceed 17 in. with units that are at least 50% greater in thickness than the units below. They can also be bonded together with corrosion-resistant metal ties which conform to those requirements for cavity walls, as previously noted. Ties at alternate courses need to be staggered, with the maximum vertical distance between ties being 18 in. and the maximum horizontal distance 36 in. Walls bonded with metal ties must then conform to the allowable stress, lateral support, thickness (excluding cavity), height, and mortar requirements for cavity walls. Since this material is not reinforced, the maximum height/thickness ratio is 18, with a minimum thickness of 8 in.

Solid masonry (UBC Section 2412)

Solid masonry consists of brick, concrete brick, or solid load-bearing concrete masonry units laid up contiguously in mortar. Refer to Figure 1-12 for details. All units are laid with full shoved mortar joints, and the head, bed, and wall joints have to be solidly filled with mortar. In each wythe, at least 75% of the units in any vertical transverse plane must lap the ends of the unit above and below a distance not less than $1\frac{1}{2}$ in., nor less than one-half the height of the units, whichever is greater. Other-

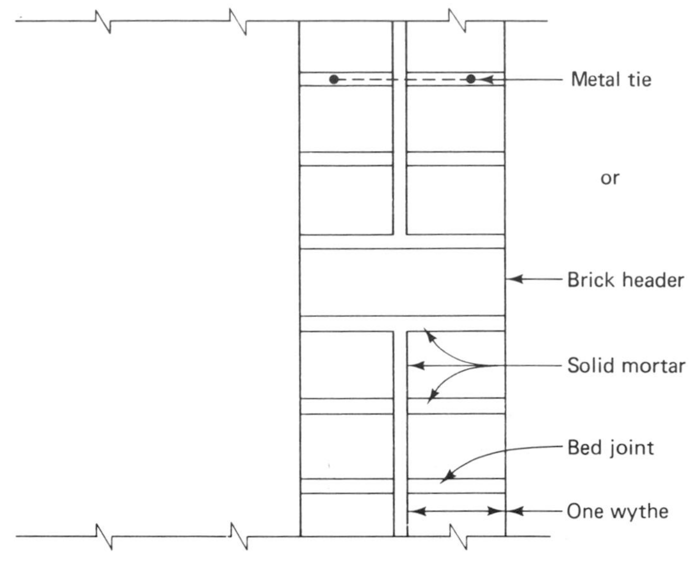

FIGURE 1-12. Solid masonry.

wise, the masonry is to be reinforced longitudinally to provide for a loss of bond, as in the case of masonry laid in stack bond. The longitudinal reinforcement amounts to a minimum of two continuous wires in each wythe, with a minimum total cross-sectional area of 0.017 in.^{2} being provided in the horizontal bed joints, with the spacing not to exceed 16 in. center to center vertically. Considerable dispute has arisen over this arbitrarily selected amount of reinforcement. For example, if one uses 6-in.-high units, the horizontal reinforcement may be spaced at 18 in. instead of 16 in. so that it conforms to the module of three 6-in. courses. This alternative of replacing the masonry unit bond by reinforcing steel is more significant with concrete units than with clay units, simply because the mortar bond to clay units is generally better than the mortar bond to concrete units. Even more important, the clay units do not have the very considerable drying shrinkage characteristic that the concrete units possess. On the contrary, clay units undergo a slight expansion due to moisture content rather than demonstrating any tendency to shrink.

When the facing and backing wythes are bonded by headers, not less than 4% of the exposed face area must be composed of solid headers extending at least 4 in. into the backing. The distance between adjacent full-length headers cannot exceed 24 in. vertically or horizontally. Where the backing consists of two or more wythes, the headers shall extend at least 4 in. into the most distant wythe, or the backing wythes have to be bonded together with separate headers whose area and spacing conforms to the requirements listed above.

Facing and backing can be bonded with corrosion-resistant unit metal ties or cross wires conforming to the cavity wall requirements previously noted. The unit ties have to be long enough to engage all wythes, with the ends embedded no less than 1 in. in mortar, or they can consist of two lengths, with the inner embedded ends hooked and lapped not less than 2 in. When the space between the metal tied wythes is solidly filled with mortar, the allowable stresses and other provisions for bonded masonry walls apply. However, where the space is not filled, they must meet the requirements for cavity walls.

Grouted masonry and reinforced grouted masonry (UBC Sections 2413 and 2414)

Grouted masonry is made with two wythes of clay brick or solid concrete brick or stone units in which the interior joint, sometimes called the "collar joint" or interior wythe, is filled with grout. This bonds the two wythes together as well as providing a space wherein the reinforcement can be located and bonded to the surrounding masonry. The thickness of grout or mortar between masonry units and reinforcement is not to be less than $\frac{1}{4}$ in., except that steel wire reinforcement may be laid in horizontal mortar joints which are at least twice the thickness of the wire diameter. Refer to Figure 1-13 for an example of reinforced grouted brick masonry construction.

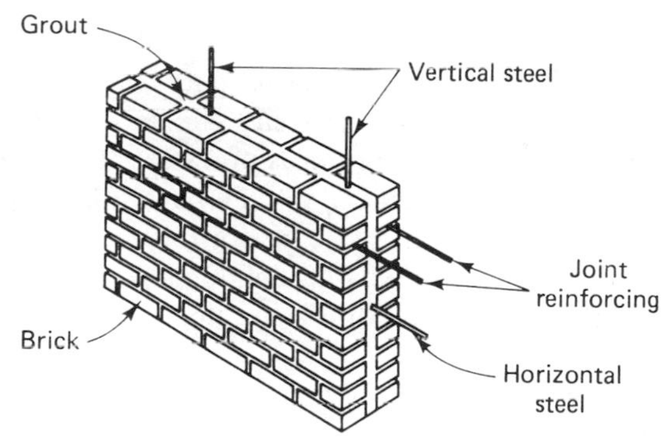

FIGURE 1-13. Reinforced grouted brick masonry.

Only Type M or S mortar can be used in grouted masonry construction, since it was developed to be worked at the higher stresses and greater spans of engineered masonry construction. This type of construction was developed in California after the disastrous Long Beach earthquake of 1933, when the use of brick masonry construction fell into disrepute because of the poor performance of unreinforced brickwork. This behavior, even in light of that day's technology, could have been predicted. But as so often happens in the course of human events, it takes a disaster to force us to recognize the reality of our negligence. Because of this poor performance, kilns were closed and the brick industry languished. However, the structural engineering profession, through its professional society (Structural Engineers Association of California), awakened to its responsibility to the public, and in conjunction with the then California State Division of Architecture developed reinforced grouted bonded walls, initially for school masonry construction. As this school building type of construction improved and evolved, grouting became recognized as a means of overcoming the inherent tensile weakness in the material by providing a superior bond between reinforcing and brickwork, making it a more homogeneous material. By so doing, it permitted the wall or beam element to readily resist flexural tensile stresses of considerable magnitude. This was considered to be a rather novel concept at that time.

There are two basic types of grouting procedures: low-lift and high-lift grouted construction. Low-lift grouting was the original method developed for reinforced masonry; high-lift grouting developed later as an improvement in the practice. The latter is recognized by many, including the OSA, as a generally superior method, since it produces a higher quality wall. The OSA regards it as the preferred method of construction, with the low-lift procedure being offered simply as an alternative in cases where the layout is not particularly suited to high-lift grouting. This might be the case, for instance, on a small job on which scheduling is difficult. Both procedures are described in a later chapter.

Reinforced hollow unit masonry (UBC Section 2415)

Reinforced hollow unit masonry is laid up as described in UBC Section 2411, but the reinforcing is grouted within the vertical cells, and horizontal reinforcing consists of joint reinforcing in the bed joints or of horizontal steel placed in the bond beam units, as described in Chapter 3 (see Figure 1-14). This method was originally developed for reinforced concrete block walls, but it is now also used in reinforced hollow clay brick construction. Here too, both low-lift and high-lift grouting methods are employed, depending upon the type of structure being built.

Masonry veneer (UBC Chapter 30)

Masonry veneer is a nonstructural installation of facing material or decorative surfacing attached to a previously constructed structural element. Two basic types exist presently: adhered and anchored veneers. These are more completely described in Chapter 15.

Alternative materials and construction methods (UBC Section 106)

It should be noted that special or unusual construction methods are permitted, on an individual-case basis, through special Research Committee recommentations. This is the way new materials or updated construction methods appear in use, even though they may not be specifically spelled out in the current code. These evolve after the committee has made a thorough evaluation of the proposal for the alternate material or new construction method and has determined that it is satisfactory. The material, method, or work offered must, for the purpose intended, be at

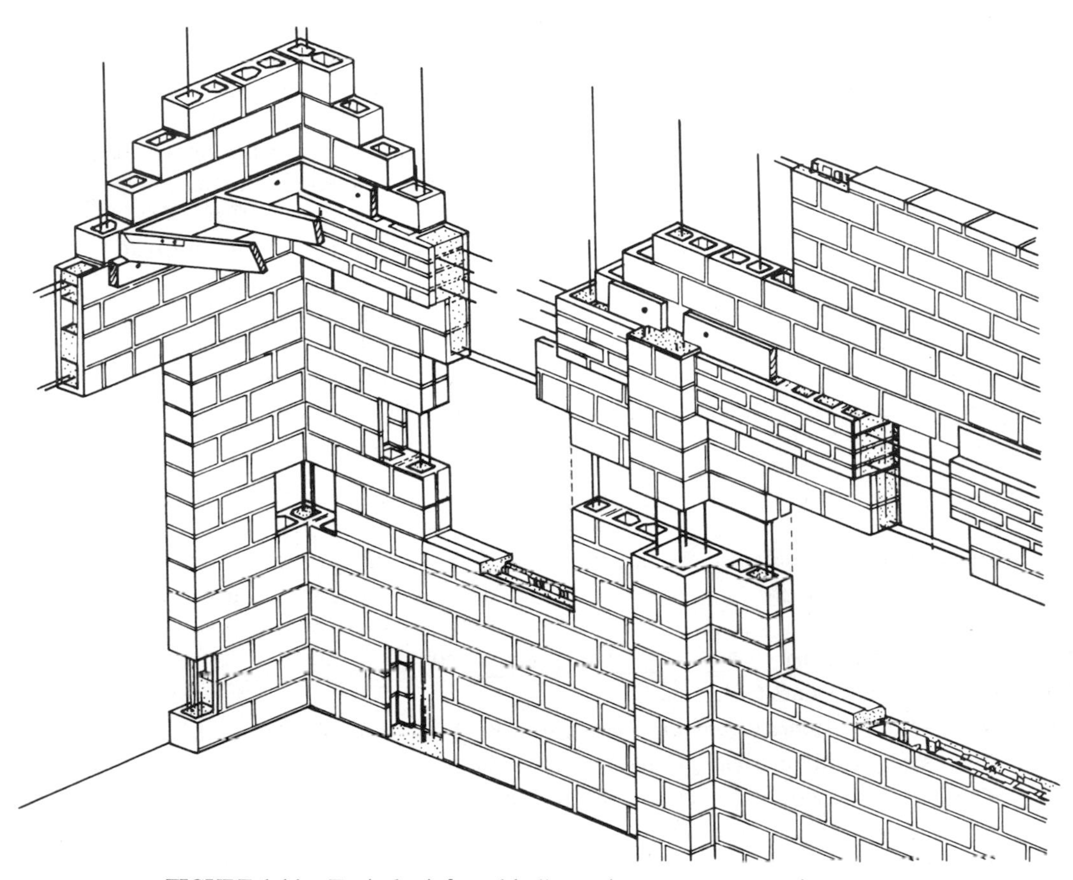

FIGURE 1-14. Typical reinforced hollow unit masonry construction.

least equivalent to that prescribed by the UBC in quality, strength, effectiveness, fire resistance, durability, and safety. Also, the proposed alternate must conform to the provisions of UBC Chapter 23, General Design Requirements.

EMPIRICAL CODE LIMITS

Building codes of one sort or another have been developing worldwide, probably even before Hammurabi's time, perhaps to protect the designer's health. For in those days, designers lost their lives if one of their structures collapsed and caused someone's death. To a lesser degree, that practice has continued to the present day. In some countries, Greece, for example, should a serious structural collapse occur, the designer or project engineer is jailed until liability has been established.

Most of these empirical relations may not have theoretical basis in fact, but they produce results that work. They have evolved from a "cut-and-try" process over the years. Now we call the process "successive approximation" or "iteration."

These terms sound more sophisticated, and besides they are less informative—a characteristic of much of our modern technological language. Call it what you will, a successive approximation is no more scientific than a cut-and-try approach. As an example of such limits, note that the height/thickness ratio (h/t) of masonry walls is limited to an abritrarily selected ratio. There are many instances of these empirical limits in the Code, and we shall encounter many of them in reference to masonry construction. Some of these limits are pulled out of the air, whereas other have a historical background. Consider, for example, the h/t limit of 14 for unreinforced stonework. Believe it or not, this dates back to somewhere around the 1200s. When building the old city walls of Rothenburg, the primitive masons on the project somehow simply sensed that this was a safe limit to be imposed. They knew virtually nothing about material strength or any conditions of wall restraint. But, it turns out, this is a very reasonable limit, albeit a conservative one.

QUESTIONS

1-1. Why shouldn't unreinforced masonry be used in high-rise masonry construction?

1-2. What are some of the limitations on the use of unburned clay masonry? (UBC Section 2404)

1-3. Where may glass masonry be used? (UBC Section 2408)

1-4. What are bond stones?

1-5. When are bond stones required in stone masonry? (UBC Section 2409)

1-6. What is cavity wall masonry? (UBC Section 2410)

1-7. What are the limitations on the size of the cavity?

1-8. What is hollow unit masonry? (UBC Section 2411)

1-9. What types of mortar are allowable for grouted masonry?

1-10. What is reinforced hollow unit masonry? (UBC Section 2415)

1-11. What two factors placed formidable limitations on the heights of masonry buildings as they were constructed around the turn of the century?

1-12. Describe the present concept of reinforced masonry.

1-13. What two kinds of load-carrying elements were the mainstays of ancient construction?

1-14. Why were these early elements most commonly used?

1-15. Distinguish between hollow unit masonry as it is defined in UBC, Section 2411, and the solid masonry of UBC, Section 2412.

1-16. Describe grouted masonry and reinforced grouted masonry per UBC, Sections 2413 and 2414.

2

Clay Brick and Tile—Material Properties

The Structural Clay Products Institute defined *brick* as "a small building unit, solid or cored not in excess of 25% of the gross flat area." Its typical form is that of a rectangular prism, formed from clay or shale and hardened by heat. Brick has been variously classified by such items as place of manufacture, raw materials used, color, surface texture, size, shape, and other special characteristics. But actually, there is no systematic terminology that applies throughout the industry nationally. *Load-bearing brick* is defined as a clay unit that supports vertical loads, as contrasted to a non-load-bearing veneer, which simply acts as an architectural surface or a weather-resistant skin.

What are the properties of brick masonry that have made it such a widely used construction material for centuries? These characteristics can be divided into two major categories: (1) those physical properties that relate to the esthetics of the material, and (2) the engineering properties that govern its structural capabilities. Basic physical properties will be described first, followed by an in-depth examination of the most significant of the engineering properties as they are presently known. Before this can be done, however, we must examine what raw material characteristics we are dealing with as well as how the units are actually made.

RAW MATERIALS

The material used in the manufacture of brick and tile is a hydrated silicate of alumina in which there may occur various impurities, such as oxides of iron, calcium, magnesium, sodium, titanium, and potassium. The presence of these impurities in varying amounts will cause variations to occur in both the chemical composition and physical properties of the clay. The effects of these variations are minimized in manufacture through the mixing of clays from different locations in the clay pit or mine. Within limits, silica will reduce burning shrinkage, but increased amounts will reduce cohesiveness. Fusing temperatures can be decreased by the inclusion of carbonate fluxes. As used in this sense, a *flux* is a substance that aids in the fusing together of materials by preventing oxidation. Iron oxide improves strength but, like the fluxes, has a strong effect on color. The variations in the raw material can be compensated for by controlling the manufacturing process; however, certain variations are bound to be reflected in the finished product.

Natural clay, which abounds in the earth's surface to a greater extent than any other material used in the production of any building product, occurs in three principal forms, all of which have similar chemical compositions but different physical properties:

Surface clays: found near the surface of the earth; sedimentary in character.

Shales: clays that have been subjected to high pressures in the earth until they have nearly hardened to a form of slate.

Fire clays: this material occurs at greater depths than either surface clays or shales. As a rule, fire clays contain fewer of the metallic oxides than do shales or surface clays, and have higher resistance to vitrification under heat.

Among the most significant of the physical properties of a clay suitable for the manufacture of brick and tile are plasticity, tensile strength, fusibility, and shrinkage. Clays must be plastic enough to permit them to be shaped or molded when mixed with water, and they must have sufficient tensile strength to maintain their shape after forming and drying. Further, clays must readily fuse together to form the final product when subjected to certain temperature ranges. For clays, unlike metals, soften slowly and melt or fuse gradually when subjected to rising temperatures. It is this fusibility that results in a hard, solid, and substantially strong unit. A fusing process can take place in three stages: (1) *incipient fusion*—the stage at which the clay particles become soft enough for the mass to stick together; (2) *vitrification*—the stage at which extensive fluxing occurs and the mass becomes rather solid and nonabsorbent; and (3) *viscous fusion*—the stage at which the clay mass breaks down completely and tends to become molten. The kiln temperature must be carefully controlled so that both the incipient and partial vitrification stages are completed, but the viscous fusion stage must not occur, or a useless clinker will result.

Shrinkage is a property inherent to a greater or lesser degree in all clays, and those possessing the least tendency to shrink are preferred for clay products. Two types of shrinkage take place: (1) *air shrinkage*, which takes place after the unit has been formed but before it is kiln-dried, and (2) *fire shrinkage*, which occurs during the burning process. Either of these, if excessive, may cause cracking or warping, both of which are to be avoided if a sound product is to be obtained.

Chemical properties of the clay are also highly instrumental in determining the color and fusibility characteristics of the finished product. For example, iron oxide tends to color the brick in a range of reds to brown; whereas its absence produces a white or cream color unit. Lime, if present in sufficient amounts, will cause a buff or cream color to predominate, since it overcomes the effect of any iron oxide present.

Carbonaceous matter causes defects and therefore must be completely eliminated. This can be accomplished through combustion, whereby oxygen combines with these carbonaceous impurities. The necessary oxygen may occur in the kiln atmosphere, or it may come from a ferric oxide within the clay itself. Burning or firing of the clay then converts the ferric oxide into ferrous oxide and it, in combination with the silica in the clay, yields the ferrous silicate.

METHODS OF MANUFACTURE

The ancient art of brick making seems to have originated on the Mesopotamian plains. It spread westward to Egypt, Asia Minor, Greece, and Rome, and from Rome to Europe and the Western world. In this long evolvement over the centuries, one point should be observed. As one country discovered brick as it was used by its neighbor, it in turn improved upon that work, until the state of the art became highly advanced and developed.

An example of this relentless advancement may be seen in the early use of clay brick by the Roman builders in England, who made and used the first brick there. Brick making on an industrial scale began in England in the thirteenth century, but it was not until the days of Henry VIII (1509–1547) that the craft became highly developed, probably owing to Flemish influence. After the fire of 1666, London became a town of brick instead of wood. But even at that time, the art of making brick was still developing. Many advances occurred through the reigns of Queen Anne and the Georges, until brick achieved its high point in the grand old manor homes of eighteenth-century England.

Stiff mud process

The modern way in which clay products are made is determined primarily by the quality and properties of the raw material. In Southern California, for instance, brick is produced by the *stiff mud* process, also referred to as the *wire-cut* process. In this process the brick clay contains the minimum amount of moisture needed for workability (12 to 15% by weight). Figure 2-1 shows the flow process involved in

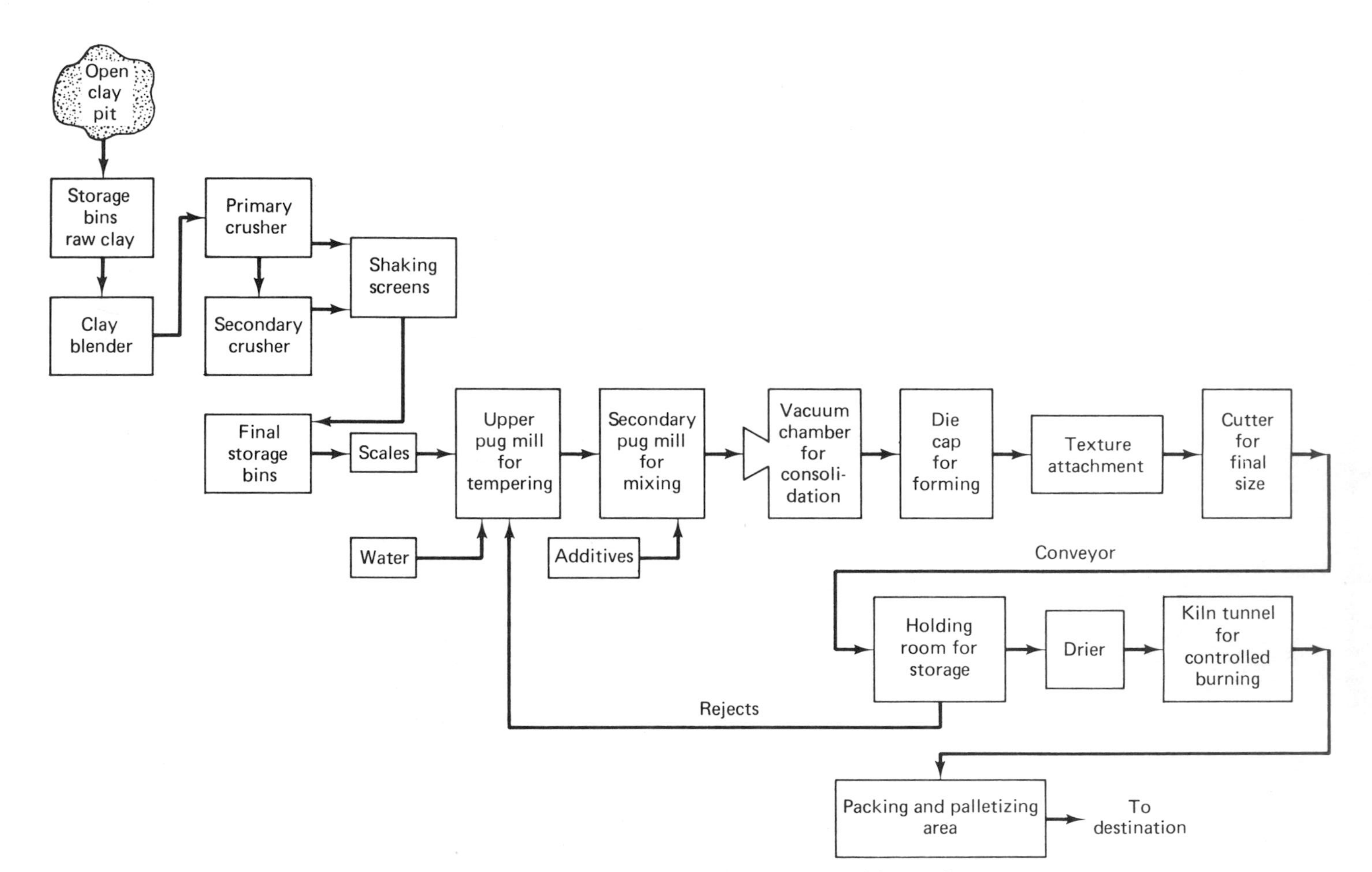

FIGURE 2-1. Flow chart: stiff mud process—clay brick manufacture.

this method of manufacture. In addition, all structural clay tile is produced in this manner. The plant site is generally located adjacent to the primary clay source, although frequently clays are transported from elsewhere to supplement the local material, when certain properties for a specific clay unit are unobtainable if the local material is used alone. The material is generally obtained by surface digging from an open clay pit. It is then transported to storage bins at the plant. If needed, the clays can be readily blended as the raw material is conveyed from the storage bins to the primary crusher, where the large chunks of clay are pulverized and the clayey materials are mixed. After passing through the primary crusher, the clay is transported to the shaking screens, where it is sized to the desired fineness. Material that was not ground finely enough by the primary crusher is reconveyed back through a secondary crusher and reused. This step is very important, for it is the fineness and uniformity of grading which, to some extent, determines the quality and properties of the finished product. From the screens, the finely divided clay is elevated to final storage bins from which it is eventually drawn.

After its initial preparation and storage, the clay is ready for tempering, a process that is accomplished in a *pug mill.* The basic objective of this procedure is to render the clay sufficiently plastic that it may be readily molded into its final shape. On the way to the pug mill, the amount of material passing a given point in a specified period of time may be weighed, in order to control the quantity of clay being fed into the pug mill at any given time. In the newer plants, the pug mill consists of two separate screw-type augers. In the upper pug mill, water is added to the dry clay, forming a heavy paste, which is forced by a continuous screw auger through the sluice and out the end, where it drops into a second mill for further mixing. As the material passes through the second mill, other ingredients may be added where called for in the specifications. These could include such additives as manganese, which might be injected to impart a particular brownish color to the brick. Or the additive could be barium carbonate, which controls scumming, a discoloration appearing on brick surfaces. At the other end of this lower pug mill box, the paste is then forced through a venturi-like throat into a vacuum chamber, where excess water and air are removed. This procedure results in greater "green" brick strength as well as in increased workability and plasticity because of the increased density obtained.

In passing through the vacuum chamber, the clay has become highly compressed in order that it may be readily forced through the die, where the brick is formed into any number of preselected shapes and sizes, the next stage in the process (Figure 2-2). After being extruded through the die like toothpaste out of a tube, the clay has assumed the shape and approximate size of the finished product. Various surface-texture effects can be achieved through the use of an attachment placed between the die and the cutter. For instance, a special surface texture is produced by a motor-driven coarse wire which travels across the top and bottom of the brick as it emerges from the die. In other variations, a stationary wire located just below the top and just above the die opening peels off the skin, producing a wire-cut texture on the top and bottom of the brick units. A core plug unit, consisting of several circular poles attached to a steel plate, placed behind the die shaper cap produces the cored brick.

FIGURE 2-2. Extruded clay column emerging from the die. The two vertical wires at the edge of the column provide accurate length, and a wire-cut texture for the ends. This will be similar to the wire-cut texture of the side of the brick so that textures will be uniform at lapped corners. (Courtesy W. L. Dickey.)

Drying and burning is more uniform on such a unit, thereby resulting in more homogeneous properties as compared with the solid units. Although shipping weight is considerably reduced with the lighter unit, no loss in strength is experienced.

As the consolidated clay emerges from the die in a continuous ribbon onto a run-out belt, it passes through a cutting reel, which locks into place and moves with the ribbon until it has rotated sufficiently to cut several bricks (12 to 15) at once (Figure 2-3). Clays shrink during both the drying and burning processes, so allowance must be made for shrinkage in both the size of the die opening and the length of cut. This shrinkage varies for different clays, thus accounting for the differences in the final size of the same nominal-size brick. The total shrinkage can vary between 4.5 and 15%. This is why the clay for any given run must be as uniform as possible. Farther down the line, the green units are removed from the conveyor and stacked onto a kiln cart (hacking), which, when loaded, is moved into the holding room. This is simply a space where the brick is stored, protected from the weather, until such time as the load of brick can be scheduled into the dryer. Rejects are returned from the end of the line by means of the conveyor belt to the upper pug mill and re-formed. At this point, the green brick contain between 7 and 30% moisture by weight. This moisture occurs in three forms: free water within the pore spaces; water clinging to the pore walls after free water has been removed; and hydroscopic, colloidal, and chemically combined water.

The kiln cart, with its load of green brick, passes from the holding room to the drier. Here free water in the pore spaces and water clinging to the pore walls is

FIGURE 2-3. The rotary frame with the steel wires moves with the clay column as it is extruded and simultaneously rotates to cut the wet clay, several units at a time.

evaporated. The remaining moisture is removed in the first stages of burning. The time required for the drying stage varies, depending upon the type of clay, but it generally ranges from 24 to 48 h. The dryer itself is heated with waste hot air which has been drawn from the kiln tunnel beyond, the temperature being about 350°F, thereby achieving an energy-saving recycle. The drying process is quite critical, since most of the shrinkage takes place during this period. Unless carefully controlled, excessive cracking can occur, resulting in an unusable brick.

In modern plants, the burning process, which takes place in a *kiln tunnel*, requires between 2 and 5 days, depending on the type of kiln and the clay characteristics. In older plants, the brick is simply fired in a field kiln, wherein the brick being burned is enclosed within a shell or walls composed of old reusable brick, which are sealed with clay mud. In the modern kiln tunnel, time of exposure at various temperature stages is carefully controlled by computer.

The degree of control is somewhat governed by the length of the tunnel. In one modern plant in Corona, California, for example, the kiln tunnel is about 380 ft long (Figure 2-4). In this type of kiln, the first two stages in the burning process subject the brick to preheat conditions. The initial preheating temperature is about 1345°F, and the second is set at about 1600°F. Preheat drives off any free water left in the units. Next come the fire zones. The kiln at the Corona plant has essentially four fire zones, where the heat is carefully taken from about 1600°F up to about 1960°F. The burning process has to be very closely controlled, for the brick must be heated only to an early stage of incipient vitrification. In this condition, the clay has begun to flow

(a)

(b)

FIGURE 2-4. Tunnel kiln. (a) Green or unfired brick on the cars entering the long tunnel. (b) The doors at the cool end from which the cars of fired and cool brick will emerge to be distributed by the car-handling equipment at the right. Note the cool-air input fan and duct at the door. (Courtesy W. L. Dickey.)

and then consolidate, but it has not yet reached the temperature of viscous fusion. If the latter stage is reached, the brick simply becomes a pile of glass, rendering it utterly useless. In the fire zone, the brick can be subjected to flashing, if desired. Thus, as the white hot brick passes through the flashing zone, it is hit with a jet of raw gas, which ignites immediately. This process produces a unique color range and degree of shading, varying with different types of clay.

After burning, the brick moves automatically into the cooling-zone portion of the kiln, where the temperature is drawn down in stages to about 100°F. This is an important stage in finishing certain classes of clay, for the rate of cooling has a direct effect on color. Besides, too rapid cooling may cause cracking and checking of the surfaces. This entire passage through the kiln is all sequenced automatically; thus each stage can be carefully governed to achieve the optimum results. From the kiln, the final product is diverted to a packaging and palletizing area, where the brick is prepared for shipment.

Soft mud process

There are other methods of manufacturing clay products, including the soft mud process, and the dry press process. The *soft mud* process is used only for the production of brick, and it is well suited for clays that contain too much water in their natural state (20 to 30%) to be forced through a die as is done in the stiff mud process. Essentially, the soft mud method of manufacture consists of mixing together clays containing 20 to 30% water. Following the initial preparation process of crushing and screening, the units are formed by molding. The oldest means of forming the brick was accomplished by hand with the use of wood molds. However, in large modern plants, the bricks are molded under pressure in a soft mud brick machine. Flow through this machine takes place in essentially five stages: (1) tempering of the clay in the pugging chamber, (2) lubricating the molds with sand or water, (3) forming the units under pressure, (4) bumping the molds for compaction, and (5) dumping the formed units onto pallets. Following this forming operation, they are dried, fired, and cooled in a manner similar to that used in the stiff mud method. If sand is used as the lubricant, the bricks are called "sand-struck," whereas if water is used, the term applied is "water-struck." Each method results in a characteristic surface texture.

Dry press process

The *dry press process* is particularly adaptable for clays possessing low elasticity. The clay is usually prepared in a grinder, following which it is processed through a pug mill, where 7 to 10% water is added. From the pug mill, the moist clay is dropped into dry press forming machines, where the brick is molded under operating pressures of between 500 and 1500 psi (Figure 2-5). After the units have been molded in this fashion, they are removed to the dryer, where they are subjected to temperatures that range from 110 to 300°F for a period of 24 to 48 h. Upon completion of this process, they are sent through burning kilns similar to those described previously.

FIGURE 2-5. A dry press machine with the removal conveyor dismantled. This is approximately 20 ft high and very rugged in order to apply adequate pressures to compact the moistened clay in the brick mold shapes. (Courtesy W. L. Dickey.)

TYPES AND PROPERTIES OF FIRED CLAY UNITS

Fired clay units include building and face brick, hollow clay tile, terra-cotta, and ceramic tile; however, the latter two are not considered to be structural materials. *Solid units* are defined as those whose net cross-sectional area in any plane parallel to the bearing surface is not less than 75% of the gross area. Units whose net area is less are referred to as *hollow units.*

Building brick

The term *building brick* refers to the common or standard basic unit of clay covered by ASTM specification C62. Of prime interest to the engineer or builder are its physical and engineering properties. The most basic of the physical characteristics include color, texture, form, size, and dimensional stability. Included among the primary engineering properties are absorption, durability (weather resistance), axial compressive resistance, flexural strength, thermal conductivity, accoustical characteristics, and fire resistance.

PHYSICAL PROPERTIES

Color The esthetic appeal of clay masonry stems from its color, texture, and form. The true color and texture of burned clay depends, as previously noted, upon its chemical composition, the intensity of burning, and the method of burning control. Of all the oxides or fluxes commonly found in clays, iron has the greatest effect on color. All clay, regardless of its basic color, containing iron in almost any form, will turn red when oxidized. This results from the formation of ferrous oxide, produced by the decomposition of ferrous silicate. The lighter colors, such as salmon brick, are generally associated with underburning, whereas overburning produces "clinker" brick, which is dark red to black in red clays and dark spotted brown in buff clays. Should the clay be burned in a reducing atmosphere (i.e., one having an oxygen deficiency), the final product takes on a purple cast. The creation of a reducing atmosphere is known as "flashing," a process referred to previously.

The many natural colors in which structural clay products are produced throughout the United States range from pure tones of pearl grays or creams, through buff, golden, and bronze tints, to a descending scale of reds and on down to the purples, maroons, and gun-metal blacks. Indeed, this is quite a variety. However, the standard common building brick occurs in terra-cotta red, with units of buff, salmon, red, or brown occasionally being supplied. If a specific color is desired, a sample must be furnished for approval prior to delivery. Table 2-1 shows some typical color ranges and texture patterns available in some commonly used architectural building brick in Southern California.

TABLE 2-1

Southern California Building Brick Color Ranges and Textures

Name	*Color*	*Texture*
Palos Verdes	Off-White to gray	Flat to uneven
Bouquet Canyon	Tan to rust	Flat to uneven
Ora Verde	Green	Flat to uneven
Santa Maria	Cream to rust	Flat
Drift Stone	Brown to black	Rough to rugged
Black Lava	Black	Rough to rugged
Kaibab Mtn.	Green to pink	Irregular
Featherock	Gray-black	Rough
Slate	Gray-green-plum	Smooth
Texas Shell	Cream	Irregular
Texas Lime	Cream	Smooth
Whitewater Canyon	Red with yellow	Irregular
Grimes Canyon	Red with yellow	Irregular

Occasional attempts are made to relate the color or tone of brick to some of its other physical properties. This is poor practice, however. For instance, higher burning causes darker tones to appear in a given clay. It also results in an increase in the compressive strength of the unit, and a reduction in its absorbtion qualities. But the chemical composition of the clay also influences the color, and certain clay combinations, because of this factor, can produce darker colors without having any significant effect on strength or absorbtion. So it can be safely stated that one should not try to generalize even on the relative values of strength or absorbtion by merely observing color or shading.

Texture *Texture* is a surface effect or appearance of the unit apart from its color, resulting from the way the unit is made. The degree of textured effect ranges from fine through medium to coarse. The natural brick texture produced in the soft mud or dry press process of manufacture is due to pressing of the clay against the sides of the steel mold. Building brick units can also be supplied with a matt face, rug face, or a wire-cut face. The *matt-face* texture consists of very light vertical lines, scored on the surface, as the unit emerges from the die in the stiff mud process. It presents an attractive pattern when left exposed. *Rug face* is a ruffled texture that is produced by making a heavy scratch on the brick surface and rolling the excess material back on the face. This brick shows an interesting wall surface of lights and shadows that are caused by these irregular scratch patterns. A *wire-cut* texture appears naturally on two faces of all brick units produced by the stiff mud process. It is characterized by an exposure of the aggregate and a slight ruffling of the surface, developed as the wire cuts through the ribbon of a green clay column to divide it into the individual brick units. A scored finish may be produced by grooving the brick face as it emerges from the die, or a combed finish may be attained by penetrating the face with parallel scratches. Other textured finishes include a stippled effect, a barklike surface, a stoneface texture which resembles the moon's surface, or the sand- or water-struck surfaces. Figure 2-6 illustrates some of these texture patterns.

Form and Dimension Building bricks are manufactured in a wide variety of sizes and shapes to permit their selection for proper scale within a wall. Some of the shapes more commonly used in the western United States are referred to as common or standard, harlequin, commercial, norman, modular, jumbo, oversize, roman, king size, Padre, Imperial, and Royale (a clay block). Some of these shapes, along with their actual dimensions, are shown in Figure 2-7. Figure 2-8 and Table 2-2 show shapes

TABLE 2-2

Nominal Modular Sizes of Eastern Brick

	Dimensions (in.)			
Unit designation	*Thickness*	*Height*	*Length*	*Modular coursing*
Standard modular	4	$2\frac{2}{3}$	8	3C = 8
Engineer	4	$3\frac{1}{5}$	8	5C = 16
Economy	4	4	8	1C = 4
Double	4	$5\frac{1}{3}$	8	3C = 16
Roman	4	2	12	2C = 4
Norman	4	$2\frac{2}{3}$	12	3C = 8
Norwegian	4	$3\frac{1}{5}$	12	5C = 16
Utility[a]	4	4	12	1C = 4
Triple	4	$5\frac{1}{3}$	12	3C = 16
SCR brick	6	$2\frac{2}{3}$	12	3C = 8
6-in. Norwegian	6	$3\frac{1}{5}$	12	5C = 16
6-in. jumbo	6	4	12	1C = 4
8-in. jumbo	8	4	12	1C = 4

[a]Also called Norman Economy, General, and King Norman.

and dimensions of typical brick units manufactured in the eastern part of the United States. As previously noted, since most clays shrink between 4 and 15% during drying and burning cycles, a dimensional allowance must be made for this shrinkage when the units are formed in order to obtain the final dimensions as shown. The actual amount of shrinkage will depend, among other things, upon the composition of the clay, the fineness to which it is ground, the amount of water mixed with it, and the firing temperatures. Further, the dimensional stability of the individual clay unit is generally governed by two factors: (1) thermal expansion, which is reversible, and (2) moisture expansion, which is not.

Final actual sizes are less than the nominal dimensions to allow for the thickness of the mortar joint, generally $\frac{3}{8}$ to $\frac{5}{8}$ in. This provides for even-dimensional modular coursing. The UBC limits the difference between actual and nominal dimensions to a maximum of $\frac{1}{2}$ in. Observe that the nominal size is always the dimension used when checking the thickness and height/thickness ratio limits specified in the UBC for various types of masonry walls.

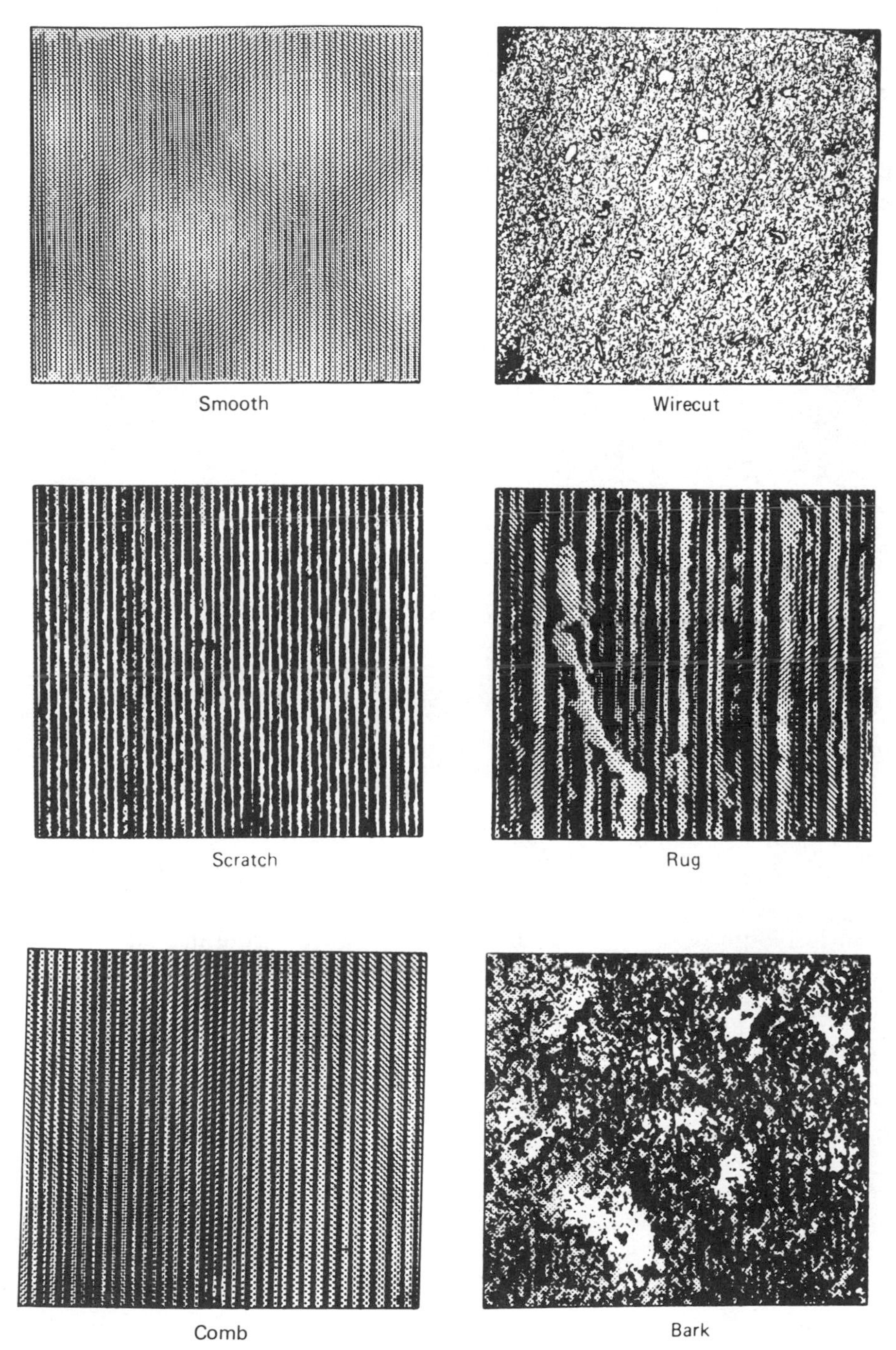

FIGURE 2-6. Some typical examples of common textures and the names that have been adopted for these by the Brick Manufacturers of Southern California.

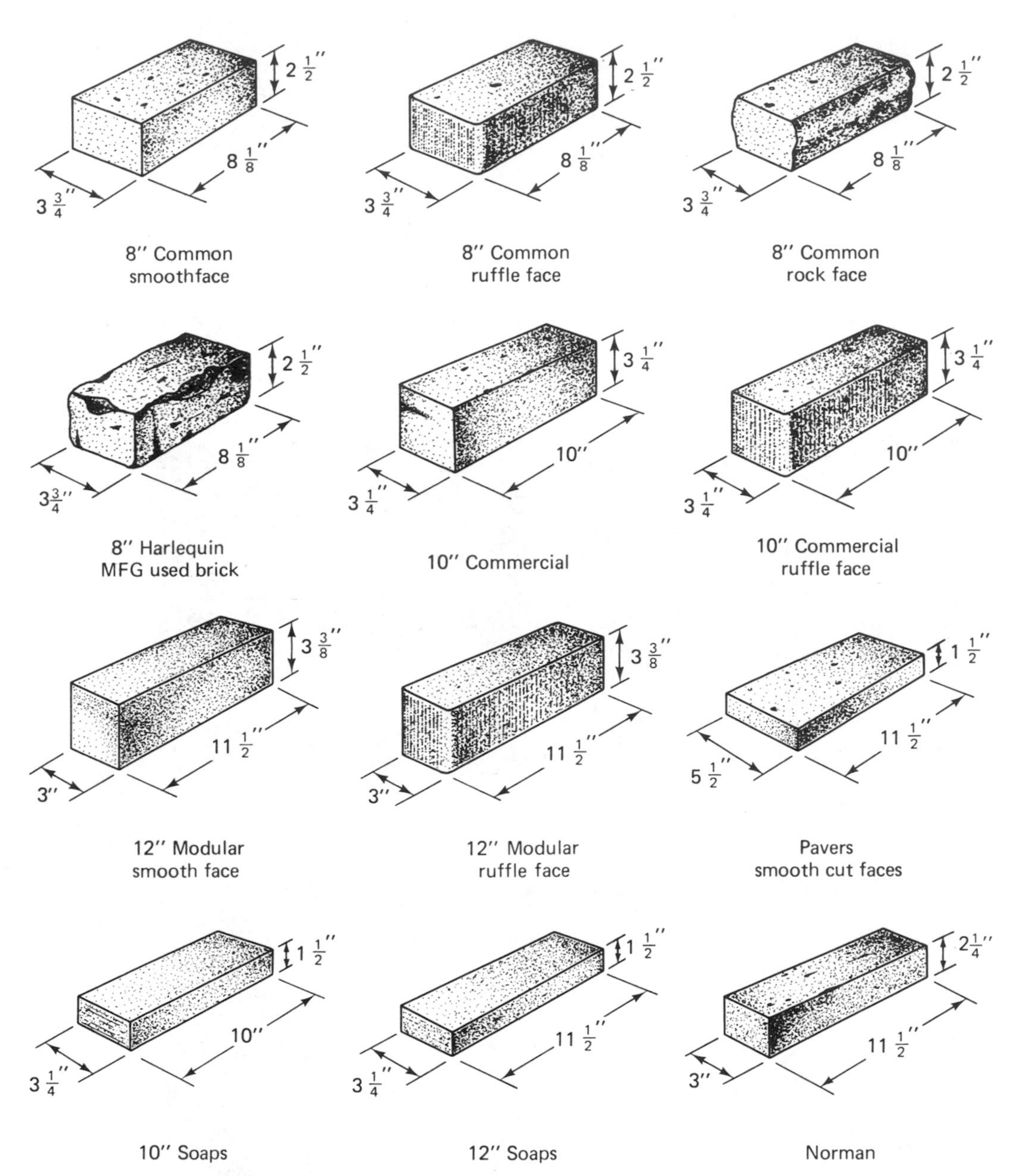

FIGURE 2-7. Typical western building brick shapes and dimensions.

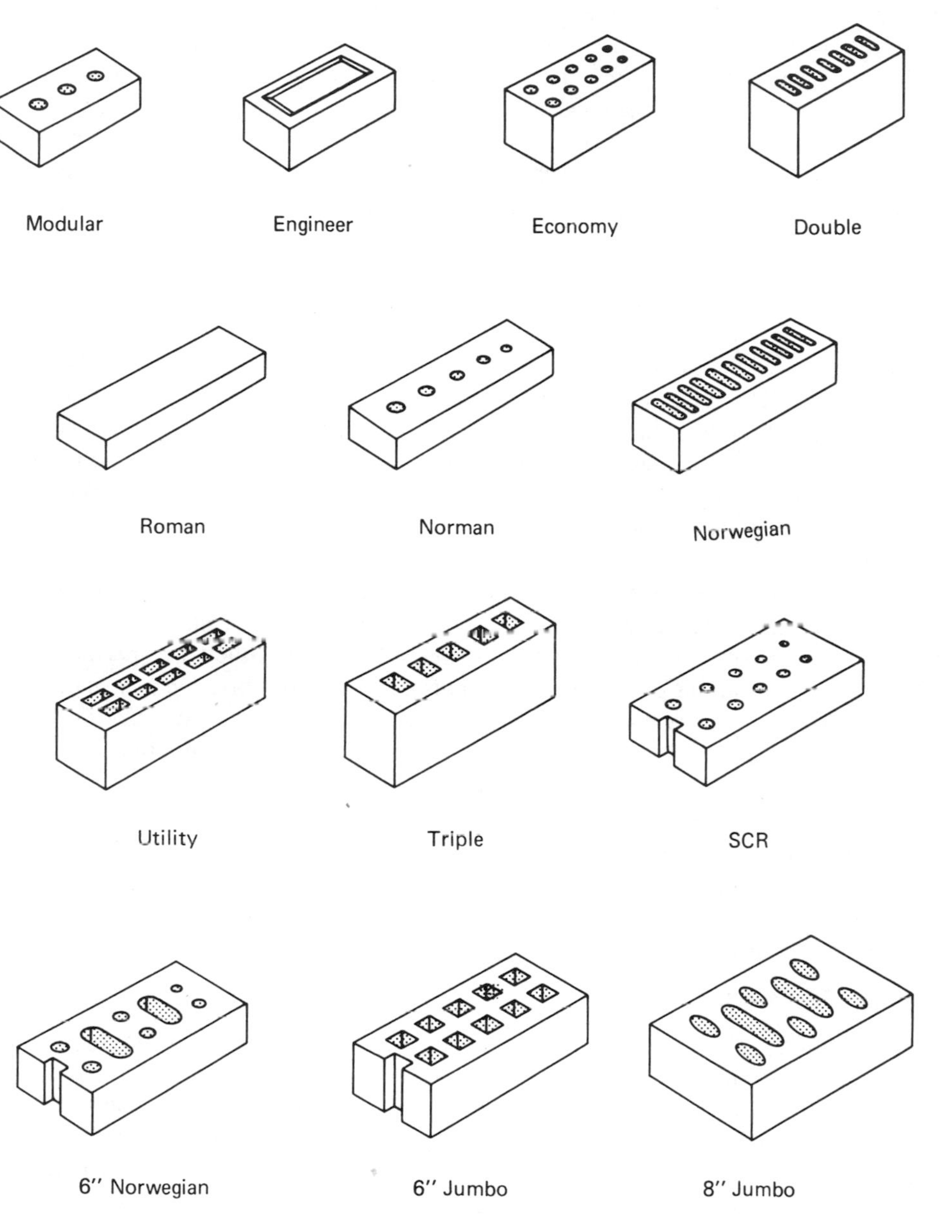

FIGURE 2-8. Typical eastern modular brick shapes.

The maximum permissible variation in the dimensions of the individual units are spelled out in ASTM C62. Often these tolerances can be too great for a precise job requirement. For instance, stack bond wall patterns require closer dimensional tolerances, since the units are placed one directly over the other. In such cases the supplier must be informed before the job starts so that remedial action may be taken to furnish the required dimensions. Observe that there are a number of factors which must be considered when one is choosing a basic unit size and shape. These would include the effects of scale on building appearance, as well as the integration of the masonry unit size with other elements in the structure.

The final weight of a burned clay unit depends upon the specific gravity of the unburned material and the various processes employed in both the production of the green unit and its subsequent burning. Weight variations within a given type may be assigned mostly to the method of manufacture alone, since the difference in specific gravity of clays and shales is quite small, varying between 2.6 and 2.8. Increased density and weight of the burned material result from a combination of fine grinding, uniform mixing of materials, pressure exerted upon the clay as it is extruded, and the completeness of the burning process. As a result of all these factors, there exists a close relationship between total absorption by the unit and its final weight. The actual weight of the average unit attains a value of about 120 lb/ft^3 for a solid unit.

To minimize weight, bricks are often cored; that is, portions of material are removed leaving circular holes within the body of the unit. As a matter of fact, the brick may be solid or cored, at the option of the seller, unless otherwise specified. According to ASTM C62, the net cross-sectional area of the cored brick in any plane parallel to the bearing surface must be no less than 75% of the gross cross-sectional area measured in the same plane. Further, no part of any hole shall be less than $\frac{3}{4}$ in. from any edge of the unit. Occasionally, a question arises as to the relative strengths of solid versus cored brick. As it turns out, coring the brick does not detract in any way from its strength. To the contrary, cored brick demonstrates higher strengths because of the fact that the cores provide for more uniform drying and burning. In addition, a better mechanical mortar bond is obtained since the mortar will press into the holes to provide an excellent mechanical "key."

Engineering Properties

Durability Designations The *physical properties* of building brick, as reflected by its color, texture, and form have now been delineated, and it has been seen how these characteristics do indeed reflect upon the esthetics of the structure. Durability grade, total absorption, rate of absorption, compressive and flexural strength, thermal conductivity, and acoustical factors of clay brick constitute those characteristics loosely grouped together and categorized as its *engineering properties*.

For durability classification purposes, building brick is divided into three grades: Severe Weathering (SW), Moderate Weathering (MW), and No Weathering (NW). The *SW grade* is intended for use where a high degree of resistance to both frost action and general disintegration by weathering is desired. The *MW grade* is generally specified if a moderate degree of resistance is required, and where the brick is not likely to be permeated with water when exposed to freezing temperatures. The *NW grade* is only used as a backup for exterior masonry, for it would rapidly disintegrate if subjected to freezing and thawing cycles. This grade can also be used for interior walls or partitions.

The effect of weathering on brick is related to the weathering index, which for any locality is the product of the average annual number of freezing cycle days times the average annual winter rainfall in inches. A freezing cycle day occurs on any day when the air temperature passes either above or below 32°F. The average number of freezing cycle days in a year equals the difference between the mean number of days during which the minimum temperature was 32°F or below and the mean number of days during which the maximum temperature was 32°F or below. Winter rainfall is the sum, in inches, of the mean monthly corrected precipitation occurring during the period between and including the normal date of the first killing frost in the fall and the normal date of the last killing frost in the spring. The winter rainfall for any period is equal to the total precipitation less one-tenth of the total fall of snow, sleet, and hail. Rainfall for a portion of a month is prorated.

A reference to Figure 2-9 will show the general areas of the United States where brick masonry is subjected to various degrees of weathering. The SW region has a weathering index exceeding 500, the MW region an index of 100 to 500, and the NW region an index of less than 100. More precise data for any specific locality,such as mountain areas, may be estimated from the tables of Local Climatological Data published by the United States Weather Bureau. Table 2-3 gives the grade requirements for brick exposures, keyed to these weathering indices limits. Note that, unless otherwise specified, SW or MW grades may be substituted for grade NW. When the grade is not specified, grade MW will be provided.

TABLE 2-3

Grade Requirements for Brick Exposures

	Weathering index		
Exposure	*Less than 100*	*100 to 500*	*500 and greater*
In vertical surfaces:			
In contact with earth	MW	SW	SW
Not in contact with earth	MW	MW	MW
In horizontal surfaces:			
In contact with earth	SW	SW	SW
Not in contact with earth	MW	SW	SW

The weathering action that has an immediate and significant disintegrating effect upon burned clay products consists of alternate freezing and thawing cycles, in the presence of moisture. Therefore, in those sections of the country where the weathering index exceeds 100, and where alternate freezing and thawing occurs frequently, the ability of the brick to withstand these effects without disintegrating becomes very

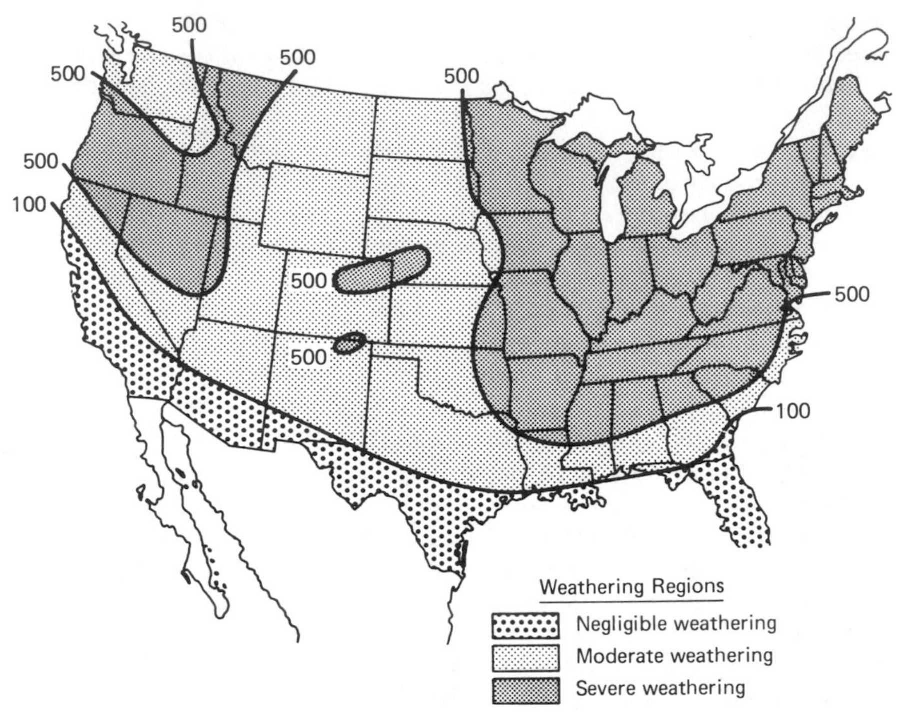

FIGURE 2-9. U.S. weathering index map.

important. The best measure of the clay product durability would be a freeze-and-thaw test in which units are subjected to from 50 to 100 alternate cycles of freezing and thawing in the presence of moisture. Such a test, however, requires 10 to 12 weeks to complete, thus rendering it impractical as a standard acceptance test. For this reason extensive research has been carried on to correlate other more measurable physical properties of clay products with their resistance to freezing and thawing.

For brick and tile produced from the same raw material and by the same manufacturing process, either *compressive strength* or *total absorption* constitute reasonably accurate indices of their resistance to freezing and thawing. This stems from the fact that the durability of clay products results primarily from the firing or burning which fuses clay particles together by incipient vitrification. Further, compressive strength and absorption are also related to firing temperatures and porosity. However, values obtained for the properties of one product may not apply to another produced from different raw materials. That is why a third property, *saturation coefficient* (C/B ratio), has been devised, and it, along with absorption and compressive strength, has been found more reliable in predicting the durability of brick than absorption and compressive strength alone. This coefficient is defined as the ratio of the absorption of the brick by a 24-h submersion in cold water to the absorption after 5-h submersion in boiling water.

Water Absorption The absorption qualities of clay brick (related to the ability to form a good bond with mortar) are taken as a part of the acceptance measurement of the durability of the material. ASTM specification C67 describes a method by which absorption of building brick can be measured. First, the samples are immersed for 24 h in cold water. The amount of water absorbed is then recorded as a percentage of the total weight of the dry unit. Following the cold water immersion, the unit is immersed in boiling water for 5 h. Again the amount of water absorbed is recorded as a percentage of the total weight of the dry unit. The ratio of these two is the *cold water/boiling water ratio*, or C/B ratio. Note that the 24-h submersion alone will not fill all the pore spaces in a brick, so more will become filled during the boiling stage. This ratio is thus that proportion of total pore space that is easily filled with moisture. Since only a part of the total pore space of the brick is naturally filled with water in place, it is thus known that room for expansion exists during freezing, without disrupting the clay body. This is the reason why this property becomes a partial measure of durability. In the UBC and ASTM standards, both the 5-h boil (%) and the C/B ratio show maximum allowable limits for each grade. Table 2-4 lists these physical requirements

TABLE 2-4

Physical Requirements for Building Brick

	Minimum compressive strength (brick flatwise), gross area (psi)		*Maximum water absorption by 5-h boiling (%)*		*Maximum saturation coefficient*	
Designation	*Average of five brick*	*Individual*	*Average of five bricks*	*Individual*	*Average of five bricks*	*Individual*
Grade SW	3000	2500	17.0	20.0	0.78	0.80
Grade MW	2500	2200	22.0	25.0	0.88	0.90
Grade NW	1500	1250	No limit	No limit	No limit	No limit

for building brick. In accordance with UBC Standard 24-1, in any area where the average annual precipitation is less than 20 inches, even with the occurrence of frost action, the 5-h-boil requirement and consequently the C/B ratio limit, may be waived. In that case the minimum average compressive strength of 2500 $lb/in.^2$ applies. Also, if the average compressive strength exceeds 8000 $lb/in.^2$, or the average water absorption is less than 8% after 24-h cold water submersion, the saturation coefficient may be waived. Should a five-brick sample comply with the requirements stated in ASTM C62 (50 cycles of freeze thaw per the procedures set forth in ASTM C67), the maximum absorption after 5-h boiling and the saturation coefficient may again be waived. Perhaps it should be noted here that over the years, through improved procedures and equipment, manufacturers have been able to produce a brick of much greater density, thereby considerably reducing the amount of water absorption and, at the same time, increasing the compressive strength. Additional research is being carried out to revise these durability factors.

Initial Rate of Absorption The pores or small openings in burned clay products function as capillaries and draw water into the unit. This action in a brick is referred to as its rate of absorption or *suction*. Although suction has little bearing upon the transmission of water through the brick, it does have an important effect on the tensile bond between the brick and the mortar. Note that water absorption is concerned with a total quantity, whereas initial absorption rate denotes a time-related function. Thus, there is no consistent relationship between the two factors. For instance, some bricks have a relatively low absorption value and a high suction rate, but the reverse can also hold true. Many bricks of high suction will soak through from end to end within a few minutes after the brick is placed on end in water. However, capillary action, as such, is not the source of wall leakage, since external water is not transmitted through the brick as much as through openings between mortar and brick.

The laboratory method for determining this initial rate of absorption is specified in ASTM C67. It calls for the partial immersion of the unit in water to a depth of $\frac{1}{8}$ in. for 1 min, following which it is removed, weighed, and the final weight compared to its dry weight. The factor, which must not exceed 0.025 oz/in.^{2} (UBC 2403 (v)) during the 1-min period, is calculated as follows:

$$X = \frac{(W_1 - W)30}{A}$$

where X = initial rate of absorption, expressed as a gain in weight, corrected on the basis of 30 in.^{2} flatwise area

W = weight, prior to immersion

W_1 = weight, after immersion

A = net cross-sectional area of immersed surface, in.^{2}

A rather rough, but effective, field test for initial rate of absorption involves drawing a circle on the brick (using, say, 25-cent coin). Then using a medicine dropper, quickly squeeze out 20 drops of water within the circle, to cover the entire area, taking care to ensure that none of the drops fall outside the circle. Note the time that it takes for all the water to be absorbed. If the time exceeds $1\frac{1}{2}$ min, the brick need not be wetted prior to laying. If the period is less than $1\frac{1}{2}$ min, they probably should be prewetted to reduce the water absorption from the mortar.

Another method has been used for a quick but rather accurate field check of the initial rate of absorption. It consists of pouring water from a graduate into a tray or pan, inserting the brick base about $\frac{1}{8}$ in. into the water for one minute, then pouring the water back into the graduate and noting the amount of water absorbed. The cc (or grams) absorbed is corrected by the ratio of the brick face area to the 30 in.^{2} standard to give an approximate IRA or suction value.

Tests indicate that the ideal initial rate for maximum bond strength falls within the 10 to 12 g/min range with a preferable maximum of about 40 g/min. If for some reason, the initial absorption rate does exceed 40 g/min, it is standard practice to wet the units, prior to laying, in order to lower this excess initial rate of absorption below the maximum stipulated. This wetting is most effective when done about 24 h before the brick units are laid, so that the water will have adequately reduced the initial absorption rate and yet not leave the surface wet.

This particular characteristic of brick is exceedingly important for several reasons:

1. If the absorption rate is too high, the brick units will be more difficult to lay in the wall, because water is removed rapidly from the mortar bed causing it to lose workability before the brick units are laid on it.
2. Should the brick exhibit an excessively high absorption rate, a bricklayer may have a tendency to tap the brick as he shoves it into place, thereby disrupting the bond between brick and mortar.
3. A brick with an extremely high initial absorption rate tends to dry out the mortar so quickly that it will not retain the proper amount of water needed for high strength and good bond.
4. Since suction has a bearing upon the water tightness of the wall, it also becomes another measure of weather resistance or wall durability.

The curve in Figure 2-10 depicts the variation in tensile bond strength with brick suction or initial absorption rate. Note that peak strength obtains at about 20 g/min; however, values up to about 40 g/min may be acceptable.

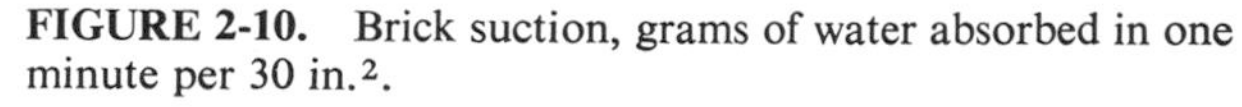

FIGURE 2-10. Brick suction, grams of water absorbed in one minute per 30 in.^{2}.

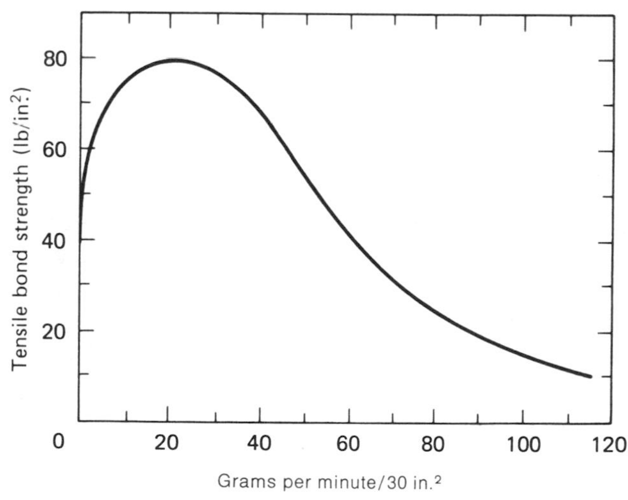

Compressive Strength The *compressive strength* of building brick may be defined as the maximum stress to which the unit can be subjected by a gradually increasing load applied perpendicular to the bedding plane or normal position. ASTM specification C67 describes the procedure for determination of the ultimate compressive strength, a value of extreme importance to the designer. As noted previously, the compressive strength of brick is significantly affected by the (1) physical properties of the clay, (2) method of manufacture, and (3) degree of burning. Other compression factors which are important to the designer are the compressive strengths of the mortar, grout, and composite prism assemblage (f'_m). Design allowable stresses are keyed to the latter value. These will be discussed in subsequent sections. Note that the ultimate compressive strength of a prism, which is a composite of mortar, brick, and grout, is less than that of the clay unit itself. This stems from the fact that the strength and stiffness properties of the brick, mortar, and grout are all different and, when acting as a composite, demonstrate a lower combined strength. In accordance with ASTM specifications, five brick specimens must be tested and they have to yield the ultimate compressive strength minimums previously given in Table 2-4. When brick must have greater strengths than those prescribed in this table, they should be classified by compressive strength rather than by durability grade. These are listed in Table 2-5. Actually, the compressive strength of most brick is much higher than that of good concrete, but few designers are aware of this fact.

TABLE 2-5

Brick Classification by Compressive Strength

	Minimum compressive strength—lb/in.²	
Designation	*Average of five units*	*Individual unit*
2,500	2,500	2,200
4,500	4,500	4,000
6,000	6,000	5,300
8,000	8,000	7,000
10,000	10,000	8,800
12,000	12,000	10,600
14,000	14,000	12,300

Stiffness An adjunct to compressive strength is the *modulus of elasticity*, a stiffness index. Tests performed at Watertown Arsenal yielded values for brick ranging between 1,400,000 and 5,000,000 lb/in.^{2}. In general, as with concrete, modulus of elasticity increases with compressive strength up to about 5000 lb/in.^{2}, after which little change is noted. As with compressive strength, the modulus of a composite prism (E_m) is somewhat less than that of the clay unit by itself. Actually, few valid test results have been published to date to establish a viable value for E_m. As a result, the current UBC specifies an extremely arbitrary value of 1000 f'_m. A change is being considered that would reduce this value to $600 f'_m$. If deflection becomes a design factor, it is recommended that a more accurate determination of E_m be made by laboratory tests. Another stiffness index, *Poisson's ratio*, or the ratio of lateral expansion to longitudinal deformation, ranges between 0.04 and 0.11 for clay masonry units.

Flexural Strength The *modulus of rupture* refers to the transverse bending capacity of brick. It is determined by supporting the brick at each end and subjecting it to a beam loading. The value for modulus of rupture, obtained from the flexure formula, Mc/I, is simply an index of flexural resistance and nothing more; ultimate strength obviously cannot be determined from this homogeneous elastic stress formula.

In concrete, recent trends have tended to relate the flexural tensile strength to the *splitting strength* as a more reliable indicator. The latter is established by loading a cylinder perpendicular to its longitudinal axis. Perhaps a similar approach would be appropriate for masonry, so that a more reliable indication of tensile strength may be had than what is currently in use.

Thermal Conductivity and Expansion The *thermal conductivity* of a material increases directly with its density, the amount of moisture present, and the mean temperature at which the coefficient is measured. However, it must be observed that the thermal conductance of individual clay units is not the same as that of the masonry assemblage because of the effect of the mortar and grout. Thermal, acoustical, and fire-resistive properties of walls are described in detail in Chapter 16.

Thermal expansion of clay products depends to a large extent upon the raw material itself, although the temperature range over which the coefficient is taken has some effect. National Bureau of Standards tests show values of between 2.8 and 3.9 $\times 10^{-6}$ in./in./°F. The thermal expansion of the individual units is different from that of the masonry assemblage, again because of the existence of the mortar joints and the grout.

Acoustical Characteristics The *acoustical characteristics* of structural clay products vary, depending upon the type of unit. Sound transmission is a function of mass; consequently, bricks, because of their high density, are most effective in controlling sound transmission rather than in absorbing sound, since most commonly used clay products exhibit rather low sound-absorption rates. However, units of special design, such as SCR Acoustile, are produced with sound absorption capacities above 60%. These are described in detail in a later section of this chapter.

Facing brick

DURABILITY GRADES

Facing brick, composed of clay or shale, or a mixture of these, must comply with ASTM C216. They are manufactured for the express purpose of forming the exposed face of a masonry wall. Thus, these units must meet more strict tolerances for color, size, warpage, and chippage than do building bricks. Only the SW and MW grades exist; however, these durability grades are each further subdivided into three texture types: FBS, FBX, and FBA. Type FBS is specified for general use in exposed exterior and interior masonry walls and partitions, since a wider variation in color and size is permitted within this classification than in Type FBX. Unless otherwise specified. Type FBS would normally be furnished. Type FBX is recommended for use in exposed faces of exterior or interior masonry walls where a high degree of mechanical perfection, narrow color range, and minimum size variation are all needed. In contrast, Type FBA is manufactured to produce those characteristic architectural effects which result from an intentional nonuniformity of color, size, or texture of the individual units. ASTM Designation C216 further calls for the faces of the brick units to be free of those imperfections detracting from the appearance of a sample wall when viewed from a distance of 20 ft for Types FBS or FBA or 15 ft for Type FBX.

COLOR–TEXTURE–FORM

Facing brick comes in a variety of colors and textures, ranging from chalk or paper white to charcoal or jet black. In between these extremes lie such colors as yellow, orange, red, blue, purple, brown, and gray, with even certain intermediate hues and shades being available. Consequently, the desired color must be carefully spelled out in the specifications so that there will be no misconception of the designer's intent.

Textures may be referred to as Norman/Ruffle, Kord Norman/Velour, Norman/Stone Face, Roman/Ruffle, Standard/Ruffle, and Economy Norman/Velour, among many others. See Figure 2-11, which shows some of the texture appearances available. To further define the limits of size, ASTM C216 lists the permissible deviations from specified dimensions and the warpage limits. It also calls for the brick to be free of cracks and other defects that will interfere with the proper setting of the brick or will impair the strength of the construction. Actually, this latter requirement applies to all structural clay products.

COMPRESSIVE STRENGTH

The compressive strength requirement of facing brick is similar to that of ordinary building brick (i.e., 2500 lb/in.^{2} minimum for MW and 3000 lb/in.^{2} for SW grades). Similar total absorption limits and saturation coefficients, where applicable, are called for also by ASTM (Table 2-4). Because of their similarities, the UBC groups both building and facing brick under the same Standard, 24-1.

Hollow brick

Hollow brick are hollow clay units very similar in size and shape to hollow concrete block masonry except that they are made of fired clay or shale, conforming to ASTM C652, which permits coring up to 40% maximum area compared to the solid brick limit of 25%. A Western version, defined by the UBC (Research Recommendation 2730), is similar, but it has certain other characteristics more appropriate to reinforced masonry construction. Western States Clay Products Association (WSCPA) has established a standard for these hollow reinforceable units. The additional requirements beyond ASTM C62, C652, and C216 for this modified version are:

FIGURE 2-11. Facing brick textures.

1. Net area strength rather than gross area strength is specified. In this way, the actual unit material strength is defined.
2. Face shell thickness is equivalent to that of hollow concrete block. The cross-web requirement, also similar to concrete block, imparts stability to the unit.
3. The cell size and area are established to permit the proper placement of reinforcing steel within the unit.
4. The tables of physical properties are related to f'_m, the composite strength, since design values stem from this property.
5. Provisions are made for a fire rating.

The WSCPA standard specifications for minimum compressive strength, water absorption by 5-h boil, and maximum saturation coefficient are shown in Table 2-6. Further, Table 2-7 provides the classification of high-strength units by compressive strength. ICBO Research Report 2730 approves f'_m values for the various H designations and also fire ratings for different hollow brick thicknesses (4, 6, 8, and 10 in.), both without plaster and with a $\frac{5}{8}$-in. plaster coat on one or both sides.

TABLE 2-6

WSCPA Physical Requirements: Hollow Brick

	Minimum compressive strength lb/in.² (brick flatwise) (net area)		*Maximum water absorption by 5-h boiling (%)*		*Maximum saturation coefficient*[a]	
Designation	*Average of five brick*	*Individual*	*Average of five brick*	*Individual*	*Average of five brick*	*Individual*
Grade I brick (exposed)	3000	2500	17.0	25.0	0.78	0.80
Grade II brick (not exposed)	2500	2000	No limit	No limit	No limit	No limit

[a]The saturation coefficient, or C/B ratio, is the ratio of absorption by 24-h submersion in cold water to that after 5-h submersion in boiling water.

TABLE 2-7

WSCPA Classification by Compressive Strength: Hollow Brick

Designation	*Compressive strength lb/in.² of five brick (net area)*	*Individual minimum (lb/in.², net area)*	f'_m *lb/in.²*
Grade II only (not exposed)	2,500–3,000	2,000	1,500
Grade I (exposed)			
H 3,000	3,000–3,999	2,500	1,800
H 4,000	4,000–4,999	3,200	2,000
H 5,000	5,000–5,999	4,000	2,300
H 6,000	6,000–6,999	4,800	2,600
H 8,000	8,000–8,999	6,500	3,300
H 10,000	10,000–11,999	8,500	4,000
H 12,000	12,000–13,999	10,600	—
H 14,000	14,000 and above	12,300	5,300

Structural clay load-bearing wall tile

Structural clay tile was developed to satisfy a need for lightweight masonry, especially when used in filler panels and spandrel walls. Also it can be utilized when the finished surface is a plaster or some other type of finish. Raw materials are the same as that of building brick, and they are all manufactured by the stiff mud process. However, the size of the basic unit is much larger, as is the percentage of voids. The result is a building material which has varying degrees of the desirable characteristics of brick (strength and durability), while reducing such undesirable qualities as weight.

Although certain types of tile do not have high strength values, others possess very high strengths and make excellent structural units. Unfortunately, the earlier and weaker units, with the accompanying weak sand-lime mortar, behaved very poorly in earthquakes. This characteristic imparted a poor image to all tile units, regardless of their quality, a sort of guilt by association indictment. Furthermore, since no reinforcing was used in those early structures, they offered little in the way of moment and perpendicular or in-plane shear resistance, often failing dramatically and suddenly (Figure 2-12).

FIGURE 2-12. Structural clay tile units (8 in.).

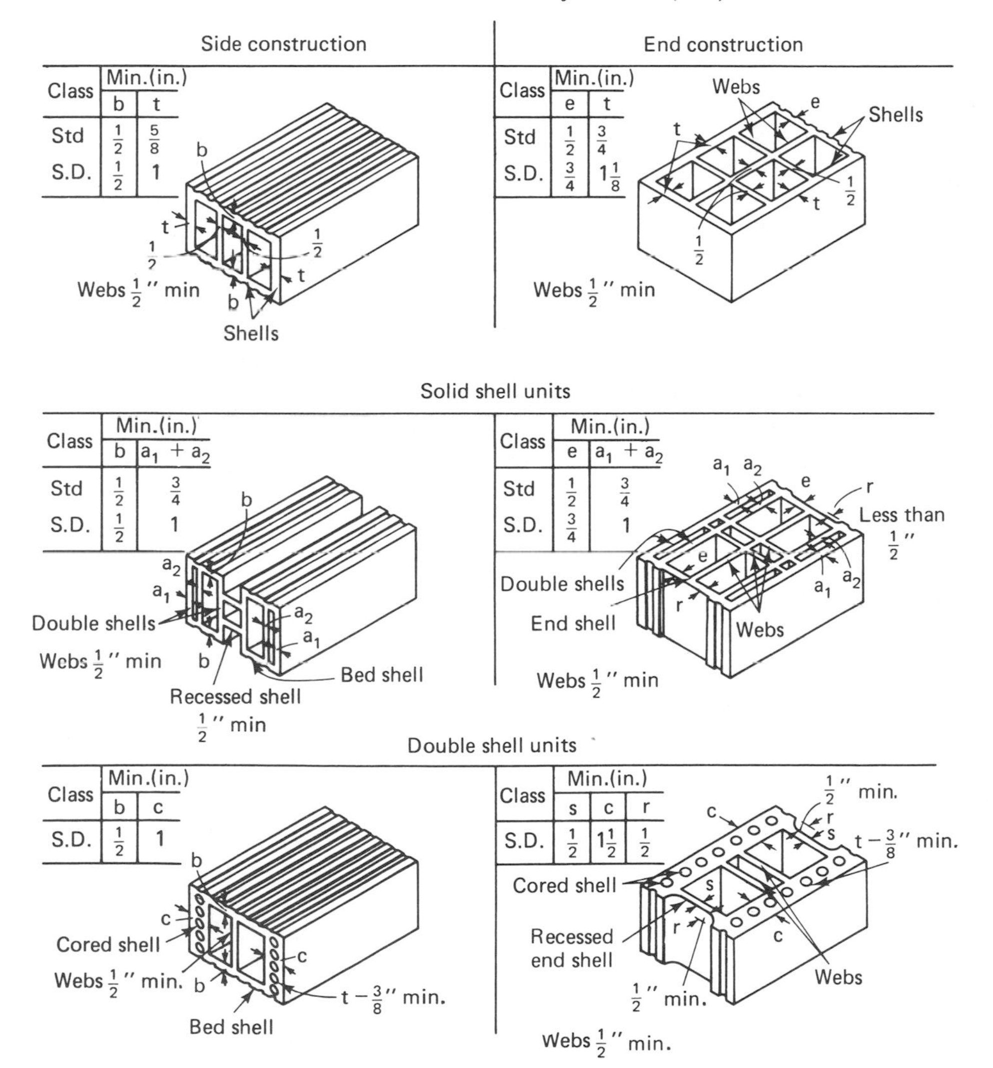

GRADE–COLOR–SIZE

This unit is covered by ASTM designation C34, which defines two grades of structural clay tile: LBX and LB. Also, UBC Standard 24-8 covers these products. Grade LBX tile is suitable for general use in load-bearing walls and adaptable for use in severe weather conditions, provided that they are burned to the normal maturity of the clay. They are also suitable for the direct application of stucco. Grade LB tile must be used only when the wall is not exposed to severe weathering action, or for exposed masonry protected with a facing of 3 in. or more of other masonry. Tile of grade LBX can be accepted under all conditions in lieu of grade LB.

Structural clay tile comes in a wide range of colors in both rough and smooth textures. However, most structural clay tiles are either red or buff, the color not being too significant, since they are usually covered with a plaster or stucco finish anyway. The units range in thickness of from 4 to 12 in. in the following nominal dimensions: 12 × 12 in., 8 × 12 in., 6 × 12 in., 8 × 8 in., and $5\frac{1}{3}$ × 12 in.

ENGINEERING PROPERTIES

Table 2-8 lists the absorption and compressive strength requirements of structural clay tile, both load-bearing (C34) and non-load-bearing (C56), as well as for facing tile (C212). Glazed units (C126) are also listed.

TABLE 2-8

Physical Requirements for Clay Tile

	Absorption %, 1-hr boiling		Minimum compression strength[a] (lb/in.^{2})			
			End constr. tile		Side constr. tile	
Type and grade	*Average of five tests*	*Indiv.*	*Min. average of five tests*	*Indiv. min.*	*Min. average of five tests*	*Indiv. min.*
Load-bearing (C34)						
LBX	16	19	1400	1000	700	500
LB	25	28	1000	700	700	500
Non-load-bearing (C56)						
NB	—	28				
Facing tile (C212)						
Types						
FTX	9	11				
FTS	16	19				
Classes						
Standard			1400	1000	700	500
Special duty			2500	2000	1200	1000
Glazed units (C126)			3000	2500	2000	1500

[a]Compression strength based upon gross area (obtained as a product of horizontal face dimension as placed in the wall times its thickness).

Miscellaneous Requirements

Other requirements for structural clay wall tile include such items as minimum number of cells (e.g., 4-in. and 6-in. horizontal thickness tiles require at least one cell, 8-in. and 10-in. require a minimum of two cells, and 12-in.-thick tiles require at least three cells in the direction of wall thickness), minimum number of crosswebs, the minimum shell thicknesses (outer face, $\frac{3}{4}$ in.; web, $\frac{1}{2}$ in.), and the permissible overall dimensional variation of $\pm 3\%$ for any form or size of tile.

The non-load-bearing units (C56) may be used for nonstructural partitions, fireproofing, furring. They are excluded from use in earthquake zones 2, 3, and 4, because all masonry there must be reinforced masonry, which requires a structural unit. Since they are nonstructural in function, no strength requirement is listed in Table 2-8.

Structural clay facing tile (unglazed)

Type

Structural clay facing tiles are made from clay, shale, fire clay, or mixtures thereof. They are described as tile designed for use in interior and exterior unplastered walls and partitions. The manufacturing processes for structural clay facing tile fall into broad categories. One type of facing tile is referred to as *unglazed*, and it possesses a smooth or rough-textured finish. In another process, the unit is produced from fire clay, shale, or a mixture thereof to which a ceramic glaze has been applied. This is described in the section "Ceramic glazed facing tile." The unglazed unit must comply with ASTM C212, which covers unglazed tile designed for use as exposed facing material in either exterior or interior walls and partitions.

There are two types produced: FTX and FTS. Strength and asborption requirements are listed in Table 2-8. Note that where Type FTS tile is not exposed to frost action, or is used in exposed masonry walls with an outside facing of 3 in. or more of stone, brick, or other masonry, the absorption limits may be waived. The FTX type is a smooth-faced tile, low in absorption, easily cleaned, and resistant to stain. Further, it possesses a high degree of mechanical perfection with a narrow color range and is subject to minimum variations in face dimensions, whereas Type FTS may come in either a smooth or rough-textured face and a medium color range. Moderate dimensional variations are permitted, as are minor surface-finish defects. A higher maximum absorption rate is also specified. Both types are adaptable for general use in exposed exterior and interior masonry walls and partitions. Both types are produced in a variety of colors, ranging from pearl grays or creams, through the buffs and browns, and descending down a scale of reds, purples, and maroons to gun-metal black.

Classes

In addition, structural clay facing tiles are classified as Standard or Special Duty. *Standard* tiles are suitable for use in both exterior or interior walls and partitions. *Special Duty* tiles, with their heavier shells and webs, are designed for superior resistance to impact and moisture penetration. They will also support greater lateral and compressive loads than the Standard tile units. When a class is not specified, Standard tile will be provided. Special Duty tile, of course, can be accepted in lieu of Standard tile except where the extra weight becomes a load factor on the supporting members.

Since these tiles are to be used as facing material, where appearance is of paramount importance, they must have special qualities which are more stringent than for the wall tile, including those for workmanship, finish, texture, color, size variation, and distortion. ASTM C212 also places requirements on size and number of cells (different from C34) as well as shell and web thicknesses (e.g., one cell for 4-in.-thick tile, two for 6-in. and 8-in. thickness, and three for 10-in. and 12-in. units).

Ceramic glazed facing tile

The ASTM C126 specifications embrace those structural clay facing units made from shale, fire clay, or a mixture thereof, which are produced with a glaze finish (Figure 2-13). Ceramic glazing consists of a number of ingredients which, after being sprayed on a clay body, fuse together to form a glasslike coating during burning at above 1500°F, making them inseparable. This coating may occur as a clear glaze or is available in either solid colors or in various multicolor arrangements. The glazed tiles are especially useful where a decorative, impervious, and easily cleaned surface is required, besides which they can also function as a high-quality structural element. They are also used in nonbearing partitions, as facings over other supports and as components of composite walls. One of their main features stems from the fact that they exhibit little tendency to shrink. Glazed tiles are highly reflective to sound. Also scored, combed, or otherwise roughened tile surfaces are produced for the direct application of plaster. Smooth tile surfaces are also acceptable as a plaster base.

Grades and Types

There are two grades and two types within each grade covered in these specifications as follows:

Grade S (select): for use with comparatively narrow mortar joints ($\frac{1}{4}$ in.).

Grade SS (ground edge): for use where face dimension variation must be extremely small.

Type I (single-faced units): for use where only one finished face will be exposed.

Type II (two-faced units): for use where two opposite finished faces will be exposed.

The basic ASTM requirement specifies compressive strength only (Table 2-8). Other ASTM conditions covered include such items as properties of the ceramic finish, color and texture, number of cells, shell and web thicknesses, and dimensional tolerances. Typically, glazed structural units are 6×12 in. in nominal face size, with the maximum permissible variation in these dimensions set at 2%. The minimum end and side face shell thickness is set at $\frac{3}{4}$ in.

6M-Series

Glazed structural wall units are modular dimensioned to lay in the wall with a nominal dimension of 6″ x 12″ including $\frac{1}{4}$″ joint. Net dimensions are shown.

Stretcher group

M-1
M-2
2 Face
M-3
M-4
M-6
M-7 1 Face
M-8 2 Face
M-7K
Kerf
M-1
M-2
6″ Wall
(2-Face)

Kerfing: Pieces marked "Kerf" are shipped to be separated at the job, or used as full pieces. A tap with a hammer will split the pieces at the kerf line.

If unglazed face is to be exposed, specify smooth-backed units.

Group I shapes

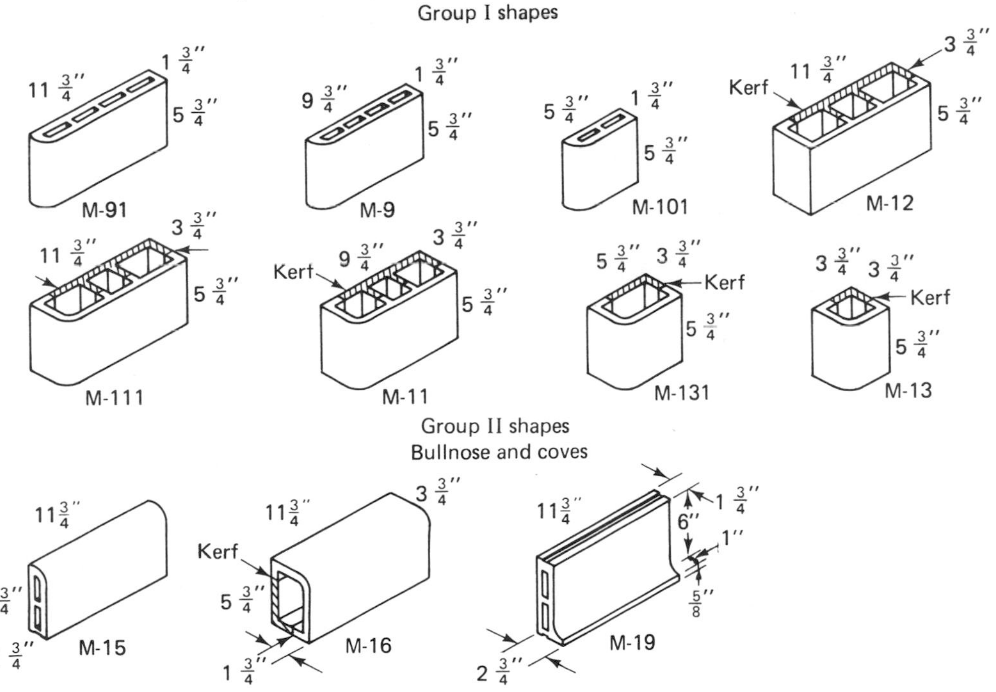

FIGURE 2-13. This is a drawing showing how western users have adapted the glazed structural units for use with reinforcing.

Some confusion has existed as to what allowable design stresses to use. Both those assumed for "Hollow Units—Grade LB" and those for solid units (grade MW brick) have been used. It would seem appropriate to permit the higher stresses, since the hollow units need develop only 1000 lb/in.^{2} on the gross area, whereas clay facing tile units must, according to ASTM C126, develop at least 2000 lb/in.^{2} on the gross area. Since the net area is a fraction of the gross area, the material strength is actually extremely high.

TERRA-COTTA

Architectural *terra-cotta* (referred to as *ceramic veneer*) also falls into the ceramic glazed classification, although it has only nonstructural uses. It is largely a custom-made product, used widely as an ornamental veneer for both exterior and interior walls.

Terra-cotta may be either machine-extruded or hand-molded, and the individual pieces are generally larger in face dimensions than brick and most facing tile. Terra-cotta is essentially made to order.

There are three general types of terra-cotta: anchored, adhesion, and hand-molded. The first two are masonry slabs, which are used as the finished faces of walls, and the terms "anchored" and "adhesion" indicate the methods used in attaching the ceramic veneer to the structural wall. Both anchored and adhesion slabs are machine-extruded. Adhesion-type slabs are $1\frac{5}{8}$ in. maximum thickness and have a maximum face area limit of 720 in.^{2}. Anchored-type slabs are usually thicker (2 to $2\frac{1}{2}$ in.) and are available in larger face dimensions than the adhesive type.

Gypsum tile

Gypsum tile (ASTM C52) has long been used as partition tile, for it has served very satisfactorily as loft or office partitions. Walls composed of gypsum tile can be reinforced for lateral forces normal to the plane of the wall by the application of mesh and plaster, but they lack adequate strength to be used in bearing walls and other structural applications. A typical tile size might be 12 in. high by 30 in. long, available in 4, 6, 8, and 12 in. thicknesses. Often the unit consists of gypsum combined with sawdust to lighten the weight of the unit. They are economical not only to produce, but also to lay up, because of their light weight. Further, this tile possesses good sound-deadening and heat-insulating properties. They can be readily removed during building alterations and, being relatively soft, the units can be easily cut and fitted around openings or other obstructions. For this reason gypsum tile will undoubtedly continue to be used, to some extent, in alteration work and in some forms of new construction.

A disadvantage that they possess stems from one of their basic physical properties (i.e., undue capacity to absorb moisture). Because of this characteristic, they are readily subjected to a loss in strength which obviously precludes their exterior use. Even where used for interior partitions, the first course often consists of clay tile, with gypsum being used above this floor course. In these applications, the gypsum tile is laid up with a gypsum mortar, similar to that used in plaster. The mix generally consists of three parts sand to one part gypsum with sufficient water to make for good workability.

Paving brick

Paving bricks are currently used in steps, sidewalks, and platforms, where the wear normally stems from light traffic only, in contrast to their function in carrying the heavy iron-wheeled equipment of earlier times. See Figure 2-14 for some typical examples of these. This present usage dictates that either ASTM C62 or ASTM C216 be used as a paving brick specification, with the following modifications on grade, size, and color:

Grade: this should stipulate only the SW grade as defined under either the building brick or facing brick specifications. The strength and durability of the SW grade is most acceptable under most conditions of use. The strength of the SW specification has been upgraded for some types of pavers in use today.

Size: the size of units to be used can be spelled out to fit the desired pattern. It is also important to state the acceptable dimensional tolerances, particularly if the paving bricks are to be used in tight pattern.

Color: the color will generally be a terra-cotta red. However, the color must be agreed upon prior to delivery by both the buyer and the seller.

Sound-absorbing clay masonry

SCR Acoustile is a relatively recent form of structural clay facing tile. It is a structural material that combines a relatively high sound absorption capacity with low sound transmission, both factors being desirable acoustical properties. Furthermore, strength has not been sacrificed in favor of these acoustical properties. Typical units are shown in Figure 2-15. The distinguishing characteristics of these structural clay facing tile units are the numerous perforations extending through the lightweight porous faces, and the fibrous glass pads behind the faces, which are excellent absorbers of sound. The perforations may be circular or slotted, uniform or variable in size, and regular or random in pattern. Sound is thus absorbed by mechanically converting it into heat. The acoustile unit is highly resistant to dirt, abrasion, fire, and ordinary wear and tear. Its strength qualities are such that it can be used as a load-bearing structural wall tile. The fibrous glass pad serves an additional purpose in that it also acts as an insulating material, thereby reducing the transmission of heat.

Observe that sound absorption and sound reflection are directly related. Sound that is not absorbed is reflected. If at a given frequency, a particular material absorbs 65% of the sound striking it, it will reflect the remaining 35%. Sound reflection is therefore extremely important in acoustical design. If an excess of absorptive

San Bernardino City Center Mall

FIGURE 2-14. Paving brick types.

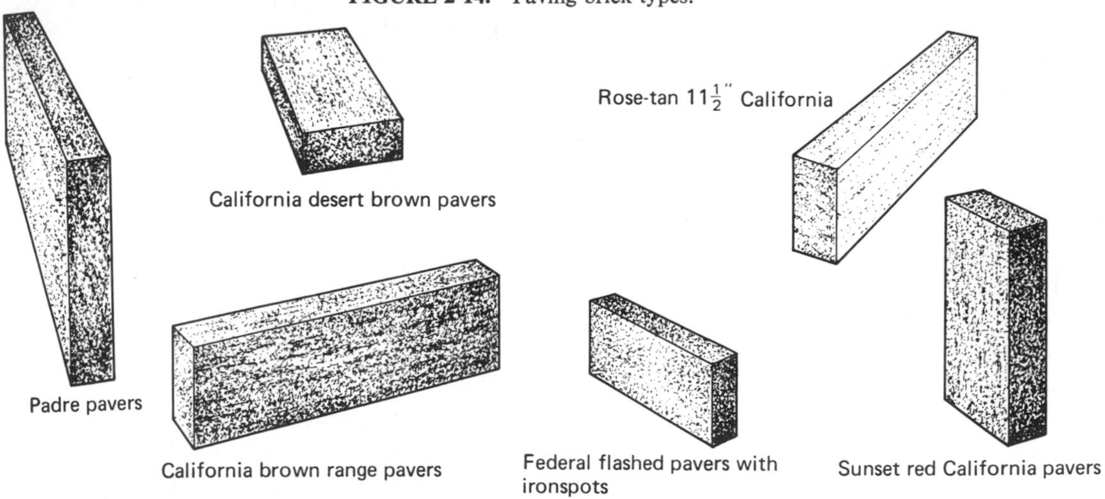

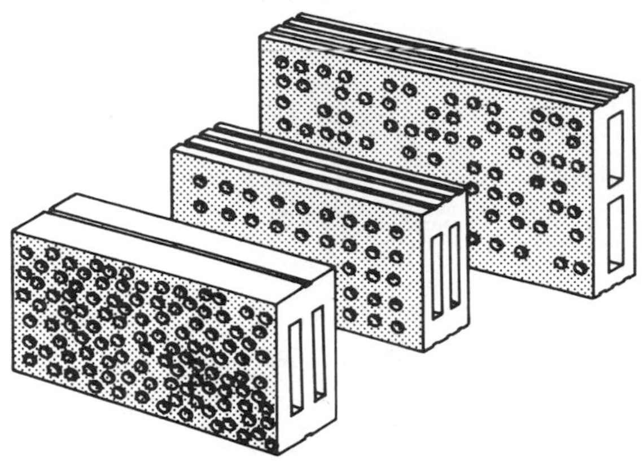

FIGURE 2-15. SCR Acoustile.

materials is used, or if they are improperly concentrated, the resultant arrangement will tend to deaden the sound. So it would seem that brick or structural tile units which have low absorption qualities or high reflective qualities can be successfully combined with the SCR Acoustile to achieve optimum results in acoustical design. Sound transmission tests at the River Bank Laboratory in Geneva, Illinois, for example, show that a 4-in. Acoustile wall will have a sound transmission of at least 53 db when plastered on the back with $\frac{1}{2}$-in. sand gypsum plaster.

Special shapes

The previously described units are typically found, with minor modifications for unique local requirements, throughout the country. In addition to these, however, a myriad of nonstandard types, shapes, and sizes may be encountered in most areas, in such nonstructural applications as fences, grilles, and sun-screens.

Details as to size, color, and texture must therefore be obtained from the local manufacturer of these specialty products.

WALL PATTERNS

Pattern bonds refer to the varied arrangements of the masonry unit's position in the wall, as well as to the texture or color used in the wall face to achieve the desired architectural effect. The six basic brick positions may be described by the terms stretcher, header, soldier, shiner, rowlock, and sailor. These are pictured in Figure 2-16. Figure 2-17 shows the terms used to describe the various parts of a brick wall. Pattern arrangement modifications may be introduced by the treatment of the mortar joint, or by the projection or recession of various bricks from the face of the wall to create a distinctive wall texture that is not entirely dependent upon the texture of the individual masonry units. Or units can be left out completely in random locations to form perforated walls or screens.

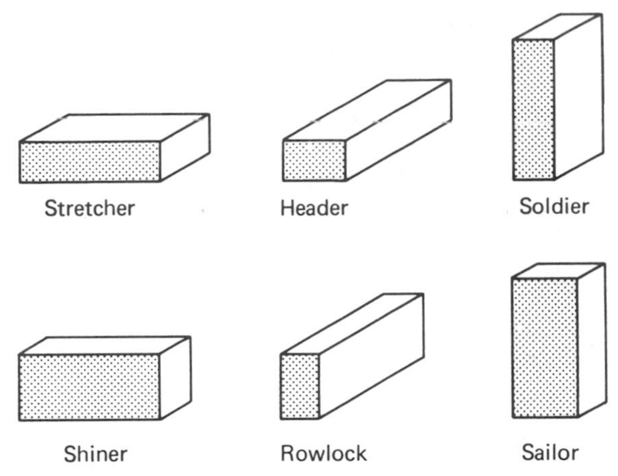

FIGURE 2-16. Terms applied to various brick positions.

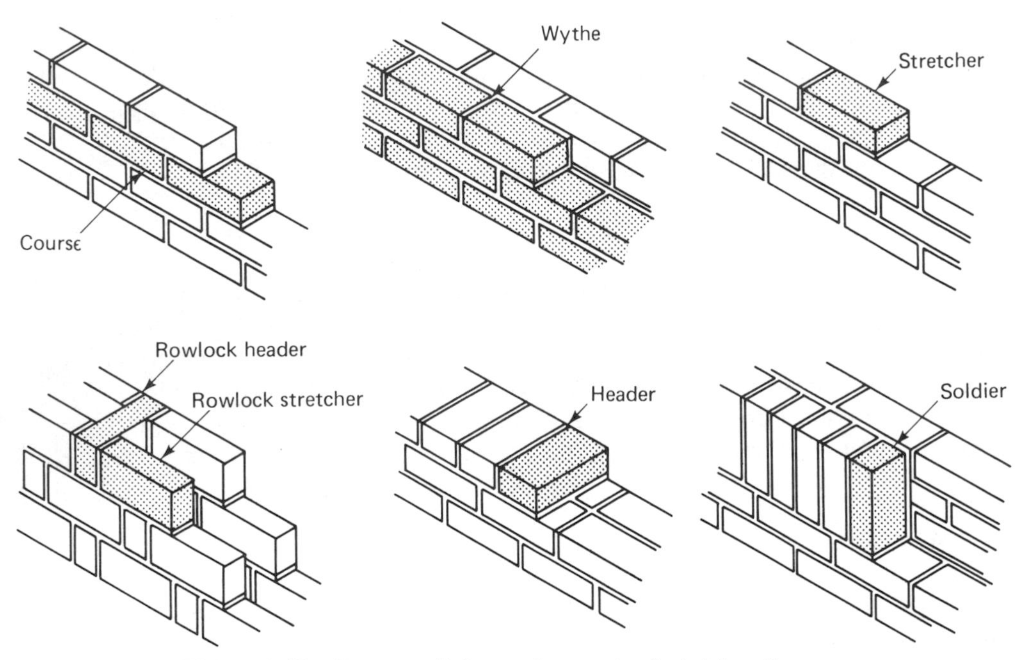

FIGURE 2-17. Terms applied to various parts of a brick wall.

Types

The most commonly used pattern bonds are running, common or American, English, Flemish, and stack, illustrated in Figure 2-18. Use of these bonds, or variations thereof, combined with different color and texture arrangements, will produce an unlimited number of wall patterns that will achieve almost any architectural effect desired.

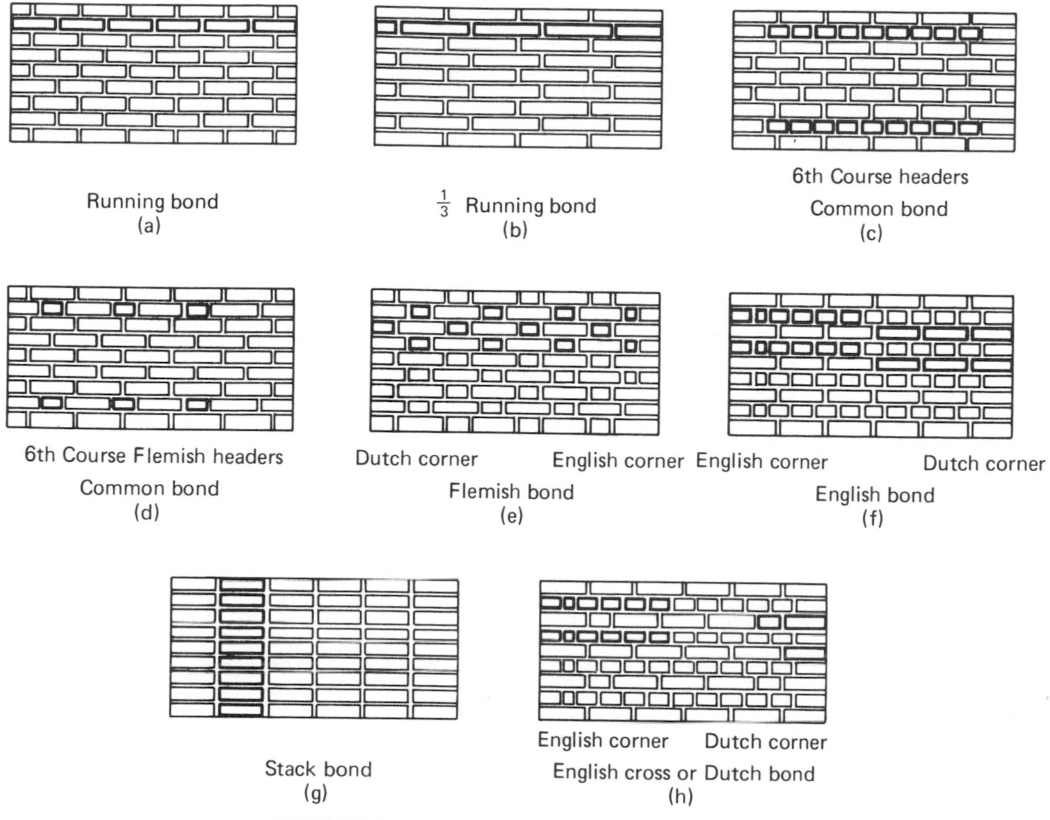

FIGURE 2-18. Traditional pattern bonds.

Running bond is the simplest of the basic patterns. Its variations include $\frac{1}{2}$ bond, $\frac{1}{3}$ bond, or $\frac{1}{4}$ bond, depending upon the amount of lap. It consists of all stretchers, wherein the head joints in adjacent courses are staggered. Running bond is probably the most commonly used of the traditional pattern bonds. *Common or American bond* is a variation of running bond, with a course of headers at regular intervals. These headers provide both a structural bond and a pattern variation. Also the common bond may be varied by using a Flemish header course. *Flemish bond* consists of courses of brick made up of alternate stretchers and headers, with the headers in alternate courses centered over the stretchers in the intervening courses. Variations in Flemish bond may be obtained by increasing the number of stretchers between headers in each course. "Snap or clipped" headers consist of half-brick substituted for full headers.

English bond is composed of alternate courses of headers and stretchers. The headers are centered on the stretchers and the joints between stretchers in all courses line up vertically. "Snap" headers may also be used. *English cross* or *Dutch bond* comprises a variation of English bond which differs only in that the vertical joints between the stretchers in alternate courses do not line up vertically. These joints center on the stretchers themselves in the courses above and below.

Stack or *block bond* is a purely patterned bond. There are no overlapping units, since all head joints line up vertically. This vertical alignment therefore requires dimensionally accurate units, or carefully prematched units for each vertical joint alignment. A manufacturer, for instance, may deliver brick from a particular batch located in a specific spot in a kiln in order to minimize dimensional variations between batches where a stack bond pattern is utilized. Say that a 12-in.-long unit is permitted a $\pm\frac{3}{8}$-in. tolerance; this could produce a difference of as much as $\frac{3}{4}$ in. between one unit and another. In stack bond this could not be tolerated, since the mortar joint is usually somewhere between $\frac{3}{8}$ and $\frac{5}{8}$ in. in thickness.

Whereas pattern bond refers to a distinctive architectural effect, *structural bond* is the term used when the brick units are interlocked or positioned, either in the plane of the wall or transverse to it, to achieve a structural tie. This may be accomplished in several ways, such as with an overlapping of the brick units, or in the use of metal ties embedded in the joints or grout core. The overlapping simply consists of alternating courses of headers and stretchers (English bond) or of alternating headers and stretchers within each course itself (Flemish bond). Thus, the stretchers develop longitudinal (in-plane) bonding strength, while the headers, being laid across the width, provide transverse bonding. The use of ties is generally permitted by most building codes and may consist of small-diameter wire placed across the width of the wall at a specified spacing.

MORTAR JOINTS

Classes

The treatment of mortar joints in the face of the wall also affects the wall pattern and texture. Figure 2-19 shows the basic terms applied to mortar joints. Mortar joint finishes fall into three classes; cut, troweled, and tooled. In the *cut joint*, the mortar is merely cut off flush with the trowel. In the *troweled joint*, the excess mortar is cut off and finished with a trowel, or "struck." In the *tooled joint*, a special tool is used to compress and shape the mortar in the joint.

FIGURE 2-19. Basic terms applied to mortar joints.

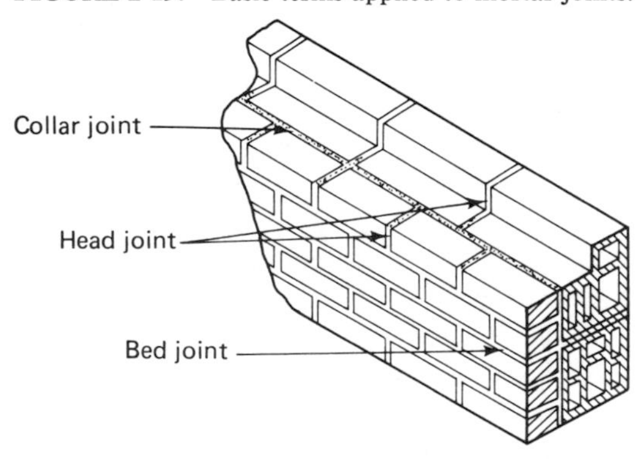

Shapes

Figure 2-20 shows a cross section of some of the typical mortar joint shapes. *Concave* and *V-shaped joints* are formed by the use of a jointing tool. Because they have been compressed in tooling, these joints are rather waterproof, and are thus recommended for use in areas subjected to heavy rains and high winds. The weather joint requires care, since it must be undercut. It sheds water readily, making it fairly watertight.

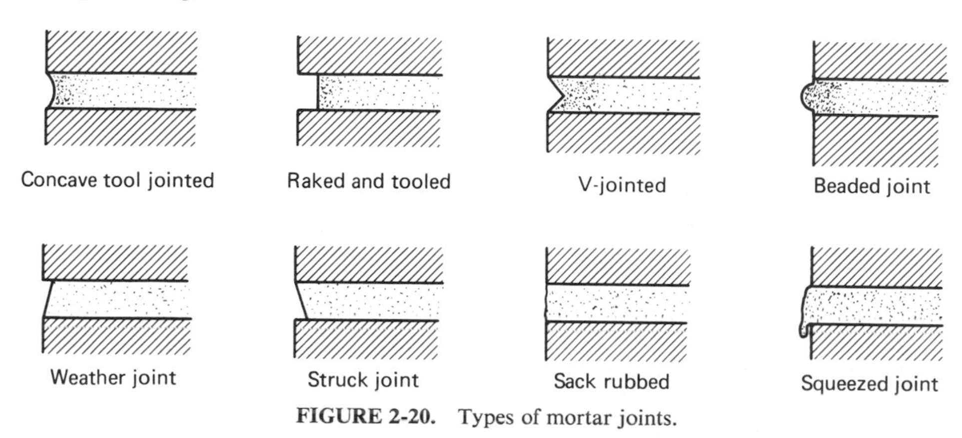

FIGURE 2-20. Types of mortar joints.

The *flush joint* is the simplest of all the joints to form, since it is made by simply holding the edge of the trowel against the brick and cutting off the excess mortar. This procedure does produce an uncompacted joint however, and further, a small hairline crack may form where the mortar is pulled away from the brick by the cutting action. It should be noted that this joint is not always watertight.

The *struck joint* is a commonly used one in brick masonry. It is a relatively easy one to strike with a trowel. Some compression occurs, but the small ledge resulting from this technique does not readily shed water; thus, it produces a more permeable joint.

The *raked joint* is formed by removing the mortar, while it is still soft, with a raking tool and then troweled with a narrow trowel. Although the joint is compacted by this latter action, it is still somewhat difficult to make it watertight. This joint, with its marked shadows, tends to darken the overall appearance of the wall, thereby achieving a distinct architectural character.

The *squeezed joint* is formed by simply forcing some of the excess mortar out beyond the face of the wall. This gives a rough-hewn appearance and is reminiscent of masonry joinery of an earlier day. It is sometimes called a "weeping joint."

Colored mortars may also be used in different ways to enhance the patterns of masonry. One method simply consists of mixing the coloring admixture in the bedding mortar itself so that the entire joint is colored. Or the tooled joint may be formed by tuck pointing. In this technique, the wall is completed with a deep raked joint, and then colored mortar is filled in later with a special tool.

QUESTIONS

2-1. List the most significant physical properties of a clay suitable for the manufacture of brick and tile, and state why these properties are important.

2-2. Explain the three stages of the fusing process of clay.

2-3. Why must viscous fusion be prevented from occurring?

2-4. Does the color of finished brick vary? If so, why?

2-5. What are the various processes employed in the manufacturing of brick?

2-6. In what forms may water be found in the brick before the bricks are kiln-dried?

2-7. What factors affect the amount of shrinkage that will occur in brick during the manufacturing process?

2-8. Why are actual brick dimensions smaller than the nominal dimensions?

2-9. What are the differences in grades for the building bricks?

2-10. What is the weathering index?

2-11. Explain a freezing-cycle day.

2-12. Explain the C/B ratio and why it is a durability factor.

2-13. The rate of initial absorption has what effect on the bonding capabilities of brick?

2-14. Why is the ultimate strength of a prism (f'_m) less than the clay unit itself?

2-15. Why are bricks often cored and how much coring is allowable in both solid and hollow bricks?

2-16. What variables affect the compressive strength of brick?

2-17. Why is gypsum tile good as a partition tile, but inappropriate for exterior use?

2-18. Distinguish between running, stack, and common bond.

2-19. How do tooled and troweled joints differ?

2-20. What are some examples of watertight joints and what makes them watertight?

2-21. Why shouldn't an attempt be made to relate the color or tone of the brick to some of its physical properties, such as compression strength or absorption qualities?

2-22. Upon what does the true color and texture of burned clay depend?

2-23. What is the effect of iron on the final color of the brick masonry unit?

2-24. List the various types of clay brick textures.

2-25. Distinguish between water absorption and the initial rate of absorption for clay brick.

2-26. Why is the initial absorption rate so significant?

2-27. In what other way are brick classified besides on the basis of the weathering index?

2-28. For what purpose is facing brick produced?

2-29. What are a hollow clay bricks and how are they classified?

2-30. Where is structural clay load-bearing wall tile used?

2-31. In ASTM C34 distinguish between the two grades of structural clay tile, LBX and LB.

2-32. What types of units are covered under ASTM C212 and for what purpose are they used?

2-33. What is architectural terra-cotta?

2-34. What is covered under ASTM C62 and where is it used?

2-35. What are the significant properties of SCR Acoustile?

2-36. Regarding the various brick positions, what is a stretcher; a header; a soldier?

2-37. What is tuck pointing?

3

Concrete Masonry Units—Material Properties

Concrete masonry embraces those units molded to various sizes—from a mixture of portland cement, and normal-weight or lightweight aggregate. These aggregates can consist of sand, gravel, crushed stone, air-cooled and expanded blast furnace slag, coal cinders, expanded shale, clay or slate, volcanic cinders, scoria, or pumice. The sizes and shapes vary in different localities, but the one becoming increasingly dominant in the western United States is two-cell hollow concrete block. One reason for this, of course, is that reinforcing steel can be readily placed and grouted within the cells. Hollow load-bearing units, often referred to as concrete block (ASTM C90), one of the newer types of masonry products—solid load-bearing units (ASTM C145), and concrete building brick (ASTM C55) are produced throughout the United States. A hollow non-load-bearing unit (ASTM C129) is also available.

RAW MATERIALS

Hollow load-bearing concrete masonry units must conform to ASTM C90. They are formed from the following raw materials, which must also conform to the individual specifications noted:

1. *Portland cement:* ASTM C150 or C175 (Air Entraining Portland Cement) or C595 (Blended Hydraulic Cement).

2. *Hydrated lime:* ASTM C207 Type S.
3. *Pozzolans:* ASTM C618 (Fly Ash or Raw Calcined Natural Pozzolans)
4. *Other constituents:* air-entraining agents, coloring pigments, integral water repellents, and so on, must be previously established as suitable for use in concrete and either conform to ASTM Standards, where applicable, or be shown by test or experience not to be detrimental to the durability of the concrete.
5. *Aggregates:* normal-weight aggregates, ASTM C33; lightweight aggregates, ASTM C331.

METHODS OF MANUFACTURE

The basic production process for concrete block is somewhat different from that of normal concrete because the block mixture, of necessity, is relatively dry. It is not poured, but rather is placed in the molds. Modern block manufacture is highly automated (Figure 3-1). First, the mold box is automatically filled with the correct predetermined mixture of solid ingredients combined with just enough water to achieve complete hydration. The material in the mold is vibrated under pressure for a controlled period to ensure solid compaction. Higher-strength blocks are produced

FIGURE 3-1. Block manufacturing machine. Photo shows a typical automated block machine. The hoppers on the lift provide material which is placed in the molds and vibrated. The molds are withdrawn and the block moved out onto a steel pallet. The pallet moves on a conveyor (pressed down for slumped block) and onto a lift that places the pallet in a rack to be removed for curing.

by subjecting the mix to increasing compaction periods. The block is then stripped from the mold and delivered in multiple units for curing either under normal atmospheric conditions or by autoclaving or steam curing. This type of operation results in a product of uniform quality, texture, size, and shape. Slumped blocks are produced in a similar manner except that after being removed from the mold, the block is squashed down with a steel plate. This procedure gives the faces of the unit a very irregular bulging appearance, presumably simulating an adobe-type texture.

ASTM DESIGNATIONS FOR CONCRETE MASONRY UNITS

The ASTM designations for concrete masonry units incorporate the following categories:

1. Hollow load-bearing concrete masonry units: ASTM C90 (and UBC Standard 24-4).
   Grade N—for general use in walls above and below grade which may or may not be exposed to moisture or weather.
   Grade S—limited to above-grade exterior walls with weather-resistant protective coatings or walls not exposed to weather. Note that in C90-59 the grades were designated as A and B and in C90-64 and 66 as U and P.
   The two types of concrete masonry units within each of the two grades are:
   Type I—Moisture-controlled: units must conform to the moisture requirements given in Table 3-1.
   Type II—Non-moisture-controlled: units need not conform to the moisture requirements specified in the table.
   Also, there are three weight classifications for all the units described herein: lightweight (105 lb/ft^3 maximum oven-dry weight), medium weight (105 to 125 lb/ft^3), and normal weight (125 lb/ft^3 minimum).
2. Solid load-bearing concrete masonry units: ASTM C145.
   At least 75% of the gross cross-sectional area of these units must be solid.
   Grades N and S—same as above. Note that each grade may be further classified as Type I or II, as described above.
3. Concrete building brick: ASTM C55.
   This is a completely solid unit, similar in both size and shape to the clay building brick, having a considerably higher compressive strength requirement than the C145 units.
   Grade N—used as architectural veneer and facing units in exterior walls and in walls where high strength and resistance to moisture penetration or severe frost action is desired.

Grade S—for general use where only moderate strength and resistance to frost action or moisture penetration is desired.

Note that each of the two grades is further classified as Type I or II, as described above.

4. Non-load-bearing concrete masonry units: ASTM C129.

   Both hollow and solid non-load-bearing units are covered under this specification. They are intended primarily for use in non-load-bearing partitions, but they can be used, under certain conditions, in non-load-bearing exterior walls above grade, where effectively protected from the weather. Further, these are classified as Type I or II, as previously described.

TABLE 3-1

Moisture-Content Requirements—Concrete Masonry

	Moisture content, maximum percent of total absorption (average of three units)		
	Humidity conditions at job site or point of use		
Linear shrinkage (%)	*Humid*[a]	*Intermediate*[b]	*Arid*[c]
0.03 or less	45	40	35
From 0.03 to 0.045	40	35	30
0.045 to 0.065, max.	35	30	25

[a]Average annual relative humidity above 75%.

[b]Average annual relative humidity 50 to 75%.

[c]Average annual relative humidity less than 50%.

HOLLOW LOAD-BEARING UNITS

Physical properties

COLOR AND TEXTURE

Regular concrete masonry units are cement gray in color. However, if so specified, they can be furnished in a myriad of colors, ranging from black to white. The desired color is achieved at the time of manufacture through the addition of the proper color pigment. One of the most desirable features of concrete block, from an architect's or owner's point of view, is the variety of colors, plus the form and texture patterns featured by concrete masonry units. Ranging from the virtually smooth finish of glass block to the rough-hewn projections of slumped block, almost any desired type of surface texture in between these can be had to achieve the desired architectural effect.

In addition, different aggregates will offer texture variations, or these can be achieved also by sand blasting. A *split-face block*, on the other hand, provides an entirely different type of surface, as shown in Figure 3-2. It is formed by splitting the face plane of the block. This technique provides a most distinctive character because of a striking multicolored matrix of aggregate particles. The blocks are usually made in pairs and split apart, forming two finished faces. Another type is produced with a very thick face shell, and the face is then split off to provide the surface with its distinctive texture. The solid piece split off may be used as a veneer facing unit.

(a) (b)

FIGURE 3-2. Split-face units: (a) end of split corner block; (b) example of half units after splitting. (Courtesy W. L. Dickey.)

Form and Dimension

Today the most commonly used concrete masonry units are manufactured with two full cores in several different overall dimensions. Common block thicknesses are 4, 6, 8, and 12 in. Also available on special order is 10- and 16-in. block. The common length is 16 in., but 12- and 24-in. lengths are available on special order. The units shown in Figure 3-3 represent the most commonly used forms, and their dimensions are generally taken as the basic standards. The 8-in.-thick units are shown, but the others have similar dimensions. Special pilaster and column block are also produced, as shown in Figure 3-4. When specifying a block size, it is common practice to give the width, height, and length, in that order. For instance, the nominal dimensions of

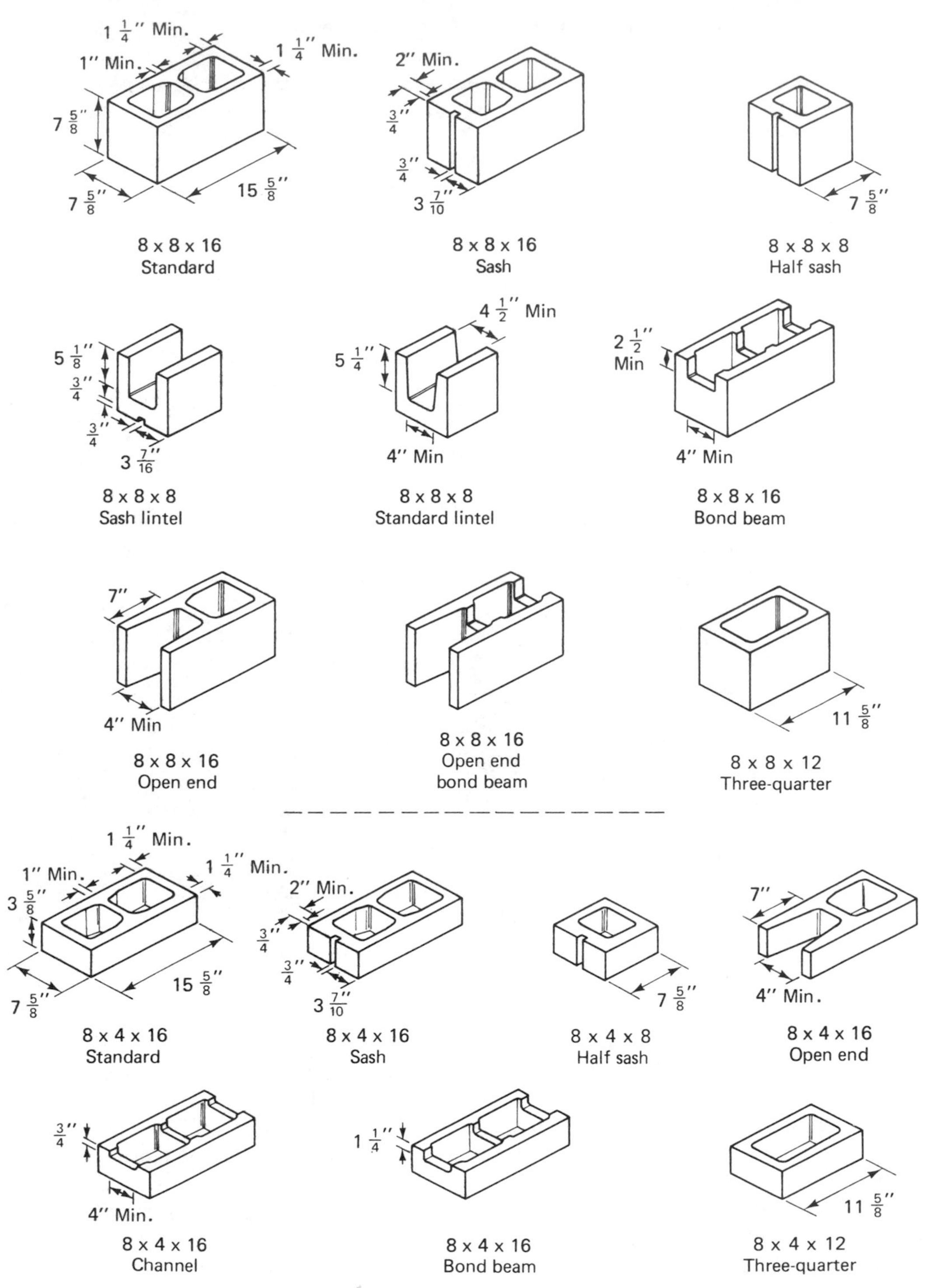

FIGURE 3-3. Concrete block dimensions (8-in. thickness).

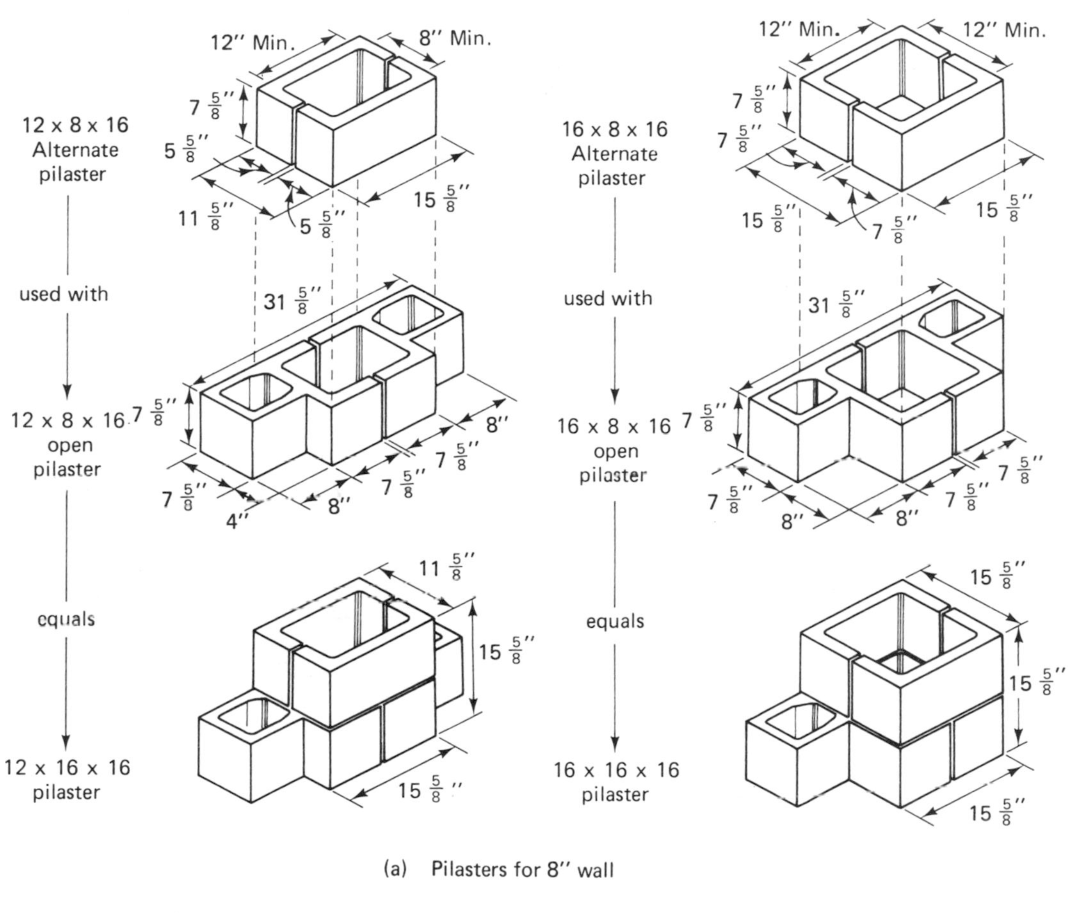

(a) Pilasters for 8" wall

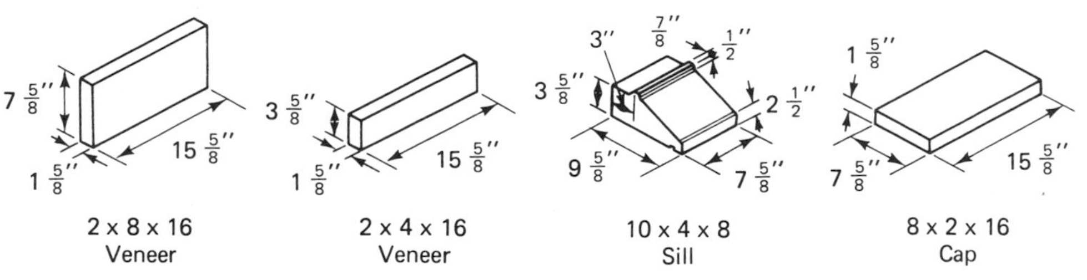

(b) Accessory blocks

FIGURE 3-4. Pilaster and accessory block units.

an 8 × 4 × 16 standard block are 8 in. wide × 4 in. high × 16 in. long. Observe, however, that the actual height, width, and length dimensions are $\frac{3}{8}$ in. less, to allow for the typical mortar joint thickness, thereby producing the modular dimensions so necessary for convenient wall layout. However, some special units, such as fence block, are made to even dimensions, such as 18 in. long. Except for units with special features, such as slump or split face, no overall dimension may differ by more than $\frac{1}{8}$ in. from these specified standard dimensions. However, typically, most manufacturers adhere to closer tolerances; for if the maximum were to continuously occur, it would be difficult to achieve good workmanship. The minimum face-shell and web thicknesses as are shown in Table 3-2. An exception to these requirements occurs in open-end or

TABLE 3-2

Minimum Thickness of Block Face Shell and Webs

		Web thickness	
Nominal width of units (in.)	*Face-shell thickness*[a] *(min. in.)*	*Webs*[a] *(min. in.)*	*Equivalent web thickness*[b] *(min. in./lin. ft)*
3 and 4	$\frac{3}{4}$	$\frac{3}{4}$	$1\frac{5}{8}$
6	1	1	$2\frac{1}{4}$
8	$1\frac{1}{4}$	1	$2\frac{1}{4}$
10	$1\frac{3}{8}$	$1\frac{1}{8}$	$2\frac{1}{2}$
	$1\frac{1}{4}$[c]		
12	$1\frac{1}{2}$	$1\frac{1}{8}$	$2\frac{1}{2}$
	$1\frac{1}{4}$[c]		

[a]Average of measurements on three units taken at the thinnest point.

[b]Sum of the measured thickness of all webs in the unit, multiplied by 12, and divided by the length of the unit.

[c]This face-shell thickness (FST) is applicable where allowable design load is reduced in proportion to the reduction in thickness from the basic face-shell thickness shown.

bond-beam units, since they cannot comply with the web-dimension requirements. However, since these units will be reinforced, the grout will more than replace the deficient webs, thereby complying with the intent of the specification.

Note in Figure 3-5 the channel and bond-beam block units. The bond-beam unit provides for the placement of horizontal steel and free flow of grout because of the $2\frac{1}{2}$ in. or more of cross-web depressions. The channel block is not recommended for use in partially grouted walls, since, although the horizontal steel can be accommodated, it does not permit proper lateral flow of grout because of the shallow cross-web depression. It could be used where all cells are grouted, however, since no lateral grout flow occurs. The 4-in.-high bond-beam units do not always provide adequate depth for lateral grout flow either, so they are frequently placed in the wall with one turned up and the other turned down, thereby providing twice the

FIGURE 3-5. Channel and bond beam block. (Courtesy W. L. Dickey.)

grout flow area. Observe also that open-end block are needed when placing them around previously positioned reinforcement.

There are also special shapes to accommodate sash or to develop lintels. Another form has a semi-interlocking head joint which is laid up without mortar in the head joints. These will, in effect, function as two solid concrete bricks with a spacer, rather than as a hollow concrete block and are so considered by UBC. Another special shape consists of two face shells of concrete with wire cross ties instead of concrete.

Surprisingly enough, there is a top and a bottom to a concrete block, because of the fact that the cores of the block mold are tapered for easy removal. This makes for thicker face shells and cross webs at one cross-sectional surface, thereby providing for a wider bed joint at that surface, and facilitating picking up the block. This surface is considered to be the top of the block (Figure 3-6).

Engineering properties

ABSORPTION

As with brick, the absorption qualities of concrete masonry units are frequently taken as an indication of their durability. ASTM specification C140 describes the method by which the water absorption quality of block can be measured, for a determination of compliance with the moisture content and absorption limits. Actually, it is questionable whether absorption is a measure of any desirable block property. At any rate, total water absorption is observed following a 24-h period of immersion. The average of three full-size units is taken for each sampling. Table 3-3 provides the maximum UBC and ASTM limits for this property.

FIGURE 3-6. The top and bottom of a concrete block. The block on the left shows the bottom and the right side of the top of the block, wider for better mortar placement, and greater ease of lifting. (Courtesy W. L. Dickey.)

TABLE 3-3

Strength and Absorption Requirements for Concrete Masonry Units

	Compression strength ($lb/in.^2$), *gross area*		*Water absorption* (*max.* lb/ft^3), *average of three units*		
			Oven-dry weight (lb/ft^3)		
Type of masonry unit	*Average of three units*	*Indiv.*	*Over 125*	*105–125*	*105 or less*
Hollow load-bearing units (C90)					
Grades N-I, N-II	1000	800	13	15[a]	18
S-I, S-II	700	600			20[b]
Solid load-bearing units (C145)					
Grades N-I, N-II	1800	1500	13	15[a]	18
S-I, S-II	1200	1000			20[b]
Concrete building brick (C55)					
Grades N-I, N-II	3500[a]	3000[a]	10	13	15
S-I, S-II	2500[a]	2000[a]	13	15	18
Hollow non-load-bearing unit (C129)					
Types I, II	600	500			

[a]Concrete brick tested flatwise.

[b]This figure applies to lightweight units weighing less than 85 lb/ft^3.

Linear Shrinkage

The *linear shrinkage* is the change in length of a unit from the wet to the dry condition. The limits are listed in tables of ASTM and UBC Standards and range from 0.03 to 0.065%. The intent of the tables is to relate the facts that greater moisture content will cause more subsequent shrinkage, but that more-humid job sites will cause less shrinkage than will occur in arid sites. The limits of shrinkage are set numerically in the tables for varying moisture contents and varying conditions of humidity, or potential amount of evaporation represented by the relation of those factors.

The amount of potential shrinkage will be a factor in the potential amount of cracking that may occur (i.e., the more shortening, the more tendency for cracking). One factor would be variation in tensile strength (i.e., the weaker the masonry units, the more cracking). Another factor, directly under the control of the designer, is the length of masonry that is restrained. Specifically, longer lengths, permitting greater accumulation of shortening, will cause greater stress. Related to the length is wall height, the greater the height or cross-sectional area, the less the unit stress due to shortening. Also, of course, there is the additional variable of temperature change, which may be additive or subtractive. These factors of amount of water evaporation, length of wall shortening, tensile strength, cross-section of wall, and amount of restraint are currently not well enough evaluated for linear shrinkage to be a sole measure of the cracking potential of masonry. However, the values in the table can be achieved, more or less easily in different areas, and will reduce the cracking tendency from that which would occur with no limits on water content. R.E. Copeland has covered this subject in an excellent article, noted in the Bibliography, which is recommended for further reading on this subject.

Moisture Content

Test procedure must conform to ASTM C140. In addition to moisture content (%), the oven-dry weight of the block units in lb/ft^3 must be reported.

Concrete units contain a gel that expands when moist and then shrinks upon drying out. For this reason, the concrete masonry units should not be wet prior to being placed in the wall. The only time that moisture should be applied would be in an extremely hot, dry climate. Even then, only a light spray should be used, because the intent is simply to cool them down enough so that not only can they be handled comfortably, but, more important, so that the mortar does not become dehydrated too rapidly. Further, it is important that the moisture content be low at the time of laying so as to minimize subsequent drying shrinkage.

Compressive Strength

In order to establish the quality of the masonry unit, its ultimate compressive strength must be determined. The procedure follows ASTM C140. Table 3-3 gives the minimum compressive strength requirements for various types of concrete masonry units, based upon the gross area. It is desirable for the designer to seek the net area strength of the units as well, since design is based upon the net or actual strength of the material.

The ultimate strength of these individual units is sometimes mistakenly taken by designers as the ultimate compressive strength of the masonry assemblage (f'_m), upon which design stresses are based. They err in so doing, for the ultimate compressive strength of the composite block and mortar (and grout if used) assemblage must of necessity be less than that of the strength of any of the individual components. For one thing, the process of mixing mortar, "buttering" the block, and shoving it into place is bound to produce zones of variable strength, where the strength of the assemblage does not come up to that of the individual blocks constituting the wall segment. Also, these individual constituents possess different moduli of elasticity and Poisson's ratios; characteristics that are bound to reduce the strength of the composite assemblage. The procedure for evaluating the ultimate strength (f'_m) of the concrete masonry assemblage is outlined in Chapter 6.

STIFFNESS

Modulus of elasticity is not generally determined for the individual masonry units and is therefore not specified in most standards. In design, it is the modulus of elasticity of the integral masonry assemblage which is a significant parameter (E_m). Very little formal research seems to have been performed in an attempt to determine this important property either, a rather peculiar circumstance. For this reason, building codes have set forth an empirical value that has become an accepted design parameter. For example, the Uniform Code specifies that the modulus of elasticity be taken as $1000f'_m$, with a 3,000,000 lb/in.^{2} maximum, regardless of the type of masonry unit or mortar used. Current deliberations may result in this figure being revised downward to $600f'_m$. This value was established statistically, although somewhat arbitrarily, by observing the stress/strain properties of masonry assemblage test prisms. In a master stroke of overkill, some very precise laboratory evaluations and theoretical calculations, carried out to several decimal points, have been based upon this arbitrary and as yet undefined stiffness characteristic.

At present, the shear modulus for masonry, E_v, is set at $400f'_m$ or $0.4E_m$. This ratio was arbitrarily assumed to be the same as that of concrete. Some designers seem to assume, however, that this value is $400.0000f'_m$ in carrying out their computer solutions. This in spite of the fact that there exists no accepted test procedure to measure this parameter. So even a reasonable approximation of its true value is as yet virtually unknown.

TENSILE STRENGTH

Of significance in evaluating the diagonal tensile resistance of masonry is the ultimate tensile strength of masonry units. Since no ASTM standard exists for this test, the following procedure has been adopted by the California Concrete Masonry Technical Committee (CCMTC). A pair of $\frac{1}{8}$-in. steel pull plates is glued to the end web of each unit with an epoxy-type adhesive. The lap of each pull plate must be at least 3 in. Each pair of plates, in turn, is symmetrically attached to a yoke for fastening to the testing machine. In this fashion, the load is applied along the centroid of the masonry-unit face shells. A load of up to one-half of the anticipated maximum load is slowly applied, following which the remainder of the load is applied at a uniform rate within a period of not less than 1 nor more than 2 min. The ultimate tensile resistance is calculated by dividing the ultimate tensile load by the minimum cross-sectional area of the specimen lying perpendicular to the direction of the load. The results of three specimens are averaged to obtain the final ultimate tensile strength. Present information indicates that the ultimate tensile strength of grade N concrete masonry units ranges somewhere between 50 and 200 lb/in.^{2}. CCMTC requires 135 lb/in.^{2} minimum.

CONCRETE BUILDING BRICK AND SOLID LOAD-BEARING UNITS

As used currently, concrete brick construction is one of the newer types of masonry construction. Its crude origin was to be seen in the use of solid chunks of concrete sidewalk which were used in retaining walls and paving during the WPA make-work of the 1930s. But the concrete brick or solid concrete units of modern masonry arc vastly different both in appearance and in material properties. The shapes of concrete brick are similar to clay brick, and because they are cast or molded, provide the opportunity for developing a variety of patterns and designs. Note that a dual set of ASTM specifications governs the quality of these units. Thus, ASTM C55 provides for Concrete Building Brick and ASTM C145 governs the Solid Load-Bearing Concrete Masonry Units. ASTM C55 covers concrete building brick and similar solid units made from portland cement, water, and suitable mineral aggregates. On the other hand, the scope of ASTM C145 extends over solid load-bearing concrete masonry units (those with 75% or more net area, similar to a cored brick) made from portland cement, water, and suitable mineral aggregates. The three weight classifications are normal, medium, and lightweight. But these two specifications are actually very similar, with most paragraphs reading identically. The most significant difference lies in the fact that C55 requires an average strength of 3500 lb/in.^{2}, whereas C145 requires a strength of only 1800 psi. This is why it is essential that the job inspector verify that the type of unit designed into the structure be delivered to the job site.

Grades and types

In both cases there are two grades, N and S, the classification having the same meaning as with the standard concrete block. There are two types under each grade: Type I, which is moisture-controlled, and Type II, which is not. Refer to Table 3-1.

Absorption

The absorption limits in both these specifications are very similar, as can be seen in Table 3-3. This item of absorption has remained a disputed point regarding its effectiveness as a measure of quality control in the manufacture of the product. The moisture content at the time of laying is probably a much more important factor, as it is with standard concrete block. The concrete bricks must also not be laid up wet, because otherwise, upon drying, they will shrink, causing unsightly cracking later.

Shrinkage

Since these are concrete products having a cement gel, they are subject to moisture volume change (i.e., they expand when moist and shrink upon drying). Although this is not spelled out in the applicable UBC Standards, shrinkage is limited in the ASTM specification and should be controlled accordingly.

NEW CONCRETE MASONRY PRODUCTS

New and varied types of concrete masonry units have emerged recently in response to a demand for lower cost combined with high-quality masonry. One example consists of a unit with both bearing surfaces ground to eliminate the mortar bed joints. When laid, the units are butted together to eliminate the head joints as well. Another unit utilizes mortared bed joints in the traditional sense, but the beveled head joints provided on the product require no mortar. This beveled face permits the grout to flow to within $\frac{5}{8}$ in. of the face-shell surface. Still other types of construction involve the use of interlocking lugs for alignment and bond in lieu of mortar. Regardless of the type of individual concrete masonry unit utilized, however, the wall must be reinforced and grouted in the typical fashion.

Also, a surface bonding material has appeared on the market recently. It has been approved for use by ICBO. This method of construction involves stacking the block wall in a running bond pattern, with no mortar either in the head or bed joints. Both sides of the wall are plastered with about a $\frac{1}{8}$-in.-thick coating of the bonding cement, which consists of a mixture of cement, sand, and glass fibers.

ARCHITECTURAL FEATURES

The wide variety of concrete masonry units available today reflects a resurging emphasis on the textured pattern in architectural design, veering away from a severity of plainness which typified many designs in the past. Modern architectural concepts in masonry now reflect warmth and glamor, and sometimes even the exotic. This versatility in design can be achieved through the clever use of the many types of concrete masonry products on the market today, either with a single type of unit or with combinations of units.

Underscoring this design versatility are the numerous architectural units, with their wide range of textured patterns. Examples of these include scored or multiscored units which are marked by a shadow line or lines extending vertically or horizontally along the face. In the offset units, one half of the face extends beyond the other half. This breaks up the sterility and monotony of long walled surfaces by producing a "reveal" pattern. A natural "rough stone" texture is obtained with split block units. The resin face units offer a smooth satin surface. "Hi-lite" units are characterized by their pyramidal projections molded into the face of the unit. "Shadowal" blocks have depressions on the face shell which create a triangular shadow effect. They can be laid up in countless ways to form either panel patterns or continuously flowing wall sweeps. The "Norman" block is a brick-shaped concrete masonry unit having a ripplelike texture. It consists essentially of two dense solid concrete masonry units, 2 in. in thickness, with a depressed cross spacer, which separates the units far enough apart to form a grout core. The cross spacer assures accurate alignment and provides a positive tie between the two units. "Calbrick" is a split-faced, brick-sized veneer approximately $2\frac{1}{4} \times 4 \times 8\frac{1}{4}$ in. in size. It produces a highly unique architectural effect. "Viking stone" is a precast concrete stone which causes a craggy rough-hewn surface effect. "Slump stone," with its dimensional variations and uneven surfaces, results in a rough, uneven texture with attractive shadow patterns.

The most recent additions to the concrete masonry family are screen blocks and grilles. These have been developed for decorative or nonstructural use. They are manufactured in standard face or sculptured design. Sizes vary over a wide range of dimensions to meet almost any decorative need. They are available in a wide variety of colors as well.

The list above represents only a few of the wide variety of concrete masonry products produced in the United States at this time. A few examples of these are seen in Figure 3-7.

WALL PATTERNS

Architectural variations can also be created by laying up the block in different ways. The most widely used wall patterns are shown in Figure 3-8, and these include running or common bond, coursed ashlar, random ashlar, and stack bond. A running bond pattern, as with brick, is formed by staggering the head joints of the masonry units. In the coursed ashlar pattern, the alternate layers consist of different height units, such as 8×16 combined with 4×16. A variation of this would place a pair of 4×16 units adjacent to the primary size, 8×16. In either case, the head joints are staggered as in running bond. As the name implies, in the random ashlar pattern, the various sizes, such as 8×16, 8×8, 4×16, and 4×8 units, are mixed and laid up in a random pattern. In any case the masonry units must always have direct vertical core alignment to permit the proper placing of the reinforcement and grout. Where continous horizontal mortar joint lines are shown, it is possible to use $\frac{7}{16}$-in. or $\frac{1}{2}$-in. mortar joints, but in the more complex ashlar patterns the standard $\frac{3}{8}$-in. modular mortar joint must be used. In the stack bond pattern, the units are all laid up with the head joints aligned. The stack bond effect may also be obtained by using a scored block laid up in a running bond pattern. Another wall pattern variation is achieved by projecting or recessing standard blocks slightly. Other effects may be produced by projecting a block at right angles to a wall so as to expose the open core, giving a shadow effect. It is possible also to run what normally would be vertical and horizontal coursing at a 45° angle to achieve a basketweave or diagonal stacking pattern.

(a)

(b)

FIGURE 3-7. Architectural textures. These are two examples of the many textures and surfaces that may be achieved with concrete block. These may be grooved, ribbed, beveled, smooth, rough, slumped, and so on, in a wide varitey. (a) A split of rough texture surface. (b) A view of a tower faced with ribbed split texture. [(a) Courtesy W. L. Dickey; (b) courtesy Masonry Institute of America.]

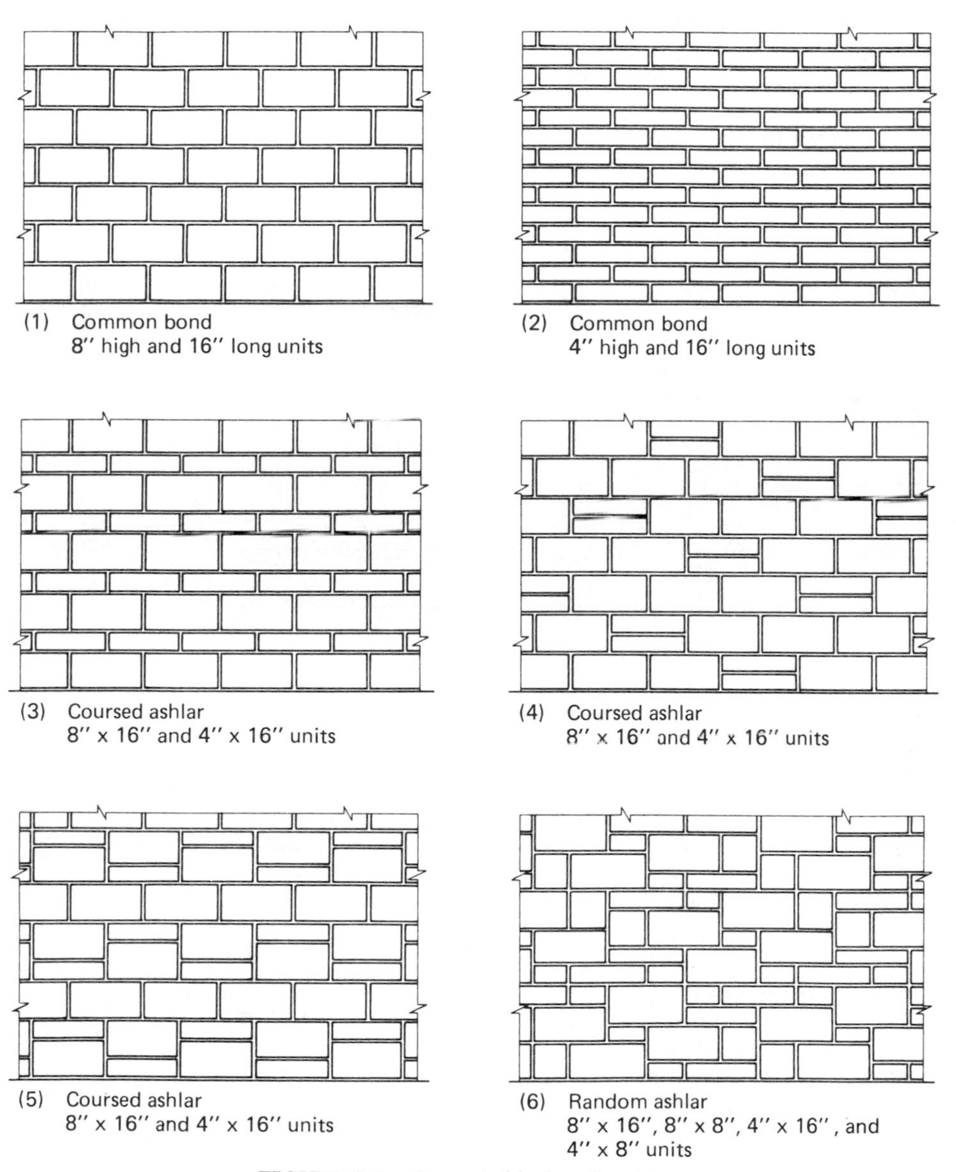

FIGURE 3-8. Concrete block wall patterns.

MORTAR JOINTS

Mortar joints are either trowel- or tool-formed. Several of the possible joint treatments are illustrated in Figure 3-9. These include the V, weather, concave or deep concave, flush, raked, squeezed or extruded, and struck joints. The choice will depend upon the function of the wall and any special architectural effect desired. V-shaped joints, commonly used, are usually neater in appearance and have very sharp

FIGURE 3-9. Mortar joint types—concrete masonry.

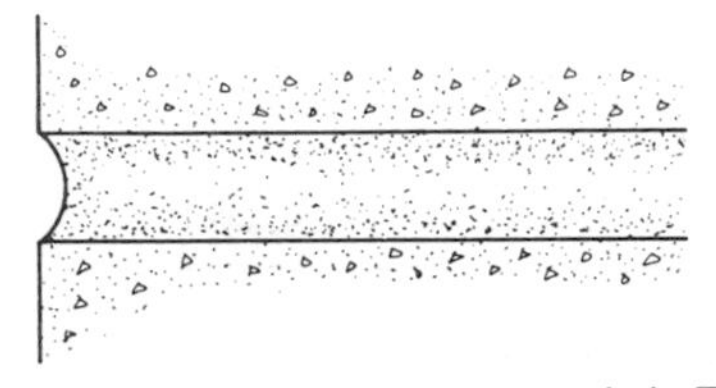

(1) Concave joint–recommended. Tooling works mortar tight to produce a good weather joint

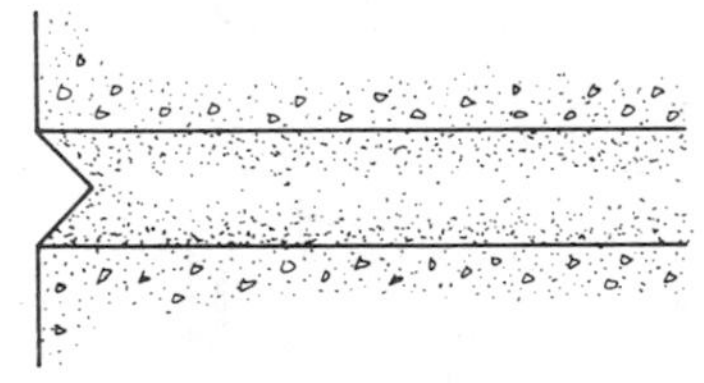

(2) V-joint–recommended. Tooling works mortar tight to produce a good weather joint

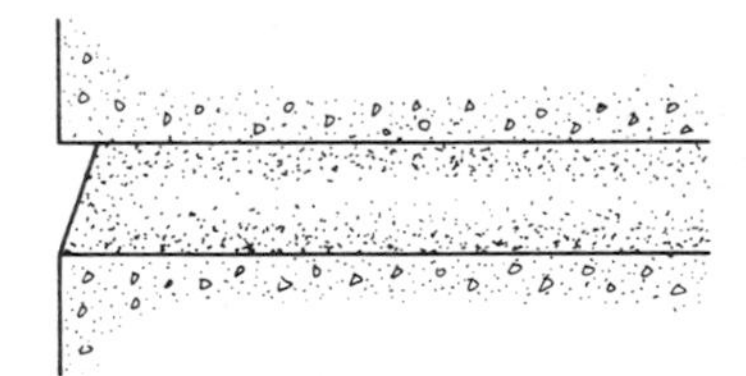

(3) Weather joint–recommended

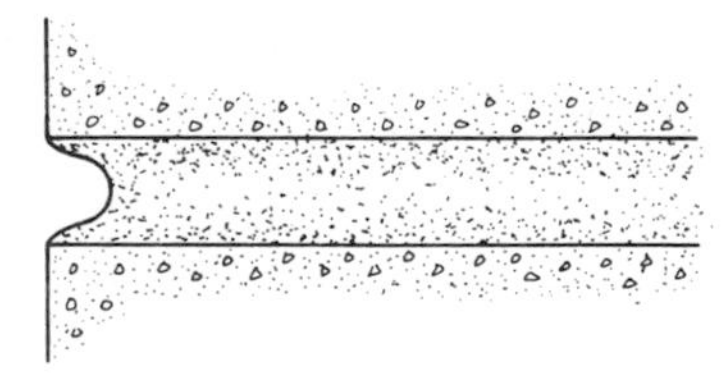

(4) Deep concave joint–recommended for special effect or to give the appearance of a raked joint

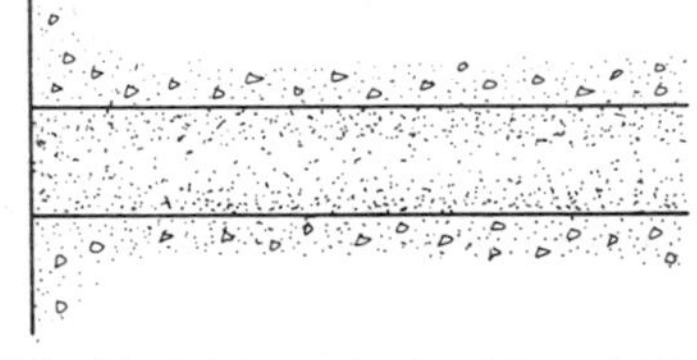

(5) Flush joint–obtained by rubbing carpet faced wood float against wall–special care required to make joint weathertight and clean

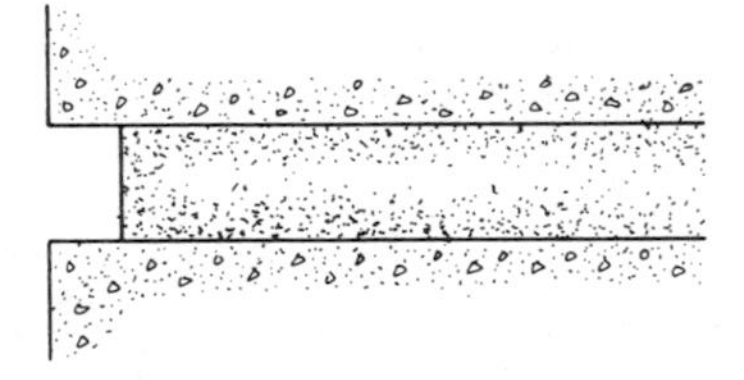

(6) Raked joint–not recommended

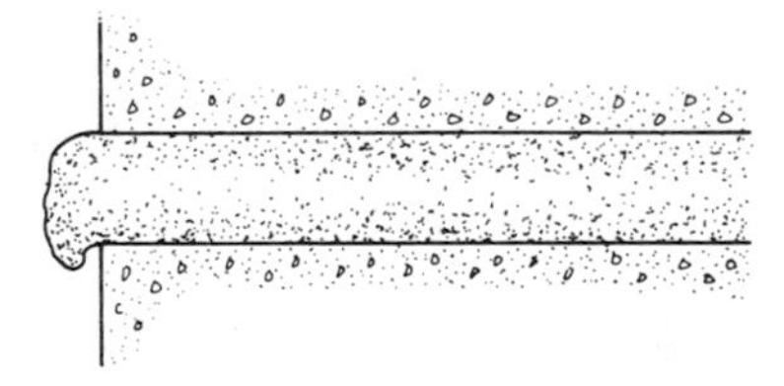

(7) Squeezed joint–not recommended this joint used to give a rustic effect

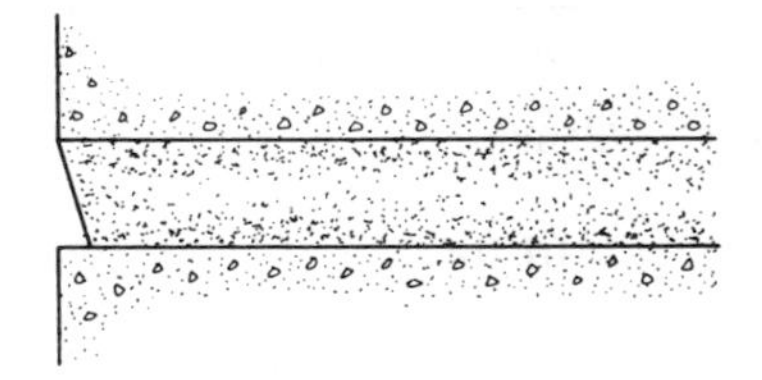

(8) Struck joint–not recommended

shadow lines. Concave joints are most popular and widely used. They have less pronounced shadows. Both of these types of joints are formed by tooling. Tooling will compress the mortar firmly against the block, thereby assuring better bond between mortar and blocks and also helping to prevent moisture infiltration into the wall. For these reasons, a tooled joint is preferred to a troweled joint. In the extruded joint, mortar that is squeezed out as the blocks are laid is not trimmed off but is left to harden in its extruded form. Neither this nor the raked joint should be used for exterior surfaces except in relatively dry climates, since they form ledges that may trap moisture, therefore having an adverse effect on the wall. A deep concave joint, which is watertight, gives the appearance of a raked joint without having any of its disavantages.

Flush joints may be used in combination with tooled joints, to accent either horizontal or vertical lines. In this treatment the joints to be emphasized are tooled. After the mortar has stiffened, it is rubbed to give it a texture similar to that of the block.

QUESTIONS

3-1. What are some of the aggregates used in concrete masonry units, and for what purposes are they used?

3-2. Describe the purpose of each type of raw material used in concrete masonry.

3-3. Under what weather conditions are the various concrete masonry grades suitable?

3-4. Why is there a discrepancy between nominal and actual dimensions of concrete masonry units?

3-5. What are the minimum face-shell thicknesses of concrete masonry?

3-6. Why is the ultimate compressive strength of the assemblage (f'_m) less than the ultimate strength of the individual units?

3-7. What are the advantages of tooled over raked mortar joints?

3-8. How does the basic production process differ between normal concrete and concrete block?

3-9. Why is it inadvisable to wet concrete masonry units prior to placing?

3-10. What is the most significant difference between the two ASTM specifications governing concrete brick?

3-11. How are the higher strengths achieved in the manufacture of concrete block?

3-12. List the various ASTM designations for concrete masonry units and the corresponding grades and types under each.

3-13. Distinguish between a channel block and a bond beam block. Where may each be used?

3-14. Absorption qualities of concrete masonry units are frequently taken as an indication of what desirable property?

3-15. What is the modulus of elasticity of the various types of masonry units? Also what is the relationship between the shear modulus and the compression modulus as specified by code?

3-16. Should the concrete building brick be layed up wet? Why?

3-17. How is a surface bonding material comprised of cement, sand, and glass fibers used in wall construction?

4

Mortar, Grout, and Steel Reinforcement

MORTAR

The use of mortar was the first logical addition to the act of simply piling rock or cut stone one on top of others as was done in early Egypt. The original purpose of the mortar was to fill the irregularities between masonry units because they were unsightly. It further served to provide resistance to the penetration of light, wind, and water, as well as to bond the units together, and possibly to add strength. However, now, in addition to the important purpose of sealing the units against the elements, the modern function includes that of bonding for strength to resist the actions of earthquakes and wind upon masonry walls. One of the earliest bonding materials, used after mud was first utilized for that purpose, was bitumen or tar, found in early clay and stone masonry. A subsequent development involved the use of lime in mortar. The early masons probably recognized what would happen when lime was used with mortar by observing the reaction of the lime rock or oyster shells used in their fireplaces. The lime calcined and then became firm again after a wetting and drying cycle. In those early days there were other more exotic admixtures employed. One frequently used mortar in some of the early European stone construction included egg whites; also, mixes with clay, urine, or oxblood combined with sand were utilized.

In modern masonry construction, the newer mortars, as specified by most of the current building codes, provide the workability needed for proper placement of the masonry unit, while producing a well-bonded weather-resistant masonry wall. These qualities were enhanced eventually by the use of portland cement. In the 1920s in Southern California, a mix of lime and sand in proportions of about 1:4 or 1:6 was delivered to the job site. Portland cement in the amount of 1:10 or 1:12 (cement to sand–lime mix by weight) was added before the mortar was applied to the wall. The portland cement produced a stronger mortar, yet one which retained the workability of lime mortar. It is those old lime mortar unreinforced buildings which are often associated with all masonry construction in the mind of the public today. They no more resemble modern reinforced masonry construction than does the first Wright brothers plane resemble a modern Boeing 747 jet. But it is extremely difficult to dispell "old wives' tales" sometimes. Following the disasterous 1933 Long Beach earthquake, a 1:3 cement/sand mix was tried, without success. Without lime, the mortar proved unworkable. So following this attempt, the so-called M,S,N,O,K (*MASONWORK*) cement and lime mortars appeared in the ASTM specifications. Later, the UBC standards (24-20) established mortar requirements by two alternative specifications, by either property or proportion standards.

Ingredients

CEMENT

The cements provided for use in mortar and grout in the UBC are portland cement and masonry cement. Portland cement is specified as either Type 1 or Type 2, in accordance with ASTM C150. Usually, "low alkali" is called for to reduce the soluble content of the cement, which may contribute to efflorescence, an unsightly deposit of soluble salts left on the surface as the water evaporates. Masonry cements must comply with ASTM C91 and portland blast-furnace slag cement with ASTM C595. There is one approved proprietary material which uses a combination of lime and an air-entraining additive that complies with ASTM C175. Note, then, that there is a "portland cement" and a "masonry cement" and that they are not interchangeable. Masonry cement cannot be used for mortar in reinforced masonry. The use of certain adhesives or bonding agents or cements of other types may be considered, but they must be approved by such agencies as the UBC under their Research Recommendations.

WATER

The requirements here are only that it be from a domestic supply and that it be clean and free from injurious amounts of oil, acid, alkali, organic matter, or any other harmful substances. This would apply to both mixing water and to the cleaning or curing water.

LIME

Lime, calcium oxide (CaO) or hydrated calcium oxide $Ca(OH)_2$, is probably one of the oldest chemicals made and used by man. It could have been discovered when he built a fire on limestone rock or in oyster shells and observed what happened. It is one of the oldest mortar ingredients used in our modern masonry. In an early mortar, it provided needed workability for one of the functions of mortar, to fill up the irregularities between units. It is still used in mortar to provide workability for watertight joints. It actually serves more as a plasticizing agent, while the portland cement provides the basic strength. Lime also increases the water retentivity, or water-holding capacity, of the mortar. This action decreases the tendency of the mortar to lose water, a phenomenon known as "bleeding," and reduces separation or segregation of the sand aggregate. While the mineral, lime, is present in many combinations, its combination with carbon dioxide, which forms limestone (calcium carbonate, $CaCO_3$), is an important source of lime as we use it in mortar. The lime in this form has the marblelike hardness of rock.

Limestone rock, as it comes from the quarry, is changed into a lime suitable for use in mortar by crushing, screening, selecting, washing, and grading. The selected stone is then placed in kilns, where it is heated to 2500°F. This drives off the moisture, including the water of crystallization, and also removes certain gasses from the stone, such as carbon dioxide. This calcining, referred to as *lime burning*, is done generally in rotating kilns. The result of the calcining and burning is *quicklime*, which is calcium oxide (CaO). Quicklime is a very caustic material; when it comes into contact with water, a violent reaction occurs, and enough heat is generated to boil the water. It can also cause severe burns, both because of the heat and the chemical action. For this reason the lime should be carefully added to the water rather than water being added to the lime. In the past, quicklime was delivered to the job for use in the masonry mortar. In that case the first step in using it was to slake it or hydrate it. This was done by adding enough water so that the oxide became a hydroxide $Ca(OH)_2$. During this slaking process, the caustic quicklime was hot and very hazardous to use. After slaking, the quicklime was set aside and stored in containers after being screened through a fine mesh. The result became known as lime putty, and this was what was added to the mortar mix. This quicklime conformed to ASTM C5 (UBC Standard 24-15).

However, use of quicklime has given way to a newer substance known as *hydrated lime*. This product is produced by slaking it in large tanks at the manufacturing plant, where it is converted to hydrated lime without necessarily saturating it with water. This hydrated lime is actually a dry powder containing just enough water to complete the chemical reaction. After hydration, the lime is pulverized, bagged, and delivered to the job site. This material must conform to ASTM C207, Type S (UBC Standard 24-19).

Sand

The sources of sand can be several. Natural sand is formed by the erosive action of rivers. Such particles are somewhat rounded in shape. They are deposited on flats and on ocean fronts as beach sand. The latter, being subjected to more abrasive wave action, are more rounded. This would make for better workability but probably have a deleterious effect on the strength. Other natural sand sources occur by wind transport, such as blow sand. On the other hand, sand may be manufactured. In this case it would be a product resulting from crushing stone, gravel or blast-furnace slag. This manufactured sand is sharper and more angular and thus may require different amounts of fine cementitious particles to provide the lubrication needed for proper workability. The deleterious substances in sand may be such items as friable particles, lightweight particles, organic impurities or excess amounts of clay or loam. These materials must be removed at the plant before the sand is sent to the job site.

Unfortunately too little concern is often shown for the quality of the sand as expressed in its grading. However the properties of sand have considerable impact upon the workability as well as the strength of the mortar. To provide some sort of guide in this respect, grading limits for mortar sand are spelled out in both ASTM C 144 and UBC Standard 24-21. The grading limits from the latter are listed in Table 4-1.

TABLE 4-1

Sand Grading Requirements (UBC Standard 24–21)

Item	*Limit*
% Passing sieve no. 4	100
〃 8	95–100
〃 100	25 max.
〃 200	10 max.

Actually, it is recommended that the sand for the mortar be on the finer side of the ranges permitted in the ASTM and UBC specifications. Those limits are rather broad, so if gradings on the coarse side of the range are used, the mortar will be harsh and not as satisfactory as it could be.

Admixtures

In general, admixtures are neither necessary nor desirable in masonry construction and therefore should not be included unless specifically approved and recommended by the designer or the building official. There are many different types of admixtures. They may be used for water reduction, for water retention to prevent bleeding, and for waterproofing. Whatever their purpose, they must be used in the proper manner and proportions.

There are numerous materials that must not be used in portland cement mortars under any circumstances and one of these is gypsum. Gypsum has sometimes been used to accelerate the set, but this introduces the hazard of later deterioration. Calcium chloride has sometimes been used for the same purpose. It is especially undesirable in reinforced masonry because calcium chloride tends to cause corrosion in the reinforcing bars or joint reinforcing.

Mortar mixes

Mortars have evolved in the past more or less by "rule of thumb." Much of the research has been performed by organizations with the intention of promoting a specific proprietary ingredient. The present specifications for engineered masonry contain provisions which are the least scientifically developed of all the controlled construction materials used therein. However, recently, because of the extreme concern for the effects of lateral forces, particularly those caused by earthquakes, on masonry construction, much greater emphasis is being placed on developing adequate standards for the control of masonry quality.

MORTAR MIX PROVISIONS

The ASTM and UBC requirements contain two alternatives for control: property specifications and proportion specifications. The requirements set forth in ASTM C270 Tables 1 and 2 are similar to those in UBC Standard 24-20.

The property specifications are those wherein the acceptability of the mortar is based upon (1) conformance of the individual ingredients with appropriate ASTM requirements, and (2) properties of a proposed mortar mix proportion with a specified flow or stiffness, as mixed in the laboratory, and tested for water retention and cube compressive strength. Should that particular laboratory mix conform to the strength value shown in ASTM Table 1, this would become the field mix.

Proportion specifications are governed by the material properties and those proportions listed in ASTM Table 2, which call for certain cement/lime/sand ratios. This is the specification generally followed because otherwise data must be presented to prove that the mortar meets the requirements of the physical property specifications in laboratory tests. Sometimes this is not always a feasible alternative. Where the proportion specifications are followed, the following conditions thereby obtain:

> The mortar will consist of a mixture of cementitious material and aggregate within the limits defined in ASTM Table 2, outlining the mix proportions, along with sufficient water to bring the mixture to a plastic state. The hydrated lime must also conform to either one of the following two conditions:
>
> 1. The total free (unhydrated) calcium oxide (CaO) and magnesium oxide (M_gO) shall not exceed 8% by weight.
>
> 2. When the hydrated lime is mixed with portland cement in the proportion set forth in ASTM Table 2, it must show an autoclave expansion of not more than 0.50%.

Now consider the UBC Code requirements and Standards. They provide for an optional choice of either (1) a mix of materials and proportions as set forth in

Table 4-2, or (2) the use of mixes, producing the properties as established by laboratory tests, and having mortar mix proportions as set forth in UBC Standard 24-20. These are reproduced as Tables 4-3 and 4-4. Table 4-2 is somewhat confusing because it combines the strength tables and the proportion tables given in Standard 24-20. It should be observed that Table 4-2 is taken from the basic code; whereas Tables 4-3 and 4-4

TABLE 4-2

Mortar Proportions (Parts by Volume)
(*UBC Table 24-A—Chapter 24*)

Mortar type	*Minimum compressive strength at 28 days (lb/in.²)*	*Portland cement*	*Hydrated limes or lime putty*[a] *Min.*	*Max.*	*Masonry cements*	*Shovel count*[b] *(at 7–8/ft.³)* *2¼*	*to 3*	*Parts*[b] *(ft.³)* *2¼*	*to 3*	*Damp loose aggregate*
M	2500	1	—	¼	—	21	28	2.81	3.75	Not less than 2¼ and not more than 3 times the sum of the volumes of the cement and lime used
		1	—	—	1	34	45	4.5	6	
S	1800	1	¼	½	—	21–25	28–34	2.8–3.4	3.7–4.5	
		½	—	—	1	25	34	3.4	4.5	
N	750	1	½	1¼	—	25–38	34–51	3.4–5.1	4.5–6.7	
		—	—	—	1	17	22	2.25	3	
O	350	1	1¼	2½	—	38–59	51–79	5.1–7.9	6.7–10.5	

[a]When plastic or waterproof cement is used as specified in Section 2403(p), hydrated lime or putty may be added but not in excess of one-tenth the volume of cement.

[b]Not a part of Table 24-A.

TABLE 4-3

Compressive Strength of Cubes for Mortar Types
(*UBC Standard 24–20, Table A*)

Mortar type	*Average compressive strength at 28 days (lb/in.²)*
M	2500
S	1800
N	750
O	350
K	75

TABLE 4-4

Mortar Proportions by Volume
(*UBC Standard 24–20, Table B*)

Mortar type	*Parts by volume of portland cement*	*Parts by volume of masonry cement*	*Parts by volume of hydrated lime or lime putty*	*Aggregate measured in a damp, loose condition*
M	1	1 (Type II)	—	Not less than $2\frac{1}{4}$ and not more than 3 times the sum of the volumes of the cements and lime used
	1	—	$\frac{1}{4}$	
S	$\frac{1}{2}$	1 (Type II)	—	
	1	—	Over $\frac{1}{4}$ to $\frac{1}{2}$	
N	—	1 (Type II)	—	
	1	—	Over $\frac{1}{2}$ to $1\frac{1}{4}$	
O	—	1 (Type I or II)	—	
	1	—	Over $1\frac{1}{4}$ to $2\frac{1}{2}$	
K	1	—	Over $2\frac{1}{2}$ to 4	

are taken from UBC Standard 24-20. It would appear that these two tables were combined to form Table 24-A into one table, as if they were directly related instead of constituting two different methods. This confusion is being clarified by proposed code changes.

At any rate, when the proportion specifications are used, the UBC Standards call for a mortar consisting of a mixture of cementitious material and aggregate, conforming to the individual ingredient property specifications, which are mixed for a minimum period of 3 min with an amount of water required to produce the desired workability in a drum-type batch mixer. Proportions must fall within the limits given in Table 4-4 for each mortar type specified.

One final note with respect to the origin of the UBC mortar-mix proportions, it has been observed that sand contains approximately one-third voids. For this reason, if sand amounts to approximately three times the combined volumes of cement and lime, then, theoretically at least, the voids in the sand should be completely filled, thereby providing a dense mix. This is a necessity if shrinkage and porosity are to be minimized. So the typical mortar mix specification of 1 part cement, $\frac{1}{4}$ to $\frac{1}{2}$ part lime, and $4\frac{1}{2}$ parts sand ($1:\frac{1}{2}:4\frac{1}{2}$), based upon damp, loose volume, has evolved. Actually, the specifications permit a range of between $2\frac{1}{4}$ and 3 for the sand volume to allow for differences in the grading of the sand and the roundness of its particles, which in turn affect workability. These proportions can be adjusted in the field to achieve the desired workability.

As previously noted, the lime is a cementing agent whose presence improves the plasticity, workability, and water retention of the mix, all of which are highly important for the attainment of maximum bond strength as well as water tightness in the masonry mortar. Further, it should be noted that the strength of the masonry assemblage in shear and flexure depends upon the mortar bond strength plus the contact area between the masonry unit and the mortar, and not directly upon the compressive strength of the mortar itself. Thus the deleterious effect of lime in reducing compression strength must not be viewed as a fatal flaw, provided that the amount of lime used in the mix is carefully controlled. Many authoritative tests have shown that masonry laid up with a $1:\frac{1}{2}:4\frac{1}{2}$ mortar mix produces up to 100% greater masonry strength in flexure and shear than does masonry laid up with a $1:\frac{1}{4}:3$ mix, primarily because of the effect of the lime on the bonding strength. For this reason it is common practice to use the upper limit of the lime quantity permitted by the UBC (i.e., $\frac{1}{2}$ part by volume).

Mixing

The basic requirement for mixing is simply that the material be thoroughly mixed. Machine mixing is ordinarily called for and is in common use; however, hand mixing is permitted for small batches. The order of placing the ingredients in the mixer is important. Should the cement be added first it is likely to lump or cling to the damp sides of the paddle blades and therefore not be uniformly dispersed throughout the mix. For this reason, the water and sand are put into the mixer first, to scour the blade surfaces and prevent any lumping in the cement. Lime is frequently added last, supposedly to prevent it from coating the sand particles before the cement gel can envelop them. No proof seems to exist that this really occurs, so this factor is often neglected.

Mixing is generally accomplished simply by shovel count in commercial work. This is deemed adequate in view of other existing variables, such as the harshness of field aggregates, which are uncontrollable to a large extent. The shovel count variation in Table 4-2 permits the tender to adjust the mix for workability. However, the California OSA requires that box measurements be used for accurate proportioning on school projects.

Mixing time, according to UBC, is set at 3 min minimum with whatever amount of water is required to produce the desired workability.

Workability

At one point, slump tests were recommended as a measure of mortar workability. However, it was found that the slump did not provide a proper measure of this mortar property, and the method was quickly abandoned. The best measure of workability is the mason's own judgment in this regard. In general, however, the mortar should be as wet as possible, while still remaining stiff enough to support the masonry units.

Tempering

Mortar that has stiffened due to water evaporation may be tempered as frequently as necessary to restore the desired consistency. UBC standards do not place limits on the length of time a mortar may be used after its initial mixing; however, ASTM C270 and many other specifications require that the mortar must be in place within $2\frac{1}{2}$ h after initial mixing. This is not a valid limitation, in itself, for the simple reason that it makes no reference to the temperature at the time of placing. If the day is cool and if the mortar is always kept plastic, it may be used satisfactorily for a much longer period than $2\frac{1}{2}$ h. But if the day is hot and dry, $2\frac{1}{2}$ h might be too long. A better limit on retempering could be when the mortar has begun its initial set and hardened enough to become harsh. Further tempering may then result in a weakened mortar which should not be used.

The tempering should not be done by simply splashing water on the surface because it will immediately run off, washing away the fines with it. Rather, the water should be dropped into a basin formed in the mortar and then thoroughly worked into the mix.

Compressive strength

Since the individual masonry units are bound together by the mortar, the strength of the resulting assemblage is somewhat dependent upon the physical properties of the mortar. As previously noted, the strength of masonry assemblages in shear and flexure depends primarily upon the mortar bond strength and not upon the compressive strength of the mortar itself; however, there are no code requirements for bond strength. Furthermore, the compressive strength requirements for laboratory specimens of mortar cubes are used only when one desires to confirm compliance of a proposed mortar mix with ASTM specifications. Further, it should be observed that the strength values specified do not actually represent the strength of the mortar as it exists in place in the field. These compressive strengths (1800 lb/in.^{2} for Type S and 2500 lb/in.^{2} for Type M) are actually developed by testing laboratory specimens (2-in. cubes) formed of the predetermined mix proportions. The mortar flow specified in these laboratory tests could not be used in the field, for the mortar would not only be too stiff to achieve a proper bond, but it would also prove unworkable. The laboratory test is covered by ASTM C270, but there are no ASTM specified field tests for mortar. Consequently, the mortar mixed to a flow suitable for use in laying masonry need not necessarily be required to meet the laboratory strength so specified.

UBC now does have an additional provision for field control adopted from the then California State Division of Architecture (now OSA) procedure. One function of this state agency is assurance of quality schoolhouse and hospital construction. So to this agency, control of field execution is of paramount importance; for example, the assurance of actual mortar strength in a wall. OSA recognized that (1) different mixes and units of different absorption qualities would result in different strengths of unknown degree; (2) different masonry units will absorb different amounts of water, influenced by variable wetting of the unit; and (3) the moisture content of mortar as tested in the laboratory for ASTM was not of a consistency suitable for field use.

Therefore, to meet OSA's desires for field control of actual construction, the agency developed a field sampling technique to approach a measure of the condition of the mortar as finally placed in the wall. This consists of: (1) mixing the mortar to proper fluid consistency for use, (2) placing a mortar bed for 1 min on units intended for the masonry construction, (3) placing that mortar sample into a 2 × 4 in. cylinder mold, and (4) obtaining the compressive strengths of the cylinders. The principle presumes that the mortar used, workable and wet, is subjected to the absorption of the masonry units in place, and thus resembles to some extent the actual mortar in the wall. It is noted that the exposure of the test sample to loss of water by absorption is rather short compared to the long time exposure in the wall. Consequently, the water/cement ratio of the sample is subjected to less water reduction than the actual mortar bed joint placed in the wall. Therefore, the mortar in the wall will have a lower water/cement ratio and correspondingly higher strength than the specimens. For this reason, the specimen was not required to have as high a compressive test strength (thus the 1500 psi minimum) as the desired or anticipated wall mortar strength. Regardless, in essence, this test specimen is influenced in the same manner as the wall mortar, although to a lesser degree, such that the sampling and test method serves as an indicator of the condition of the in-the-wall mortar.

Adhesives

Various adhesives have been proposed and used as substitutes for portland cement mortar. These should not be used, however, unless they do have a good test or service record, because there are many factors that affect the quality of the masonry, such as weathering, adhesive strength, durability, heat resistance, and fire resistance. Some of these adhesives have been approved for use in mortars by ICBO Research Committee; however, their usage is not widespread at present. Their methods of application and design provisions are specifically outlined in these Recommendations, since their performance is different from that of conventional masonry.

Fire clay

Fire clay is a mortar used for fire brick and generally is a type that develops its strength during firing in a furnace. It has no place in the mortar used in reinforced masonry. Some have used it as an additive because of the workability it adds to the mortar, but it does not add to the strength. As a matter of fact, it may even weaken the mortar. In time it may wash out because it will not be subjected to the high refractory temperatures that are needed to make it set and achieve an adequate strength.

GROUT

Grout is simply a high-slump concrete made with small-size aggregate, such as pea gravel. It derives its name from a Swedish term meaning "groot" or "porridge," terms that provide some indication of its pouring consistency. It must be fluid enough to fill all voids in the grout space and to completely encase the reinforcement. It serves the dual functions of: (1) bonding the wythes together into a composite element of masonry construction, and (2) bonding the reinforcement to the masonry so that the two materials will act together as something of a homogeneous material. The latter is absolutely essential if the desired flexural resistance is to be obtained, since the flexural tensile forces must be developed by the reinforcement.

Ingredients

CEMENT

Portland cement, Types 1 or 2 (ASTM C150), are typically used. Also portland blast-furnace slag cement (ASTM C595) may also be employed in lieu of portland cement.

FINE AGGREGATE

The same specifications apply as for masonry mortars.

COARSE AGGREGATE

The basic ASTM specification for coarse aggregates to be used in masonry is C404, which in turn refers to C33.

The UBC Standard 24-23 covers grout aggregates also. In that specification, a table is provided establishing grading requirements for this material. This is shown in Table 4-5. The coarse grout aggregate normally used, referred to as "pea gravel," has a maximum size of about $\frac{3}{8}$ in. in diameter. It should be noted, however, that in large grout spaces it is permissible to use up to a $\frac{3}{4}$-in. aggregate. As with the sand, it is desirable to keep the grading on the finer side of the range rather than on the coarser side. Also, the grading of the combined sand/gravel mix should be considered.

TABLE 4-5

Aggregate Grading Requirements
(*UBC Table 24-23-A*)

	Amounts finer than each laboratory sieve, square openings (wt. %)				
	Fine aggregate			*Coarse aggregate*	
	Size 1	*Size 2*		*Size 8*	*Size 89*
Sieve size		*Natural*	*Manufactured*		
$\frac{1}{2}$ in.	—	—	—	100	100
$\frac{3}{8}$ in.	100	—	—	85–100	90–100
No. 4 (4.76 mm)	95–100	100	100	10–30	20–55
8 (2.38 mm)	80–100	95–100	95–100	0–10	5–30
16 (1.19 mm)	50–85	60–100	60–100	0–5	0–10
30 (595 μm)	25–60	35–70	35–70	—	0–5
50 (297 μm)	10–30	15–35	20–40	—	—
100 (149 μm)	2–10	2–15	10–25	—	—
200 (74 μm)	—	—	0–10	—	—

There has been some use of lightweight aggregate in grout to reduce the weight of the structure. It has also been used because it improves the insulating properties of the grout. This material is governed by ASTM C330 and UBC Standard 26-3. It should be noted that lightweight aggregates from different sources will perform differently, for some may contain fines obtained by crushing larger particles, producing a more harsh aggregate. Lightweight aggregate should be prewet before use, to minimize absorption of water from the mix. This is especially important for grout that is to be pumped into the wall. Pumping pressures will force water from the matrix into the pores of the aggregate, tending to stiffen the mix.

There are two basic types of lightweight aggregates in use: natural and manufactured. The natural lightweight aggregates are mined and are found in natural deposits, usually volcanic in nature. Pumice is the most common and it consists of a volcanic glass full of minute cavities, which cause its light weight. Scoria is a frothy, basaltic magma or lava, or possibly a burned clay or clinker deposit. It may also be slag refuse produced from the reduction or smelting of ores. Volcanic ash consists of porous particles roughly the size of peas or shot, which look like ashes. Tuff is a porous, finegrained volcanic rock of light weight.

Manufactured lightweight aggregates consist of expanded slag or expanded clay or shale. Expanded slag is a frothy, solidified rock and waste that contains a large number of cavities. Expanded clay or shale has been fired in a kiln. The heat caused the material to bubble and then to become firm by vitrification. Some are fired in large lumps and then broken and graded to size. Others are formed into pellets of various sizes and fired.

LIME

The same specification as noted for mortar applies here as well. Lime has occasionally been used in grout to increase fluidity and to provide some water retentivity. It has not been used extensively in Western regions, but its use is permitted up to 10% by volume.

Mix proportions

The grout used in masonry construction may be either fine or coarse, depending upon the size of the grout space. Fine grout generally consists of one part portland cement to $2\frac{1}{2}$ to 3 parts sand, with lime possibly being added in an amount not to exceed $\frac{1}{10}$ of a part. The coarse grout is composed of 1 part portland cement to which may be added not more than $\frac{1}{10}$ part lime and 2 to 3 parts sand and not more than 2 parts coarse aggreggate. A commonly used mix in this instance would be 1 : 3 : 2.

Coarse grout can be used in grout spaces 2 in. or more in width in two-wythe masonry or in grout spaces in filled-cell construction 4 in. or more in both horizontal directions. An exception to this stems from the use of mortar in lieu of grout in chimney and fireplace construction. In that case, additional water must be provided for a pouring consistancy.

The $\frac{3}{8}$-in aggregate in coarse grout may be increased to $\frac{3}{4}$ in. where the grout space is 3 in. or more wide in two-wythe construction. The larger aggregate is preferred, since it will help to minimize later shrinkage of the wet, rich mix.

Grout used in high-lift masonry construction, where fluidity is especially important, must be richer and wetter than grout used in low-lift operations. As an alternative, it may have an approved admixture to provide this fluidity and water retentivity, and also possibly to minimize volume loss due to absorption.

Mixing

Grout may be mixed in two ways. On smaller jobs, it is done with job mixers. In that case, the ingredients must be added in such a way that there is not a "balling" of the cement, for the cement powder may form into damp balls and not be distributed uniformly into the grout mass. However, on larger jobs and where high-lift grouting operations are employed, the mixing of the grout is done in transit-mix trucks, because better control is provided. In that case, the grout is pumped directly from the truck into the grout space in the wall. With both methods care must be taken to ensure that there is no segregation of materials during placing.

One of the additives used in high-lift grouting is a proprietary material called Grout Aid. The volume of wet grout used in high-lift operations is composed partially of excess water, and when that water is removed by suction or absorption by the adjacent masonry units, the loss of water volume causes a reduction in grout volume. This, in turn, produces settlement and numerous cracks. Further, there may be an arching effect caused as the material tends to bridge between the two wythes. Also, there may be a separation at the center, as the grout bonds to the two wythes and separates at the center, or the separation may occur at one wythe as the grout in the grout core becomes smaller. Thus, a reconsolidation must be made to prevent this. To overcome this need for reconsolidation, it was necessary to discover an admixture that could be used even with a highly absorbent brick. Further, one was needed that would provide greater fluidity.

Compressive strength

There is no ASTM specification for grout testing; however, UBC Section 2403(s) 3 stipulates that grout must achieve a minimum compressive strength of 2000 lb/in.^{2} at 28 days. It also refers to a field strength test of grout from UBC Standard 24-22. UBC Standard 24-22 further establishes a way of obtaining samples that will be somewhat representative of the strength of the grout as it exists in the wall, where it has been subjected to absorption by the masonry units, thereby lowering its water/

cement ratio. This procedure entails the making of a mold comprised of the type of units (i.e., clay brick or concrete units) used in the wall. The present method is shown in Figure 4-1.

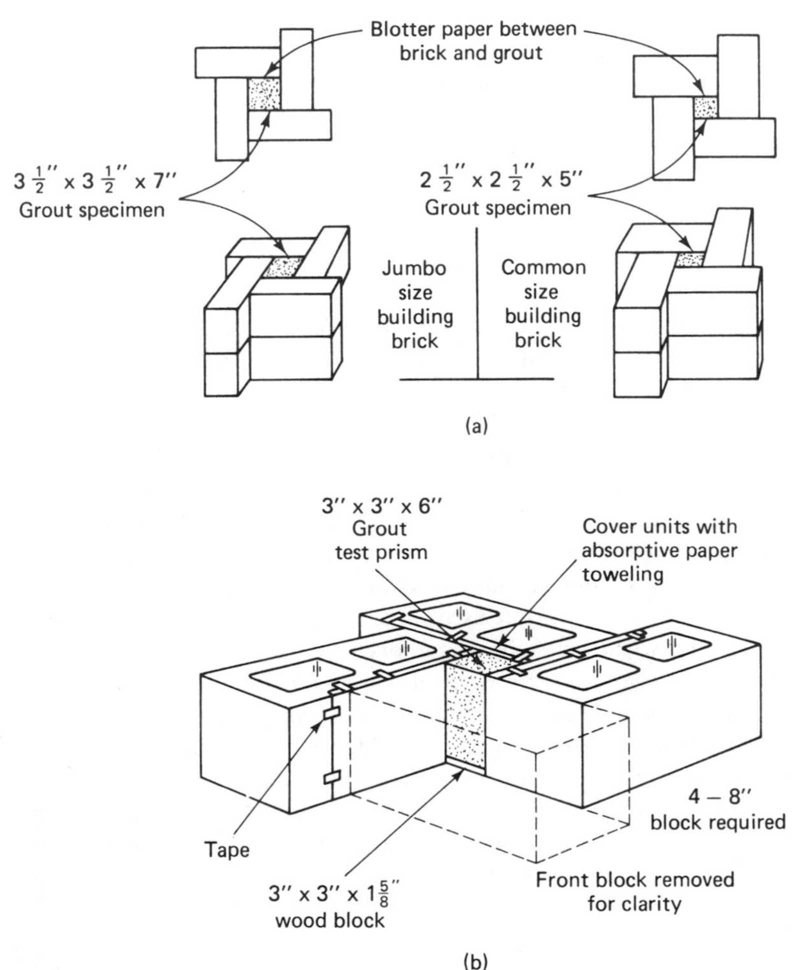

FIGURE 4-1. Grout field test specimens. (a) The present general method using brick, showing two sizes of specimens, both with a height/width ratio of 2:1. (b) Use of concrete block mold to provide a specimen of 2:1 ratio.

STEEL REINFORCEMENT

Reinforced masonry is engineered construction. This means that the structural element, be it a beam, column, or wall, must be designed to withstand specific loads. Reinforcement must be placed in the proper amount and location within

the structural elements, to resist those longitudinal and diagonal tensile forces induced by various load conditions. In addition, reinforcement may be provided to resist other forces, such as a portion of the compressive load carried by columns and pilasters, the axial tensile force in a diaphragm chord member, flexural tensile stresses developed in bond beams, or the tensile stresses in walls or columns caused by an eccentric vertical load.

Limits

Reinforced masonry, as described in the UBC, contains many arbitrary requirements covering the amount and location of the reinforcing steel, to improve its field performance in major earthquakes. These are not needed for other zones; thus provision is made for a "partially reinforced masonry" in which only the reinforcing required by design is necessary. The UBC specifies the minimum arbitrary amount of wall reinforcement for reinforced masonry as $0.002bt$ (wall width times thickness), no less than one-third of which must be placed in either the vertical or horizontal direction, whether or not it is actually needed to resist stress. The maximum bar spacing in the field of the wall is 48 in. on-centers either vertically or horizontally. The minimum bar diameter is $\frac{3}{8}$ in. except for approved wire joint reinforcement. The OSA raises this arbitrary minimum total to $0.003bt$. Such empirical Code specifications for the amount of reinforcement required in certain seismic areas is not based upon sound engineering calculations but merely represents a "feeling." Actually, to some extent, the limits apparently stem from observations of the performance of unreinforced, unanchored, and unengineered masonry—an extremely different kind of material assemblage than the modern engineered reinforced systems that we encounter today.

Bearing walls may also have to be reinforced for combined bending caused by the horizontal loads, plus direct compression caused by vertical loads. Beams and shear walls may require "shear steel," or web reinforcement, to prevent diagonal tensile failures. Steel is arbitrarily provided around all openings, whether required for stress or not, simply to prevent the diagonal cracks that tend to radiate from the corners of these openings. Examples of some of these conditions may be seen in Figure 4-2.

One disadvantage accruing from the presence of an excessive amount of reinforcement is the practical one of placement. When the bar spacing is too small, say closer than the length of a mason's trowel and forearm, it becomes difficult to lay up the masonry courses with good workmanship. The mason cannot keep from jarring the bars while the grout is still wet, or he may not be able, because his movements are inhibited, to place the mortar evenly. In any event, a good bond between masonry units, or between grout and reinforcing, becomes less likely under these circumstances. For example, it might be preferable, where No. 4 at 24 in. center to center (cc) are called for, to place pairs of No. 4 at 48 in. cc, the optimum spacing, or use No. 6 at 48 in. cc.

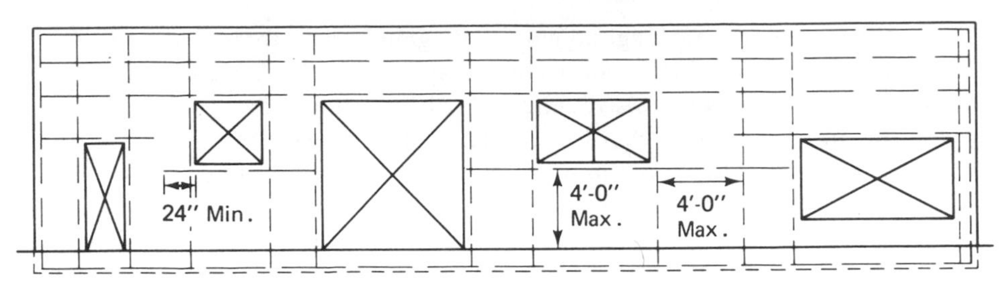

FIGURE 4-2. Typical wall reinforcing requirements for reinforced masonry.

Types of reinforcement

The type of steel used to reinforce masonry is the same as that used in reinforced concrete (i.e., the bars must comply with ASTM Standard A615-Grade 40, 50, or 60). The allowable design stress for grades 40 and 50 is 20,000 lb/in.^{2} in flexural tension (walls and beams), increasing to 24,000 lb/in.^{2} for steel with a yield of 60,000 lb/in.^{2} or more. The allowable axial compressive stress in columns is set at 0.4 of the minimum yield strength, with a 24,000 lb/in.^{2} maximum. The unknown factor of the relationship between the moduli of elasticity of the masonry and reinforcing ($n = E_s/E_m$), plus the fact that higher steel stresses, accompanied by greater elongations, might result in undesirable cracking in the masonry, tend to discourage the use of the higher-strength steels. Generally grade 40 is recommended for its greater ductility. However, in special circumstances where there are very heavy loads on high-rise bearing walls or masonry columns, a high-strength steel (A615-grade 60) might be used. Maximum size reinforcing must be limited to No. 11 bars. Sizes are specified in terms of the number of eighths of inch of bar diameter.

Prefabricated joint reinforcing (ASTM A-82) can be used in the masonry bed joints, either as a part of the required minimum horizontal reinforcing or as flexural tensile reinforcing. The allowable stress may be taken as 50% of the minimum yield, with a 30,000 lb/in.^{2} maximum. The longitudinal wires in the ladder type are joined with intermittent perpendicular cross wires called "spacers" (Figure 4-3). Another type has diagonal cross members forming a sort of truss.

Joint reinforcing possesses certain advantages. Since it has a greater surface area, it will develop a better bond with the masonry than will the larger reinforcing bars. Further, since it is closer to the outer fibers, it will begin to function much earlier in the loading process, with less cracking of the masonry taking place.

QUESTIONS

4-1. When mortar was originally used, what purposes did it serve?

4-2. What are some of the modern functions of mortar?

4-3. Why do you add lime to the mortar mix?

4-4. What are the significant properties of mortar? Why are they important?

4-5. Describe the OSA recommended field method for mortar test specimens.

4-6. What is the purpose of the field test for mortar strength?

4-7. Why is it not too detrimental that lime be used, even though it may reduce the compressive strength of the mortar?

4-8. What is tempering and when is it unadvisable to temper mortar?

4-9. What is grout?

4-10. What are the two main functions of grout?

4-11. What is the arbitrary minimum steel requirement for reinforced masonry?

4-12. Describe the method of preparing and testing grout prisms.

4-13. Explain the two alternative methods for mortar specifications.

4-14. Which specification of mortar usually governs? Why?

4-15. May adhesives be used in place of portland cement mortar?

4-16. Why must steel be included in reinforced masonry?

4-17. Why is steel provided around wall openings?

4-18. Why might joint reinforcing be more desirable than reinforcing bars?

4-19. Why is the mortar flow specified for the laboratory compression tests on mortar not suitable for field use?

4-20. What should be done to a lightweight aggregate before it is used in a grout mix, and why?

4-21. Name the two basic types of lightweight aggregates in use today and cite some examples of each.

FIGURE 4-3. Wall joint reinforceing.

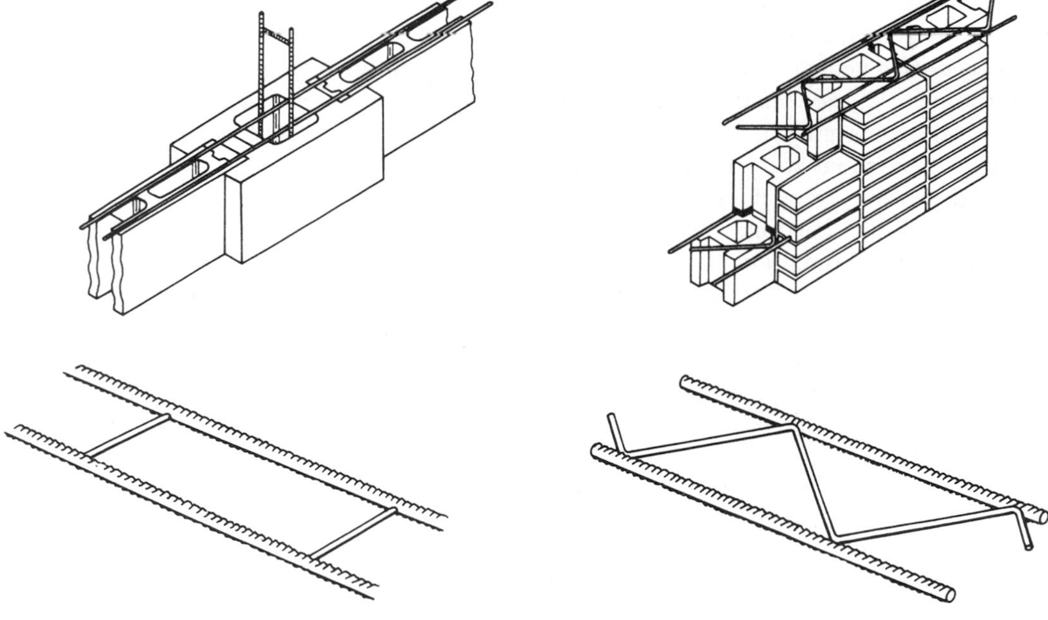

(a) Ladder-type joint reinforcing

(b) Truss-type joint reinforcing

5

Load Types and Intensities

The sizes of structural members and their connections are primarily governed by the requirements that these elements must adequately carry all the loads imposed upon them without reaching their usable strength or undergoing excessive deformations. It is essential that various load combinations be considered, but it probably should be pointed out that the structure is not expected to resist each of these various combinations with the same margin of safety. Obviously short-term loads, such as high winds or earthquakes, are going to have to be resisted at a lower safety level than will the long-time vertical load impositions, if the structure is to be economically viable. For instance, one might measure a desired margin of safety under a certain dead load/live load combination, whereby the usable strength might be taken as the yield strength of the material. This property would be utilized in precluding the development of excessive deformations which might impair or even destroy the useful function of the element, long before a complete collapse became imminent. Yet conceivably this limit would prove to be too conservative, and therefore untenable, when considering the relative short-term effects of wind or seismic action, wherein the design load combination might be dead plus lateral. Obviously, the design limits for that load group must be raised, thereby lowering the safety margin. Also, perhaps the limit should be measured against the ultimate capacity of the material. For, under the most severe natural disaster conditions, we are concerned only with preventing a total collapse of the structure so as to allow the occupants to escape to safety. Post-yield deformations and drift displacements are expected and tolerable under these circumstances. Thus we are no longer as concerned about any functional impairment within the building itself, unless perhaps it is classified as an essential facility [UBC 2312(k)]. Codes crudely recognize this distinction by arbitrarily raising the allowable stresses for such short-term loadings by 33%. Or one might set the limiting capacity at a tolerable deflection, to minimize the possibility of cracking a nonstructural attached element such as plaster or other materials.

Statistical and probablistic evaluations of these random-load-intensity occurrences, and the structural response to a specific load pattern under these loads, are often possible. Thus, an attempt is made in the various codes to consider these statistical probabilities by the use of *load factors*, such as those found in current ultimate-strength design procedures for reinforced concrete structures. Actually, these load factors are based to some extent on the relations similar to those formulated by Lind in his publication "Consistent Partial Safety Factors," *ASCE Proc.*, Vol. 97, 1971. Here if $\bar{L}$ represents the mean load, and $\bar{S}$ the mean strength, then, according to Lind:

$$a\bar{L} \leq b\bar{S}$$

where a = partial safety coefficient larger than 1
b = partial safety coefficient smaller than 1

The magnitude of each partial safety coefficient depends upon the variance of the quantity to which it applies, $\bar{L}$ or $\bar{S}$, and on the chosen value of the safety index of the structure. The ACI load factors are designated by a and the material quality reduction factors (Φ in ACI 318) by the coefficient b. The ACI Code relates these to design values rather than to mean values of $\bar{S}$ and $\bar{L}$. It assigns different values of a to different kinds of loads and different values of b to different kinds of structural functions, this assignment being based to some extent upon statistical information.

At any rate, the student at this point probably has not had much experience in originating the loads on a structure, so it might be appropriate to look at the sources of the loads and to see how they are placed on a structure. Live-load values, for example, are generally spelled out in governing building codes and these are considered legal minimums. But they really should serve only as guides. In some critical cases, such guides might prove insufficient, and therefore it becomes the prerogative of the structural engineer to make his own evaluation. This might occur, for example, where snow may accumulate on a particular shape of roof surface in a geographical area where loads of this type, because of their variability, may not be specified in the local building code. In this case it would be up to the local building official to define the unit load limits. Also, new types of structures are continually being developed for which there may be no specified legal load limits. In cases where unique loadings are imposed on such structures, determination of the critical or controlling loads is not a routine matter, and perhaps would require a certain amount of research and often even physical testing to establish the loading requirements.

For common structures, however, the building codes do specify minimum design loads, depending upon use and occupancy of the structure. Generally, these have been established through years of experience and in the main have proven to be adequate without being too excessive. Actually, what is the degree of intensity of this load that you take from the building code? Is it the peak load intensity to which the structure might ever be subjected? If it were, the cost of building would be prohibitive. It is apparent, then, that some kind of a probabilistic consideration must enter the picture, be it a direct evaluation or an intuitive one. A probability model for the maximum load can be devised by building it up from a probability density function for the projected loading conditions, as shown in Figure 5-1. In this curve, the area under the curve between the two abcissas, such as loads L_1 and L_2, represents the probability of occurrence of loads of magnitude $L_1 < L < L_2$. Correspondingly, the probability of occurrence of loads larger than L_d is given by the shaded area to the right of L_d. Thus even these design loads, which are not the maximum possible attainable values, are rarely, if ever, reached during the life of most structures, since they are based upon rather conservative estimates. For instance, it is not likely that, even if a crowded corridor were completely full of people, the specified design load of 100 lb/ft^2 would be exceeded, or even attained.

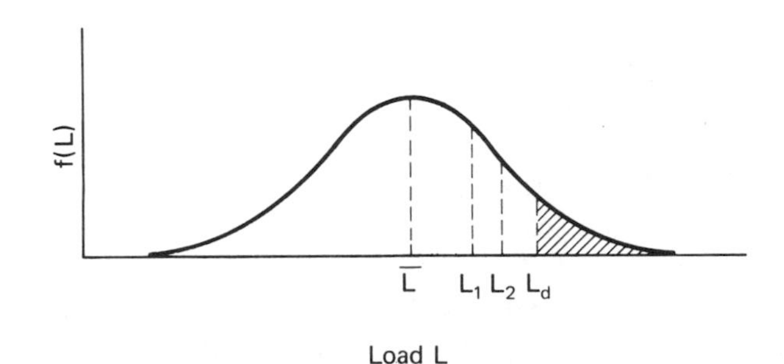

FIGURE 5-1. Load frequency curve.

In essence, loads can be placed into three broad categories: dead, live, and lateral, such as wind, blast, or earthquake. That combination of those three load sources which produces the highest controlling stress or deflection condition then becomes the final criterion for design. Current design practice then goes well beyond a simple consideration of vertical loads, dead and live, only. To ensure the stability of the modern building against overturning effects of high winds, blast, or seismic ground motion, attention must also be paid to other complex load patterns. One of the most critical would be that of miminum dead load combined with these dynamic lateral force effects.

DEAD LOADS

Dead loads may be categorized as those which remain essentially constant during the life of a structure, such as the weight of the building components or any permanent or stationary construction, such as partitions or equipment, which may be present. A definition for dead load is given in the UBC Section 2302. There it is referred to as "the vertical load due to the weight of all permanent structural and non-structural components of a building, such as walls, floors, roofs, and fixed service equipment." In the main, dead loads are relatively simple to determine, since they may be computed from component sizes and known material densities. In most cases, a preliminary estimate must be made prior to sizing the structural members, to obtain their contribution to the total design load. However, with a bit of experience, one can readily make a reasonably accurate estimate for preliminary design purposes. Generally, these will not be far off from the final dead-load values.

The unit weights of various common building materials may be found from any number of sources. The densities of some of the more commonly used construction materials are listed in Table 5-1. Tables 5-2 and 5-3 provide needed dead-load information for reinforced grouted concrete block and brick walls. Observe that Table 5-2 provides an average hollow block wall weight in lb/ft^2, depending upon the spacing of the reinforcing (and thus the grouted cores) and the aggregate weight of the masonry unit. This is especially useful when one is calculating the seismic weight of the wall component, as will be explained later. The equivalent solid thickness is also useful for determining the axial compressive stress on a partially grouted wall section. It is an equivalent thickness for the solid material in a wall, as if there were no hollow cores. Values may be found in Table 7-4.

TABLE 5-1

Approximate Weights of Building Materials

Material	*Weight (lb/ft^2)*	*Material*	*Weight (lb/ft^2)*
Floors		Ceilings	
Concrete finish, per inch of thickness	12	$\frac{3}{4}$-in. plaster on metal lath furring	8
Lightweight concrete fill, per inch of thickness	9	$\frac{3}{4}$-in. gypsum plaster on metal lath and channel suspended ceiling construction	10
$1\frac{1}{2}$-in. Terrazzo floor finish directly on slab	19	Acoustical fiber tile directly on concrete blocks or tile	1
$\frac{3}{4}$-in. ceramic or quarry tile on 1-in. mortar bed	22		
$\frac{3}{4}$-in. mastic floor	9	Walls	
Hardwood flooring, $\frac{7}{8}$-in. thick	4	Windows, glass, frame and sash	8
Subflooring (soft wood), $\frac{3}{4}$-in. thick	$2\frac{1}{2}$	Structural glass, per inch of thickness	15
$\frac{1}{2}$-in. Douglas fir plywood	3	Stone 4-in. thick	55
		Glass block 4-in. thick	18
Roofs		Partitions[a]	
Five-ply felt and gravel (or slag)	$6\frac{1}{2}$	4-in. clay tile	18
Three-ply felt and gravel (or slag)	$5\frac{1}{2}$	6-in. clay tile	25
Five-ply composition roof, no gravel	4	8-in. clay tile	31
Three-ply felt composition roof, no gravel	3	4-in. gypsum block	13
Concrete tile	16	2-in. solid plaster	20
Sheathing, $\frac{3}{4}$-in. thick, yellow pine	$3\frac{1}{2}$		
Gypsum, per inch of thickness	4	2×4 studs, or metal studs, lath and $\frac{3}{4}$-in. plaster	18
Metal deck (18 gauge)	3		
Corrugated metal (20 gauge)	$1\frac{1}{2}$	Steel partitions	4

[a]UBC Section 2304 (d) specifies that on floors where movable partitions are used a uniformily distributed load of 20 lb/ft^2 shall be included in design.

TABLE 5-2

Average Weight of Grouted Hollow Unit Walls[a]

	Concrete units						*Hollow clay block (lb/ft²)*
	Medium weight 125–105 (lb/ft²); average 120 (lb/ft²)			*Normal weight 125 (lb/ft²) or more; average 138 (lb/ft²)*			
Wall thickness (in.):	*6*	*8*	*12*	*6*	*8*	*12*	*8*
Solid grouted wall	*56*	*77*	*118*	*68*	*92*	*140*	*88*
Vertical cores grouted at:							
16 in. oc	46	60	90	58	75	111	71
24 in. oc	42	53	79	53	68	99	64
32 in. oc	40	50	73	51	65	93	61
40 in. oc	38	47	70	50	62	89	58
48 in. oc	37	46	68	49	61	87	55
No grout in wall	31	35	50	43	50	69	45

[a]Average weights of completed walls of various thickness in pounds per square foot of wall face area. A small quantity has been added to these values to include the weight of bond beams and reinforcing steel. Grout and mortar use sand-gravel aggregate.

TABLE 5-3

Average Weight of Two-Wythe-Reinforced Grouted Brick Walls[a]

Wall thickness (in.)	*Weight (lb/ft²)*
8	86
8½	93
9	100
9½	106
10	112
10½	118
11	125
11½	131
12	138
13	151
14	164

[a]Based upon average brick weight at 10 lb/ft² of vertical surface per inch of thickness and grout core at 13 lb/ft²/in. of thickness.

Most dead loads are established on the high side, but then it is desirable that the estimates do err somewhat on the conservative side. Unless the initial estimate varies grossly from the final weight, it usually is not necessary to refine the preliminary dead-load values selected. However, there are certain unique situations wherein the dead load should not be overestimated, and, if anything, should be underestimated.

A case in point stems from the situation where the dead load is depended upon to provide stability against overturning. Obviously, if the dead load is overestimated, resistance to overturning is also overestimated and the factor of safety against overturning could be reduced to an intolerable level. Another example in this regard occurs in reinforced concrete or masonry columns. It may be that the overturning effect is great enough, and the dead load small enough that certain columns in the structure could be subjected to tensile stresses over a portion of their cross section. If so, these tensile stresses must be considered in the design and adequate reinforcing provided. Overestimating the dead load would obviously result in an underevaluation of these critical tensile stresses, possibly to the extent that serious cracking or even failure might occur. It is common practice, for instance, to consider a load combination of 0.9 dead plus lateral to preclude any overestimation of the dead-load effect in an earthquake.

LIVE LOADS

Live loads are also referred to as *occupancy loads*. Specifically, Section 2302 of the UBC defines live load as "the load superimposed by the use and occupancy of the building, not including the wind load, earthquake load, or dead load." Occupancy loads include those loads which are directly caused by people, machines, or other movable objects. They may be considered as short-duration loads, since they act only intermittently during the life of a structure. Table 5-4 specifies minimum floor live-load requirements for various types of occupancies or uses. These therefore become

TABLE 5-4

Uniform and Concentrated Floor Live Loads
(*UBC Table 23-A*)

Use or occupancy			
Category	*Description*	*Uniform load* (lb/ft^2)[1]	*Concentrated load* (lb/ft^2)[1]
1. Armories		150	0
2. Assembly areas[4] and auditoriums and balconies therewith	Fixed seating areas	50	0
	Moveable seating and other areas	100	0
	Stage areas and enclosed platforms	125	0
3. Cornices, marquees and residential balconies		60	0
4. Exit facilities, public[5]		100	0
5. Garages	General storage and/or repair	100	[3]
	Private pleasure car storage	50	[3]
6. Hospitals	Wards and rooms	40	1000[2]
7. Libraries	Reading rooms	60	1000[2]
	Stack rooms	125	1500[2]
Manufacturing	Light	75	2000[2]
	Heavy	125	3000[2]

TABLE 5-4 (Continued)

Use or occupancy		Uniform load (lb/ft²)[1]	Concentrated load (lb/ft²)[1]
Category	Description		
8. Offices		50	2000[2]
9. Printing plants	Press rooms	150	2500[2]
	Composing and linotype rooms	100	2000[2]
10. Residential[6]		40	0
11. Rest rooms[7]			
12. Reviewing stands, grand stands and bleachers		100	0
13. Schools	Classrooms	40	1000[2]
14. Sidewalks and driveways	Public access	250	[3]
15. Storage	Light	125	
	Heavy	250	
16. Stores	Retail	75	2000[2]
	Wholesale	100	3000[2]

[1]See Section 2306 for live load reductions.

[2]See Section 2304 (c), first paragraph, for area of load application.

[3]See Section 2304 (c), second paragraph, for concentrated loads.

[4]Assembly areas include such occupancies as dance halls, drill rooms, gymnasiums, playgrounds, plazas, terraces and similar occupancies which are generally accessible to the public.

[5]Exit facilities include such uses as corridors and exterior exit balconies, stairways, fire escapes and similar uses.

[6]Residential occupancies include private dwellings, apartments, and hotel guest rooms.

[7]Rest room loads shall be not less than the load for the occupancy with which they are associated but need not exceed 50 lb/ft^2.

the live-load design criteria for a structure having a given use. The minimum roof live loads are shown in Table 5-5. They are keyed to the roof tributary area in square feet and slope, for method 1. Method 2 begins with a basic uniform load, in psf, and then permits a specified % rate of reduction, per square foot of roof area, up to a specified maximum. Thus, the total vertical loading on a beam, truss, or wall becomes a uniformly distributed load (lb/foot) calculated from the unit live-load requirement (lb/ft^2) and the dimension contributing to the loading of that element (w lb/ft^2 × tributary width) plus the weight of any components such as slabs or partitions being supported, plus the weight of the structural element itself. Occupancy loads, as specified by the Code, prove to be extremely conservative; consequently, floor failure due to overloading is extremely rare. To take into account the fact that the full live-load density will not occur simultaneously over large floor or roof areas, most codes provide for a live-load reduction factor keyed to the floor area. For example, UBC Section 2306 permits a floor live-load reduction when the tributary area loading on a column, pier, wall, foundation, truss, beam, or slab exceeds 150 ft^2. The live load may be reduced by 0.08 %/ft^2 of area for floors supported by the member. However, the total percentage reduction (R) is limited to the lesser of two values: 40 % for horizontal

TABLE 5-5

Minimum Roof Live Loads[a]
(*UBC Table 23-C*)

	Method 1				*Method 2*	
	Tributary loaded area for any structural member (ft^2)					
Roof slope	*0–200*	*201–600*	*Over 600*	*Uniform load*[b]	*Rate of reduction r* (%)	*Maximum reduction R* (%)
1. Flat or rise less than 4 in./ft. Arch or dome with rise less than one-eighth of span	20	16	12	20	0.08	40
2. Rise 4 in./ft to less than 12 in./ft. Arch or dome with rise one-eighth of span to less than three-eighths of span	16	14	12	16	0.06	25
5. Rise 12 in./ft and greater. Arch or dome with rise three-eighths of span or greater	12	12	12	12	No reduction permitted	
4. Awnings except cloth covered[a]	5	5	5	5	No reduction permitted	
5. Greenhouses, lath houses, and agricultural buildings	10	10	10	10	No reduction permitted	

[a]Where snow loads occur, the roof structure shall be designed for such loads as determined by the building official. See UBC Section 2305(d). For special purpose roofs, see Section 2305(e).

[b]See Section 2306 for live-load reductions. The rate of reduction r in Section 2306 Formula (6-1) shall be as indicated in the table. The maximum reduction R shall not exceed the value indicated in the table.

members or vertical members receiving load from one level only, increasing to 60% for other vertical members, or

$$R = 23.1 \times \left(1 + \frac{DL}{LL}\right)$$

No reduction is permitted in places of public assembly or where the live load exceeds 100 lb/ft^2.

Occasionally, a live load may actually be distributed over a relatively small area, in which case it really should be treated as a concentrated load rather than as a uniformly distributed one. This might occur in a storage area where an occasional heavy load, such as safes, storage racks, or the like, might be permanently located within a relatively small area. Concentrated loads may be considered sometimes as an alternative loading to a uniform load. Whichever of the two types produces the most critical loading would then become the design criterion. Refer back to Table 5-4 to observe how UBC handles this possibility.

Sometimes a dynamic live load occurs due to the presence of moving vehicles or machinery. These loads can produce very large dynamic stresses, if their frequency is in tune with the natural frequency of the structure, even though the actual load magnitude itself may be relatively small. To simplify the analysis, it is often possible to consider this magnification effect simply by assuming some sort of equivalent static load on the floor or girder. For instance, moving cranes in industrial facilities certainly create dynamic live loads on the building. To provide for this dynamic effect, the combined crane dead plus static lift load is increased by anywhere from 10 to 25% to represent the design criterion for the crane loading. This increase also provides consideration of the effect of suddenly applied loads. To account for crane de-acceleration, a horizontal force of 10% of the total vertical load is specified by UBC. This load is applied horizontally along each runway. However, in more complex dynamic loadings, a precise method of analysis must be employed to calculate the dynamic response of the structure. This might occur for example in suspension bridges, towers, or elevator supports.

Snow and rain also represent types of live loads. Snow effect may often be obtained from local building officials or such sources as the American National Standards Institute (ANSI), which has developed a snow-load map, Figure 5-2, for the purpose of determining typical snow loads in any given geographical area. They represent minimum values, however, and in cases where high local accumulation of snow may be possible, say at higher elevations, even larger values must be used. These can be obtained usually from the local building official who has jurisdiction of the area in which the structure is to be built. For instance, 1 in. of newly fallen snow is roughly the equivalent of about 0.5 lb/ft^2. But this value can be even larger at higher elevations, where the snow is often denser; e.g., the density of packed snow is about 12 pcf. In addition, the snow load may have to be adjusted by applying corrections to the basic value to account for such factors as (1) degree of exposure, (2) roof slope, (3) existence of ice, and (4) possible uneven snow deposition. A miminum of about 10 lb/ft^2 of roof area is recommended, although snow and ice loads of up to 60 lb/ft^2 on flat roofs in some areas are not uncommon. In addition, the possibility of uneven loading must be considered with snow, since it is often possible that only part of the roof would be covered at any given time due to rapid melting or to wind action on the exposed surface. Thus, this kind of uneven load distribution can produce critical stresses and deflections that would exceed those obtained when the loading is spread over the entire roof. Failures of this sort occur more often than we care to admit.

Rain does not usually constitute a separate loading condition because it does not exceed the design snow load. Further, because roof drains are provided, rainwater usually does not accumulate in sufficient amounts to become a significant load factor. However, failures have occurred when rainwater did accumulate on a roof because of inadequate or plugged drains, causing an excessive local deflection. Such ponding can produce increasing accumulations of water and therefore progressive

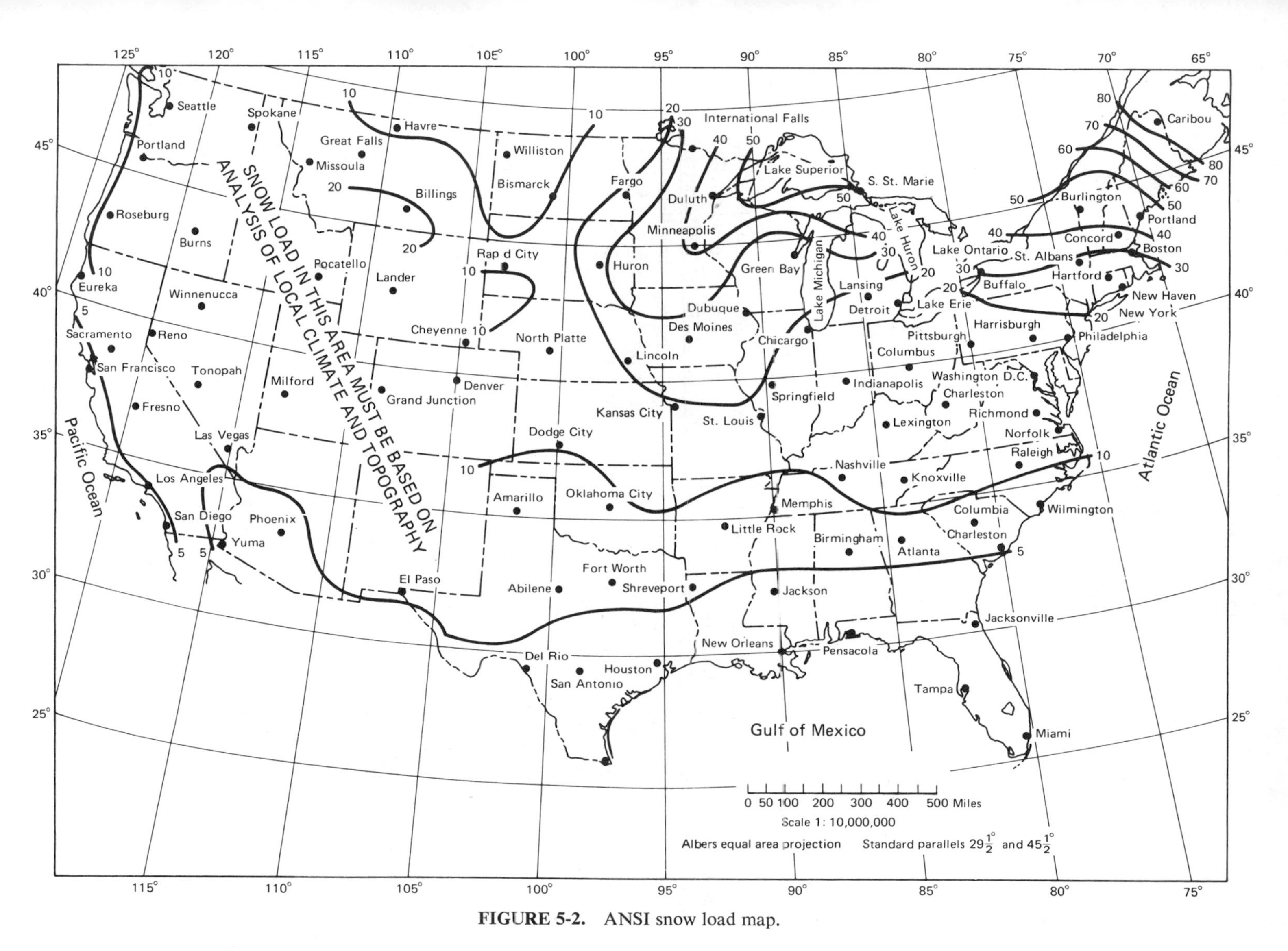

FIGURE 5-2. ANSI snow load map.

loading increases on the roof. If the flexural stiffness of the roof is relatively small compared to the span, a ponding failure could occur. To account for this possibility, the American Institute of Steel Construction (AISC) specifications (1970) contain provisions for ponding by specifying minimum stiffnesses for primary and secondary roof beams.

LATERAL LOADS

Lateral loads may be generated by various natural phenomena. These would include, of course, wind or seismic activity. Also, lateral loads may be caused by blast pressure or by the dynamic effects of moving vehicles or equipment. Water or retaining-wall pressures are, of course, lateral loads, too, but they fall into the live-load category.

Wind loads

Wind loads exert pressure (or suction) on exposed surfaces and therefore constitute an extremely important factor in the design of multistory structures. The magnitude, frequency, and distribution of wind loads depends upon a number of factors. Extensive research of recent origin (ASCE among others) regarding these has been conducted. Some of the results have been codified to permit the use of rather simplified formulas in design practice. Design forces are based on the maximum wind velocity occurring on a standard anemometer some 30 ft above the ground at a particular location. These maximum velocities are obtained from weather station data and are expected to occur only once in about 50 years. An ASCE report (1961) contains a considerable amount of information regarding how this wind velocity varies with elevation above the ground.

To codify wind-load values that may be readily used in design, the kinetic energy of wind motion must first be converted into a dynamic pressure. This dynamic pressure then may be converted into an equivalent static pressure as follows:

Kinetic energy of moving air:

$$\mathrm{KE} = \tfrac{1}{2}mV^2$$

where m is the mass and V the velocity. Taking weight of air at 0.07635 lb/ft^2, the kinetic energy or velocity pressure, q_0, is obtained as

$$q_0 = 0.002558V^2$$

where V is the design wind velocity, mph, taken 30 ft above ground. Applying a shape factor to q_0 to obtain the equivalent static pressure in lb/ft^2, p:

$$p = 0.002558C_sV^2$$

where C_s is the building-shape coefficient, which for most buildings lies between 1.3 and 1.5. The basic static wind pressure based upon design wind velocities in various geographical locations occurring at a height of 30 ft above grade may be obtained from a resultant wind pressure map such as appears in the UBC (Figure 5-3).

Variation in wind velocity with height depends upon the surface roughness of the terrain around the structure, and on the temperature distribution of the interacting air masses. One expression describing the velocity at a given height is $V_z = V_{30}(h_z/h_{30})^{1/7}$, where V_{30} is the datum air velocity at elevation $h_{30} = 30$ ft. If the terrain is particularly rough, as in urban areas, the increase with height may be much less, since the turbulence reduces the velocity, and the $\frac{1}{7}$ exponent may increase to as much as 0.3. Generally, it is not necessary to compute the velocity at a particular height in this fashion, however, for most building codes contain a table giving the velocity or pressure values which govern at various heights above the ground. One such table is UBC Table 24-F, reproduced here as Table 5-6. This list provides the design pressure for various height zones and geographical locations in the United States, with wind pressures varying between 20 and 50 lb/ft². For example, the basic design wind pressure (elevation below 30 ft) for a datum wind pressure intensity of 20 lb/ft² (geographical location, Southern California) is 15 lb/ft². This force would act inward or outward on the gross area of the vertical projection of that portion of the building or structure measured above the average level of the adjoining ground.

TABLE 5-6

Wind Pressures for Various Height Zones Above Ground[a, b]
(UBC Table 23-F)

Height zones (ft)	*Wind-pressure-map areas (lb/ft²)*						
	20	*25*	*30*	*35*	*40*	*45*	*50*
Less than 30	15	20	25	25	30	35	40
30 to 49	20	25	30	35	40	45	50
50 to 99	25	30	40	45	50	55	60
100 to 499	30	40	45	55	60	70	75
500 to 1199	35	45	55	60	70	80	90
1200 and over	40	50	60	70	80	90	100

[a]See Figure 5-3. Wind pressure column in the table should be selected which is headed by a value corresponding to the minimum permissible resultant wind pressure indicated for the particular locality.

[b]The figures given are recommended as minimum. These requirements do not provide for tornadoes.

The basic expression for pressure loading does not take into account rapid pressure variations. Gust effect, if needed, may be handled approximately by the use of a separate gust factor, C_g. For instance, for short gusts of about 1-s dura-

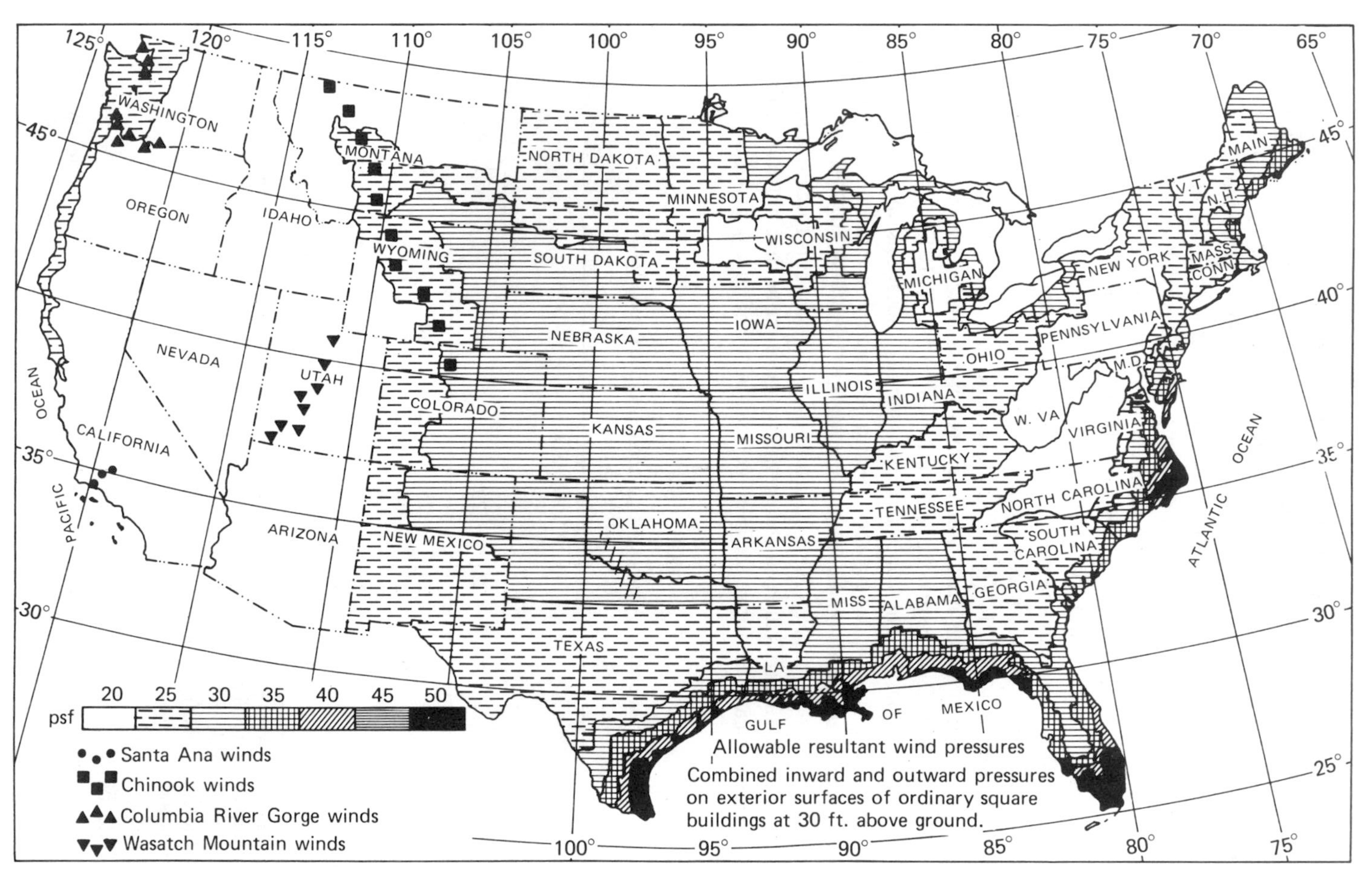

FIGURE 5-3. Wind pressure map.

tion, a factor of $C_g = 1.3$ is recommended for narrow structures. For longer gusts, say 10 s, C_g could be taken as 1.1. Actually, the C_g factor should be applied to the wind velocity, V, before calculating the equivalent static pressure loading, p.

Wind forces may also exert considerable uplift pressures, which could occasionally reach as much as 40 lb/ft^2. This effect must be considered when designing structures, particularly open ones; for failure to consider such effects could lead to structural collapse. According to the UBC, the roofs of all enclosed buildings or structures must be designed and constructed to withstand pressures acting upward normal to the surface equal to 0.75 times values given in Table 5-6 for the height zone under consideration. It defines an enclosed building as "a building enclosed at the perimeter with solid exterior walls." Roofs of unenclosed buildings, roof overhangs, architectural projections, eaves, canopies, cornices, marquees, or similar structures unenclosed on one or more sides must be designed to withstand upward pressures of 1.25 times values set forth in this table. The upward pressures are assumed to be acting over the entire roof area. If the roof slope exceeds 30°, the roof area must be designed to withstand pressures, acting inward normal to the surface, equal to those specified for the height zone in which the roof is located, and applied to the windward slope only. The importance of this uplift effect can be appreciated when one recognizes the fact that wind blowing across an airfoil provides sufficient lift for an aircraft weighing many tons. Adequate anchorage of the roof to walls and columns must be provided to resist this uplift pressure.

The overturning effect of wind forces must also be considered when designing structures. This becomes most pronounced for those structures which have a relatively narrow dimension in the direction of the wind loading. The overturning moment must be balanced by the stabilizing moment of the dead load only; live load may not be considered active in this particular case, and adequate anchorage of walls and columns to the foundation must be provided to resist this effect. It is required that the factor of safety against overturning be at least 1.5 (i.e., the overturning moment from the wind pressure must not exceed two-thirds of the dead-load stabilizing moment).

Earthquake forces

The February 1971 San Fernando, California, earthquake was probably the most thoroughly documented earthquake in history with regard to measurements of ground-motion behavior, acceleration intensities, and building response and damage incurred. Since that time, the structural engineering profession has examined and refined earlier assumptions regarding the ultimate design of earthquake-resistant structures. This reexamination has resulted in three revisions (1971, 1973, 1974) of the "Recommended Lateral Force Requirements and Commentary" by the Seismology Committee of the Structural Engineers Association of California (SEAOC). Actually, this is not a code, per se, but a series of recommendations. However, most, if not all, of the recommended seismic design changes that have appeared in these publications have been adopted, with certain modifications, both by local jurisdictions and ICBO in their Uniform Building Code.

To properly interpret these seismic codes, and their many revisions, it is becoming of paramount importance that the design engineer understand the basic theory substantiating, to some degree, these various code provisions. The following discussion represents a very brief and somewhat incomplete treatment of these considerations. Much of the following discussion has been condensed from an article by Edward J. Teal, appearing in the *Engineering Journal of the AISC* (Fourth Quarter, 1975), "Seismic Design Procedure for Steel Buildings." The authors express their appreciation to AISC for granting permission to reproduce portions of the article herein. It has been reduced to as simple terms as possible to facilitate understanding by those readers who possess a limited knowledge of the dynamics of structures. Since this text is concerned only with the design of reinforced masonry structures, those aspects of seismic analysis that bear directly or indirectly upon this type of construction are emphasized. Those items that relate primarily to moment resisting frame buildings are, in the main, omitted from this discussion, except where they are needed for clarification. The reader is referred to the SEAOC Recommendations and other publications, such as *Seismic Design Practice for Steel Buildings*, published by the AISC, and the *Lateral Force Design of Small Reinforced Concrete Buildings*, published by the Portland Cement Association, for further insight into seismic design procedures. Actually, the intent herein is merely to mention some of the more fundamental aspects of seismic design as they may apply to masonry, in preparation for handling, with some facility, certain design cases. Certainly, it is not expected that a student will become a seismic design expert upon completion of this simple treatise. That becomes a matter of further study, accompanied by a considerable amount of sound engineering experience, both in the office and in the field. This study must be an ongoing one, since so much of the material is undergoing constant change and revision.

Seismic Design Philosophy

Before any further discussion is offered regarding the latest seismic design provisions, the following points should be noted, because they represent the general SEAOC philosophy underlying the latest seismic codes.

1. Code seismic provisions apply to all buildings, high or low, whether they utilize shear walls, braced frames, or moment resisting frames to stabilize the structure laterally. Parenthetically, it should be pointed out that, in the total presentation, it would appear that more emphasis is placed on tall buildings and moment frames. This is probably true, inasmuch as the available dynamic data source stems mainly from such types of structures, since they are the ones most frequently instrumented. The sections dealing with drift control, for instance, apply strictly to tall buildings, for drift is almost always critical to the design of any moment resisting space frame (steel or concrete), whereas drift is less critical in shear wall or braced frame structures.
2. The vibratory response of buildings to ground motion falls primarily within the elastic range, except when the ground-motion intensity is violent enough

to cause the system to move into the inelastic range. Thus, the dynamic response analysis almost always involves elastic vibration. That applies to the considerations in this discussion, unless otherwise noted.

3. The Code may seem to imply that response forces due to lateral ground motions apply to only one axis at a time. Obviously, horizontal ground motions are not unidirectional. For any given site, there will be ground motions oriented radially to the source event in random directions and perpendicular to those radial motions. These motions, of course, can be resolved into components parallel to the two building axes. Thus, the Code addresses itself to a consideration of only one of these components at a time, to simplify the design approach. It is tacitly assumed that the amplification of the forces in the building due to simultaneous vibration on the two axes can be generally represented by some sort of amplification of the uniaxial forces.

4. The Code does not require design consideration of vertical ground motions. This stems from the fact that the design for vertical loads includes safety factors that will generally provide for the instantaneously added dynamic forces imposed by vertical ground motions. It is recognized that this premise may not apply when large overturning forces are involved, but this factor is generally included in the total probability of occurrence rather than as a separate variable.

DYNAMIC FACTORS IN EARTHQUAKES

Dynamic forces

In an earthquake, all parts of a structure exert dynamic loads, because of the inertial effect of motion. Static loads (dead and live) are entirely independent of the supporting structure, whereas a dynamic loading is dependent upon the dynamic characteristics of the supporting structure. Furthermore, static loads are independent of any loads preceding the loads being considered. On the other hand, dynamic loads vary with every change in motion and, at any given time, generally depend upon the characteristics of the preceding motion, as well as the motion at the instant considered.

Dynamic systems

A *single-degree-of-freedom* (SDF) *system* possesses one lumped mass, which can be simulated with a single weight set at the top of a slender, weightless cantilever rod. The period of vibration (defined later) will depend upon the weight plus the stiffness of the rod. A structure with several lumped masses constitutes a *multidegree-of-freedom* (MDF) *system* and its vibration becomes a combination of the vibrations due to several lumped masses, there being as many vibration modes as there are lumped masses. Each mode has a period of its own and can be represented by a SDF system of the same period. The mode with the longest period is referred to as the *first* or *fundamental mode* and subsequent modes, having shorter periods or higher frequencies, are called *higher modes*. It is to be observed that the period relations depend upon the mode shapes.

The deflected shape of the structure for any single mode of vibration is constant for that structure regardless of the vibration magnitude. This really means that, although the displacement amplitude varies with time, the distribution of the displacement magnitudes throughout the height remains constant. So the acceleration distribution, for a single mode of vibration, will also remain constant. If the mode shape and maximum vibration at the top are known, the maximum vibration at any level can be obtained for that mode. Figure 5-4 simulates some modal shapes (no scale) for, say a three-floor (lumped-masses) building.

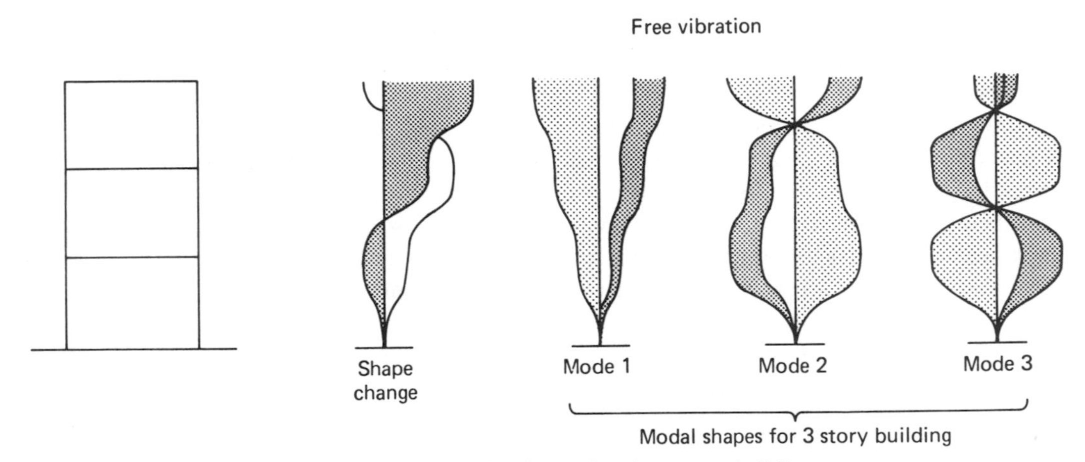

FIGURE 5-4. Modal shapes for three-story building.

Ductility

Ductility typically refers to the ability of an element or a system to withstand a considerable amount of postyield distortion, without exhibiting any significant loss of strength. This is an extremely important structural property, particularly in the case of the lateral-load-resisting system, since it does measure the ability of that structural system to adjust to any local inelastic behavior, by simply shifting additional load increments to some other section or component which has yet to exhibit a postyield condition. Furthermore, this property is extremely important, because it gives visual evidence of impending distress due to the large distortions taking place, and thereby precludes the possibility of any sudden failure occurring without due warning. In earthquakes, the property is doubly important because it allows the structure to absorb considerable energy, thereby withstanding more intense ground motions, after its yield strength has been exceeded. Most assuredly, brittle behavior, with its sudden explosive-type failure mode, has no place in earthquake-resistant structures.

For static loadings, if a system is ductile enough, the yield strength of any one element can be exceeded, but failure will not occur until the entire system reaches the yield capacity. For dynamic loading, failure will not even occur after the entire system reaches yield as long as the $P\Delta$ moments are not excessive. However, the ductility may have to be maintained through a number of small inelastic strain reversals and some large strain reversals.

Damping

A perfectly elastic system, set into vibratory motion, would continue to vibrate forever if the vibration were not stopped by some exterior force. However, no system is perfectly elastic, and therefore the vibratory motion will die out as a result of energy losses resulting from internal strains. This energy absorption is referred to as *damping*. This characteristic is generally expressed as a percentage of *critical damping*. The latter is the damping effect that would cause the vibratory motion to cease in one swing after free vibration starts. The first small percentages of damping greatly reduce peak responses, because peak responses are generally associated with short response-time durations and therefore involve little energy. Damping represents energy losses from many sources.

This property is extremely critical in seismic considerations, as the higher the percent critical damping, or energy absorption, the lower the value of the seismic response coefficient C. This lowered coefficient, since it reflects the acceleration of the structure, will result in a reduced horizontal force being exerted on the building by the ground motion. Refer to Figure 5-5, which illustrates this damping effect. This spectrum analysis of an El Centro earthquake, 1940, indicates, for instance, that an undamped response of 500%g at a $\frac{1}{4}$-s period, is reduced to about 60%g with 20% critical damping.

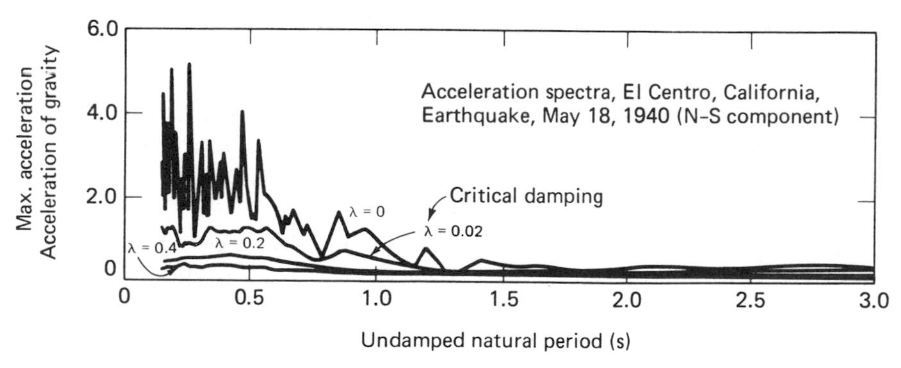

FIGURE 5-5. Effect of critical damping on acceleration.

Period of vibration

For the purpose of seismic analysis, all structures are assumed to vibrate in accordance with the laws of harmonic motion, their response being governed by the dynamic characteristics of the structure itself. These characteristics are a function of the weight and stiffness of the structure; thus, the building's response to ground motion at its base will be measured by these dynamic properties. One of the parameters of building response is the period of vibration, T, which is the time required to complete one cycle of oscillation and is the reciprocal of the natural frequency of vibration f. The natural frequency equals the circular frequency ω divided by 2π. Furthermore, the circular frequency of a SDF structure is

$$\omega = \sqrt{\frac{\lambda}{m}} \qquad \text{and} \qquad f = \frac{\omega}{2}$$

where m, the mass, $= W/g$ and λ, the stiffness, $= F/\Delta$ (force $\div$ deflection). Then the period becomes

$$T = \frac{1}{f} = \frac{2\pi}{\omega} = 2\pi\sqrt{\frac{m}{\lambda}} \qquad \text{or} \qquad T = 2\pi\sqrt{\frac{W\Delta}{gF}}$$

where T is the period for a SDF system. Expressing F as $C_1 W$, the period can also be defined as

$$T = \frac{2\pi}{\sqrt{g}}\sqrt{\frac{\Delta}{C_1}} = 0.32\sqrt{\frac{\Delta}{C_1}}$$

where Δ, the drift, is expressed in inches.

As can be seen, the building period is directly related to the square root of the horizontal displacement at the roof (drift) caused by a dynamic horizontal force equal to the weight of the building above the base. Since this is a dynamic force, the total horizontal force due to the weight of the building must be distributed throughout the height of the building in direct proportion to the variation in the dynamic response acceleration of the building. This acceleration, relative to the base, increases approximately uniformly with the height for most buildings. This concept accounts for the triangular force distribution assumption, with the maximum lateral force occurring at the top, as currently defined in the SEAOC Recommendations. It is also the basis for UBC formula 12-3, which closely approximates the fundamental or first-mode period of the building when the design drift is known.

A much simpler formula than UBC 12-3, obtained by referring to the previously derived basic period formula, formulated by Teal in his discussion, expressed in general terms becomes

$$T = C_T\sqrt{\frac{\Delta}{C_1}}$$

where Δ is the horizontal displacement at the top, in inches, and C_1 is the force coefficient used to derive Δ. When the total building weight is used as the horizontal dynamic force, $C_1 = 1$. Otherwise, $C_1 =$ ZICS (refer to base shear formula in a later section), if the code drift-force coefficient is used. The period constant, C_T, depends upon the deflected shape of the structure, but it can be reasonably approximated as 0.25 to 0.30 for the typical MDF constant-drift building. This simplified version of Code formula 12-3 can be used to approximate the period when only the drift coefficient is known. By definition, this story drift coefficient equals the ratio of interstory horizontal displacement to story height, usually expressed as a percentage (e.g., 0.5% for frame buildings).

Since the majority of masonry buildings, however, are relatively low, rather stiff structures, it is not appropriate in most cases to use the drift coefficient as a parameter in evaluating their periods. Thus, the UBC period formula 12-3A, which states that $T = 0.05h_n/\sqrt{D}$ (h_n = height of building and D = building dimension in direction of lateral force), is provided for those circumstances. This empirical formula is based upon recorded periods for shear-wall buildings. This formula seems to give a close empirical estimate for T in this type of building.

Response Coefficient

The vibration of a SDF system due to a continuously varying base motion will, at any time, be the summation of the effects of the base motion impulses up to that time. The maximum vibration magnitude reached during any length of time after the base motion starts is defined as its spectral value. If a series of SDF systems are subjected to the same base motion, there will be a series of maximum values related to SDF system periods, which will form a spectral curve for that base motion. Thus, any given irregular motion will produce an individual response curve or response spectrum. Knowing the base motion, SDF period, and percent critical damping, one can obtain, from the appropriate curve, the maximum vibration at the top of a SDF system, measured in terms of acceleration, velocity, or displacement relative to the base. These relationships are represented in two different manners in Figures 5-6 and 5-7. The log scale in Figure 5-6 permits the graphing of these different curves on one coordinate system, but it does tend to radically condense the important areas of the period scale while expanding those areas of little concern. The standard coordinate plot in Figure 5-7 does not exhibit this distortion; however, three separate curves would be required to portray the information conveyed in Figure 5-6.

For a MDF system, if any fixed relation between the periods of the vibration modes is assumed, a computer can be programmed to obtain the response of all the considered modes simultaneously. Thus, a complete MDF response spectrum for any structure, with the given modal relationship, can be obtained for a given input motion. MDF spectra for the most typical modal period relationship ($1, \frac{1}{3}, \frac{1}{5}, \frac{1}{7}$, etc.) have been run for many of the key design earthquake motions.

The Code provides a "static equivalent" seismic response formula:

$$C = \frac{0.067}{\sqrt{T}} \qquad \text{(maximum value of 0.12)}$$

This formula is actually expressed in the form of a generalized response spectrum, it being based upon certain assumptions and theory. Earthquake ground motions are erratic, random, and constantly changing throughout the duration of strong motion, because they are generated in and transmitted through extremely complex ground structural arrangements. The determination of the dynamic response to this very erratic motion is a complex mathematical problem that can only be solved practically by a modern computer. The entire record of response for a large number of different SDF systems involves an immense volume of data, so it is generally reduced to simplify the analysis to that of a history of only the maximum response reached at any given time during an earthquake for each SDF system. This, then, produces a spectrum of maximum responses (spectral acceleration, velocity, or displacement) to a given ground-motion sequence for a range of dynamic systems, as defined by their dynamic periods of vibration.

FIGURE 5-6. Log tripartite plot of spectra. Damping values are 0, 2, 5, 10, and 20% of critical.

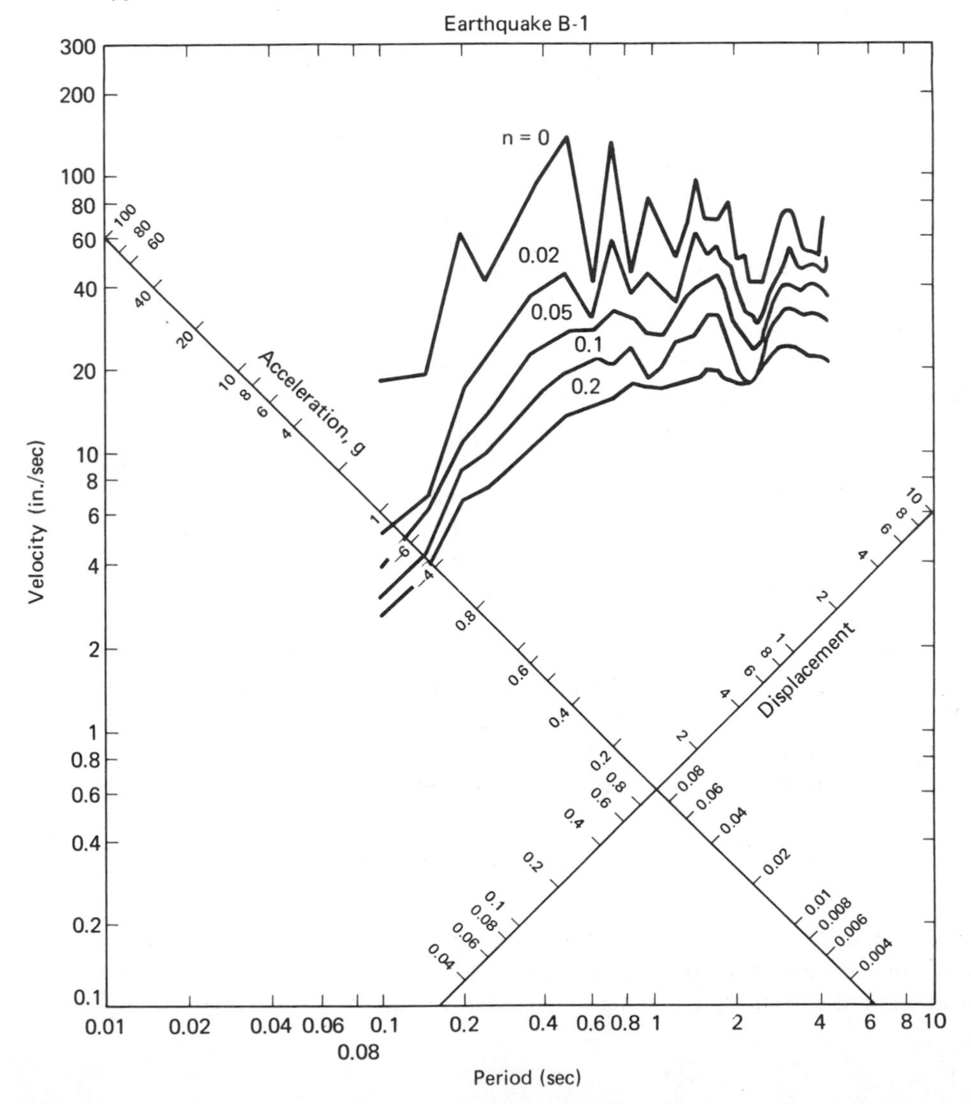

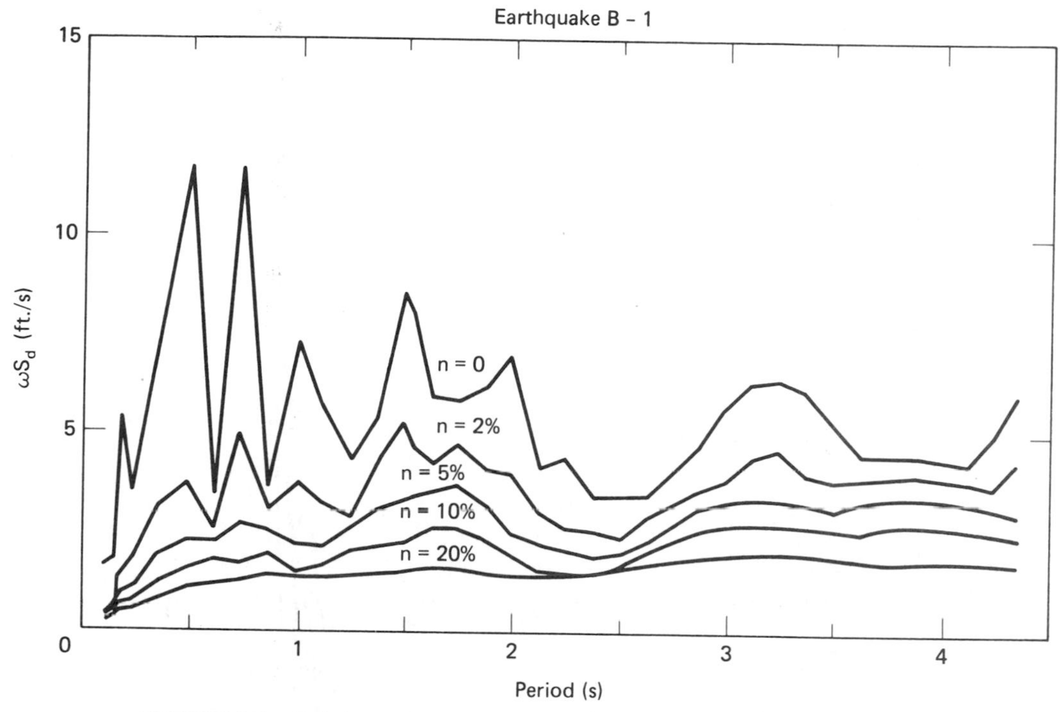

FIGURE 5-7. Relative velocity response spectrum. Damping values are 0, 2, 5, 10, and 20% of critical.

Now each individual site ground motion will have a unique response spectrum, and as such there will be a dominant period. For that ground motion, the maximum response (S_a, S_v, or S_d) of each SDF system can be picked off the spectrum and, further, the effect of a change in period can be exactly obtained. These will be modified by the degree of damping in the foundation and structure. However, any individual erratic random ground motion produces an erratic response curve, so that the effect on response of small period changes is also erratic and does not necessarily reflect a general trend for other ground motions. Thus, to obtain a meaningful design spectrum, one must examine a large number of both actual and statistically generated ground motions, and from these produce a smooth curve that would represent envelopes of all spectra. This envelope curve does not represent a true relationship between response and system-period for a specific ground motion, but it does indicate that any given motion response lies within the bounds of the envelope produced. Only those responses that reach the envelope limits at some point are critical to the development of the envelope. Figure 5-8 portrays an acceleration spectra for an actual strong earthquake (El Centro) and several simulated ones, thus demonstrating how envelope spectra can show the trend of the relation between acceleration and building period for a SDF system. The MDF spectra for a typical modal period relationship are shown in Figure 5-9.

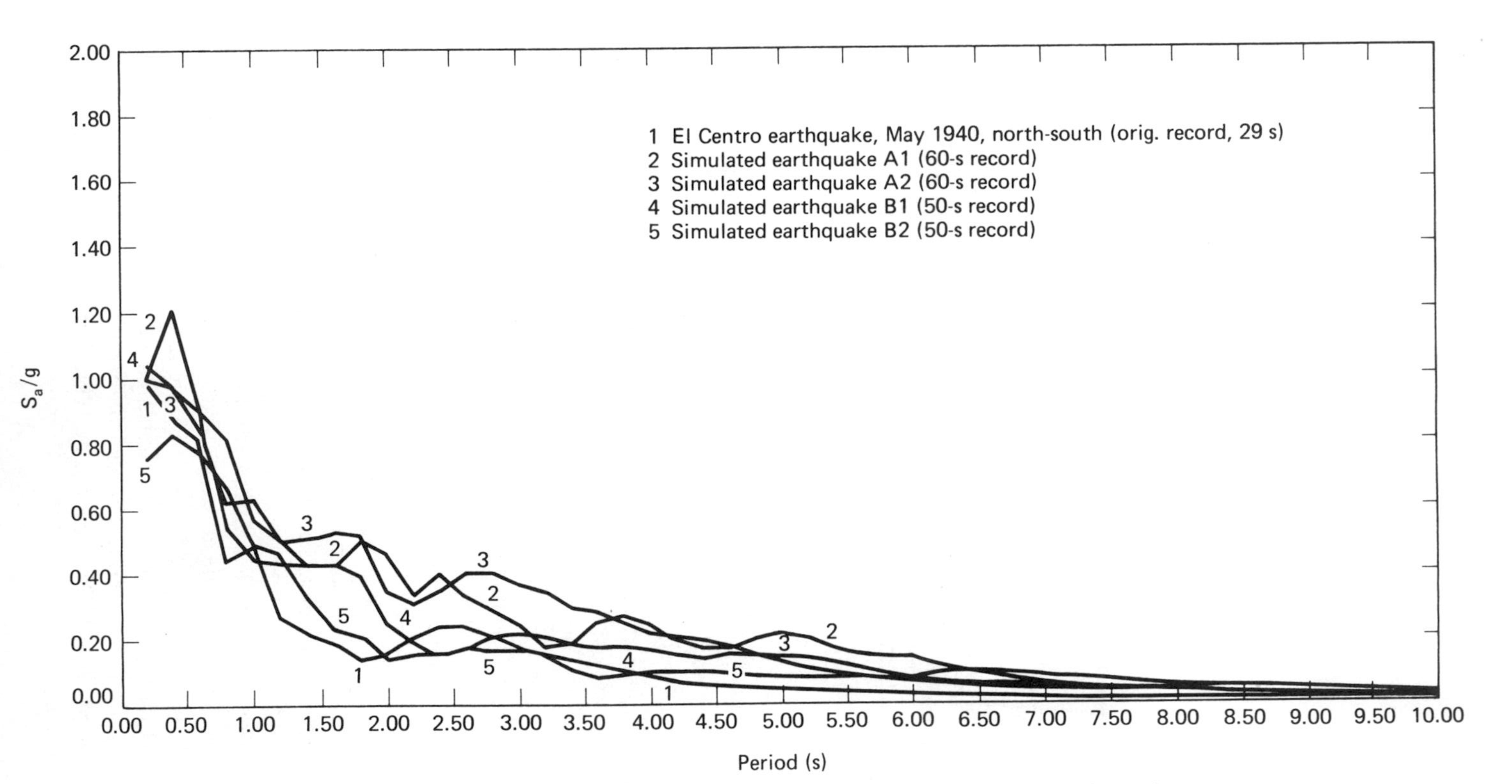

FIGURE 5-8. Single degree of freedom absolute acceleration spectra, 5% of damping.

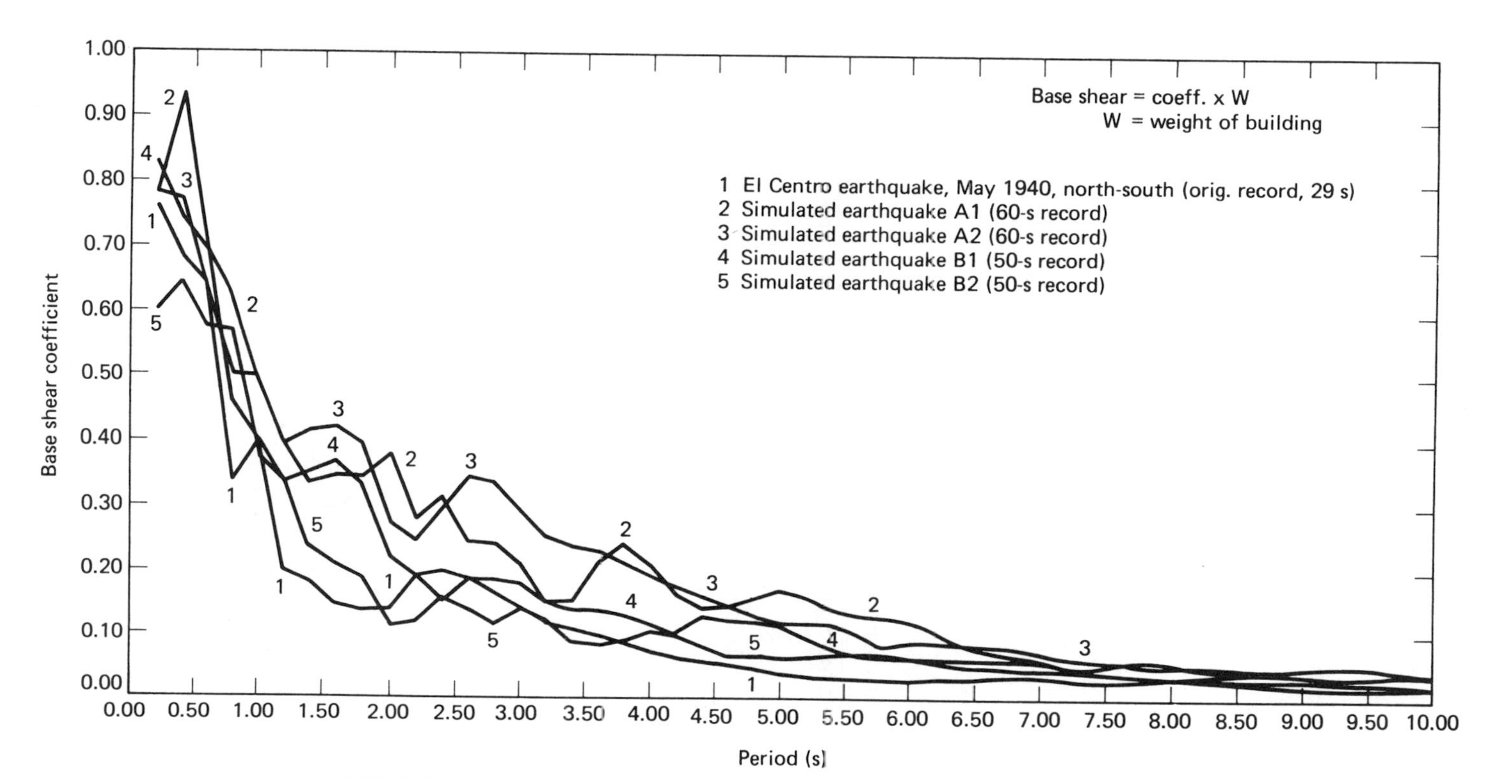

FIGURE 5-9. Multi-degree-of-freedom base shear coefficient spectra. Computed for eight modes, 5% damping. Modal period relations 1, $\frac{1}{3}$, $\frac{1}{5}$, $\frac{1}{7}$, $\frac{1}{9}$, $\frac{1}{11}$, $\frac{1}{13}$, and $\frac{1}{15}$.

It can be shown from SDF response analysis that the purely elastic response of a SDF system for a range of periods with a constant spectral velocity, is given by the formula $a/g = C'/T$, where $C' = S_v(2\pi/g)$. This basic formula would apply over the period range, where the S_v is a constant value or up to the peak acceleration period. For a uniform-mass and uniform-drift-coefficient building, the purely elastic MDF response, assuming a constant spectral velocity, is given by the formula $a/g = C'/T^{3/4}$. If it is assumed that the spectral velocity value in the design range is increasing somewhat with period, rather than remaining constant, the exponent for T would be less than $\frac{3}{4}$. The Code formula ($C = 0.067/T^{1/2}$), which uses an exponent of $\frac{1}{2}$ for T, could reflect this assumption, or it could merely represent a subjective empirical design response adjustment. The effect of decreasing the exponent (say from $\frac{3}{4}$ to $\frac{1}{2}$) is to decrease the formula response for shorter building periods, and to increase the response values for longer period buildings. Since this expression is used for design, with modifiers and working stresses, the constant in the formula does not actually directly relate to the response to any given level of ground-motion intensity.

STATIC LOAD EQUIVALENCE (SEAOC)

Base shear formula

This total horizontal seismic shear force, acting at the base of a structure, is a function of the acceleration of each of the masses of the structure relative to the base, the effects being added algebraically at any instant. In formulating a "static design" equivalent, since force equals total building mass times an assumed relative acceleration, UBC formula 12-1 states that the base shear V (kips) $= ZIKSCW$. It has four empirical modifiers against the response formula C: Z, I, K, and S. The value Z can be determined from a seismicity map that shows the various areas of the United States where the probabilities of earthquakes of varying degrees of intensities are likely to occur. This map, taken from Chapter 23 of the UBC, is shown in Figure 5-10. In zone 0, there is no likelihood of earthquake damage. Zone 1 has a Z value of $\frac{3}{16}$. This is described as an area where minor damage could occur. Here distant earthquakes may cause damage to structures with fundamental periods in excess of 1.0 s. It corresponds to locations where earthquake intensities of V and VI on the modified Mercalli (MM) Intensity Scale of 1931 may occur. Zone 2 ($Z = \frac{3}{8}$) lies within an area where moderate damage could occur. It corresponds to an intensity of VII on the MM Scale. Zone 3 ($Z = \frac{3}{4}$) could experience major property demage, and it corresponds to an intensity of VIII and higher on the MM Scale. Zone 4 ($Z = 1$) embraces those areas as in zone 3 that lie in close proximity to certain prescribed major fault systems.

The I value represents an *occupancy importance coefficient*. According to Table 5-7, essential facilities would have a factor of 1.5; any building where the primary occupancy is for assembly use with more than 300 persons per room has a factor of 1.25; and all other occupancies are assigned a value of 1.0. Essential facilities refer to those buildings or structures that must be safe and usable for emergency purposes after a major earthquake has occurred, in order to preserve the peace, health, and safety of the general public. Such facilities could include hospitals, fire and police stations, municipal government centers, public utility service centers, civilian emergency centers, communication centers, and primary and secondary school buildings.

TABLE 5-7

Values for Occupancy Importance Factor I
(*UBC Table 23-K*)

Type of occupancy	*Value of I*
Essential facilities	1.5
Any building where the primary occupancy is for assembly use for more than 300 persons (in one room)	1.25
All others	1.0

K is the *framing coefficient*, which is supposed to indicate, in general, the dynamic characteristics of the lateral-force-resisting system. To some extent it is a measure of the system's ultimate damping and ductility characteristics which were described earlier. Table 5-8 shows values of *K*, which range from 0.67 for buildings with a ductile moment resisting frame through 1.33 for buildings with a box system.

TABLE 5-8

Horizontal Force Factor *K* for Buildings or Other Structures[a]
(*UBC Table 23-I*)

Type or arrangement of resisting elements	*Value[b] of K*
1. All building framing systems except as hereinafter classified	1.00
2. Buildings with a box system as specified in Section 2312(b)	1.33
3. Buildings with a dual bracing system consisting of a ductile moment resisting space frame and shear walls or braced frames using the following design criteria: a. The frames and shear walls shall resist the total lateral force in accordance with their relative rigidities considering the interaction of the shear walls and frames. b. The shear walls, acting independently of the ductile moment resisting portions of the space frame, shall resist the total required lateral forces. c. The ductile moment resisting space frame shall have the capacity to resist not less than 25% of the required lateral force.	0.80
4. Buildings with a ductile moment resisting space frame designed in accordance with the following criterion: the ductile moment resisting space frame shall have the capacity to resist the total required lateral force.	0.67
5. Elevated tanks plus fill contents, on four or more cross-braced legs and not supported by a building	2.5[c]
6. Structures other than buildings and other than those set forth in UBC Table 23-J	2.00

[a]Where wind load as specified in Section 2311 would produce higher stresses, this load shall be used in lieu of the loads resulting from earthquake forces.

[b]See Figures 1, 2, and 3, UBC Chapter 23 and definition of *Z* as specified in Section 2312(c).

[c]The minimum value of *KC* shall be 0.12 and the maximum value of *KC* need not exceed 0.25.

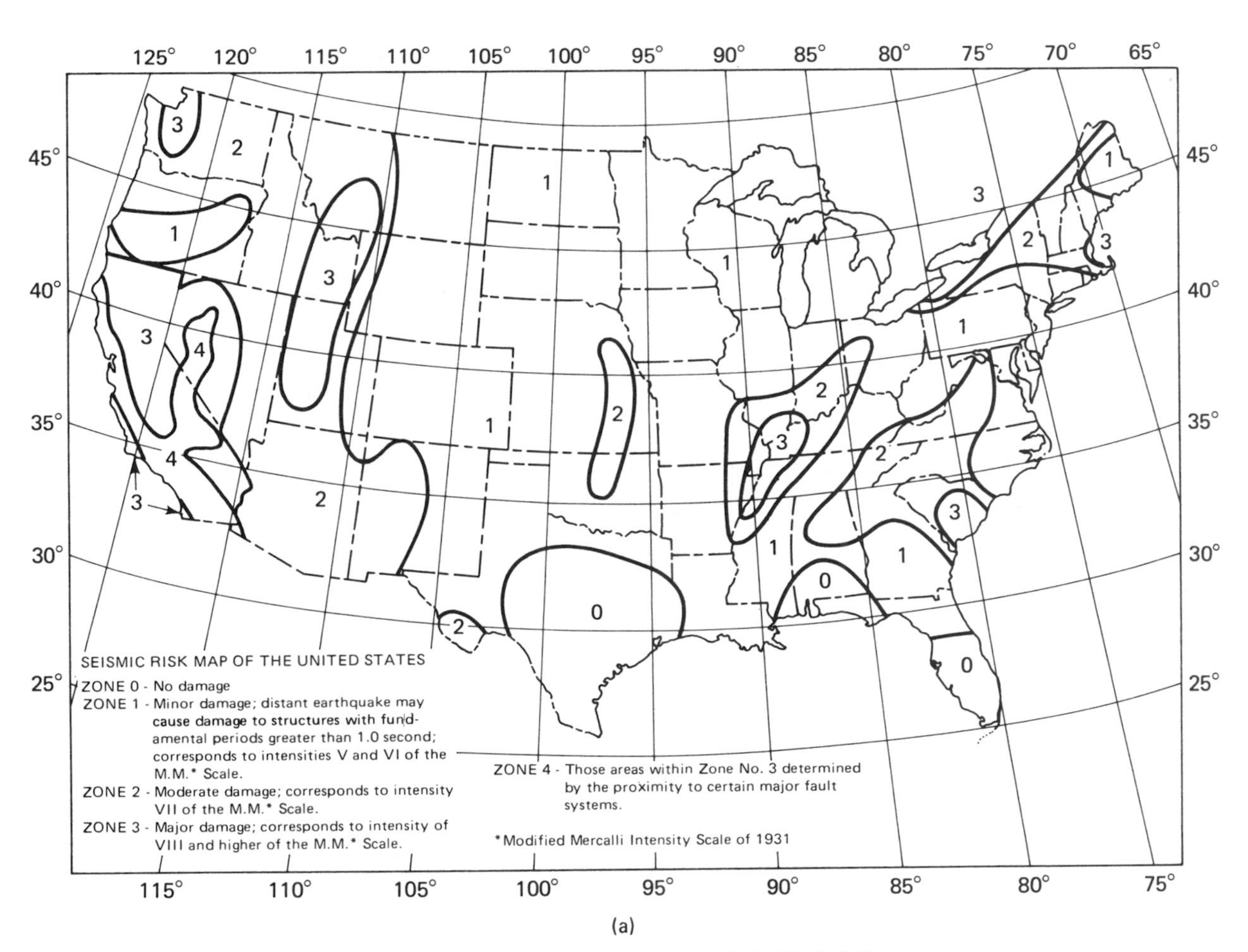

(a)

FIGURE 5-10. Seismic zone map of the United States.

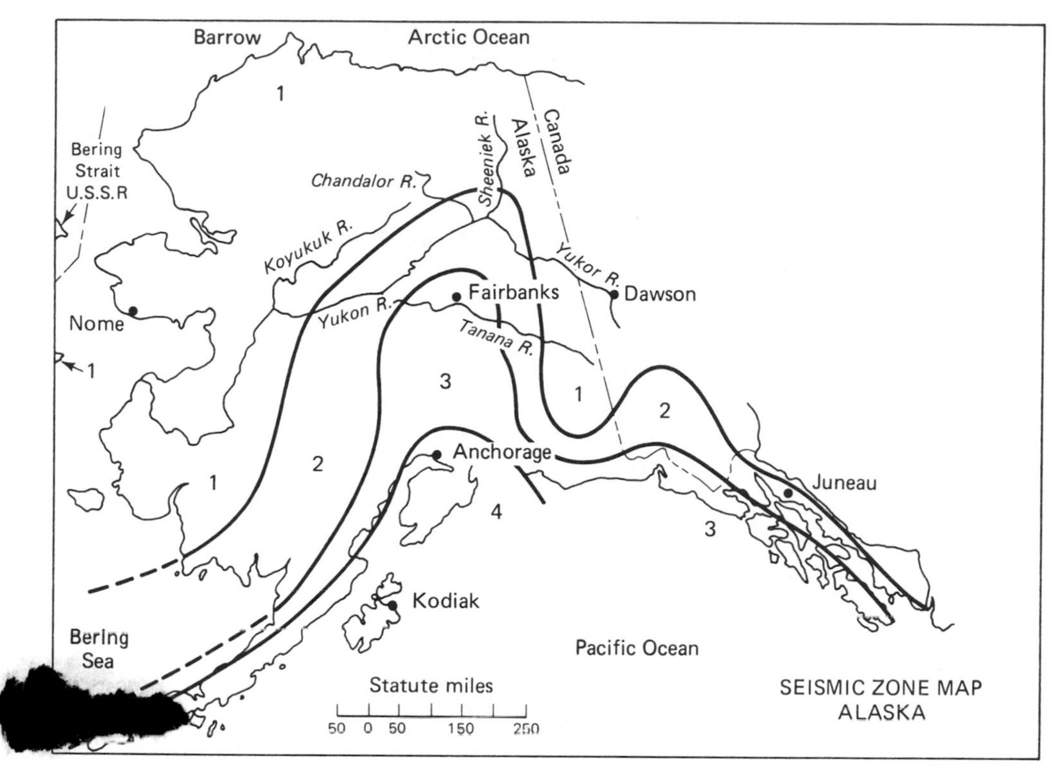

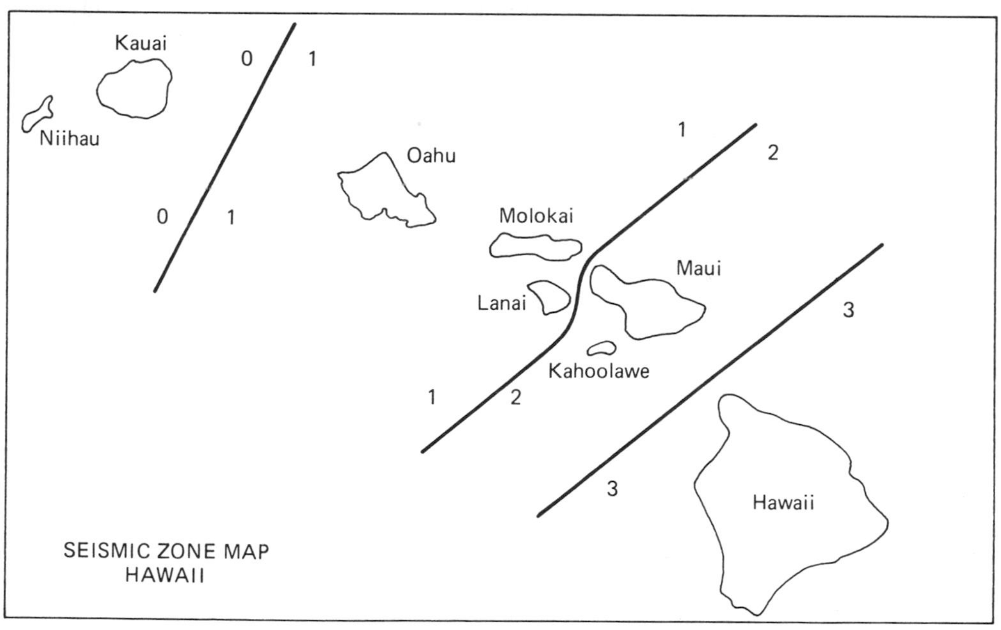

(b)

FIGURE 5-10 (cont.) Seismic zone map of Hawaii and Alaska.

Masonry building design will generally be covered under Types 1 or 2. They are arbitrarily limited to a maximum height of 160 ft. These types may be described as follows:

> *Item 1:* "All building framing systems except as hereinafter classified," having an assigned $K = 1.00$, include building systems consisting of complete vertical load-carrying frames, with the lateral resistance provided by shear walls. In effect this recognizes that, if the lateral system fails due to overstress from unpredictable brief lateral forces, the building would not immediately collapse. Therefore it would be somewhat less hazardous than a building which depended upon the lateral resisting element (e.g. bearing-shear wall) for vertical support also.
>
> The example shown in Chapter 13 (Figure 13-26) combines a multistory complete precast concrete frame to carry the vertical loads, with shear walls to provide the lateral stability.
>
> *Item 2:* "Buildings with a box system," with $K = 1.33$, is the most common masonry system. The shear walls provide the lateral stability as well as vertical support. These are used in low, medium, and high rise, as shown in the examples depicted in Chapter 13. (Figures 13-18, 13-20, 13-22, etc.)

The modifier S is a *site structure resonance factor*, based upon the assumption that a direct relation exists among the site period, the ground motion that may be expected to be generated at the site, and the response of different per[illegible] buildings on that site. It may be evaluated from UBC formula 12-4, wherei[illegible] expressed in terms of T, the building period, and T_s, the characteristic site period. S cannot be taken as less than 1.0. T is calculated as previously described, being limited to a 0.3-s minimum. The range of T_s values would be obtained from properly substantiated geotechnical data. T_s will be taken as that value within the range of site periods that is nearest to T, with a 2.5 maximum value. Figure 5-11 shows how the S factor relates to the ratio T/T_s. Figure 5-12 gives the relation between minimum and maximum S and the building period, T. Finally, Figure 5-13 plots both minimum and maximum C and $C \times S$ factors against T. Actually, Figure 5-13 can serve as a very useful design aid inasmuch as CS and T are interdependent and cannot be directly evaluated.

The last value in the base shear formula, W, is simply the total dead weight of the building itself, including any partition loading. In storage and warehouse occupancies, 25% of the floor live load must be included in the W value. The dead load, as defined herein, includes the vertical load due to the weight of all permanent structural and nonstructural components of a building, such as walls, floors, roof, and fixed service equipment. Movable partitions, per Section 2304(d), are considered as a uniform dead load of 20 lb/ft^2.

Base shear distribution

If a building is assumed to vibrate predominantly in its fundamental mode, with the deflection curve a straight line (uniform dynamic drift), the amplitude of vibration is proportional to the height. Since the fundamental period of vibration

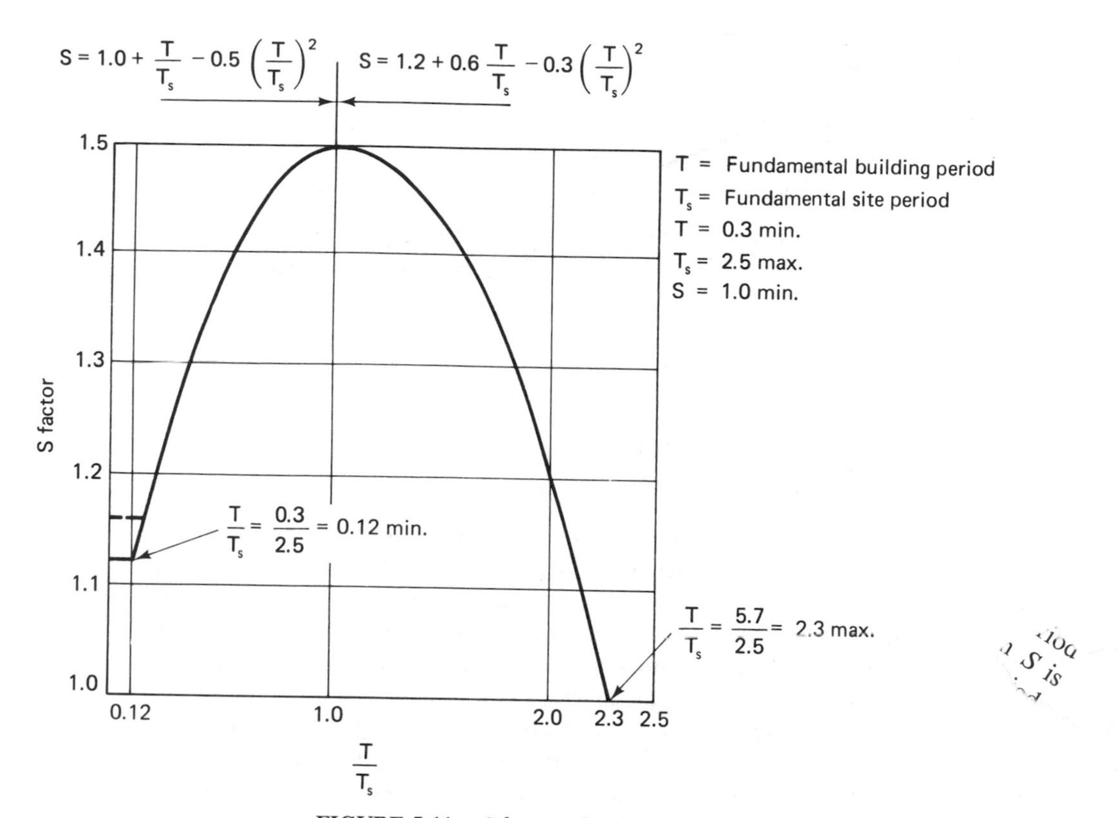

FIGURE 5-11. *S* factor related to *T*/*T*.

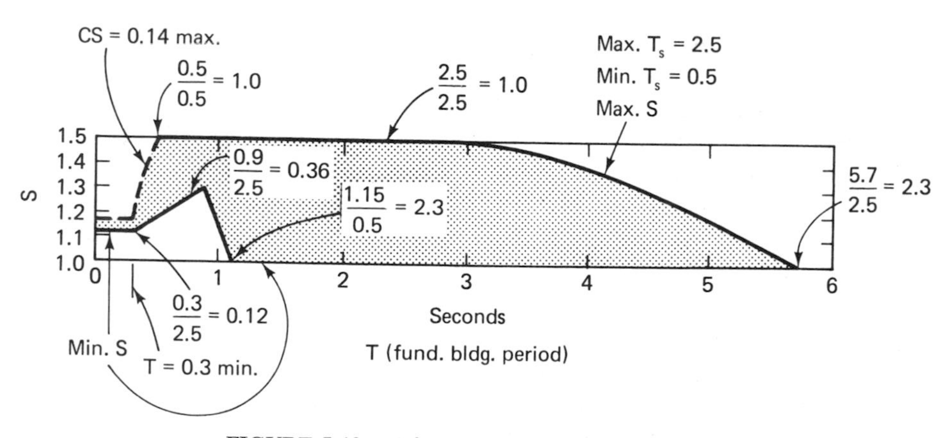

FIGURE 5-12. *S* factor related to building period.

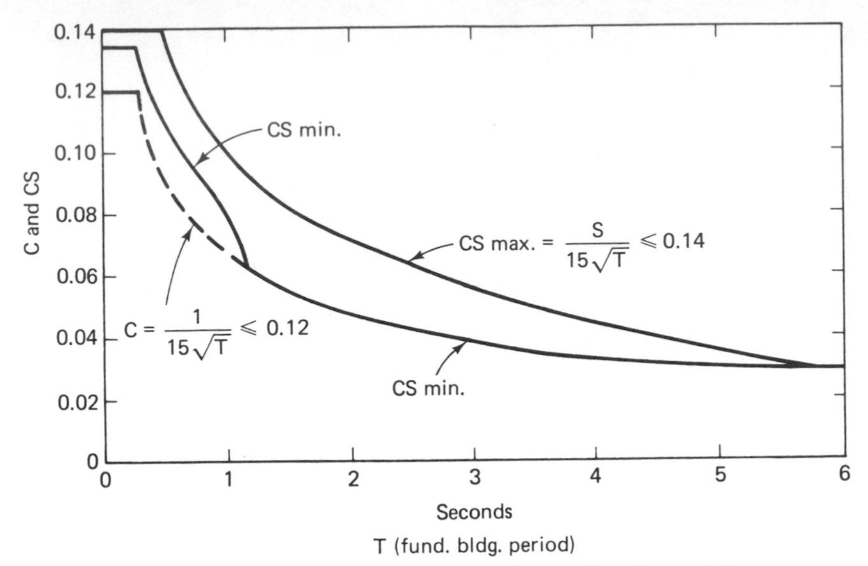

FIGURE 5-13. *C* and *CS* factors related to building period.

applies to the entire building, the acceleration at any level must also vary directly with the elevation above the base. The lateral forces acting on the structure will therefore vary linearly from zero at the base to a maximum at the top. The Code defined triangular force distribution, as illustrated in Figure 5-14, is based upon this general assumption. Accordingly, when the forces are accumulated from the top down at any floor level, a parabolic shear distribution envelope is produced.

Code formula 12-5 separates out of the base shear a horizontal force, F_t, (UBC 12-6), which is assumed to act directly at the top of the building. This latter force ostensibly represents a sort of "whiplash" effect, and it has a marked effect on design only at the top of the building, where the frame is light, and also on the overturning moment. The expressions are

$$V = F_t + \sum_{i=1}^{n} F_i$$

where

$$F_t = 0.07TV \leq 0.25V$$

Note that F_t may be considered as zero where T is 0.7 s or less. The remainder of the base shear is distributed to all levels above the base, including the roof according to Code formula 12-7, which states that

$$F_x = \frac{(V - F_t)w_x h_x}{\sum_{i=1}^{n} w_i h_i}$$

At each level designated as x, the force F must be applied over the height of that story in accordance with the mass distribution at that level.

What this formula does, then, is to provide for that basic triangular force distribution which occurs in a uniform dynamic stiffness and uniform dynamic drift building. Moderate variations of the triangular distribution, which would be caused

FIGURE 5-14. Loading and shear diagrams.

by dynamic stiffness variations related to mass distribution, are accounted for by including the actual weight at each level and the height above the base for each level. Actually, the envelopes of maximum dynamic shears, obtained by time-history dynamic computer analysis for many buildings subjected mathematically to several earthquake ground motions, confirm this general MDF envelope distribution of the horizontal forces.

Overturning moment

The overturning moment, imposed on the building, equals the algebraic sum of the moments of all the forces above the base multiplied by their heights above that base. If the forces were represented by an envelope of maximum responses which are reached at different times, the overturning moment would be considerably overestimated. However, since the first mode is dominant for these overturning moments, and the forces for the first mode do all reach an algebraic maximum at the same time, the Code-applied effects do not attain an excessive influence. It could perhaps be recognized that overturning moments generally do not pose a threat to actually overturn the building anyway, because the transitory nature of the loading does not allow enough time to bring the building to a position past its center of rotation. However, this transitory phenomenon cannot be assumed to dispense with a look at the generation of axial forces in columns or piers within shear walls due to overturning, with its resulting magnification of the preexisting vertical load (dead and live) stresses. The imposition of these instantaneous forces, transitory though they may be, can cause serious consequences unless provisions are made in the design to overcome their effects. This magnification of the axial loads is demonstrated in the analysis of a two-story building provided in Chapter 12.

Distribution of total horizontal force

The Code specifies that the total horizontal force that applies at any level be distributed to the shear walls in proportion to their relative rigidities where the floor or roof diaphragms can be classed as rigid elements (see Chapter 10). Furthermore, at any level, the incremental changes of design overturning moment in the story under consideration must be distributed to the various shear resisting elements in that same proportion. Thus, the Code, in a general way, requires that an analysis be made for those axial forces induced on columns and piers by the design horizontal forces, distributed throughout the height of the building, as it cantilevers from its base. A rational stiffness-related distribution of horizontal forces, and the effect of the resulting induced axial forces, is thus required at each story level, from the top down to the foundation. Also, when checking in-plane shear stresses in masonry shear walls, one must multiply the design shear force acting on each wall by a factor of 1.5. Furthermore, it is generally assumed that the effects of uplift, caused by the overturning moment, are resisted by the least dead load that could possibly exist.

Horizontal torsional moments

In addition to the direct shears, distributed as noted above, torsional shear may be transmitted to vertical elements due to the existence of any horizontal torsional moment generated by the eccentricity between the center of mass and the center of vertical element stiffness (center of rigidity), when they do not coincide. This actual eccentricity, per UBC, cannot be taken as less than 5% under any circumstances. These torsional shear forces can only be added to the direct in-plane shear forces to determine the total shear force being carried by each of the vertical shear resisting elements, since the Code requires that negative torsional shears be neglected. Provision for neglecting negative shears is conservative, but it defies rational explanation.

Lateral forces on elements of structures

Under this heading are included parts or portions of a structure or elements which are attached to the structural system. The attached elements respond dynamically to the motion of the system, not to the motion of the ground. Since the system has fixed modal periods, resonance between the attached element and the structural sytem may pose serious problems. Thus, Code formula 12-8 reads $F_p = ZIC_pSW_p$. Values of C_p are given in Table 5-9, and these coefficients represent an attempt to provide a static force equivalent for the various elements listed, by considering both their possible response and the consequences of a failure. For example, C_p for an exterior bearing or nonbearing wall which must span between floors or pilasters is 0.20. Note that where C_p is 1.0 or more, C and S may each be taken as 1.0.

TABLE 5-9

Horizontal Force Factor C_p for Elements of Structures
(*UBC Table 23-J*)

Part or portion of buildings	*Direction of force*	*Value of* C_p[1]
1. Exterior bearing and nonbearing walls, interior bearing walls and partitions, interior nonbearing walls and partitions. Masonry or concrete fences	Normal to flat surface	0.20
2. Cantilever parapet	Normal to flat surface	1.00
3. Exterior and interior ornamentations and appendages	Any direction	1.00

TABLE 5-9 (Continued)

Part or portion of buildings	Direction of force	Value of C_p
4. When connected to, part of, or housed within a building:		
a. Towers, tanks, towers and tanks plus contents, chimneys, smokestacks, and penthouse	Any direction	0.20
b. Storage racks with the upper storage level at more than 8 ft in height plus contents	Any direction	0.20
c. Equipment or machinery not required for life safety systems or for continued operation of essential facilities	Any direction	0.20
d. Equipment or machinery required for life safety systems or for continued operation of essential facilities	Any direction	0.50
5. When resting on the ground, tank plus effective mass of its contents	Any direction	0.12
6. Suspended ceiling framing systems (applies to seismic zones 2, 3, and 4 only)	Any direction	0.20
7. Floors and roofs acting as diaphragms	Any direction	0.12
8. Connections for exterior panels or for elements complying with Section 2312(j) 3C	Any direction	2.00
9. Connections for prefabricated structural elements other than walls, with force applied at center of gravity of assembly	Any direction	0.30

[1]These C_p values have been modified somewhat in the 1979 UBC. See comments in this regard at the beginning of Chapter 12.

QUESTIONS

5-1. What does a load-frequency curve show?

5-2. What does the code mean by dead load?

5-3. How are dead loads generally determined?

5-4. Considering a normal-weight aggregate, what is the average weight of a hollow unit block wall 8 in. thick having a vertical steel spacing of 48 in?

5-5. What is meant by equivalent solid thickness, and where is it used?

5-6. What is live load in terms of code definition?

5-7. What are the two methods for determing roof live loads, per UBC recommendations?

5-8. Describe how the live-load percentage reduction works.

5-9. Describe how wind-load values are codified so that they may be readily used in design.

5-10. What is the datum level at which the wind velocity is generally measured?

5-11. What does a wind pressure map show?

5-12. How does the UBC treat uplift pressures in design?

5-13. What are some other factors affecting wind velocity and its resulting pressure on a building?

5-14. The overturning moment from wind pressure must not exceed what factor of the dead-load stabilizing moment?

5-15. How does the Code handle the fact that for any given site there will be ground motions oriented radially to the source event in random directions and perpendicular to those radial motions?

5-16. What is the first or fundamental mode?

5-17. How does the displacement amplitude vary with time and building height?

5-18. What is ductility and why is it highly significant or desirable in a building structure?

5-19. What is damping, and what is meant by percent critical damping?

5-20. Why is damping an extremely critical factor in seismic considerations?

5-21. One of the parameters of building response is period of vibration. To what does this refer? What is Teal's version of the expression for period in terms of horizontal drift of the top of the building?

5-22. What is the code formula that gives a close empirical estimate for the period of shear-wall buildings?

5-23. What is meant by a spectral value with reference to acceleration, velocity and displacement of a building frame?

5-24. To what does the response coefficient refer, and what is the static equivalent seismic response formula given by the Code?

5-25. Express the SEAOC base shear formula and describe what the loading diagram would look like.

5-26. Upon what assumption is the load distribution of Question 5-25 based. Sketch the shear distribution envelope for it.

5-27. The base shear formula stems from what fundamental law of motion?

5-28. To what does Z refer, and how is it determined?

5-29. To what does I refer, and how is it determined?

5-30. What is K, and how is it determined?

5-31. What is the modifier S in the base shear formula? What are the minimum and maximum values for S?

5-32. If S is not evaluated by an accepted geotechnical exploration method, what factor is assigned to it? Also, the product C times S is assigned a maximum value. What is it?

5-33. The last value in the base shear formula, W, refers to what?

5-34. Code formula 12-6 separates out of the base shear a horizontal force F_t. To what does it refer and what is its value?

5-35. Write the expression for the base shear distribution at each floor level above the ground, and define the terms.

5-36. In a multistory building, where would the overturning effects be significant?

5-37. Where rigid diaphragms exist, how does the Code specify that the total horizontal force at any level be distributed?

5-38. How are torsional shear stresses generated in a building? What minimum accidental eccentricity does the Code specify, and how are negative torsional shears handled?

5-39. To what does Code formula 12-8 refer? Write the expression for it.

5-40. To what does the coefficient C_p refer, and how is it obtained?

6

Engineering Design

It is the express intent of this chapter to discuss those basic concepts which form the basis for the analysis and design of reinforced masonry and further to develop rational design procedures for the various types of individual structural elements as well as for complete building systems. The precepts involved here are those of the elastic theory of *working stress design* (WSD), long utilized in designing reinforced concrete elements. As a matter of fact, most of the design or analysis formulas are similar to those for reinforced concrete except that the ultimate strength of the masonry, f'_m, and the allowable stresses are reduced to reflect the properties of masonry instead of concrete. At present, reinforced masonry design is based upon this theory entirely. The state of the art has not progressed sufficiently to embrace the more sophisticated precepts of ultimate strength design, principally because of the lack of knowledge of many of the fundamental material properties (e.g., ultimate strain of the masonry assemblage), the performance characteristics of reinforced masonry systems, and the wide scatter of variable values reflected in much of the test data.

It would be well at this stage to review the theory of working stress design for those students who may not have been exposed to it in their courses in reinforced concrete. In this theory, allowable stresses are established as a fraction of the usable strengths of the materials, which are taken as the yield strength of the reinforcing steel and the ultimate compressive strength of the masonry assemblage. Working stress design is based upon the concept that a straight-line stress distribution obtains under the application of the working load. Members are then proportioned so that these allowable stresses are not exceeded when working loads are imposed. The working load may be defined as the sum of the actual dead load of the structure, the maximum live load that must be superimposed, and the governing lateral load, if any (i.e., wind or seismic forces). The live load is generally designated by a cognizant building code authority, as noted in Chapter 5.

FUNDAMENTAL CONCEPTS

The forces acting at a section of a beam may be resolved into components normal and tangential to this section (Figure 6-1). The components normal to this section, the flexural or bending forces, are tensile on one side of the neutral axis (usually the centroidal axis of the cross section) and compressive on the opposite side. Internal equilibrium conditions dictate that the total compressive force, C, equals the total tensile force, T. They therefore form a couple whose function is to resist the external bending moment at the section. The tangential components are known as the *shear stresses*, and they in turn resist the external transverse shear forces. The fundamental assumptions underlying this straight-line theory are briefly described as follows:

1. A plane section before bending remains a plane section after bending. This means then that the unit strains in the cross section both above and below the neutral axis remain directly proportional to their distance from that axis.
2. The bending stress, f_m, at any point is a function of the strain at that point and is described by the stress/strain characteristics of the masonry. Since a straight-line stress/strain curve is presumed within the working stress range, stress is presumed to be linear also.
3. The masonry is assumed to offer no tensile resistance whatsoever; consequently, the reinforcing steel must carry all tensile stresses.
4. Because the reinforcing steel becomes thoroughly bonded within the grout, or so it is assumed, the strain in the steel at a given point is the same as that of the surrounding grout. Thus, the stress in the equivalent transformed masonry section is f_s/n.
5. The modulus of elasticity of the masonry/mortar/grout assemblage remains constant throughout the working load range.
6. The shearing force is averaged over the entire cross section; thus, the shearing stresses are assumed to be uniformly distributed over the cross section.

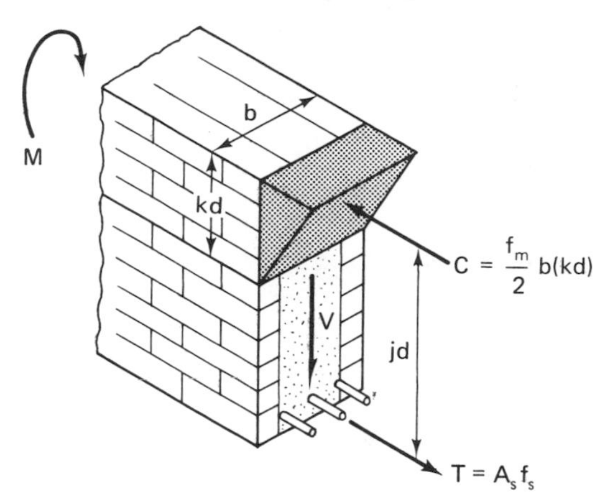

FIGURE 6-1. Internal force system—cracked system.

7. The member is a straight prismatic section.
8. The external force system is in equilibrium with the internal resistance; internal and external shears and moments are balanced.
9. The span of the member is considered large compared to its depth, i.e., shear deformations are neglected.

Now, the assumption that the masonry carries no tensile stresses implies a cracked section. This sometimes leads to some erroneous conclusions. For instance, where the steel is located in the center of a wall, the wall would necessarily have to crack through more than one-half the wall thickness for the steel to resist the tensile stresses. It is not likely that cracking occurs to this extent. Therefore, it would seem somewhat more logical to assume that the masonry does take some tensile stress. However, owing to the uncertainty involved in measuring whatever tensile strength masonry may possess, it is not depended upon in the design process.

The principle of the transformed section is resorted to in analyzing the internal mechanics at a section. Figure 6-2 shows the straight-line stress and strain distribution patterns assumed to exist in the cracked section. From the internal force system shown in Figure 6-1, assuming that equilibrium obtains throughout, the necessary design and analysis formulas are developed. As previously noted, it is assumed that all the masonry on the tensile side of the neutral axis has cracked and is therefore effectively absent structurally. Thus, the transformed section consists of masonry in compression on one side of the neutral axis, and n times the steel area on the other; since for the same deformation, the force in the steel would be n times that in the masonry at the same level in the cross section. The symbols, and their meaning, used in subsequent expressions developed are listed in the Appendix. (Note that where masonry strength, f'_m, is specified, this refers to the composite material: the masonry unit, mortar, and grout assemblage.)

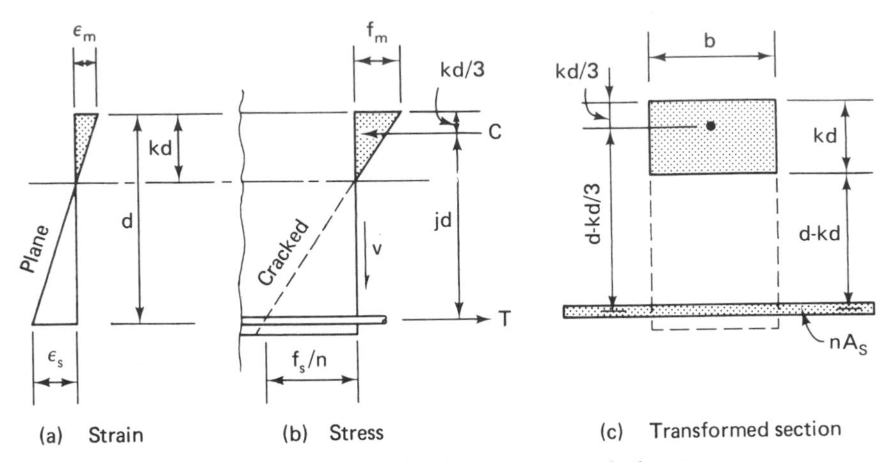

FIGURE 6-2. Stress/strain patterns—cracked system.

FLEXURAL BEHAVIOR OF REINFORCED MASONRY

Plain masonry beams are obviously inadequate as flexural members because they offer virtually no tensile resistance in bending. This stems from the fact that the tensile strength of masonry is only a small fraction of its compression strength. Thus, unless some way is found to reinforce this inadequate tensile resistance, the full benefits of the strength of the masonry will never be achieved. There certainly is no practical reason not to reinforce the masonry with steel bars in the same manner that concrete is reinforced. The joint action of the masonry and steel is assured, provided that the steel is adequately bonded to the grout, which in turn bonds to the masonry unit, be it brick or concrete block. This does occur, of course, because the deformed bars have both a chemical adhesion and a high mechanical bond at the steel/grout interface. The bars can be anchored additionally by the use of 90° bends or standard hooks.

The structural behavior of a reinforced masonry beam can best be described in three stages: (1) the uncracked condition, (2) the cracked state, and (3) the ultimate capacity region.

Stage 1: Uncracked-section condition

If a reinforced masonry beam is subjected to rather low loads, the maximum tensile stress in the masonry and grout may be less than the ultimate tensile resistance of the materials, in which case the entire cross section will be effective in resisting the moments produced by those low load levels (Figure 6-3). That is, tensile stresses will be developed within the masonry itself. In addition, the reinforcing, which is assumed to be deforming at the same rate as the adjacent masonry, is also subjected to low tensile stresses. Thus, at this stage, all stresses in the masonry are relatively small in magnitude and are therefore proportional to the strain, the section thus being entirely in the elastic condition. The distribution of stresses and strains over a cross section are seen in Figure 6-3, along with a transformed section.

Stage 2: Cracked-section state

Let us now assume that the load is increased to the extent that the tensile strength of the masonry has been reached and passed, in which case tension cracks will develop on the bottom of the masonry section, in the case of a simply supported beam. These tension cracks will tend to propagate upward as the load increases, causing the neutral axis to shift slightly upward, thereby reducing the area of the compression zone. If the beam is properly designed, the width of these cracks will be hairline in size and actually will not even be observed under the action of service load magnitudes.

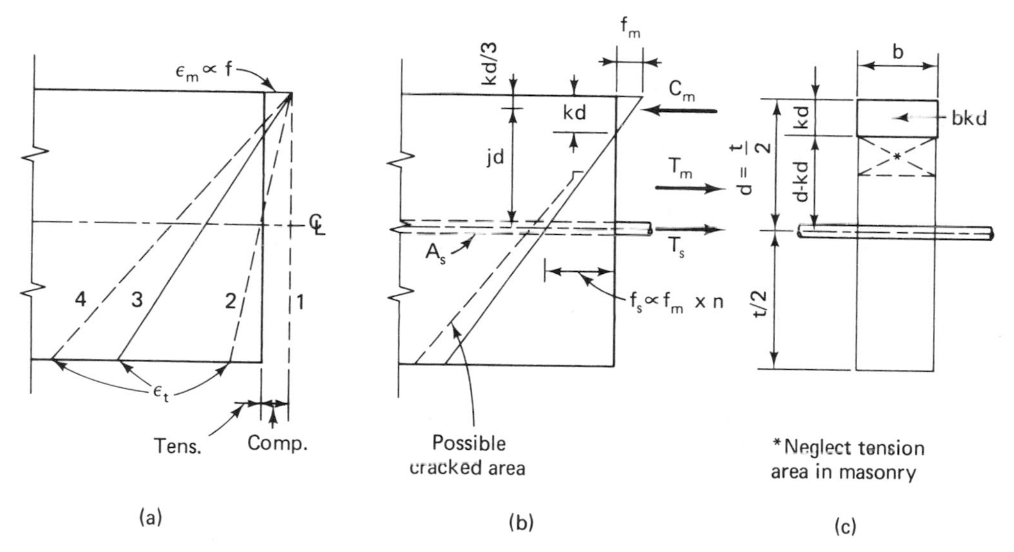

FIGURE 6-3. (a) Shows a typical wall with the plane of deformation as assumed in elastic design. It is shown at an f_m which might be taken as a design value, and is also shown by dashed lines as it might be varied for different consequent stress distribution, proportional to the strain: (1) uniform compression over the area; (2) flexural tension and compression over total uncracked section, as unreinforced masonry is designed and as the section functions until tension cracks form; (3) a design stress strain assumption; (4) a plane that would cause extreme cracking. (b) Shows the probable true condition of the stress: some of the section cracked, with zero consequent tension, and a portion of the uncracked section carrying some tension contributing to moment capacity. It shows the steel being deformed to a stress proportional to n times the f_m that would have occurred at that point. (c) The transformed section, as assumed conservatively, in reinforced masonry design: neglecting the tension of the masonry, which will be contributing to the capacity by an indeterminate and uncertain amount.

However, they do significantly affect the performance of that beam. In a cracked section the masonry cannot be called upon to transmit any tensile stresses, so the reinforcing steel will have to resist the entire tensile force. Even at the relatively moderate service loads, the masonry stresses and strains will probably remain in this elastic state. Certainly, the reinforcing steel stresses and strains continue to remain proportional, since the yield condition in this material is far from being reached within this stage, which was portrayed in Figures 6-1 and 6-3.

Stage 3: Ultimate capacity region

Finally, the ultimate load-carrying capacity of the masonry beam will eventually be reached. In that event, failure may be caused in either one of two ways. If the beam is underreinforced, the steel will reach its yield strength prior to the achievement of the compression capacity of the masonry. At this point the steel will yield and stretch a considerable amount, thereby causing the hairline tension cracks to widen considerably and propagate upward a significant distance into the compression zone. This deformation also causes a considerable deflection of the beam. Eventually, then, the compression zone is decreased in area to the extent that the masonry can no longer develop any increased flexural resistance and it will crush, producing a *secondary compression failure*. In these cases it is the reinforcing steel that limits the capacity, and thus the usable strength is measured by the yield stress of the steel. It should be pointed out that in masonry, as in concrete, if a failure must occur, it should take place in this manner. For prior to any actual collapse, there are extremely visible signs of distress, such as spalling of the masonry or excessive deflections. There is also a finite time factor involved between the time that these evidences of impending failure show up and the time that total collapse occurs. This would enable anyone in the proximity of such a failure to remove themselves before total collapse finally takes place.

However, if reinforcing is present in sufficiently large amounts, so that the yield strength of the reinforcing is not achieved before the masonry reaches its ultimate compression capacity, then a disastrous brittle failure occurs, i.e., the masonry will fail very suddenly by crushing and completely dissolve the structural integrity of that beam. This kind of failure must be avoided, for it is explosive in nature and it occurs with virtually no warning. Therefore, it is recommended that masonry beams be proportioned under at least the philosophic premise that should a failure occur, it will be initiated by yielding of the steel rather than by crushing of the masonry.

Masonry beams are to be proportioned and reinforced assuming that the second stage obtains, i.e., the section has cracked but remains elastic in behavior. Whether this actually occurs under various working load levels is somewhat problematical. For this reason, the margin of safety built into the design is rather high and undefinable. Assuming the second stage of performance, then, the following design and analysis formulas are all predicated on the working stress theory.

RECTANGULAR SECTIONS

Tensile reinforcement only

The neutral axis must first be located. This is found by equating the moment of the transformed tensile steel area about the centroidal axis of the cross section to the moment of the compression area producing the following, as noted in Figure 6-2:

$$b(kd)\left(\frac{kd}{2}\right) = nA_s(d - kd)$$

Therefore,

$$\frac{b(kd)^2}{2} - nA_s(d - kd) = 0$$

Substitute $p = A_s/bd$ to obtain

$$\frac{b(kd)^2}{2} - npbd(d - kd) = 0$$

Now divide through by bd^2 to obtain

$$\frac{k^2}{2} - pn(1 - k) = 0$$

from which

$$k = \sqrt{2pn + (pn)^2} - pn \qquad \boxed{\text{Location of neutral axis}} \qquad (1)$$

This expression for k can be used in analyzing the moment capacity of a beam with a given cross section for which the modular ratio, n, and the steel ratio, p, are known, or for calculating unit stresses in the steel and masonry when a known moment is applied. It cannot be used in design, since p is usually unknown.

The total compression force is computed as $C = b(kd)(f_m/2)$ and the total tensile force $T = A_s f_s$. Equilibrium dictates, of course, that these two forces must be numerically equal, and the moment of the couple which they comprise can be expressed in terms of the tensile force T; thus,

$$M_s = T(jd) = A_s f_s jd \qquad \boxed{\text{Steel moment}} \qquad (2)$$

and the steel stress then becomes:

$$f_s = \frac{M}{A_s jd} \qquad \boxed{\text{Steel stress}} \qquad (2a)$$

Taking moments about T gives the following:

$$M_m = Cjd = b(kd)\frac{f_m}{2}(jd) = \frac{f_m}{2}jkbd^2 \qquad \boxed{\text{Masonry moment}} \qquad (3)$$

from which the masonry stress then becomes

$$f_m = \frac{2M}{bd^2jk} = \frac{M}{bd^2}\frac{2}{jk} \qquad \boxed{\text{Masonry stress}} \qquad (3a)$$

The value of $2/jk$ is a readily determinable one in terms of pn and is often tabulated as an analysis parameter. Note that the basic allowable $F_b = 0.33f'_m$ with a 900 lb/in.^{2} maximum, whereas f_s allowable = 20,000 lb/in.^{2} for grade 40 reinforcing steel. Note then that the maximum valve of f_m becomes F_b.

For computational convenience, the term $f_m/2\ jk$ can be expressed as K. Then the value $K = M/bd^2$ indicates the masonry stress level. The basic design equation for proportioning the masonry section then becomes

$$bd^2 = \frac{M}{K} \qquad \text{where } K = \frac{F_b}{2}jk \qquad \boxed{\text{Masonry size}} \qquad (4)$$

It can be readily seen that

$$jd = d - \frac{kd}{3} \qquad \text{or} \qquad j = 1 - \frac{k}{3} \qquad \boxed{\text{Lever arm}} \qquad (5)$$

The foregoing, except for equation (4), are generally considered as analysis expressions when investigating the ability of a given cross section to resist a specified service moment.

It is often convenient to express k in terms of the allowable stresses in both the masonry and steel, for design purposes. From the geometry of the strain diagram (Figure 6-2):

$$\frac{\epsilon_m}{\epsilon_s} = \frac{kd}{d - kd}$$

Substituting $\epsilon_m = f_m/E_m$ and $\epsilon_s = f_s/E_s$ (since at the working stress level, stress and strain are assumed proportional to each other), one obtains

$$\frac{f_m}{E_m} \cdot \frac{E_s}{f_s} = \frac{k}{1 - k}$$

Since the modular ratio $n = E_s/E_m$ and $f_m(\text{max.}) = F_b$:

$$\frac{n}{f_s/F_b} = \frac{k}{1 - k}$$

The Code specifies an arbitrary value of $1000f'_m$ for E_m with a 3000-kips/in.² maximum. E_s is generally taken as 30,000 kips/in.² A solution for k at this balanced state (i.e., where these arbitrary allowable stresses in both the steel and the masonry are achieved simultaneously) produces the value

$$k_b = \frac{n}{n + f_s/F_b} \qquad \boxed{\text{Location of neutral axis-balanced condition}} \qquad (6)$$

In this case, k_b is expressed in terms of the modular ratio, n, and the ratio of the allowable stresses in the steel and masonry. This expression must, of necessity, be applied in design cases only, since while both stresses are specified, they must have a strain compatibility. It could be stated that at this stage the *balanced steel ratio*, p_b, had been achieved.

To determine the value of p_b at which a balanced design obtains, again equate the tensile and compressive forces at the section:

$$\frac{F_b}{2} bk_b d = (p_b bd)f_s$$

Dividing both sides by bd and solving for p_b,

$$p_b = \frac{F_b}{2}\frac{k_b}{f_s} = \frac{k_b}{2f_s/F_b}$$

With the desired simultaneous stress values known, substitute the value for k_b in the expression above, obtaining

$$p_b = \frac{n}{2f_s/F_b(n + f_s/F_b)} \qquad \boxed{\text{Balanced Steel Ratio}} \qquad (7)$$

This ficticious ratio p_b is called the *balanced steel ratio*, and it too can be used only for design. If the actual steel percentage, p, is less than p_b, the steel stress will reach its allowable value before the masonry reaches its allowable limit; thus, the design moment will be governed by the permitted steel stress. If p is greater than p_b, the masonry will reach its allowable stress first, and it therefore will dictate the value of the permissible moment on the beam.

Certain of the preceding relations can be manipulated in terms of K such that the percentage of steel, p, can be conveniently determined for a given masonry section. For instance, should f_s (allowable) govern, then

$$K = \frac{M}{bd^2} = pjf_s \quad \text{or} \quad p = \frac{K}{f_s j} \qquad \boxed{\text{Steel Ratio}} \qquad (8)$$

↑ assume (say 0.9), then verify

where

$$p = \frac{A_s}{bd}$$

$$k = \sqrt{2pn + (pn)^2} - pn$$

$$j = 1 - \frac{k}{3}$$

$$f_m = \frac{M}{bd^2} \times \frac{2}{jk} < F_b \qquad \text{and} \qquad f_s = \text{allowable}$$

When F_b (allowable masonry stress) governs,

$$K = \frac{M}{bd^2} = \frac{F_b}{2} jk$$

Also from the strain compatibility requirement, Figure 6-2, where f_m reaches F_b:

$$\frac{F_b}{kd} = \frac{f_s/n}{d - kd} \qquad \text{or} \quad F_b = \frac{kd}{d - kd}\frac{f_s}{n} = \frac{k}{1-k}\frac{f_s}{n}$$

then

$$f_s = nF_b\left(\frac{1-k}{k}\right) \qquad \text{with } f_s < \text{allowable} \tag{8a}$$

where k is obtained from the following computations:

$$M = \tfrac{1}{2}F_b(jk)bd^2 \qquad \text{or} \qquad jk = \frac{2M}{F_b bd^2} = \text{numerical value}$$

and $j = 1 - k/3$, so

$$jk = k - \frac{k^2}{3}$$

Then $k^2 - 3k + 3(jk) = 0$, using the above numerical value for jk. Solve this quadratic for k; then evaluate j, from which the steel requirement is found as

$$p = \frac{K}{f_s j} \qquad \text{or} \qquad A_s = \frac{M}{f_s jd} \qquad \text{where} \quad f_m \text{ reaches } F_b \text{ (allowable masonry stress)}$$

↑ from (8a) but below allowable f_s (8b)

These derivations provide good demonstrations of the fundamentals involved in an evaluation of this sort. However, they would be somewhat cumbersome to apply in actual practice, particularly where F_b governed. Thus, if a curve were plotted with $K/F_b = M/bd^2F_b$ as the ordinate and np as the abcissa, one could readily obtain the correct p, regardless of which stress governed. Examples demonstrating the use of this and other design aids are presented in the Appendix as well as in Chapters 7, 8, 9, and 10.

Examples illustrating the use of fundamental expressions developed in the prior discussion follow.

EXAMPLE 6-1: Obtain Steel and Masonry Stresses

A simply supported beam with a 12-ft span sustains a working live load of 50 lb/ft². The adjacent beams are 10 ft cc. The beam effective depth $d = 24.5$-in.; 8-in.,-wide, 4-in.-high concrete blocks are used. No inspection is to be provided. The steel reinforcement consists of two No. 4's. Determine the stresses, f_m and f_s. Assume $f'_m = 1500$ lb/in.².

Solution:

tributary width = center-to-center distance of beams = 10 ft

live load = 50 lb/ft² × 10 ft = 500 lb/ft

dead load = 77 lb/ft² × 28 in./12 = 180 lb/ft (Table 5-2 for medium-weight block)

total load = 500 + 180 = 680 lb/ft

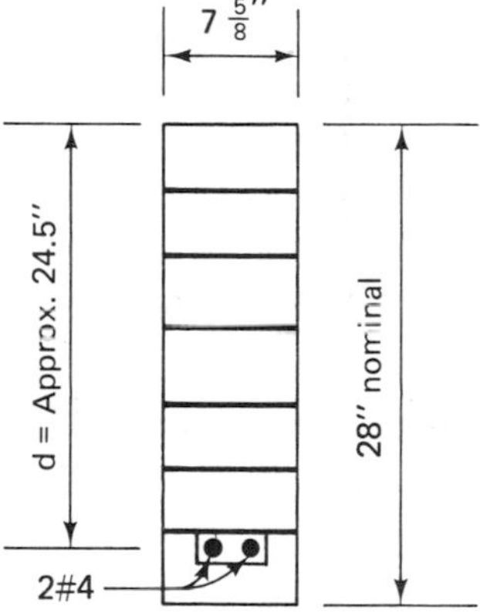

$$M = \frac{wl^2}{8} = \frac{(680 \text{ lb/ft})(12 \text{ ft})^2}{8} = 12.2 \text{ ft-kips}$$

$$A_s = 0.40 \text{ in.}^2 \quad (2 \text{ No. 4's})$$

$$b = 7.63 \text{ in.}$$

$$p = \frac{A_s}{bd} = \frac{0.40 \text{ in.}^2}{(7.63 \text{ in.})(24.5 \text{ in.})} = 0.00214$$

$$f'_m = 1500 \text{ lb/in.}^2 \qquad \therefore F_b = 250 \text{ lb/in.}^2 \text{ for } \tfrac{1}{2} \text{ stress}$$

$$n = \frac{30{,}000}{\frac{1}{2}f'_m} = \frac{30{,}000}{750} = 40 \quad \text{(no inspection)}$$

$$k = \sqrt{2pn + (pn)^2} - pn$$

$$= \sqrt{2(0.00214)(40) + [0.00214(40)]^2} - 0.00214(40)$$

$$= 0.337$$

$$j = 1 - \frac{k}{3} = 0.89$$

$$f_m = \frac{2M}{jkbd^2} = \frac{2(12.2) \times 12{,}000}{(0.89)(0.337)(7.63 \text{ in.})(24.5 \text{ in.})^2} = 213 \text{ lb/in.}^2 < 250 \text{ lb/in.}^2 \qquad \therefore \text{OK}$$

$$f_s = \frac{M}{A_s jd} = \frac{12.2 \times 12{,}000}{0.40 \text{ in.}^2 (0.89)(24.5 \text{ in.})} = 16{,}785 \text{ lb/in.}^2 < 20{,}000 \text{ lb/in.}^2 \qquad \therefore \text{OK}$$

Or the allowable moment may have been sought, in which case the determination would be made on the basis of the allowable stresses in both the steel and the masonry. Thus,

$$M_s = A_s f_s jd = 0.40 \times 20{,}000 \times 0.89 \times 24.5 = 174{,}440 \text{ in.-lb.}$$

and

$$M_m = \frac{f_m}{2} jkbd^2 = \frac{250}{2} \times 0.89 \times 0.34 \times 7.63 \times (24.5)^2 = 173{,}235 \text{ in.-lb } \textit{governs}$$

In this case, the masonry stress governs, although this condition is very close to the balanced state [i,e., the allowable $f_m (= F_b)$ and f_s being reached simultaneously].

EXAMPLE 6-2: Design the Lintel

A reinforced brick lintel, 20 ft long, must carry a live load of 1000 lb/ft in addition to its own weight. Special inspection is provided. Grade 40 steel and MW grade brick are used. The actual width of the lintel is 9 in. Determine the minimum depth and reinforcement required.

Solution:

$$b = 9 \text{ in.}$$

$$f'_m = 2500 \text{ lb/in.}^2 \qquad \text{(MW grade but with special strength units)}$$

$$F_b = \tfrac{1}{3}(2500) = 833 \text{ lb/in.}^2 \qquad (n = 12)$$

$$f_s = 20{,}000 \text{ lb/in.}^2 \qquad \text{(grade 40)}$$

Assume beam depth $\sim$24 in. to estimate dead load:

$$\text{DL} = (100 \text{ lb/ft}^2)(24 \text{ in.}/12) = 200 \text{ lb/ft}$$

$$\text{LL} = 1000 \text{ lb/ft}$$

$$\text{total load} = 1000 + 200 = 1200 \text{ lb/ft}$$

$$M = \frac{\text{w}l^2}{8} = \frac{(1200 \text{ lb/ft})(20 \text{ ft})^2}{8} \times 12 = 720 \text{ in.-kips}$$

Base the initial design on achievement of balanced condition, thus:

$$k = \frac{1}{1 + (f_s/nF_b)} = \frac{1}{1 + [20{,}000/12(833)]} = 0.333$$

$$j = 1 - \frac{k}{3} = 0.89$$

$$M_m = \frac{F_b}{2} jkbd^2$$

$$d = \sqrt{\frac{2M_m}{F_b jkb}} = \sqrt{\frac{2(720{,}000 \text{ lb/in.})}{(833 \text{ lb/in.}^2)(0.89)(0.333)(9 \text{ in.})}} = 25.5 \text{ in.}$$

∴ Overall depth would be about 30 in.—close enough to assumed 24 in. for dead load.

$$M = A_s f_s jd$$

$$A_s = \frac{M}{f_s jd} = \frac{720{,}000 \text{ lb/in.}}{(20{,}000 \text{ lb/in.}^2)(0.89)(25.5)} = 1.59 \text{ in.}^2$$

$$p = \frac{A_s}{bd} = \frac{1.59 \text{ in.}^2}{(9 \text{ in.})(25.5 \text{ in.})} = 0.0069$$

Compare with the balanced steel ratio:

$$p_b = \frac{n}{2f_s/F_b(n + f_s/F_b)} = \frac{12}{2(20{,}000/833)(12 + 20{,}000/833)} = 0.0069$$

So this is a "balanced" beam.

If the effective depth, d, had been increased to fit a certain modular coursing, or for some other reason, then p actual would be less than p_b and the beam will be considered as underreinforced. Then determine p from $A_s = M/f_s jd$. Shear and bond should also be checked. See examples in Chapter 9.

T-beams (partially grouted hollow unit masonry)

In the T-section of a partially grouted masonry wall section, the face shells on the compressive side are assumed to comprise the flange, and the web is formed by the reinforced solid grouted cell. That portion of the wall cross section which acts as a T-beam is shown in Figure 6-4. Shown in Figure 6-5 is the internal resisting couple, which must maintain equilibrium with the outside force system. It should be observed that a T-beam condition, as depicted, does not usually occur in a wall thickness under about 10 in. when the steel is located in the center. So if the wall thickness is less than 10 in., kd will generally be less than the face-shell thickness, calling for a rectangular beam analysis. Note that the total compressive force C consists of the compression in the flange, C_f, and compression in the web, C_w. This value of C may be developed from a consideration of the stress conditions in the flange and web. The average stresses in the flange and web are:

$$\text{Flange:}\quad f_f = \frac{1}{2}\left(f_m + \frac{kd - t_f}{kd} f_m\right) = f_m \frac{2kd - t_f}{2kd}$$

$$\text{Web:}\quad f_w = \frac{f_m}{2}\frac{kd - t_f}{kd}$$

Then the compressive forces in both the flange and web become

$$C_f = f_m \frac{2kd - t_f}{2kd} bt_f \quad \text{and} \quad C_w = f_m \frac{kd - t_f}{2kd}[b'(kd - t_f)]$$

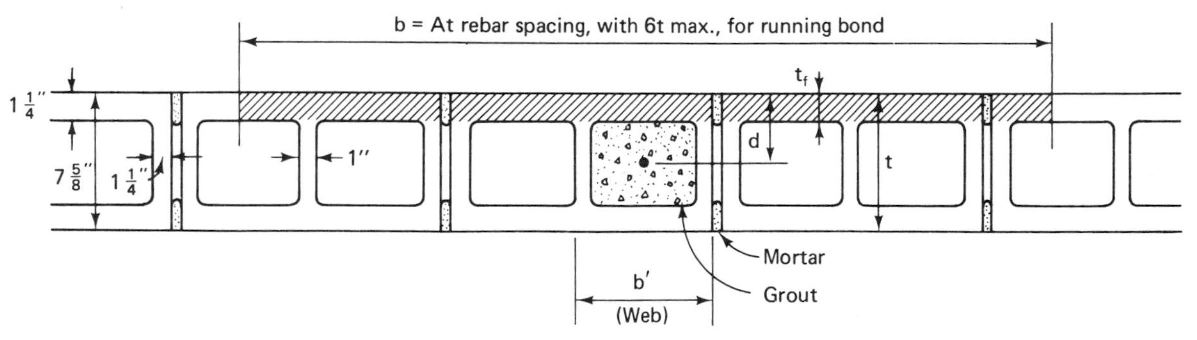

FIGURE 6-4. T-beam in partially grouted wall.

The lever arm $j_f d$ for the flange component, C_f, may be found from the location of the centroid of the trapezoidal stress pattern occurring on the flange cross section. The

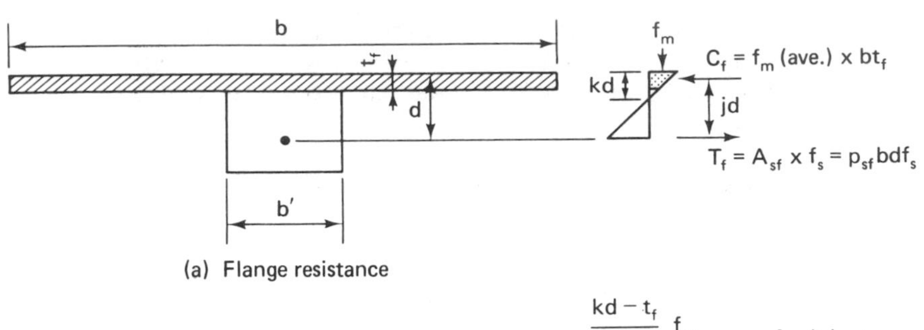

(a) Flange resistance

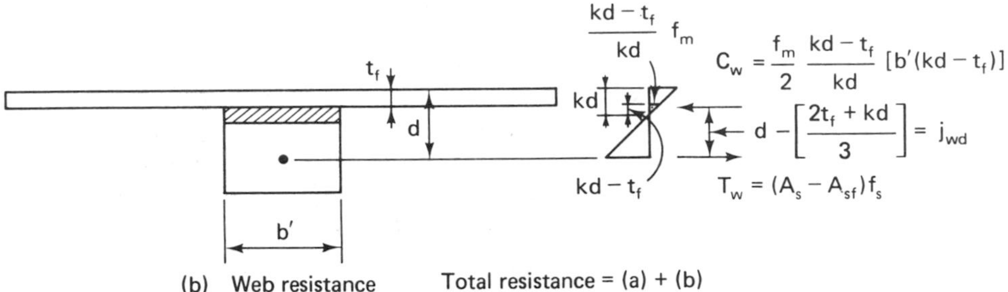

(b) Web resistance Total resistance = (a) + (b)

FIGURE 6-5. T-beam internal force system.

expression would be similar to the value of j given later, where the web component, C_f, is neglected except that the value of p_f (Figure 6-5a) would be substituted for p in equations (9) and (10). The lever arm $j_w d$ for the web component would be obtained on the basis of the triangular stress pattern existing over that compressive area (Figure 6-5b). The total resisting moment becomes:

$$M_T = \overbrace{f_m \frac{2kd - t_f}{2kd}(bt_f)j_f d}^{\text{flange}} + \overbrace{f_m \frac{kd - t_f}{2kd} b'(kd - t_f)j_w d}^{\text{web}}$$

Generally, however, the contribution of the web component is small enough to be neglected, thereby simplifying the preceding expression considerably. In that case,

$$T \sim C_f$$

or

$$pbdf_s = f_m \frac{2kd - t_f}{2kd} bt_f$$

Also, from the strain compatibility relation:

$$k = \frac{n}{n + f_s/f_m} \quad \text{or} \quad f_m = f_s \frac{k}{n(1 - k)}$$

Substituting this value of f_m into the previous equation to eliminate the unit stresses,

$$k = \frac{np + \frac{1}{2}(t_f/d)^2}{np + (t_f/d)} \tag{9}$$

and the distance to the resultant compressive force C_f from the top of the beam is

$$Z = \frac{3kd - 2t_f}{2kd - t_f}\frac{t_f}{3} \qquad \text{and} \qquad jd = d - Z$$

yielding a value for the lever arm of

$$j = \frac{6 - 6t_f/d + 2(t_f/d)^2 + (t_f/d)^3(\frac{1}{2}pn)}{6 - 3t_f/d} \tag{10}$$

The resisting moment becomes

$$M_s = A_s f_s jd \tag{11}$$

or (whichever is smaller)

$$M_m = f_m\left(1 - \frac{t_f}{2kd}\right)(bt_f)jd \tag{11a}$$

The foregoing equations may be readily used for analysis purposes; that is, for determining the allowable resisting moment for a T-beam section as demonstrated in Example 6-3. However, they become somewhat cumbersome and impractical in design. So if one assumes, conservatively, that (1) the lever arm jd is never less than $d - t_f/2$ (actually the center of gravity of the trapezoidal stress pattern is always above the mid-depth of the face-shell thickness), and (2) the average unit compressive stress in the trapezoidal pattern, $f_m(1 - t_f/2kd)$, is never as small as $f_m/2$, the moment equations conservatively simplify to

$$M_s = A_s f_s\left(\frac{d - t_f}{2}\right) \tag{12}$$

and

$$M_m = \frac{f_m}{2}bt_f\left(\frac{d - t_f}{2}\right) \tag{12a}$$

Observe that these equations should be used only in design cases (Example 6-4), never for review problems. Wall T-beam design simply involves the determination of the steel area required.

Note that a T-beam, as used in the context herein, can exist only where the wall is partially grouted (only reinforced cells filled) in hollow unit masonry construction. Where solid grouted hollow unit masonry (all cells filled) occurs, one simply analyzes it as a rectangular beam in accordance with the principles developed in the previous section.

EXAMPLE 6-3: Determine Allowable Moment

An 8-in.-thick block wall has No. 6 bars at 48 in. cc. The wall is partially grouted, and the steel reinforcement is located at the center of the wall. Assume no special inspection and that $f_s = 20{,}000$ lb/in.². Determine the resisting moment.

Solution:

$$f'_m = 1350 \text{ lb/in.}^2 \text{ (partially grouted)}$$

$$F_b = \tfrac{1}{3}(\tfrac{1}{2})(1350) = 225 \text{ lb/in.}^2 \; (n = 44)$$

$$b_{\text{flange}} = 48 \text{ in. (max.} = 6t)$$

$$t_{\text{flange}} = 1\tfrac{1}{4} \text{ in.}$$

$$b_{\text{web}} = \text{grout space} + \text{cross webs} = 6 \text{ in.} + 1\tfrac{1}{4} \text{ in.} + 1 \text{ in.} = 8\tfrac{1}{4} \text{ in.}$$

$$d = \frac{7.63 \text{ in.}}{2} = 3.8 \text{ in.}$$

$$A_s = 0.44 \text{ in.}^2 \text{ (1 No. 6)}$$

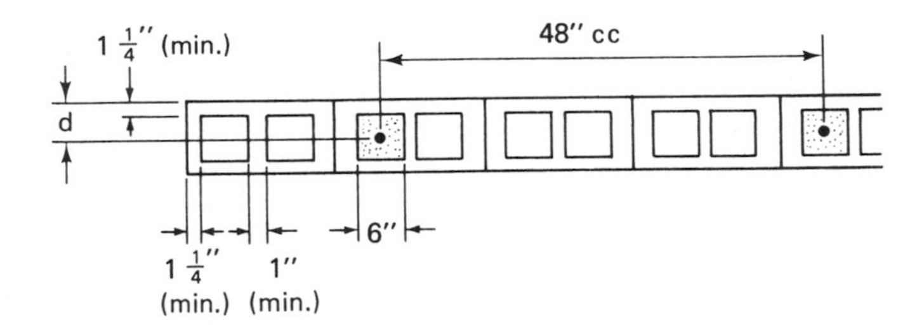

At this point it is not apparent whether the wall acts as a rectangular or a T-beam.

Assume that the section will act as a *T*-beam:

$$p = \frac{A_s}{bd} = \frac{0.44}{48 \times 3.82} = 0.0024 \quad \text{and} \quad pn = 44 \times 0.0024 = 0.106$$

$$t_f/d = 1.25/3.82 = 0.327$$

$$k = \frac{pn + 0.5(t_f/d)^2}{pn + (t_f/d)} = \frac{0.106 + 0.5(0.327)^2}{0.106 + 0.327} = 0.370 \quad \text{and} \quad kd = 0.370 \times 3.82$$

$$= 1.41 \text{ in.} > 1.25 \text{ in.} \qquad \therefore \text{acts as T-beam [use Eq. (10)]}$$

$$j = \frac{6 - 6(t_f/d) + 2(t_f/d)^2 + (t_f/d)^3(1/2pn)}{6 - 3(t_f/d)}$$

$$= \frac{6 - 6(0.327)^2 + 2(0.327)^2 + (0.327)^3(1/2 \times 0.106)}{6 - 3(0.327)} = 0.88 \quad \text{and} \quad jd = 3.36 \text{ in.}$$

The allowable bending moment becomes:

$$M_s = A_s f_s jd = 0.44 \times 20{,}000 \times 3.36 = 29{,}570 \text{ in.-lb.}$$

$$M_m = F_b(1 - t_f/2kd)bt_f jd = 225\left(1 - \frac{1.25}{2 \times 1.41}\right)(48 \times 1.25 \times 3.36)$$

$$= 25{,}300 \text{ in.-lb. } \textit{governs}$$

At half-stresses, this combination approaches that of a balanced state. Full stress condition ($F_b = 450$ lb/in.²) would suggest a rectangular beam action in which the steel stress at 20,000 lb/in.² would govern the permissible moment (= 30,500 in.-lb.). Consequently little would be gained in this case by specifying full stresses in the masonry.

EXAMPLE 6-4: Select Reinforcement Required for a Given Moment

A partially grouted wall is to be constructed of 12 in. concrete block, without inspection. What reinforcement is needed to carry a moment of 1000 ft-lbs? $f'_m = 1350$ lb/in.², $F_b = \frac{1}{2} \times \frac{1}{3} \times 1350 = 225$ lb/in.² ($n = 44$), $f_s = 20{,}000$ lb/in.², $t_f = 1.25$ in., $d = 5.82$ in. Try steel at maximum spacing (i.e., $b = 48$ in.).

Solution:

Trial steel area from equation (12):

$$A_s = \frac{M \text{ per ft}}{f_s(d - t_f/2)} = \frac{12 \times 1000}{20{,}000(5.82 - 1.25/2)} = 0.115 \text{ in.}^2/\text{ft}$$

Minimum for 12 in. wall (0.0007 bt) = 0.097 in.²/ft < 0.115 in.²/ft required for moment. Try No. 6 at 48 in. ($A_s = 0.11$ in.²/ft)
Check trial steel area:

$$p = \frac{A_s}{bd} = \frac{0.11}{12 \times 5.82} = 0.0016 \quad \text{and} \quad pn = 0.0016 \times 44 = 0.0704$$

$$t_f/d = \frac{1.25}{5.82} = 0.215$$

$$k = \frac{0.0704 + 0.5(0.215)^2}{0.0704 + (0.215)} = 0.33 \times 5.82 \text{ in.} = 1.91 \text{ in.} \qquad \therefore \text{figure as a T-beam.}$$

$$j = \frac{6 - 6(0.215) + 2(0.215)^2 + (0.215)^3\left(\frac{1}{2 \times 0.0704}\right)}{6 - 3(0.215)} = 0.90$$

$$A_s(\text{revised}) = \frac{12 \times 1000}{20{,}000 \times 0.90 \times 5.82} = 0.113 \text{ in.}^2 \qquad \therefore \text{No. 6 at 48 in. OK}$$

$$f_m = \frac{12 \times 1000}{\left(1 - \frac{1.25}{2 \times 0.33 \times 5.82}\right)(48 \times 1.25)(0.90 \times 5.82)} = 56 \text{ lb/in.}^2 \qquad \therefore \text{OK}$$

Note that solid grouting the wall would serve no useful structural purpose, since the steel governs here when spaced at the maximum allowable distance.

Compressive reinforcement

From the expression previously developed for a rectangular section comes the evaluation of the resisting moment of that section in terms of the maximum masonry stress:

$$M_m = \left(\frac{F_b}{2}\right) bd^2 (j_b k_b)$$

Should the maximum design moment for that section, M, exceed M_m, compression reinforcement could be provided to bolster that section's compressive resistance. This does not often become necessary in reinforced masonry design. However, since it could occur, the condition should be explored. From the strain distribution pattern in Figure 6-6, we obtain

$$\frac{\epsilon_s}{\epsilon'_s} = \frac{d - kd}{kd - d'} = \frac{f_s}{f'_s}$$

$$f'_s = f_s \frac{kd - d'}{d - kd} = f_s \frac{k - d'/d}{1 - k} \leq f_{s\,(\text{allow})} \qquad \boxed{\text{Compressive steel stress}} \qquad (13)$$

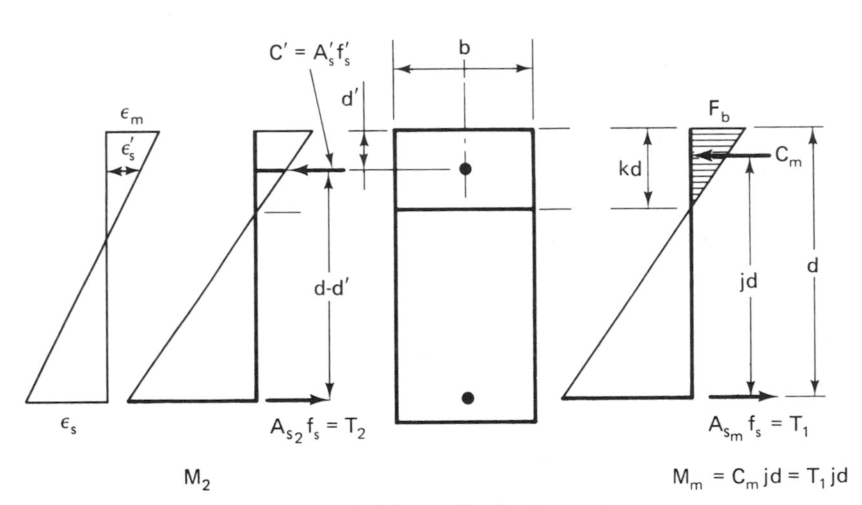

$M = M_m + M_2$
M_m = Moment governed by masonry stress, at balanced state
A'_s = Compressive steel
A_{s_m} = Steel required to balance allowable masonry compression
A_{s_2} = Remainder of tensile steel

FIGURE 6-6. Beam moment, with compressive reinforcement.

Now, if both the steel and masonry remain perfectly elastic, the stress in compression steel, f'_s, would be n times the compression stress in the masonry at the same location, since unit strains in the steel and surrounding masonry at the same points are presumed equal. However, this stress ratio for f'_s remains true only at the lower strain levels, since at the higher strain levels, the stresses do not increase proportionally in the masonry. The strains in the compression steel and adjacent masonry do remain the same, and are proportional to the distance from their neutral axis, regardless of the strain condition (if the concept that a plane section remains plane holds). This requires that at higher strain levels, the unit stress in the steel, behaving according to its elastic stress/strain characteristics, will actually be larger than it would be if the masonry had continued to react in an elastic fashion, so as to accept its proportionate share of the load. But this the masonry ceases to do, because the stress is increasing at a lesser rate than before, when masonry stress and strain were somewhat proportional. As a consequence, the masonry tends to shift a portion of its burden to the adjacent reinforcing. Thus because of this behavior, the compressive steel stress actually becomes greater than elastic stress computations would indicate.

To compensate for this effect, the ACI Code specified that, in working stress design, the stress in compression reinforcement in concrete beams be taken as two times the value obtained from the strain compatibility relation, this resulting value not to exceed the allowable steel stress in tension. This approach represented an attempt to correct for a fundamental deficiency in the elastic theory, and presumably it could also be followed in masonry design. However, it is not clear just what multiplying factor to use in masonry, because no test data exist to substantiate one. At the extremely low allowable stress levels permitted by the Code, this nonlinearity may not be too significant anyway. So this factor will be neglected in subsequent computations.

Also note that the compressive steel stress may be expressed in terms of the elastic stress in the extreme masonry fiber:

$$\frac{F_b}{kd} = \frac{f'_s/n}{kd - d'} \quad \text{by} \sim \triangle \text{ from stress distribution pattern}$$

Therefore,

$$f'_s = nF_b\frac{kd - d'}{kd} = nF_b\frac{k - d'/d}{k} \leq f_{s\,(\text{allow})} \tag{13a}$$

Figure 6-6 shows that, for analysis purposes, the resisting moment consists of:

M_m = moment that can be developed by a given masonry section without compression reinforcement; this requires a portion of the total tensile reinforcement to form the resisting couple, the amount being A_{s_m}, with the value of f_m reaching its specified allowable stress value (F_b)

M_2 = excess moment developed by the compressive reinforcement, acting with the remainder of the tensile steel, A_{s_2}

$$\text{total resisting moment } M = M_m + M_2$$

$$\text{total steel area } A_s = A_{s_m} + A_{s_2}$$

$$M_m = \frac{F_b}{2} j_b k_b b d^2$$

where

$$k_b = \frac{n}{n + f_s/F_b} \quad \text{and} \quad j_b = 1 - \frac{k_b}{3}$$

Thus,

$$A_{s_m} = \frac{M_m}{f_s j_b d} \tag{14}$$

and

$$A_{s_2} = A_s - A_{s_m} \tag{14a}$$

Then $M_2 = A'_s f'_s(d - d')$ or $M_2 = A_{s_2} f_s(d - d')$, where f_s is the allowable tensile steel stress and f'_s is obtained from equation (13). Use the smaller value of $A_{s_2} f_s$ or $A'_s f'_s$ to govern the allowable moment, then

$$M_{\text{total}} = M_m + M_2 \qquad \boxed{\text{Total resisting moment}} \tag{15}$$

Note that M_2 must be computed from both the compressive and tensile forces in that couple. This is necessary, since it is not known whether the compressive steel force ($A'_s f'_s$) or the tensile steel force ($A_s f_{s_2}$) places the limit on the magnitude of M_2 in an analysis computation.

In some design situations, the masonry section might be limited in size for some reason other than structural—perhaps an architectural consideration, such that it is not capable of sustaining the applied working load without the additional compressive steel. In that case the following design procedure could be followed:

$$M_m = \frac{F_b}{2} j_b k_b b d^2 = K_b b d^2 \quad \text{and} \quad A_{s_m} = \frac{M_m}{f_s j_b d} \tag{16}$$

The deficient moment, for which compressive steel is needed, is $M_2 = M_{\text{design}} - M_m$, and therefore,

$$A_{s_2} = \frac{M_2}{f_s(d - d')} \tag{17}$$

The total tensile steel requirement becomes

$$A_s = A_{s_m} + A_{s_2}$$

Tensile reinforcement (18)

The compressive steel requirement is found from

$$f'_s = f_s \frac{k - d'/d}{1 - k} \leq f_s \quad \text{and} \quad A'_s = \frac{M_2}{f'_s(d - d')[(n - 1)/n]}$$

Compressive reinforcement (19)

Of perhaps academic interest only is an expression for the actual location of the neutral axis for the doubly reinforced beam:

$$k = \sqrt{[np + (n - 1)p']^2 + 2[np + (n - 1)p'd'/d]} - [np + (n - 1)p']$$

Neutral axis—compression steel (20)

Also the distance of the resultant compressive force (sum of compression resistance in masonry and compressive resistance of steel) below the extreme compression fiber is

$$Z = \frac{\frac{1}{6} + \frac{(n - 1)A'_s}{kbd} \times \frac{d'}{kd} \times \left(1 - \frac{d'}{kd}\right)}{\frac{1}{2} + \frac{(n - 1)A'_s}{kbd} \times \left(1 - \frac{d'}{kd}\right)}$$

Lever arm—compression steel (21)

and

$$j = 1 - Zk \qquad (\text{for } A'_s = 0; \text{ then } Z = \tfrac{1}{3}) \tag{21a}$$

The following example will illustrate the foregoing principles. Others will be found in Chapters 7, 8, and 9.

EXAMPLE 6-5: Determine the Allowable Moment Capacity

An 8 in. concrete block beam is solid-grouted. Special inspection is provided. The beam is doubly reinforced, with $d = 24.5$ in. and $d' = 4.5$ in. Tensile reinforcement consists of two No. 6's, and compressive reinforcement three No. 5's. Determine the moment capacity of the beam.

Solution:

$A_s = 0.88$ in.^{2} (two No. 6)

$A'_s = 0.93$ in.^{2} (three No. 5)

$b = 7.63$ in.

$$F_b = \tfrac{1}{3}(1500) = 500 \text{ lb/in.}^2 \quad (n = 20)$$

$$f_s = 20{,}000 \text{ lb/in.}^2$$

$$p = \frac{A_s}{bd} = \frac{0.88 \text{ in.}^2}{(7.63 \text{ in.})(24.5 \text{ in.})} = 0.0047$$

$$p_b = \frac{n}{2(f_s/F_b)(n + f_s/F_b)} = \frac{20}{2(20{,}000/500)(20 + 20{,}000/500)} = 0.0042$$

$p\,(= 0.0047) > p_b\,(= 0.0042)$ $\therefore$ Beam is overreinforced, and masonry stress would limit capacity. Thus, the compressive steel will provide additional capacity along with all or part of p in excess of p_b. Solve as doubly reinforced beam.

The total moment capacity is divided into two portions: M_m, as governed by the masonry stress, and M_2, as developed by the effective compression steel.

$$k_b = \frac{n}{n + f_s/F_b} = \frac{20}{20 + 20{,}000/(500)} = 0.333$$

$$j_b = 1 - \frac{k}{3} = 0.89$$

$$M_m = \frac{F_b}{2} j_b k_b b d^2 = \frac{500 \text{ lb/in.}^2}{2}(0.89)(0.333)(7.63 \text{ in.})(24.5 \text{ in.})^2 = 339.3 \text{ in.-kips}$$

$$A_{s_m} = \frac{M_m}{f_s j d} = \frac{339{,}300}{(20{,}000 \text{ lb/in.}^2)(0.89)(24.5 \text{ in.})} = 0.78 \text{ in.}^2$$

Therefore,

$$A_{s_2} = A_s - A_{s_m} = 0.88 \text{ in.}^2 - 0.78 \text{ in.}^2 = 0.10 \text{ in.}^2$$

To check k for a double reinforced section, use the following expression:

$$k = \sqrt{[np + (n-1)p']^2 + 2[np + (n-1)p'(d'/d)]} - [np + (n-1)p']$$

$$p = 0.0047, \; np = 0.094$$

$$p' = \frac{A_s'}{bd} = \frac{0.93 \text{ in.}^2}{(7.63 \text{ in.})(24.5 \text{ in.})} = 0.0050$$

$$k = \sqrt{[20(0.0047) + (19)(0.005)]^2 + 2[20(0.0047) + (19)(0.005)(4.5/24.5)]} - [20(0.0047) + (19)(0.005)] = 0.32 \text{ (showing that } k = 0.33 \text{ is close enough)}$$

$$f_s' = f_s \frac{k - d'/d}{1 - k} = 20{,}000\left(\frac{0.32 - 4.5/24.5}{1 - 0.32}\right) = 4120 \text{ lb/in.}^2 < f_s = 20{,}000 \text{ lb/in.}^2$$

$$\left.\begin{aligned} A_{s_2} f_s &= 0.10 \times 20{,}000 = 2000 \text{ lb} \\ A_s' f_s' &= 0.93 \times 4120 = 3830 \text{ lb} \end{aligned}\right\} \text{(use smaller value)}$$

Thus,

$$M_2 = A_{s_2} f_s (d - d') = 2000(24.5 - 4.5) = 40 \text{ in. kips}$$

$$\text{Total moment capacity} = M_m + M_2 = 339.3 + 40 = 379.3 \text{ in.-kips}$$

It appears that there is an excess of compressive reinforcement in this design.

INFLUENCE OF WEB REINFORCEMENT

Beam behavior without web reinforcement

The behavior of reinforced masonry beams without web reinforcement probably resembles in some respects the behavior of reinforced concrete. Once longitudinal tension cracks form in the masonry, the flexural tensile strength must be developed by the reinforcing steel. Thus, higher loads can be sustained than would be possible if no longitudinal steel whatsoever were present. At these higher load levels diagonal tension stresses of considerable intensity would be created where the shear forces are the highest, primarily close to the supports or near concentrated loads. These are produced by the combined effect of longitudinal tension and tangential shear stresses, as noted in Figure 6-7. A simple combined stress analysis will indicate the maximum diagonal tensile stress occurring at that point. However, the magnitude and direction of this state of stress is different at every elevation in the cross section, as well as at every point longitudinally along the span. At any rate, this variation renders it almost impossible to establish a rational definition of the effects of these stresses in reinforced masonry beams. To further complicate this condition, the diagonal tensile value will also be influenced by the relative strength of the masonry units versus the strength of the head and bed joints. Modes of shear failure may be seen in Figure 6-8.

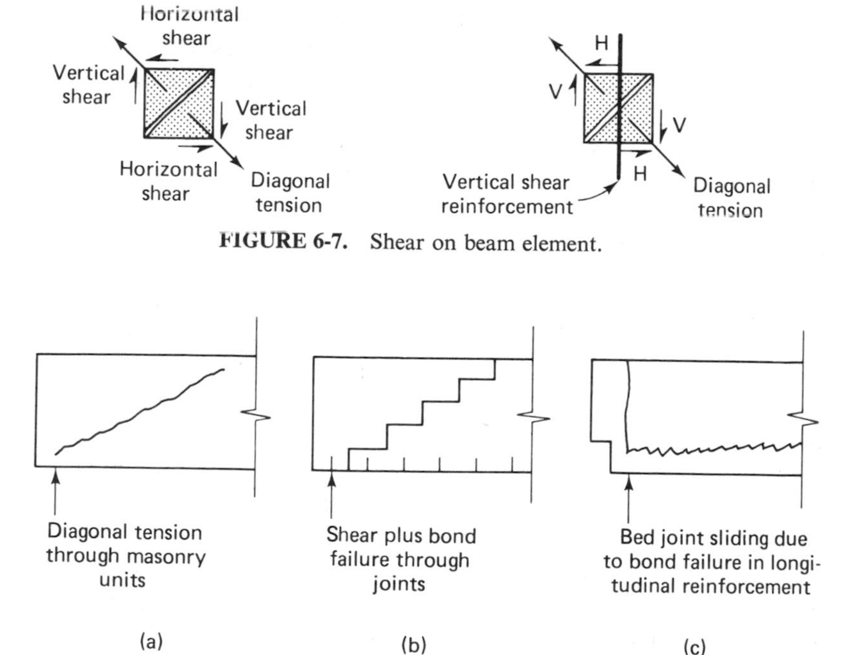

FIGURE 6-7. Shear on beam element.

FIGURE 6-8. Shear failures in masonry beams: (a) diagonal tension through masonry units; (b) shear plus bond failure through joints; (c) bed joint sliding due to bond failure in longitudinal reinforcement.

Longitudinal tensile reinforcement is effective in resisting these tensile stresses occurring near it; however, this tension steel cannot offer much resistance to those diagonal tensile stresses occurring higher up in the cross section anywhere along the span. Eventually, then, somewhere along the span, these diagonal tensile stresses will reach magnitudes that exceed the tensile strength of the masonry, and consequently will open up diagonal cracks oriented at various angles with the longitudinal axis of the beam (Figure 6-9). By examining the formation of these diagonal tensile cracks a little more closely, one finds that the relative values of moment and shear will affect both the magnitude and the direction of the diagonal tensile stresses that develop.

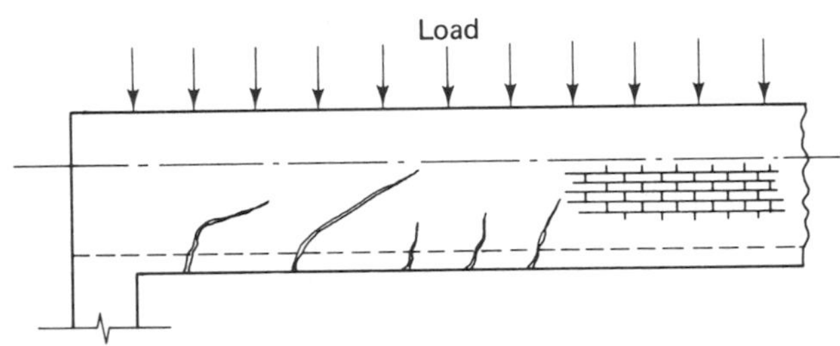

FIGURE 6-9. Diagonal and flexural tension cracks in a masonry beam.

This phenomenon has been verified experimentally in masonry piers subjected to various bending moment-to-shear ratios. No significant research has been done on masonry beams to substantiate this phenomenon, however. Because of this paucity of information, the present Code simply specifies a single allowable shear stress value for flexural members, in contrast to a variable allowable shear stress, depending upon the M/Vd ratio, for masonry shear panels. Certainly, a similar type of relationship needs to be established for flexural members as well.

Observe that there is a distinct difference between diagonal tensile and longitudinal tensile cracks. It has already been observed that the hairline flexural cracks occurring on the tension side of the beam are assumed to produce no structural deficiencies in the reinforced flexural member. This is due to the fact that adequate longitudinal reinforcement is available to resist all tensile forces. In contrast, diagonal tensile cracks are not resisted to any great extent by the longitudinal reinforcement. Unless web reinforcement (shear steel) is provided to offset the effects of this diagonal cracking, serious consequences may be incurred. For once the diagonal crack is formed, it tends to spread, and will continue to propagate into the compression zone under even slight load increases, ultimately reducing the compression area to an ineffective amount. Note then that the presence of adequate web reinforcement produces a dual effect. It tends to inhibit the growth of these diagonal cracks beyond the initial cracking phase; and, even more important, it increases the ductility of the beam, thereby enabling it to absorb a considerable amount of energy before failure occurs.

Thus, when diagonal cracks form in a masonry beam, they cause the following changes in stress patterns:

1. Depending upon the extent of the penetration of these cracks above the neutral axis, the amount of uncracked cross-sectional area remaining decreases. This reduction, in turn, results in an increase in the actual unit shear acting on the vertical face. For a computational comparison with an allowable value, it is averaged as $v = V/bd$, although the actual value would be V/by (Figure 6-10).

2. Since the remaining compression zone has been reduced, the flexural compression stresses increase by an indefinable amount.

3. Dowel action, also numerically indefinable, can become critical, for it can cause the grout to split along the reinforcing bar back toward the support. This action opens up the possibility of a bond failure should this distance become long enough, since the anchorage of the bar within the grout will eventually be destroyed.

Unless the combined action of the masonry and longitudinal reinforcing can adjust to these necessary stress increases in maintaining equilibrium, a failure could occur due to one or more of these stress adjustments.

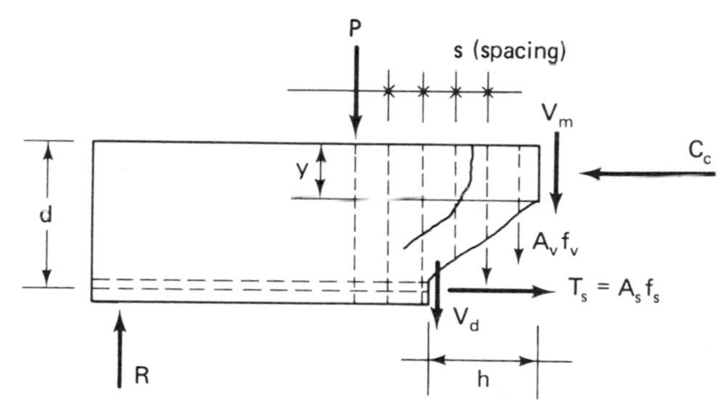

FIGURE 6-10. Beam shear resistance with web reinforcement.

On the other hand, when the proper amount of web reinforcement is provided, the flexural member will then be able to develop its full moment capacity, rather than having its strength limited by premature shear failure. So even though the procedure for placing web reinforcement in masonry beams does not have a very solid theoretical basis, tests have shown that when certain empirical rules are followed, the steel will be adequate to prevent these diagonal tension failures from occurring before the beam achieves its full moment capacity, thereby producing a more ductile-like behavior pattern throughout these upper loading stages.

Beam behavior with web reinforcement

Web reinforcement does not appear to have any particular effect on the behavior of the masonry beam prior to the formation of diagonal cracks. This was suggested in some early flexural tests conducted on reinforced brick beams by Schneider at USC in 1951. However, after diagonal cracks have developed, the beam behavior becomes drastically altered. The shear steel contributes to the change in the diagonal tensile resistance of the beam in several ways:

1. Obviously, a portion of the total shear is resisted by the bars crossing a particular crack. In reinforced concrete, therefore, the amount of web reinforcement needed becomes a function of the difference between the ultimate shear on the section and that shear resistance assigned to the concrete. However, in masonry, perhaps because its behavior is less well understood, the UBC assumes that the shear steel must resist all the shear at any section, should the allowable shear resistance assigned arbitrarily to the masonry by the Code be exceeded. There is no experimental basis for this limitation (see Figure 6-11).

2. These vertical bars tend to inhibit the growth of the cracks, thereby minimizing their penetration into the compression zone and also perhaps cutting down on the number of cracks forming during any particular overload condition. Thus, more uncracked masonry is available to resist flexural compression and shear stresses.

3. The stirrups tie the longitudinal reinforcement into the main bulk of the masonry, providing some restraint against splitting. This also has the effect of constraining the masonry within a steel cage, thereby effecting something in the nature of ductile behavior. Thus, even though there is not a great deal of theoretical basis for this incorporation of shear steel in masonry, the end effect produces a beam that does provide satisfactory performance under service-load conditions.

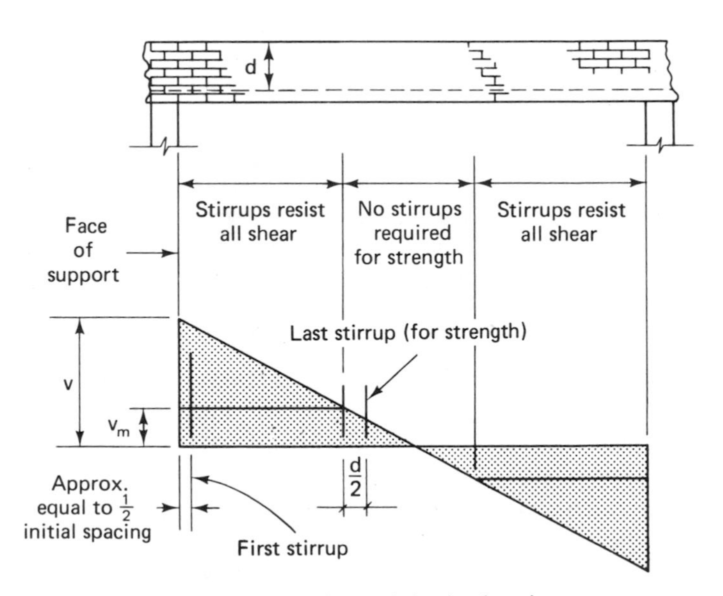

FIGURE 6-11. Shear reinforcing locations.

Design of web reinforcement

In reinforced masonry, shear analysis may arise either in reference to (1) the determination of the adequacy of a flexural member in resisting transverse shear, with or without web reinforcement; or (2) the case involving the capability of a shear wall to resist in-plane shear forces. Flexural shear seldom governs in the design of masonry walls, although it is often critical in masonry beams. Regarding flexural shear, the force system shown in Figure 6-12 applies; therefore, the unit horizontal shear at the neutral axis on an element of length dx is found from:

$$\sum F_x = 0 \qquad \text{thus} \qquad v(b\,dx) + T = T + dT$$

where

$$T = \frac{M}{jd} \qquad \text{and} \qquad T + dT = \frac{M + dM}{jd}$$

thus,

$$v = \frac{1}{bjd}\frac{dM}{dx} = \frac{V}{bjd} \qquad \boxed{\text{Flexural shear stress}} \qquad (22)$$

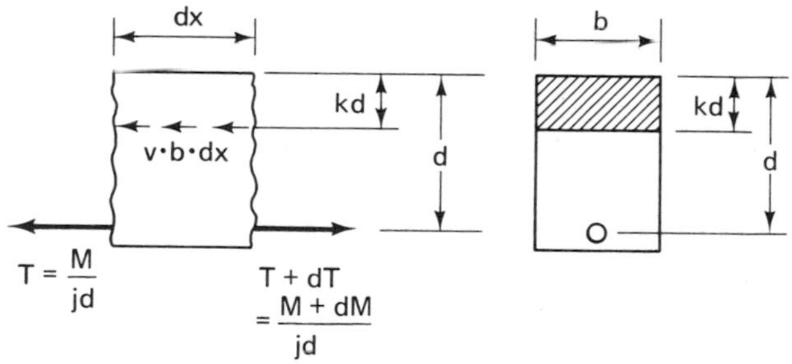

FIGURE 6-12. Flexural shear resistance.

The flexural shear area for partially grouted hollow unit masonry is evaluated as shown in the section "Wall Design Dimensions," Chapter 7. The Code-assigned value for the allowable flexural shear in masonry without web reinforcement is $1.1\sqrt{f'_m}$ (50 lb/in.^{2} max.). Thus, if v as determined from equation (22) does not exceed this value, theoretically no web reinforcement is needed; if v is greater, web reinforcement must be designated as shown in the next paragraph. However, as pointed out in the previous discussion, it is highly desirable that a nominal amount of web reinforcement be placed in all masonry beams anyway, mainly to make for a more ductile behavior pattern.

For the in-plane shear stress of solid grouted masonry (load parallel to the wall), the unit shear stress is found as

$$v = \frac{V}{t \times L}$$ In-plane shear (23)

where t = actual wall thickness
L = length of wall

The in-plane shear area for partially grouted hollow unit masonry is computed as shown in the section "Wall Design Dimensions," Chapter 7.

When web reinforcement is required, it may be selected in the following manner. Since for ease in placing and forming, the shear steel consists generally of a straight No. 3, No. 4, or No. 5 bar (with a hook or 90° bend at each end), the actual web reinforcement design consists simply of determining the location and spacing of these stirrups. Refer to Figure 6-10 for the free body diagram. Equilibrium in the vertical direction dictates that internal shear resistance equal external shear force, or

$$V_m + \sum_n A_v f_v + V_d = V_{\text{ext}}$$ where n is the number of stirrups traversing any diagonal crack

If s equals the stirrup spacing and h the horizontal projection of the crack, then

$$n = \frac{h}{s} \sim \frac{d}{s}$$ (for diagonal crack flatter than 45°)

Prior to failure, part of the load is carried by the uncracked masonry at the head of the crack (V_m). It would be assumed that this value is equivalent to the actual shear stress occurring at the instant the diagonal crack forms, whatever that value happens to be. The dowel action V_d is neglected as being of uncertain value. Then

$$V_{\text{ext}} = V_m + nA_v f_v = V_m + A_v f_v \frac{d}{s}$$

$$V_{\text{ext}} - V_m = A_v f_v \frac{d}{s}$$

Divide both sides by bd, and if

$$\frac{V_{\text{ext}} - V_m}{bd} = v - v_m \quad \text{(unit stresses)}$$

then

$$s = \frac{A_v f_v}{b(v_{\text{ext}} - v_m)}$$

where A_v = area of shear steel bar
v_{ext} = unit shear caused by the service loads
v_m = unit shear resistance assigned to the masonry

But if web reinforcement must arbitrarily take all the shear as required by the Code, then v_m becomes zero; thus,

$$s = \frac{A_v f_v}{b v_{\text{ext}}} \qquad \boxed{\text{Stirrup spacing}} \qquad (24)$$

The Code permits a maximum flexural shear value of $3.0\sqrt{f'_m}$ (150 lb./in.^{2} max.) where web reinforcement is utilized.

Essentially, checking the adequacy of a masonry beam or lintel for shear resistance simply becomes a matter of performing the following computations:

1. Calculate the average shear index, $v_{\text{ext}} = V/bjd$.
2. Compare v_{ext} with v_{allow}, which equals $1.1\sqrt{f'_m}$ (50 lb/in.^{2} max.), where special inspection is called for, or, in its absence, $v_{\text{allow}} = 25$ lb/in.^{2}.
3. If $v_{\text{ext}} < v_{\text{allow}}$, theoretically no stirrups are required. If $v_{\text{ext}} > v_{\text{allow}}$, stirrups are required.
4. Where stirrups are required, select a bar size, No. 3, 4, or 5, typically. Then spacing over the distance where stirrups are required (i.e., until v_{ext} drops below v_{allow}; see Figure 6-10) is taken as

$$s = \frac{A_v f_v}{b v_{\text{ext}}}$$

Note that s can be increased where the distance is great enough to justify it, to reflect the decreasing v_{ext} toward the center of the span. Also, observe that if $v_{\text{ext}} > 3.0\sqrt{f'_m}$ (150 lb/in.^{2} max.) or 75 lb/in.^{2} (no inspection), then b or d of the masonry section must be increased, since this is the maximum shear permitted under any circumstances.

Numcrical examples illustrating these concepts are given in Chapter 9.

FLEXURAL BOND

To develop the expression for bond stress required to maintain the composite action of the masonry and reinforcement, consider the force system shown in Figure 6-13. A short length of beam, dx, is shown, and the moment at one section differs from that at the other by the amount dM. Therefore, this incremental moment produces a change in the bar force amounting to

$$dT = \frac{dM}{jd}$$

To maintain equilibrium, this incremental tensile force is resisted by the bond force between the bar and the masonry. Thus, if U equals the bond force per unit bar

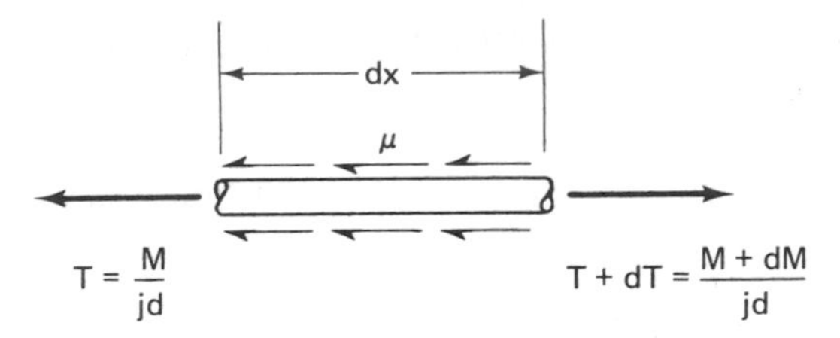

FIGURE 6-13. Flexural bond stress.

length,

$$U\,dx = (T + dT) - T = dT$$

from which

$$U = \frac{dT}{dx} = \frac{dM}{jd}\frac{1}{dx} = \frac{V}{jd}$$

The unit bond stress can be defined as

$$u = \frac{U}{\Sigma_0} \quad \text{or} \quad u = \frac{V}{\Sigma_0 jd} \qquad \boxed{\text{Flexural bond stress}} \qquad (25)$$

Note that, unlike concrete, where the bond emphasis is placed upon anchorage and development length, in masonry only a flexural bond check is actually called for at the present time. Adequate anchorage is generally achieved by conforming to UBC Section 2418 requirements for bond embedment lengths or special anchorage. These are summarized in Chapter 9.

Properly designed and placed grout will develop sufficient bond strength with the masonry units. This strength is developed in two ways: (1) the chemical adhesion produced by the portland cement paste, and (2) the mechanical interlock of the rebar. Allowable bond stresses are as follows:

Plain bars	30 lb/in.^{2}—no special inspection 60 lb/in.^{2}—special inspection required
Deformed bars	100 lb/in.^{2}—no special inspection 140 lb/in.^{2}—special inspection required

COMBINED BENDING AND DIRECT STRESS

Interaction formula

Reinforced masonry walls are often subjected to a combination of axial and flexural stress. Exterior masonry bearing walls, for example, resist the compressive stresses induced by the vertical live and dead loads. In addition, they often must span vertically between floor diaphragms when subjected to the lateral effects of either wind or seismic forces acting normal to the wall. These forces, therefore, impose flexural compressive and tensile stresses on the wall in addition to the axial compressive stresses caused by the gravity loads. Should the vertical loads not be axial, additional flexural stresses will be superimposed on the wall section due to this loading eccentricity.

To determine the adequacy of such sections, present practice stipulates that the basic interaction formula be utilized (Figure 6-14). It evolves as follows from the combined stress condition:

$$f_m = \frac{P}{A} + \frac{Mc}{I} = \frac{P}{A} + \frac{M}{S}$$

Divide by f_m to obtain

$$1 = \frac{P}{Af_m} + \frac{M}{Sf_m}$$

Let f_m in the first term be the allowable masonry stress in axial compression, where only the load P is acting. This then becomes F_a, which is $0.20f'_m$ for walls, or $(.18f'_m + .65p_g f_s)$ for columns, modified by a height/thickness reduction factor as defined in Chapter 7. In the second term, f_m, becomes F_b, the allowable masonry stress in flexure $(0.33f'_m)$, where only the moment M is applied. This value may be increased by one-third when bending occurs by virtue of wind or seismic load effects. Also let $P/A = f_a$, the actual axial stress. For walls, $f_a = P/bt$ and for columns $f_a = P/[bt + (n - 1)A_s]$. Also, the term M/S is symbolized as f_b, the actual bending stress. Should the bending stress exceed the axial compressive stress, as shown in Figure 6-14, a cracked section occurs and f_b equals $M/bd^2 \times 2/jk$. The analysis form of the interaction equation then becomes

$$\frac{f_a}{F_a} + \frac{f_b}{F_b} \leq 1.00 \qquad \boxed{\text{Interaction formula}} \qquad (26)$$

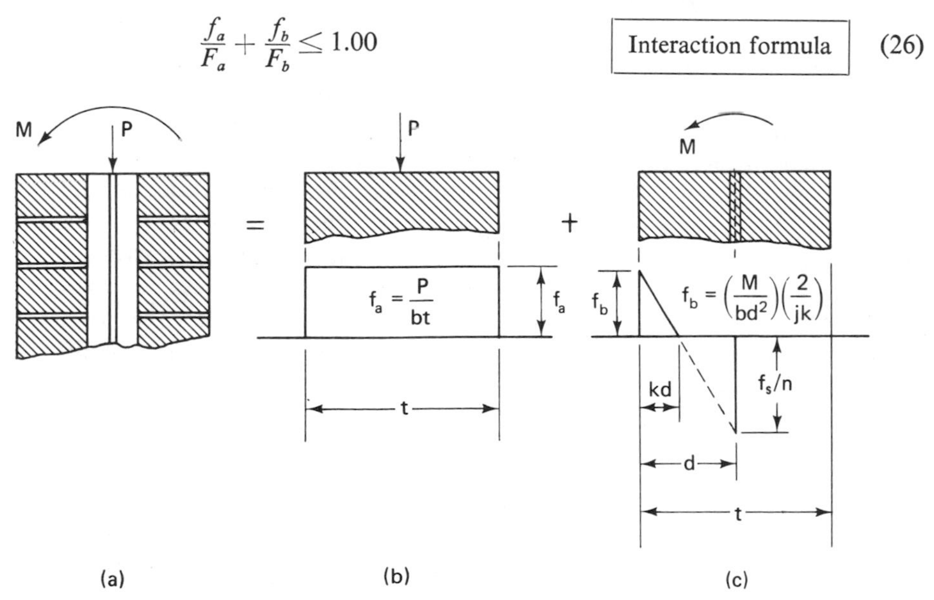

FIGURE 6-14. Combined bending plus direct stress distribution.

Measuring the Code adequacy of a given section simply involves checking compliance with the foregoing unity expression. Note that this expression does not reflect the effect of the axial compressive load upon the flexural tensile stress.

To use the interaction formula as a design expression, it may be modified into the form

$$f_b = \left(1 - \frac{f_a}{F_a}\right)F_b \tag{26a}$$

In this case, f_b, represents a sort of modified allowable masonry stress, reduced from the permitted F_b to take into consideration the effect of the applied axial load. Generally, the procedure here would be to assume a reasonable amount of reinforcement, then to check the actual bending stress to determine if it is less than the value of f_b as found in equation (26a).

The foregoing approach may be modified somewhat for the condition whereby the axial compressive stress overcomes the flexural tensile stress to the extent that the entire cross section remains under a compressive stress condition and therefore does not crack in tension. For instance, assume that the section does not crack; then compute f_a and f_b. Note that the bending stress, f_b, is calculated on the basis of an uncracked transformed section, as is the axial stress if that section is a column. When $f_a \geq f_b$, compressive stress dominates the entire cross section and only minimum arbitrary vertical steel is required (i.e., $0.007bt$). When $f_a < f_b$, the section is assumed to have cracked. Then proceed as described previously, using the interaction formula.

Interaction diagrams

Interaction diagrams portray a reasonably accurate delineation of the overall column or wall capacity when it is subjected to combined bending plus axial compression. However, they only have real meaning when expressed in ultimate strength terms, P_u and M_u. Reducing these to working stress values with the arbitrarily specified Code factors of safety can at best provide only a qualitative indication of wall or column capacity. This is why it is imperative that any future realistic appraisal of reinforced masonry wall or column strength be based upon accurate limiting stress or strain capacities of the masonry and steel. However, we do not possess sufficient information at this time to establish such a strength level; consequently, we must temporarily remain with working stress factors. Figure 6-15 might represent a qualitative indication of column capacity under these conditions.

The simple interaction equation in the UBC is:

$$\frac{f_a}{F_a} \pm \frac{f_b}{F_b} \leq 1$$

It is a workable relationship that was taken from an early concrete design theory. This could be plotted as a simple linear function as in the solid line of Figure

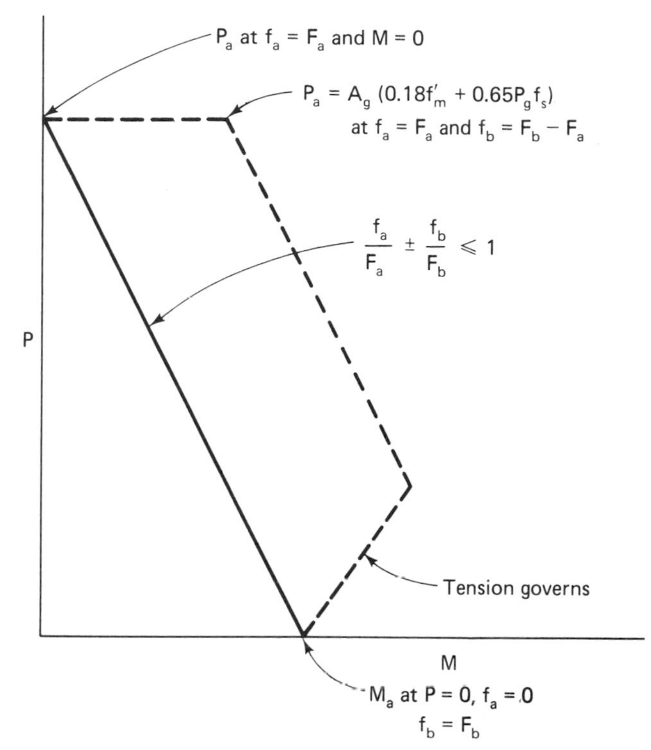

FIGURE 6-15. Working-stress design, compression plus bending. Masonry column.

6-15 if considered as a simple mathematical expression. However that may be varied by recognition of the fact that the average axial compressive stress would be $0.18f'_m$ or $0.20f'_m$, but the maximum bending compressive stress is $0.33f'_m$. Therefore moment can be increased by an amount of $(F_b - F_a) \times I/c$ without changing the compressive load and the resulting curve could be similar to the dashed line on Figure 6-15. Refer to the example shown in Chapter 7.

There are other refinements that may be made, but great precision is really not warranted in view of the wide variations in the material capacities and large factors of safety.

DETERMINATION OF MASONRY DESIGN STRENGTH (f'_m)

Procedure

In order to establish the ultimate design strength of the masonry assemblage, f'_m, UBC Section 2404(c) provides for two options: assumption or prism test. This value is needed since the allowable design values for reinforced masonry are expressed as functions of f'_m, as noted in Chapter 7.

Assumed ultimate compressive strength, f'_m

Where prism tests are not made, the f'_m may be assumed in accordance with the values listed in Table 6-1. When the assumed f'_m exceeds 2600 lb/in.^{2}, field tests in accordance with UBC Section 2404(c) 2, as described in the next section, are required.

TABLE 6-1

Assumed f'_m Values for Various Types of Masonry Units[a]

	Value of f'_m lb/in^2
Solid clay units—14,000 lb/in.^{2} gross	5300
Solid clay units—10,000 lb/in.^{2} gross	4000
Solid clay units—6,000 lb/in.^{2} gross	2600
Solid units—3000 lb/in.^{2} gross	1800
Solid units—2500 lb/in.^{2} gross	1500
Hollow concrete units—grade N	1350
Hollow concrete units—grade N grouted solid	1500
Hollow clay units—grade LB (1¼ in. minimum face shell)	1350
Hollow clay units—grade LB (1¼ in. minimum face shell) grouted	1500
Hollow clay units—Type I 5000 lb/in.^{2} net	2500

[a]For solid units, intermediate values may be interpolated. Compressive tests of solid clay units shall be conducted in accordance with UBC Standard 24-24.

Prism tests to establish f'_m

When the ultimate compressive strength is to be established by test, the test specimens consist of prisms made from the same materials with the same bonding arrangements as the structure being built, although they may be arranged in stack bond, in any case. Further, the moisture content and consistency at the time of laying, as well as the mortar joint thickness, should approximate the actual structure conditions as closely as possible. The established value of f'_m will then be the average of all specimens tested, but it cannot exceed 125% of the minimum value obtained on any test specimen. The unit stress, f'_m, is obtained by dividing the ultimate load by the net area of the masonry. In this case, this testing must be done prior to starting construction. In addition, one field test must be made for every 5000 sq ft of wall during construction, with a minimum of three tests being required per building.

Test prisms of hollow unit masonry should consist of one masonry unit in plan for a height/depth (h/d) ratio of 2.0, which is the basic h/d ratio (Figure 6-16).

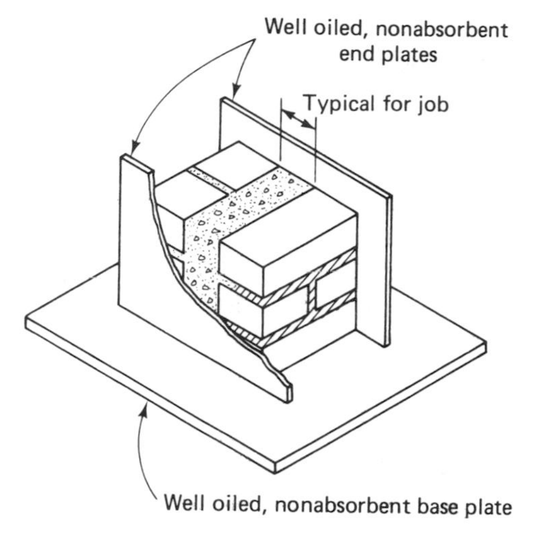

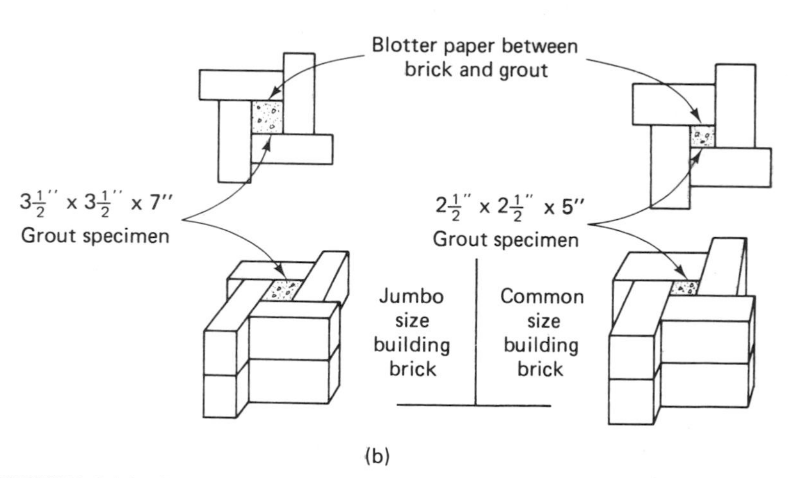

FIGURE 6-16. Masonry test specimens. (a) Shows a recommended procedure for grout specimens. The use of blotter paper or other porous material when pouring grout test prisms will prevent the grout from sticking to the brick. This method of grout test yields laboratory strengths which are closer to field conditions than is the case with test cylinders poured in nonabsorbent molds. This may also be used for brick wall specimens when used without the bond breakers. (b) An alternate method of field sampling grout.

Brick prisms should conform to this dimensional ratio also; although shorter lengths for these and solid grouted hollow unit prisms are now permitted (4 in. minimum), so that smaller-capacity testing equipment may be used. These will probably lead to more conservative results. A correction factor must be applied for an h/d ratio other than 2.0, as follows:

h/d ratio	1.5	2.0	3.0	4.0	5.0
Correction factor	0.86	1.00	1.20	1.30	1.37

The UBC specifications also call for special conditions of sampling, storage, curing, and testing at 28 days. There may be found in Section 2404(c), C and D.

PROBLEMS

When the material used is not specified, assume that the steel is of grade 40 and the brick is of grade MW and the concrete block is Grade LB, mediumweight units.

6-1. A simply supported brick beam with a 14-ft span sustains a working live load of 1000 lb/ft. The width of the beam is 10 in. and it has an overall depth of 36 in. The tensile reinforcement consists of two No. 5's. No special inspection is provided. Compute working stresses and the allowable moment. Is the beam adequate?

6-2. A solid grouted 8-in. concrete masonry beam is reinforced with two No. 4's. It has an effective depth of 28.5 in. Compute the allowable moment capacity (with and without special inspection). Is the beam over- or underreinforced? Why?

6-3. Same as Problem 6-2, but change the reinforcing steel to two No. 5's.

6-4. A simply supported brick beam with a 15-ft span sustains a working live load of 700 lb/ft. The width of the beam is 9 in., $f'_m = 1500$ lb/in.^{2}, and special inspection is provided. Determine the required beam depth and steel area assuming a balanced design.

6-5. Given a brick beam 10 in. wide which carries a moment of 22,000 ft-lb. Determine the required beam depth and steel area (with and without inspection).

6-6. Given a 12-in. partially grouted block wall with No. 7 at 48 in. cc placed at the center and no inspection is provided. Determine the allowable moment capacity. Remember to check the T-beam action and to include a one-third increase in the allowable stresses.

6-7. Determine the required reinforcing steel for a 12-in. partially grouted block wall subjected to a moment of 1900 ft-lb/ft. The steel is placed at the center at 32 in. cc. Special inspection is provided and include a one-third increase in the allowable stresses.

6-8. An 8-in. concrete block beam has A_s = two No. 6's and A'_s = two No. 5's, with $d' = 3.5$ in. and $d = 21.5$ in. Special inspection is provided. Determine the allowable moment capacity.

6-9. A 9-in.-wide brick beam is limited in depth to 20 in. It carries a moment of 16 ft-kips. Special inspection is provided. Find A_s and A'_s, assuming that $d' = 4$ in.

6-10. Given a 10-in.-wide brick beam with a 20-ft span, where

$f'_m = 3000$ lb/in.^{2} $\qquad n = 10$

$LL = 3500$ lb/linear foot $\qquad$ shear reinforcement = No. 4

$f_v = 20{,}000$ lb/in.^{2}

Determine the adequacy of the shear reinforcement shown by drawing the curve of shear capacity provided versus the capacity required for the entire beam.

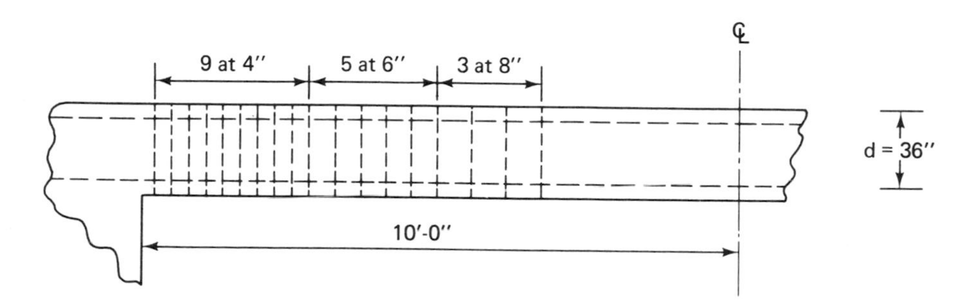

6-11. Given an 8-in. solid grouted concrete block beam on a 20-ft span with a d of 24 in. (no inspection is provided). It supports a live load of 1200 lb/ft. Determine the required stirrup spacings (assume that $j = 0.9$). Use a No. 4 stirrup.

7

Reinforced Walls

It is the intent of the next four chapters to tie together the previously developed design and analysis expressions with the essentially empirical rules, dictated by the Code and by experience, to form a basic, yet readily usable design procedure. It is felt that the approach delineated in these chapters will prove rigorous enough to satisfy the theoretical demands of an analytical analysis, and yet not be so cumbersome and time consuming that it cannot be readily utilized by the practicing structural engineer. Therefore, the dual objective sought herein will be (1) to present and clarify the principles of reinforced masonry for students, to familiarize them with current design practice; and (2) to meet the needs of structural engineers in their everyday problem solving. To facilitate application of the design procedures, the highly condensed but extremely versatile design aids provided in the Appendix should be helpful. These comprise charts and tables to facilitate the use of many of the design and analysis equations. For clarity, these chapters will divide and proceed along the lines of the basic structural elements involved: bearing walls, shear walls, deep-wall beams, nonbearing and filler walls, parapets, columns and pilasters, beams and lintels, followed by lateral-load-resisting systems such as diaphragms and box-type structures. To facilitate use of these procedures, UBC Table 24-H is expanded herein into Table 7-1 to give the specified maximum working stresses for reinforced solid and hollow unit masonry having various ultimate strength values, f'_m. This chapter will be devoted to a discussion of bearing walls, nonbearing walls, shear walls, parapet walls, and deep-wall beams.

CODE DESIGN LIMITATIONS FOR BEARING WALLS

Height/thickness limits

The UBC places arbitrary limitations on the sizes of various walls, and these are given in Table 7-2 (UBC Table 24-1). These limits are expressed in the form of a specified minimum wall thickness and a maximum unsupported wall height/thickness (h/t) ratio for various types of masonry walls. It might be observed initially that the minimum wall thicknesses and h/t ratios evolved strictly from the practice, of an earlier time, of constructing unreinforced walls. Eventually, or perhaps belatedly, limits were set forth for reinforced masonry. But actually the latter had no real basis in fact; rather the limits probably grew out of a compromise over what the members of a "Code committee" felt was appropriate. For instance, it has been suggested that this h/t ratio of 25 was set by a committee to limit the tension on a wall joint to a maximum value of 50 lb/in.^{2} when the wall is subjected to a wind load of 15 lb/ft^2! Such arbitrary establishment of design limits sometimes happens during the selection process for Code-promulgated empirical rules or limits. Unfortunately, as the state of the art advances, in the form of better material qualities stemming from improved production methods and construction practices, as well as through the development of more precise design procedures, many of the Code requirements remain unchanged. Examples abound, both in the laboratory and in the field, where reinforced walls not conforming to these regulations function quite adequately under actual service-load conditions. The Code belatedly tries to overcome this lack of validity by providing an "exception":

> The h/t ratio may be increased and the minimum thickness decreased when data is submitted which justifies a reduction in the requirements specified.

One way to achieve an effective increase in the h/t limits would be to make some provision, regarding the end restraint of the wall, which would have the actual effect of decreasing the value of h. This practice is described in more detail later in this section. Observe that for reinforced masonry bearing walls (solid grouted or hollow), this h/t limit equals 25, and the minimum wall thickness is 6 in. A subsequent approval permits the use of a 4-in.-thick wall at an h/t maximum of 27. Of significance also is the h/t limit of 30 for exterior, reinforced, nonbearing walls.

This requirement, of course, does not preempt the need to maintain the structural integrity of the wall insofar as dead, live, and lateral loads are concerned; that is, it must also be adequate to resist these loads regardless of the h/t ratio. Actually, separate considerations must be made for (1) the compressive resistance to total vertical loads, including those imposed by the overturning effect of the floors above

TABLE 7-1

Maximum Allowable Working Stresses for Reinforced Solid and Hollow Unit Masonry (lb/in.^{2})
(*UBC Table 24-H*)

Type of stress	*Allowable stress or stress coefficients*		*Hollow clay units,[a] grade LB; Hollow concrete units, grade N or S*		*Grouted solid hollow units; Concrete, grade N or S; Clay, grade LB; Solid units, 2500 lb/in.^{2} on gross*		*Solid units 3000 lb/in.^{2} on gross*	
Ultimate compressive strength:	f'_m		1350		1500		1800	
Special inspection required:	No	Yes	No	Yes	No	Yes	No	Yes
Compression—axial								
Walls	$\frac{1}{2}(0.2f'_m)$	$0.2f'_m$	135	270	150	300	180	360
Columns	$\frac{1}{2}(0.18f'_m)$	$0.18f'_m$	122	243	135	270	162	324
Compression—Flexural	$\frac{1}{2}(0.33f'_m)$ 450 lb/in.^{2} max.	$0.33f'_m$ 900 lb/in.^{2} max.	225	450	250	500	300	600
Shear								
No shear reinforcement			(Net section)					
Flexural[f]	25	$1.1\sqrt{f'_m}$ 50 lb/in.^{2} max.	25	40	25	43	25	47
Shear walls[g] $M/Vd \geq 1$	17	$0.9\sqrt{f'_m}$ 34 lb/in.^{2} max.	17	33	17	34	17	34
$M/Vd = 0$[h]	25	$2.0\sqrt{f'_m}$ 50 lb/in.^{2} max.	25	50[e]	25	50	25	50
Reinforcing taking all shear			(Net section)					
Flexural	75	$3.0\sqrt{f'_m}$ 150 lb/in.^{2} max.	75	110	75	116	75	127
Shear walls[g] $M/Vd \geq 1$	35	$1.5\sqrt{f'_m}$ 75 lb/in.^{2} max.	35	55	35	58	35	64
$M/Vd = 0$[h]	60	$2.0\sqrt{f'_m}$ 120 lb/in.^{2} max.	60	73	60	77	60	85
Modulus of elasticity[i]	$\frac{1}{2}(1000f'_m)$ 1,500,000 lb/in.^{2} max.	$1000f'_m$ 3,000,000 lb/in.^{2} max.	675,000	1,350,000	750,000	1,500,000	900,000	1,800,000
Modular ratio, $n = E_s/E_m$	$30{,}000/\frac{1}{2}f'_m$	$30{,}000/f'_m$	44	22	40	20	33	16.7
Modulus of rigidity[i]	$\frac{1}{2}(400f'_m)$ 600,000 lb/in.^{2} max.	$400f'_m$ 1,200,000 lb/in.^{2} max.	270,000	540,000	300,000	600,000	360,000	720,000
Bearing on full area[j]	$\frac{1}{2}(0.25f'_m)$ 450 lb/in.^{2} max.	$0.25f'_m$ 900 lb/in.^{2} max.	169	338	187	375	225	450
Bearing on $\frac{1}{3}$ or less of area[j]	$\frac{1}{2}(0.30f'_m)$ 600 lb/in.^{2} max.	$0.30f'_m$ 1200 lb/in.^{2} max.	203	405	225	450	270	540
Bond								
Plain bars	30 lb/in.^{2}	60 lb/in.^{2}	30	60	30	60	30	60
Deformed	100 lb/in.^{2}	140 lb/in.^{2}	100	140	100	140	100	140

[a]Stresses for hollow unit masonry are based on net section.
[b]Special testing shall include preliminary tests conducted as specified in Section 2404(c) to establish f'_m and at least one field test during construction of walls per each 5000 ft^2 of wall, but not less than three such field tests for any building.
[c]Ultimate compressive strength may be assumed up to $f'_m = 2600$ lb/in.^{2} if solid clay units have an ultimate strength of at least 6000 lb/in.^{2} on gross area. No prism test would be required, but the strength of the brick would have to be established in accordance with UBC Standard 24-24.
[d]Ultimate compressive strength may be assumed up to $f'_m = 5300$ lb/in.^{2} in accordance with Section 2404(c) 3, but field prism tests would have to be made to verify strength, consisting of at least one prism per 5000 ft^2 of wall nor less than three prisms per building.
[e]A heavy vertical line indicates a limit of allowable stresses by the UBC. Values may be exceeded if local jurisdiction permits.

Special testing[b]; f'_m established by prism tests–lb/in.²

2000	2500[c]	2700	3000	3500	4000	4500	5000	5300[d]	6000
Yes	Yes	Yes	Yes	Yes	Yes	Yes	Yes	Yes	Yes
400	500	540	600	700	800	900	1000	1000	1200
360	450	486	540	630	720	810	900	954	1080
667	833	900	[e] 900	900	900	900	900	900	900
49	50	50	50	50	50	50	50	50	50
34	34	34	34	34	34	34	34	34	34
50	50	50	50	50	50	50	50	50	50
134	150	150	150	150	150	150	150	150	150
67	75	75	75	75	75	75	75	75	75
89	100	104	110	118	120	120	120	120	120
2,000,000	2,500,000	2,700,000	3,000,000	3,000,000	3,000,000	3,000,000	3,000,000	3,000,000	3,000,000
15	12	11	10	10	10	10	10	10	10
800,000	1,000,000	1,080,000	1,200,000	1,200,000	1,200,000	1,200,000	1,200,000	1,200,000	1,200,000
500	625	675	750	875	900	900	900	900	900
600	750	810	900	1050	1200	1200	1200	1200	1200
60	60	60	60	60	60	60	60	60	60
140	140	140	140	140	140	140	140	140	140

[f]Web reinforcement shall be provided to carry the entire shear in excess of 20 lb/in.² whenever there is required negative reinforcement and for a distance of one-sixteenth the clear span beyond the point of inflection.
[g]When calculating shear or diagonal tensile stresses in shear walls that resist seismic forces, use 1.5 times the force required by Section 2312(d) 1.
[h]M is the maximum bending moment occurring simultaneously with the shear load V at the section under consideration. Interpolate by straight line for M/Vd values between 0 and 1 (see Figure 7-8).
[i]Where determinations involve rigidity considerations in combination with other materials or where deflections are involved, the modulii of elasticity and rigidity under columns entitled "yes" for special inspection shall be used, or tests shall be made to determine the correct values.
[j]This increase shall be permitted only when the least distance between the edges of the loaded and unloaded areas is a minimum of one-fourth of the parallel side dimension of the loaded area. The allowable bearing stress on a reasonably concentric area greater than one-third, but less than the full area, shall be interpolated between the values given.

TABLE 7-2

Minimum Thickness of Masonry Walls

Type of masonry	*Maximum ratio unsupported height or length to thickness*	*Nominal minimum thickness (in.)*
Bearing walls		
Unburned clay masonry	10	16
Stone masonry	14	16
Cavity wall masonry	18	8
Hollow unit masonry	18	8
Solid masonry	20	8
Grouted masonry	20	6
Reinforced grouted masonry	25	6
Reinforced hollow unit masonry	25	6
Nonbearing walls		
Exterior unreinforced walls	20	2
Exterior reinforced walls	30	2
Interior partitions unreinforced	36	2
Interior partitions reinforced	48	2

the wall in question; (2) the capacity of the wall to resist lateral loads (wind or earthquake) acting both normal to (bending) and in the plane of the wall (shear), and (3) resistance to in-plane bending due to the overturning effect of lateral loads, under all or even partial dead load with no live load acting. For the lateral load combinations, stress increases of one-third are permitted. When the lateral load acts normal to the wall, it must span between supports, either vertically, diaphragm to diaphragm, or horizontally, between pilasters. In both cases the h/t limit would be 25. However, in the latter case, should an upper portion of the wall act as a beam to pick up the roof or floor loads, the h/t limit would increase to 30, since the wall itself then becomes nonbearing in performance. Note also that the limiting h dimension, when spanning between pilasters, becomes a horizontal distance. Actually, the Code intent here was for the limiting h dimension to be either a vertical or a horizontal one. An editorial clarification defines h now as "the distance between supporting members (vertical or horizontal stiffening elements)."

Additional moments in the wall can be caused by the eccentric application of any vertical dead and live load imposed upon the wall. This eccentricity exists because the center of the floor or roof connection generally cannot be made coincident with the wall center line, owing to the manner in which the connection must be achieved. The manner in which this eccentricity may be determined is described in the section "Assumed Location of Beam and Truss Reactions" in Chapter 8. In that event, the eccentric moment at the top of the wall would equal the live plus the dead load times the eccentricity. Also the, combined effect of dead-load eccentric moment at the vertical midspan, in combination with the moment caused by the lateral out-of-plane forces, must be considered. Later examples will illustrate these conditions.

Axial stress limits

The allowable unit axial stress in reinforced masonry walls (F_a) is defined in UBC Section 2418(j) 2 as

$$F_a = 0.20f'_m\left[1 - \left(\frac{h}{40t}\right)^3\right]$$

where t = nominal wall thickness, in.
h = clear unsupported distance between wall support, in.
max. f'_m = 6000 lb/in.^{2}

Note that no load capacity allowance is made for the vertical steel in bearing walls. The factor $[1 - (h/40t)^3]$ represents a sort of stress reduction factor to account for the $P\Delta$ effect (Figure 7-1) or the increase in moment due to lateral buckling. You may recall that the AISC steel column formulas, which are considerably more precise, account for this effect through an amplification factor.

Figure 7-2 draws a comparison between the allowable stress for f'_m = 2000 lb/in.^{2} and what is perceived as the expression for ultimate buckling capacity of a wall. For the long elements, this capacity is defined by the *Euler formula*,

$$\frac{P}{A} = \frac{\pi^2 E_m}{(kh/r)^2}$$

In these cases the element buckles before any of the material anywhere in the cross section reaches a yield state. However, should any portion of the material reach a yield condition before buckling actually occurs, then a modification of the tangent modulus formula,

$$\frac{P}{A} = \frac{\pi^2 (E_m)_t}{(kh/r)^2}$$

might be more appropriate. The term $(E_m)_t$ represents the reduced modulus obtained as the tangent to the stress/strain curve, corresponding to the actual stress condition at the instant the element buckles. But this value becomes somewhat difficult to define in a practical design case. The problem is further complicated by the fact that the masonry section is composed of two materials of very different stiffnesses, creating a distinct nonhomogeneity. A more reasonable approach would follow along the lines of a modified inelastic buckling formula which would conform to the various test data generated to develop it. This would then represent the transitional expression between the Euler elastic buckling formula and the ultimate compressive strength of the material itself. Perhaps it could be reasonably formulated by an expression such as the Johnson–Euler, the curve for which is sketched in Figure 7-2. This expression could thus reflect the performance of a nonhomogeneous reinforced masonry section.

Note, for example, that at an h/t ratio of 25, the Code-allowed f_m reduces to about 230 lb/in.^{2} ($R \sim 0.77$), while the Johnson–Euler curve would set the buckling strength at about 1350 lb/in.^{2} still yielding a rather high safety margin, nearly 6. These

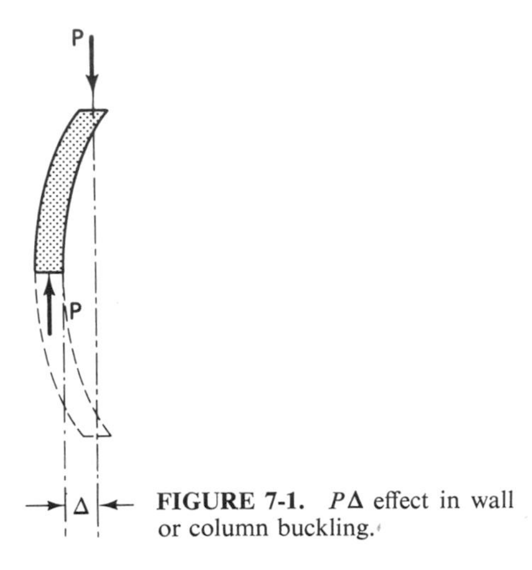

FIGURE 7-1. $P\Delta$ effect in wall or column buckling.

FIGURE 7-2. Wall or column buckling curves. The shape of the reduction curve is not correct; it goes to 0 at an h/t of 40 and to minus values at values greater than 40 (i.e., the capacity would increase). A few cities have recognized this and have extended the curve beyond the h/t of 25 by a curve of the general shape of 1/5 of an appropriate Euler curve. Others should be similarly enlightened!

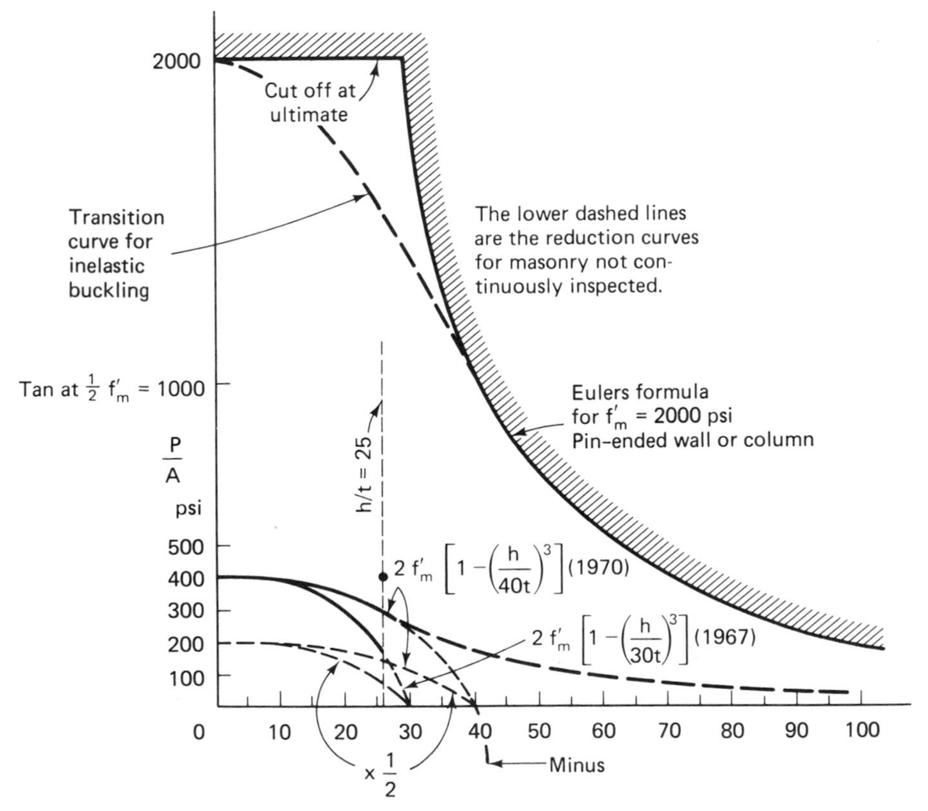

curves clearly demonstrate that this Code expression for the reduction factor R does not represent a valid curve. For instance, we must recognize that if a wall with $h/t = 25$ has a capacity of about three-fourths that of $h/t = 0$, the capacity does not suddenly reduce to a minus value (or zero) at an h/t of about 40, as the reduction formula would indicate. Consequently, an expression based upon both the Johnson–Euler and the elastic Euler formula would seem to be much more appropriate. However, considerably more test data must be generated before these can be accurately evaluated. It was just such test data that substantiated an increase in the present stability factor over that of the 1967 factor, which was set at $1 - (h/30t)^3$.

The value of the unsupported height can be modified by a proper consideration of wall-end restraint, the k factor in the Euler formula. Table 7-3 lists both

TABLE 7-3

k Factors for Walls and Columns

	(a)	(b)	(c)	(d)	(e)	(f)
Buckled shape of column is shown by dashed line	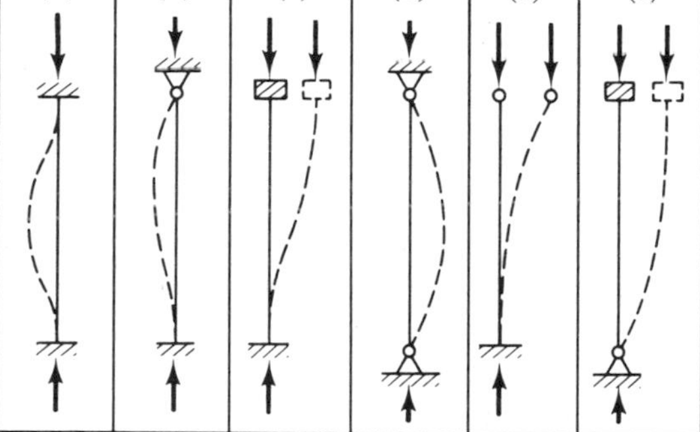					
Theoretical K value	0.5	0.7	1.0	1.0	2.0	2.0
Recommended design value when ideal conditions are approximated	0.65	0.80	1.2	1.0	2.10	2.0
End condition code	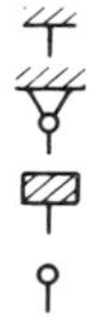 Rotation fixed and translation fixed Rotation free and translation fixed Rotation fixed and translation free Rotation free and translation free					

the theoretical and recommended values of k for various restraint conditions in either walls or columns. This is similar to the end-restraint factors sometimes used for concrete or steel columns.

The restraint referred to can be achieved in several ways. For instance, if the wall can be fixed at the footing in some fashion and simply supported at the top, say by the roof, then the recommended effective design height, h', becomes $0.8h$. The presence of a knee brace from roof beam to wall, with a fixed base, permits an h' of about 0.65 times the distance from knee brace to floor. When the wall is designed to span horizontally between pilasters across several bays, it is to be noted that some *fixity*, or resistance to rotation, is provided at the pilaster, both by the torsional resistance of the pilaster and by the continuity of the wall section over several spans. Therefore, the effective h' can become the distance between points of inflection or about 0.6 times the clear distance between pilasters. In effect, the h/t limit then becomes actually $30/0.6 = 50$ instead of 30, assuming of course, that a beam at the roof or floor level carries the vertical loads. Refer to Figure 7-3, which illustrates these conditions. In another scheme, the walls may span between footing pads, in which case the wall may be assumed to be a "deep beam" carrying the vertical loads between those pads. The behavior and design of these deep-wall beams are described in a later section. With this system the vertical height becomes the limiting h dimension, but a greater h/t limit is permitted under a special Code provision for this type of wall.

Combined axial and flexural stresses

Where combined bending and axial forces are encountered, the previously developed interaction formula is to be used [per UBC Section 2418(g)]:

$$\frac{f_a}{F_a} + \frac{f_b}{F_b} \leq 1$$

where F_a = allowable axial wall stress per UBC Section 2418(j)(2), including any stress increase permitted in Section 2418

f_a = computed axial wall sttess = P/A

F_b = maximum allowable bending stress = $0.33f'_m$ (900 lb/in.^{2} max.), including any stress increase permitted in Section 2418

f_b = computed compressive bending stress in the masonry

Recognize that this approach (interaction formula) is very conservative, for it in no way accounts for the influence of the axial compressive force in reducing the flexural tension developed at the pier boundary (jamb). Nondimensional interaction curves need to be developed for masonry walls and columns, much like the ACI interaction curves for reinforced concrete elements. Once these are generally available and verified, then rather than using the simplified interaction formula above, one can observe whether or not all design load combinations fall within the pier section's interaction envelope curve. This would indeed provide a more realistic appraisal of wall or pier adequacy under working-load conditions. These curves can be readily nondimensionalized simply by dividing the P ordinate by btf'_m and the M abcissa by $bt^2f'_m$. Otherwise, an interaction curve must be plotted for each different type of masonry unit and size, as well as for various probable reinforcing areas per lineal foot of wall.

For example, refer to Figure 7-4, which is a plot of P versus M for a specific unit. It is for an 8-in. clay unit of high strength which develops a masonry assemblage

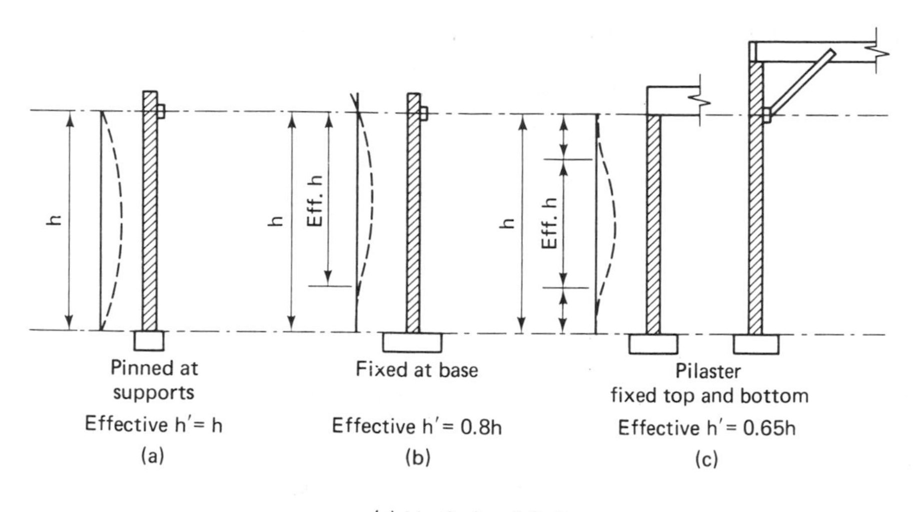

(a) Vertical wall fixity

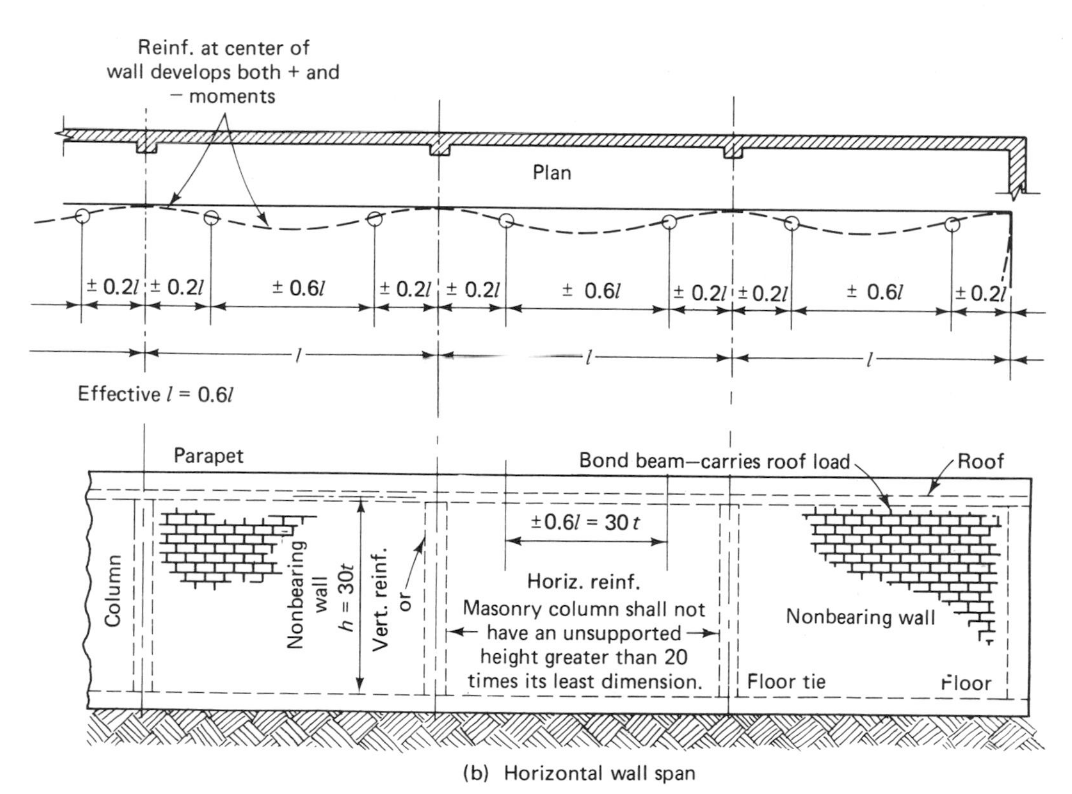

(b) Horizontal wall span

FIGURE 7-3. Conditions of vertical and horizontal wall restraint.

strength, f'_m, of 2500 lb/in.². The data considered in the curve are:

$$f'_m = 2500 \text{ lb/in.}^2$$

$$F_b = 0.333 \times 2500 = 833 \text{ lb/in.}^2$$

$$F_a = 0.20 \times 2500 = 500 \text{ lb/in.}^2$$

$$n = \frac{30{,}000{,}000}{2500} = 12$$

$$A = 90 \text{ in.}^2/\text{ft}$$

$$A_s = 0.20 \text{ in.}^2/\text{ft}$$

The accompanying calculations (page 183) indicate the assumed conditions that were used to develop the curve by varying the deformation of a plane section and determining the consequent stress magnitude and distribution and the corresponding values of P and M.

The assumption of zero steel stress in the compression phases is because the equation for bearing-wall capacity does not consider the contribution of wall steel. This approximation is valid in practice since the amount of steel in bearing walls is relatively small and would not influence the capacity appreciably. The stress in the steel, which is assumed to be tensile only, is computed as n times the stress that would

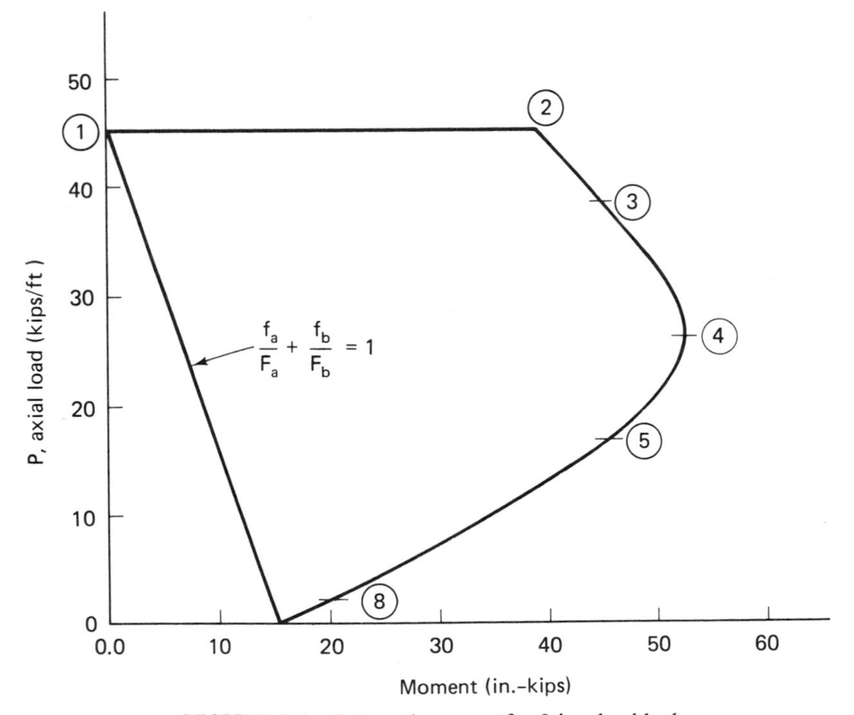

FIGURE 7-4. Interaction curve for 8-in. clay block.

$f'_m = 2500$; $F_b = 833$; $F_a = 500$; $n = 12$; $A'_m = 90\ \text{in.}^2/\text{ft}$; $A_s = 0.20\ \text{in.}^2/\text{ft}$.

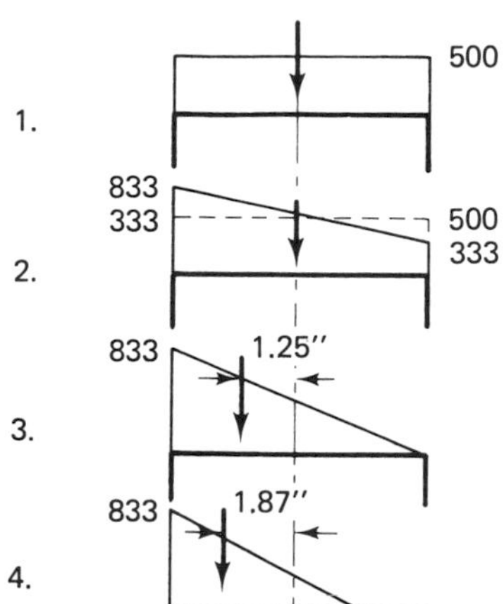

1.

$$P_m = 90 \times 500 = 45{,}000 \text{ lb}$$
$$M = 0$$

2.

$$P_m = 90 \times 500 = 45{,}000 \text{ lb}$$
$$M = 333 \times \frac{12 \times 7.5^2}{6} = 38{,}461 \text{ in.-lb}$$

3.

$$P_m = \frac{833}{2} \times 90 \times 1 = 37{,}485 \text{ lb}$$
$$M = \quad \times \frac{7.5}{6} = 46{,}855 \text{ in.-lb}$$

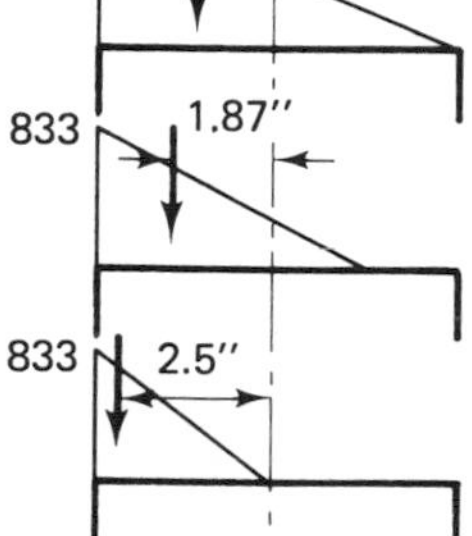

4.

$$P_m = \frac{833}{2} \times 90 \times 0.75 = 28{,}113 \text{ lb}$$
$$M = \quad \times 1.87 = 52{,}572 \text{ in.-lb}$$

5.

$$P_m = \frac{833}{2} \times 90 \times 0.5 = 18{,}740 \text{ lb}$$
$$M = \quad \times 2.5 = 46{,}856 \text{ in.-lb}$$

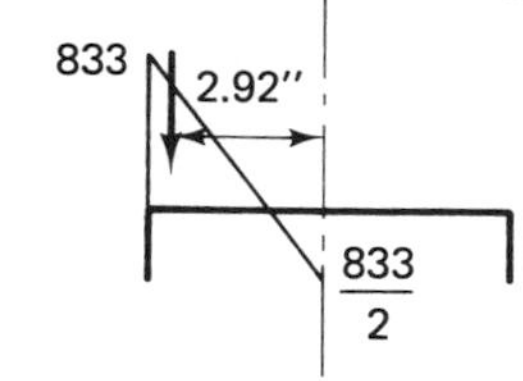

6.

$$P_m = \frac{833}{2} \times 90 \times 0.33 = 12{,}370 \text{ lb}$$
$$P_s = \frac{833}{2} \times 12 \times 0.20 = \frac{-1{,}000}{11{,}370}$$
$$M_m = \quad \times 2.92 = 36{,}120 \text{ in.-lb}$$

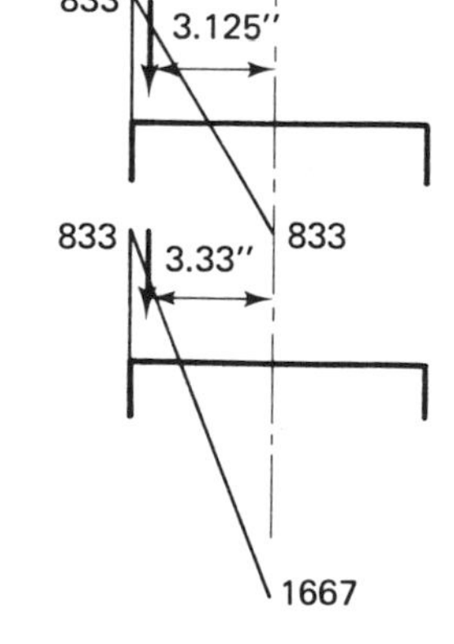

7.

$$P_m = \frac{833}{2} \times \frac{90}{4} = 9{,}371 \text{ lb}$$
$$P_s = 833 \times 12 \times 0.20 = \frac{-1{,}999}{7{,}372 \text{ lb}}$$
$$M_m = \quad \times 3.125 = 29{,}285 \text{ in.-lb}$$

8.

$$P_m = \frac{833}{2} \times \frac{90}{6} = 6{,}247 \text{ lb}$$
$$P_s = 1667 \times 12 \times 0.20 = \frac{-3{,}998}{2{,}249 \text{ lb}}$$
$$M_m = \quad \times 3.33 = 20{,}804 \text{ in.-lb}$$

have occurred in masonry at that point due to the deformation of the plane section. The assumption that masonry takes no tension is not true but is conservative for design of items with high flexural loadings.

Observe the variations of stress and consequent P versus M values as the deformations are imposed.

1. At point 1 the assumption is that a uniform F_a of 500 lb/in.^{2} average stress over the total section occurs.
2. Keeping the average stress at 500 lb/in.^{2} but imposing bending until the maximum allowable flexural stress, $F_b = 833$ lb/in.^{2} occurs at point 2, it is obvious that the total section is working, increasing the compressive stress on one side with decreased value on the other. The moment M increases, while average load P remains constant.
3. Between points 2 and 3, additional deformations increase moment but decrease the compressive stress to 0 at one face.
4. Between points 4 and 5, increased deformation would impose varying degrees of tension (assumed not to occur until the steel at the center line would come into action, as shown in 5).
5. As the plane deforms further, the compression area becomes smaller and the steel stress becomes greater, until the full capacity of the steel would be reached at $P = 0$.

The lower portion of the curve would have been modified considerably if the steel had been a greater amount, as in a highly stressed beam, or in one in which d is much greater than $t/2$. However, this discussion does emphasize that the simple interaction formula, in the same figure, is quite conservative, and also that the calculated capacity is not a precise consideration requiring use of many significant digits in the final design.

Another interesting set of curves are plotted in Figure 7-5. They show the maximum axial load P that may be imposed upon 10-in.-thick walls of different heights (h) built with various masonry strengths (f'_m). The dashed boundary line shows where the tension stress due to lateral-load bending produced by a 15-lb/ft^2 wind load is overcome by the axial compressive stresses produced by vertical loads. Combinations of P and h to the right of the boundary produce compression over the entire cross section; therefore, no steel is needed for moment resistance. Where tension does occur, minimum steel is still adequate for most conditions, except those which lie in the shaded area. In such cases the minimum amount of steel (0.077 in.^{2}/ft) may need increasing to carry the lateral-load moment.

Minimum reinforcement

Experience (reinforced by Code requirements) dictates that reinforced masonry walls must contain a certain minimum amount of reinforcement to provide for safe performance during catastrophic earthquakes. This is arbitrarily stipulated

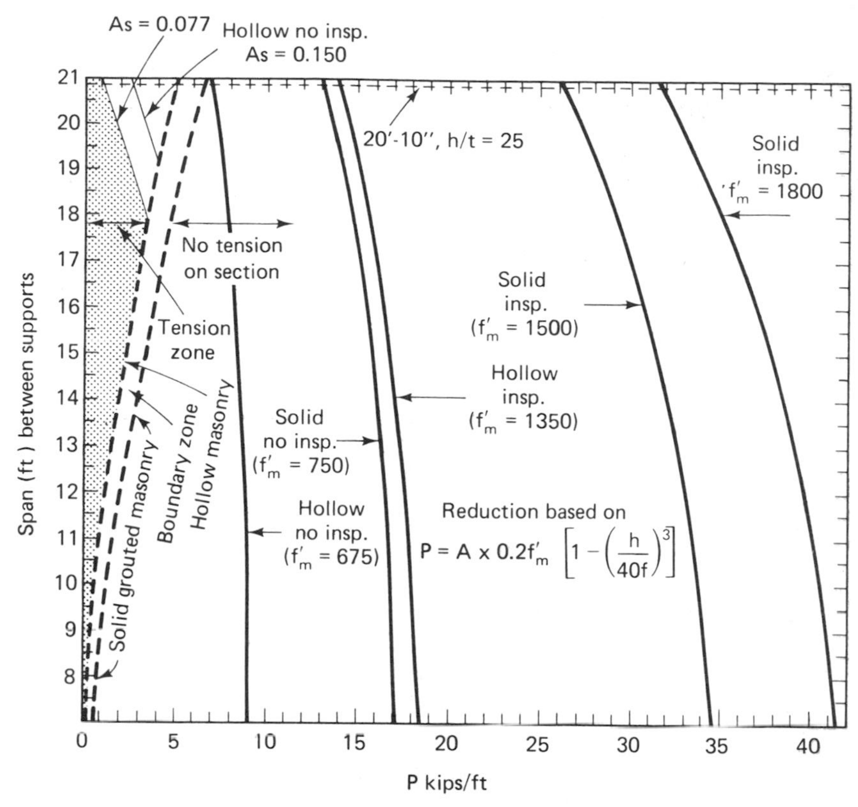

FIGURE 7-5. Axial concentric load *P* for 10-in. bearing walls.

presently as 0.002 times the gross cross-sectional area of the wall, not less than one-third of which is required in either the vertical or the horizontal direction. Furthermore, maximum spacing for the principal wall steel in either direction is set at 48 in. cc, with the minimum bar diameter being $\frac{3}{8}$ in. Also, joint reinforcement may be considered as contributing to the required horizontal minimum. Horizontal bars must be placed at the top of all footings, adjacent to all wall openings, at intersecting roof and floor levels, and at the top of parapet walls. Any or all of these may also be considered as contributing to the minimum horizontal steel requirement, provided that the bars are continuous along the entire length of the wall. If a wall consists of more than two wythes or 12 in. of thickness, the minimum vertical reinforcement requirement must arbitrarily be divided into two mats, one at each wall face, except in retaining walls. Further, in bearing walls, at least one No. 4, or two No. 3 bars, must be placed on all sides of and adjacent to every opening of more than 24 in. in size in either direction. Such bars must extend at least 40 bar diameters (24 in. minimum) beyond the opening corners. These are called for in addition to the specified minimum reinforcement (refer to Figure 4-3).

An example of how this minimum reinforcement requirement may be satisfied is shown in the following:

1. A reinforced masonry wall has No. 4's at 24 in. cc vertical steel and

five No. 4 horizontal bars in a 10-ft-high, 8-in.-thick wall. Is this adequate?
Horizontal steel (consider vertical wall section):

$$5 \text{ No. 4's give } \frac{5 \times 0.20 \text{ in.}^2 \times 100}{10 \text{ ft} \times 12 \text{ in.} \times 7.6 \text{ in.}} = 0.110\% > 0.07\% \text{ min.} \qquad \therefore \text{OK}$$

Vertical steel (consider horizontal wall section):

$$\text{No. 4's at 24 in. cc gives } \frac{0.20 \text{ in.}^2 \times 100}{7.6 \text{ in.} \times 24 \text{ in.}} = 0.110\%$$

and

$$0.110 + 0.110 = 0.22\% > 0.20\% \text{ min.}$$

$\therefore$ OK for total minimum steel

2. Select the minimum steel required for an 8 in. concrete block wall, 12 ft high.

$$\text{total area required} = 0.002 \times 12 \text{ in.} \times 7.62 \text{ in.} = 0.183 \text{ in.}^2/\text{ft}$$

with not less than 0.061 in.^{2}/ft in either direction. Thus, No. 5's at 48 in. oc (max. spacing) is

$$\frac{0.31 \text{ in.}^2}{4} = 0.077 \text{ in.}^2/\text{ft vertical} \qquad \therefore \text{OK}$$

The horizontal steel required $= 0.183 - 0.077 = 0.106$ in.^{2}/ft. Thus, 0.106×12 ft $= 1.27$ in.^{2} total horizontal steel area is required.

With one No. 4 at the top of the parapet, two No. 5's in the bond beam, and one No. 5 at the top of the footing, the total area would be 1.13 in.^{2}. The remainder ($1.27 - 1.13 = 0.14$ in.^{2}) would be supplied by intermediate horizontal No. 4 bars in the wall field (max. bar spacing = 48 in.) or No. 9 wire joint reinforcing at 24 in. cc (= 0.017 in.^{2}/ft).

Partially reinforced masonry

Partially reinforced masonry is a form of masonry construction that is reinforced to function structurally under specific gravity or wind loads to resist flexural tension or shear stresses. It need not adhere to any of the arbitrary reinforcing requirements specified for reinforced grouted masonry [UBC Section 2418(j)3] located in regions of high seismicity (see "Types of Masonry Walls," Chapter 1). However, the h/t limits given in Table 7-1 for reinforced masonry walls also apply to partially reinforced masonry walls. Other differences, as spelled out in the Code, include:

1. Only Type M or S mortar may be used.
2. The maximum spacing of the vertical reinforcement in exterior walls may be increased to 8 ft.

3. Horizontal reinforcement, not less than 0.2 in.^{2} in area, is called for at the top of all footings, at the top and bottom of all openings, near roof and floor levels, and at the top of all parapet walls.

In-wall columns

Where the bearing wall supports a concentrated load, it may still be designed either as a load-bearing wall or as an in-wall column (nonprojecting pilaster). As a load-bearing wall, the vertical wall steel cannot be counted on to contribute to the load-carrying capacity, however. On the other hand, if a portion of the wall is considered to act as a column, the allowable column stresses and reinforcing requirements apply, and the vertical reinforcement carries its proportionate share of the vertical load. Additional reinforcement may be needed, either as column steel to help carry the reaction, or to reinforce the wall for the additional bending caused by an eccentrically applied reaction if one exists. When the wall is considered to act as a column, the effective design length may not exceed the smaller of the following: (1) center-to-center distance between loads, or (2) width of bearing plus four times the wall thickness. The bearing stress on the masonry caused by the beam or truss reaction must not, in any case (column or wall section), exceed $0.25f'_m$ (900 lb/in.^{2} max.) for bearing on the full area, and $0.30f'_m$ (1200 lb/in.^{2} max.) for bearing on one-third or less area.

WALL DESIGN DIMENSIONS

Effective depth

The nominal thickness of concrete masonry units is typically 6, 8, 10, 12, or 16 in., with the actual thickness being $\frac{3}{8}$ in. less. The effective *d*, for steel at the center, would then equal the actual thickness divided by 2; for example, 7.625 divided by 2 = 3.81 in. for the *d* in a nominal 8-in. wall. On the other hand, if the steel were placed in two layers, to achieve a greater *d* value, the effective depth could be about 5.3 in. for 8-in. block, 9.0 in. for 12-in. block, and 13.0 in. for 16-in. block.

It is actually more economical to place the steel in the center of the wall, rather than placing it in two layers, if no other factors require the latter. Remember that this is the tension reinforcement in the wall span, which carries the lateral wind or seismic load to the supporting members. For example, consider No. 4's at 24 in. cc placed at the center of the wall versus two No. 3's at 24 in. cc placed in two layers.

$$M = A_s f_s j_d = \frac{0.2 \times 20{,}000 \times 0.9 \times 3.81}{2} = 6858 \text{ in.-lb of moment developed per horizontal foot of wall}$$

using No. 4's at 24 in. cc (0.10 in.^{2}/ft). Now with steel in two layers,

$$M = \frac{0.11 \times 20{,}000 \times 0.9 \times 5.3}{2} = 5250 \text{ in.-lb/ft}$$

with two No. 3's at 24 in. cc (0.11 in.^{2}/ft). Obviously, the greater moment of 6858 in.-lb with the single mat is achieved with slightly less steel area.

Now consider clay brick walls. Because of their double- or triple-wythe construction, there is not a standard wall thickness per se. However, a typical $3\frac{1}{4}$-in.-thick brick with a standard $2\frac{1}{2}$-in. grout space produces a 9-in. wall thickness, wherein d could then be simply $\frac{9}{2}$, or $4\frac{1}{2}$ in. Thus, different effective depths will be achieved by using different brick thicknesses and grout spaces. If the steel were placed in two layers in the grout space in order to achieve a maximum d, a 9-in. wall thickness would have a d of about 5 in., whereas a 10-in.-thick wall produces a d of approximately 6 in., and beyond that the effective depth simply increases by about $\frac{1}{2}$ in. for every $\frac{1}{2}$-in. increase in wall thickness. Actually, the minimum grout space is $\frac{3}{4}$ in., so a wall conceivably could be 3 in. + 3 in. + $\frac{3}{4}$ in. = $6\frac{3}{4}$ in. thick. However, since the reinforcing must have $\frac{1}{4}$ in. cover, the grout space width must be increased beyond the minimum, depending on the amount of steel placed therein.

Effective width

The effective width of a wall in flexure (loading perpendicular to wall) may be horizontal or vertical, depending upon which direction the wall spans. At any rate, the UBC requires that the effective b used in computing flexural stresses or moment capacity shall not be greater than six times the wall thickness for running bond, nor more than three times the wall thickness for stack bond [UBC Section 2418(f)] or the bar spacing (refer to Figure 7-6a).

In 1971, the Masonry Institute of America carried out a test program which attempted to determine the effective width (b) in flexure for concrete block wall panels (8 ft 8 in. wide × 20 ft 0 in. high). They concluded that in both the 6- and 8-in.-thick panels, laid up in running bond, the vertical steel functioned as effectively at an 8-ft spacing as it did for a 2-ft spacing. This would represent an effective b for a spacing/thickness ratio of 96:8 = 12. Further, they found that the 8-in.-thick stack bond panels functioned as if the effective width were about one-half that of the running bond panels. These results would certainly suggest that taking the effective width, in running bond, at six times the block thickness, provides an extremely conservative structural design.

Equivalent solid thickness

The axial compression area may be obtained as the product of the actual width × the length of the wall. This applies for grouted brick and solid grouted hollow unit masonry. However, if partially grouted hollow unit masonry is used, the effective compression area may be computed from a value for the *equivalent solid thickness*. This refers to the actual solid area present; that is, the volume of solid material in the wall divided by the face area of the wall. These values are shown in Table 7-4.

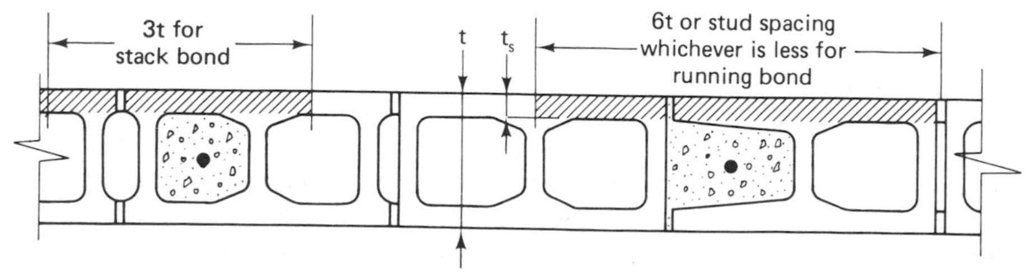

(a) Area assumed effective in flexural compression, force normal to face

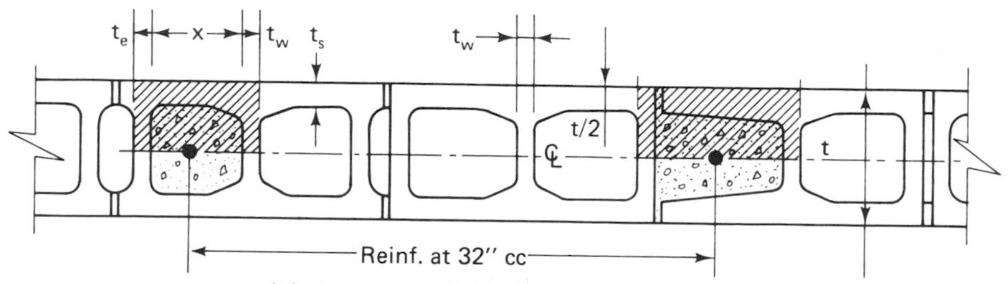

(b) Area assumed effective in shear, force normal to face

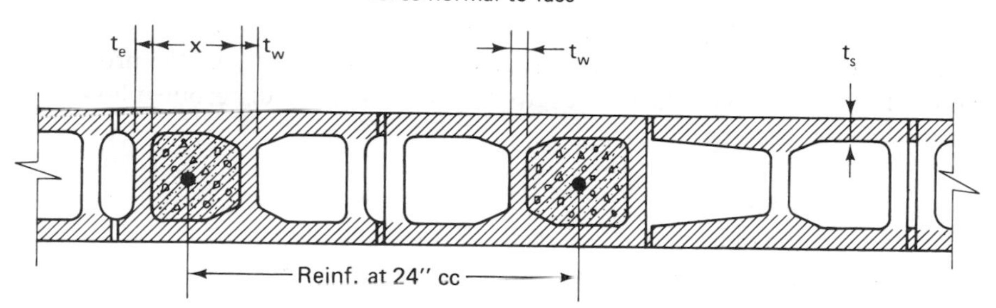

$x = 6''$ $t_w = 1''$
$t_s = 1.25''$ $t_e = 1.25''$

(c) Area assumed effective in shear, force parallel to face

FIGURE 7-6. Compressive and shear areas.

TABLE 7-4

Equivalent Solid Thickness—Hollow Unit Masonry Walls[a]

Grout spacing	*Nominal unit thickness (in.)* 6	8	12	*In-plane shear area—in.²/ft (8 in. block)*
Solid grouted	5.6	7.6	11.6	88.5
Grout at 16 in. oc	4.5	5.8	8.5	60.9
24	4.1	5.2	7.5	50.5
32	3.9	4.9	7.0	45.3
40	3.8	4.7	6.7	42.3
48	3.7	4.6	6.5	40.5
No grout	3.4	4.0	5.5	

[a]Some special shapes will very from the figures shown.

Shear areas

The flexural shear width for solid grouted wall sections, loaded out of plane, is governed by the dimensions specified for a flexural compression determination. However, for partially grouted, hollow unit masonry construction, as can be seen in Figure 7-6b, the flexural shear width for a 32-in. grouted cell spacing for steel would be determined as follows:

Flexural Shear (8-in. block):

$$\text{shear width per 32 in.} = b = \text{grouted cell width} + \text{web and end wall}$$
$$= 6 \text{ in.} + 1 \text{ in.} + 1\tfrac{1}{4} \text{ in.} = 8\tfrac{1}{4} \text{ in.}$$

Then

$$8\tfrac{1}{4} \times \tfrac{12}{32} = 3.1 \text{ in./ft}$$
$$d = 3.81 \text{ in.}$$
$$v = \frac{V \text{ per foot}}{bjd} = \frac{V}{3.1 \text{ in.} \times 0.9 \times 3.81 \text{ in.}} \text{ lb/in.}^2$$

Seldom is the flexural shear stress critical, however. The *in-plane direct shear area* (load parallel to wall), as on a solid grouted shear wall, would simply be the product of the actual wall thickness times the wall length. However, in partially grouted hollow unit masonry, the area becomes:

in-plane direct shear area
= total length of mortared face shells × face-shell thickness
+ grouted core area + (total mortared web and end wall thicknesses adjacent to grouted core × distance between face shells)

Thus, for a 24-in.-cc spacing (Figure 7-6c):

$$(2 \times \overset{t_s}{1\tfrac{1}{4}} \times 24) + 6(5) + (\overset{t_w}{1} + \overset{t_e}{1\tfrac{1}{4}})(5) = 60 + 30 + 11$$
$$= 101 \text{ in.}^2/24 \text{ in. or } 50.5 \text{ in.}^2 \text{ average per foot of wall length}$$

For a 48-in.-cc spacing:

$$(2 \times 1\tfrac{1}{4} \times 48) + 6(5) + (1 + 1\tfrac{1}{4})(5)$$
$$= 161 \text{ in.}^2/48\text{-in. or } 40.5 \text{ in.}^2 \text{ average per foot of wall length.}$$

For raked joints, deduct $\frac{1}{4}$ in., for each side raked, from the face-shell width. Refer to Table 7-4 for the average in-plane shear values for various wall steel spacings in terms of square inches per foot of 8 in. wall length.

MISCELLANEOUS WALL DETAILS

Wall patterns

Wall strength may be somewhat affected by the type of unit or wall-pattern arrangement, depending upon the kind of loading to which the wall is subjected. In direct compression vertically, no significant difference in wall strength has been observed. This is no doubt due to the fact the direct compressive strength of a wall depends primarily upon the strength of the individual masonry units and is not appreciably affected by mortar strength. For this reason, mortar strength is not a primary specification criterion in most codes.

That bond patterns do have some influence on flexural strength and cracking resistance is evidenced by the UBC provisions for unreinforced masonry (UBC Table 24-B). These provide for a tension allowance across a bed joint, a condition that exists when the wall spans vertically. However, should the wall span horizontally between pilasters, it is the head joint that becomes subjected to flexural tensile stress. For this state, the allowable stress in tension in the head joint becomes twice the tensile allowance for the bed joint. Further, no tension is permitted in the stack bond across the head joints, which precludes the use of a stack-bond wall pattern in horizontal wall spans. On the other hand, these strength differentials are not reflected, currently, in the assumptions for reinforced masonry. The Code design stresses permitted for this type of construction are valid for stack as well as running bond, so that, at this point, any added strength achieved by the latter simply becomes a design bonus.

Grout space width

The following discussion relating to the grout-space width requirements is intended primarily to assist designers in establishing a brick wall thickness because that will determine their design dimensions, such as the effective depth, *d*. For unlike masonry of single-wythe thickness with its standard unit thickness, a two-wythe or more reinforced wall thickness can vary over a wide range, for it depends upon both the thickness of the masonry unit and the grout space width. In low-lift grouted construction, the minimum grout space width is $\frac{3}{4}$ in., where only joint reinforcement is utilized, but typically this space varies between 2 and $3\frac{1}{2}$ in. in two-wythe construction, depending upon the size of the reinforcing provided in the grout space. In the low-lift grouting procedure, one exterior wythe may be carried up 18 in. before grouting, but the other exterior wythe has to be laid up and grouted in lifts not to exceed six times the width of the grout space to a maximum height of 8 in. The lift-height requirement is inserted to ensure ease of pouring, with a maximum of 8 in. being specified to limit the fluid pressure of the wet grout upon the freshly laid brick. In walls consisting of three wythes or greater, the bricks making up the interior wythes must be embedded in the grout so that at least $\frac{3}{4}$ in. surrounds the ends and sides of each brick unit.

Where high-lift grouting is called for, the grout space must be at least 3 in. in width, but not less than that thickness needed to locate the reinforcement (vertical and horizontal) in it, with a required clearance all around of $\frac{1}{4}$ in. minimum. This minimum width may be reduced to 2 in. if no horizontal steel is located in the grout space to impede the pour. In the high-lift grouting of two-wythe construction, continuous pours for the full height of the wall are permitted in lifts of 4 ft.

Because of the limited working space within the grout core, it may be advantageous to use bundled groups of smaller bars spaced closer together, rather than placing separate larger bars. To determine the equivalent perimeter of the bundle (Σ_0), obtain the diameter of a single bar whose area equals that of the bundled group, and base the perimeter on that diameter. This is needed when one is checking the bond stress, which in masonry is still based upon a flexural bond concept (i.e., $u = V/\Sigma_0 jd$), as derived in Chapter 6.

PARAPET WALLS

A parapet wall is simply that portion of a bearing or a nonbearing wall that extends above the roof for fire protection or for some architectural reason. This portion of the wall must be designed to resist 100% of gravity ($C_p = 1.0$), in acting as a cantilever above the roof level. The projecting wall is therefore designed as follows:

$$M = \frac{wh^2}{2}$$

where M = moment, ft-lb/ft of wall
w = weight of wall, lb/ft² of surface area
h = unsupported parapet height, ft

Then

$$A_s = \frac{M}{f_s jd}$$

Here j is assumed to be 0.9. However, the assumed lateral load of 100% g need not be carried past the base into the supporting structure. It has been said that the 100% factor is not truly valid, but was introduced hysterically after many poorly anchored parapets and cornices were shaken off during earthquakes, with dramatic and disastrous results.

EXAMPLES OF WALL DESIGN and ANALYSIS

EXAMPLE 7-1: Determine the Steel Requirement for a Concrete Masonry Wall

A building is to be constructed using 8-in. concrete block. The total wall height is 20 ft. The parapet is $3\frac{1}{3}$ ft of the 20-ft height. Assume that the wall is partially grouted and that it is located in seismic zone 4.

Determine the steel requirements for a wall with maximum $h/t = 25$ subjected to the following loads and conditions:

(*a*) *Lateral force on end wall—wind at 15 lb/ft² or seismic zone 4 half-stresses (i.e., no continuous inspection). Assume that the wall is nonbearing.*

(*b*) *Vertical dead load and moment due to $e = 4$ in., combined with moment due to lateral loads on wall. Use full stresses (i.e., by providing continuous inspection).*

(*c*) *Consider the parapet wall. Check for moment, bond, and shear using full stresses.*

(*d*) *Calculate the allowable moment for steel furnished in part (b).*

(*e*) *Reduce wall in part (b) to 10 ft 8 in. high and check.*

LOADS (ROOF):

Dead load:	5-ply roofing	2.0 lb/ft²
	2 × 4 at 24 in. oc	0.8
	1/2-in. plywood	1.5
	purlins	1.5
	miscellaneous	1.7
	total roof DL	7.5 lb/ft²

Solution:

(*a*) *Lateral force only on end walls (nonbearing):*

$$f'_m = 1350 \text{ lb/in.}^2 \qquad \tfrac{1}{2}f'_m = 675 \text{ lb/in.}^2 \text{ (no inspection)}$$

$$f_s = 20{,}000 \text{ lb/in.}^2$$

$$n = 22 \qquad n = 44$$

$$F_b = \tfrac{1}{3}f'_m = 450 \text{ lb/in.}^2 \qquad \tfrac{1}{2}F_b = 225 \text{ lb/in.}^2$$

$$t = 7.625 \text{ in.}$$

From Table 5-6, the minimum wind force = 15 lb/ft².

$V = ZIKCSW$

$w = 75$ lb/ft² From Table 5-2, for 8-in. block, assuming vertical cells grouted at 16 in. oc. This figure is quite conservative. In a practical situation one would assume the maximum spacing of 48 in. cc; then $w = 61$ lb/ft². Actually, this spacing was selected here to demonstrate the procedure to be followed when the allowable masonry stress, F_b, governs the requirement for A_s.

$C_p = 0.2$ (Table 5-9)

Therefore,

$$F_p = ZIC_pSW_p = (1.0)(1.0)(0.2)(1.5)(75) = 22.5 \text{ lb/ft}^2 > 15 \text{ lb/ft}^2$$

Thus, seismic governs over wind forces.

$$w = (22.5 \text{ lb/ft}^2)(1 \text{ ft width of wall}) = 22.5 \text{ lb/ft}$$

$$h = 16 \text{ ft 8 in. (distance from floor to bond beam)}$$

Assume that the wall acts as a simply supported vertical beam:

$$M = \frac{wh^2}{8} = \frac{(22.5\ \text{lb/ft})(16.67\ \text{ft})^2}{8} = 781.6\ \text{ft-lb/ft of wall}$$

To estimate the steel required, assume that the allowable steel stress (f_s) governs and solve for A_s:

$$M_s = f_s A_s jd$$

$$f_s = (1\tfrac{1}{3})(20{,}000\ \text{lb/in.}^2) = 26{,}700\ \text{lb/in.}^2\ \text{(allow for one-third increase due to seismic load}$$

$$j = 0.9\ \text{(assumed)}$$

$$d = \frac{t}{2} = \frac{7.625}{2} = 3.8\ \text{in.} \qquad \text{(steel at center of 8-in. nominal block)}$$

$$A_s = \frac{M_s}{f_s jd} = \frac{(781.6\ \text{ft-lb/ft})(12\ \text{in./ft})}{(26{,}700\ \text{lb/in.}^2)(0.9)(3.8\ \text{in.})} = 0.103\ \text{in.}^2/\text{ft of wall}$$

Check to determine whether allowable f_m should govern. Or one could calculate f_m from $M/bd^2 \times 2/jk$ to determine if F_b were exceeded.

$$M_m = \tfrac{1}{2} f_m kjd^2$$

$$F_b = 1\tfrac{1}{3} \times 225 = 300\ \text{lb/in.}^2$$

$$kj = \frac{2M_m}{F_b bd^2} = \frac{2(781.6\ \text{ft-lb/ft})(12\ \text{in./ft})}{(300\ \text{lb/in.}^2)(12\ \text{in.})(3.8\ \text{in.})^2} = 0.361$$

Since $j = 1 - k/3$, multiply each side by k to obtain

$$kj = k\left(1 - \frac{k}{3}\right) = k - \frac{k^2}{3} = 0.361$$

This forms a quadratic equation:

$$k^2 - 3k + 3(0.361) = 0$$

Therefore, $k = 0.42$. Then, from equation (8a), Chapter 6:

$$f_s = nF_b \frac{1-k}{k} \leq f_{s(\text{allow.})}$$

where F_b = allowable masonry stress $= 225\ \text{lb/in.}^2 \times 1\tfrac{1}{3} = 300\ \text{lb/in.}^2$

$$f_{s(\text{actual})} = 44(300)\frac{1-0.42}{0.42} = 18{,}230\ \text{lb/in.}^2 \qquad j = 1 - \frac{k}{3} = 1 - \frac{0.42}{3} = 0.86$$

$$A_s = \frac{M}{f_s jd} = \frac{781.6 \times 12}{18{,}230 \times 0.86 \times 3.8} = 0.157\ \text{in.}^2/\text{ft}$$

Use larger A_s, observing that F_b governs the design rather than the allowable f_s. The larger A_s provides a larger k, and hence reduces the induced f_m in the masonry.

Compare with results given in the Appendix:

FIRST, TRY CURVE B-2:

$$M = Kbd^2$$

$$K = \frac{M}{bd^2} = \frac{(781.6 \text{ ft-lb})(12 \text{ in./ft})}{(12 \text{ in.})(3.8 \text{ in.})^2} = 54.1$$

to allow for $\frac{1}{3}$ stress increase in chart, enter with $\frac{3}{4} \times 54.1 = 40.6$

Thus, use

$$\frac{K}{F_b} = \frac{40.6}{0.225} = 180.4$$

With this value as the ordinate, find $pn = 0.153$ or $p = 0.0035$. $A_s = pbd = 0.0035(12 \text{ in.})(3.8 \text{ in.}) = 0.158 \text{ in.}^2/\text{ft}$ of wall, as before. From curve B-2, it can be readily seen that masonry stress governs, so $f_m = 300 \text{ lb/in.}^2$ Also, f_s can be taken off as $13{,}700 \times 1\frac{1}{3} = 18{,}300 \text{ lb/in.}^2$.

TRY CURVE B-3:

With $K = 54.1$, find $pn = 0.153$ for $F_b = 300 \text{ lb/in.}^2$. Thus, since F_b governs,

$$f_s = \frac{M}{A_s jd} = \frac{781.6 \times 12}{0.158 \times 0.86 \times 3.8} = 18{,}200 \text{ lb/in.}^2$$

Determine bar size and spacing for vertical steel: Try for maximum spacing of 48 in. (4 ft) to minimize the number of grouted cells and to provide for better laying of the masonry.

No. 7 at 48 in. oc ($A_s = 0.150 \text{ in.}^2/\text{ft}$) areas from Table D-2.

No. 6 at 32 in. oc ($0.166 \text{ in.}^2/\text{ft}$)

No. 5 at 24 in. oc ($0.154 \text{ in.}^2/\text{ft}$)

Check minimum steel requirements:

$$\text{minimum steel} = 0.0007bt = 0.0007(12 \text{ in.})(7.625 \text{ in.}) = 0.064 \text{ in.}^2 < 0.150 \text{ in.}^2 \quad \therefore \text{OK}$$

This is also given in Table D-1.

$$\text{minimum total wall steel} = 0.002bt = 0.002(12 \text{ in.})(7.625 \text{ in.}) = 0.183 \text{ in.}^2/\text{ft of wall}$$

$$0.183 - 0.150 = 0.033 < 0.064 \quad \text{(Use minimum for horizontal reinforcement)}$$

0.064×20 ft of wall height $= 1.28 \text{ in.}^2$ total horizontal steel needed in height of wall. Typically, one would provide at least

$$\text{No. 5 at 4 ft} = 6 \times 0.3 = 1.8 \text{ in.}^2$$

which provides required horizontal steel area. On the basis of the 48-in.-cc spacing, $w = 61$

lb/ft² and F_p reduces to 61/75 × 22.5 = 18.3 lb/ft. This would result in a revised $A_s = 0.084$ in.² based upon f_s allowable. A check of f_m discloses that it does not exceed F_b [see part (b)].

(*b*) *Combined vertical dead load and moment due to e = 4 in. and moment due to lateral loads:*

Continuous inspection is provided; thus, $f'_m = 1350$ lb/in.²; $n = 22$.

$$\text{wall load} = 1/2(16.7\text{ ft})(61\text{ lb/ft}^2) + (3.3\text{ ft})(92\text{ lb/ft}^2) = 813\text{ lb/ft}$$

↑ assuming solid-grouted parapet wall

$$\text{roof load} = 7.5\text{ lb/ft}^2 \times 10\text{ ft} = 75\text{ lb/ft (10 ft tributary loading width)}$$

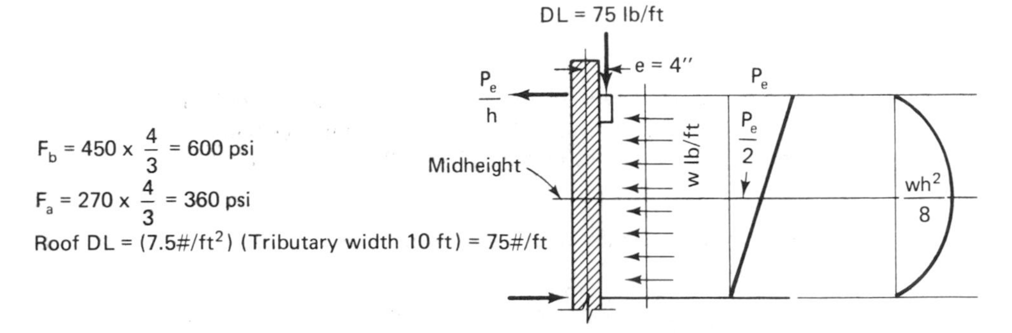

Maximum seismic moment occurs at the midheight: assume "pinned" or simple span.

$$M_1 = \frac{Pe}{2} = \frac{(75\text{ lb/ft})(4\text{ in.})}{2 \times 12} = 12.5\text{ lb-ft/ft} \quad \text{(due to eccentricity)}$$

$$M_2 = \frac{wh^2}{8} = \frac{(18.3\text{ lb/ft})(16.7\text{ ft})^2}{8} = 638\text{ ft-lb/ft} \quad \text{(due to lateral load, assumed steel spacing at 48 in. cc)}$$

$$\text{total } M = M_1 + M_2 = 12.5 + 638 = 650.5\text{ ft-lb/ft}$$

AXIAL STRESS:

$$F_a = 0.2f'_m R = 0.2(1350)(0.756) = 204\text{ lb/in.}^2 \text{ where } R = \left[1 - \left(\frac{h}{40t}\right)^3\right] = 0.756$$

for $h/t = 25$ (also see Diagram E-1)

$$f_a = \frac{P}{bt} = \frac{75 + 813}{12 \times 4.6} = 16.1\text{ lb/in.}^2$$

Note that the equivalent solid thickness for axial compression for an 8-in. block with reinforcing at 48 in. cc is 4.6 in. (Table 7-4).

INTERACTION EQUATION:

$$\frac{f_a}{F_a} + \frac{f_b}{F_b} \leq 1.33 \text{ (}\tfrac{1}{3}\text{ stress increase allowed for lateral loads)}$$

$$f_b = \left(1.33 - \frac{f_a}{F_a}\right)F_b = \left(1.33 - \frac{16.1}{204}\right)450 = 563.0\text{ lb/in.}^2$$

where the allowable masonry stress $F_b = \frac{1}{3}f'_m$. This would permit the masonry stress to reach 563.0 lb/in.^{2} under this combined loading condition.

STEEL AREA:

$$M = Kbd^2 \qquad \text{or} \qquad K = \frac{650.5 \text{ ft-lb} \times 12}{12 \times (3.8)^2} = 45.0$$

Try $f_{s(\text{allow.})}$; then

$$A_s = \frac{M}{f_s jd} = \frac{650.5 \times 12}{26{,}700 \times 0.86 \times 3.8} = 0.090 \text{ in.}^2/\text{ft}$$

If $F_{b(\text{allow.})}$ governs, solve for k.

$$jk = \frac{2M}{f_b bd^2} = \frac{2 \times 650.5 \times 12}{563 \times 12 \times (3.8)^2} = 0.160 \quad \text{and} \quad k^2 - 3k + 3(0.160) = 0 \quad \therefore k = 0.17$$
$$j = 0.94$$

$$f_s = nF_b\frac{1-k}{k} = 22(563)\frac{1-0.17}{0.17} = 60{,}473 \text{ lb/in.}^2 > 26{,}700 \text{ lb/in.}^2$$

Thus, steel governs and we use $A_s = 0.090$ in.^{2}

Actually, the wall could be designed as a cantilever span supported at the ground and roof levels, the cantilever reducing the design moment near the midheight of the wall as shown in sketch. Thus, seismic design moment in the span could be reduced to 562 ft-lb from 638 ft-lb. The eccentric moment due to the vertical load, 12.5 ft-lb, must be added making the total design moment equal to 574.5 ft-lb. This would reduce A_s slightly, but not significantly. Provide No. 6's at 48 in. cc ($A_s = 0.110$ in.^{2}/ft > min.).

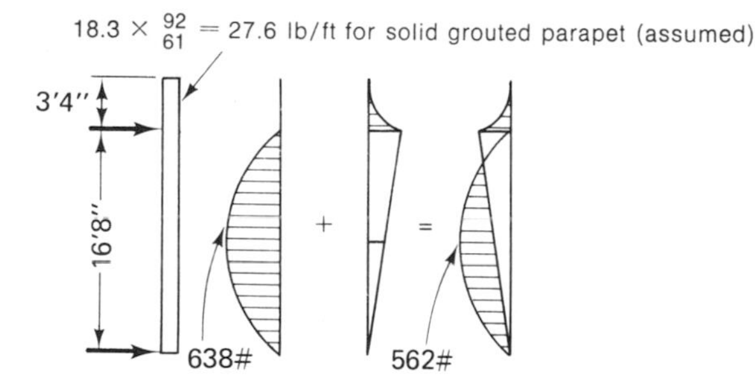

USE OF DESIGN AIDS:

From Curve B-4, for $nK = 990$ and $f_s = 26{,}700$ lb/in.^{2}, find $np = 0.044$ or $p = 0.0020$.

$$A_s = 0.0020 \times 12 \times 3.8 = 0.091 \text{ in.}^2/\text{ft}$$

as before.

(c) *Parapet wall design:*

Note from Table 5-9 that C_p for cantilever parapets is 1.00 rather than 0.20. This 1.00 factor is for the cantilever itself, and need not be considered in the rest of the structure as a distributed moment or load. The seismic force on the wall element,

$$F_p = ZC_pISW_p = (1.0)(1.0)(1.5)(61 \text{ lb/ft}^2) = 92 \text{ lb/ft}^2,$$

governs, where W_p is assumed here to be 61 lb/ft². If the parapet wall were solid grouted, the bond stress would become excessive, calling for smaller bars at closer spacing in the parapet.

$$M = \frac{wh^2}{2} = \frac{92}{2} \times (3.33)^2 = 511 \text{ ft-lb/ft of wall}$$

By comparison with the wall moment below the parapet, A_s (0.110 in.²/ft) is adequate for the cantilevered parapet wall.

BOND:

$$V = 92 \text{ lb/ft}^2 \times 3.33 = 306 \text{ lb/ft and } \Sigma_0 = 0.59 \text{ in. for No. 6's at 48 in. cc}$$

$$u = \frac{V}{\Sigma_0 jd} = \frac{306}{0.59 \times 0.91 \times 3.8} = 150 \text{ lb/in.}^2 < 140 \times \tfrac{4}{3} \text{ lb/in.}^2 \qquad \therefore \text{OK (usually is)}$$

FLEXURAL SHEAR:

$$v = \frac{V}{bjd} = \frac{306 \text{ lb/ft} \times 4 \text{ ft}}{8.25 \times 0.91 \times 3.8} = 42.9 \text{ lb/in.}^2 < 40 \text{ lb/in.}^2 \times 1.33 \text{ allowed} \qquad \therefore \text{OK}$$

(*d*) *For illustration purposes, calculate the permitted moment for steel furnished:*

$$p = \frac{0.110}{12 \times 3.8} = 0.0024$$

$$pn = 22 \times 0.0024 = 0.053$$

$$F_b = 0.33 \times 1350 \times 1.33 = 600 \text{ lb/in.}^2$$

$$k = \sqrt{2pn + (pn)^2} - pn = \sqrt{2(0.053) + (0.0028)} - 0.053 = 0.277$$

Therefore, $j = 1 - k/3 = 0.91$ and $2/jk = 7.93$ (also from Table C-1).

$$M_m = \frac{F_b bd^2}{2/jk} = \frac{600 \times 12 \times (3.8)^2}{7.93} = 13{,}111 \text{ in.-lb/ft as governed by } F_b$$

$$M_s = A_s f_s jd = 0.110 \times 26{,}700 \times 0.91 \times 3.8$$
$$= 10{,}156 \text{ in.-lb/ft (846 ft-lb/ft) as governed by } f_s.$$

USING CURVE B-2:

$M_s = Kbd^2 = 58.2 \times 12 \times (3.8)^2/12 = 841$ ft-lb/ft (from curve B-2 for $pn = 0.053$, $K/F_b = 97$; $\therefore K = 97 \times 0.450 \times 1.33 = 58.2$)

(*e*) *Reduce* h/t *to 16 (height = 10 ft-8 in.):*

Then the A_s required for moment due to lateral forces only (half-stresses) becomes 0.041 in.²/ft; thus minimum steel governs. No. 5's at 48 in. cc meets the criterion. When the combined eccentric ($e = 4$ in.) vertical load and the lateral loads are considered, minimum steel still governs. This is a graphic demonstration of the results obtained by holding the h/t below 25.

EXAMPLE 7-2: Determine the Steel Requirement for a Brick Wall

Using the information provided in Example 7-1, determine the steel requirement for the wall with the loads given in parts (a) and (b), using an $8\frac{1}{2}$-in. brick wall and maximum $h/t = 25$. Thus, $f'_m = 1500$ lb/in.^{2}; $\frac{1}{2} = 750$ lb/in.^{2} for no inspection; and $F_b = \frac{1}{3} \times 750 = 250$ lb/in.^{2} with $n = 40$.

Solution:

(*a*) *Lateral force on end wall:*

$$w = 93 \text{ lb/ft}^2 \text{ (Table 5-3) for } 8\tfrac{1}{2}\text{-in. grouted brick wall}$$

$$F_p = ZIC_pSW_p = 1.0 \times 1.0 \times 0.20 \times 1.5 \times 93 = 27.9 \text{ lb/ft}^2 > 15\text{-lb/ft}^2 \text{ wind}$$

so seismic governs. Maximum $h = 25t = 25 \times 8.5$ in. $= 17.7$ ft > 16.67 ft actual, so OK. Actually, the $8\frac{1}{2}$-in. represents a nominal 9-in. wall, so the h/t calculation *could* be $25 \times \frac{9}{12} = 18.75$.

Required steel area:

$$M = \frac{wh^2}{8} = \frac{27.9 \text{ lb/ft}}{8} \times (16.67)^2 = 969 \text{ ft-lb/ft of wall}$$

$$M = Kbd^2 \qquad \text{or} \qquad K = \frac{M}{bd^2} = \frac{969 \times 12}{12 \times (4.25)^2} = 53.6$$

where $d = t/2 = 4.25$ in.

From Curve B-2: for $K/F_b = (\frac{3}{4} \times 53.6)/0.250 = 161$ find $pn = 0.123$ (F_b governs). Thus, $p = 0.0031$ and $A_s = 0.0031 \times 12 \times 4.25 = 0.158$ in.^{2}/ft. For $K/F_b = 161$, find $f_s = 15{,}700 \times \frac{4}{3} = 20{,}880$ lb/in.^{2} and $f_m = 250 \times \frac{4}{3} = 333$ lb/in.^{2}; OK.

From Curve B-3: for $K = 53.6$, find for $F_b = 333$ lb/in.^{2} a value for $pn = 0.124$, as before.

From Curve B-4: for $nk = 40 \times 53.6 = 2140$ and $f_s = 26{,}700$ lb/in.^{2} a lesser value of np is obtained, indicating that F_b governs. This is immediately apparent in Curve B-2 also. For any value of $K/F_b > 148.2$ (for basic allowable $f_s = 20{,}000$ lb/in.^{2}), F_b governs and the value of f_s is obtained directly from the curve.

(*b*) *Combined vertical dead load and moment due to e = 4 in. and moment due to lateral load:*

$$\text{roof DL} = 75 \text{ lb/ft}$$

$$\text{wall load} = \tfrac{1}{2}(16.67)(93) + 3.33(93) = 1085 \text{ lb/ft}$$

The maximum moment at midheight is

$$M_1 = \frac{Pe}{2} = \frac{75 \text{ lb/ft}}{2}\,\frac{4}{12} = 12.5 \text{ ft-lb/ft}$$

$$M_2 = \frac{wL^2}{8} = \frac{(27.9 \text{ lb/ft})(16.67 \text{ ft})^2}{8} = 969.0 \text{ ft-lb/ft}$$

$$\text{total } M = 12.5 + 969.0 = 981.5 \text{ ft-lb/ft}$$

AXIAL STRESS:

From Diagram E-1 for $h/t = 23.5$, $R = 0.80$ and $F_a = 0.2f'_mR = 120$ lb/in.^{2}.

$$f_a = \frac{P}{bt} = \frac{75 + 1085}{(12 \text{ in.})(8.5 \text{ in.})} = 11.4 \text{ lb/in.}^2$$

INTERACTION EQUATION:

$$\frac{f_a}{F_a} + \frac{f_b}{F_b} = 1.33 \text{ where } f_b = 250 \text{ lb/in}^2$$

$$f_b = \left(1.33 - \frac{f_a}{F_a}\right)F_b = \left(1.33 - \frac{11.4}{120}\right) \times 250 = 309 \text{ lb/in.}^2$$

$$M = Kbd^2 \qquad K = \frac{M}{bd^2} = \frac{(981.5 \text{ ft-lb/ft})(12 \text{ in./ft})}{(12 \text{ in.})(4.25 \text{ in.})^2} = 54.3$$

From Curve B-3: for $K = 54.3$ and $f_b = 309$ lb/in.^{2}, find $pn = 0.140$ or $p = 0.0035$.

From Curve B-4: for $f_s = 26{,}700$ lb/in.^{2} and $nk = 2172$, find $pn = 0.092$. Thus, the steel area is governed by the permitted masonry stress of 309 lb/in.^{2}

$$A_s = pbd = 0.0035(12)(4.25) = 0.18 \text{ in.}^2\text{/ft of wall}$$

EXAMPLE 7-3: Obtain Maximum h/t Ratio

Refer back to Example 7-1, part (a). What is the maximum h/t permitted for the 8-in. block wall when minimum reinforcing is provided, say No. 5's at 48 in. cc? Check with both half- and full stresses.

Solution:

Check by use of Table C-1.

Continuous Inspection:

$$M = 18.3 \text{ lb/ft.} \times \frac{h^2}{8} \qquad h^2 = \frac{8M}{18.3}$$

$$f'_m = 1350 \text{ lb/in.}^2 \qquad F_b = 450 \times \tfrac{4}{3} = 600 \text{ lb/in.}^2 \qquad n = 22$$

$$p = 0.077/(12 \times 3.8) = 0.0017;\ np = 0.374;\ j = 0.92;\ 2/kj = 9.1$$

(from Table C-1)

$$M_m = 600 \times 12 \times \frac{3.8^2}{9.1} = 11{,}425 \text{ in.-lb} = 952 \text{ ft-lb}$$

$$M_s = 26{,}700 \times 0.077 \times 0.92 \times 3.8 = 7187 \text{ in.-lb} = 600 \text{ ft-lb governs}$$

$$h = \sqrt{8 \times 600/18.3} = 16.2 \text{ ft}$$

maximum $h/t = 16.2 \times 12/8 = 24.3$ (governed by A_s)

No Inspection:

$$f'_m = 675 \text{ lb/in.}^2 \qquad f_m = 225 \times \tfrac{4}{3} = 300 \text{ lb/in.}^2 \qquad n = 44$$

$$p = 0.0017 \qquad np = 0.0748 \qquad j = 0.89 \qquad \frac{2}{kj} = 7.1$$

$$M_m = 300 \times 12 \times \frac{3.8^2}{7.1} = 7322 \text{ in.-lb} = 610 \text{ lb-ft}$$

$$M_s = 26{,}700 \times 0.077 \times 0.89 \times 3.8 = 6953 \text{ in.-lb} = 579 \text{ ft-lb governs}$$

$$h = \sqrt{8 \times 579/18.3} = 15.9 \text{ ft}$$

$$\text{maximum } h/t = 15.9 \times 12/8 = 23.8 \text{ (governed by } A_s\text{)}$$

Since the allowable steel stress governs, little is gained by going to the continuous inspection (full) stresses.

EXAMPLE 7-4: Design Portion of Wall Adjacent to Large Openings

Design the section of wall between the 3 ft 4 in. and 8 ft 0 in. openings shown in the front elevation. At the windows, the wall above and below the opening generally must span to the sides of the opening, then vertically to the roof diaphragm and to the floor. This places a concentration of loads on the wall portion between the openings.

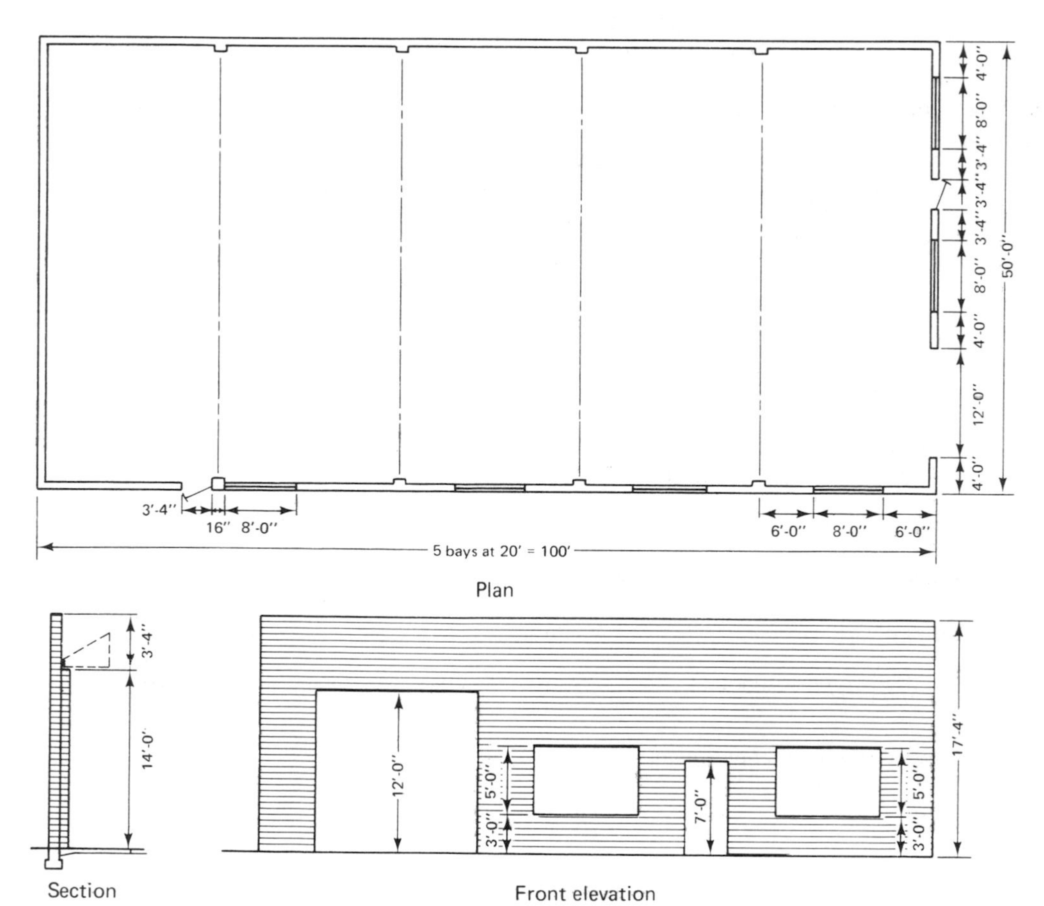

Solution:

Using a wind load of 15 lb/ft², the load transferred horizontally to be resisted by the portion of wall between the openings is

$$15[8 \text{ ft}/2 + 3.33 \text{ ft} + 3.33 \text{ ft}/2] = 135 \text{ lb/vertical foot of wall (40 in. wide)}$$

Then the design moment becomes the lateral-load moment plus the eccentric vertical load moment at midheight of wall, or:

$$M_1 = \frac{wh^2}{8} = 135 \times \frac{14^2}{8} = 3307 \text{ ft-lb (wall height is 14 ft)}$$

$$M_2 = 280 \text{ lb/ft (roof dead load to wall)} \times 5 \text{ in.}/12 \text{ (eccentricity)} \times \tfrac{1}{2} \text{ (wall mid-height)} \times (8 \text{ ft}/2 + 3.33 \text{ ft} + 3.33 \text{ ft}/2) = 525 \text{ ft-lb}$$

$$\text{Design load} = M_1 + M_2 = 3307 + 525 = 3835 \text{ ft-lb}$$

Estimate amount of steel, assuming that $j = 0.89$:

$$A_s = \frac{3835 \times 12}{26{,}700 \times 0.89 \times 3.8} = 0.51 \text{ in.}^2$$

Choose three No. 5 bars vertical in wall between openings, one to be placed at each jamb and one between.

$$p = \frac{3 \times 0.31}{40 \times 3.8} = 0.0061 \qquad pn = 0.0061 \times 44 = 0.27$$

$$t_f/d = \frac{1.25}{3.8} = 0.328 \qquad b = 40 \text{ in.}$$

From equations (9) and (10), Chapter 6:

$$k = \frac{0.27 + 0.5(0.328)^2}{0.27 + 0.327} = 0.54$$

$$j = \frac{6 - 6(0.328) + 2(0.328)^2 + (0.328)^3(1/2 \times 0.27)}{6 - 3(0.328)} = 0.86$$

$$kd = 0.54 \times 3.8 = 2.05 \text{ in.} > t_f \qquad \therefore \text{acts as T-beam, and lever arm } jd = 3.28 \text{ in.}$$

The allowable T-beam moment becomes [per equation (11), Chapter 6]:

$$M_s = A_s f_s jd = 0.93 \times 26{,}700 \times 3.28 = 81{,}400 \text{ in.-lb}$$

$$M_m = f_m\left(1 - \frac{t_f}{2kd}\right)bt_f jd = 300\left(1 - \frac{1.25}{2 \times 2.05}\right)(40)(1.25)(3.28) = 34{,}300 \text{ in.-lb} \leftarrow \text{governs}$$

Consider the interaction equation, $f_a/F_a + f_b/F_b \leq 1.00$; thus,

$$\text{axial load} = [(75 \times 9.33 \text{ ft}) + 280] \times 9 = 8820 \text{ lb} \quad \text{(9.33 ft is wall height above openings)}$$

$$\text{axial stress} = \frac{8820}{40 \times 5.8} = 38 \text{ lb/in.}^2$$

where 5.8 is the equivalent solid thickness and 75 lb/ft² is weight of partially grouted section (Tables 5-2 and 7-4).

$$F_a = 0.2(1350) \times \tfrac{1}{2} \times 0.855 = 115 \text{ lb/in.}^2$$
$$F_b = \tfrac{1}{3} \times 1350 \times \tfrac{1}{2} \times 1\tfrac{1}{3} = 300 \text{ lb/in.}^2$$

Check the combined stress: (38/115) + (3835 × 12/34,300) = 1.7 > 1.0. (In this example, the ratio of the actual to allowable moment was used for f_b/F_b and the $\frac{1}{3}$-stress increase was taken on this term only. Since the result is more than 1.0, the wall is overstressed, if reinforced as noted. Now try solid-grouting the wall and place six No. 5's three in each face. The section is now designed as a solid rectangular beam. Then $d = 5.3$ in. and

$$pn = \frac{0.93 \times 40}{40 \times 5.3} = 0.175 \qquad K = \frac{3835 \times 12}{40 \times (5.3)^2} = 41.0$$

From Table C-1: $j = 0.846$, $2/kj = 5.14$. Then

$$f_m = \frac{M}{bd^2} \times \frac{2}{jk} = \frac{3835 \times 12 \times 5.14}{40 \times 5.3^2} = 211 \text{ lb/in.}^2$$

Also, from Curve B-3 for $K = 41.0$ and $pn = 0.175$, find f_m about 215 lb/in.²

$$\text{axial stress} = \frac{8820}{40 \times 7.6} = 29 \text{ lb/in.}^2$$

$$\text{allowable axial stress, } F_a = 0.2(1500) \times \tfrac{1}{2} \times \left[1 - \left(\frac{h}{40t}\right)^3\right] = 128 \text{ lb/in.}^2$$

$$\text{allowable masonry stress, } F_b = \tfrac{1}{3} \times 1500 \times \tfrac{1}{2} \times 1\tfrac{1}{3} = 333 \text{ lb/in.}^2$$

Now check the combined stress formula,

$$\frac{29}{128} + \frac{211}{333} = 0.86 \qquad \therefore \quad \text{OK}$$

This demonstration illustrates the increase in wall strength obtained by placing bars next to each face of wall; of course, the steel area is greatly increased (i.e., six No. 5's versus three No. 5's). Because of the importance of these bars, extra care should be taken to see that they are properly detailed and correctly positioned in the wall and grouted in place. In the section between openings, check the shear stress for forces perpendicular to the wall.

$$v = \frac{V}{bjd} = \frac{135 \times 7}{40 \times 0.90 \times 5.3} = 5 \text{ lb/in.}^2$$

$$\text{bond stress } u = \frac{135 \times 7}{5.3 \times 0.9 \times 5.9} = 34 \text{ lb/in.}^2$$

Shear and bond stresses very seldom govern design in masonry walls.

Another solution to this problem would call for special inspection, thus permitting the use of higher allowable stresses.

SHEAR WALLS

A reinforced masonry shear wall is essentially a deep vertical cantilever beam. By resisting the in-plane shear and bending moments caused by the horizontal force brought to it by a floor or roof diaphragm, it imparts lateral stability to the structure. In modern masonry construction, this element often functions as a bearing wall as well, thereby being subjected, in addition, to vertical compressive stresses. This concept is portrayed in Figure 7-7. So this system becomes a composite of masonry bearing walls combined with lateral stiffening diaphragms and shear walls which, in essence, comprise the box-type lateral force resisting system, whose behavior is described in considerable detail in Chapter 10. Thus, the need for a costly stabilizing moment resisting frame is eliminated. Also, the presence of stiff shear walls means that there will be less nonstructural damage, such as that incurred at partitions, windows, doors, ceilings, furnishings, and so on, during an earthquake than would occur in the more flexible frame building.

FIGURE 7-7. Shear-wall concept.

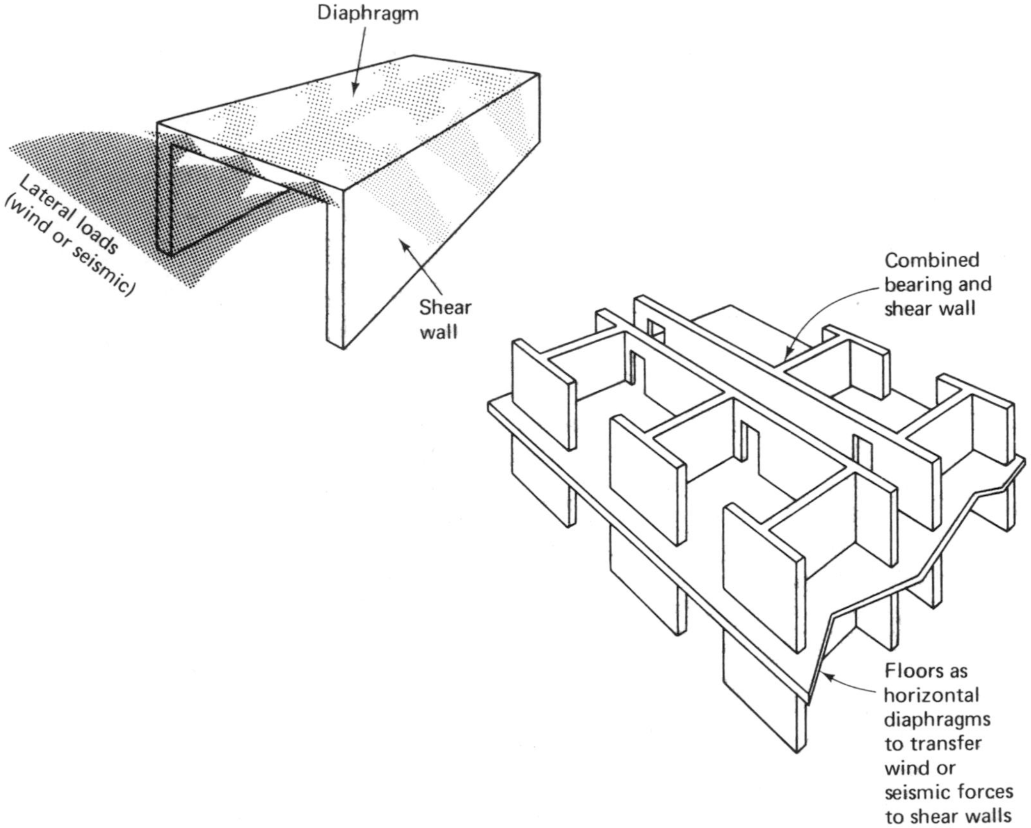

In the box system, columns may exist, either within the bearing walls as pilasters, or as distinct interior members, but in either case they almost always sustain vertical loads only. An exception to this condition would occur when the wall, subjected to a lateral load acting normal to it, spans horizontally between pilasters. In that event the pilaster or column must be reinforced to carry these wall reactions in flexure, as well as the vertical loads, by spanning vertically between diaphragms.

The roof or floor, functioning as a sort of deep flanged girder, distributes the lateral wind pressure or the inertial effects of those masses, accelerated by earthquake ground motion, to the shear walls. These "diaphragms" consist of concrete or gypsum slabs, timber panels, steel deckings, or horizontal bracing trusses. How these forces are distributed to the various shear walls is described in detail in Chapter 10. But briefly stated, the lateral forces are brought to the walls on the basis of tributary areas when the diaphragm is flexible (timber), or by a relative wall rigidity distribution when the diaphragm is rigid (concrete).

Allowable design stresses

An examination of Table 7-1 discloses that the allowable shear stress for a shear wall with no web reinforcement is

$$\frac{M}{Vd} \geq 1 \qquad f_v = 0.9\sqrt{f'_m} \text{ (34 lb/in}^2\text{ max)}$$

$$\frac{M}{Vd} = 0 \qquad f_v = 2.0\sqrt{f'_m} \text{ (50 lb/in}^2\text{ max)}$$

where M = the maximum moment applied to the pier by the lateral shear force V. Thus, if the pier is considered fixed top and bottom (a multistory wall), M/Vd becomes $h/2d$ (pier height/effective depth). For cantilever piers (single-story wall), $M/Vd = h/d$ (Figure 7-8a). Also, these represent full design stresses (i.e., they are permitted where special inspection occurs). Otherwise, these values must be reduced to 17 and 25 lb/in.^{2}, respectively.

A straight-line interpolation is permitted between the M/Vd limits as shown in Figure 7-8b. Should these limits be exceeded, the masonry is no longer assigned any shear value, and special horizontal shear steel must be provided to carry the entire lateral shear force. In that event the maximum stresses permitted (with special inspection) are raised to

$$\frac{M}{Vd} \geq 1 \qquad f_v = 1.5\sqrt{f'_m} \text{ (75 lb/in}^2\text{ max)}$$

$$\frac{M}{Vd} = 0 \qquad f_v = 2.0\sqrt{f'_m} \text{ (120 lb/in}^2\text{ max)}$$

With no inspection, these values reduce to 35 and 60 lb/in.^{2}, respectively.

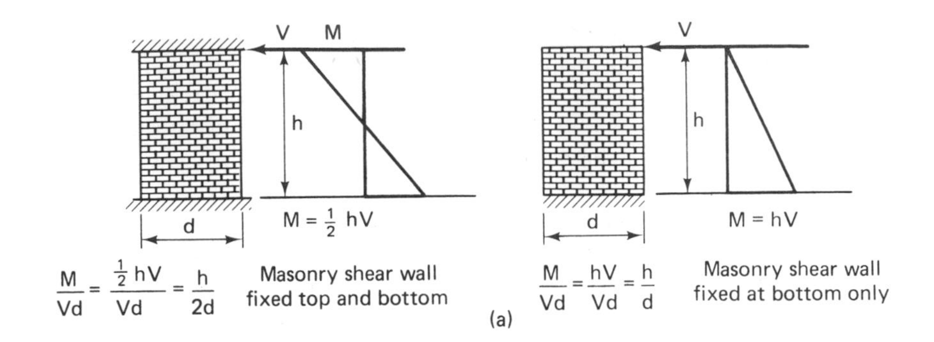

(a)

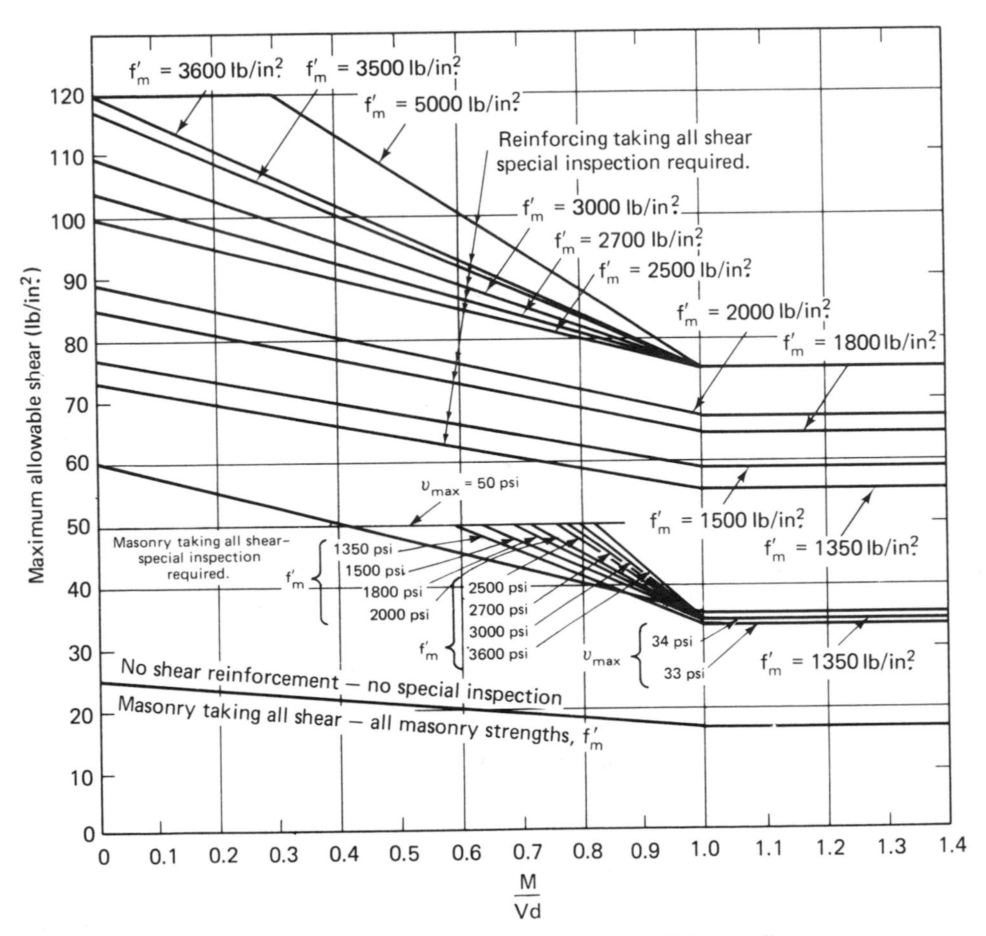

(b) Maximum allowable shear stress, psi, all shear walls

FIGURE 7-8. Allowable shear stress for shear wall.

Experimental shear wall behavior

A one- or two-story shear wall functions essentially as a deep stiff cantilever beam, wherein the primary critical stress is direct or in-plane shear. However in multistory walls, high in-plane flexural stresses may also develop. This condition becomes further complicated when openings occur uniformly throughout the height of the wall, creating what is often referred to as "coupled shear walls." In these cases, they really do not act as independent cantilevers, owing to the coupled bending action of the floor slabs. One approach toward evaluating this type of behavior is described in Chapter 10.

Actually, it is the intent of this section to describe the influence of certain significant parameters, such as the M/Vd ratio and reinforcement arrangement, upon the behavior of the type of shear walls that are often encountered in low-rise masonry buildings. The influence of various parameters on the behavior of single-story concrete masonry piers was firmly established by results obtained in a series of tests devised by Walter Dickey for the Masonry Institute of America and carried out at Cal Poly, Pomona, in 1969 under the direction of Robert R. Schneider. The tests were made primarily to evaluate the structural performance of full-size concrete masonry piers, having variable dimensions, upon which horizontal loads of various magnitudes were imposed. There were a number of test parameters injected into this extensive investigation, among which were the following:

1. *The h/d or M/Vd ratio.* With this particular group of piers, dimensional variation was obtained by varying the width/height dimensions. Further, both fixed-end piers and cantilever piers were utilized. The former were obtained simply by constructing a heavy spandrel beam above and below the pier in question; whereas with the cantilever group, only a base spandrel was constructed, the top remaining free and unrestrained. Jamb steel was kept constant and was designed to represent a typical amount for a pier of this type (i.e., two No. 5's or one No. 7). Typical specimen types are shown in Chapter 17. The findings obtained herein showed that the shear strength of the pier as measured by the expression "V (total shear) divided by $b \times L$ (shear area)" definitely increases with a decrease in the h/d ratio. This rate of increase changes rapidly from an h/d of about 3:1 until an h/d of about 1:1 is reached, after which the rate of change is drastically less. It is interesting to note that the results of these tests are reflected in the current UBC, as was seen in the earlier section on allowable design stresses. The test-data variation between the h/d ratio and shear stress may be seen in Figure 7-9.

When looking at stiffness in a qualitative manner, it was observed that those piers, heavily restrained top and bottom by spandrel beams and having an h/d ratio of 1 or less, experienced very small deflections while sustaining relatively high loads. The steep load deflection curves produced in those instances pointed to the existence of a rather stiff pier element, at least until attainment of the ultimate load.

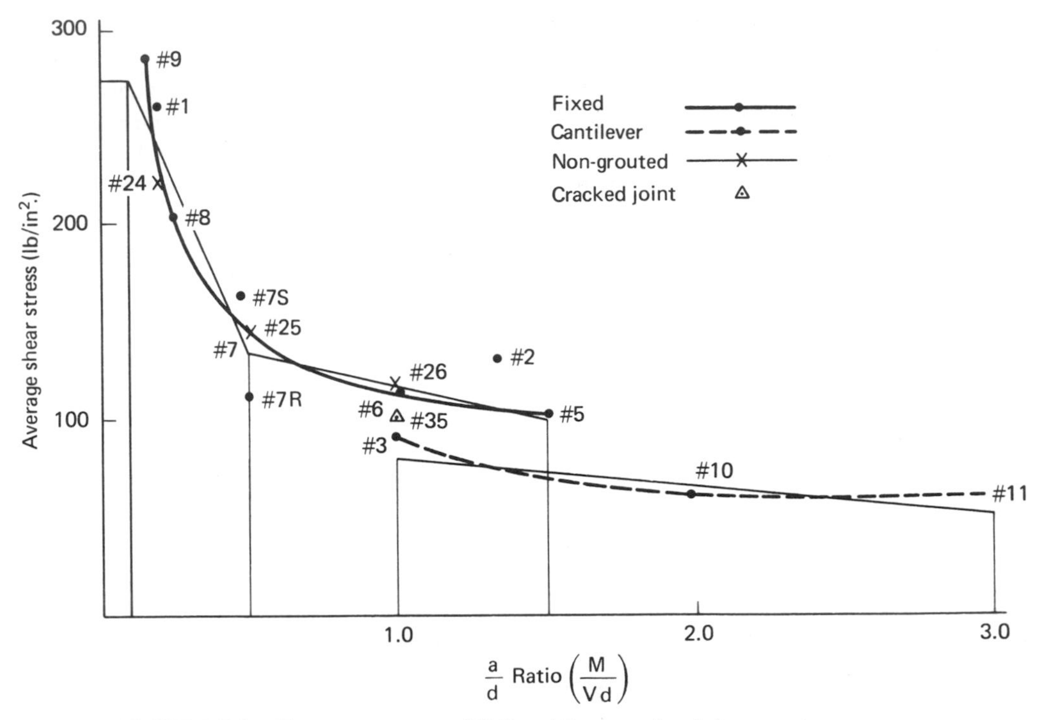

FIGURE 7-9. Shear stress versus $h/2d(=a/d)$—no web reinforcement.

The cracking pattern throughout was typically diagonal tensile in nature. Post-ultimate load performance then became one of continuing deflections, during which time the pier continued to sustain a significant portion of the maximum capacity previously reached. This behavior would definitely suggest a ductile type of performance. On the other hand, because of the predominant moment deflections, the cantilever pier deflected very rapidly from the outset. The nature of the load deflection curves here signifies a similar capacity for ductile behavior, although to a lesser degree. In all these cases, failure was initiated in the masonry. Measurements indicated that in no case did the steel reach a yield stress value.

2. *Web reinforcement: horizontal.* The variable built into this group consisted of horizontal web reinforcement in three amounts, that is, No. 5's at 8 in. oc (noted as a heavy amount of shear steel), No. 5's at 16 in. oc (moderate), and No. 4's at 24 in. oc (light). It was observed that there was a significant increase in the ultimate shear stress value for both fixed and cantilever types over that exhibited by the previous group, which contained no shear steel. This differential became more pronounced as the h/d ratio decreased. The upper h/d ratio for which any appreciable difference in strength existed appeared to be somewhere between $3\frac{1}{2}$ and 4.

It would appear that here, too, there is a definite increase in shear strength with a decrease in the h/d ratio, as delineated in Figure 7-10. This rate of change is not nearly as great as that exhibited by the piers with no web reinforcement. For the lightly reinforced piers, a slight increase in shear strength over that of the control group developed at an h/d ratio below about 2. Above that ratio, however, there appeared to be little change.

A second distinct event relating to the behavioral characteristics of this group occurred in the failure mechanism. The load-deflection curves were moderately flat and extended through a very wide deflection range. Thus, each pier continued to sustain a significant proportion of its ultimate load-carrying capacity while undergoing rather large deflections in the immediate post-ultimate region. Certainly, these energy-absorbing qualities of properly reinforced concrete masonry are of paramount importance in seismic regions. Other tests clearly show similar behavior by reinforced masonry. These tests indeed demonstrate the ability of this type of construction to absorb a considerable amount of energy without completely collapsing, thereby creating a definite ductile behavior pattern.

The higher loads sustained by the piers within this group naturally resulted in increased jamb steel stresses. So when the shear capacity is increased, this may also call for additional jamb steel to compensate for this added shear strength. However, it is preferable that the flexural strength, developed by the jamb steel, remain the limiting factor in establishing the pier capacity, for this condition

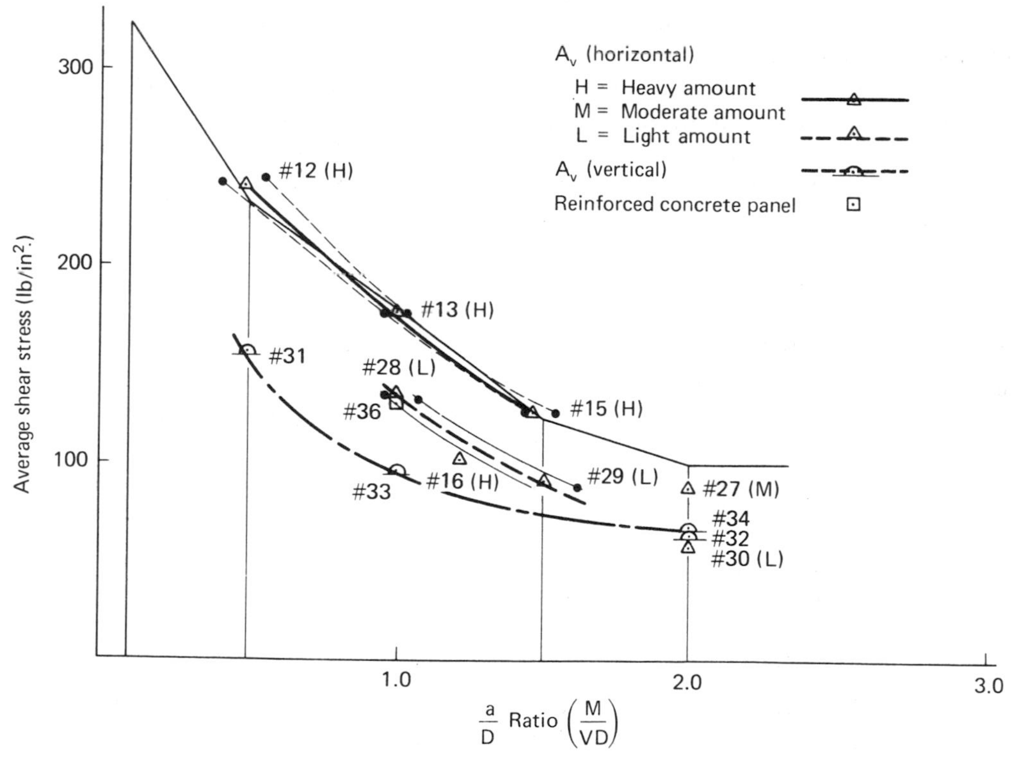

FIGURE 7-10. Shear stress versus $h/2d(=a/D)$—with web reinforcement.

presages a ductile type of failure mode rather than the brittle type of collapse usually associated with shear failure, as was exhibited by test group 4.

3. *Web reinforcement: vertical.* The presence of vertical steel in the field of the pier did not significantly increase the pier resistance, in terms of shear stress, over that of comparable piers in group 1. Further, the failure mechanism and the cracking patterns exhibited by piers with this type of reinforcement suggested the possibility that vertical web reinforcement does not restrain the propagation of diagonal cracks nearly as effectively as does horizontal shear steel. Load deflection characteristics also indicated that the vertical web reinforcement does not impart the stiffness to the pier that the horizontal stirrups do. Thus, these test results leave little doubt that horizontal shear steel is far more effective than is vertical.

4. *Increased jamb reinforcement.* In this group, the jamb reinforcement was doubled over that provided in the control group. It appears that the shear resistance increases only slightly, if at all, under such conditions. The presence of heavy jamb reinforcement, however, altered the failure pattern significantly. Prior to attainment of the ultimate load, the piers behaved as rather stiff elements and exhibited only small deflections. At failure, however, they behaved in a rather brittle manner, as evidenced by a sudden drop in the load being sustained.

It can be seen from these test results that the shearing resistance of a reinforced masonry pier is definitely affected by the h/d ratio, as well as by the amount and location of the horizontal web reinforcement present, when the latter is needed. It is on the basis of tests described above that the UBC has moved from the practice of assigning constant values to the allowable shear stress in a masonry shear wall, to a recognition that a variable value is a more reliable indication of performance and therefore should be used, as has been done for reinforced concrete beams and walls for several years.

Design considerations

IN-PLANE SHEAR STRESS

Once the magnitude of the in-plane lateral shear force, brought to the shear wall by the diaphragm, has been obtained, a check on the shear stress is readily made from

$$v_{\text{actual}} = \frac{V_{\text{total}}}{t \times L}$$

where L = effective length of shear wall

t = actual thickness of wall section for solid grouted masonry or, in the case of partially grouted hollow unit masonry, use the in-plane shear area given in Table 7-4.

Thus the actual v, shear stress, must not exceed the allowable shear stress, previously designated in an earlier section of this chapter, if shear steel is not to be provided. However if it is called for (i.e., $v_{\text{actual}} > v_{\text{allowable}}$), the amount can be determined as denoted in the following section. Observe, too, that there is a maximum stress value specified by the UBC even where this steel is provided.

Spacing of Shear Steel

As noted in the previous section, when allowable stress in the masonry is exceeded, horizontal web reinforcement is required. In that event, the shear steel must be designed to assume the entire in-plane shear force, none being assigned to masonry as opposed to reinforced concrete design.

The web reinforcement can be spaced according to the following relation, as derived in Chapter 6:

$$s = \frac{A_v f_v}{b v_{act}}$$

where, f_v = allowable shear stress in bar (i.e., 20,000 lb/in.^{2})
A_v = area of stirrup

This spacing is called for over the entire story height of the wall. (Remember that the spacing must conform to the modular height dimensions of hollow unit masonry in order that the shear steel can be located within the block dimensions.

Pier Axial and Flexural Stresses

Although the in-plane shear seems to be the primary stress condition considered, other factors must be accounted for if the pier is to remain structurally sound. The lateral force brought to the top of the pier by the diaphragm produces an in-plane moment of $Vh/2$ for the case of restrained ends and Vh for a cantilever condition, as demonstrated in Figure 7-11. In single-story construction, the vertical load consists of the roof live and dead load carried by that pier, and the design axial load, P_{total}, then becomes

$$P_{total} = P_{live} + P_{dead}$$

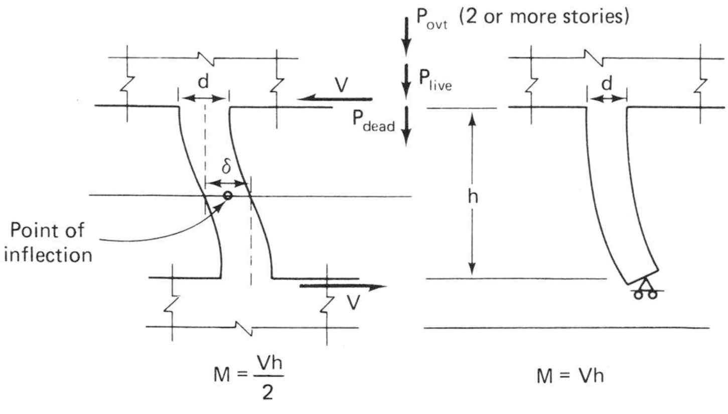

FIGURE 7-11. In-plane pier moments.

In multistory construction, an additional axial load increment may be imposed upon each pier, caused by the overturning effect of the stories above the one in question. This may be evaluated in the following manner:

1. Overturning moment at second-floor level (Figure 7-12):

$$(M_{ovt})_2 = V'_r(h_2 + h_3) + V'_3h_2$$

2. Axial load and moment on piers in first-story wall: Determine the proportion of total lateral force at the second-floor level, $V = V'_r + V'_2 + V'_3$, distributed to each pier 1, 2, 3, 4, 5 on the basis of relative rigidity distribution. Then, total overturning moment on piers in the first story, $M_{ovt} = (M_{ovt})_2 +$ total $V \times$ distance to the second-floor level from the critical level of the pier in the first story (e.g., at the sill height of piers). Thus the axial load on a pier due to overturning, P_{ovt} (from Mc/I) is:

$$P_{ovt} = \frac{M_{ovt}(\bar{l})(A)}{I_n}$$

where $\bar{l}$ = distance from the center of gravity of the net wall section in the first story to the centroid of the pier in question
A = cross-sectional area of pier
I_n = moment of inertia of net wall section in first story.

The moment on pier (in-plane) is:

$$M = V_{pier} \times \frac{\text{pier height}}{2} \text{ (for pier fixed at each end)}$$

3. $P_{total} = P_{dead} + P_{ovt} + P_{live}$ where P_{ovt} caused by wind or $P_{total} = P_{dead} + P_{ovt} +$ partial P_{live} where P_{ovt} caused by seismic (see paragraph following item 5).
4. The jamb steel at the pier boundary is

$$A_s = \frac{M}{f_s \times 0.9d}$$

5. The following UBC load combinations should be examined, among others, using the most critical:
(a) $P_{dead} + P_{live}$ with no stress increase permitted.
(b) Where wind governs, use

$$(P_{dead} + P_{live} + P_{ovt}) + M_w \text{ effect} + V_w \text{ effect}$$

1/3 stress increase permitted.

(c) Where seismic governs, use

$$(P_{\text{total}}) + M_e \text{ effect} + V_e \text{ effect}$$

or

$$0.9P_{\text{dead}} + P_{\text{ovt}} + M_E \text{ effect} + V_E \text{ effect}$$

1/3 stress increase permitted in both cases.

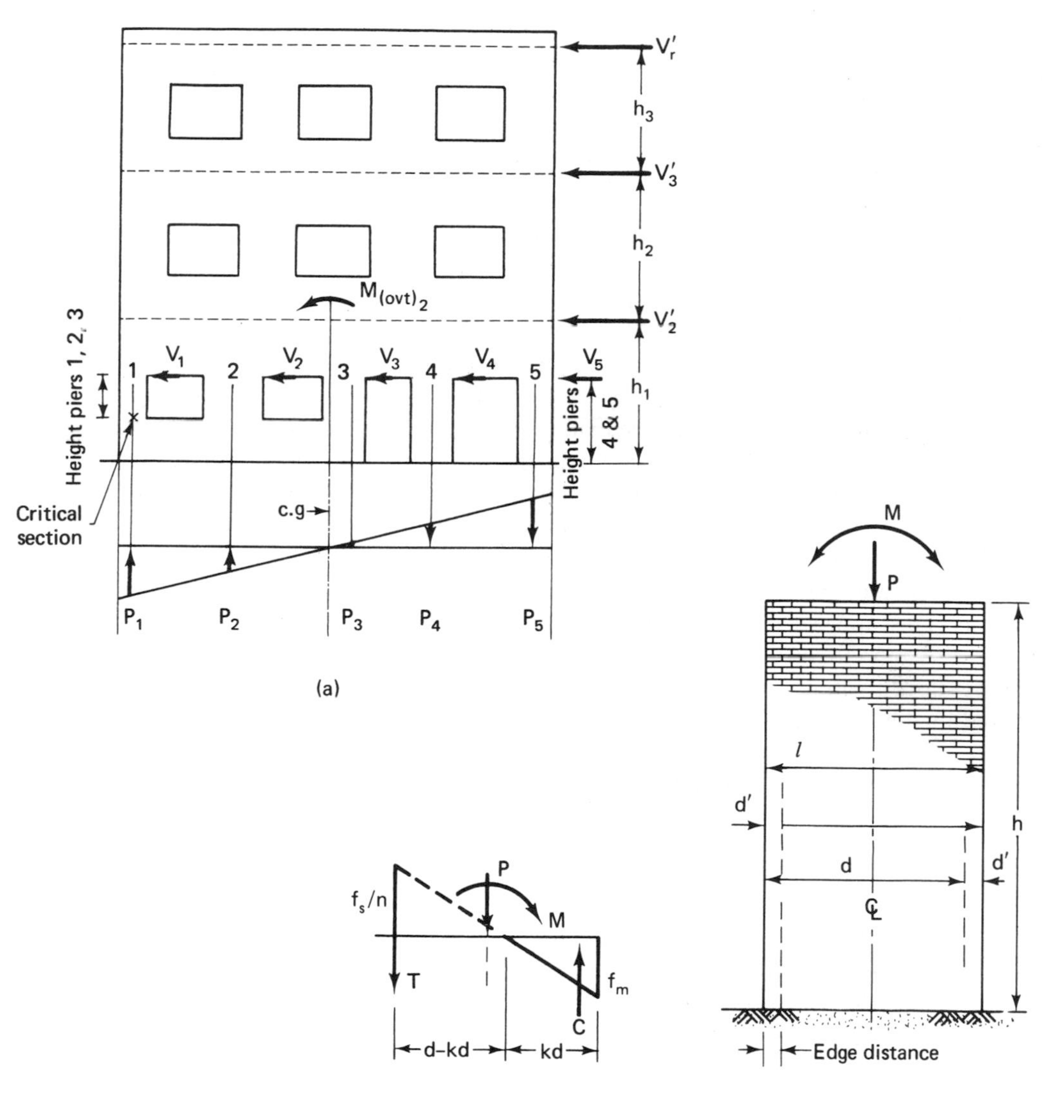

FIGURE 7-12. Axial load on pier due to overtuning.

Unique circumstances may warrant other load combinations. Check the adequacy of the pier, reinforced as in part 4, using the interaction formula:

$$f_a/F_a + f_b/F_b \leq 1.33$$

Where seismic effects control, the value of P_{total} could be either $(P_{dead} + P_{ovt})$ or $(P_{dead} + P_{ovt} + P_{partial\ live})$, depending upon those assumptions made regarding the loading probabilities. One of the assumptions made in seismic design is that the lateral loads are based on the probable real loads that may be in place at the time a quake occurs. This will certainly include the dead load of the structure. Since this may not have been determined exactly, the magnitude used in calculations may be adjusted, say $\pm 10\%$ (i.e., increased slightly if the dead-load stress would be additive on the section, or decreased if it would relieve a pier uplift tension stress). The live load is not generally assumed to contribute to the lateral or overturning load, since experience has shown that most of the time only a small fraction of the live load will be in place during a major earthquake. For this reason the live load, for either lateral or vertical influence, usually need not be combined with $P_{ovt} + P_{dead}$. However, if the live load is largely an equipment load that could be in place during an earthquake, all or part of it should be included. Also, if the live load is storage, such that at least a portion of it could be present, perhaps 25% might be included. This accounts for the exception noted in UBC Section 2312(c) in defining W for seismic calculations. Or, if the live load is a storage type that might probably be in place most of the time, 100% of LL could justifiably be added to the total load both for the overturning effect and also in determining the W for base shear values. One example consists of the load in the bins of processing plants, which may all be kept nearly full during plant operations.

The amount of jamb steel at the pier boundary was determined in step 4 as $M/f_s jd$. This approach ignores the axial compressive load effect on the tensile steel stress. A more refined solution might involve an analysis of the combined force system developed within a pier subjected to both axial load and in-plane moment as skectched in Figure 7-12(b).

From the force diagram, $C = \frac{1}{2} f_m (kd) t$ where the numerical value of $f_m = f_a + f_b$ in which $f_a = P/tl$ and $f_b = (1.33 - f_a/F_a)F_b$.

$$F_a = .2f'_m[1 - (h/40t)^3] \qquad F_b = f'_m/3 \text{ (900 lb/in.}^2 \text{ max.)}$$

For non-inspected work use $\frac{1}{2}F_a$ and $\frac{1}{2}F_b$.

Since $\sum M = 0$ about T, $C(d - kd/3) = P(l/2 - \text{edge dist.}) + M$. Solve this expression for (kd) and k.

Since $\sum F = 0$, $P + T = C$; thus find the tensile force in the jamb steel, T, and the resulting stress, $f_s = [(1 - k)/k]nf_m$. The required jamb steel area can then be found as

$$A_s = T/f_s$$

A negative value for T would indicate that theoretically no additional jamb steel would be required and the minimum steel area would then suffice.

One approximate, and perhaps more usable, method assumes $j = 0.9$ and corresponding $k = 0.3$, i.e., C (the bending compressive force resultant in the masonry) will act at $0.1d$ from the compressive face, and the axial compression P will act at the centroid. The effect of the overturning moment M to be resisted by the tension steel may then be reduced by the value of $P \times (0.5d - 0.1d = 0.4d\pm)$, i.e., the resistance to overturning provided by the axial load P. Thus $T = (M - P \times 0.4\text{d})\ 10.9\text{d}$.

DEEP-WALL BEAMS

Deep-wall beams, as they are designated by the International Conference of Building Officials Research Committee, are permitted a higher h/t ratio than that of brick masonry walls supported on continuous footings. The deep-wall-beam concept provides an alternative to conventional bearing-wall construction, since the wall is designed to span between spread footing pads as a beam, instead of having a continuous foundation support as shown in Figure 7-13. This concept has been used in reinforced concrete construction for many years. The technique was probably introduced in the early 1930s by the International Association for Bridge and Structural Engineers, where concrete walls were designed to span between supports consisting of concrete caissons. Then in recent years this concept has been used extensively in the "tilt-up" method of concrete construction often used for commercial and industrial buildings. It was not until the late 1960s, however, that the Masonry Institute of America, under the direction of Walter L. Dickey, explored the feasibility of using the technique in reinforced masonry construction. A research program was devised by them, and it was on the basis of these test results that the ICBO Research Committee issued a recommendation in their reports 2727 (Reinforced Brick Wall Panels) and 2868 (Concrete Masonry Wall Panels). The basic provision of this report permits the h/t ratio to be increased from 25 to 36, for load-bearing or shear walls and to 48 for other uses. The effect of these changes may be seen in Figure 7-14. In this type of

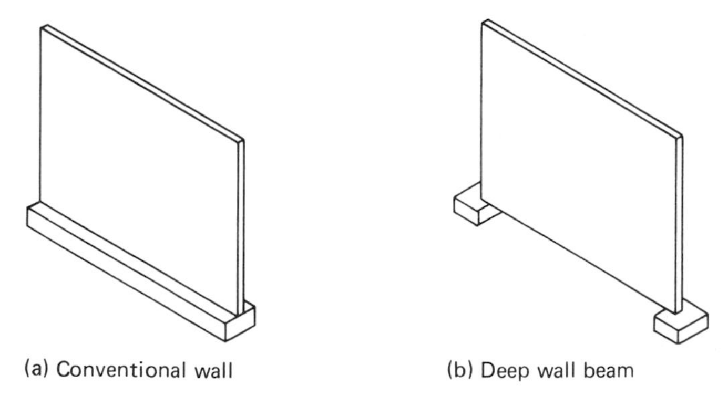

FIGURE 7-13. Deep-wall-beam concept.

construction the wall is actually considered as a flexural member, and therefore must carry the loads in bending instead of in direct compression. This is particularly useful for one-story applications, as in warehouses, banks, and other types of industrial and commercial buildings. For instance, a 16-ft-high wall supported continuously on a wall footing would have a minimum arbitrary thickness requirement of 8 in. to conform to the $h/t = 25$ limitation. However, if the wall is designed to span between isolated spread footing pads, the minimum wall thickness could be reduced to 6 in., in conformance with the increased h/t ratio of 36.

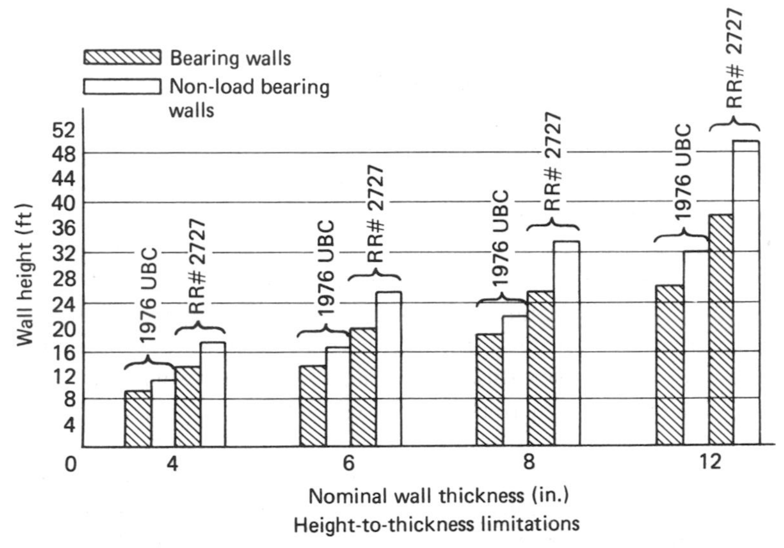

FIGURE 7-14. h/t limits for conventional bearing walls and deep-wall beams.

Basic advantages

The basic advantage of this design concept lies in its lower cost. One cost reduction results from the reduced wall thickness permitted, and this factor, further magnified by a reduced wall weight, in turn leads to smaller foundation requirements. The use of the deep-wall-beam concept often reduces wall thicknesses by a nominal 2 in. less than that for conventional design. It has been estimated that reducing the wall thickness from 8 in. to the 6 in. permitted by the deep-wall design concept would result in a labor and material cost saving of about 12%. Furthermore, this thickness reduction can be even greater in nonbearing walls. For example, if the unsupported height is 16 ft, the conventional nonbearing 8-in. wall can be reduced to 4 in. if the deep-wall-beam design were utilized. Even the reinforcement costs can be lowered. Where the amount of steel required is governed by the UBC minimum of $0.002 \times$ the cross-sectional area, it can be readily seen that considerably less steel is needed in the thinner wall. Reducing the thickness from 8 in. to 6 in., for example, will produce a steel cost saving of about 27%.

Code requirements

As with the conventional masonry wall, deep-wall beams must be reinforced to carry both in-plane lateral forces and those seismic or wind forces acting normal to the face. In addition, the previously specified minimum steel for walls remains unchanged for deep-wall beams (i.e., 0.0007*bt* minimum in either the vertical or horizontal direction, with the minimum total steel ratio remaining at 0.002). The actual wall thickness should be utilized in these calculations.

Since the deep-wall beam is considered to be a flexural member, the length of wall supported on the pad must not exceed one-tenth of the center-to-center spacing between supports. The bearing stress allowed, then, is given in UBC Table 24-H, which calls for $0.25f'_m$, with a 900 lb/in.^{2} maximum. The calculated bearing stress equals the beam reaction divided by the bearing area, which is defined as the net area of the section. Thus, in hollow unit masonry, only the area of the mortar and the grouted cells may be counted as bearing area. Cross web areas (when full bed joints are not used) and empty cells must not be included in these bearing area values.

The ICBO research report specifies that the axial compression stress shall not exceed that obtained from the formula

$$F_a = 0.20f'_m\left[1 - \left(\frac{h}{48t}\right)^3\right]$$

where h = unsupported vertical or horizontal height, in., whichever is applicable
t = nominal wall thickness, in.

Thus, it is assumed that the wall acts as a column at its vertical boundaries, having an in-plane dimension equal to the bearing width plus twice the nominal thickness, as shown in Figure 7-15. The reaction at the midheight of the equivalent column is thus

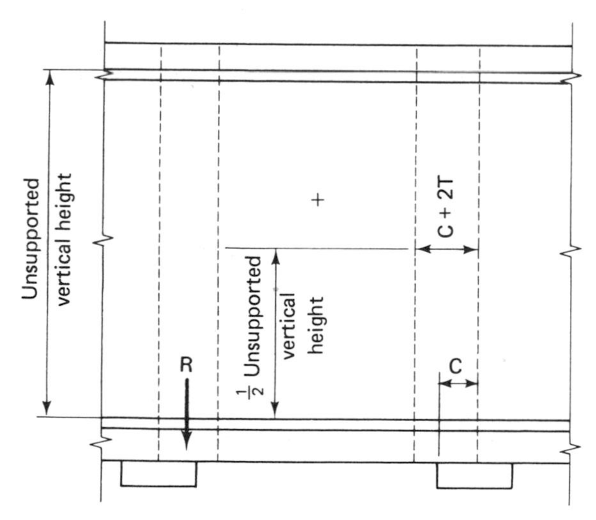

FIGURE 7-15. Deep-wall-beam edge columns.

divided by the net cross sectional area to obtain the applied column compression stress. Observe that the allowable compression stress, as defined by the Code, still incorporates a conventional buckling reduction factor.

Design procedure

A deep-wall beam can still serve as a shear wall. To do so, it must carry any in-plane shear forces incurred. Furthermore, as noted before, any seismic forces (not wind) transmitted by the diaphragm to the shear wall must be increased by 50% when checking the applied shear stress on it. The allowable shear stresses remain the same as specified in Table 7-1 for shear walls, either with or without shear steel, as the case may be. These values, too, may be increased by one-third for lateral loads.

Deep-wall beams are designed to resist flexural stresses in the same way that conventional masonry beams are analyzed. Working stress design is used to determine the amount of in-plane flexural steel required. The applied loads consist of roof live and dead load plus the panel weight itself. This flexural reinforcement is typically placed at the bottom on the tension side of the wall, as shown in Figure 7-16. ICBO Research Report 2727 further stipulates that if the height/length ratio of the deep-wall beam exceeds 4:5, a special bending stress analysis must be made. To comply with this stipulation, one approach often used assumes that the deep-wall beam is divided into two parts by a horizontal plane located usually 4 to 8 ft above the foundation. The lower portion is considered to act as a beam supporting all loads, including the dead weight of the upper and lower portions of the wall beam, as shown in Figure 7-12. The lower portion is then viewed as if it were a conventional masonry beam to obtain the needed steel requirement (refer to Chapter 6). This approach usually leads to conservative results. Another method of design entails application of the procedures developed by the PCA for deep girders.

The wall must also be designed to resist the wind or seismic forces acting perpendicular to it as it spans the floor and the roof diaphragm. The same procedure is followed as that utilized in analyzing conventional bearing or nonbearing walls.

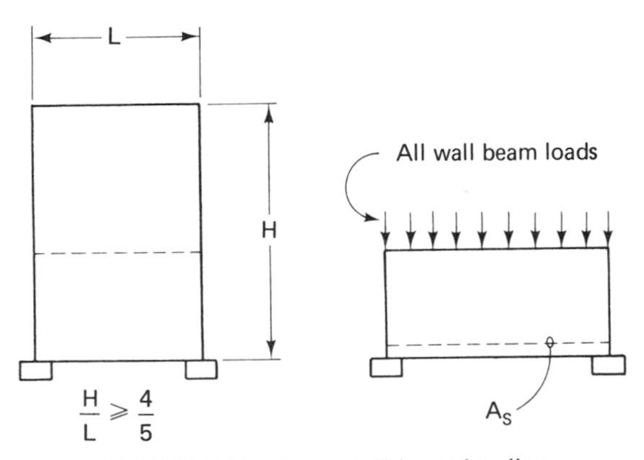

FIGURE 7-16. Deep-wall-beam bending.

Where axial compression forces exist, in addition to the normal forces perpendicular to the wall, the interaction relation must be reviewed. In this case f_a equals the vertical load stress at the midheight of the wall distributed over an area of $t(c + 2t)$ at each end of the panel (Figure 7-15). The f_b is based upon an effective b of the total panel length.

To check the shear stress caused by the forces normal to the wall, the shear area equals the depth of the tensile steel times the combined widths of the grouted cell, plus the adjacent brick cross webs, as seen in Figure 7-17.

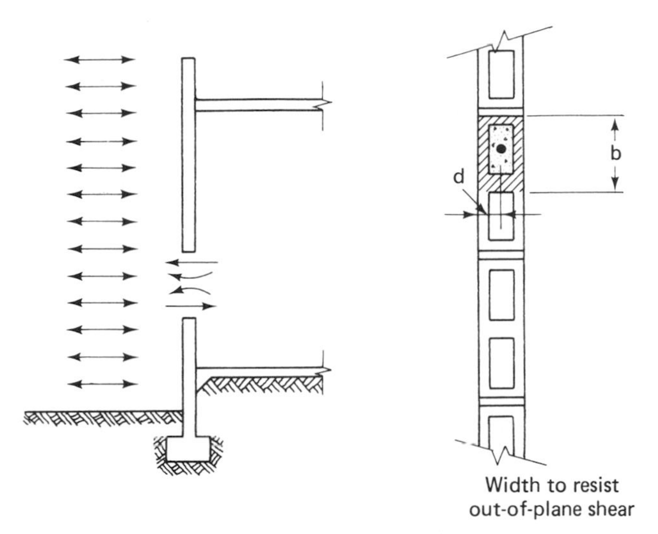

FIGURE 7-17. Out-of-plane shear area.

EXAMPLE 7-5

Using the deep-wall-beam concept, design the walls for the bank building shown in Figure 7-18. The 4-in. brick bearing wall uses H 8000 brick (ICBO Research Report 2730) and design conforms to ICBO Research Report 3118. The unsupported height is 10 ft, and the h/t ratio equals 30. Section A shows the wall supported on two piers, 21 ft apart, and projecting from the foundation wall. This concept allows for temporary wood framing, later removed, to be placed under the first course of brick before laying. Section A′ represents an alternative solution, wherein a slot filled with Styrofoam is made in the concrete floor slab. The roof is fastened to the 4-in. wall with $\frac{1}{2}$-in. ϕ ⌋ bolts embedded 2 in., with a 2-in. vertical projection. Note that ICBO Report 3118 allows 640-lb bolt shear for this configuration.

Solution:

Loads:

Wind = 15 lb/ft^2

In-plane shear: Wind = 73 lb/ft — Taken from the original design
Seismic = 118 lb/ft

Roof: DL = 135 lb/ft
LL = 225 lb/ft

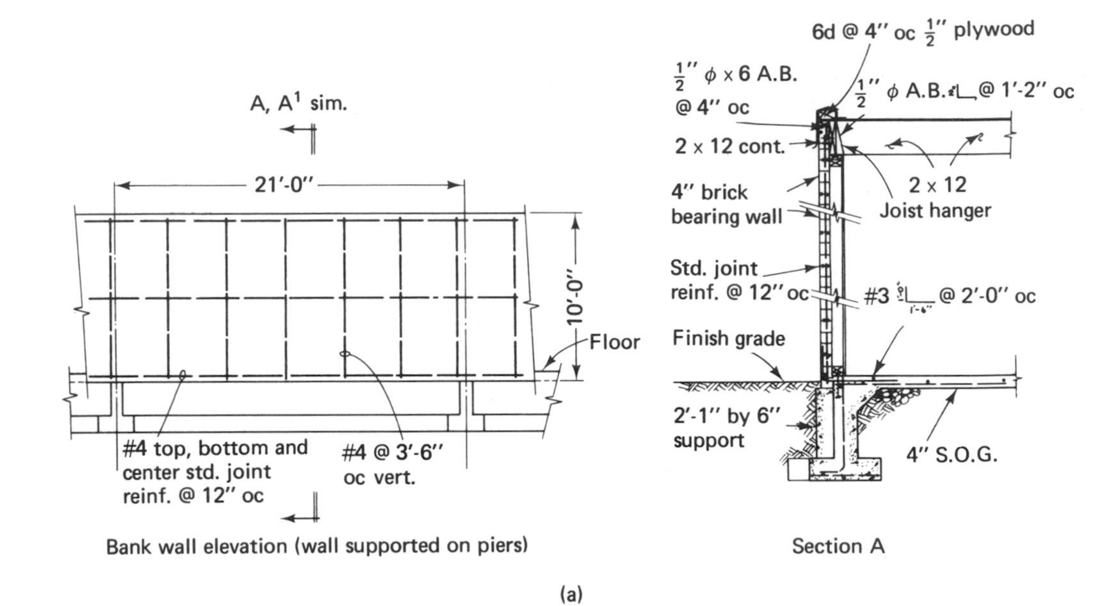

(a)

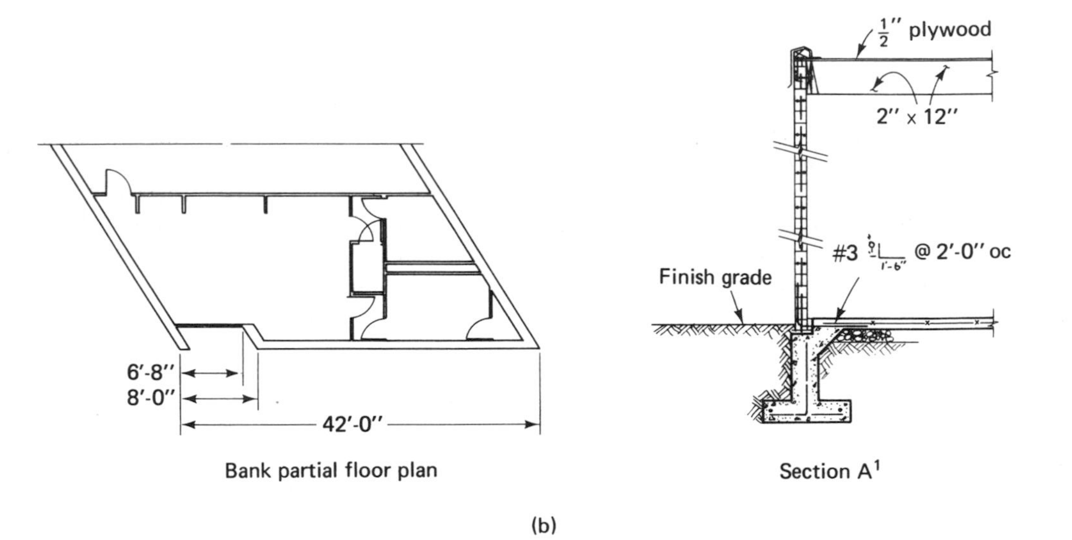

(b)

FIGURE 7-18. Bank building details.

Panel dead load

$3.5/12 \times 140 \times 10 \text{ ft} = 408 \text{ lb/ft}$

Assumes 140 lb/ft^3

Geometry:

$h = 10 \text{ ft} \rightarrow 120 \text{ in.} \quad t = 4 \text{ in.}$

$h/t = 120/4 = 30 < 36$ OK

Per ICBO Research Report 2727

Minimum steel:

Try 4′-0″ oc

$0.0013 \times 48 \times 3.5 = 0.22$ horiz.

$0.0007 \times 48 \times 3.5 = 0.12$ vert.

Use No. 4, bottom, top, and center plus stnd. wt. joint reinf. horizontal

No. 4 at 4′-0″ oc vert.

UBC Section 2418(j)3

Bearing:

$\text{Area} = 2 \times 0.75 \times 12 + 2 \times 4 = 26 \text{ in.}^2$

$\text{reaction} = (135 + 225 + 408) \times 2\,1/2 = 8064 \text{ lb}$

$F = 0.25 f'_m = 0.25 \times 2600 = 650 \text{ lb/in.}^2$—allow

$f = 8064/26 = 310 \text{ lb/in.}^2$ actual $\quad \therefore$ OK

0.75-in. bed joints, each face plus one grouted cell

Table 7-1

Compression:

$\text{Area} = 2 \times .75(12 + 2 \times 4) + 2 \times 4 = 38 \text{ in.}^2$

$\text{Load} = (135 + 225 + 204) \times 2\,1/2 = 5922 \text{ lb}$

$F_a = 0.20 \times 2600[1 - (120/48 \times 4)^3] = 393 \text{ lb/in.}^2$

$f_a = 3922/38 = 155.8 \text{ lb/in.}^2$ actual $\quad \therefore$ OK

Column width = C + 2t plus one grouted cell

Load at midheight

Equation Report 2727

$f'_m = 2600 \text{ lb/in.}^2$ H8000 brick with reduced design stress to avoid prism testing

UBC Section 2404

In-plane-shear:

1. Seismic

$v = 1.75\sqrt{f'_m} = 89 \text{ lb/in.}^2$

Table 7-1

$v_{\text{allow}} = 89 \times 1.33 = 119 \text{ lb/in.}^2$

$V_{\text{applied}} = 118 \text{ lb/ft}$

$V_{\text{design}} = 118 \times 1.5 = 177 \text{ lb/ft}$

UBC Section 2418(h)

$v = V_{\text{design}}/12 \times 2 \times 3/4$

$= 177/12 \times 2 \times 0.75 = 10.0 \text{ lb/in.}^2$

$\frac{1}{3}$ stress increase

$A_{s(\text{req.})} = 10.0 \times 1.5 \times 48/(20{,}000 \times 0.9)$

$= 0.04 \text{ in.}^2/4'0''$ OK

2. Wind

$v = 1.75\sqrt{f'_m} = 89 \text{ lb/in.}^2$

$v_{\text{allow}} = 89 \times 1.33 = 119 \text{ lb/in.}^2$

$V_{\text{applied}} = 73 \text{ lb/ft}$ OK

Table 7-1

UBC Section 2303

3. Beam

$v = 3.0\sqrt{f'_m} = 153 \text{ lb/in.}^2 \geq 150 \text{ lb/in.}^2$

$V = 8064 \text{ lb}$

Table 7-1

Max. allowable = 1500 lb/in.^2

$v = V/bjd = 8064/1.5 \times 0.9 \times 120 = 50$ lb/in.^{2}
$A_s = 50 \times 1.5 \times 48/20{,}000 \times 0.9 = 0.20$ — UBC Section 2418(h)
No. 4 at 48″ oc vert. OK

In-plane-bending:

$h/l \cong 10/21 \leq 4/5$ — Deep beam analysis is not required

$M_{\text{applied}} = 768 \times (21 - 12.6'')^2/12 = 25{,}472)$ ft-lb — Assumes a continuous span
$b = 1.5$ in., $d = 118$ in., $n = 11.54$
$f_s = 20{,}000$, lb/in.^{2} $f_m = 858$ lb/in.^{2} — Working stress design
$A_s = 1$ No. $4 = 0.20$ in.^{2}

Out-of-plane:

1. Bending and compression

$M = 15 \times 10^2/8 = 187$ ft-lb/ft — Span, floor to roof
Try No. 4 at $3' - 6''$ oc, $b = 24$ in., $d = 1.75$ in., — Width $= 6t$ UBC Section 2418(f)
$M_{\text{applied}} = 187 \times 3.5 = 654$ ft-lb/ft.
$n = 11.54, f_s = 26{,}666$ lb/in.^{2}, $f_m = 1144$ lb/in.^{2} — $\frac{1}{3}$ increase, UBC Section 2303
$K = 0.281$, $M_s = 705$ ft-lb
$f_a = (135 + 225 + 204)/(12 \times 1.5) = 31.3$ lb/in.^{2}
$F_A = 0.20 \times 2600(1 - (120/40 \times 4)^3) \times 1.333$ — $\frac{1}{3}$ increase, UBC Section 2303
$= 400$ lb/in.^{2}
$f_b/F_B + f_a/F_A \leq 1.00$ — UBC Section 2418(g)

$$\frac{654}{705} + \frac{31.3}{400} = 1.0 \quad \text{OK}$$

2. Shear

$v_{\text{allow}} = 50 \times 1.33 = 66.5$ lb/in.^{2} — Table 7-1
$V_{\text{applied}} = 15 \times 10/2 = 75$ lb/ft — Max. allowable $= 50$ lb/in.^{2}
$v = V/bjd$ — UBC Section 2418(h)
$b = 4$ in. $+ (2 \times \frac{3}{4}) = 5\frac{1}{2}$in./3′-6″ — One grouted cell plus two cross webs
$d = 1.75$ in. $j = 0.93$
$v = 75 \times 3.5/5.5 \times 0.93 \times 1.75 = 29.2$ psi lb/in.^{2}

Connections:

Roof: 2×12 ledger with joist hanger $\frac{1}{2}''$ ϕ bolts — $\frac{1}{2}''$ ϕ UBC bolt allowable in wood
$V_A = 135 + 225 = 360$ lb/ft
$V_{\text{bolt}} = 430$ lb in wood — UBC allowable
$= 640$ lb in masonry — ICBO R.R. #3118
$s = 430/360 = 1.2$ ft
$\frac{1}{2}''$ ϕ J-bolts at $1' - 3''$ oc

Floor: $V = 118$ lb/ft
$A_s \approx 1.5 \times 118/20{,}000 = 0.01$ in.^{2}/ft — Shear friction, UBC Chapter 26
Use No. 3 dowels at $2' - 0''$ oc

Roof tie: 200 lb/ft — UBC Section 2313
6d nails at 84 lb/nail, 3-6d per foot

High-strength 4-inch hollow reinforced brick

With the advent of such developments as the ICBO's Research Report 2727 on the deep-wall-beam concept, higher h/t ratios were permitted, thereby leading, in effect, to thinner walls. To further satisfy this need, a high-strength 4-in. hollow clay brick was developed through the resources of the Western States Clay Products Association (WSCPA). These grade I fired clay units have compressive strengths of designation H8000 or H10,000 (as called for in ICBO Research Report 3118). These units must comply with WSCPA specifications for hollow brick for reinforced brick masonry (hollow masonry units made from clay or shale) as defined in ICBO Research Report 2730.

In design, f'_m becomes 3300 lb/in.^{2} for H8000 and 4000 lb/in.^{2} for H10,000 brick. Structural sizing and reinforcing requirements must still conform to UBC Section 2418. Allowable J-bolt ($\frac{1}{2}$-in. diameter) shear is 640 lb, with a $2\frac{1}{2}$-in. minimum embedment and a 2-in. vertical projection. The maximum h/t ratio permitted for bearing walls is 27. Only a running bond pattern is permitted (see Figure 7-19 for construction details).

STUD-TYPE WALL DESIGN

The grouting of hollow unit masonry units offers an alternative method of wall design from that previously described, when lateral loads (wind, seismic, earth pressure) act normal to the wall face. In this view the solid grouted reinforced cell can be construed to be a kind of "concrete stud," which becomes the actual structural member resisting these forces. In this concept the block merely serves as a form to contain the concrete grout. The remaining portion of the wall, nonstructural in function, merely acts as the wall enclosure (Figure 7-20).

An advantage to this method stems from the fact that it can be built without special inspection, while basing the design on the higher concrete stresses allowed (f_c). For comparison, observe that for solid grouted masonry, with no inspection, the allowable flexural compressive stress $F_b = 250$ lb/in.^{2}. On the other hand, by considering only the grouted stud as the structural element, the comparable allowable stress, f_c, becomes 900 lb/in.^{2} (0.45×2000 lb/in.^{2} concrete). This method is especially suited for the analysis of masonry assembled without mortar, such as surface-bonded or interlocking block walls.

EXAMPLE 7-6

Perhaps the following simple calculations for a 12 in. wall will illustrate the use of this method as compared to the "conventional" approach.

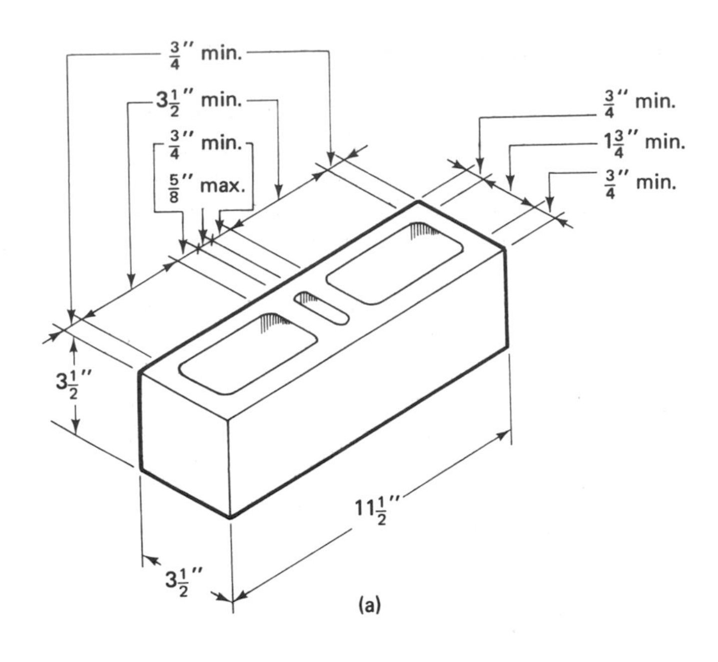

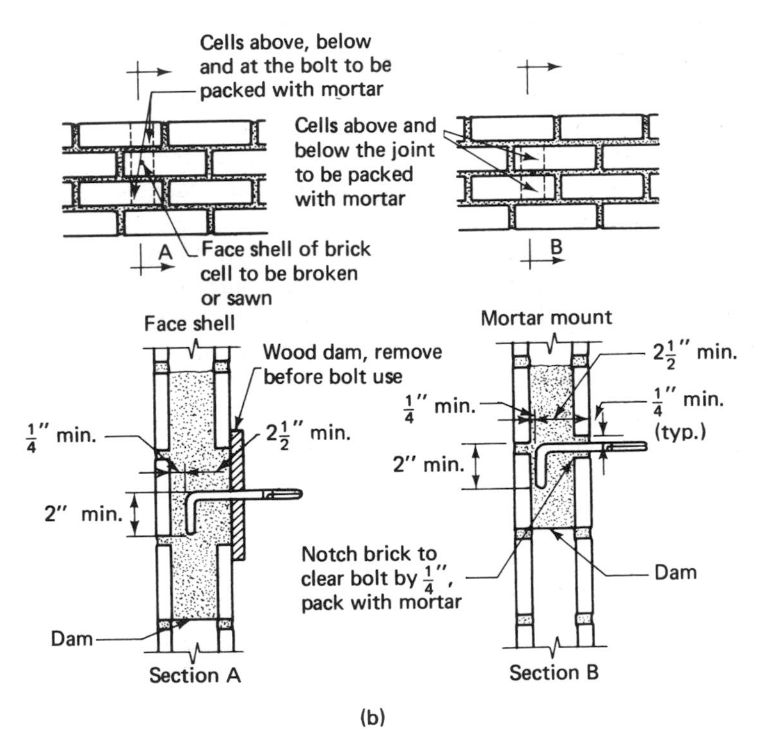

FIGURE 7-19. High-strength 4-in. hollow reinforced brick construction.

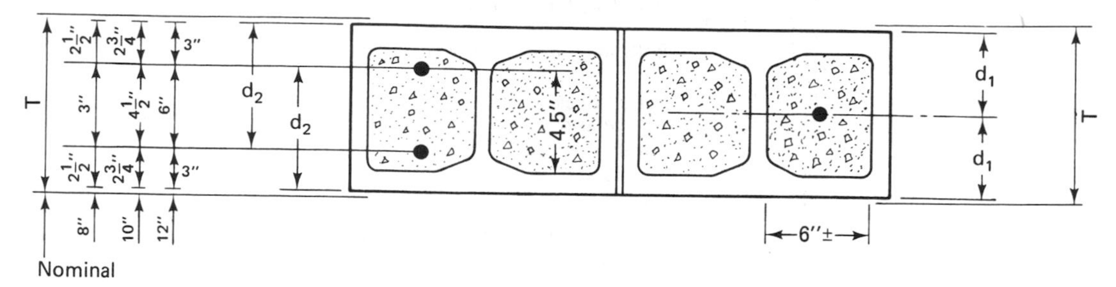

FIGURE 7-20. Stud-type wall concept.

Conventional reinforced masonry method of analysis

$b = 12$ in., $d = 11.5/2 = 5.7$ in.
$f_m' = 1500$ lb/in.^{2}, $F_b = 250$ lb/in.^{2}, $n = 40$
$A_s = 0.18$ in^2/ft (for No. 7 at 40 in. cc)

$$p = \frac{0.18}{12 \times 5.7} = 0.026,\ pn = 0.105$$

$j = 0.87$ and $2/jk = 6.24$ (Table C-1)

$$M_s = A_s f_s jd$$
$$= 20{,}000 \text{ lb/in.}^2 \times 0.18 \times 5.7 \times 0.87$$
$$= 17{,}850 \text{ in.-lb}$$
$$M_m = f_m \times bd^2/2/jk$$
$$= \frac{250 \times 12 \times 5.7^2}{6.24} = 15{,}620 \text{ in.-lb} \quad \text{(governs)}$$

Stud-type method of analysis

$$b = 2 \text{ cells} \times 6 \text{ in.} \times \frac{12}{16 \text{ in./ft}} = 9 \text{ in. concrete}$$

$d = 4.5$ in. from grout interface to rebar
$f_c' = 2000$ lb/in.^{2}, $f_c = 900$ lb/in.^{2}, $n = 11$
$A_s = 0.18$ in./ft^2

$$p = \frac{0.18}{9 \times 4.5} = 0.0044,\ pn = 0.049$$

$j = 0.91$ and $2/jk = 8.2$

$$M_s = 20{,}000 \times 0.18 \times 4.5 \times 0.91$$
$$= 14{,}740 \text{ in.-lb} \quad \text{(governs)}$$
$$M_c = \frac{900 \times 9 \times 4.5^2}{8.2} = 20{,}000 \text{ in.-lb}$$

Note that the capacity in the stud-type wall design would have been increased by adding reinforcing to the point where a balanced design would have been approached, in which case the allowable moment would reach something in the neighborhood of 20,000 ft-lb. Then this approach would have permitted a higher design moment than that determined by the conventional reinforced masonry method. In that case, any additional reinforcing would have contributed nothing, for the 15,620 in.-lb permitted moment is governed by the masonry stress.

WALL CONSTRUCTION DETAILS

Concrete masonry

The following section is provided to demonstrate what properly reinforced masonry walls look like in section. A typical concrete block wall section appears in Figure 7-21. Often, only the reinforced cells need to be grouted, at least for structural purposes. On the other hand, a completely solid grouted wall may be called for, if, for example, a higher fire rating is needed to conform to the Code. Note that the horizontal reinforcing in the field of the wall may consist of rebars placed in channel or bond beam block at 48 in. cc maximum (see Chapter 3). Where wire

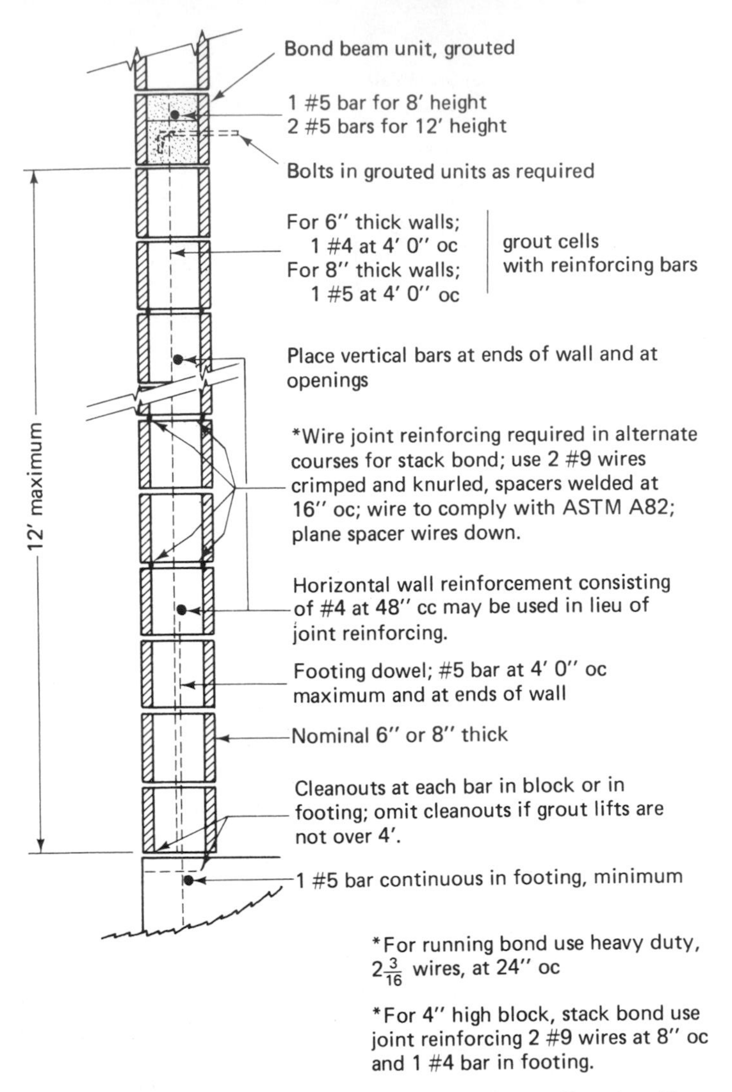

FIGURE 7-21. Typical reinforced concrete masonry wall section. *Note:* This provides for minimum temperature reinforcing—additional reinforcing may be required for wind, seismic, and other loads. Use of joint reinforcing permits ungrouted cells to be filled easily with perlite, vermiculte, or foam for insulation, thus reducing the *U* factor.

reinforcement is embedded in the joints to serve as part of the minimum horizontal steel, the detail would appear as in Figure 7-21. The basic principle of joint reinforcement presumes that the wall spans horizontally from intersection to intersection, or from support to support.

The reinforcing details for various types of wall intersections can be seen in Figure 7-22. Note the continuity maintained by the reinforcement around the corner in details (a) and (b). Finally, refer back to Figure 1-14, which presents an isometric view of typical commercial concrete block construction. Observe, among other things, how the tensile reinforcement fits within the lintels over the openings.

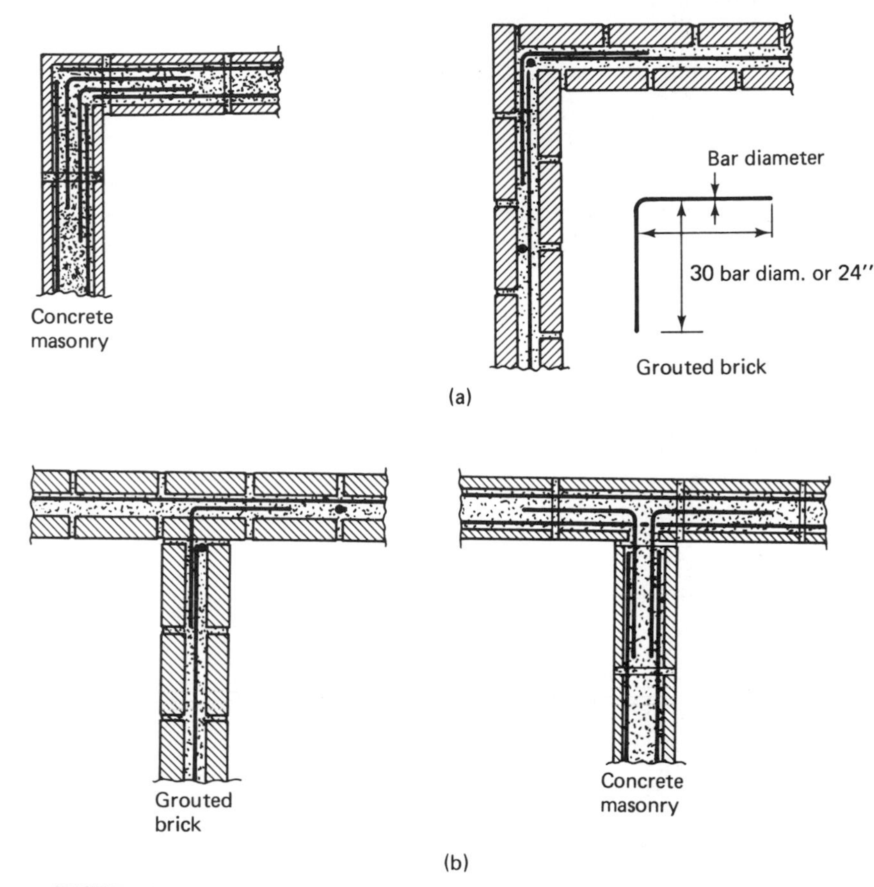

FIGURE 7-22. Typical masonry wall intersections and corner details. (a) Corner connection—exterior bond beams. Lap all bars minimum of 30 bar diameters or 24 in. (b) Wall intersections—minimum lap of 30 bar diameters or 24 in.

Grouted brick masonry

A typical reinforced grouted brick wall section, two wythes thick, is shown in Figure 7-23. Note here the location of the vertical and horizontal reinforcement, within the grouted core between the exterior wythes.

The figures shown in this section demonstrate what sound engineered reinforced masonry construction looks like. It should become apparent to anyone, after thoroughly examining these details and others throughout this text, that reinforced masonry, when properly designed and constructed, is actually no more complicated than is reinforced concrete, or even nonengineered masonry, for that matter. But there is indeed a vast difference in performance between engineered and nonengineered masonry, insofar as its ability to resist the destructive forces of nature is concerned.

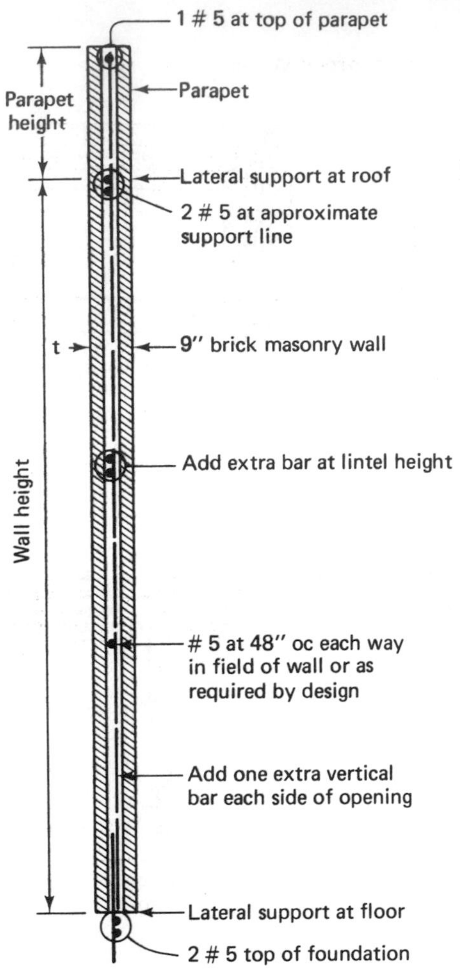

FIGURE 7-23. Typical reinforced brick wall sections.

WALL DESIGN AND ANALYSIS AIDS

The fundamentals of masonry design have been thoroughly examined and the basic design and analysis formulas were derived in Chapter 6. These have then been extended so that they can be used in proportioning specific structural elements such as the various types of masonry walls described in this chapter and the columns, beams, and lateral-load-resisting systems as discussed in the following three chapters. In addition to those fundamental equations that may be used in some of the simpler design cases, it is often convenient to make use of certain design aids, such as tables and charts, to speed the process toward a solution. Typical design aids in the form of charts, tables, and curves, appear in the Appendix.

They are presented primarily to introduce design aids to students and to familiarize them with the use of such aids. However, it is recommended that these not be used indiscriminantly or depended upon exclusively in problem solving. Rather, one should continue to return to the fundamentals. However, practitioners will find these invaluable when seeking economical and efficient solutions for various types of reinforced masonry structures.

PROBLEMS

For the following problems, wind pressure = 15 lb/ft², $Z = 1$, $I = 1$.

7-1. Determine the minimum steel required for a 10-in. brick wall, 16 ft high. The parapet extends 3 ft above the roof line. Use joint reinforcement between the footing and bond beam. Assume two No. 4's in bond beam and top of footing.

7-2. Determine the amount of reinforcement needed for a nonbearing 9-in. brick wall spanning 15 ft vertically. The steel is grade 40 and the brick is MW. Solve with the basic formula, then check the result with design aids (with and without inspection). Check the minimum steel requirements. Steel is placed at the center.

7-3. Determine the amount of reinforcement needed for a brick bearing wall, 9 in. thick, spanning 15 ft vertically. The steel is of grade 40 and the brick is of MW. The 200-lb/ft roof load is applied with an eccentricity of $\frac{1}{2}t$. Solve with the basic formula, then check the result with design aids (with and without inspection). The steel is placed at the center of the wall section.

7-4. Same as Problem 7-2, but change the wall to an 8-in. partially grouted concrete masonry wall.

7-5. Same as Problem 7-3, but change the wall to an 8-in. partially grouted concrete masonry wall.

7-6. A $3\frac{1}{2}$-ft high partially grouted concrete masonry parapet wall 8 in. thick is reinforced with No. 6's at 40 in. cc. Determine the allowable load. Check the shear and bond. Use full stresses.

7-7. Given an 8-in. concrete masonry bearing wall with No. 6's at 48 in. cc, find the maximum h/t permitted (with and without inspection).

7-8. Same as Problem 7-7, but change the steel to No. 4's at 30 in. cc.

7-9. Given a masonry wall with the loading as shown, determine the axial load on piers A and C due to overturning effect.

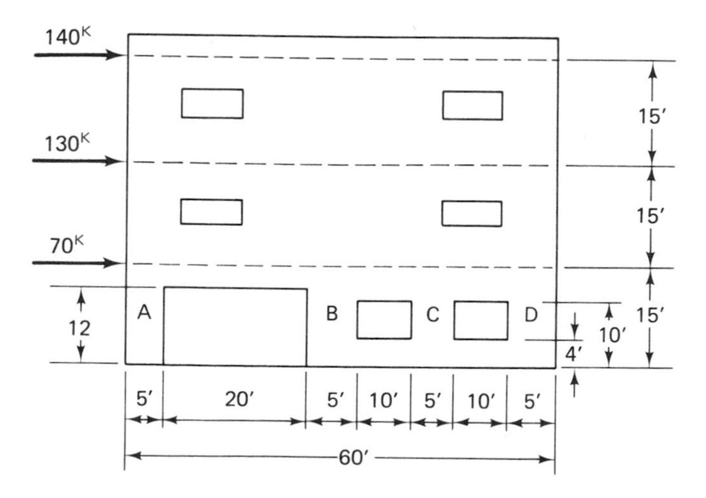

7-10. Determine the allowable moment per foot of the given 12-in. retaining-wall section by the conventional approach and by the stud method. No special inspection is provided.

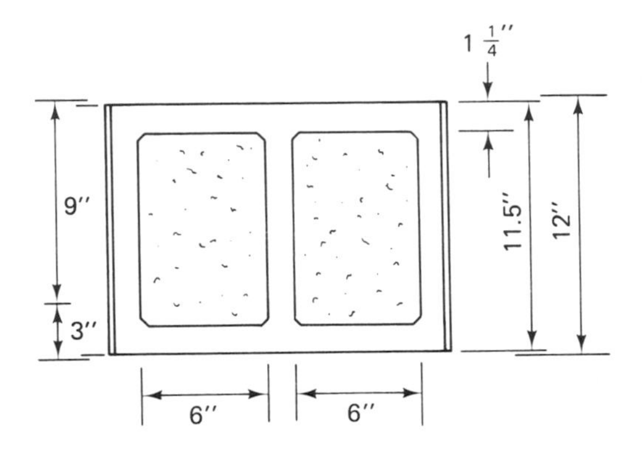

$f'_m = 1500$ lb/in^2. $n = 40$ (masonry)
$f'_c = 2000$ lb/in^2. $n = 11$ (concrete)
$f'_s = 20{,}000$ lb/in^2.
$A_s = 0.09$ in^2/ft

7-11. Given a 9-in. brick masonry wall, determine the reinforcement required for section C under a wind pressure of 15 lb/ft^2. Special inspection is provided.

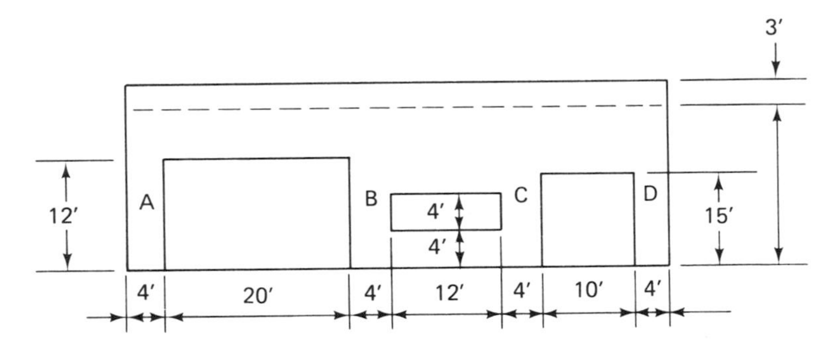

8

Masonry Columns and Pilasters

A *masonry column* is a vertical compression member designed to support a load along its longitudinal axis. Depending upon its location and function, a masonry column may be built as a separate supporting member, or it may be contained entirely within a wall, acting as a *flush wall column*. A pilaster functions as a column within a wall, projecting from either or both faces of the wall. It also serves to stiffen the adjoining wall panels (see Figure 8-1). The vertical loads usually consist of concentrated reactions imposed by girders or trusses. In addition, it may have to sustain lateral loads in bending, when the wall panel between pilasters is designed to span horizontally between these pilasters. This may be done to improve the h/t ratio, as was pointed out in Chapter 7. The pilaster in this case must then have the capacity to span vertically between diaphragms, in carrying the wall panel reactions caused by wind or seismic effects. Additional bending stresses may result should the supported member be eccentrically located with respect to the centroidal axis of the column or pilaster.

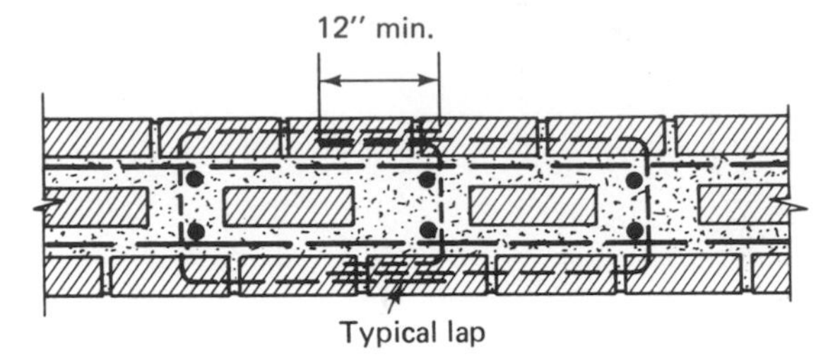

(a) Flush wall column-brick; Ties in mortar joint

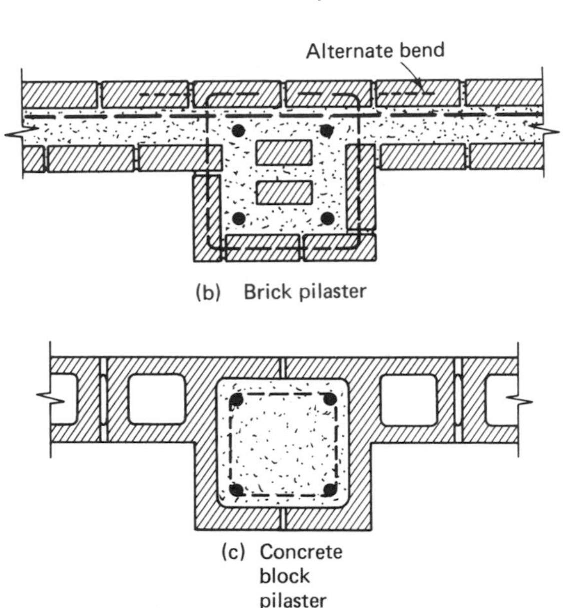

(b) Brick pilaster

(c) Concrete block pilaster

FIGURE 8-1. Wall columns and pilasters.

COMBINED LOADING EFFECTS

In the latter two loading conditions cited in the preceding section, the column or pilaster must be designed to sustain the combined effects of axial compressive plus flexural stresses. Thus, several load combinations usually must be examined, with the most severe stress condition governing the final design. Often, the direct compressive stresses resulting from the vertical loads, which are generally considerable in magnitude, will exceed the flexural tensile stresses, so that all segments of the column remain in compression. The *interaction* (or *unity*) *design equation* is reasonably adequate for the analysis or design of such a structural element. However, in some exterior columns and pilasters, flexural tensile stresses do exceed the axial compressive stresses, so that tension will be produced on some portion of the column section and the neutral axis will lie somewhere within the cross-sectional boundaries, as demonstrated by the stress distributions in Figure 8-2. This may occur where (1) vertical dead and live loads act at large eccentricities, (2) flexural members are connected to the columns to the extent that some degree of fixity occurs, (3) lateral thrusts are transmitted to the columns as a result of wind or earthquake forces, or (4) connected horizontal members expand and contract due to a thermal or moisture change. The design then will be governed either by the flexural stresses developed in the tensile reinforcement or by the combined axial and flexural compressive stresses occurring within the masonry section. The precise analytical analysis in this case becomes rather complex, and the use of certain types of proposed interaction curves would seem to greatly simplify the computations. Without these, however, the basic interaction formula has been used in such instances, even though it fails to account for the effect of the axial compression in reducing the tensile steel stress.

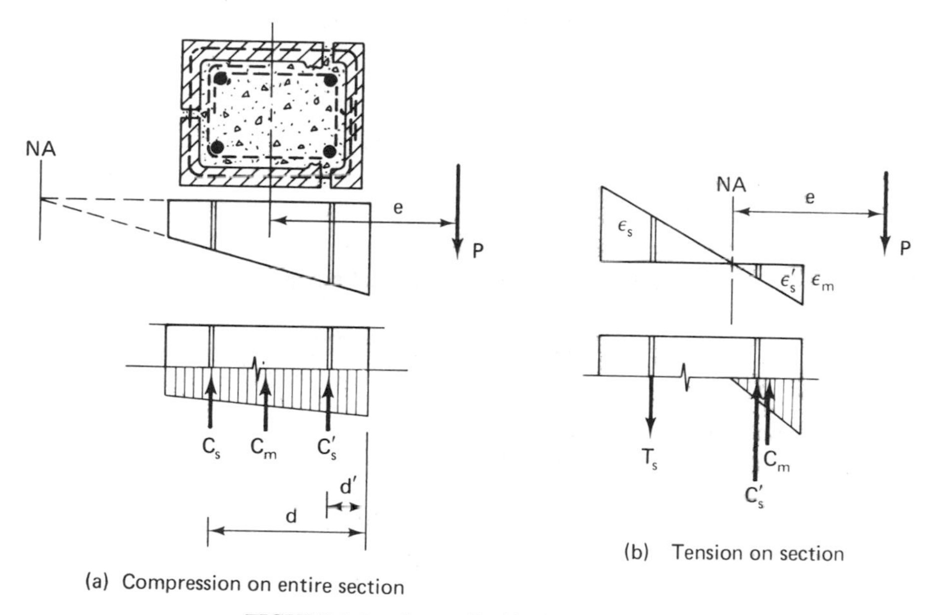

FIGURE 8-2. Stress distribution on columns.

The National Concrete Masonry Association (NCMA) has developed a group of design equations based upon the basic force system developed under combined loading. Appreciation is expressed for their permission to reproduce the resulting expressions here.

Starting with the interaction formula expressed in the form $f_a = (1 - f_b/F_b)F_a$, wherein $F_a = (P/A_g)/[1 + p_g(n - 1)]$ and $F_b = 0.33f'_m$ the NCMA derivations lead to the following results:

1. Section in compression (interval a-b, Figure 6-15):
   (a) Load governed by allowable masonry stress:

$$P_e = \frac{F_a A_g}{\dfrac{F_b - F_a}{F_b[1 + p_g(n-1)]} + X_1 \dfrac{F_a}{F_b}}$$

where P_e = allowable equivalent column load on short columns subjected to combined axial and bending loads

$$\text{where } X_1 = \frac{2k}{k^2 + p_g[(3n-1)k - (d'/t)(n-1) - n]}$$

k is obtained from

$$k^3 + ak^2 + bk - c = 0$$

$$\text{where } a = 3\left(\frac{e}{t} - \frac{1}{2}\right)$$

$$b = 3p\left[2\frac{e}{t}(3n-1) + \left(2\frac{d'}{t} - 1\right)(n+1)\right]$$

$$\text{with } p_g = \frac{A_{st}}{bt};\ p = \tfrac{1}{2}p_g$$

$$c = 3p\frac{d'}{t}\left[2\frac{e}{t}(n+1) + 2\frac{d'}{t}(3n+1) - 5n - 1\right] + 3pn\left(\frac{2e}{t} + 1\right)$$

(b) Load governed by allowable compressive steel stress:

$$P_e = \frac{f_m A_g}{X_1} \qquad \text{where } f_m = \frac{20{,}000}{(2n-1)(1 - d'/kt)}$$

2. Section in partial tension (interval b-c, Figure 6-15):
The load governed by allowable steel stress is

$$P_e = \frac{f_m A_g}{X_1} \qquad \text{where } f_m = \frac{20{,}000}{n\left(\dfrac{t - d'}{kt} - 1\right)}$$

3. When the entire section remains uncracked (i.e., $e/t < \frac{1}{3}$):

$$P_e = \frac{A_g(0.18f'_m + 0.65f_s p_g)}{1 + X_2(0.54 + 3.9pf_s/f'_m)}$$

where

$$X_2 = \frac{e}{t}\,\frac{6}{[1 + 3p_g(n-1)(1 - d'/t)^2]} \qquad \text{and} \qquad p = \tfrac{1}{2}p_g$$

The National Concrete Masonry Association in their publication "Design and Construction of Reinforced Concrete Masonry Columns and Pilasters," presents tabulated values of X_1, X_2, and k evaluated from the parameters d'/t, f'_m, p_g, and e/t.

In contrast, interior columns normally need be designed to carry only vertical loads, since the lateral effects of wind or seismicity are negligible. All horizontal forces are taken by the lateral-force-resisting system, usually a box system with shear-wall elements. Masonry lateral-load-resisting moment frames are virtually nonexistent. Presently, at least, not enough is known about the continuity performance of reinforced masonry frames, so it would be somewhat imprudent to depend upon such a system for lateral stability. As research continues to reveal more about the behavior of this composite material, and its ductile performance has been thoroughly established, we may eventually expect to see safe economical reinforced masonry moment frames used in certain locales.

ASSUMED LOCATION OF BEAM AND TRUSS REACTIONS

When designing columns and pilasters, or walls for that matter, one must give careful attention to the manner in which the horizontal members are supported and how they bear on that element (Figure 8-3). When these members are not restrained against rotation by the connection, they will produce a vertical reaction on the supporting member that acts approximately parallel to its vertical axis at an eccentricity determined by the type of connection provided. For instance, should a beam, which may rotate under load, be supported on a steel bearing plate, the resultant reaction will move outward toward the edge an indeterminate amount, depending upon the relative stiffness of the wall or column. In such instances it is generally satisfactory to consider that the vertical component of the beam reaction is applied at the third point of the bearing plate, as shown in Figure 8-3b. On the other hand, a truss or deep beam will usually exhibit little rotation under load. For this reason, it might be more realistic to assume that the reaction is uniformly distributed over the length of bearing, and therefore the resultant reaction would occur at the center of bearing, as shown in Figure 8-3c.

GENERAL COLUMN BEHAVIOR

Column buckling loads

For those long columns that reach a condition of instability and buckle before the material yields, the elastic buckling load for a homogeneous section is simply defined by the Euler column formula:

$$P_{cr} = \frac{\pi^2 EI}{(kh)^2} \quad \text{or} \quad \left(\frac{P}{A}\right)_{cr} = \frac{\pi^2 E}{(kh/r)^2}$$

Generally, these elements are referred to as "long slender columns," since theirs is an *elastic* buckling failure, wherein no portion of the material reaches a yield state. How-

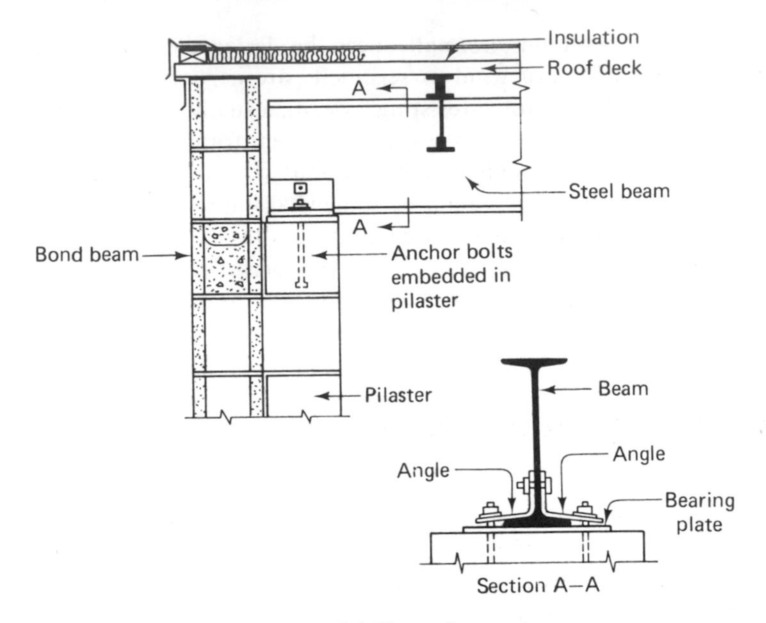

(a) Type of support

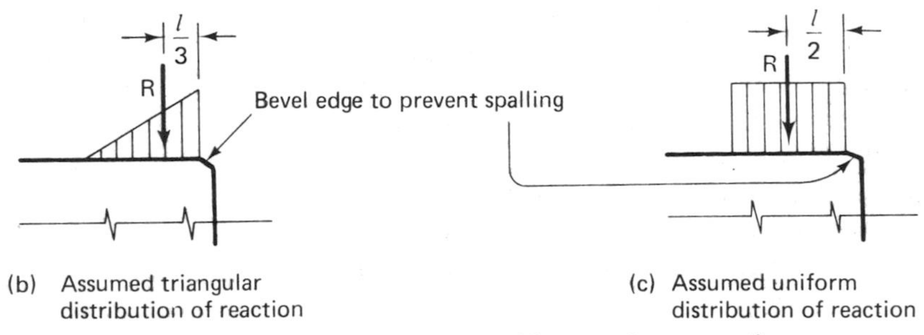

(b) Assumed triangular distribution of reaction

(c) Assumed uniform distribution of reaction

FIGURE 8-3. Assumed location of beam and truss reactions.

ever, the majority of columns in actual practice fall into the *inelastic* range. For these *intermediate members*, some part of the material reaches the inelastic state prior to buckling. Thus, we have what is referred to as the inelastic buckling failure mode. The inelastic buckling load for a homogeneous section is given by the tangent-modulus relation,

$$\left(\frac{P}{A}\right)_{cr} = \frac{\pi^2 E_t}{(kh/r)^2}$$

where E_t = modulus of elasticity (tangent modulus) corresponding to the critical stress at buckling, and, thus, a function of the stress strain characteristics of the material.

Note the qualitative limits of these buckling modes in Figure 8-4.

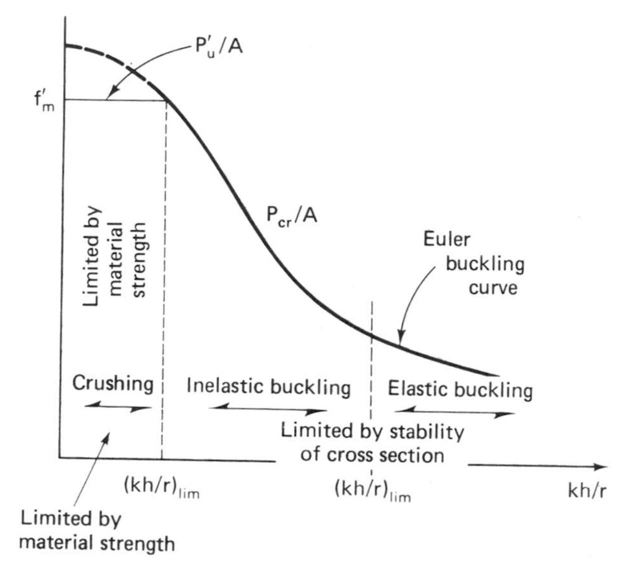

FIGURE 8-4. Load-carrying capacities of columns versus kh/r.

At the other end of the compression spectrum lie those short compression "blocks" wherein instability does not occur at all, since failure in these cases results strictly through reaching the ultimate compression strength of the material itself.

Actually, for a homogeneous material such as structural steel, the distinction between elastic and inelastic performance is sharply defined. In a nonhomogeneous material such as reinforced concrete or masonry, this distinction becomes somewhat blurred, since no well-defined limiting stress (i.e., yield) occurs; thus, the ultimate load capacity of the column, regardless of the limiting material factors (i.e., yielding of the tensile steel or crushing of the masonry), really becomes the significant design criterion.

Extensive research has been conducted on both steel and concrete columns in an attempt to define the ultimate load capacity when the inelastic buckling failure mode obtains. The modified Euler formula (tangent modulus), although theoretically sound, would be impractical to apply under actual design circumstances, since the tangent modulus, E_t, is a function of the critical buckling stress being sought.

Many attempts have been made to formulate practical column design equations for most materials; i.e., ones that will reasonably reflect the buckling behavior seemingly indicated by the curves produced from extensive laboratory test data. The recent AISC column design deliberations produced some rather precise evaluations of this buckling performance. The ACI Code also provides for evaluation of the ultimate capacity of reinforced concrete columns. In this approach, one must establish at the outset the "balanced state"; that is, where the tensile steel yields at the same time that the compression concrete reaches its limiting crushing strain, defined by the ACI Code as 0.003 in./in. Knowing this physical boundary, then, the designer proceeds to project whether the actual column being analyzed or designed is governed by the limiting yield stress of the tensile reinforcing steel or the limiting compressive strain of the concrete, before the final strength capacity or size determination can be made. The use of such design aids as the computerized ACI interaction diagrams greatly facilitates this procedure.

Unfortunately, this degree of precision is unavailable to the designer of reinforced masonry, since not enough in the way of performance research has been conducted on full-size reinforced masonry columns to enable anyone to formulate the needed rational ultimate capacity expressions upon which a proper design could be based. So presently we have to rely solely upon empirical Code design formulas which ostensibly employ sound but arbitrarily selected safety factors to define a "safe" working load for a short column. Then, to account for the added moment introduced by the lateral displacement of the deflected column shape (the $P\Delta$ effect), the Code inserts a load-reduction factor against this permissible working load, as was done with masonry bearing walls. As with walls, the simplified interaction (unity) formula is often applied to check the combined stress condition.

Effective column height

Another factor that must be considered stems from both the effect of column end restraint (connection fixity) and whether or not lateral deflection (sidesway) occurs at the top of the column. In the Euler expression defining column buckling load, the basic longitudinal dimension h' is the distance between pinned ends, so the effective column height becomes the distance between points of inflection (equivalent to pinned ends), as their locations are modified by the particular conditions of end restraint. If the column is laterally supported, top and bottom, in the direction of both principal axes and is not restrained by the connection, it is considered pin-ended and the value of h', the effective column height, becomes the actual distance between supports. On the other hand, if the column is somehow connection-restrained at either, or both, ends, then h must be converted into an equivalent value, $kh = h'$. Figure 8-5 gives the effective column heights for various restraint conditions. Should a column be braced from another member, such as a stiff roof girder, the unsupported height becomes the distance from the braced point to the bottom of the column. Note that the conditions portrayed serve to illustrate the fact that compression members, free to buckle in a sidesway mode, are always considerably weaker than those braced against sidesway. To simplify the analysis, the recommended values for k, given in Table 7-3, may be used for most masonry columns, as well as walls, unless one can reliably spell out the actual conditions of end restraint.

CODE REQUIREMENTS

Dimensional limits

The dimensional limitations of masonry columns specified in Section 2418(k), relate to the following items: (1) the least cross-sectional dimension is 12 in., except that this value may be reduced to 8 in. minimum, provided that the design is

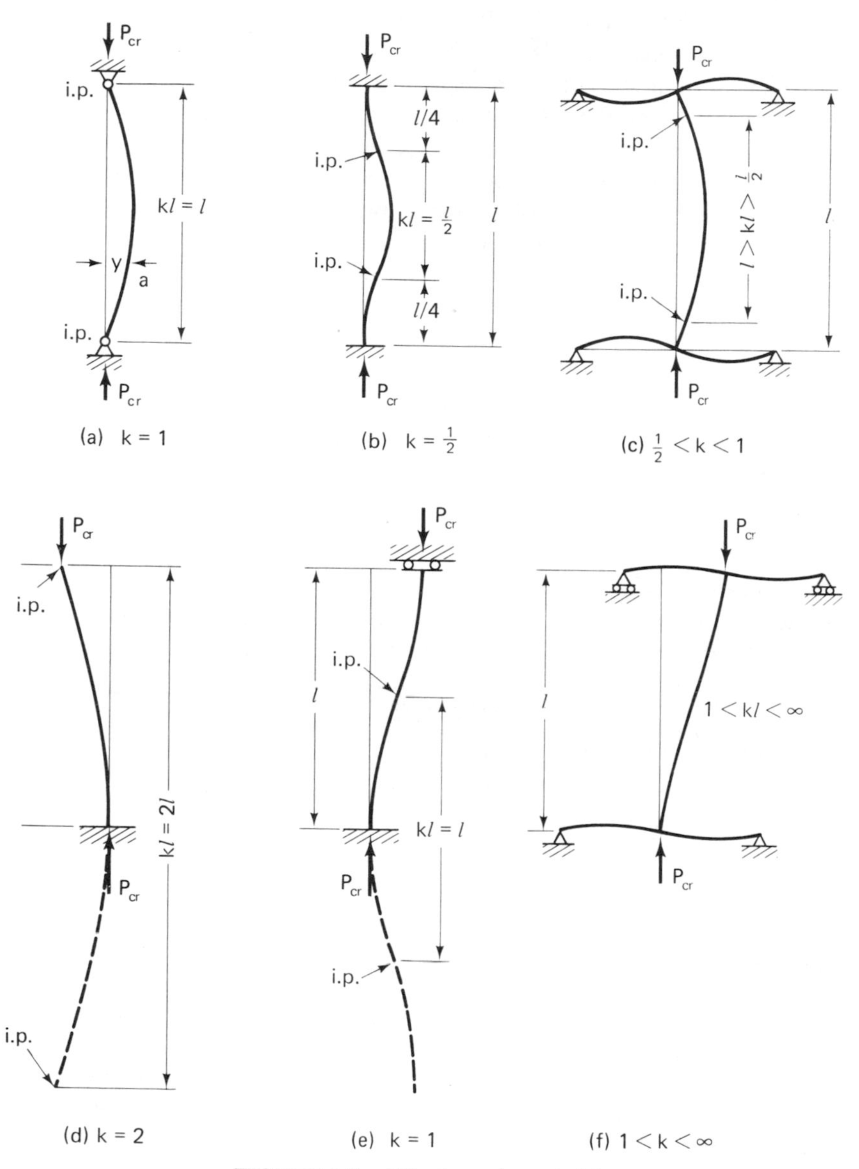

FIGURE 8-5. Effective column heights.

then based upon one-half the allowable stress for axial loads only (i.e., the allowable bending stresses need not be so reduced), (2) the unsupported column height must not exceed 20 times its least cross-sectional dimension, and (3) any length of bearing wall that is shorter than three times its thickness must be designed as a column rather than as a pier or length of wall. The latter ratio is not a real nor an intrinsic limit but has been established as an arbitrary limit between a column and a pier, and it may be varied by valid engineering considerations.

UBC Section 2419(c) defines the effective height (h') of a column as its actual height when it is provided with lateral supports in the directions of both principal axes at the upper and lower column ends. Further, the actual height is to be taken as the clear distance between the floor surface and the underside of the deeper beam framing into the column from either direction at the upper end of the column. Where a column is provided with lateral support in the directions of both principal axes at the bottom, and in the direction of only one principal axis at the top, its effective height, relative to the direction of the top support, should be taken as the height between supports. But its effective height at right angles to this direction must be taken as twice the height above the lower support. This rule has its basis in the theoretical case cited in the preceding section, wherein a column, restrained at the bottom end and laterally unsupported at the top, has an effective height equal to twice the actual height (Figure 8-5d). Also, if there is no support at the top, the effective height of the column with respect to both principal axes must be taken as twice its height above the lower support.

Load capacity

The Code presumes that the load resistance of columns, unlike walls, is proportioned between the gross area of the masonry and the area of the vertical reinforcing provided. The allowable axial load (P) is therefore prorated as follows:

$$P = A_g(0.18f'_m + 0.65p_g f_s)$$

where A_g = gross area of masonry column section
p_g = ratio of area of vertical reinforcement to A_g (max. = 4%, min. = 0.5%)

The same stress-reduction factor used for walls, to account for the buckling effect, applies to columns as well [i.e., $1 - (h/40t)^3$]. Refer back to the section "Axial Stress Limits" in Chapter 7 for a discussion regarding the reliability of this factor.

Reinforcement

Regarding Code requirements for vertical column reinforcement, Section 2418(k) stipulates arbitrarily that the ratio p_g shall not be less than 0.5% or more than 4%. The number of bars must not be less than 4, with the minimum bar diameter being $\frac{3}{8}$ in. and the maximum a No. 10. Where the vertical reinforcement is lapped, the length of the lap must be sufficient to transfer the working stress by bond, but in no case shall the length of the lap splice be less than 30 bar diameters. Welded splices are permitted, but they must be full butt welds.

Ties must be placed not less than $1\frac{1}{2}$ in. or more than 5 in. from the surface of the column. They may be placed directly against the vertical bars, if feasible, or placed in the horizontal bed joints. All vertical bars must be stayed laterally by the corner of a complete tie having an included angle of not more than 135°, or by a hook at the end of the tie. The corner longitudinal bars must have such support provided by a complete tie enclosing the longitudinal bars, as shown in Figure 8-6a. Additional

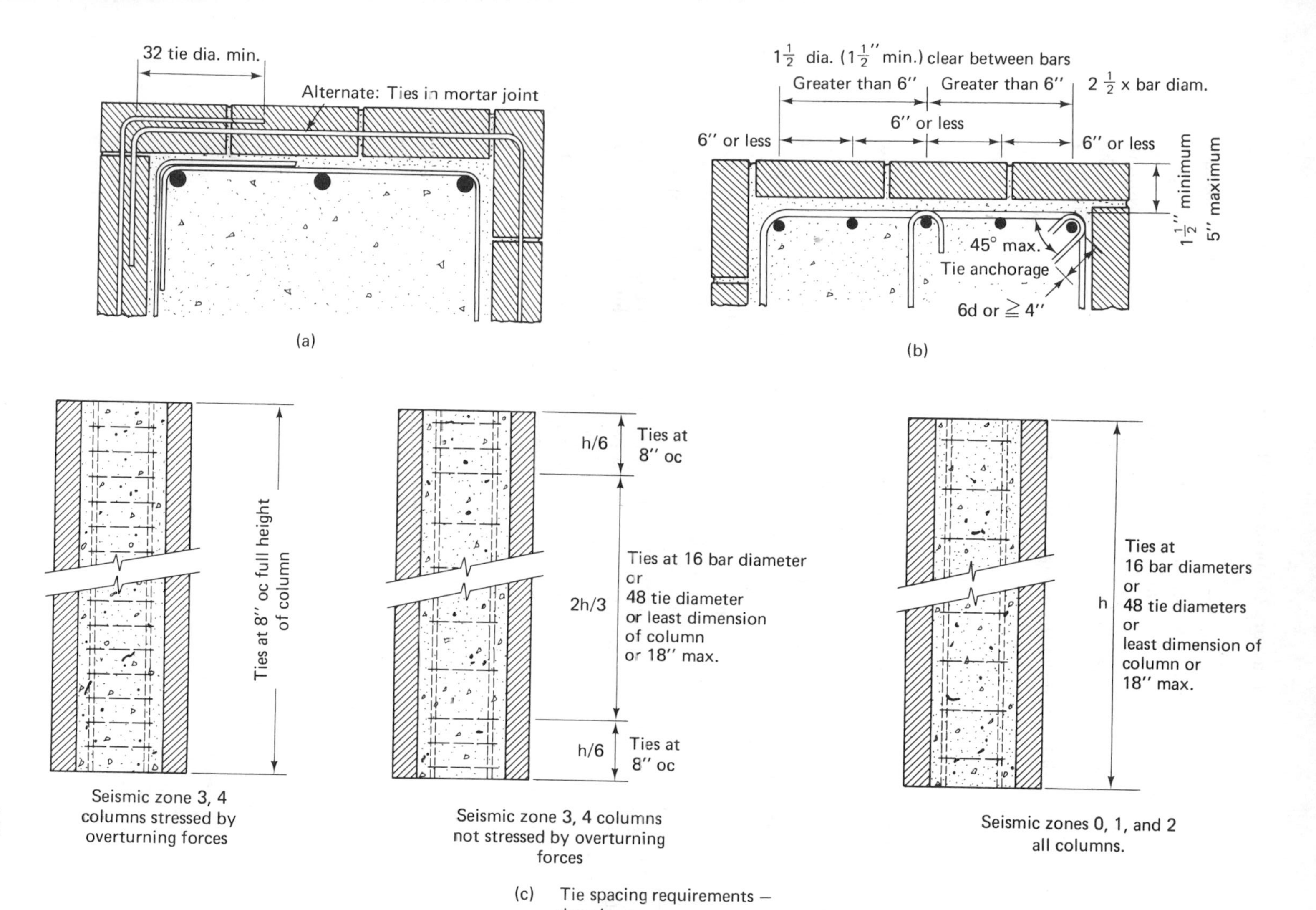

FIGURE 8-6. Column reinforcing details: (a) column reinforcing details—sections, seismic zones 0, 1, 2; (b) column reinforcing details—sections, seismic zones 3 and 4; (c) tie spacing requirements—elevations.

arbitrary requirements in Seismic Zones 3 and 4 state that alternate longitudinal bars must have such lateral support provided by ties, and no bar shall be further than 6 in. from such laterally supported bars, as noted in Figure 8-6b. Furthermore, in these zones, the maximum tie spacing is limited to 8 in. cc for the full height of any column subjected to overturning forces caused by seismic loads. For all other columns, the maximum spacing of 8 in. cc must extend over a distance of one-sixth of the clear height at the top and bottom of the column, but not less than 18 in. or the maximum column dimension. Tie spacing for the remaining height is not to exceed 16 bar diameters, 48 tie diameters, or the least dimension of the column, whichever is the least value, with 18 in. being maximum spacing in any case. The latter spacing limitation also applies to the full height of columns in seismic zones 1 and 2. These spacing limits are similar to the ACI values for reinforced concrete columns, and were established for masonry simply for that reason, not because of any valid test data. See Figure 8-6c, which illustrates these spacing requirements. The minimum tie size is $\frac{1}{4}$ in. in diameter for No. 7 or smaller longitudinal bars, with a No. 3 minimum being specified for No. 8, No. 9, or No. 10 longitudinal bars. An exception to these tie size limits is permitted where they are located in the bed joints. When permitted per Section 2414(b), they may be smaller, but not less than $\frac{1}{4}$ in. in diameter, provided that the total cross-sectional area of such smaller ties crossing a vertical plane is equal to the area of the larger ties at their required spacing. Additional ties (two No. 4 or three No. 3) must be provided around all anchor bolts which are set in the top of columns for buildings located in seismic zones 2, 3, and 4. Such ties must engage at least four bolts or four vertical column bars, and they are to be located within the top 5 in. of the column.

To assist one in determining how much space is required to accommodate different bar combinations in concrete masonry columns, refer to Table 8-1. It gives a

TABLE 8-1

Spacing Requirements to Accommodate Longitudinal Column Steel

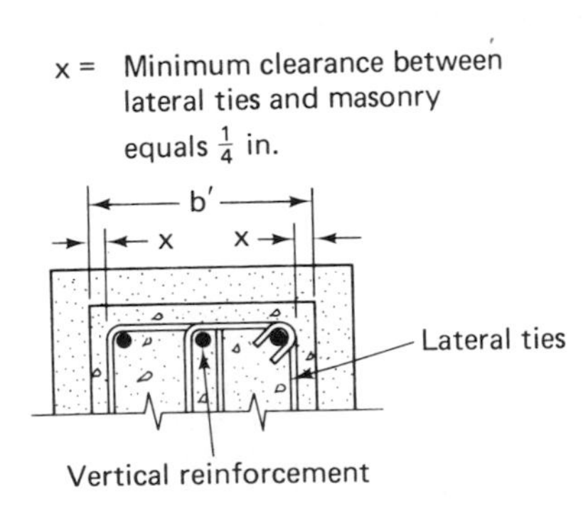

Size of vertical bars	*b′. Minimum width (in.) Number of bars at one face*				
	2	*3*	*4*	*5*	*6*
No. 4	4.0	6.0	8.0	10.0	12.0
No. 5	4.25	6.375	8.5	10.625	12.75
No. 6	4.75	6.75	9.0	11.25	13.5
No. 7	4.75	7.125	9.5	11.875	14.25
No. 8	5.0	7.5	10.0	12.5	15.0
No. 9	5.45	8.25	11.10	13.90	16.75
No. 10	5.96	9.125	12.30	15.50	18.65
No. 11	6.45	9.95	13.50	17.00	20.55

Note: The values in the table are based on the assumption that the maximum size of coarse aggregate will not exceed one inch and lateral reinforcement consists of No. 3 bars.

minimum width needed in a vertical plane parallel to one face of a column or pilaster for various bar quantities.

DESIGN CONSIDERATIONS

Columns

Often, masonry column dimensions are established by other than structural needs, such as architectural or construction features or connection dimensional limits and their necessary clearances. Whatever the circumstances, the design of masonry columns is not a very precise procedure, nor is it a correct one. It merely consists of solving some arbitrary equations with the appropriate f'_m values, f_s values, physical size dimensions, and modifying the assumed design load by the appropriate reduction factor. It is conservative regarding masonry compressive strength; for example, the use of an allowable masonry stress of $0.18f'_m$ or 270 lb/in.^{2} for 1500-lb/in.^{2} masonry yields a high factor of safety of 5.6. It is not precise regarding the calculation of area of reinforcing, because the changes in amount of reinforcing available vary in rather large incremental steps. Also, the requirement for number of bars is at least 4, not smaller than No. 3, with an understanding that the distribution of bars will be symmetrical and that they all will be the same size. Furthermore, any great degree of precision in design calculations is not justified by any correctness of theory regarding the strength and function of the steel, since the assumption of $0.65f_s$ for the working steel stress is a violation of the assumed elastic behavior. For example, if column steel, with a modulus of elasticity of 30,000,000 lb/in.^{2}, is strained sufficiently to produce a stress of $0.65 \times 24{,}000 = 15{,}600$ lb/in.^{2}, the same strain in the adjacent masonry would produce a stress of $15{,}600/n$, which might vary from $15{,}600/10 = 1560$ lb/in.^{2} to $15{,}600/44 = 355$ lb/in.^{2}, compared with the corresponding 540 to 121 lb/in.^{2} permitted by the Code factor of $0.18f'_m$. The value of $n = E_s/E_m$ varies from 10 to 44, and is based on $E_m = 1000f'_m$, an assumption that may be considerably in error. In view of these uncertainties, it would be inappropriate to carry out the calculations to more than two significant figures, until further research defines truer values.

The calculations for the Code-permitted column capacity can be quite simple. One method uses an equivalent load, by increasing the design load with the reduction factor, and then checking a combination of masonry capacity plus steel capacity to verify adequacy. The masonry capacity is ($0.18f'_mA_g$) and the steel capacity is ($0.65f_sA_s$), with A_s limited between $0.005A_g$ and $0.04A_g$. The f_s permitted is 40% of the yield, but not to exceed 24,000 lb/in.^{2}.

There are some simple design aids which facilitate a selection. These may assume the form of load tables which provide the separate load-carrying capabilities of both the masonry section (for various f'_m values) and the minimum and maximum amounts of vertical steel. Tables F-1 and F-2 and Curve F-3 in the Appendix constitute useful examples for column design aids. Tables F-1 and F-2 consist simply of a tabulation of the loads carried by the masonry (for various f'_m values) and the reinforcing (for various bar combinations) based upon the UBC-permitted load-capacity expression for columns. Curve F-3 was devised by the NCMA to evaluate this same expression graphically in terms of a given f'_m and f_s, providing a very convenient way to use this expression. Note that steel requirements are expressed in terms of the steel ratio, p_g.

Where combined bending plus axial compression occurs, the basic interaction formula, previously set forth for walls will apply:

$$\frac{f_a}{F_a} + \frac{f_b}{F_b} = 1.00$$

Typically for columns, this case would arise when the vertical load, P (e.g., a truss reaction), acts eccentrically with the longitudinal axis of that member. The stress conditions should be examined at both the top of the column and at its midheight as shown in Figure 8-7, by using the following load and moment combinations:

1. Top of column: consider the combination of the axial load, P, plus the eccentric moment Pe. Here the design value of P need not be increased by the load-reduction factor, since at the end of the column no $P\Delta$ effect, as such, exists.
2. Midpoint of column: since the $P\Delta$ effect is maximized somewhere near the column midheight, increase the design load by dividing P by the load-reduction factor. Consider this load in combination with the eccentric moment at the center line, which approaches $Pe/2$.

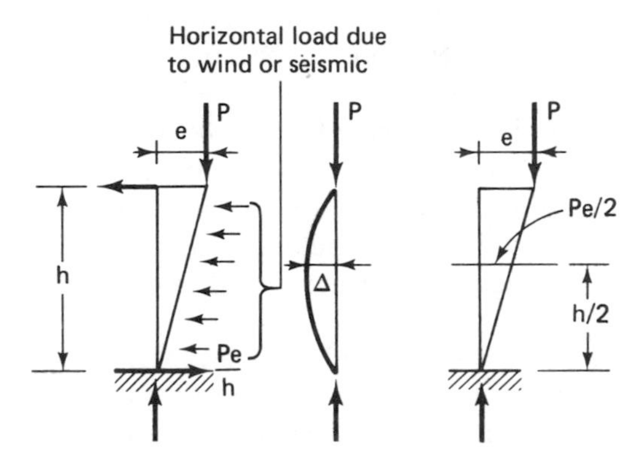

FIGURE 8-7. Eccentrically loaded column.

Pilasters

Opening Effect

As noted previously, a pilaster is essentially a stiffening element within a wall that can be considered laterally supported in the plane of the wall. Thus, since the critical buckling dimension is that perpendicular to the wall, the governing slenderness ratio, kh/r, becomes the one with respect to the axis parallel to the wall. Essentially, then, in a case of oversimplification, the pilaster can be designed or analyzed similarly to an isolated masonry column. But special reinforcement considerations must be given to those pilasters located adjacent to any wall openings of significant size, even when the actual wall spans vertically. In that event, the portion of the wall above and below the opening must of necessity, because of the opening discontinuity, transfer all the tributary lateral loads to the pilaster by spanning horizontally as a one-way slab.

For instance, consider the case portrayed in Figure 8-8. Because of the wall discontinuities caused by the large openings on either side of the pilaster, the wall above and below the openings will have to span horizontally. The lateral reactions will therefore be carried by the pilaster in spanning vertically between floor and roof. The tributary width for the lateral load (say, wind at 20 lb/ft²) will equal 18 ft (= 20/2 + 1.33 + 13.33/2).

$$\text{moment at center of pilaster span due to wind} = 20\ \text{lb/ft}^2 \times 18 \times \frac{16^2}{8} = 11{,}520\ \text{ft-lb}$$

$$\text{moment at center of pilaster due to eccentricity of truss reaction} = 12{,}000 \times \frac{4}{12} \times \frac{1}{2} = 2000\ \text{ft-lb}$$

$$\text{design moment for pilaster} = 13{,}520\ \text{ft-lb}$$

$$\text{design load on pilaster} = (18\ \text{ft} \times 300\ \text{lb/ft}) + 12{,}000 = 17{,}400\ \text{lb}$$

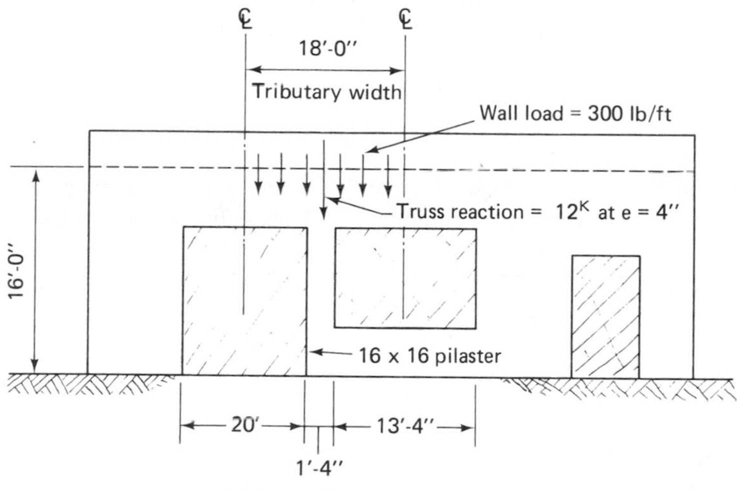

FIGURE 8-8. Pilaster adjacent to opening.

Horizontal Wall Spans

When a wall is specifically reinforced to span horizontally between pilasters, it is usually assumed to act as a one-way slab, spanning in that direction. However, in reinforced masonry, because of the Code requirement for minimum reinforcement, there will be a significant amount of vertical wall steel present. When this condition does occur, it may be appropriate to consider that the wall panel functions structurally as a two-way slab, transmitting the lateral loads to both vertical and horizontal edge supports. The amount of lateral load transmitted in either direction in that case will be a function of (1) the flexural rigidity in either direction expressed in terms of the horizontal/vertical span ratio, (2) the fixity or restraint developed at the edges, and (3) the distribution of the loads applied to the wall panels.

Curves have been developed by NCMA (Figure 8-9) which give coefficients that ostensibly approximate the proportion of the wall panel load that is transmitted horizontally to concrete masonry pilasters. Appreciation is extended to NCMA for granting permission to reproduce their approach for handling this special loading condition. For more complete details, refer to their publication "Reinforced Concrete Masonry Columns and Pilasters." Curve expressions were derived by equating the maximum deflection values for a strip of wall in both directions. The coefficients (k) are then used in conjunction with a formula that was devised to calculate the approximate proportion of lateral load carried by the pilaster expressed as w_p in lb/ft of height. They are a function of the support restraint conditions at the panel edges and the appropriate panel length/height ratio. The value of w_p is then determined approximately from the formula

$$w_p = kw_dA$$

where w_d = design wind or seismic lateral load perpendicular to wall panel, lb/ft²
A = horizontal span, center to center of pilasters, ft
w_p = approximate lateral load transferred to the pilaster by the wall panel, lb/ft of pilaster height
k = coefficient varying with horizontal to vertical span ratio and support conditions

Note that in this procedure (see Figure 8-10), there are three degrees of panel edge restraint at the vertical pilaster boundary: (1) walls fixed at the pilasters,

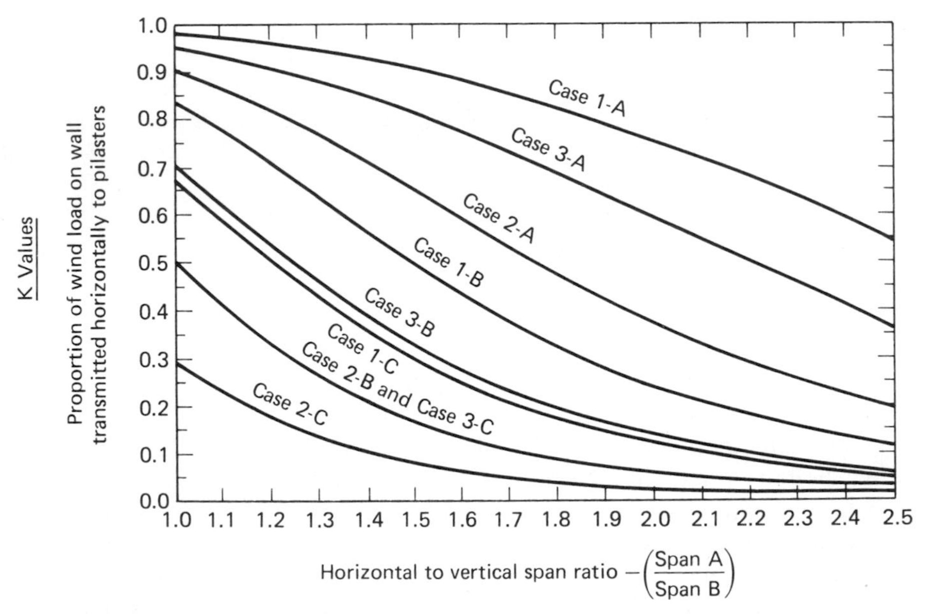

FIGURE 8-9. Proportion of horizontal panel load transmitted to pilasters. Approximate method of calculating wind load to be carried by pilasters.

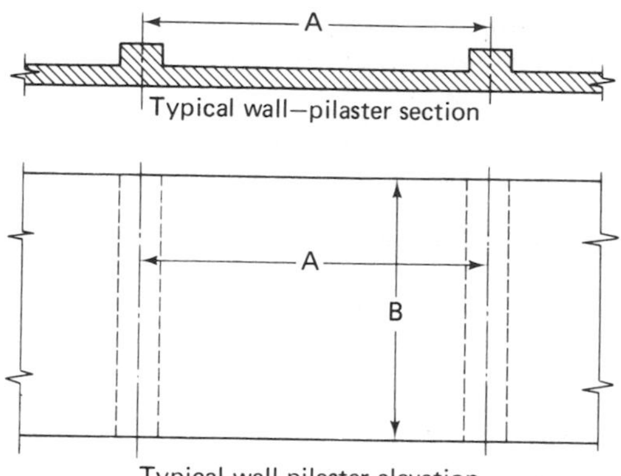

Typical wall-pilaster elevation

Case 1: Walls fixed at pilasters

A. Fixed at bottom, free at top
B. Supported top and bottom
C. Fixed at bottom, supported at top

Case 2: Walls supported at pilasters

A. Fixed at bottom, free at top
B. Supported top and bottom
C. Fixed at bottom, supported at top

Case 3: Walls fixed at one end. Supported at other

A. Fixed at bottom, free at top
B. Supported top and bottom
C. Fixed at bottom, supported at top

FIGURE 8-10. Wall-panel edge restraints.

(2) walls simply supported at the pilasters, and (3) walls fixed at one end and supported at the other end. Under each of these three cases are three degrees of restraint at the other horizontal panel edges: (A) panel fixed at bottom, free at top; (B) panel supported at both top and bottom; and (C) panel fixed at bottom and supported at the top.

The following example might demonstrate this approach.

EXAMPLE 8-1: Find the Shear and Moment on Pilaster Due to Lateral Load on Wall Panels

Assume a building with concrete masonry exterior walls extending 12 ft between the foundation and the roof diaphragm, subjected to a wind pressure of 20 lb/ft². The walls span 18 ft cc between pilasters, which are built as an integral part of the concrete masonry wall. The roof loads will be carried by trusses simply supported on the pilasters; thus, the wall edges may be considered supported at the top and fixed at the bottom, and at the pilasters the continuous panel edges are assumed fixed. Find the value $k = 0.3$ from the curve for case 1-C with a wall length/height ratio 18:12, or 1.5. Then the value of w_p will be determined as

$$w_p = 0.3 \times 20 \text{ lb/ft}^2 \times 18 \text{ ft} = 108 \text{ lb/ft}$$

The value of 108 lb/ft then becomes the uniformly distributed load applied over the full height of the pilaster. The moment and shear developed in the pilaster as a result of this load will depend upon assumptions made as to the manner in which the pilaster is supported at its top and bottom. In one-story construction, for example, typically one might consider that it is supported at the top and fixed at the bottom, in which case the design moment and shear become

$$M_{\max} = \frac{w_p h^2}{8} = \frac{108 \times 12^2}{8} = 1944 \text{ ft-lb}$$

$$V_{\max} = \frac{5 w_p h}{8} = \frac{5 \times 108 \times 12}{8} = 810 \text{ lb}$$

COMBINED LOAD EFFECTS

Should the pilaster be required to provide vertical support for a truss or beam, it may be subjected to combined axial load and bending. Say for example, that the preceding pilaster must carry a vertical concentrated load of 25 kips. If the load transmitted to the pilaster by the supported beam or truss is eccentrically applied, which is generally the case, the moment due to that eccentricity must be added to the lateral-load moment to obtain the total design moment. These values are additive, since it is generally considered that the lateral loads can act on either side of the wall.

EXAMPLES OF COLUMN DESIGN AND ANALYSIS

EXAMPLE 8-2: Design a Brick and a Concrete Masonry Column

An interior column supports an axial load of 200 kips. Determine the size of the column, vertical reinforcing steel and the tie spacing, if the column is constructed of:

(*a*) *Reinforced brick*, $f'_m = 1800$ *lb/in.*².

(*b*) *Reinforced concrete masonry* $f'_m = 1500$ *lb/in.*².

The height of the column is 30 ft and grade 40 steel is used. Assume special inspection.

Solution:

(*a*) *Reinforced brick*, $f'_m = 1800$ *lb/in.*²

This method is basically a trial-and-error solution, involving an attempt to find reasonable column dimensions and an economical amount of steel. Try 18 in. × 18 in., actual size.

$$f_s = 0.40 f_y = 0.40(40{,}000 \text{ lb/in.}^2) = 16{,}000 \text{ lb/in.}^2$$

$$R = 1 - \left(\frac{h}{40t}\right)^3 = 0.875 \text{ (Diagram E-1)}$$

$$\text{corrected } P' = \frac{P}{R} = \frac{200 \text{ kips}}{0.875} = 228.6 \text{ kips}$$

$$\text{load on masonry } P_m = 0.18 f'_m A_g = 0.18(1800 \text{ lb/in.}^2)(18 \text{ in.} \times 18 \text{ in.}) = \left(\frac{1}{1000}\right)$$

$$= 105 \text{ kips}$$

$$\text{balance of load on steel} = P' - P_m = 228.6 - 105 = 123.6 \text{ kips}$$

$$P_s = 0.65 p_g f_s A_g = 0.65 f_s A_{st}$$

$$p_g = \frac{P_s}{0.65 f_s A_g} = \frac{123600}{0.65(16{,}000 \text{ lb/in.}^2)(18 \text{ in.} \times 18 \text{ in.})} = 0.0367$$

$$\therefore\ A_{st} = 0.0367 \times 18 \times 18 = 11.9 \text{ in.}^2 \text{ provide } 10 \# 10$$

limits for p_g: $0.5\% \leq 4.0\%$

$$0.005 \leq 0.0367 \leq 0.04 \qquad \therefore \text{ OK}$$

Assume Zone 3 or 4 (column not subjected to overturning):

Maximum tie spacing: 16 bar diameters $= 16 \times 1.25 = 20$ in.

48 tie diameters $= 48 \times 0.375 = 18$ in. ← will govern for central 20 ft

column dimension $= 18$ in,

ties at 8 in. cc for top and bottom 5 ft

Try for more economical amount of steel; thus, increase the masonry section to 20 in. × 20 in. and use design Tables F.

For 20 × 20, actual size: from Diagram E-1:

$$\frac{h}{t} = \frac{30(12)}{20} = 18 \qquad \therefore\ R = 0.909$$

$$P' = \frac{200}{0.909} = 220 \text{ kips}$$

From Table F-1, load on masonry = 129 kips

balance of load on steel = 220 − 129 = 91 kips

From Table F-2, for $f_s = 16{,}000$ lb/in.², provide 12 No. 8's (load taken by steel = 98.6 kips). Note that an even number of bars was selected in order to provide a symmetrical section (four bars on a side), with only a slight excess of computed capacity.

(*b*) *Reinforced concrete masonry* — $f'_m = 1500$ *lb/in.*²

USE OF EQUATIONS AND COLUMN LOADING TABLES:

$$f_s = 16{,}000 \text{ lb/in.}^2$$

$$P = 200 \text{ kips}$$

Solution:

Use 8 in. block: $b = 7\frac{5}{8}$ in. Try 24 × 24 in. (actually, $23\frac{5}{8} \times 23\frac{5}{8}$ in.)

$$A_g = 23\tfrac{5}{8} \times 23\tfrac{5}{8} = 558 \text{ in.}^2$$

$$R = 1 - \left(\frac{h}{40t}\right)^3 = 0.945$$

$$P' = \frac{P}{R} = \frac{200}{0.945} = 211.7 \text{ kips}$$

$$P_m = 0.18 f'_m A_g = 0.18(1500)(558)\left(\frac{1}{1000}\right) = 150.7 \text{ kips} \qquad \text{(also see Table F-1)}$$

$$P_s = P' - P_m = 211.7 - 150.7 = 61.0 \text{ kips}$$

$$p_g = \frac{P_s}{0.65 f_s A_g} = \frac{61{,}000}{0.65(16{,}000)(558)} = 0.0105$$

$$0.005 \leq 0.0105 \leq 0.04 \qquad \therefore \text{ OK (may also use range shown in left of Table F-1)}$$

$$A_s = p_g A_g = 0.0105 \times 558 = 5.87 \text{ in.}^2$$

USE NCMA COLUMN LOAD TABLES:

Refer to column load Curve F-3. Find axial $P = 212$ kips on the ordinate of the chart. Intersect the A_g curve for 560 in.². Read the abcissa $p_g = 0.011$ for A_{st} of about 6.0 in.². The results are similar to those obtained with the basic UBC strength formula.

EXAMPLE 8-3: Design an Eccentrically Loaded Brick Column

A brick column supports a vertical load of 70 kips with an eccentricity of 8 in. The $f'_m = 1800$ lb/in.² and $f_s = 16{,}000$ lb/in.²; $n = 16.7$, with special inspection provided. The height of the column is 36 ft. Determine the column size and vertical reinforcement.

$$P = 70 \text{ kips}$$

$$M = Pe = (70)(8) = 560 \text{ in.-kips}$$

Solution:

The procedure here simply involves the selection of (1) a column size, related somewhat to the magnitude of the axial load, and (2) an economical p_g, perhaps between 1 and 2%. Then check this selection for load capacity.

Try 24 in. × 24 in. and six No. 9 bars ($A_s = 6.00$ in.²):

$$A_g = 576 \text{ in.}^2$$

$$p_g = \frac{A_s}{A_g} = \frac{6.00}{576} = 0.0104 \qquad \therefore \text{ OK}$$

$$R = 0.909$$

$$F_a = [0.18(1800) + 0.65(0.0104)(16{,}000)]0.909 = 392.8 \text{ lb/in.}^2$$

$$F_b = 600 \text{ lb/in.}^2$$

$$f_a = \frac{70(1000)}{576 + (6.00)(16.7 - 1)} = 104.4 \text{ lb/in.}^2$$

$$f_b = \left(1 - \frac{104.4}{392.8}\right)600 = 440 \text{ lb/in.}^2 \text{ remaining for bending}$$

$$b = 24 \text{ in.} \qquad d = 24 - 4 = 20 \text{ in.}$$

Assume that two No. 9 bars are located at each face of the column parallel to the axis of bending (see the sketch on page 251)

$$A_s = 2 \times 1.00 = 2.00 \text{ in.}$$

$$p = \frac{A_s}{bd} = \frac{2.00}{24(20)} = 0.0042$$

$$np = 16.7(0.0042) = 0.0696 \qquad \text{and} \qquad K = \frac{560{,}000}{24 \times 20^2} = 58.3$$

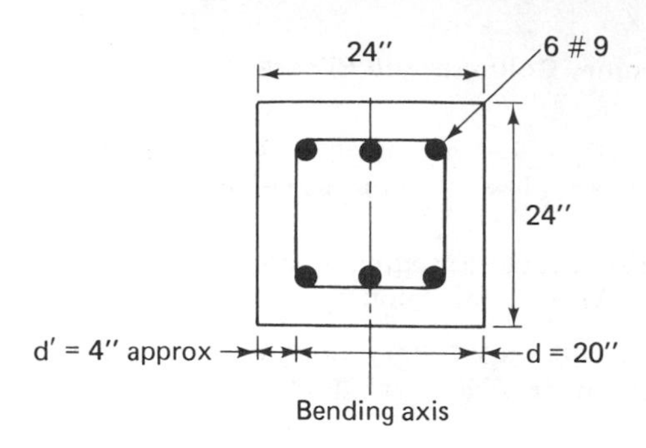

$$k = \sqrt{2(0.0696) + (0.0696)^2} - 0.0696 = 0.309 \qquad \text{(also from Table C-1)}$$

$$j = 1 - \frac{k}{3} = 0.897$$

$$f_b = \frac{2(560)(1000)}{(0.897)(0.309)(24)(20)^2} = 420.9 \text{ lb/in.}^2 < 440 \text{ lb/in.}^2 \text{ allowable}$$

$$f_s = \frac{560(1000)}{(2.00)(0.897)(20)} = 15{,}600 \text{ lb/in.}^2 < 16{,}000 \text{ lb/in.}^2 \qquad \therefore \text{ OK}$$

Thus, it would appear that as economical a column size and reinforcement ratio have been selected as is permitted under the present design rules.

Or refer to Curves B-1 and B-3:

Curve B-3, for $K = 58.3$ and $np = 0.07$, find $f_m = 425$ lb/in.^{2}.
Curve B-4, for $nK = 974$ and $np = 0.07$, find $f_s = 15{,}500$ lb/in.^{2}.

EXAMPLE 8-4: Determine the Allowable Axial Load on a Brick Column

A 10 in. × 18 in. brick column is reinforced with four No. 9's. The allowable steel stress is $f_s = 16{,}000$ lb/in.^{2}. Assume that the column height is the maximum allowable. Assume full inspection. Determine the vertical load if axially loaded. Use $f'_m = 1500$ lb/in.^{2}, $n = 20$.

Solution:

Assume the maximum $h/t = 20$; therefore, $h = 20t = 20(10) = 200$ in. $= 16.67$ ft.

$$f'_m = 1500 \text{ lb/in.}^2$$

$$f_s = 16{,}000 \text{ lb/in.}^2$$

$$A_{st} = 4.00 \text{ in.}^2 \quad \text{(four No. 9's)}$$

$$A_g = (10)(18) = 180 \text{ in.}^2$$

$$p_g = \frac{A_{st}}{A_g} = \frac{4.00}{180} = 0.0222$$

$$R = 1 - \left(\frac{h}{40t}\right)^3 = 1 - \left[\frac{200}{40(10)}\right]^3 = 0.875$$

$$F_a = (0.18f'_m + 0.65pgfs) \times R = [0.18(1500) + 0.65(0.0222)(16{,}000)]0.875 = 438 \text{ lb/in.}^2$$

Since the minimum dimension permitted for full stresses is 12 in., then, per UBC, reduce the F_a above to $\frac{1}{2} \times 438 = 219$ lb/in.^{2}.

$$P = F_a A_g = (219)(180) = 39.4 \text{ kips}$$

From Curve F-3: For $A_g = 180$ in.^{2} and $p_g = 0.0222$, find an axial load of 90 kips. Allowable $P = 0.875 \times 90 \text{ kips} \times \frac{1}{2} = 39.4$ kips, as before.

EXAMPLE 8-5: Compute the Allowable Eccentric Load on a Concrete Masonry Column

Consider a 16 in. × 16 in. concrete masonry column reinforced with four No. 8's. Use full allowable masonry stresses to determine what vertical load it can carry at an eccentricity of 8 in.

Solution:

1. Solve using only the basic interaction formula. $d' = 3.5$ in., so $d'/t = 3.5/15.62 = 0.224$; $e/t = 8/15.62 = 0.51$; $d = 15.62 - 3.5 = 12.1$ in. $f'_m = 1500$ lb/in.^{2} $\therefore\ n = 20$

$$p_g = \frac{4 \times 0.79}{15.62 \times 15.62} = 0.0130 \qquad p = \frac{2 \times 0.79}{15.62 \times 12.1} = 0.00836 \qquad pn = 0.167$$

$$F_a = 0.18 f'_m + 0.65 p_g \times 16{,}000 = 0.18 \times 1500 + 0.65 \times 0.0130 \times 16{,}000$$
$$= 405 \text{ lb/in.}^2$$

$$F_b = \tfrac{1}{3} f'_m = 500 \text{ lb/in.}^2$$

$$f_a = \frac{P}{bt + (n-1)A_s} = \frac{P}{244 + 19 \times 3.16} = 0.00329P$$

For $pn = 0.167$, $j = 0.86$, $k = 0.43$, and $2/jk = 5.39$ from Table C-1,

$$f_b = \frac{M}{bd^2} \times \frac{2}{jk} = \frac{8P}{15.62 \times (12.1)^2} \times 5.39 = 0.0189P$$

$$\frac{0.00329P}{405} + \frac{0.0189P}{500} = 1 \qquad \therefore\ 1.65P + 7.65P = 202{,}500$$
$$\text{or} \quad P = 21.80 \text{ kips}$$

2. Use the more precise NCMA design equations which reflect the influence of the compressive force upon the steel stress.
   (a) Consider a cracked section first, since $e/t > \frac{1}{3}$. Allowable load limited by masonry compressive stress.

$$P = \frac{F_a A_g}{\dfrac{F_b - F_a}{F_b[1 + p_g(n-1)]} + X_1 \dfrac{F_a}{F_b}} = \frac{325 \times 244}{\dfrac{500 - 325}{500[1 + 0.013(20 - 1)]} + 3.5\left(\dfrac{325}{500}\right)}$$
$$= 31.02 \text{ kips} \leftarrow \text{governs for this method}$$

$$\text{where } F_a = \frac{P/A_g}{1 + p_g(n-1)} = \frac{[A_g(0.18 f_m + 0.65 P_g)(16{,}000)]/A_g}{1 + p_g(n-1)}$$
$$= \frac{0.18 \times 1500 + 0.65 \times 0.0130 \times 16{,}000}{1 + 0.0130(20 - 1)} = 325 \text{ lb/ft}^2$$

$$X_1 = \frac{2k}{k^2 + p_g[(3n-1)k - (d'/t)(n-1) - n]} = 3.5$$

A determination based upon either the allowable compressive or tensile stresses in the longitudinal reinforcement reveals that neither of these factors governs.

(b) Check the uncracked section condition as well:

$$P = \frac{A_g(0.18f'_m + 0.65p_g f_s)}{1 + X_2[0.54 + 3.9p(f_s/f_m)]}$$

$$= \frac{244(0.18 \times 1500 + 0.65 \times 0.0130 \times 16{,}000)}{1 + 2.5\,(0.54 + 3.9 \times 0.0065 \times [16{,}000/1500])}$$

$$= 32.66 \text{ kips}$$

where $X_2 = e/t \dfrac{6}{1 + 3p_g(n-1)[(1 - d'/t)^2]} = 2.50$

and in this case $p = \frac{1}{2}p_g = 0.0065$

CONSTRUCTION DETAILS: TYPICAL SECTIONS

Concrete masonry pilasters may be constructed with either special or standard shapes. Typical sections are portrayed in Figure 8-11a and b. An isolated concrete masonry column is seen in Figure 8-12. Figure 8-13 shows an isometric of a typical concrete masonry pilaster and bond beam. Note location of horizontal steel

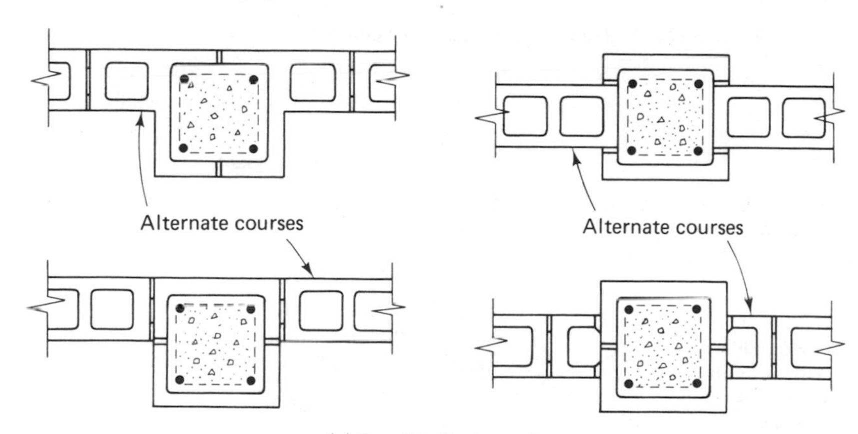

(a) Special pilaster units

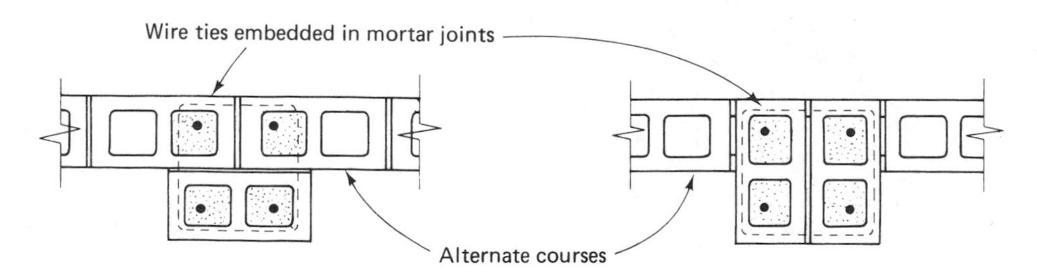

(b) Built with two core standard masonry units

FIGURE 8-11. Typical concrete masonry pilaster sections.

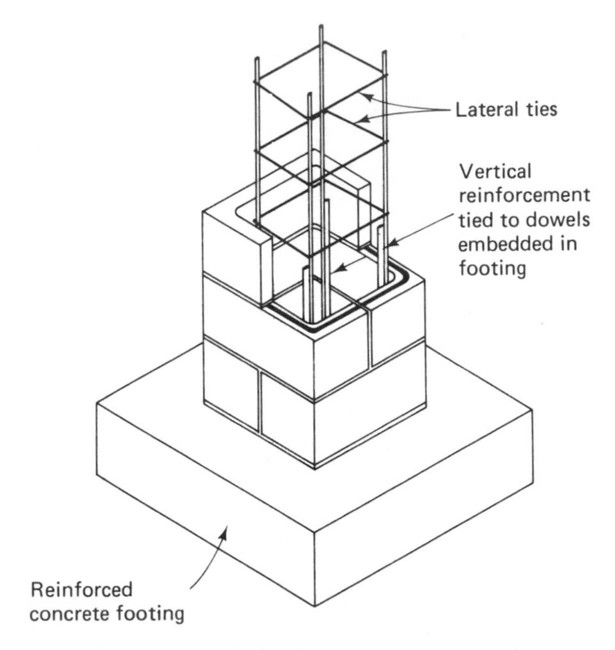

FIGURE 8-12. Isolated concrete masonry column.

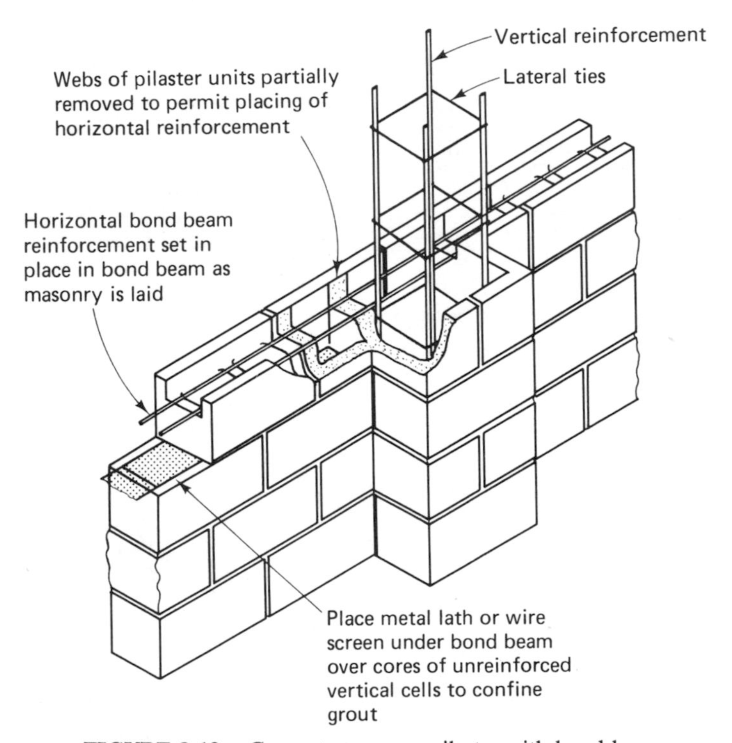

FIGURE 8-13. Concrete masonry pilaster with bond beam.

in the bond beam. Figure 8-14 shows how a vertical splice could be achieved in a continuous concrete masonry column.

Typical reinforced brick column sections are seen in Figure 8-15, with a pilaster section shown in Figure 8-16. As in concrete masonry, the lateral ties may be located in either the mortar joints between the courses or within the grout core, tied against the vertical steel, depending upon the type of masonry unit used.

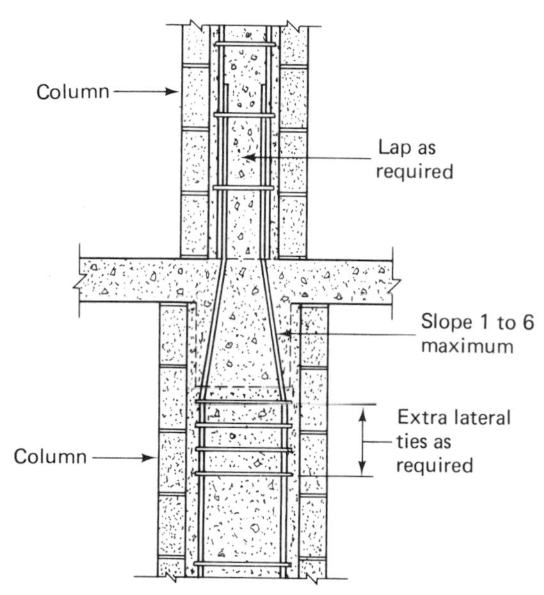

FIGURE 8-14. Vertical splice detail for concrete masonry column.

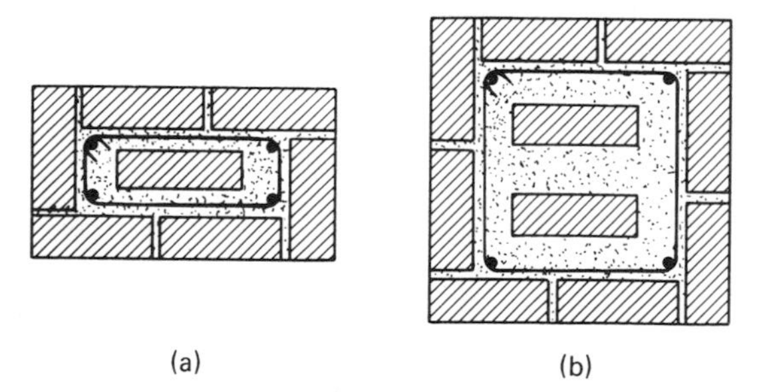

FIGURE 8-15. Reinforced brick column sections.

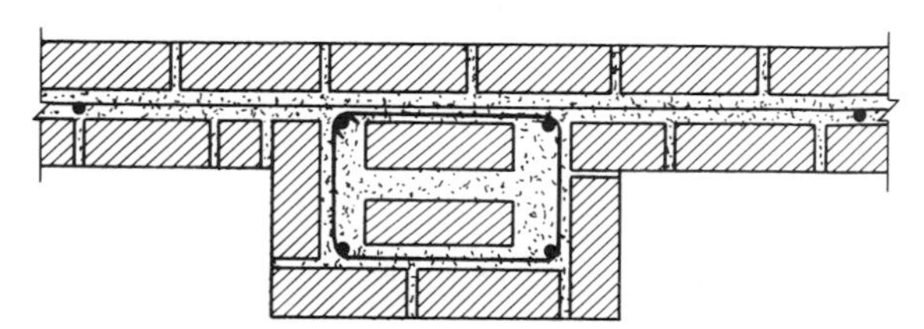

FIGURE 8-16. Reinforced brick pilaster section.

PROBLEMS

Solve the following problems using basic formulas, then check the result with the design aids.

8-1. Given a 20 in. × 20 in. brick column reinforced with four No. 10's $f'_m = 1500$ lb/in.^{2}, $f_s = 16{,}000$ lb/in.^{2}. Assume that column height is the maximum allowed. Determine the allowable load (with and without special inspection).

8-2. Same as Problem 8-1 except that the load is applied at an eccentricity of 4 in.

8-3. Given a 10 in. × 18 in. × 16 ft high brick column with four No. 8's. $P = 15$ kips; $e = 3$ in. parallel to the longer side, $f'_m = 1500$ lb/in.^{2}, $f_s = 16{,}000$ lb/in.^{2}. Is the column adequate (with and without inspection)?

8-4. Consider a square interior brick column with $h = 20$ ft, $P = 160$ kips, $e = 0$, $f'_m = 1500$ lb/in.^{2}, $f_s = 16{,}000$ lb/in.^{2} Determine:

(a) Column size and steel reinforcement. (Assume that $p_q \sim 1.5\%$.)

(b) Specify the ties.

8-5. Same as Problem 8-4 except that the column is concrete masonry, with $f'_m = 1500$ lb/in.^{2}.

8-6. Repeat Problem 8-4 with $e = 3$ in.

8-7. Repeat Problem 8-5 with $e = 3$ in.

9

Rectangular Beams

A *lintel* is simply a horizontal beam carrying vertical loads across an opening (Figure 9-1). It may be constructed of structural steel shapes, concrete, or reinforced masonry itself. Other types of beam terminology encountered in masonry construction include roof or floor beams, bond beams, and grade beams. A bond beam generally occurs near the roof or floor level. It often serves a dual function in that (1) it tends to tie the building together around its perimeter, and (2) it may also serve as the diaphragm chord member wherein the moment resistance of that element is developed. It further can be called upon to carry wall, roof, and floor loads, if the wall below is to be nonbearing in function. A grade beam, usually reinforced concrete, is simply a beam at ground level which may serve as a foundation supporting the wall above by spanning between isolated footing pads. This chapter will be concerned with the principles and procedures involved in designing the various types of masonry beams.

It might be pointed out that the primary beam design factor involved herein deals with the vertical load-carrying capability of simple or continuous beams as they bear on assumed nonyielding supports. This means that no continuity is developed between beam and column or wall supports. Thus, we are dealing typically

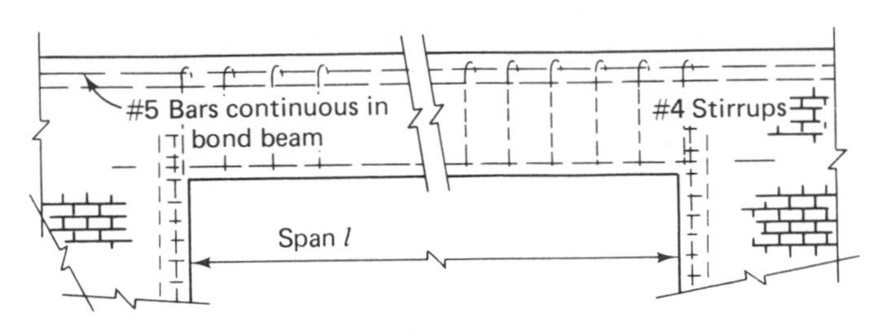

FIGURE 9-1. Lintel elevation.

with a system wherein the beams and columns are called upon to resist vertical loads only and the lateral stability and stiffness is provided by shear elements, rather than through beam-column continuity.

CODE REQUIREMENTS

The Codes have adopted certain rules for reinforcing masonry beams that are based largely upon earlier experience with concrete beam design and construction. They should be used with caution, heeding the warning that since they are arbitrary and empirical in nature, they may be different in different locales.

UBC Section 2418(i) requires that the tensile negative reinforcement in any span of a continuous, restrained, or cantilever beam must be adequately anchored by bond, hooks or mechanical anchors, or through the supporting member. Within any such span, every reinforcing bar, except in a lap splice, whether required for positive or negative moment, must extend at least 12 bar diameters beyond the point at which it is no longer needed to resist tensile stress.

No flexural steel can be terminated in a tension zone unless one of the following conditions is satisfied:

1. The actual shear does not exceed one half of that normally permitted, including the allowance for shear reinforcement, if any.
2. Additional stirrups, in excess of those required, are provided each way from the cutoff, a distance equal to the depth of the beam. The stirrup spacing shall not exceed $d/8r_b$, where r_b is the ratio of the area of bars cut off to the total area of bars at the section.
3. Continuing bars provide double the area required for flexure at that point or double the perimeter required for flexural bond.

At least one-third of the total reinforcement provided for the negative moment at the support must extend beyond the extreme position of the point of inflection a distance either (1) sufficient to develop, by bond, one-half the allowable stress in such bars, or (2) equal to one-sixteenth of the clear span length, or (3) equal to the depth of the member, whichever of these three limits is the greatest. The tension in any bar must be developed by bond on a sufficient straight or bent embedment, or by any other permitted anchorage. Table 9-1 shows the embedment length

TABLE 9-1

Length of Embedment Without Hook or Bend[a]

	$f_s = 20{,}000$ lb/in.²		$f_s = 24{,}000$ lb/in.²	
Bar no.	*No special inspection* $\mu = 100$ lb/in.² (in.)	*Special inspection* $\mu = 140$ lb/in.² (in.)	*No special inspection* $\mu = 100$ lb/in.² (in.)	*Special inspection* $\mu = 140$ lb/in.² (in.)
3	19	14	23	16
4	25	18	30	22
5	31	23	38	27
6	38	27	45	33
7	44	32	53	38
8	50	36	60	43
9	56	41	68	49
10	63	45	75	54
11	69	49	83	59

[a] $l_d = \dfrac{d f_s}{4\mu}$

needed for the longitudinal tension steel in order to fully develop the given bar size (at 20,000 or 24,000 lb/in.²). Where space is limited and the full embedment length is not available, a standard hook or bend, whose dimensions are shown in a following paragraph, may be used in lieu of a part of that length, as shown in Table 9-2. Of the

TABLE 9-2

Length of Embedment with Hook or Bend[a]

	$f_s = 20{,}000$ lb/in.²		$f_s = 24{,}000$ lb/in.²	
Bar no.	*No special inspection* $\mu = 100$ lb/in.² (in.)	*Special inspection* $\mu = 140$ lb/in.² (in.)	*No special inspection* $\mu = 100$ lb/in.² (in.)	*Special inspection* $\mu = 140$ lb/in.² (in.)
3	12	8	15	11
4	16	11	21	15
5	20	14	26	18
6	23	17	31	22
7	27	20	36	26
8	31	22	41	29
9	35	25	46	33
10	39	28	52	37
11	43	31	57	41

[a] $l_d = \dfrac{d f_s}{4\mu}$

positive reinforcement in a continuous beam, not less than one-fourth of the bar area must extend along the same face of the beam into the end support a distance of at least 6 in. In simple beams, or at the freely supported end of continuous beams, at least one-third the required positive reinforcement must extend along the same face of the beam in to the support a distance of at least 6 in. Compression steel in beams and girders must be anchored by ties or stirrups not less than $\frac{1}{4}$ in. in diameter, and spaced not farther apart than 16 bar diameters or 48 tie diameters. Such ties or stirrups are required throughout the length of the compression steel, regardless of the web reinforcement requirements. These requirements are pictorially summarized in Figures 9-2 and 9-3.

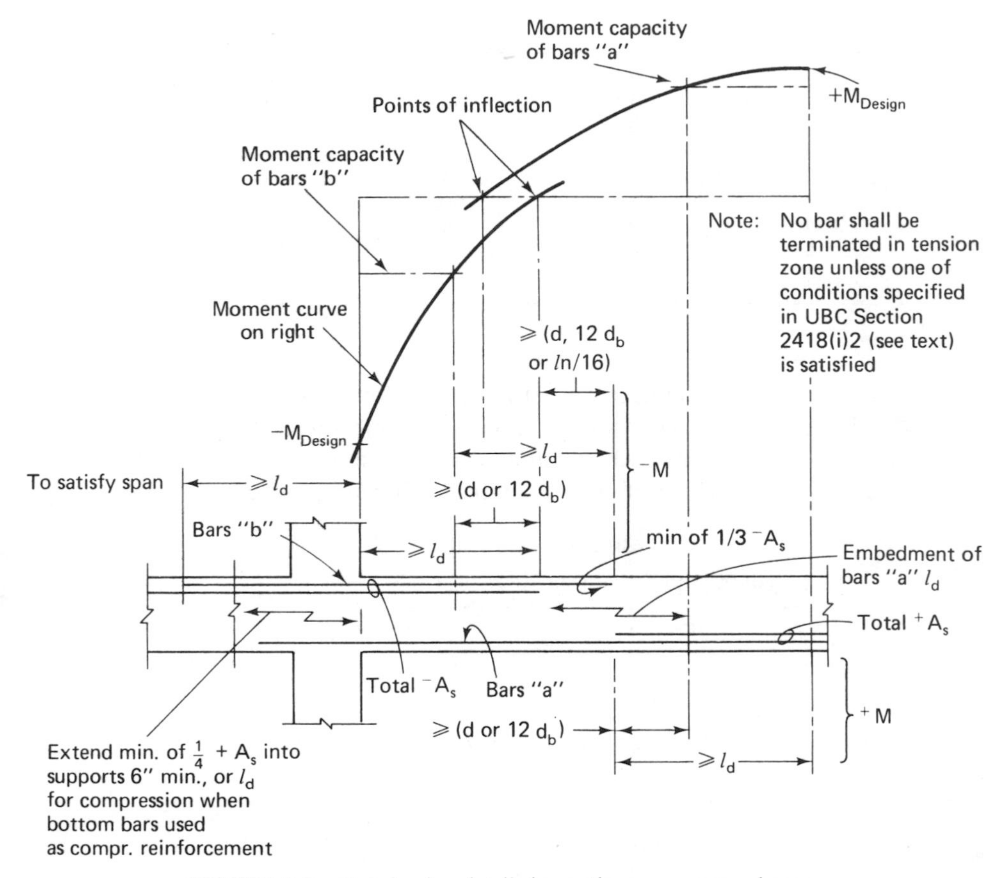

FIGURE 9-2. Reinforcing details in continuous masonry beams.

The anchorage of the web reinforcement is exceedingly important, if it is to be depended upon to develop any diagonal tension resistance. Therefore, the Code specifies that any one of the following methods is permitted:

1. Welding to the longitudinal reinforcement.

2. Hooking tightly around the longitudinal reinforcement through at least 180°.
3. Embedment above or below the middepth of the beam on the compression side, a distance sufficient to develop the stress to which the bar will be subjected at a bond stress not to exceed the allowable bond stresses permitted by the Code (i.e., 140 lb/in.^{2}, or 100 lb/in.^{2} if no special inspection provided).
4. By a standard hook considered as developing 7500 lb/in.^{2} plus embedment sufficient to develop by bond the remaining stress in the bar at the unit allowable bond stress. The effective embedment length shall not be assumed to exceed the distance between the middepth of the beam and the tangent of the hook.

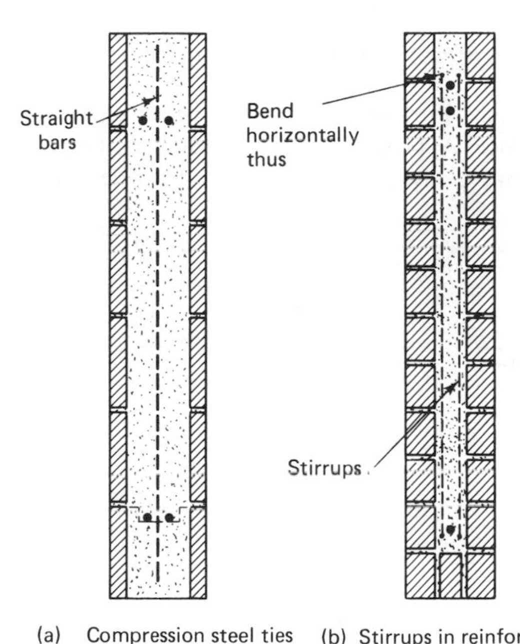

FIGURE 9-3. Typical web reinforcing in grouted masonry beams.

Figure 9-3 shows how the stirrups fit within the grouted cells of concrete masonry or in the grouted core of two-wythe construction.

The standard "hook" provides a means of bonding the reinforcement where adequate anchorage distance is not available. See Figure 9-4 and Table 9-3 for required dimensions and the tensile capacity for 180° hooks or the alternate 90° bends. If the bend radius exceeds six bar diameters, the bend length is counted as simply an extension of the bar. They are not permitted in the tension portion of any beam except at the ends of simple or cantilever beams or at the freely supported ends

TABLE 9-3

Tension Capacity and Dimensions of Bend or Hook
(at 7500 lb/in.^{2} developed stress)

Bar no.	*90° bend or 180° hook (lb)*	*D (in.)*	*J (in.)*	*A (in.)*
3	825	$2\frac{1}{2}$	3	6
4	1500	3	4	8
5	2330	$3\frac{3}{4}$	5	10
6	3300	$4\frac{1}{2}$	6	12
7	4500	$5\frac{1}{4}$	7	14
8	5920	6	8	16
9	7500	9	$11\frac{1}{4}$	19
10	9530	$10\frac{1}{4}$	$12\frac{3}{4}$	22
11	11800	$11\frac{1}{4}$	$14\frac{1}{4}$	24

of continuous or restrained beams. Further, they cannot be considered effective in adding any compressive resistance. Approved mechanical devices, capable of developing the strength of the bar, may be used in lieu of hooks, provided that tests demonstrate their adequacy.

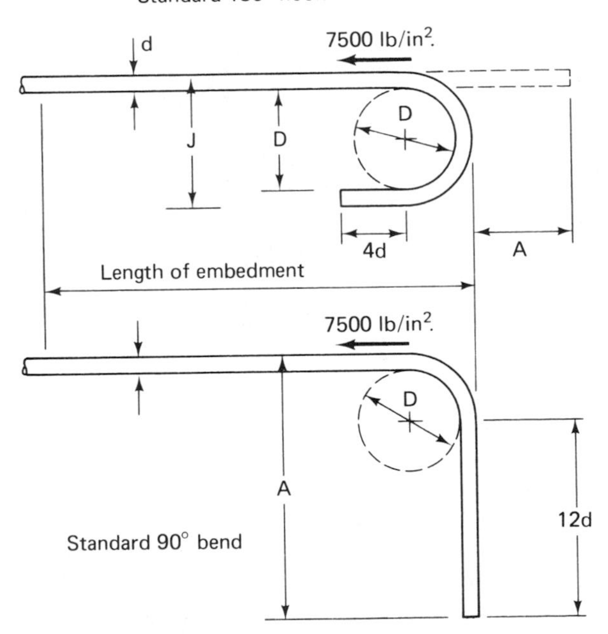

FIGURE 9-4. Standard hook and bend dimensions.

DESIGN PROCEDURES

Lintels

Loads brought to isolated beams or lintels are generally evaluated on the basis of tributary widths, where they support any roof or floor loads. For a lintel where a considerable expanse of wall extends above it, the load magnitude becomes somewhat vague, since the wall actually tends to arch over the opening, and therefore the entire wall load above the lintel may not be transferred to it. Some designers assume a 45° triangular load extending from a maximum at the center of the span to zero at the supports. Arching action of the masonry over the opening may then be counted upon to support the remaining portion of the load. It is important to recognize that the horizontal thrust resulting from any arch action must be provided for by the mass in the adjoining wall or by properly designed tension devices. When the floor loads are applied below the apex of the triangle, arching does not support the loads, so it is then assumed that all the uniform loads directly above the lintel must be carried by it. Refer to Figure 9-5.

Assuming a triangular loading ($\beta = 45°$) on the lintel, the moment at the center due to the material weight becomes

$$M \text{ at center line} = \frac{WL}{6} = \frac{L}{6}\left(\frac{w'L^2}{4}\right) = \frac{w'L^3}{24} \text{ ft-lb}$$

where W = total triangular load due to material weight, lb
w' = material weight of vertical wall surface, lb/ft²

Concentrated loads from beams and trusses above, framing into the wall, may be considered to transfer downward from the apex of a triangle whose sides make an angle of 60° with the horizontal. The load is then considered to be uniform over the base of the triangle and will probably extend over only a portion of the opening span (see Figure 9-6).

With the assumed 60° distribution (or $\alpha = 30°$ with vertical) for a concentrated load, its effect on the lintel may be measured as:

$$a = (h' \tan\alpha + 0.5L - x') = (0.577h' + 0.5L - x')$$

$$w = \frac{p'}{2h' \tan\alpha} = \frac{p'}{1.155h'}$$

where w = that part of concentrated load p' converted to an equivalent uniform load

$$\text{moment at center line} = R_L \frac{L}{2} \qquad \text{when } a < \frac{L}{2}$$

$$= \frac{wa^2}{2L}\frac{L}{2} = \frac{wa^2}{4} \quad \text{in ft-lb}$$

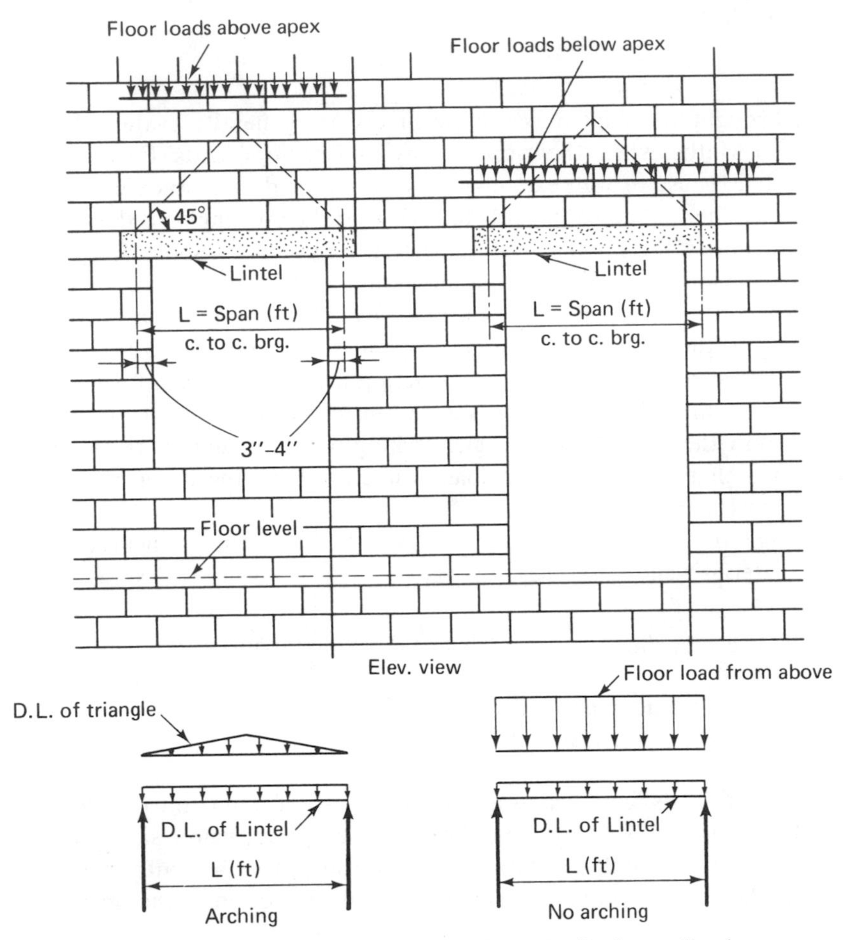

FIGURE 9-5. Approximate uniform load distribution on lintels.

Once the lintel load has been established, the procedure simply becomes a matter of selecting adequate dimensions and specifying a sufficient amount of reinforcement. See Figure 9-7 for typical brick and concrete block lintel sections.

The allowable shear on the masonry beam must be checked. In the event that it is exceeded, then shear steel must be added, in a sufficient amount to resist the entire shear, since none may be assigned to the masonry. Figure 9-8 indicates how these stirrups might be spaced. The amount of longitudinal reinforcement needed will be based upon the flexural tension requirements, which can be determined as described in Chapter 6. The effective span for moment may be taken as the clear span plus 4 in. at each end.

The beam dimensions may have already been predetermined, or they may have to be assumed. This would be the case for a beam within a wall, for example, since the width, b, would be the wall thickness and the effective depth, d, could be

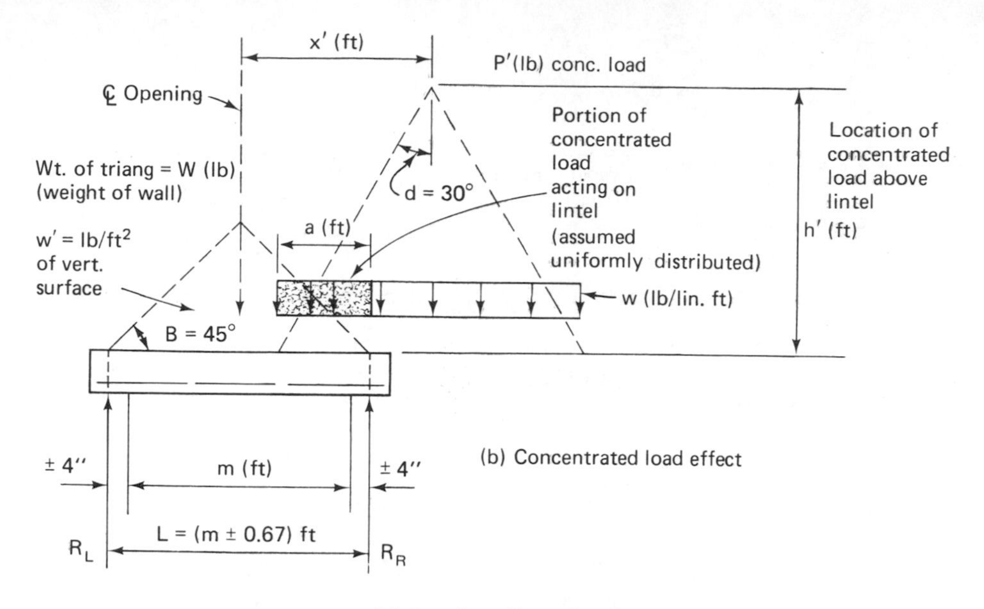

(a) Loading effect of wall weight on lb/ft² of vertical wall surface

FIGURE 9-6. Approximate concentrated load distribution on lintel.

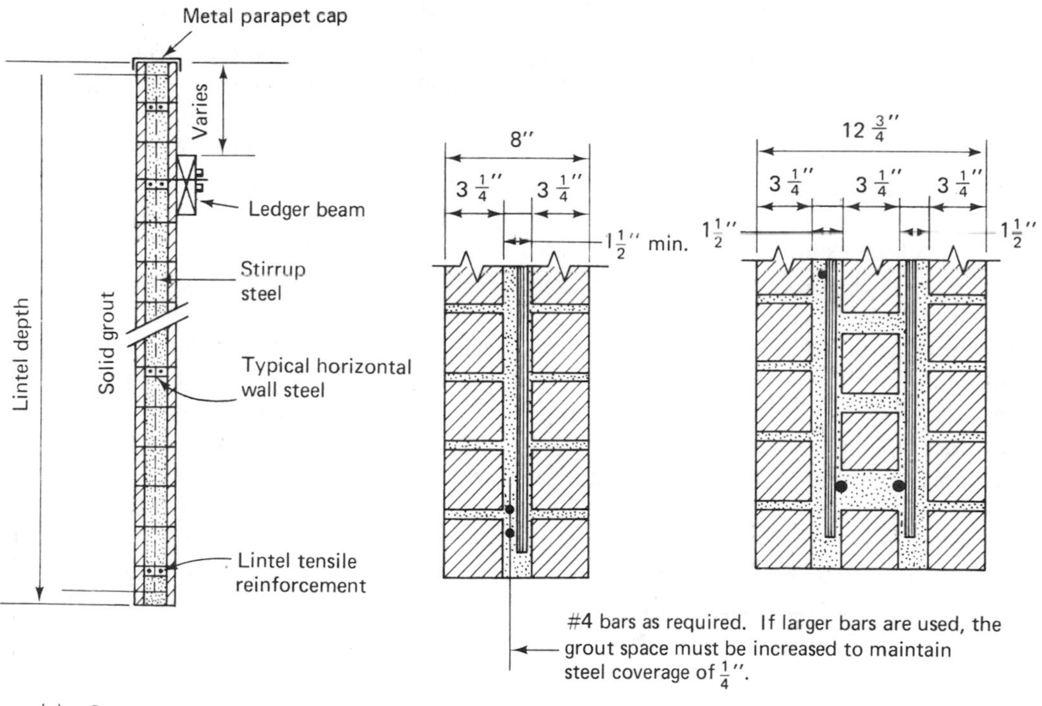

(a) Concrete masonry lintel section

(b) Grouted brick lintel section

FIGURE 9-7. Masonry lintel sections.

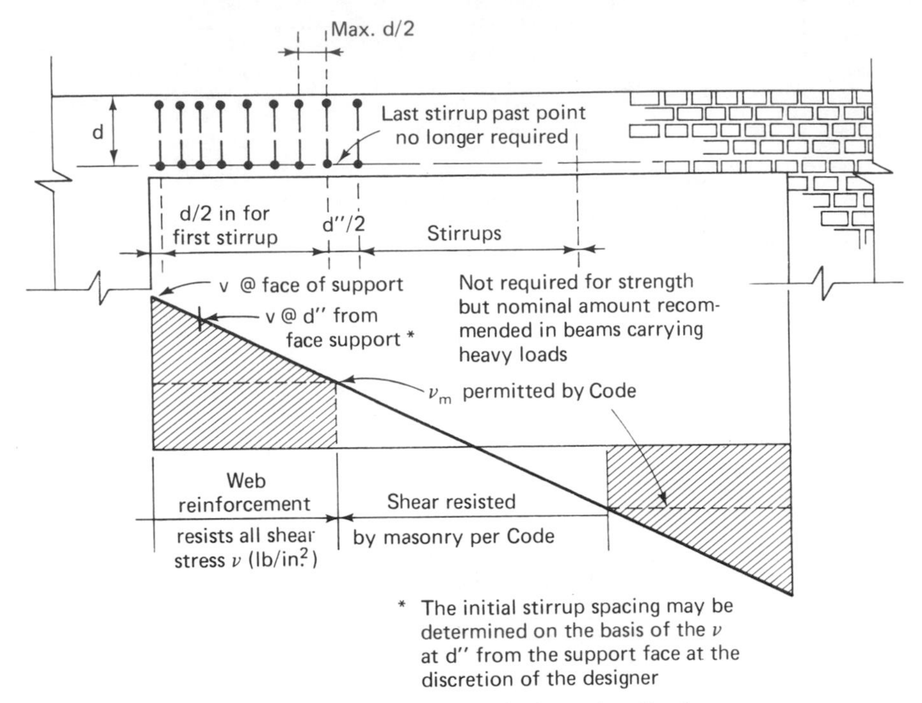

FIGURE 9-8. Location of stirrups in a lintel.

arbitrarily selected within the wall height. For the condition where the top of the wall is no more than, say, 3 ft above the soffit of the lintel, the depth might be taken as this full dimension. However, where there is considerable wall height above the lintel soffit, d would have to be arbitrarily selected. One method stipulates that the effective depth be obtained from the masonry shear requirement $V_{\text{actual}}/v_{\text{allow}}jb$. This d is then used as the effective depth of the lintel to determine the longitudinal tensile steel amount. A later example will demonstrate this simple but reliable procedure.

There are certain other design considerations that may have to be observed as well. For instance, it is often necessary to control the deflection of the beam. There are times when this becomes a more severe limitation than allowable stress considerations. It could become the design criterion, for instance, where a beam supports some sort of brittle nonstructural element, such as a plastered ceiling, or over door openings where any appreciable deflection might cause malfunction of the door operators. Presently, there does not seem to be much in the way of theoretical or experimental investigations to shed much light on the establishment of any rational deflection limits. Thus, as is often the case, the Code injects "rule of thumb" limits that are based upon past performance and experience. For example, one UBC empirical rule simply states that the live-load deflection shall be limited to the span/300 and the dead plus live-load deflection to the span/240. The Tri-Services Manual, on the other hand, recommends arbitrarily a value of L/50 for long-time dead load and L/30 for combined dead and live loads. They further recommend that the live-load deflection be based upon the full E_m value, but that E_m for dead load be reduced to one-third of that value.

Another Code stipulation states that the clear distance between lateral supports for a masonry beam must not exceed 32 times the least width of the compression flange or face. The intent here is to provide adequate lateral support so that the compression flange will not buckle or deflect laterally.

Roof and floor beams

If the beam is called upon to carry the vertical roof (or floor) live and dead loads, the steel area must be determined for this loading also. This condition occurs when the wall below is nonbearing and spans horizontally between pilasters, as described in Chapter 7. In that event the design moment would be based upon the conditions of continuity. The steel area would simply amount to:

$$+A_s = \frac{+M}{f_s jd}$$

$$-A_s = \frac{-M}{f_s jd}$$

Assume that

$$j = 0.9$$

$$d = \text{depth required for flexural compression or shear}$$

$$\left.\begin{aligned} +M &= wl^2/16 \\ -M &= wl^2/10 \end{aligned}\right\} \text{reasonable approximations}$$

where w = live- and dead-load combinations; see the discussion in the next paragraph

Refer back to Figure 9-2 for the location of positive and negative beam reinforcing.

Note that where uniform floor loads act on continuous beams not subjected to any lateral loads, design practice calls for an application of several possible vertical load combinations. In every case, however, the full dead load remains on all spans. To produce the maximum positive moment, for instance, load (with full live load) that span in question and all alternate spans. Or for maximum negative moment over a support, and maximum shear in the span, load the spans on either side of that support plus all other alternate spans. The influence of these loading possibilities must be thoroughly analyzed in order to properly reinforce a continuous masonry beam.

Bond beams

As previously noted, bond beams in masonry construction serve two important functions by (1) tying the structure together around its perimeter, and (2) acting as the diaphragm flange, thereby developing its moment resistance. Although typically one thinks of a bond beam occurring at a roof or floor level, it may also occur at the top of a parapet or masonry foundation wall. Typically, the reinforcement would consist of a pair of No. 4 or No. 5 bars (see Figures 9-9 and 9-10).

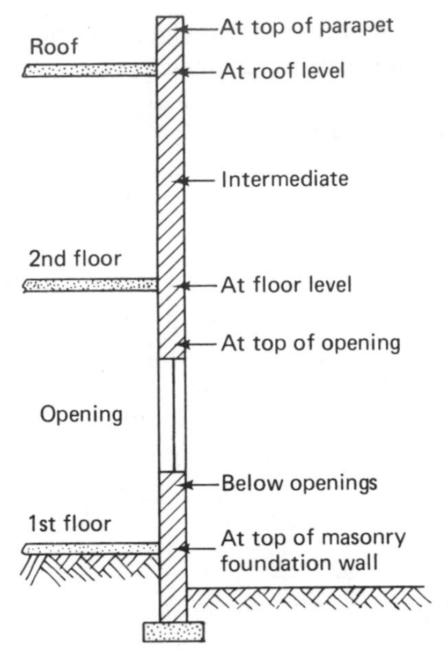

FIGURE 9-9. Location of bond beams.

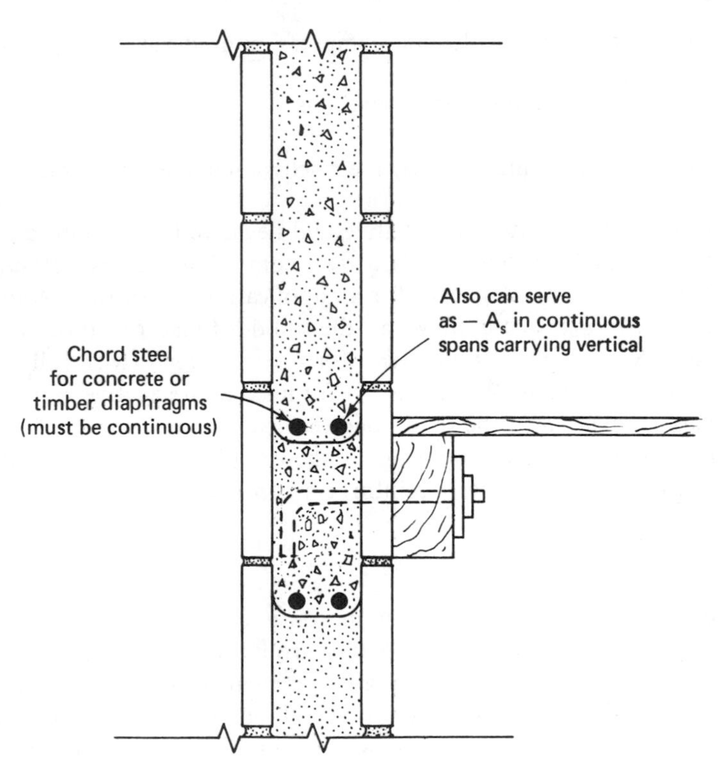

FIGURE 9-10. Bond-beam reinforcing details.

However, the bond beam at the roof or floor level must be carefully detailed. Reinforcement must be lapped 40 bar diameters or 24 in., whichever is greater, at splices, intersections, and corners. Bar splices must be staggered. Further, where it acts as the chord of a roof or floor diaphragm, it must be designed to carry the lateral loads as shown in Figure 9-11. The axial tensile force, T, would be obtained as

$$T = \frac{wl^2}{8h}$$

The steel required to develop this force would amount to

$$A_s = \frac{T}{20{,}000\ \text{lb/in}^2 \times \frac{4}{3}}$$

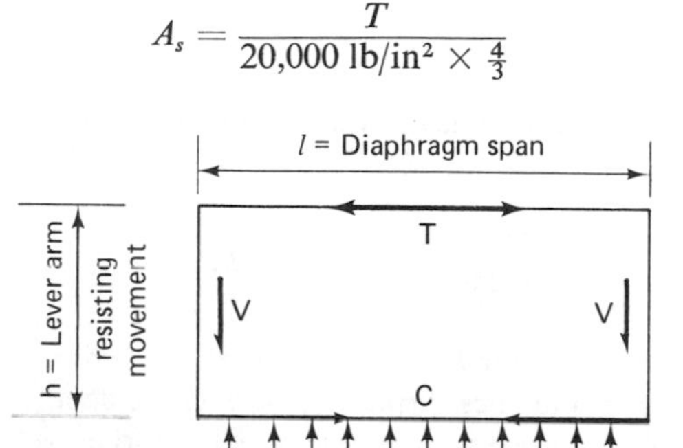

FIGURE 9-11. Bond-beam loading when serving as a diaphragm chord.

EXAMPLES OF BEAM DESIGN AND ANALYSIS

The fundamental beam design and analysis formulas were completely developed in Chapter 6, utilizing the working stress theory of elastic design. The student is referred to the appropriate design formulas and relations in that chapter for their use in the following examples, which were selected to illustrate the various design procedures involved in designing lintels and bond beams.

EXAMPLE 9-1: Determine the Required Reinforcement in a Lintel

A lintel over an opening has a clear span of 12 ft and carries a load $w = 770$ lb/ft. The lintel is constructed of 8-in. concrete block. Therefore, $f'_m = 1500$ lb/in.². Use half-stresses for no continuous inspection. Determine the reinforcement required.

Solution:

Assume that the lintel acts as a simply supported beam. Therefore,

$$M = \frac{wL^2}{8} = \frac{(770\ \text{lb/ft})(12\ \text{ft})^2}{8} = 13{,}860\ \text{ft-lb} = 166{,}300\ \text{in.-lb}$$

$$n = \frac{30{,}000}{f'_m \times \frac{1}{2}} = \frac{30{,}000}{1500 \times \frac{1}{2}} = 40 \quad \text{(no inspection)}$$

$b = 7.63$ in. (for 8-in. nominal block); assume that $j = 0.89$

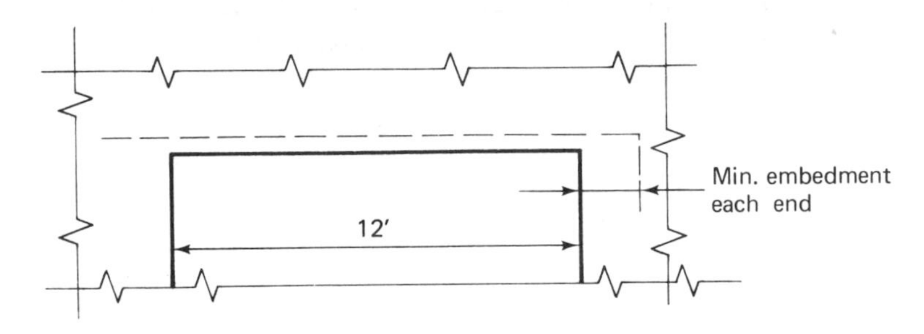

Select a value of d based upon a shear requirement, assuming that adequate height above the lintel soffit is available:

$$d = \frac{V}{vjb} = \frac{770 \times \frac{12}{2}}{25 \times 0.89 \times 7.63} = 27.2 \text{ in.}$$

Nearest block module gives $d = 28.5$ in., for $h = 32$ in. Note that the overall depth, h, for concrete masonry units occurs in 4-in. modules. Typically, the effective depth d may be taken as $h - 3.5$ in. (bottom channel block turned up, and one over bar turned with notch down to provide space for grout around bars).

Assume that $f_s = 20{,}000$ lb/in.² governs steel requirement:

$$A_s = \frac{M}{f_s j(d)} = \frac{166{,}300 \text{ in.-lb}}{(20{,}000 \text{ lb/in.}^2)(0.89)(28.5 \text{ in.})} = 0.33 \text{ in.}^2 \text{ Provide two No. 4's } (A_s = 0.40 \text{ in.}^2)$$

$$p = \frac{A_s}{bd} = \frac{0.33}{(7.63 \times 28.5)} = 0.0015, \; pn = 0.0015 \times 40 = 0.060$$

$$k = \sqrt{2pn + (pn)^2} - pn = \sqrt{2(0.060) + (0.060)^2} - 0.060 = 0.29$$

$$j = 1 - \frac{k}{3} = 1 - \frac{0.29}{3} = 0.90 \qquad \therefore \text{ OK as assumed}$$

Check on masonry capacity for selected A_s:

$$F_b = \tfrac{1}{3}f'_m = \tfrac{1}{3}(\tfrac{1}{2} \times 1500) = 250 \text{ lb/in.}^2$$

$$jk = \frac{M}{(F_b/2)bd^2} = \frac{166{,}300 \text{ in.-lb}}{(250 \text{ lb/in.}^2/2)(7.63 \text{ in.})(28.5)^2} = 0.215$$

$$j = 1 - \frac{k}{3} \qquad \therefore jk = k\left(1 - \frac{k}{3}\right) = 0.215$$

Therefore,

$$k^2 - 3k + 3(0.215) = 0$$

Thus, $k = 0.233$.

$$f_s = nF_b \frac{1-k}{k} = 40 \times 250 \frac{1 - 0.233}{0.233} = 32{,}920 \text{ lb/in.}^2 > 20{,}000 \text{ lb/in.}^2$$

Thus, steel stress governs as initially assumed.

Or check masonry stress directly: For $p = 0.0015$,

$$k = 0.29, \qquad j = 0.90 \qquad \therefore \frac{2}{jk} = 7.60$$

Then

$$f_m = \frac{M}{bd^2} \times \frac{2}{jk} = 26.8 \times 7.60 = 204 \text{ lb/in.}^2 < 250 \text{ lb/in.}^2$$

$\therefore$ OK, verifies statement above

Check using design aids:

CURVE B-2:

$$K = 26.8$$

For $K/Fb = 26.8/0.250 = 107$, find $pn = 0.060$ and $p = 0.0015$, as before.

CURVE B-4:

For $nk = 40 \times 26.8 = 1070$ and $f_s = 20{,}000$ lb/in.^{2}, find $pn = 0.060$, as before.

Check bond stress ($\Sigma_o = 3.1$ in. for two No. 4's):

$$\mu = \frac{V}{\Sigma_o jd} = \frac{770 \times 12/2}{3.1 \times 0.90 \times 28.5} = 58 \text{ lb/in.}^2 < \mu_{\text{allow.}} = 100 \text{ lb/in.}^2 \qquad \therefore \text{ OK}$$

Also from Table 9-1, provide 25 in. minimum embedment at each end.

SHEAR STRESS:

Since d was based upon allowable v without web reinforcement, obviously no further check is needed for shear. Note that if d had been reduced to 24.5 in., the allowable masonry stress would have called for a p of 0.0020, giving an A_s of 0.38 in.^{2}, which is virtually the same tensile steel amount required for $d = 28.5$ lb/in.^{2}. However, web reinforcement would have been needed in the former case. If the depth is available, it would be preferred to eliminate the need for stirrups. Since the h of a lintel must be solid-grouted, no larger depth should be selected, as a rule, than is necessary for shear or flexural stress, whichever happens to be critical.

EXAMPLE 9-2: Obtain the Moment Capacity of a Brick Beam

A brick beam has a total depth = 36 in. (say $d = 30$ in.), and $b = 10$ in. The tension reinforcement consists of two No. 7's. The stresses are $f'_m = 1800$ lb/in.^{2} with half-stresses called for ($n = 33$), and $f_s = 24{,}000$ lb/in.^{2}. Determine the moment capacity of the beam.

Solution:

$$A_s = 1.20 \text{ in.}^2 \quad \text{(two No. 7's)}$$

$$p = \frac{A_s}{bd} = \frac{1.20 \text{ in.}^2}{(10 \text{ in.})(30 \text{ in.})} = 0.004, \qquad pn = 0.132$$

$$k = \sqrt{2pn + (pn)^2} - pn = \sqrt{2(0.004)(33) + [0.004(33)]^2} - 0.004(33) = 0.398$$

$$j = 1 - \frac{k}{3} = 1 - \frac{0.398}{3} = 0.87$$

FOR STEEL STRESSES:

$$M = A_s f_s jd = (1.20 \text{ in.}^2)(24{,}000 \text{ lb/in.}^2)(0.87)(30 \text{ in.}) = 751{,}680 \text{ in.-lb}$$

FOR MASONRY STRESSES:

$$F_b = \frac{1}{3} f'_m = \frac{1}{3}\left(\frac{1}{2}\right)(1800) = 300 \text{ lb/in.}^2$$

$$M = \frac{F_b}{2} jkbd^2 = Kbd^2$$

$$K = \frac{300 \text{ lb/in.}^2}{2}(0.87)(0.398) = 51.9$$

$$M = (51.9)(10 \text{ in.})(30 \text{ in.})^2 = 467{,}100 \text{ in.-lb} \leftarrow \text{governs}$$

Thus, the steel stress is:

$$f_s = \frac{M}{A_s jd} = \frac{467{,}100}{1.20 \times 0.87 \times 30} = 14{,}900 \text{ lb/in.}^2, \quad \text{and} \quad f_m = F_b = 300 \text{ lb/in.}^2$$

From Curve B-3: for $pn = 0.132$ and $F_b = 300 \text{ lb/in.}^2$, find $K \sim 52$. Then $M = Kbd^2 = 52 \times 10 \times (30)^2 = 468{,}000$ in.-lb. Check $f_s = M/A_s jd = 14{,}930 \text{ lb/in.}^2$. Therefore, masonry F_b governs and allowable moment is 468 in.-kips.

From Curve B-2: for $pn = 0.132$, find $K/F_b = 172.5$; thus, $f_s = 15{,}100 \text{ lb/in.}^2$, and $K = 172.5 \times 0.3 = 51.8$, as before.

From Table C-1: for $pn = 0.132$, find $j = 0.867$, $2/kj = 5.80$

$$M_s = 1.20 \times 24{,}000 \times 0.87 \times 30 = 751{,}700 \text{ in.-lb}$$

$$M_m = \frac{300 \times 10 \times 30^2}{5.80} = 465{,}500 \text{ in.-lb (governs)}$$

EXAMPLE 9-3: Consider the Effects of Different Material Strengths on Beam Capacity

Using the beam described in Example 9-2, analyze the cases for:

(*a*) $f'_m = 1500/2 \text{ lb/in.}^2$. $n = 40$.

(*b*) $f'_m = 1500 \text{ lb/in.}^2$. $n = 20$.

(*c*) $f'_m = 1800 \text{ lb/in.}^2$. $n = 16.7$.

Is there any advantage to be gained by using grade 60 rather than grade 40 steel?

Solution:

(*a*) $p = 0.004$ *and* $pn = 0.160$:

From Curve B-2, find $K = 46$ and note that masonry allowable stress $= 250 \text{ lb/in.}^2$ governs;

thus, $M_m = 46 \times 10 \times 30^2 = 414$ in.-kips. Observe that for $f'_m = 1800/2$, $M_m = 467$ in.-kips, so very little is gained by going to a higher-strength masonry unit when half-stresses are specified for design.

(b) $p = 0.004$ *and* $pn = 0.080$:

From Curve B-2, find $K = 72$ and note that this is very close to the balanced state where $f_s = 20{,}000$ lb/in.² and $F_b = 500$ lb/in.² are achieved simultaneously.

$$M = 72 \times 10 \times 30^2 = 648 \text{ in.-kips}$$

In both instances the allowable masonry stress is reached before the $f_s = 24{,}000$ lb/in.² is achieved, so no advantage would be gained by specifying a higher-strength steel. This is usually the case, at least in the lower masonry buildings.

(c) $p = 0.004$ *and* $pn = 0.0668$:

From Curve B-2, find $K = 72$ where $f_s = 20{,}000$ lb/in.² (which governs), and $K = 82$ if $f_s = 24{,}000$ lb/in.², in which case the allowable masonry stress governs. Thus,

$$M = 72 \times 10 \times 30^2 = 648 \text{ in.-kips} \quad \text{and}$$

$$M = 82 \times 10 \times 30^2 = 738 \text{ in.-kips}$$

Contrast the situation here:

1. For $f_s = 20{,}000$ lb/in.², the allowable moment is virtually the same, whether $f'_m = 1500$ lb/in.² or 1800 lb/in.²; in the later case, it is the steel that governs the capacity.
2. For $f_s = 24{,}000$ lb/in.² and $f'_m = 1800$ lb/in.², a 13% increase in strength over that for $f'_m = 1500$ lb/in.² is achieved.

EXAMPLE 9-4: Determine Compressive Steel Requirement for a Beam of Limited Depth

A concrete block beam with a depth of 48 in., solid-grouted, and special inspection provided, sustains an applied moment of 200 ft-kips. The stresses are $f'_m = 1500$ lb/in.² ($n = 20$), and $f_s = 20{,}000$ lb/in.². Assume a cover of 3.5 in. ($d = 44.5$ in.) and $d' = 4$ in. Use 8-in. block. Determine the tensile and compressive steel required.

Solution:

The solution for the required tensile and compressive steel involves two steps:

1. Determine the moment capacity and steel required for the balanced condition, when F_b and allowable f_s are reached if no compression steel is assumed.
2. Determine the compressive steel required to resist that portion of the design moment which is in excess of that resisted under the balanced condition, and add tension steel accordingly.

STEP 1:

$$F_b = \frac{1}{3} f'_m = \frac{1}{3}(1500) = 500 \text{ lb/in}^2.$$

$$b = 7.63 \text{ in.}$$

$$p_b = \frac{n}{\frac{2f_s}{F_b}\left(n + \frac{f_s}{F_b}\right)} = \frac{20}{2\left(\frac{20,000}{500}\right)\left(20 + \frac{20,000}{500}\right)} = 0.00417$$

(from curve B-2, find $pn = 0.083$ at balanced point; $p_b = 0.083/20 = 0.00415$)

$$A_{s_m} = p_b bd = 0.00417(7.63 \text{ in.})(44.5 \text{ in.}) = 1.41 \text{ in.}^2$$

$$k_b = \frac{1}{1 + (f_s/nF_b)} = \frac{1}{1 + [20,000/20(500)]} = 0.33$$

$$j_b = 1 - \frac{k}{3} = 0.89$$

$$M_m = A_{s_m} f_s j_b d = (1.41 \text{ in.}^2)(20,000 \text{ lb/in.}^2)(0.89)(44.5 \text{ in.}) = 1116.9 \text{ in.-kips} = 93.1 \text{ ft-kips}$$

$$M_2 = M - M_m = 200 - 93.1 = 106.9 \text{ ft-kips} - \text{deficient moment}$$

$$A_{s_2} = \frac{M_2}{f_s(d - d')} = \frac{106.9(12000)}{(20,000 \text{ lb/in.}^2)(44.5 \text{ in.} - 4.0 \text{ in.})} = 1.58 \text{ in.}^2$$

$$A_s = A_{s_m} + A_{s_2} = 1.41 + 1.58 = 2.99 \text{ in.}^2$$

$$p = \frac{A_s}{bd} = \frac{2.99}{7.63(44.5)} = 0.00881$$

Note that A_s is the total tensile steel required, with A_{s_m} equal to the steel required to balance masonry strength, and A_{s_2} equal to the steel tensile needed to resist the deficient moment.

STEP 2:

To evaluate the compressive steel stress and compressive steel area, assume a k for the total beam at 0.33:

$$M_2 = 106.9 \text{ ft-kip}$$

$$A'_s = \frac{M_2}{f'_s(d - d')\left(\frac{n-1}{n}\right)}$$

$$f'_s = nF_b \frac{kd - d'}{kd} \le f_s = 20(500)\left(\frac{0.33 \times 44.5 - 4.0}{0.33 \times 44.5}\right) = 7280 \text{ lb/in.}^2 < 20,000 \text{ lb/in.}^2$$

or

$$f'_s = f_s \frac{k - d'/d}{1 - k} \le f_s = 20\left(\frac{0.33 - 4/44.5}{1 - 0.33}\right) = 7170 \text{ lb/in.}^2$$

comparable to the value above

$$A'_s = \frac{106.9 \times 12000}{(7170 \text{ lb/in.}^2)(44.5 - 4.0)\left(\frac{20 - 1}{20}\right)} = 4.65 \text{ in.}^2 \text{ and } A_s = 2.99 \text{ in.}^2$$

$$p' = \frac{A'_s}{bd} = \frac{4.65 \text{ in.}^2}{(7.63 \text{ in.})(44.5 \text{ in.})} = 0.0137$$

EXAMPLE 9-5: Design a Concrete Masonry Lintel

(*a*) *Determine the depth of solid grouting required for a lintel so that web reinforcement is not required. Also determine longitudinal steel requirements. The span is 20 ft and the load is 1100 lb/ft.*

Solution:

One very simple way to determine a lintel depth requirement is by simply solid grouting enough concrete block courses so that the resulting shear will be less than the maximum permissible on the masonry. For instance (for $f'_m = 1500$ lb/in.^{2} with no special inspection):

$$d = \frac{V}{jbv_{(\text{allow.})}} = \frac{1100 \times 20/2}{0.9 \times 7.6 \times 25} = 64.3 \text{ in.}$$

Solid grout 68 in. of masonry to provide d of about 64.5 in.

$$M = wl^2/10 = 1100 \times 20^2/10 = 44 \text{ ft-kips}$$

assuming some fixity of each end.

$$A_s = \frac{M}{f_s jd} = \frac{44 \times 12}{20 \times 64.5 \times 0.9} = 0.45 \text{ in.}^2$$

Provide two No. 5's ($A_s = 0.60$ in.^{2})

$$\text{actual } p = \frac{A_s}{bd} = 0.00122 \qquad \text{and} \qquad pn = 0.00122 \times 40 = 0.0488$$

Thus, $k = 0.267$, $j = 0.91$, $2/jk = 8.23$ (Table C-1).

$$f_m = \frac{M}{bd^2} \times \frac{2}{jk} = \frac{44 \times 12{,}000}{7.63 \times (64.5)^2} \times 8.23 = 137 \text{ lb/in.}^2 < 250 \text{ lb/in.}^2 \qquad \therefore \text{ OK}$$

(*b*) *Should sufficient depth not be available for the masonry to resist shear without stirrups, they may be placed as follows:*

Solution:

Say h limited to 60 in. then $d = 56.5$ in.

$$A_s = \frac{44 \times 12}{20 \times 56.5 \times 0.9} = 0.52 \text{ in.}^2$$

Use two No. 5's, actual $p = 0.00139$, $pn = 0.056$, $2/jk$ (Table C-1) $= 7.82$.

$$f_m = \frac{44 \times 12{,}000}{7.63 \times (56.5)^2} \times 7.82 = 170 \text{ lb/in.}^2 < 250 \text{ lb/in.}^2 \qquad \therefore \text{ OK}$$

$$v = \frac{1100 \times 20/2}{7.63 \times 0.9 \times 56.5} = 28.5 \text{ lb/in.}^2 > 25 \text{ lb/in.}^2$$

Provide stirrups to take the entire shear.

STIRRUP SPACING (USE NO. 3 STIRRUP):

$$s = \frac{A_v f_v}{bv} = \frac{0.11 \times 20{,}000}{7.6 \times 28.5} = 10.1 \text{ in.; say, 8 in. cc to fit block module.}$$

Theoretically, we could cut the stirrups off when v reaches 25 lb/in.^{2}, but it would be good practice to extend these out from each end for 2 or 3 ft. or more.

PROBLEMS

Solve the following problems using basic formulas, then check the result with the design aids. Use MW brick and grade 40 steel if they are not specified in the problems.

9-1. Consider a 9-in.-wide three-span continuous beam, having a span of 18 ft. $f'_m =$ 2700 lb/in.^{2}, $f_s = 24{,}000$ lb/in.^{2}, $LL = 2000$ lb/ft.

(a) Determine the moments due to pattern loadings (use moment coefficients; then compare your results with those obtained by an elastic beam analysis).

(b) Determine the depth required to carry the given load without using compression steel.

(c) Specify the steel required.

(d) Check the shear and bond requirements.

(e) Detail the beam, showing the reinforcement and cutoff locations.

For the following problems, assume that the lintel is simply supported.

9-2. Given an 8-in. concrete block lintel with a span of 16 ft. $f'_m = 1500$ lb/in.^{2}, $d =$ 24.5 in., $A_s =$ two No. 4's. Determine the masonry and steel stresses. Check the shear and bond stresses. Special inspection is provided. The wall above the lintel is grouted at 48 in. cc.

9-3. Given a 10-in. brick lintel with a span of 16 ft. The wall above the opening is 12 ft. high. The roof load at that height is 200 lb/ft. Determine the depth and reinforcement needed. Avoid the use of stirrups. Special inspection is provided.

9-4. Same as Problem 9-3, except that the wall height is 5 ft where the 200-lb/ft load is applied.

9-5. Same as Problem 9-4, but change the brick lintel to a 12-in. concrete block lintel. The wall above the lintel is grouted at 48 in. cc. $f'_m = 1500$ lb/in.^{2}. No inspection is provided.

9-6. Consider an 8-in. concrete block lintel with a span of 16 ft. $f'_m = 1500$ lb/in.^{2}, $d =$ 20.5 in., $d' = 3.5$ in. Determine the reinforcement required. No inspection is provided. The wall above the lintel is fully grouted.

9-7. Consider an 8-in. concrete block lintel with a span of 16 ft. A concentrated load of 2000 lb is applied at 7 ft above and 4 ft from the center line of the opening. Determine the maximum moment and shear for the given loads. The wall above the lintel is fully grouted.

9-8. For the beam in Problem 9-7, determine the lintel depth and reinforcement needed, assuming a balanced design. No inspection is provided for an $f'_m = 1500$ lb/in.^{2}.

10

Lateral-Load-Resisting Elements

Modern building structures are subjected to many different types of loadings and combinations thereof. These were detailed thoroughly in Chapter 5. As was shown there, load combinations which include live, dead, and lateral effects must be carefully evaluated when designing modern structures. Although the vertical loading effects alone must be examined in the process, the ingenuity of the structural engineer rarely becomes challenged until he considers the ability of the structure, and its various elements, to resist certain types of lateral loads, such as winds or earthquakes. Perhaps it would be an oversimplification to state that it is really not difficult to design a building to sustain only vertical live and dead loads; rather, the real problem emerges when one must provide economically also for the lateral stability of a structure subjected to the effects of wind or seismic action. Unfortunately, in some parts of the country, even in certain regions that experience high wind intensities, this consideration has not been given its just due, particularly when it comes to masonry structures. A collapse under some dynamic show of nature is often considered simply "an act of God"; when in reality the building perhaps could have been designed to withstand these forces had they been properly considered.

In design, one must determine at the outset which of the horizontal forces governs (i.e., wind or seismic). However, in seismic zones 3 and 4 there is no question but that, generally, the most destructive force to be assumed acting on a building is caused by the ground motion occurring during an earthquake. For when the ground beneath the structure moves suddenly, the building masses tend to remain in their original positions due to inertia, and as a result, the building suffers distortions of varying degrees. Unless that building has been properly designed to adjust to those distortions, collapse or serious structural damage will occur. The seismic forces specified by the UBC, which are applied as equivalent static loads to a structure (Figure 10-1), were thoroughly described in Chapter 5. Dynamic analysis considerations are important and they are generally used in the design of the larger buildings; however, the majority of construction volume involves smaller buildings, and these are often designed with the Code equivalent static loads. It will be the intent of this chapter to demonstrate how the building must be designed to resist these equivalent static lateral forces. It should be noted that although the ground motion can occur in any direction, it becomes a computational convenience to consider the effects of this ground motion as if it were to act only in directions parallel to the perpendicular axes of the building (i.e., in the X and Y directions). It is therefore customary to investigate the capacity of the building from the standpoint of these two directions only, and thereby to assume that this would account for whatever direction the particular ground motion might actually take. There are some structures, however, in which loads are more critical in a specific direction, and these must be considered accordingly (e.g., a rectangular-shaped tower supporting a water tank at the top).

If the building is to be designed by current Code practice, then these inertial forces are replaced by specified static lateral loads which are assumed to act as loads applied at different floor levels, to simulate the lateral inertia effects as indicated in Figure 10-1. There are several different types of structural systems

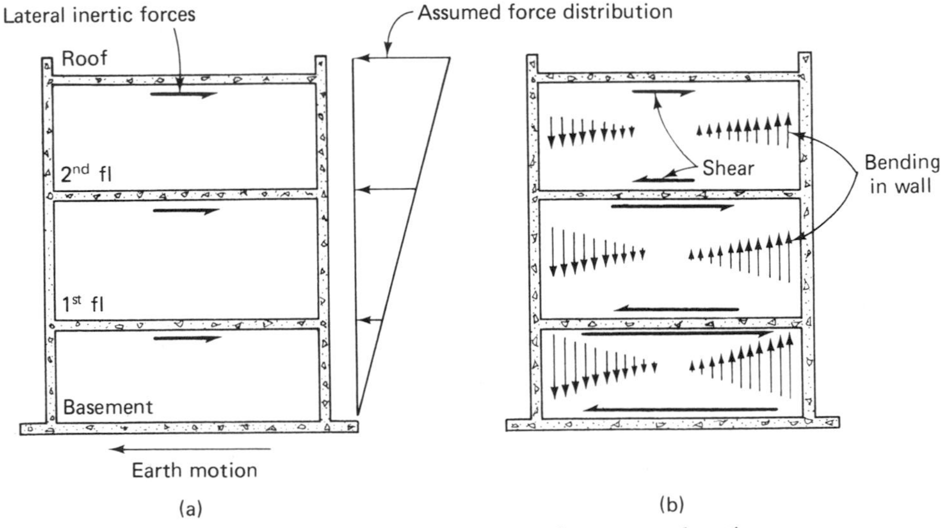

FIGURE 10-1. Forces and stresses due to ground motion.

employed to resist these horizontal forces and carry them from the various floor levels down into the foundation. The vertical structural elements used to transfer lateral forces include (1) shear walls, (2) braced frames, (3) moment-resisting space frames, and (4) some combination of these. The horizontal structural elements which distribute these forces to the vertical resisting elements are the floor and roof diaphragms, or horizontal bracing. As has been pointed out previously, it does not appear economically feasible at this point in the state of the art to construct masonry moment-resisting frames, so it will be the intent of this chapter to limit the discussion to that of shear walls and horizontal diaphragms or bracing systems.

GENERAL BEHAVIOR OF BOX SYSTEMS

An examination of Figure 10-2 will show in a general way how a box-type shear-wall structure is intended to sustain the lateral loads and transfer them to the ground. Observe that a box system does not have a vertical load-carrying frame but depends upon the walls to not only carry the vertical loads, but also to provide the necessary lateral stability. The Code assigns a value of 1.33 to K for this type

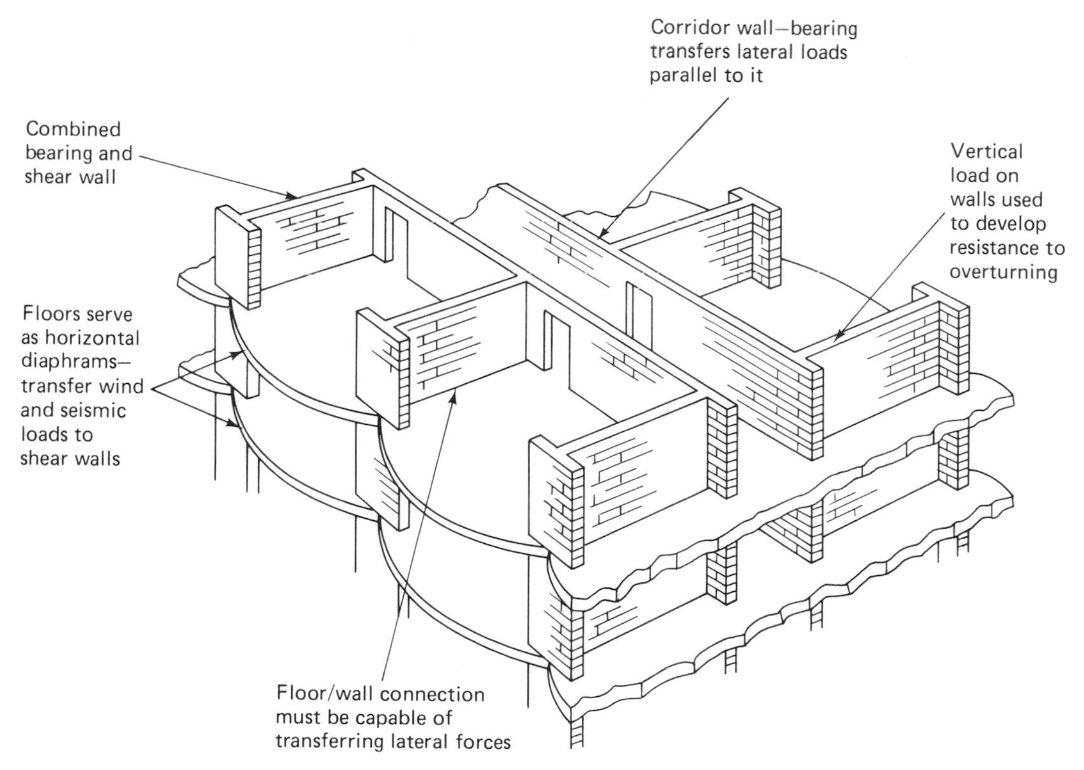

FIGURE 10-2. Box-type shear-wall structure.

of system. In contrast, a space frame, which possessses no bearing walls, may be stabilized laterally with nonbearing shear walls, in which case K decreases to 1.00. At the other end of the spectrum lies the ductile moment-resisting space frame, which depends 100% on the moment frame for lateral stability, it having a K value of 0.67. Note that the walls, which are perpendicular to the assumed direction of the ground motion, must span vertically between the floor diaphragms. Therefore, the inertial effect of one-half the wall height, both above and below the floor level in question, is considered to be transferred to that floor diaphragm. In the case of the one-story building, the load transferred to the roof diaphragm stems from the wall midheight, but it also includes all the parapet wall, if one exists (see Figure 10-3). In addition, the inertial effect of the roof dead load itself must be taken by the diaphragm. In computing these effects, it is computationly convenient to simply consider a 1-ft-wide strip extending from the midheight of one perpendicular wall, horizontally across the roof and down to the midheight of the opposite parallel wall, as shown in Figure 10-3. This produces a force, w pounds per foot, which then becomes the lateral loading on the roof diaphragm. Now the diaphragm, which will be described in detail later, behaves essentially as a horizontal plate girder, wherein the boundary members or

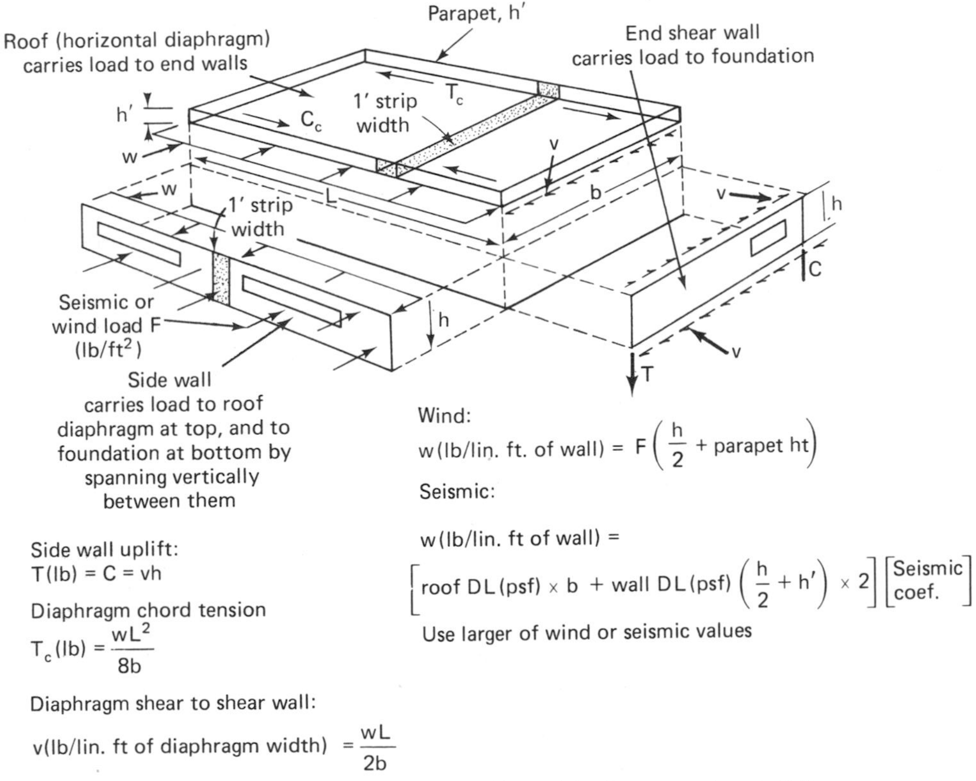

FIGURE 10-3. Distribution of lateral loads on one-story building.

chords (often a bond beam in masonry construction) serve as the girder flanges and the decking functions as the web to carry the flexural shear force. The diaphragm therefore spans between the supporting shear walls; that is, those walls which are parallel to the direction of the assumed lateral force. Through an appropriate connection detail at the diaphragm boundary, the total horizontal shear is transferred directly to the shear wall. This lateral shear transfer to any shear wall may be based on the adjacent tributary area, or on the relative rigidities of the various shear walls. These alternative procedures will be described in more detail in subsequent paragraphs. In addition to the diaphragm shear, each shear wall must resist the force produced by its own inertial effect. The sum of the diaphragm shear and this wall shear constitutes the total direct shear force which must be withstood by the shear-wall materials. If the shear wall is adequately designed, these forces will be distributed into the foundation. In addition to this direct shear force, an additional torsional shear may be transferred into the wall. This takes place when the building center of gravity and the center of rigidity of the vertical resisting elements do not coincide. The magnitude of this torsional shear force arises from the eccentricity of the diaphragm load, which is simply the distance between the center of gravity and the center of rigidity. These two forces, the total direct lateral shear and the torsional shear, are combined so that the sum becomes the total Code-designated force imposed upon the shear wall (see Figure 10-4). As explained later, the Code calls for the actual design force to be 1.5 times this computed total shear value when checking the shear capacity of the masonry wall.

A simple numerical example will demonstrate how a box-type system receives the various loads and distributes them through the diaphragm to the shear walls (refer to Figure 10-3).

1. Compare wind versus seismic forces for governing loading:

$$V = ZIKCSW = 1.0 \times 1.0 \times 1.33 \times 0.14W^{1} = 0.186W$$

Note that if the period, T, of the building were calculated from $T = 0.05h_n/\sqrt{D}$, then $T = 0.05 \times (14\text{ft})/\sqrt{50\text{ft}} = 0.10$ s. The acceleration coefficient $C = 1/15\sqrt{T}$ becomes $C = 1/15\sqrt{0.10} = 0.21$, but $C_{\max} = 0.12$.

(a) Wind: 15 lb/ft² (14 ft/2 + 3.5 ft) = 158 lb/ft.

(b) Seismic Forces:

(1) Transverse direction: Consider, for computational convenience, a 1-ft-wide strip. Then both walls normal to seismic force will span from ground to roof. This means that 7 ft of load in the span goes to the roof diaphragm, plus the 3.5 ft of parapet above it. Also,

[1] If the site response factor is not obtained from a geotechnical investigation, the max $S(1.5)$ specified in the Code must be used; thus, $C \times S = 1.5 \times 0.12 = 0.18$, but CS is limited by Code to a maximum value of 0.14. This is why the value 0.14 was used above in obtaining the base shear V. Generally, the max. CS value of 0.14 is used in one- and two-story buildings.

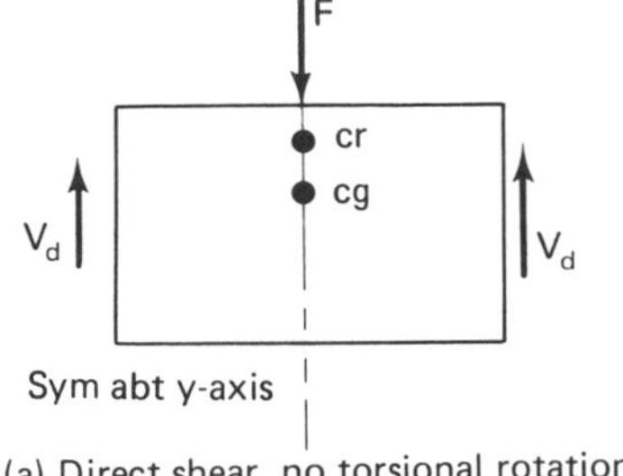

(a) Direct shear, no torsional rotation

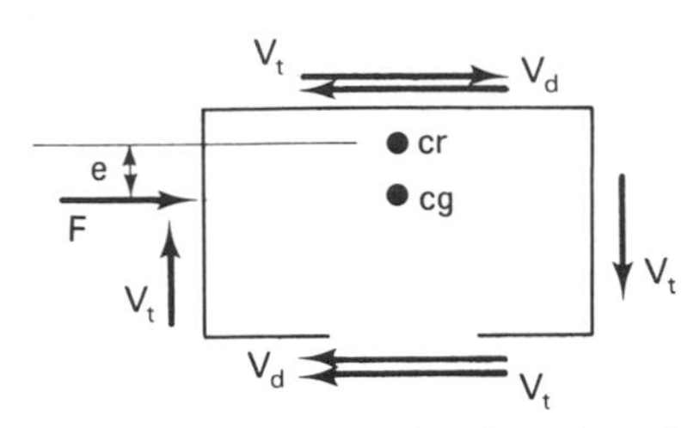

(b) Combined shear, torsional rotation, T = Fxe

Note on the S wall that direct shear V_d combines with torsional shear, V_t, to produce the total design shear; i.e., $^*V = V_d + V_t$. (Negative torsional shears are precluded from consideration by Code.)

FIGURE 10-4. Combined torsional plus direct shear. (*When checking shear adequacy for UBC, multiply this value by 1.5.)

the roof dead load (DL) will add to the load on the diaphragm, or:

$$\begin{aligned} \text{Roof DL: } 9\ \text{lb/ft}^2 \times 50\text{ft} &= 450\ \text{lb/ft} \\ \text{Wall DL: } 2(61\ \text{lb/ft}^2)(14\text{ft}/2 + 3.5\text{ft}) &= 1280\ \text{lb/ft} \\ w &= 1730\ \text{lb/ft} \end{aligned}$$

(61 lb/ft² is weight of the masonry wall in this case)

Then $w = 0.186(1730) = 322$ lb/ft $<$ 158 lb/ft due to wind, so seismic governs.

(2) Longitudinal direction: In this direction the seismic loading becomes:

$$\begin{aligned} \text{Roof DL: } 9\ \text{lb/ft}^2 \times 100\ \text{ft} &= 900\ \text{lb/ft} \\ \text{Wall DL: } 2 \times 61\ \text{lb/ft}^2 \times 10.5\ \text{ft} &= 1280\ \text{lb/ft} \\ w &= 2180\ \text{lb/ft} \end{aligned}$$

Then $w = 0.186 \times 2180 = 405$ lb/ft. Seismic governs in this direction also.

2. Roof diaphragm shear: The transverse direction often governs because of the longer span of the diaphragm (L), the lesser depth of the diaphragm, and the shorter shear wall length (b) to resist the loads.

$$V = \frac{wL}{2} = 322 \times 100\ \text{ft}/2 = 16.1\ \text{kips}$$

(based upon tributary width since wood roof is categorized as a flexible diaphragm)

$$v = \frac{V}{b} = 16{,}100 \text{ kips}/50 \text{ ft} = 322 \text{ lb/ft}$$ diaphragm shear to be transferred to shear wall through connections

3. Shear at wall midheight:

$$V = \underbrace{16.1}_{\text{diaphram shear}} \text{ kips} + \underbrace{(50 \text{ ft} \times 10.5 \text{ ft} \times 61 \text{ lb/ft}^2)(0.186)}_{\text{wall inertia force at midheight}} = 22.1 \text{ kips}$$

$$v = \frac{V}{b} = \frac{22{,}100}{50 \text{ ft}} = 442 \text{ lb/ft}$$
↑ (assumes no openings)

Actually, a masonry wall would have to be designed for in-plane shear, according to the Code, to carry 1.5 × 22.1 kips, or 33.15 kips.

HORIZONTAL STABILITY ELEMENTS

Diaphragms

As previously noted, a horizontal diaphragm in this connotation is analogous to a plate girder lying in a horizontal plane where the floor or roof deck functions as the web to resist the shear force and the boundary elements, such as bond beams, serve as the girder flanges in developing the resisting moment (see Figure 10-5).

Diaphragms may be constructed of any number of different types of materials such as concrete, masonry, wood, gypsum, or metal. Combinations of materials are sometimes used. In masonry construction, one would normally encoun-

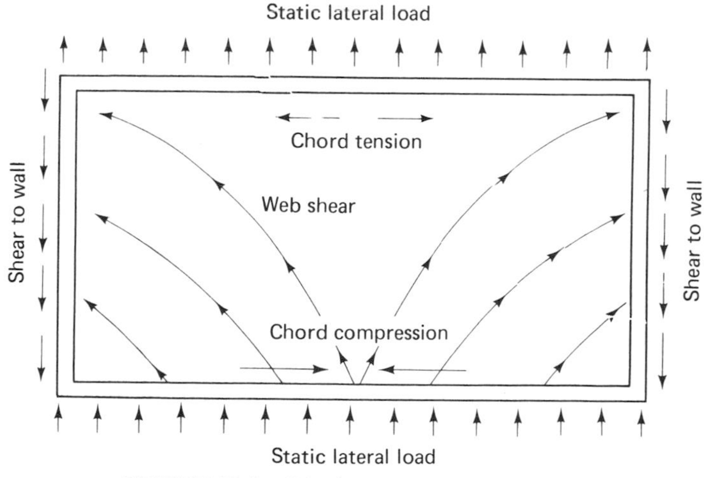

FIGURE 10-5. Diaphragm beam-like behavior.

ter either cast-in-place or precast concrete slabs, or a unitized roof system composed of timber joists supporting a plywood or diagonal sheeting deck, or a roof system stabilized with rod bracing.

The total shear force distributed to the various vertical shear-resisting elements comprising the lateral force-resisting system will depend upon how rigid they are compared to the rigidity of the diaphragm. The effect of diaphragm stiffness on the distribution of lateral forces is described in the following paragraphs. For the purpose of this discussion, the diaphragms are classified into three groups of relative flexibilities: rigid, flexible, and semirigid.

A *rigid diaphragm* is assumed to distribute the horizontal forces to the vertical resisting elements in direct proportion to the relative rigidities of those elements. This premise stems from the fact that, under a symmetrical loading, the rigid diaphragm, which in itself does not deform appreciably will cause each vertical element to deflect the same amount. Since the deflections of all these shear elements are therefore equal, then for each wall, the amount of force that it takes to cause that deflection must be directly proportional to the rigidity of the element in question (see Figure 10-6). Rigid diaphragms are considered capable of transferring torsional shear deflections and forces.

A *flexible diaphragm* may be likened to a series of simple spans extending between very rigid supports (i.e., the vertical resisting elements). It is assumed here that the relative stiffness of these nonyielding supports is very great compared to that of the diaphragm, which therefore deflects as a limber beam. This limber beam, having no appreciable continuity across the supports, thus develops no negative moment over them which would affect the distribution of load. Each simple span load therefore distributes half its load to each support. Thus, the flexible diaphragm will distribute the lateral forces to the vertical resisting elements on a tributary width rather than on a continuity reaction basis (see Figure 10-6). It should be noted here that a flexible diaphragm is not considered capable of distributing torsional stresses resulting from the condition where the center of mass and the center of rigidity do not coincide.

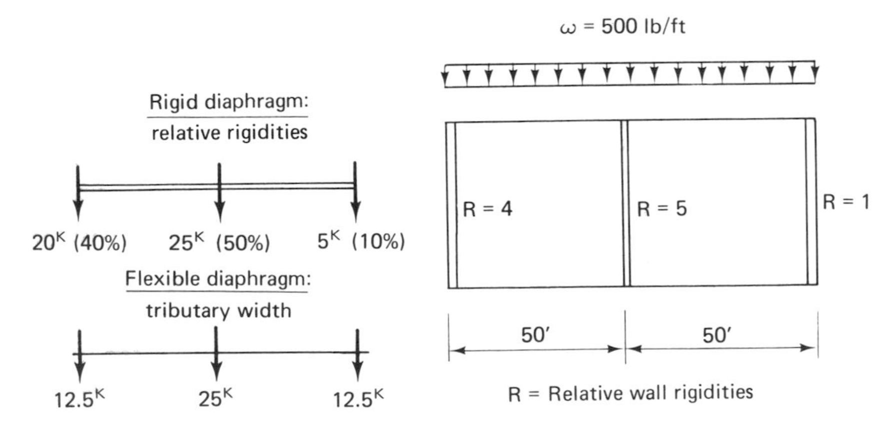

Note: Torsional shear not included

FIGURE 10-6. Rigid versus flexible diaphragms.

Semirigid diaphragms are those which exhibit significant deflection under load, but which also have sufficient stiffness to distribute a portion of their load to the vertical elements in direct proportion to the rigidities of those elements. This action is somewhat analogous to that of a continuous concrete beam bearing on yielding supports. In that case, the support reactions are dependent upon the relative stiffness of both diaphragm and vertical element. A rigorous analytical analysis for something of this type is very time consuming and usually is not justified, since a number of questionable assumptions have to be made in order to make the computations palatable.

Design Considerations

In general, certain features must be given very careful consideration when selecting and designing the components of the lateral-force-resisting system for reinforced masonry buildings of any type. Included among these are the following:

1. The type of vertical resisting element definitely affects the type of floor or roof diaphragm that can be used, both from a structural and an economical standpoint. For example, one would not expect to use wood framing to support a concrete diaphragm. Also, do not overlook the fact that these horizontal elements must be designed to carry vertical loads as well. In reinforced masonry, the selection would normally be limited to wood or concrete, since these are more readily tied into the masonry shear-wall elements, although structural steel trusses or beams supporting a concrete topping or steel decking diaphragm have been utilized successfully.
2. In high-rise masonry, most assuredly a rigid diaphragm such as concrete must be provided. Anything flexible may permit erratic vibrations in portions of the structure, which could lead to disastrous results.
3. Connection details must be given the utmost consideration, a practice often slighted. Unless a concrete deck or a unitized plywood panel system is properly joined to the walls, the shear forces cannot be distributed to them. Furthermore, boundary members at diaphragm edges must be designed to resist direct tensile or compressive (chord or flange) stresses.
4. Where openings occur, details must be devised to ensure that shear stresses can be developed around them.
5. In L- or T-shaped plan layouts, continuity in flexible diaphragms must be developed through such elements as "drag struts" or "collectors," so that the loads can be transferred through the junction of the L or T. See Figure 10-7 and also Example 11-2 for further illustrations of this component. Without such a connection, deflection incompatibilities could occur between adjacent diaphragms such as A and B in the L, thereby producing large tearing forces at their junction (point 1). These must be handled by the drag strut, which transmits these forces to a nearby shear wall, such as C. An alternative solution would provide for a structural separation so that the different areas (A and B) can deflect and function independently of each other. Also portions of buildings that have different periods of vibrations could be separated in this manner. This method actually constitutes a preferred solution, provided that it can be readily achieved.

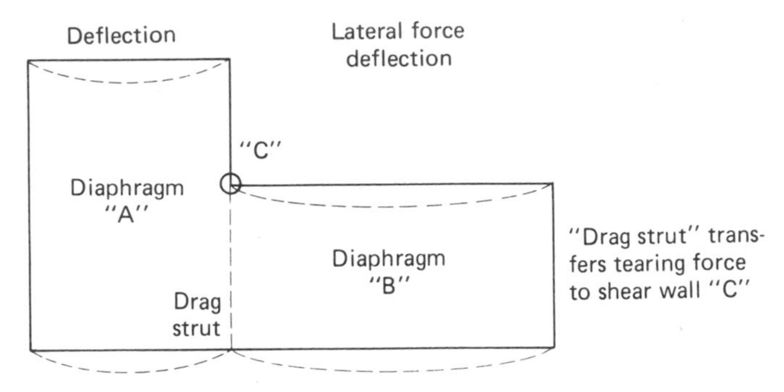

FIGURE 10-7. Drag strut in L-shaped building.

6. Deflection limitations must be imposed upon the design of a diaphragm to ensure an adequate stiffness in order that it may support the vertical wall elements without excessive and damaging deflections.

Deflections

A diaphragm must not only be designed to carry the assumed loadings to the vertical elements, but it also must possess enough strength and stiffness so that the vertical elements laterally supported by it can safely sustain the deflections induced by their response to any wind or seismic motion. The basic deflection of any diaphragm consists of the sum of two components: (1) the flexural deflection, and (2) the web deflection. The first component, the flexural deflection, may be evaluated like that of a beam deflection, wherein the shear contribution is neglected. It is assumed that all the flexural stresses are developed in the flanges of the diaphragm and none in the web.

The nature of the second component, web deflection, will vary somewhat depending upon the type of diaphragm, although the shear stress is often taken to be uniformly distributed within the web depth. Other factors may also have to be taken into account, such as the added deflection due to nail and chord splice slip in the case of wood diaphragms (see page 295).

The magnitude of the diaphragm deflection under the Code-specified lateral forces is often used as the criterion to establish the adequacy of the diaphragm stiffness when the supported walls are concrete or masonry. The numerical limitation on deflection generally consists of a maximum value prescribed for the relative deflection or drift of the walls between the level of diaphragm in question and the floor below. Any deflection in excess of this value would serve as a warning that the diaphragm may not be stiff enough to support the rigid masonry walls without subjecting them to dangerous strain levels. However, for relatively flexible diaphragms, the only flexibility limitation needed simply becomes a maximum span/width ratio as shown in Table 10-1. Those are empirical and arbitrary and may be waived if design calculations justify that no hazard will be incurred.

TABLE 10-1

Maximum Diaphragm Dimension Ratios
(*UBC Table 25-I*)

Material	*Horizontal diaphragms: Maximum span/width ratios*	*Vertical diaphragms: Maximum height/width ratios*
1. Diagonal sheathing, conventional	3:1	2:1
2. Diagonal sheathing, special	4:1	$3\frac{1}{2}$:1
3. Plywood, nailed all edges	4:1	$3\frac{1}{2}$:1
4. Plywood, blocking omitted at intermediate joints	4:1	2:1

Wood diaphragm construction

CONSTRUCTION DETAILS

Wood diaphragms can be constructed either with a plywood decking, or with 1 or 2 in. sheathing laid at 45° to the supporting rafters, called *diagonal sheathing*. With diagonal sheathing, the wood diaphragm may be used to resist a horizontal shear not to exceed 300 lb/lin. ft of width at the diaphragm edge. For 2-in. nominal material, the maximum design shear may be increased to 400 lb/lin. ft. For heavy lateral loads, a specially constructed double sheathed diaphragm of plank has been used, but plywood is more commonly installed. A shear value of up to 600 lb/lin. ft is permitted on this type of diagonal sheathed deck.

An anomaly in the UBC appears to exist regarding wood diaphragms and masonry or concrete walls. For instance, Section 2416(c) states that "masonry shall not be supported by wood members except as provided for in Section 2516." According to this latter provision, wood members may not be used to resist horizontal forces contributed by concrete or masonry members in buildings over one story in height. But the exception contained therein specifically permits wood floors or roof members to resist horizontal forces imposed by wind, earthquake, or earth pressure on concrete or masonry walls, provided that such forces are not resisted by rotation of the truss or diaphragm.

Where plywood sheathing is used, all boundary members must be proportioned and spliced where necessary to provide continuity through the chord member, while the plywood itself acts as the web in resisting the flexural shear stresses. Shear stresses have proven essentially uniform across the depth of the diaphragm, rather than demonstrating a tendency toward a parabolic variation, as exhibited in the web of a shallow beam. Similarly, the chords carry all the flange stresses in axial tension or compression rather than sharing these with the web. Framing members need be at least 2-in. in nominal thickness. Panel edges bear on the framing members and butt along their center lines. Nails have to be placed not less than $\frac{3}{8}$ in. in from the panel edge and not more than 12 in. apart along intermediate supports, and 6 in. along panel edge bearings. Closer nail spacings per Table 10-2 (UBC Table 25-J) provide for higher allowable shears permitted to be transferred to the vertical shear resisting element.

Design Capacity

The load-carrying capacity and stiffness of diaphragms depends not only on plywood thickness and nailing but also on whether or not they are blocked. Blocking consists of lightweight nailers, usually 2 × 4′s, framed between the joists or other primary structural supports, for the specific purpose of connecting the edges of the plywood sheets (see Figure 10-8). The reason for the blocking in diaphragms is to allow nailing of sheets at all edges for better shear transfer. The capacity of unblocked diaphragms is controlled by the buckling of unsupported and unnailed plywood panel edges, with the result that they possess a much lower capacity than if they were blocked. Increased edge nailing in this case will simply not increase capacity. So for the same nail spacing, design loads permitted on a blocked diaphragm vary from $1\frac{1}{2}$ to 2 times the design loads for its unblocked counterpart. In addition, the maximum loads for which a blocked diaphragm with optimum nail spacing can be designed are several times greater than those without blocking.

One must not overlook the fact that the plywood must carry vertical live and dead loads across a span that equals the spacing of the supporting joists. This flexural requirement generally dictates the plywood thickness needed. To aid in this selection, Table 10-3 gives permitted loads for various plywood spans and thicknesses, where the plywood sheathing is continuous over two or more supports.

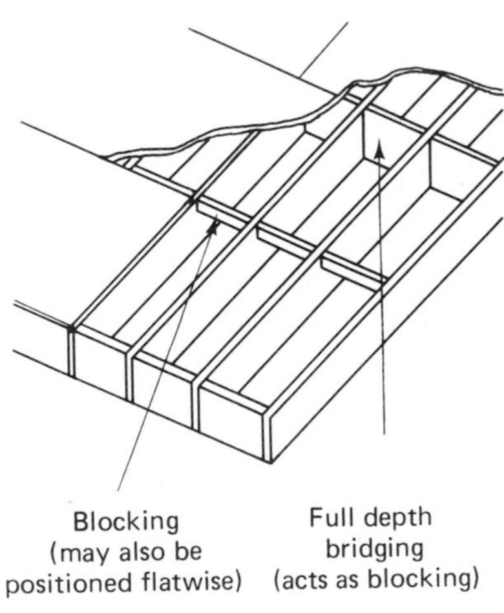

FIGURE 10-8. Blocking in plywood diaphragm construction.

TABLE 10-2

Allowable Shear in Pounds per Foot for Horizontal Plywood Diaphragms with Framing of Douglas Fir, Larch, or Southern Pine[a–c] (*UBC Table 25-J*)

					Blocked diaphragms				*Unblocked diaphragms*	
					Nail spacing at diaphragm boundaries (all cases) and continuous panel edges parallel to load (cases 3, 4, 5, and 6)				*Nails spaced 6 in. max. at supported end*	
					6	4	$2\frac{1}{2}$	2		
					Nail spacing at other plywood panel edges					
Plywood grade	*Common nail size*	*Minimum nominal penetration in framing (in.)*	*Minimum nominal plywood thickness (in.)*	*Minimum nominal width of framing member (in.)*	6	6	4	3	*Load perpendicular to unblocked edges and continuous panel joints (case 1)*	*All other configurations (cases 2, 3, and 4)*
Structural I	6d	$1\frac{1}{4}$	$\frac{5}{16}$	2	185	250	375	420	165	125
				3	210	280	420	475	185	140
	8d	$1\frac{1}{2}$	$\frac{3}{8}$	2	270	360	530	600	240	180
				3	300	400	600	675	265	200
	10d	$1\frac{5}{8}$	$\frac{1}{2}$	2	320	425	640[b]	730[b]	285	215
				3	360	480	720	820	320	240
C-D, C-C, Structural II and other grades covered in UBC Standard 25–9	6d	$1\frac{1}{4}$	$\frac{5}{16}$	2	170	225	335	380	150	110
				3	190	250	380	430	170	125
			$\frac{3}{8}$	2	185	250	375	420	165	125
				3	210	280	420	475	185	140
	8d	$1\frac{1}{2}$	$\frac{3}{8}$	2	240	320	480	545	215	160
				3	270	360	540	610	240	180
			$\frac{1}{2}$	2	270	360	530	600	240	180
				3	300	400	600	675	265	200
	10d	$1\frac{5}{8}$	$\frac{1}{2}$	2	290	385	575[b]	655[b]	255	190
				3	325	430	650	735	290	215
			$\frac{5}{8}$	2	320	425	640[b]	730[b]	285	215
				3	360	480	720	820	320	240

[a]These values are for short-time loads due to wind or earthquake and must be reduced 25% for normal loading. Space nails 10 in. on center for floors and 12 in. on center for roofs along intermediate framing members.
Allowable shear values for nails in framing members of other species set forth in Table 25-17 of UBC Standards shall be calculated for all grades by multiplying the values for nails in Structural 1 by the following factors: group III, 0.82, and group IV, 0.65.

[b]Reduce tabulated allowable shears 10% when boundary members provide less than 3 in. nominal nailing surface.

[c]Framing may be located in either direction for blocked diaphragms.

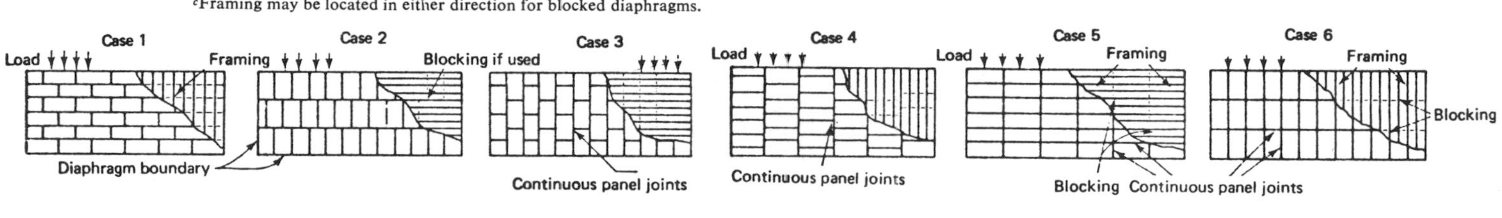

NOTE: Framing may be located in either direction for blocked diaphragms.

TABLE 10-3

Allowable Spans and Vertical Loads for Plywood Decks
(UBC Tables 25-R-1 and R-2)
(a) Face grain perpendicular to supports[a]

Panel identification index[c]	Plywood thickness (in.)	Roof[b] Maximum span (in.) Edges blocked	Edges unblocked	Load (lb/ft²) Total load	Live load	Floor maximum span[d] (in.)
12/0	$\frac{5}{16}$	12		155	150	0
16/0	$\frac{5}{16}, \frac{3}{8}$	16		95	75	0
20/0	$\frac{5}{16}, \frac{3}{8}$	20		75	65	0
24/0	$\frac{3}{8}, \frac{1}{2}$	24	16	65	50	0
30/12	$\frac{5}{8}$	30	26	70	50	12[e]
32/16	$\frac{1}{2}, \frac{5}{8}$	32	28	55	40	16[g]
36/16	$\frac{3}{4}$	36	30	55	50	16[g]
42/20	$\frac{5}{8}, \frac{3}{4}, \frac{7}{8}$	42	32	40[f]	35[f]	20[g]
48/24	$\frac{3}{4}, \frac{7}{8}$	48	36	40[f]	35[f]	24

[a]These values apply for Structural I and II, C-C and C-D grades only. Spans shall be limited to values shown because of possible effect of concentrated loads.

[b]Uniform load deflection limitation: 1/180 of the span under live load plus dead load, 1/240 under live load only. Edges may be blocked with lumber or other approved type of edge support.

[c]Identification index appears on all panels in the construction grades listed in footnote.

[d]Plywood edges shall have approved tongue-and-groove joints or shall be supported with blocking, unless $\frac{1}{4}$-in. minimum thickness underlayment is installed, or finish floor is $\frac{25}{32}$-in. wood strip. Allowable uniform load based on deflection of 1/360 of span is 165 lb/ft.

[e]May be 16-in. if $\frac{25}{32}$-in. wood-strip flooring is installed at right angles to joists.

[f]For roof live load of 40 lb/ft² or total load of 55 lb/ft², decrease spans by 13% or use panel with next greater identification index.

[g]May be 24 in. if $\frac{25}{32}$-in. wood-strip flooring is installed at right angles to joists.

(b) Face grain parallel to supports[h]

	Thickness	No. of plies	Span	Total load	Live load
Structural I	$\frac{1}{2}$	4	24	35	25
	$\frac{1}{2}$	5	24	55	40
Other grades	$\frac{1}{2}$	5	24	30	25
covered in UBC	$\frac{5}{8}$	4	24	40	30
Standard 25–9	$\frac{5}{8}$	5	24	60	45

[h]Uniform load deflection limitations: 1/180 of span under live load plus dead load, 1/240 under live load only. Edges shall be blocked with lumber or other approved type of edge support.

CONNECTIONS

The three major parts of a diaphragm are the web, the chords, and the connections. Connections are extremely critical in achieving diaphragm action stemming from the fact that (1) the individual pieces of the web must be connected to form a unit, (2) the chord members in all probability are also not single length pieces, and (3) web and chords must be secured so that they act together. Connection detailing actually becomes a major part of the diaphragm design procedure. For instance, with masonry walls, the perimeter detail for the roof could be handled in several ways. Horizontal lumber edge members, adequately bolted to the masonry wall, could serve as the chords; or the plywood web could be attached directly to a masonry bond beam at the roof line. In the latter case, the bond beam would function as the chord element. Because of the ease of nailing plywood to a lumber chord and of bolting this chord to the masonry, lumber chords are often used. Examples may be found in Chapter 11.

PLYWOOD DESIGN PROCEDURE: EXAMPLE

The following example will show how the values in Table 10-2 are used when designing a wood diaphragm. This procedure, devised by the American Plywood Association (APA) and approved by ICBO, was based essentially on a long and extensive series of tests conducted by the American Plywood Association (APA) Research Center, the Forest Products Laboratory, and the Oregon Forest Research Center. This example appears in an APA pamphlet, reproduced here with APA's permission.

The roof diaphragm for the building shown in Figure 10-9 is to be designed. It is assumed that the size and spacing of the framing members have been already determined, as has the plywood thickness, all based upon the vertical live- and dead-load requirements. Shear walls are spaced so as to comply with permissible length/width ratios (i.e., 4:1 as specified in Table 10-1).

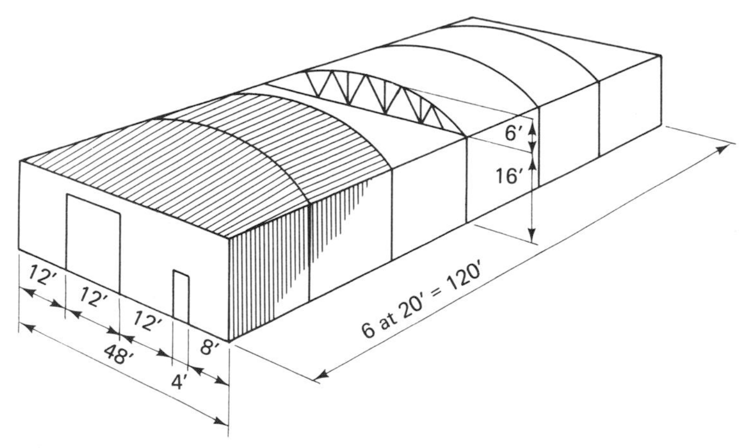

FIGURE 10-9. Diaphragm design procedure—example.

Note that the diaphragm need not be flat to resist shear. In this case the bowstring trusses act as web stiffeners. Assume in this example that the lateral wind load of 25 lb/ft² governs, since the building is located in seismic zone 1. The roof joists consist of 2 × 12's at 24 in. cc and they are solidly blocked at their ends.

Step 1: Diaphragm Loading

$$w = 25 \text{ lb/ft}^2 (6 \text{ ft} + \underbrace{16 \text{ ft}/2}_{\downarrow \text{ half-wall height}}) = 350 \text{ lb/lin. ft} \longleftarrow \text{lateral force at eave line}$$

↓ half-wall height (16 ft/2); ↑ vertical truss projection (6 ft)

Step 2: Diaphragm Shear

$$\text{total shear} = (350 \text{ lb/ft})\left(\frac{120 \text{ ft}}{2}\right) = 21{,}000 \text{ lb} \qquad \text{tributary width distribution}$$

$$\text{unit shear} = 21{,}000 \text{ lb}/48 \text{ ft} = 438 \text{ lb/ft}$$

Step 3: Plywood Panel Layout Note cases 1 to 6 in Table 10-2. They are simply illustrations of the rule expressed in the note under the table. Design stresses in all cases depend upon the direction of the continuous panel joints with reference to the load direction. In unblocked diaphragms, the direction of the unblocked edges is also significant.

As the table indicates, a $\frac{3}{8}$-in. Structural I plywood (APA grade and U.S. Product Standard Governing Construction and Industrial Plywood), with face grain perpendicular to joists, can carry the required shear of 438 lb/ft, provided the panel is blocked. The continuous panel joints will run in the long direction of the building. Since framing and construction joints are in same direction, this fits case 2 in Table 10-2 (refer to Figure 10-10).

Step 4: Plywood Nailing Schedule Table 10-2 lists allowable loads for a number of different combinations of nailing, plywood species and thickness, and framing lumber width, as well as for both blocked and unblocked diaphragms. As an example of its use, the figure 420 in the first line of the table signifies that a maximum shear stress of 420 lb/lin. ft is carried by a case 2 blocked diaphragm of $\frac{5}{16}$-in. Structural I APA grade plywood; with 6d nails spaced 2 in. oc around the diaphragm boundaries and 3 in. oc on the other panel edges, on 2-in.-nominal-width framing members. Obviously, since it contains so much information in such compact form, this table must be used with caution.

The allowable shears given in Table 10-2 have been derived strictly from tests. They are actually higher than those that would be obtained theoretically from the standard nail values. Shears higher than those listed in the table may be justified by theoretical calculations.

Interpolation from Table 10-2 indicates that nailing at $3\frac{1}{2}$ in. and 5 in. will carry the 438-lb/lin. ft maximum shears in the diaphragm under consideration, using $\frac{3}{8}$-in. Structural I C-D plywood and 8d nails. This spacing is determined as follows: $\frac{600}{438} = 1.37 \times 1\frac{1}{3}$ (increase for short-duration loading) $= 1.82$.

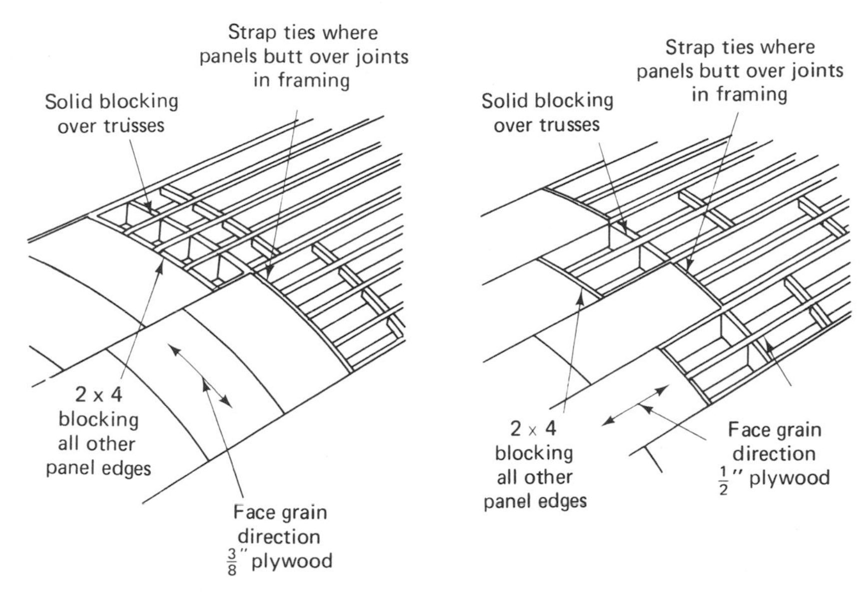

FIGURE 10-10. Plywood panel layout. Three-eighths-inch plywood with its face grain perpendicular to joists will adequately carry vertical load, but requires blocking at 4-ft centers. An alternative would be use of $\frac{1}{2}$-in. plywood with its face grain parallel to the joists, requiring blocking only at the panel ends. An analysis of blocking and labor costs for alternate 1 versus the additional cost of plywood for alternate 2 will indicate which is desirable. Note that either layout is case 2, Table 10-1.

$1.82 \times 2 = 3.6$, say, $3\frac{1}{2}$ in cc (boundary nailing)

$1.82 \times 3 = 5.5$, say $5\frac{1}{2}$ in. cc (nail spacing at all panel edges)

Nailing could be reduced to 6 in. at both the panel edges and at the boundaries where the shear becomes 270 lb/ft or less.

$$\frac{270}{438} \times \frac{120}{2} = 37 \text{ ft measured outward from the center, or}$$
$$23 \text{ ft in from the end of the diaphragm}$$

Blocking can be omitted where the shear reduces to 180 lb/ft (Table 10-2 unblocked diaphragms cases 2, 3, 4, 5, and 6).

$$\frac{180}{438} \times \frac{120}{2} = 25 \text{ ft measured outward from the center, or}$$
$$35 \text{ ft in from the end of the diaphragm}$$

The shear diagram below illustrates the points at which nail spacing may be changed and blocking omitted.

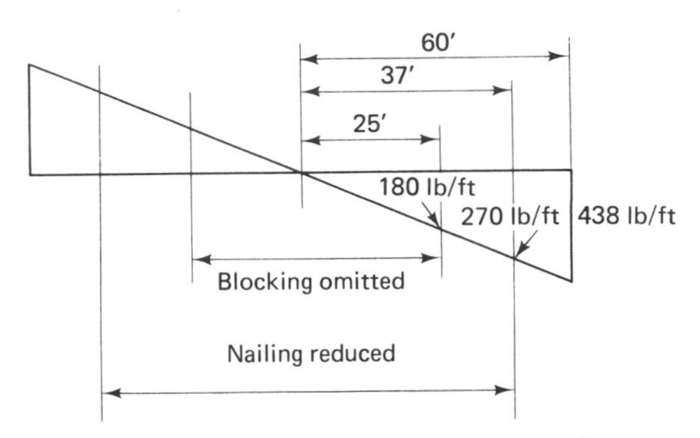

Step 5: Chord Size Chords must resist axial forces only, that is, those caused by the bending moment in the roof diaphragm:

$$\frac{wl^2}{8b} = \frac{(350 \text{ lb/ft})(120 \text{ ft})^2}{8(48 \text{ ft})} = 13{,}100 \text{ lb}$$

With an allowable tension stress of 1100 lb/in.^{2} increased one-third for short-time loads, a single No. 1 KD southern pine 2 × 8 will sustain 16,000 lb when well supported laterally.

Chord splices must be designed to resist the full load in the spliced members at the point where the joint occurs. At the actual point where one of the 2 × 8's is butted, all 13,100 lb must be carried by the other 2 × 8. Each 2 × 8 must be designed to carry the whole 13,100 lb, but only one-half of that load is transferred through bolts at the splice. In this case the bolts should be designed for 6550 lb. Refer to Figure 10-11 for a typical wood chord detail.

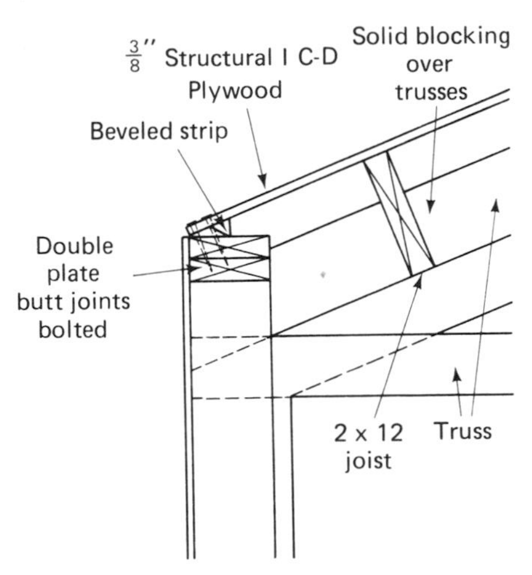

FIGURE 10-11. Wood chord detail.

Step 6: Diaphragm Deflection Ordinarily, the deflection of a timber diaphragm need not be calculated if the diaphragm length/width ratio limits noted in Table 10-1 are observed. However, should it become necessary to compute the deflection, the following approach can be adopted. Consider that the total maximum lateral deflection is comprised essentially of four factors: (1) bending deflection, (2) shear deflection, (3) nail slip, and (4) chord splice slip. Thus, the basic formula becomes

$$d = \frac{5vL^3}{8EAb} + \frac{vL}{4Gt} + 0.094Le_n + \frac{\sum(\Delta_c X)}{2b}$$

where

d = deflection, in.

v = shear, lb/ft

L = diaphragm length, ft

b = diaphragm width, ft

A = area of chord cross section, in.^{2}

E = elastic modulus of chords, lb/in.^{2}; for use in this formula, E values listed in the lumber standards should be increased by 3%, since shear deflection is separately calculated. This 3% restores the usual reduction included in tabulated E values to account for shear deflection

G = shearing modulus of webs, lb/in.^{2}

t = effective plywood thickness for shear, in.

e_n = nail deformation from Figure 10-12 at calculated load per nail, on perimeter of interior panels, based on shear per foot divided by number of nails per foot. If the nailing is not the same in both directions, use the greater spacing for calculations

$\sum(\Delta_c X)$ = sum of individual chord-splice slip values (Δ_c) on both sides of the diaphragm, each multiplied by its distance to the nearest support

In this example it was assumed that the chord splices occurred every 20 ft and that each of 10 splices (five each side) exhibited a slip of $\frac{1}{16}$ in. Also, the value of e_n was taken from Figure 10-12 as 0.05 in. The material is Douglas fir or southern pine lumber; green when nailed; tested after seasoning to 13% moisture content; average specifie gravity = 0.48.

$$\frac{5 \times 438 \times 120^3}{8 \times (1.03 \times 1.900{,}000) \times 21.8 \times 48} = 0.231 \text{ in.}$$

$$\frac{438 \times 120}{4 \times 90{,}000 \times 0.371} = 0.394$$

$$0.094 \times 120 \times 0.05 = 0.564$$

$$\frac{(4 \times \frac{1}{16} \times 20) + (4 \times \frac{1}{16} \times 40) + (2 \times \frac{1}{16} \times 60)}{2 \times 48} = 0.234$$

Total deflection, d = 1.423 in.

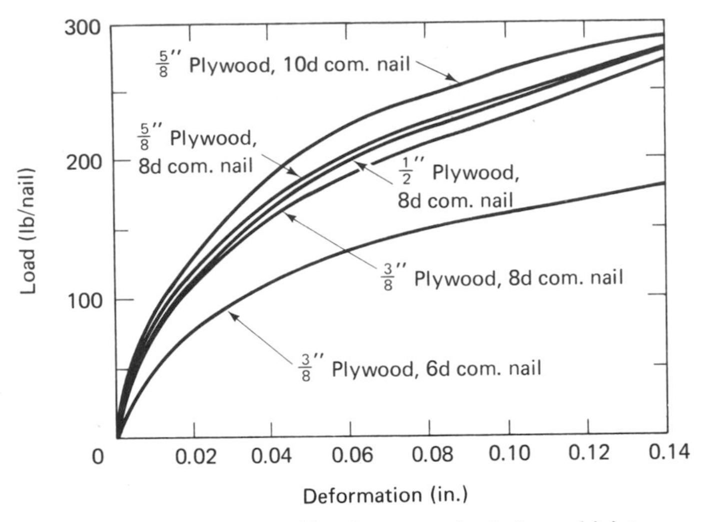

FIGURE 10-12. Lateral bearing strength of plywood joints.

Should the end-wall deflection prove significant it should be added to this total. This computed value should then be compared with the deflection magnitude which a reinforced masonry wall, with a given h/t ratio and thickness, can tolerate. If it became necessary to reduce this deflection, the most effective measure would consist of increasing the plywood thickness or decreasing the nail spacing. Placing an additional shear wall within the building would have a similar effect, since the span would be reduced. In this instance, perhaps $\frac{1}{2}$ in. plywood over 3×12 joists spaced at 32 in. oc would suffice, should the deflection need to be reduced.

Concrete diaphragm construction

ANALYSIS

Concrete diaphragms must conform to ACI 318-77. The concrete diaphragm webs are designed as concrete slabs which must carry the vertical loads between the framing members, or they must be supported on other vertical load-carrying elements. According to an analysis procedure, outlined by the Portland Cement Association, the horizontal inertial force carried by the rigid concrete diaphragm itself is calculated as $F_p = ZIC_pSW_p$, where C_p equals 0.12. (Table 5-5). Since we are dealing with reinforced concrete in this instance, the value of F_p must be converted to an ultimate design horizontal force acting on the floor. This is obtained by multiplying F_p by a Code-specified load factor, U, of $1.3W$ (wind) or $1.43E$ (earthquake). This force (F_pU) is then distributed to the shear walls on the basis of their relative rigidities. The average unit stress is computed as

$$v_u = \frac{V_u}{\phi b_w d}$$

where $d =$ net slab length

b_w = the slab thickness
$\phi = 0.85$ = material quality factor

This should be compared to the allowable shear, $v_c = 2\sqrt{f'_c}$.

If the span/depth ratio of the floor slab is less than a value of $\frac{4}{5}$, which characterizes the lower bound of that ratio for a deep and simply supported flexural member, the floor slab may be considered to act as a deep girder. Where the floor slab acts as an ordinary flexural member, assuming a homogeneous section, the approximate maximum tensile stress in the concrete at a section near the center of the span, due to a uniform load, is determined as

$$\text{uniform load (kips/ft)}, W = F_p U \div \text{diaphragm span}$$

$$\text{stress index}, f = \frac{Mc}{I}$$

where f = modulus of rupture, lb/in.^{2}
$M = W \times (\text{diaphragm span})^2/8$
$c = (\text{diaphragm depth}/2) \times 12$
$I = \text{diaphragm thickness} \times (\text{diaphragm depth} \times 12)^3/12$

Actually, since the value of f represents the modulus of rupture, the allowable value may be approximated at $7.5 \times \sqrt{f'_c}$. Normally the flexural stress due to the forces acting on the slab is easily resisted by the concrete alone without even considering the contribution of the steel reinforcement in the slab.

Reinforcing and Connection Details

There are some special considerations concerning reinforcing arrangements and location which must be observed. It is common practice to provide a minimum of two No. 5's as continuous chord reinforcing, adding more when the design calculations stipulate.

Also, it is desirable to introduce additional diagonal reinforcement in the outside corner of the floor slabs to minimize excessive cracking, as shown in Figure 10-13a. This is done by augmenting the shrinkage reinforcement with dowels located between the slab and the wall, and by extending a diagonal band of bars from the corner into the floor slab. The exact amount of reinforcement varies for each case, depending upon the configuration, slab thickness, span, and loads, among other factors.

Stress concentrations occur at points of discontinuity such as those created by large unframed openings. At such locations, an amount of reinforcement equivalent to that interrupted by the opening should be located near the sides of such openings. Further diagonal bars, as shown in Figure 10-13b, are needed.

Deflections

These concrete slabs are categorized as rigid diaphrams, and thus are dimensionally governed only by the drift permitted on masonry walls subjected to wind or seismic forces. The diaphragm deflections must be determined by using unfactored loads.

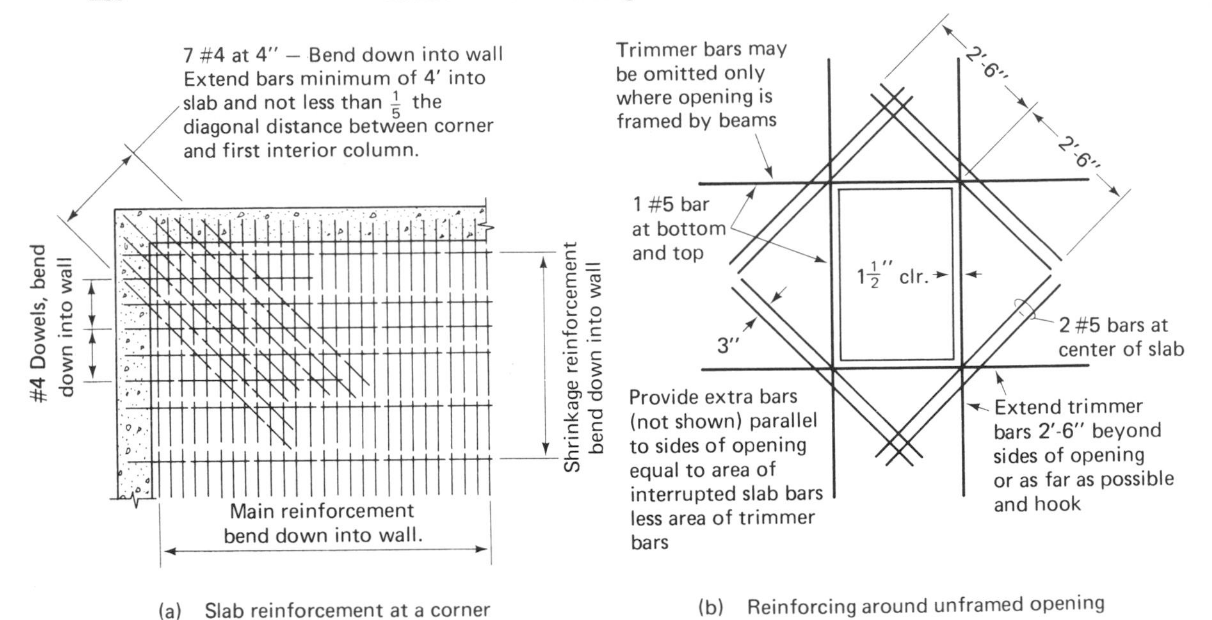

FIGURE 10-13. Special reinforcement—concrete diaphragms.

Horizontal bracing

Instead of using horizontal diaphragms, the strength and rigidity of a horizontal bracing system may be used to distribute the horizontal forces to the vertical lateral-force-resisting elements, which, in turn, could consist of vertical wall bracing in lieu of shear walls. Such a truss-type system tends to minimize lateral sidesway or drift. Lateral bracing may be constructed of any approved material, such as concrete, steel, or wood. Structural steel shapes or rods are often used because of the ease in fabricating such a system with this material. When planning the bracing system for a building, one must consider the structure as a whole. This means visualizing all ways in which the structure might become unstable and then providing bracing with the strength and rigidity needed to maintain the structure's stability in any failure mode. The wall framing towers must be such that they will carry the reactions brought to them by the horizontal bracing system. Consideration must also be given when locating the bracing so that it does not become an obstruction in doors, windows, or passageways.

DESIGN

Actually, the horizontal bracing system is nothing more than a truss made up of chords and diagonals lying in the plane of the roof or the floor. The horizontal forces to be transferred are assumed to act at the joints of the truss, and these loads are carried by truss action to the side walls where the truss reactions become the shears transferred to the side shear wall or wall bracing. So the members are simply sized on the basis of a truss analysis, with the loads stemming from the appropriate wind or seismic factors evaluated as shown previously.

LAYOUT

There are many different schemes for providing bracing in the plane of the roof or the floor, and the preferred layout must be determined from each individual requirement. Typical layouts may be seen in Figure 10-14a and b. In any of these systems, note that they are only as good as the strength of the connections at the joints. Thus, these connections must be strong enough to develop the useful strength of the member.

Where turnbuckles are used to join a rod bracing system together, care must be exercised to ensure that all turnbuckles are uniformly tight so that they will act together as a complete unit. Otherwise, some rods may carry considerably more than the design stress, while others may be very understressed. This could lead to a failure within the system, with a corresponding loss in structure stability.

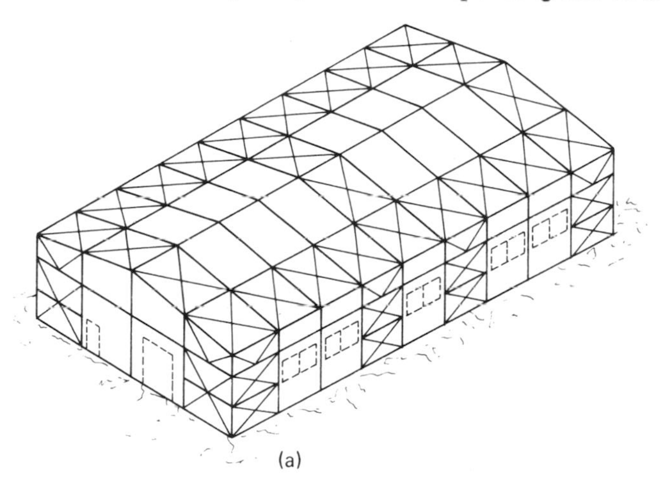

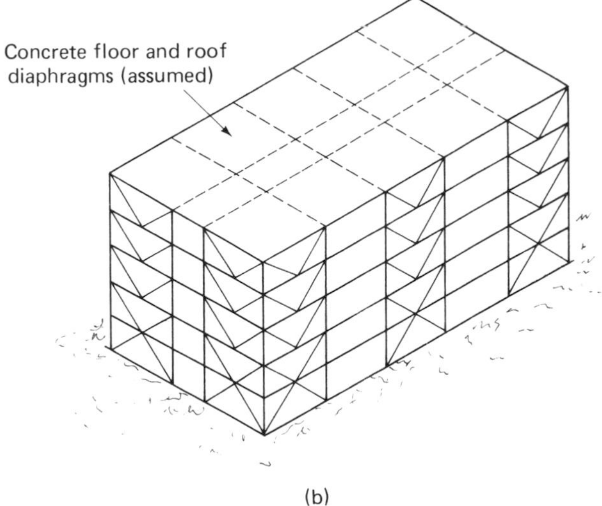

FIGURE 10-14. Bracing layout: (a) bracing for industrial building; (b) bracing for tier building.

VERTICAL STABILITY ELEMENTS

Solid shear walls

Horizontal forces at any floor or roof level, as previously pointed out, may be transferred to the foundation through the strength and rigidity of the side walls in a box-type lateral force-resisting system ($K = 1.33$). These elements are referred to as *shear walls.* The design strength of shear walls is often governed by flexure; however, in low walls, the governing design criterion may be shear. A shear wall is a sort of cantilever plate girder, standing upright, wherein the wall performs the function of the plate girder web, with the pilasters or floor diaphragms serving as web stiffeners. The tension/compression flanges are provided for by the reinforcement (jamb steel) located at the vertical boundaries at each end of the wall, or by specially designed vertical boundary elements. Refer to Figure 10-3 to view the forces acting on one of these side walls.

WALL LOADS

As may be seen in Figure 10-15, walls may be subjected to various types of forces. They may, for instance, be called upon to carry vertical dead and live loads other than their own weight, whereupon they are referred to as *bearing walls*. These gravity loads induce compressive stresses, which must be compared to the axial compressive stress permitted on reinforced masonry. Should such gravity loads act off the centroidal axis of the wall section, the combined eccentric bending and axial effects must be examined.

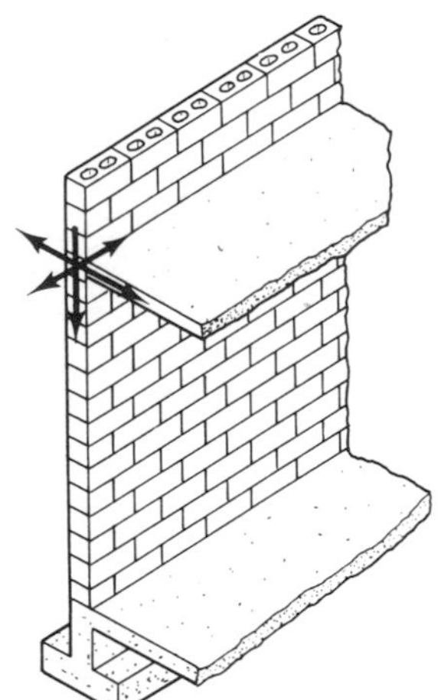

FIGURE 10-15. Forces on shear wall.

Horizontal forces, caused by either wind or seismic action, will act normal to the wall and parallel to it. When a wall is called upon to resist in-plane horizontal shear forces, it becomes a shear wall. As such, it may or may not function as a bearing wall as well, although it often does carry vertical loads. In the latter case, the combined effect of horizontal forces and vertical loads must be considered when verifying the structural adequacy of the wall. In addition, tensile forces may result from the overturning moment caused by the in-plane lateral forces, and they must be resisted by boundary reinforcement properly anchored to the foundation. A nonbearing wall, since it does not carry vertical loads, can serve as a filler, curtain wall, or partition, as well as a shear wall or a combination of these.

Loads parallel to the walls (i.e., when the wall acts as a shear wall) produce deformations which are qualitatively represented in Figure 10-16. The rigidity of a shear wall is inversely proportional to its deflection under a unit horizontal force. Where shear walls are tied together by a rigid diaphragm or bracing so that all must deflect equally, the total lateral force is shared in direct proportion to their relative rigidities. One way of calculating lateral stiffness and stresses in a single shear wall without openings involves simple flexural theory only. Wall deflection is the sum of the deflections due to bending and shear. However, for architectural or environmental reasons, openings are often introduced, thereby reducing the effectiveness and drastically altering the deformation characteristics of such *coupled shear walls*. The supporting mechanism (floor) also has an effect. Hence, a mathematical analysis of this condition can become extremely complex. Procedures for both cases will be discussed in subsequent paragraphs.

The forces that walls must resist normal to their surface stem from the action of wind or seismicity. Whichever of these two is larger will govern the design. The Code-specified seismic force normal to the wall, F_p, equals ZIC_pSW_p, where C_p (the acceleration coefficient) equals 0.20 (Table 5-9). This force is considered to act on the wall in both the inward and outward directions. The wall may be designed to carry this load by spanning vertically between floor diaphragms or between the roof diaphragm and the ground. Or it may be reinforced to span horizontally between pilaster or cross-wall supports. At any rate, the spans may be considered continuous or noncontinuous across these supports, depending upon how they are reinforced.

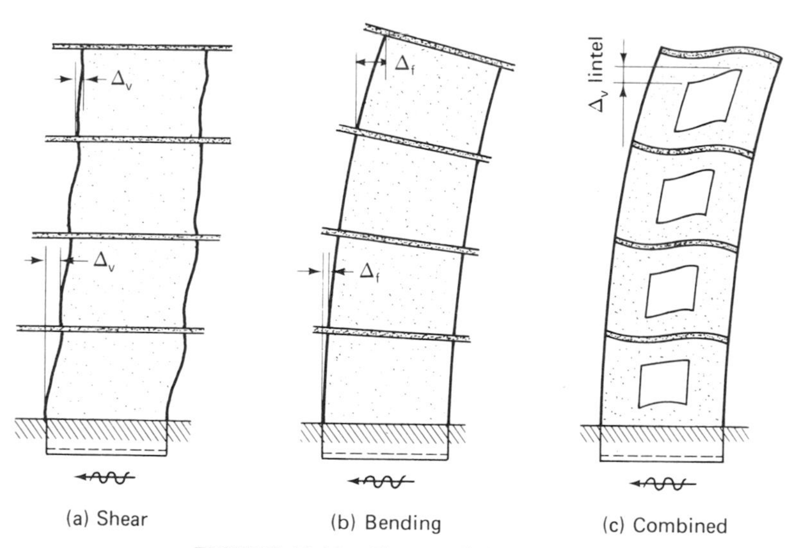

FIGURE 10-16. Shear-wall deformations.

Coupled shear walls

Multistory shear walls with openings present a number of problems. If the openings are very small, their effect on the overall stress is minor. However, larger openings have a much more pronounced effect. Openings (windows, doors, etc.) normally occur in regularly spaced vertical rows throughout the height of the wall, and the tie between the vertical wall sections is provided either by connecting beams which form a part of the wall, or through floor slabs, or a combination of both. From the structural-analysis standpoint, the problem is highly redundant, and a mathematical solution for the determination of stresses and deformations is extremely complex. The term "coupled shear wall" is often applied in such cases. One solution to this problem is described in some detail in the next section.

METHOD OF ANALYSIS

The phenomenon of coupled shear walls has evolved recently through the increase in the number of high-rise masonry buildings being erected for both residental and commercial purposes (e.g., apartments and motels). Although the loads in the walls may be determined from the assumption that all walls deflect equally due to the relatively high in-plane stiffness of the slabs, even solid high-rise walls do not really act as independent cantilevers because of the coupled bending action of the floor slabs. Walls that contain openings throughout their height behave in a similar manner (Figure 10-17), and a practical approach to the solution of this problem is needed to obtain a satisfactory analysis of such coupled-shear-wall structures. One practical answer to

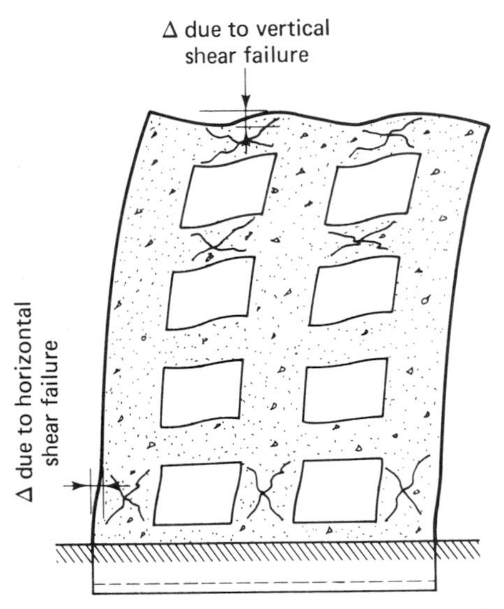

FIGURE 10-17. Deformation of shear wall with openings.

this problem has been provided by Alexander Coull and J. R. Choudbury, and the following material has been excerpted with permission from their article "Stresses and Deflections in Coupled Shear Walls" published in the *ACI* Journal in February 1967.

One previous approach to this problem was based upon the assumption that a discrete system of connections, formed by lintel beams or floor slabs, may be replaced by an equivalent continuous medium. By assuming that the cross beams have their points of contraflexure at midspan, and that the axial deflection is minimal, the behavior of this type of system may be expressed as a single second-order differential equation, which produces a general closed solution to the problem. Model results provided good verification of this theory. Figure 10-18 shows how the individual connecting beams of stiffness EI_p are replaced by a continuous medium or lamina of stiffness EI_p/h per unit height. Assuming that the connecting beams do not deform axially under the reaction of the lateral loading, both walls will deflect equally, with the point of contraflexure occurring at the midpoint of each connecting beam. Then, if the laminas are considered cut at their midpoints, the only force acting at the cut section is a shear force of intensity q per unit of height x. By considering the deformations of the cut laminas, one establishes compatability conditions which

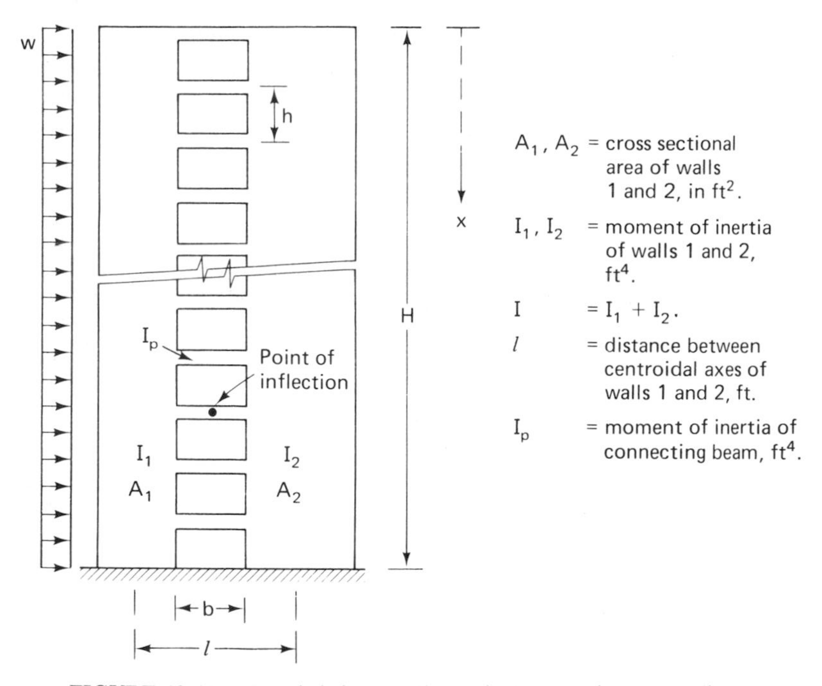

FIGURE 10-18. Coupled shear wall—equivalent continuous medium.

state that no resultant relative deformations occur at the cut. This leads to the following governing differential equation:

$$\frac{d^2T}{dx^2} - \alpha^2 T = -\beta x^2 \tag{1}$$

where
$$\alpha^2 = \frac{12I_p}{hb^3}\left(\frac{l^2}{I} + \frac{A}{A_1 A_2}\right)$$
$$\beta = \frac{1}{2} wl\left(\frac{12I_p}{hb^3}\right)\frac{1}{I}$$
$$T = \int_0^x q\, dx$$
$$= \frac{2\beta}{\alpha^4}\left(1 + \frac{\sinh \alpha H - \alpha H}{\cosh \alpha H} \sinh \alpha x - \cosh \alpha x + \tfrac{1}{2}\alpha^2 x^2\right)$$

However, the solution, based upon this differential equation defining T, which represents the total shear force in the continuous connection from top of wall to position x, proved too labo ious to be feasible for an office design procedure. So a more suitable approach to coupled-shear-wall analysis was sought by the authors, with the following results.

Wall Stresses

Expressions defining the extreme fiber stresses for the actual coupled-shear-stress distribution shown in Figure 10-19b were written in the earlier solution in terms of the wall moments M_1 (wall 1) and M_2 (wall 2), the shear force T, and the cross-sectional properties of the wall (i.e., area and moment of inertia). This actual stress distribution, although theoretically a superposition of a uniform axial stress on a linear bending stress distribution, could more conveniently be defined by using the alternative superposition of two pure bending stresses: (1) the bending stress occurring when the wall system is assumed to act as a single composite cantilever, the neutral axis being situated at the centroid of the two wall elements (Figure 10-19c): and (2) the two linear stress distributions obtained on the assumption that the walls act independently of each other, with a neutral axis located at the centroid of each wall as in Figure 10-19d.

Now, considering these two stress patterns:

1. Composite cantilever action:

$$\text{Moment at any section} = \tfrac{1}{2}wx^2 \frac{K_2}{100}$$

where K_2 = a computational convenience representing the percent of load carried by composite cantilever action

Then the composite cantilever extreme fiber stresses in the wall at edges A and B of wall 1 are obtained as:

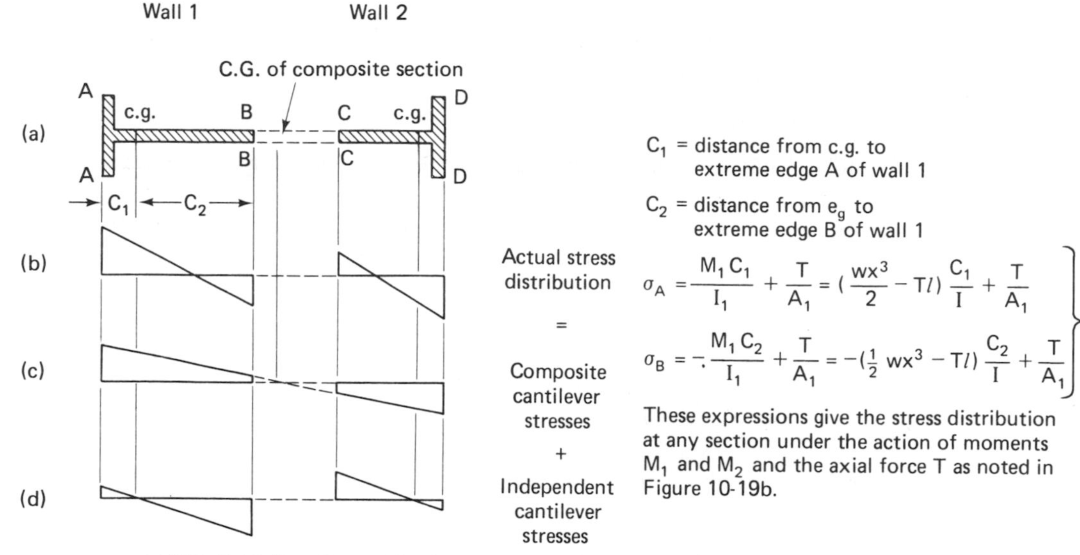

FIGURE 10-19. Stress distribution on walls 1 and 2. Superposition of stress distributions due to composite and individual cantilever action to give true stress distribution in walls.

$$\left.\begin{aligned}\sigma_A &= \frac{wx^2}{2I'}\left(\frac{A_2 l}{A} + C_1\right)\frac{K_2}{100}\\ \sigma_B &= \frac{wx^2}{2I'}\left(\frac{A_2 l}{A} - C_2\right)\frac{K_2}{100}\end{aligned}\right\} \quad I' = I_1 + I_2 + \frac{A_1 A_2}{A} l^2 \qquad (2)$$

which represents the moment of inertia of the composite cantilever

Similar expressions are obtained for wall 2.

2. Individual cantilever action:
   The wall moments, M_1 and M_2, at any level x, are expressed as:

$$M_1 = \tfrac{1}{2} wx^2 \frac{I_1}{I}\frac{K_1}{100}$$

$$M_2 = \tfrac{1}{2} wx^2 \frac{I_2}{I}\frac{K_1}{100}$$

where K_1 = another simplifying constant representing the percent of load carried by independent cantilever action

The individual cantilever extreme fiber stresses in the wall become

$$\left.\begin{aligned}\sigma_A &= \frac{M_1 C_1}{I_1} = \frac{1}{2} wx^2 \frac{C_1}{I}\frac{K_1}{100}\\ \sigma_B &= -\frac{M_1 C_2}{I_1} = -\frac{1}{2} wx^2 \frac{C_2}{I}\frac{K_1}{100}\end{aligned}\right\} \text{similar expressions hold for wall 2} \qquad (3)$$

These expressions are based upon the assumption that the axial deformations in the connecting beams may be ignored, and that since the walls deflect equally, the loads they carry will be proportional to their stiffnesses.

Connection Beam Stresses

The values of K_2 and K_1 may be obtained by equating the corresponding stresses, at the four extreme fiber locations, from the previous stress equations for wall 1 (eqs. 1, 2, and 3), and using the value of T as defined from the solution to the differential equation $d^2T/dx^2 - \alpha^2T = -\beta x^2$. The results are shown in Figure 10-20. Note that $K_1 = 100 - K_2$.

To obtain the stress in the connection beam, find

$$q = \frac{dT}{dx} = \frac{wH}{l}\frac{1}{\mu}K_3 \tag{4}$$

where q = shear force/unit height in the equivalent continuous system of connecting laminas

Note that $\mu = (A/A_1A_2)(I/l^2)$ and K_3 is obtained from Figure 10-21.

If equation (4) is differentiated, the maximum shear force intensity is found to be

$$q_{\max} = \frac{wH}{l}\frac{1}{\mu}(K_3') \tag{5}$$

where K_3' (from Figure 10-21) is the value of K_3 evaluated at the position where the maximum shear force intensity occurs.

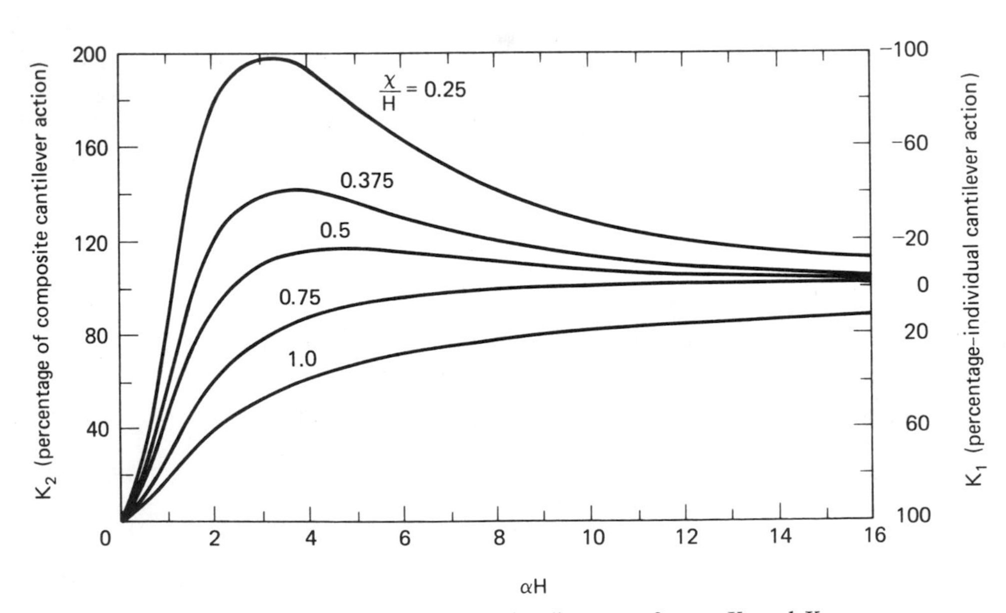

FIGURE 10-20. Variation of wall bending stress factors K_1 and K_2.

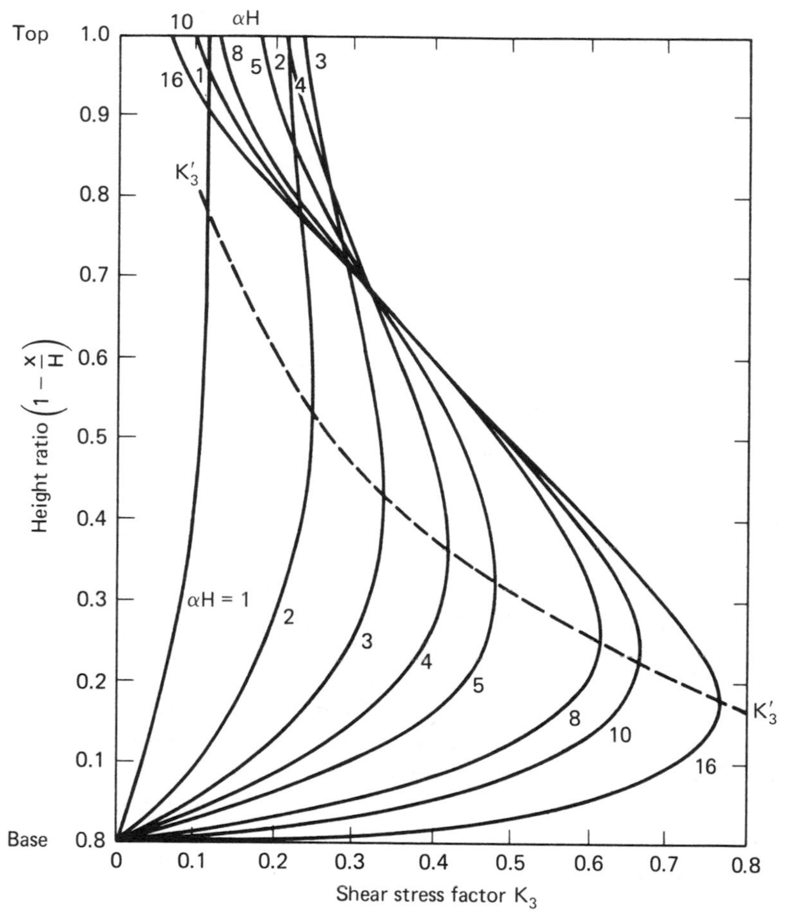

FIGURE 10-21. Variation of connecting beam stress factors K_3 and K_3'.

Deflections

The moment curvature relation for each wall is given by

$$\frac{EId^2y}{dx^2} = \frac{1}{2}wx^2 - Tl$$

where y = horizontal deflection, ft, at any height x

Integrating this expression twice produces the expression for the deflection, y, when the appropriate boundary conditions have been inserted therein. Then the maximum deflection at the top of the structure is obtained as

$$y_m = \frac{1}{8}\frac{wH^4}{EI}K_4 \tag{6}$$

where K_4 is obtained from Figure 10-22.

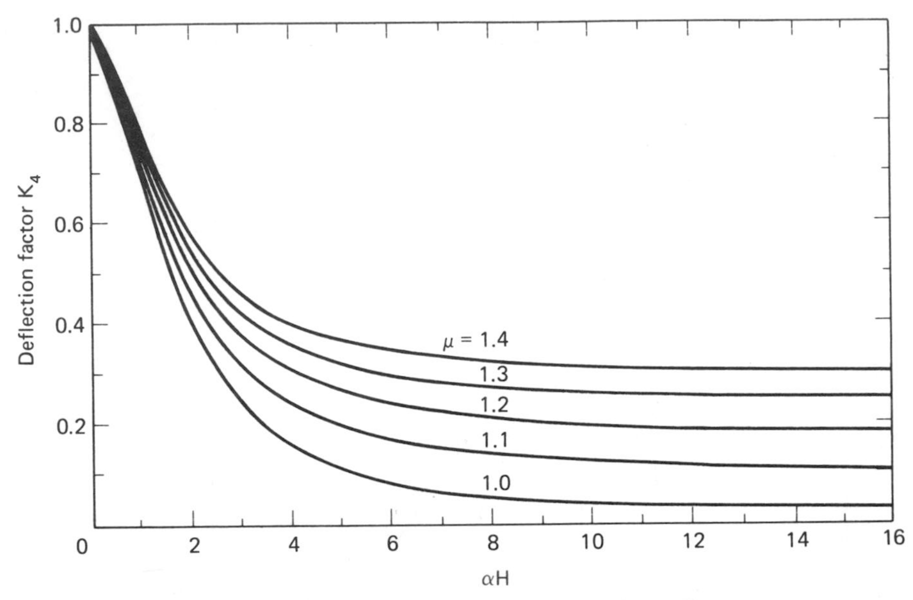

FIGURE 10-22. Variation of deflection factor K_4.

SOLUTION PROCEDURE

After the evaluation of all the stiffness parameters for the wall system, the proportions of the load carried by composite cantilever action K_2, and individual cantilever action K_1, may be determined from Figure 10-20 for any height position. The bending stresses in the walls may then be obtained from ordinary beam theory, based on the superposition of the two distinct cases described in equations (2) and (3).

The stresses in the connecting beam can be obtained next, by first computing the intensity of the shear forces in the equivalent continuous connecting medium from equation (4), the required coefficient, K_3, being taken from Figure 10-21. The shear force, Q, in any particular beam is given by the area underneath the curve between half-story-height levels above and below the beam position. The shape of the curves enables the maximum value of the shear force to be readily determined. The maximum bending moment in the connecting beam is then given by $\frac{1}{2}Qb$.

The maximum shear intensity is defined by equation (5) and the maximum shear force in any connecting beam will thus be given approximately by the product of the maximum intensity (the coefficient K_3' being determined from Figure 10-21) and the story height. This value will always be greater than the true maximum, the error involved being dependent on the number of stories as wall as the other geometrical parameters. The greater the number of stories, the smaller the error will be. The maximum deflection at the top of the structure is given by equation (6), the required coefficient K_4 being given in Figure 10-22.

It would thus appear that the curves developed in this article present a practical procedure for the rapid evaluation of stresses and deflections of coupled shear walls. They are general, in that they hold for any shape of wall; thus, cross walls acting as flanges can even be incorporated into the analysis. An example of this procedure follows in Chapter 13.

Wall deflections and rigidities

INDIVIDUAL CANTILEVER AND FIXED-END PIERS

As was pointed out in a previous section, the roof or floor diaphragm is considered to transmit wind or seismic loadings to the transverse shear walls on the basis of relative wall rigidities, if supported by a rigid diaphragm. With rigid diaphragms, such as reinforced concrete or steel decking, where the center of the mass coincides with the center-of-the-wall rigidities (the point where lateral force must act in order to produce equal deflections of the resisting elements), the supporting walls will deflect equally, as seen in Figure 10-23a. Consequently, horizontal forces transferred by the diaphragm will be distributed to the cross walls in inverse proportion to their flexibility. Thus, a very flexible wall will resist only a small portion of the total seismic force. It will take a proportionately larger portion of that force to cause a stiffer wall to deflect the same amount. In terms of stiffness (i.e., the lateral force required to produce a unit deflection) or the reciprocal of flexibility, the lateral forces will be distributed in direct proportion to the relative stiffness of the resisting elements. It should be noted here that it is not necessary to compute the absolute stiffness of these resisting walls, rather only the relative stiffness, as each one is simply being compared to the other elements.

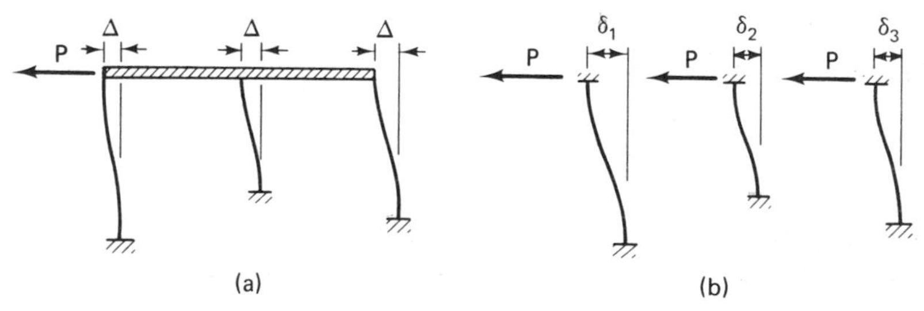

FIGURE 10-23. Deflection of wall consisting of connected piers-rigid diaphragm support.

In masonry structures, it is generally conceded that in one- and two-story buildings the walls may be considered cantilevered. On the other hand, in a multistory building, the segments of the walls between stories above the first story would be considered fixed at both top and bottom. These two idealized conditions of end support are shown in Figure 10-24. When a force is applied at the top of a pier, it will produce a deflection, Δ, which is the sum of the deflections due to bending moment (Δ_b) plus that due to shear (Δ_v). For the case of both ends fixed, that is, restrained against rotation, the total pier deflection, Δ_F, is defined as

$$\Delta_F = \Delta_b + \Delta_v = \frac{PH^3}{12E_mI} + \frac{1.2PH}{E_vA} \qquad (7)$$

where P = lateral force on pier—lbs
H = height of pier—in.
A = cross-sectional area of pier—in.^{2}
I = cross-sectional moment of inertia of pier in direction of bending—in.^{4}
E_m = modulus of elasticity in bending—lb/in.^{2}
E_v = shear modulus—lb/in.^{2}

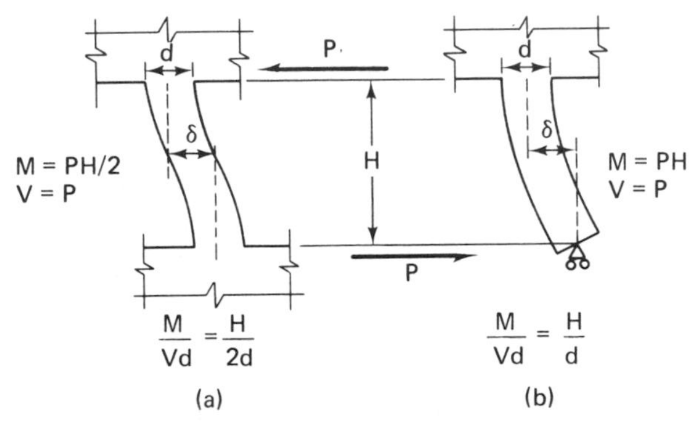

FIGURE 10-24. End-support restraints.

For the cantilever pier, that is, bottom end fixed, the total deflection, Δ_C, is:

$$\Delta_C = \Delta_b + \Delta_v = \frac{PH^3}{3E_mI} + \frac{1.2PH}{E_vA} \tag{8}$$

The rigidity of the pier is defined as

$$k = \frac{1}{\Delta_b + \Delta_v}$$

The preceding expressions (7) and (8) are general in that they apply to any homogeneous material. For those materials, such as masonry, where $E_v \sim 0.4E_m$, they can be written as

$$\Delta_F = \frac{P}{E_mt}\left[\left(\frac{H}{d}\right)^3 + 3\left(\frac{H}{d}\right)\right] \tag{7a}$$

where t = wall thickness-in.
d = effective length of wall-in.

$$\Delta_C = \frac{P}{E_mt}\left[4\left(\frac{H}{d}\right)^3 + 3\left(\frac{H}{d}\right)\right] \tag{8a}$$

Actually, since only the relative rigidity is sought, any value could be used for E_m. One set of curves, taken from "Reinforced Load Bearing Concrete Block Walls," published by the Masonry Institute of America, selects the convenient but arbitrary values $P = 1 \times 10^6$, $E_m = 1.2 \times 10^6$, and $t = 10$ in. to obtain the following expressions from (7a) and (8a):

$$\Delta_F = 0.0833\left(\frac{H}{d}\right)^3 + 0.25\left(\frac{H}{d}\right) \tag{7b}$$

$$\Delta_C = 0.333\left(\frac{H}{d}\right)^3 + 0.25\left(\frac{H}{d}\right) \tag{8b}$$

Figure 10-25 provides a graphical solution for these expressions, and they are also listed in tabular form in Table 10-4. The expressions above can be used for any wall thickness, actually, even though a value of $t = 10$ in. was used in obtaining expressions (7b) and (8b), since only relative rigidities are desired. However, where walls of different thicknesses are being compared, the values can be factored by the ratio of the different wall thicknesses. For instance, say that three walls of thicknesses

TABLE 10-4

Pier Deflections, Δ

		Δ *Fixed end*	Δ *Cantilever*
$\frac{H}{d}$	$\left(\frac{H}{d}\right)^3$	$0.0833\left(\frac{H}{d}\right)^3 + 0.25\left(\frac{H}{d}\right)$	$0.333\left(\frac{H}{d}\right)^3 + 0.25\left(\frac{H}{d}\right)$
0.1	0.001	0.0251	0.0253
0.2	0.008	0.0507	0.0527
0.3	0.027	0.0773	0.0840
0.4	0.064	0.1053	0.1214
0.5	0.125	0.1354	0.1668
0.6	0.216	0.1680	0.2221
0.7	0.343	0.2036	0.2896
0.8	0.512	0.2427	0.3710
0.9	0.729	0.2857	0.4685
1.0	1.000	0.3333	0.5840
1.1	1.331	0.3859	0.7196
1.2	1.728	0.4440	0.8772
1.3	2.197	0.5080	1.0588
1.4	2.744	0.5786	1.2665
1.5	3.375	0.6561	1.5023
1.6	4.096	0.7412	1.7368
1.7	4.913	0.8343	2.0659
1.8	5.832	0.9358	2.3949
1.9	6.859	1.0464	2.7159
2.0	8.000	1.1664	3.1720
2.1	9.261	1.2964	3.6182
2.2	10.648	1.4370	4.1064
2.3	12.167	1.5885	4.6388
2.4	13.824	1.7515	5.2172
2.5	15.625	1.9265	5.8438
2.6	17.760	2.1294	6.5818
2.7	19.683	2.3146	7.2491
2.8	21.952	2.5286	8.0320
2.9	24.389	2.7567	8.8709
3.0	27.000	2.9910	9.7500

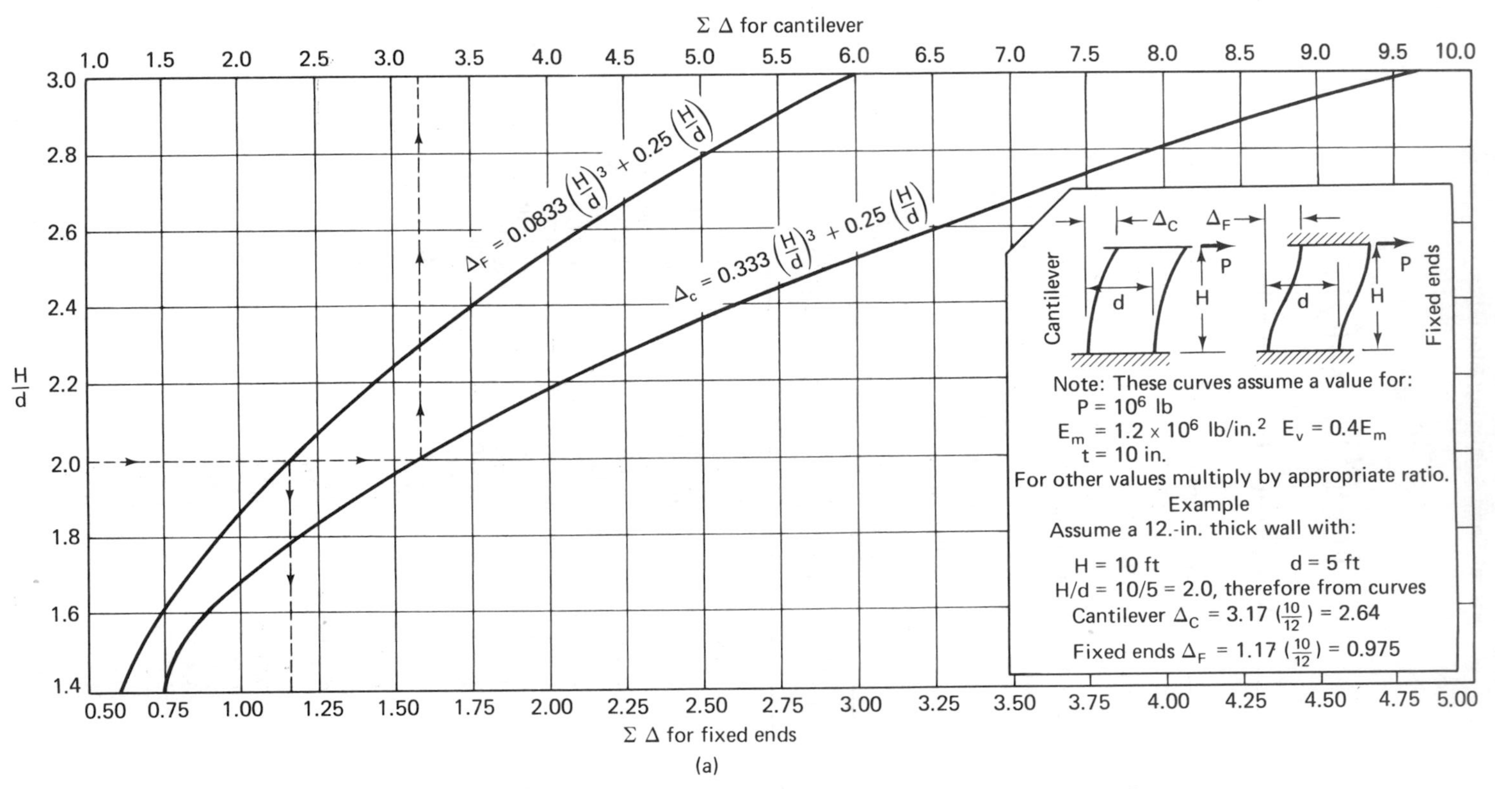

FIGURE 10-25. Wall deflections, Δ. (a) $0.0 \leq H/d \leq 1.5$. (b) $1.5 \leq H/d \leq 3.0$.

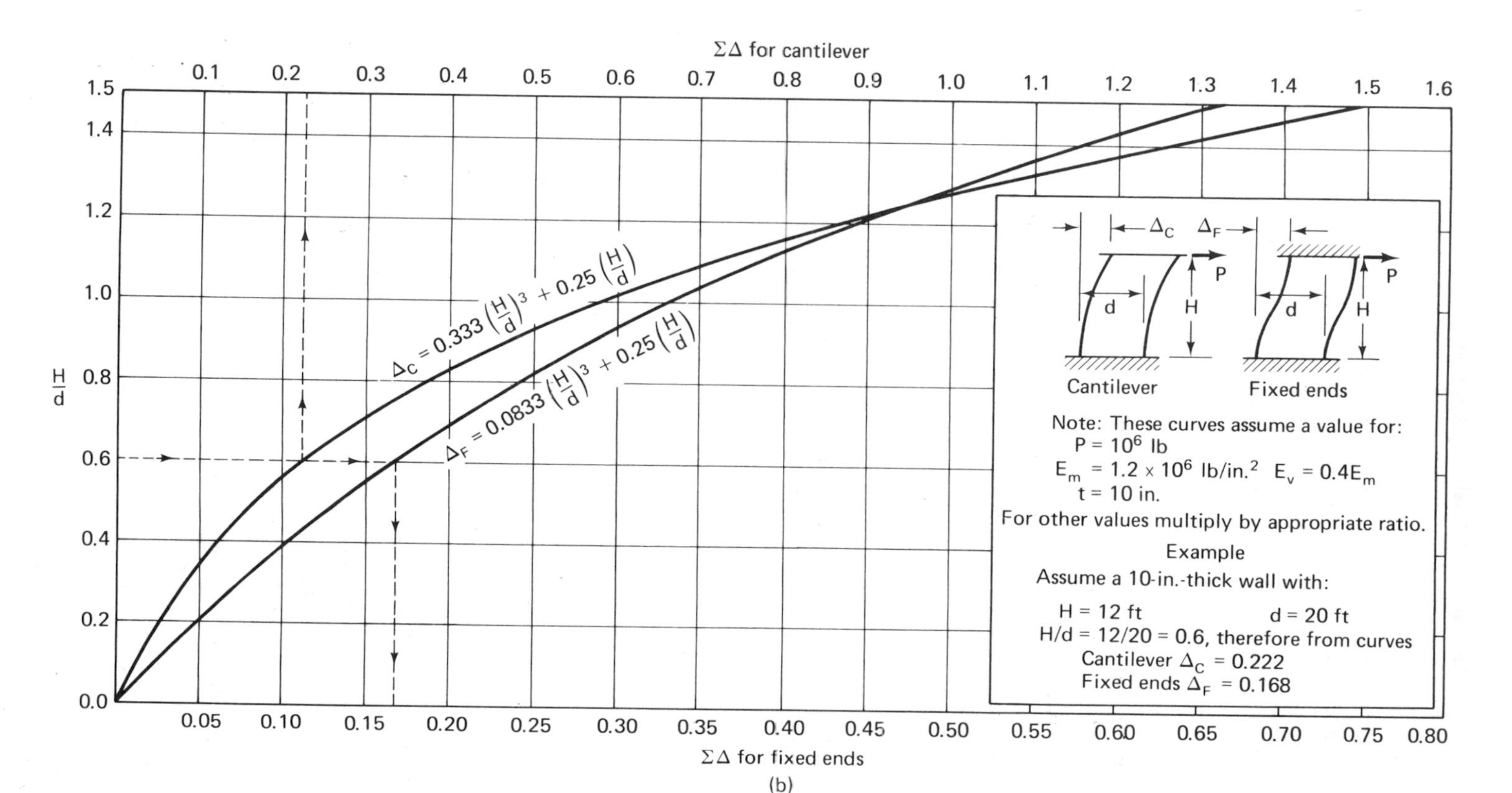

FIGURE 10-25. (continued)

9, 10, and 12 in. are supported by the diaphragm. Take the 10-in. wall as the datum thickness (actually, any thickness can be established as the basic one); then where the foregoing chart or table is used to obtain R values, multiply them by the following factors:

$$k_{12} = k_{\text{chart}} \times \tfrac{12}{10} \qquad \text{and} \qquad k_9 = k_{\text{chart}} \times \tfrac{9}{10}$$

Deflection of Wall Consisting of Connected Piers

A wall consisting of several piers connected at their tops was shown in Figure 10-23. The lateral deflection at the top of such a wall, Δ, due to a horizontal load, P, can be obtained if the deflections of the component piers when subjected separately to the load are known or calculated first (Figure 10-23b). Identical end restraint conditions are assumed for the piers in both cases. Thus, if the deflection of a component pier, i, under the horizontal load, P, is denoted by δ_i, then its stiffness, (i.e., the force required to produce a unit deflection) is given by

$$k_i = \frac{P}{\delta_i}$$

The deflection of a wall consisting of n connected piers, each having individual lateral stiffnesses $k_i = P/\delta_i$ $(i = 1, 2, 3, 4, \ldots, n)$ can then be obtained from the expression

$$\Delta = \frac{P}{\sum_{i=1}^{n} k_i} = \frac{P}{k_1 + k_2 + k_3 + \ldots k_n} \tag{9a}$$

or

$$\Delta = \frac{P}{\sum_{i=1}^{n} 1/\delta_1} = \frac{1}{1/\delta_1 + 1/\delta_2 + 1/\delta_3 + \ldots 1/\delta_n} \tag{9b}$$

$$\text{Wall rigidity } k = 1/\Delta \quad \text{(See Method III in Example 10-1)} \tag{9c}$$

Another approach to the determination of the relative rigidity of a wall consisting of several connected piers was suggested in a Concrete Masonry Association of California publication. This method also appeared in the Tri-Services publication, "Seismic Design for Buildings." In this method the deflection of the solid wall as a cantilever (assuming a one- or two-story building) is determined; then the cantilever deflection of an interior strip, having a height equal to that of the highest opening, is calculated and subtracted from the solid wall deflection. This step thereby removes the entire portion of the wall containing all the openings. Then the deflections of all the piers within that interior strip are determined from their own individual rigidities and added to the modified wall deflection to arrive at the total deflection of the actual wall with openings. The reciprocal of this value becomes the relative rigidity of that wall. This manner of calculating wall rigidities, although sometimes cumbersome to apply, probably yields a more accurate determination than any of the other practical methods. For one thing, it tends to consider more accurately the influence of the portion of the wall both above and below the openings.

To illustrate this basic approach and the use of the pier deflection chart and table, consider the 12 × 15 ft wall shown below (Figure 10-26):

1. Cantilever deflection of solid 12 × 15 ft wall:

$$\frac{H}{d} = \frac{12}{15} = 0.80 \qquad \therefore \Delta_C = 0.3710$$

2. Cantilever deflection of 4 × 15 ft midstrip:

$$\frac{H}{d} = \frac{4}{15} = 0.267 \qquad \therefore \Delta_C = 0.073$$

3. Fixed-end deflection of 4 × 5 ft piers *A* and *B*:

$$\frac{H}{d} = \frac{4}{5} = 0.80 \qquad \therefore \Delta_F = 0.2427$$

$$k_A = k_B = \frac{1}{0.2427} = 4.12$$

Then $k_{A+B} = 4.12 + 4.12 = 8.24$:

$$\Delta_{A+B} = \frac{1}{8.24} = 0.1214$$

4. Total wall deflection:

$$0.3710 - 0.0730 + 0.1214 = 0.4194$$

5. Relative wall rigidity:

$$k = \frac{1}{\Delta} = \frac{1}{0.4194} = 2.39$$

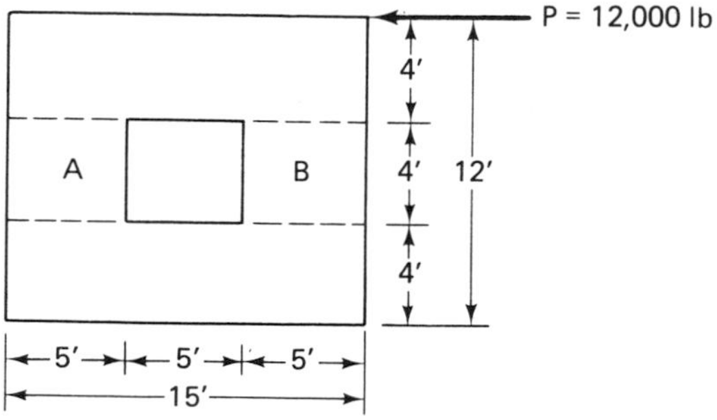

FIGURE 10-26. Rigidity of perforated wall.

The solution on page 315 is carried out to four significant figures; however, this is done merely for calculation purposes and does not imply that the deflections would actually be accurate to that degree of precision. Probably any figures past the second digit could be subject to question because of uncertainties involving the determination of a true E_m or E_v value, the actual distribution of stress, and the degree of end rotation occurrence among other factors. For another illustration refer to Method I in Example 10-1.

A rather simplified approach, although perhaps not as reliable as the previous two methods, involves simply a summation of the rigidities of only those vertical piers framed between window and door openings in the wall. Under this consideration the pier height is taken as the vertical dimension of the adjacent opening. Where different size openings occur on each side of the wall, the smaller dimension is taken as H. This approach is illustrated in Method II of Example 10-1.

Relative Stiffness of Connected Parallel Walls

Now consider the case of separate parallel walls, for instance, A, B, C, and D as shown in the plan in Figure 10-27. The walls are assumed to be tied together at the top by a horizontal roof slab which is stiff enough in its plane to be condidered rigid. In this particular case, the centers of rigidity and mass coincide so the walls will deflect equally and the force, P, will be distributed to the walls in proportion to their relative stiffnesses. For any wall, i, the relative stiffness is given by

$$R_i = \frac{k_i}{\sum_{i=1}^{4} k_i} = \frac{k_i}{k_1 + k_2 + k_3 + k_4} \tag{10a}$$

or

$$R_i = \frac{1/\delta_i}{\sum_{i=1}^{4} 1/\delta_i} = \frac{1/\delta_i}{1/\delta_1 + 1/\delta_2 + 1/\delta_3 + 1/\delta_4} \tag{10b}$$

where k_i = lateral stiffness of wall i

δ_i = wall deflection ($i = 1, 2, 3, 4$) when subjected separately to the same horizontal load, and the same end-restraint conditions as when the walls are connected

Then each wall, i, will resist that proportion of P equal to

$$P_i = \left(\frac{k_i}{\sum_{i=1}^{4} k_i}\right)P = \left(\frac{1/\delta_i}{\sum_{i=1}^{4} 1/\delta_i}\right)P \tag{11}$$

As can be seen from this expression, the distribution of the applied lateral force among the individual vertical elements depends only on the relative stiffness of the connected elements. Furthermore, it does not depend upon the magnitude of the load P. P_i thus becomes the magnitude of the direct lateral shear force acting on the wall, i.

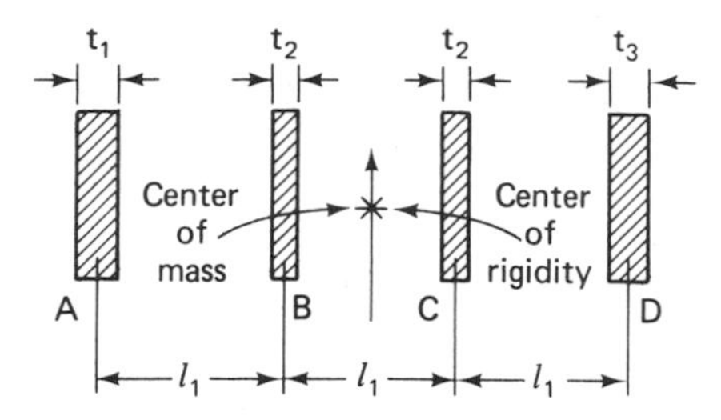

FIGURE 10-27. Connected parallel walls.

Torsional shearing stresses

TORSIONAL ECCENTRICITY

When the center of mass and the center of rigidity, as indicated in Figure 10-28b, do not coincide, torsional shear forces will be induced on the walls in addition to the direct shear forces defined in the previous paragraph. The horizontal load, P, will act at the center of mass, thus a torsional moment, M_t, is induced which is equal to $P_y \times e$, where e equals the distance between the line of force (center of mass) and the center of rigidity. Even in those instances, as indicated in Figure 10-28a, where theoretically e equals 0, the UBC requires the use of a minimum eccentricity amounting to 5% of the building dimension, this value sometimes being referred to as an *accidental eccentricity*.

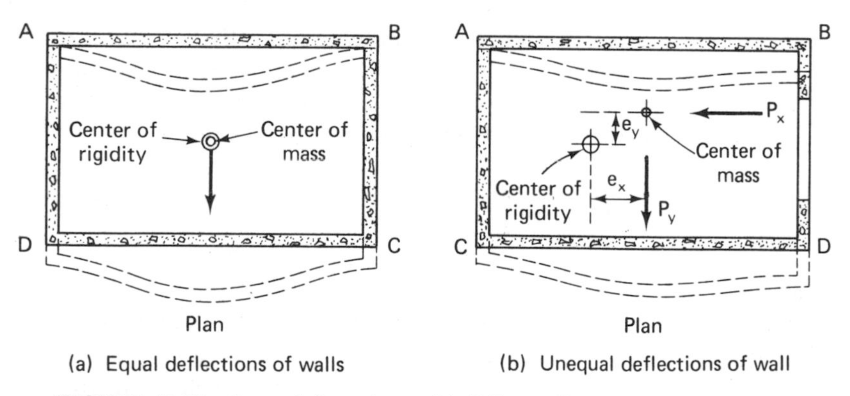

FIGURE 10-28. Lateral distortions of buildings—direct plus torsional shear.

TORSIONAL SHEAR DETERMINATION

Refer to Figure 10-29 and note that the location of the center of mass, $\bar{x}_m$, is found by taking statical moments about the center line of any wall, say wall A here, using the respective weights of the walls as forces in the moment summation. The distance, $\bar{x}_r$, from the center line of wall A to the center of rigidity is then found by taking statical moments, about the center line of the same wall, of the relative lateral stiffnesses of the component walls, determined as shown previously.

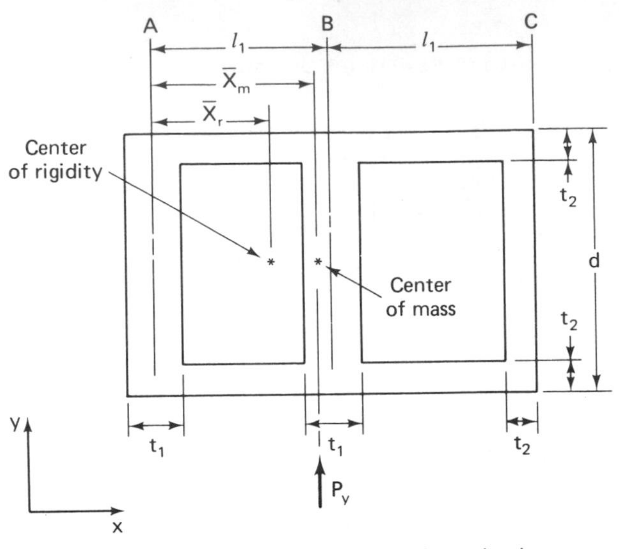

FIGURE 10-29. Torsional shear determination.

Thus, if R_y represents the relative wall rigidity of a particular wall in the y direction, the x coordinate of the center of rigidity of the wall system, $\bar{x}_r$, with respect to an arbitrary axis of reference such as the y axis of wall A, is given by the moment summation for all walls parallel to the y axis

$$\bar{x}_r = \frac{\sum R_y x}{\sum R_y} \tag{12a}$$

Since the walls parallel to the x direction do not contribute significantly to the lateral resistance in the y direction, these relative rigidity terms do not appear in this summation. On the other hand, the y coordinate of the center of rigidity, $\bar{y}_r$, entails the use of the R_x terms (in-plane lateral stiffness of the wall in the x direction) as follows:

$$\bar{y}_r = \frac{\sum R_x y}{\sum R_x} \tag{12b}$$

In the expressions above, x is the distance to the reference axis from the axis of any given parallel wall in the y direction; y is a corresponding distance for walls parallel to the x direction. Note that $e_x = \bar{x}_m - \bar{x}_r$ and $e_y = \bar{y}_m - \bar{y}_r$.

The total horizontal shear, $(P_y)_i$, resisted by a particular wall element, with an axis parallel to the y direction, due to the applied horizontal load, $(P_y)_i$, may be obtained from the expression

$$(P_y)_i = \underbrace{\left(\frac{R_y}{\sum R_y}\right) P_y}_{\text{direct shear}} + \underbrace{\left(\frac{R_y \bar{x}}{J_r}\right) P_y e_x}_{\text{torsional shear}} \tag{13a}$$

where $\bar{x}$ or $\bar{y}$ = perpendicular distance from the center of rigidity, $\bar{c}_r$ to the axis of the wall in question

Similarly, for an applied horizontal force P_x in the x direction,

$$(P_x)_i = \left(\frac{R_x}{\sum R_x}\right)P_x + \left(\frac{R_x\bar{y}}{J_r}\right)P_x e_y \tag{13b}$$

Note that J_r equals the relative rotational stiffness of all the walls in the story under consideration. It corresponds to a polar moment of inertia and may be found from the expression

$$J_r = \sum (R_x\bar{y}^2 + R_y\bar{x}^2) \tag{14}$$

Note that in the equations above, the plus sign is always used. This stems from the fact that since the horizontal load, P, is a reversible one, the codes generally require that the effect of torsional moments be considered only when they tend to increase the direct shears. The UBC, for example, specifically prohibits the use of any negative torsional shear values.

Examples of shear-wall design and analysis

EXAMPLE 10-1: One-Story Building

The four reinforced brick shear walls, all 50 ft long and 20 ft high, to be analyzed are shown in Figure 10-30. Their thicknesses and opening dimensions are given therein. The diaphragm shear loading is generated by those walls lying perpendicular to the assumed direction of the ground motion, in addition to that of the roof dead load. The manner in which this diaphragm loading is evalvated was shown in the example given in the section dealing with the general behavior of a box-type system.

(*a*) *Calculate the relative wall rigidities by three different methods discussed in Chapter 10 and compare the results in terms of the percent of lateral force each would pick up if supported by a rigid diaphragm.*

(*b*) *Of the total force assigned to wall C from* (*a*), *determine what portion of that will be assigned to piers 1, 3, 5, and 6 on the basis of relative pier rigidities.*

(*c*) *Assume that the diaphragm is plywood and that the walls are parallel to each other on 50-ft centers. Determine the total shear force to be carried by each wall. Note that a flexible diaphragm such as this cannot transfer any torsional shear forces.*

(*d*) *Then assume that the diaphragm is concrete and therefore considered rigid. With the walls arranged in the same configuration, determine the total shear to be carried by each wall. Consider torsional shear effects if present. Also determine the force distribution to the individual piers within the walls.*

Solution:

(*a*) *Wall rigidities* (*refer to the section "Wall deflections and rigidities," this chapter*):

METHOD I

To demonstrate this method, calculate the relative rigidities of walls C and D. Refer to Figure 10-25 and Table 10-4 for rigidity and deflection values.

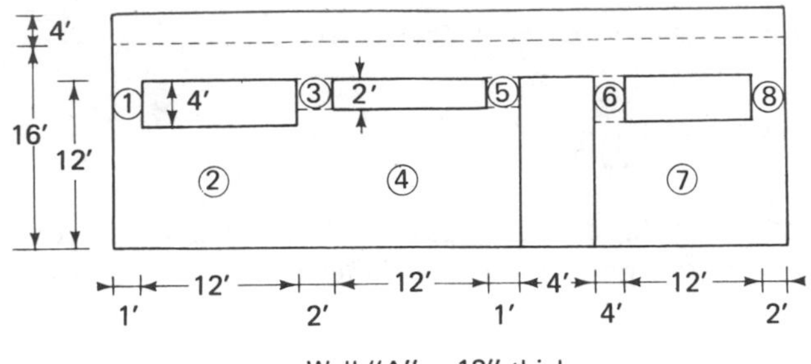

Wall "A" – 12" thick

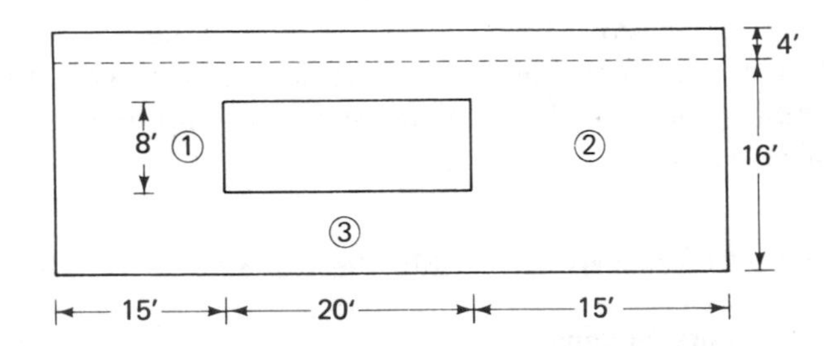

Wall "B" – 8 $\frac{1}{2}$" thick

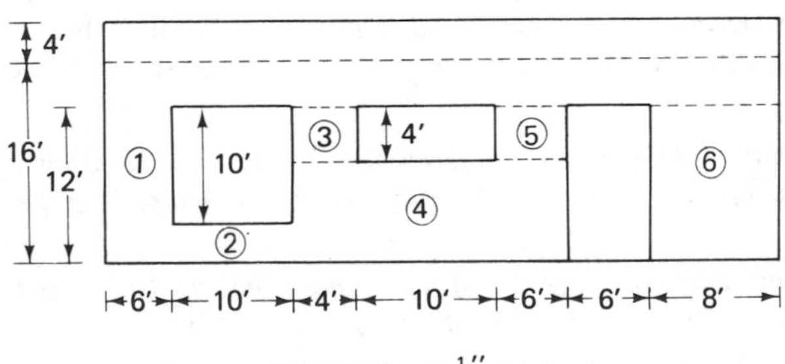

Wall "C" – 8 $\frac{1}{2}$" thick

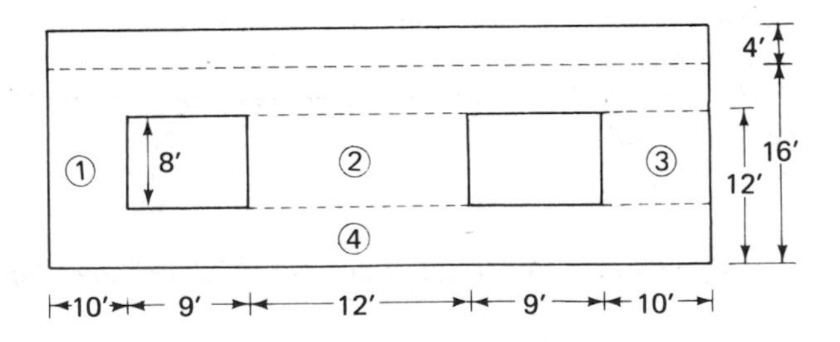

Wall "D" – 10" thick

FIGURE 10-30. One-story shear-wall building.

1. Obtain the deflection of the solid cantilever wall.
2. Subtract the deflection of a cantilever strip having a height equal to that of the highest opening in the wall.
3. Compute the deflection of all composite piers with openings lying within that strip.
4. Add these deflections of the individual piers to the modified wall deflection obtained in (2) to obtain the final wall deflection.
5. Take the reciprocal of this compound deflection to obtain the relative wall rigidity.

Wall C:

Although support will be at roof level 16 ft above floor, the h for stiffness more reasonably includes the total wall height of 20 ft.

Entire wall: $h/d = 20/50 = 0.4$, $\Delta_C = 0.121$, $k_c = 1/\Delta = 8.237$
Strip (with openings): $h/d = 12/50 = 0.24$, $\Delta_c = 0.065$, $k_c = 15.385$
Pier 1–2: $h/d = 12/16 = 0.75$, $\Delta_F = 0.220$, $k_F = 4.545$
Strip (pier 1 and opening): $h/d = 10/16 = 0.625$, $\Delta_F = 0.175$, $k_F = 5.714$
Pier 1: $h/d = 10/6 = 1.67$, $\Delta_F = 0.806$, $k_F = 1.240$
$\Delta = 0.220 - 0.175 + 0.806 = 0.851$, $k_F = 1.175$
Pier 3–4–5: $h/d = 12/20 = 0.60$, $\Delta_F = 0.168$, $k_F = 5.952$
Strip (piers 3 and 5 and opening:) $h/d = 4/20 = 0.20$, $\Delta_F = 0.051$, $k_F = 19.724$
Pier 3: $h/d = 4/4 = 1.0$, $\Delta_F = 0.333$, $k_F = 3.000$
Pier 5: $h/d = 4/6 = 0.67$, $\Delta_F = 0.193$, $k_F = 5.181$

$$\sum k_i = 3.000 + 5.181 = 8.181 \text{ and } \Delta = \frac{1}{\sum k_i} = 0.122$$

$$\Delta = 0.168 - 0.051 + 0.122 = 0.239 \text{ and } k = 1/\Delta = 4.180$$

Pier 6: $h/d = 12/8 = 1.5$, $\Delta_F = 0.656$, $k_F = 1.524$
Entire strip: $\sum k_i = 1.175 + 4.180 + 1.524 = 6.879$ and $\Delta = 1/\sum k_i = 0.145$
Entire wall: $\Delta = 0.121 - 0.065 + 0.145 = 0.201$ and $k = 1/\Delta = 4.975$
For solid wall: $k = 8.237 \times 8.5/8.5 = 8.237$ (in terms of 8.5 in. thickness)
For entire wall: $k = 4.975 \times 8.5/8.5 = 4.975$ (in terms of 8.5 in. thickness)

Note that this wall with openings as shown has a relative rigidity of 60% of that of a solid wall with the same overall dimensions.

Wall D:

For entire wall: $h/d = 20/50 = 0.40$, $\Delta_c = 0.121$, $k_c = 8.237$
For strip (with openings): $h/d = 8/50 = 0.16$, $\Delta_c = .0417$, $k_c = 24.000$
Pier 1: $h/d = 8/10 = 0.8$, $\Delta_F = 0.243$, $k_F = 4.120$
Pier 2: $h/d = 8/12 = 0.67$, $\Delta_F = 0.193$, $k_F = 5.181$
Pier 3: $h/d = 8/10 = 0.8$, $\Delta_F = 0.243$, $k_F = 4.120$

$$\sum k_i = 4.120 + 5.181 + 4.120 = 13.421,\ \Delta = 1/\sum k_i = 0.075$$

For entire wall: $\Delta = 0.121 - 0.042 + 0.075 = 0.154$, $k = 1/\Delta = 6.481$
For solid wall: $k = 8.237 \times 10/8.5 = 9.691$ (in terms of 8.5-in.-thick wall)
For entire wall: $k = 6.481 \times 10/8.5 = 7.625$ (in terms of 8.5-in.-thick wall)

The rigidity of walls A and B would be computed in a similar fashion. The table on page 322 summarizes these various rigidities obtained in this manner.

Wall	*k*	*R*
A	5.344	0.218
B	6.557	0.268
C	4.975	0.203
D	7.625	0.311
		1.000

METHOD II

Determine the rigidity by considering only the sum of the rigidities of the individual piers between openings within the wall. This is probably the shortest procedure shown in this discussion. In this example the method yields results that are comparable to the others, which are somewhat more precise, though not necessarily more correct.

Wall A:

Pier 1: $h/d = 4$, $\Delta_F = 6.331$, $k_F = 1/\Delta_F = 0.158$
Pier 3: $h/d = 1$, $\Delta_F = 0.333$, $k_F = 3.000$
Pier 5: $h/d = 2$, $\Delta_F = 1.167$, $k_F = 0.857$
Pier 6: $h/d = 1$, $\Delta_F = 0.333$, $k_F = 3.000$
Pier 8: $h/d = 2$, $\Delta_F = 1.167$, $k_F = 0.857$

$$\text{Wall } k = \sum k_F = 7.872 \times \frac{12}{8.5} = 11.113 \text{ (in terms of 8.5-in.-thick wall)}$$

METHOD III

Compute k on the basis of the summation of the various individual pier deflections.

Wall A:

Pier 2: $h/d = 0.62$, $\Delta_F = 0.172$, $k_F = 5.814$ Other pier values given in previous method.
Pier 7: $h/d = 0.44$, $\Delta_F = 0.114$, $k_F = 8.772$

Pier 1–2: $\Delta_1 + \Delta_2 = 6.331 + 0.172 = 6.503$, $k_F = 0.154$

$$\text{Pier 3–4–5: } \Delta_{3,4,5} = \Delta_4 + \frac{1}{1/\Delta_3 + 1/\Delta_5} = 0.192 + \frac{1}{3.000 + 0.857} = 0.451$$

$$k_F = 2.216$$

$$\text{Pier 6–7–8: } \Delta_{6,7,8} = \Delta_7 + \frac{1}{1/\Delta_6 + 1/\Delta_8} = 0.114 + \frac{1}{3.000 + 0.857} = 0.373$$

$$k_F = 2.679$$

$$\sum k: \quad 0.154 + 2.216 + 2.679 = 5.049 \times \frac{12}{8.5} = 7.128 \text{ (in terms of 8.5-in.-thick wall)}$$

Summary of wall rigidities by the various methods:

	Method I		*Method II*		*Method III*	
Wall	*k*	*R*	*k*	*R*	*k*	*R*
A	5.344	0.218	11.113	0.214	7.128	0.189
B	6.557	0.268	13.986	0.270	10.929	0.291
C	4.975	0.203	10.947	0.211	7.110	0.189
D	7.625	0.311	15.789	0.305	12.448	0.331
		1.000		1.000		1.000

Thus, fairly good correlation for the relative rigidity values among walls *A*, *B*, *C*, and *D* is obtained.

(*b*) *Force distribution to piers within a wall:*

Wall C (refer back to method I for pier 1, 3, 5, 6 deflections):

Pier 1–2 for $\Delta_{1,2}$: $\Delta_1 = 0.806$ and for $h/d = 0.125$, $\Delta_2 = 0.038$

$\therefore \Delta_1 + \Delta_2 = 0.806 + 0.038 = 0.844$ and $k = 1/\Delta = 1.185$

Pier 3–4–5 for $\Delta_{3,4,5}$: for $h/d = 0.40$, $\Delta_4 = 0.105$ ($k = 9.524$)

$$\Delta_{3,4,5} = \Delta_4 + \frac{1}{1/\Delta_3 + 1/\Delta_5} = 0.105 + \frac{1}{1/0.333 + 1/0.193} = 0.227 \text{ and } k = 4.401$$

Pier 6: $\Delta_6 = 0.656$, $k = 1.524$

$\sum k$: $1.185 + 4.401 + 1.524 = 7.110$

$$\left.\begin{aligned} &\text{To Pier 1–2:} \quad k/\sum k = \frac{1.185}{7.110} = 16.7\% \\ &\text{To Pier 3–4–5:} \quad k/\sum k = \frac{4.401}{7.110} = 61.9\% \\ &\text{To Pier 6:} \quad k/\sum k = \frac{1.524}{7.110} = 21.4\% \end{aligned}\right\} \text{of the shear assigned to wall } C$$

For Pier 3–4–5: Take 61.9% of the shear assigned to wall *C* and distribute it to piers 3 and 5 as follows:

$$\text{Pier 3:} \quad k/\sum k = \frac{3.000}{3.000 + 5.181} = 36.7\%$$

$$\text{Pier 5:} \quad k/\sum k = \frac{5.181}{3.000 + 5.181} = 63.3\%$$

Summary:

Pier	*% of shear on wall C going to each pier*	
1		16.7
3	36.7% of 61.9% =	22.7
5	63.3% of 61.9% =	39.2
6		21.4
		100.0

(*c*) *Determine force transferred to each wall if diaphragm is constructed of timber* (*flexible diaphragm*):

With this parallel wall layout, we have a series of three simple spans, 50 ft long, as described in Chapter 10. Thus, the lateral load will be distributed by this flexible diaphragm on a tributary width basis. This yields 25 ft × 300 lb/ft = 7500 lb on walls *A* and *D* and 50 ft × 300 lb/ft = 15,000 lb on walls *B* and *C*. These would be the forces transferred from the diaphragm into the wall. Thus, the connection must resist 7500 lb/50 ft = 150 lb/ft at *A* and *D* or 15,000/50 = 300 lb/ft at *B* and *C*.

In addition to the diaphragm shear, the piers within the wall must carry the inertial effect of the wall dead weight. This may be calculated at the bottom of the window openings, or it is sometimes taken at the midheight of the wall.

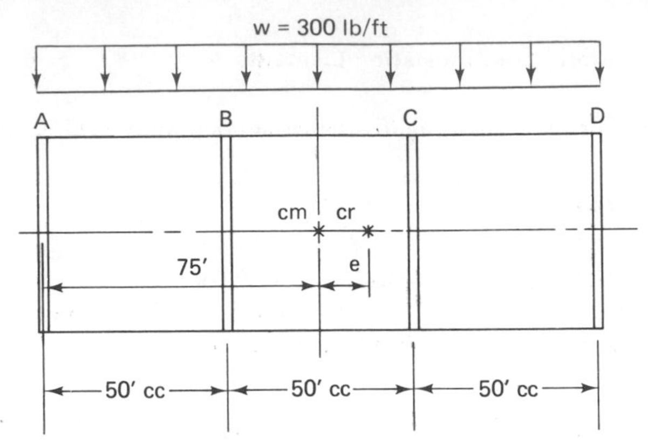

In-plane shear $V = ZIKCSW = 1.0 \times 1.0 \times 1.33 \times 0.14W = 0.186W$ (use $CS_{\max} = 0.14$)

Inertial wall load = 0.186 (wall weight, lb/ft²)[(h/2 + parapet)(wall length) − (opening areas)]

Wall A:

$$\text{Load} = 0.186\,(120)[(16/2\text{ ft} + 4\text{ ft})(50\text{ ft}) - (12 \times 4 + 12 \times 2 + 4 \times 4 + 12 \times 4)] = 10{,}350\text{ lb}$$

Summary of wall loads:

Wall	*Loads (lb)* *Diaphragm*	*Wall*	*Total for flexible diaphragm*[a] *(lb)*
A	7,500	10,350	17,850
B	15,000	8,200	23,200
C	7,500	7,850	22,850
D	15,000	9,800	17,300

[a]These forces would be multiplied by 1.5 when determining in-plane shear capacity according to UBC.

(*d*) *Determine force transferred to each wall if diaphragm is a concrete slab (rigid diaphragm):*

Direct in-plane shear forces:

Wall	*R*	*Load (lb)* *Diaphragm*	*Wall*	*Total for rigid diaphragm,*[a] V_D *(lb)*
A	0.218	9,810	10,350	20,160
B	0.268	12,060	8,200	20,260
C	0.203	9,135	7,850	16,985
D	0.311	13,995	9,800	23,795

[a]See Table above.

Diaphragm shear = 300 lb/ft × 150 ft = 45,000 lb

e.g., 0.218 × 45,000 = 9,810 lb

Torsional Shear Forces:

Since concrete slab is considered rigid in comparison with the one-story walls, it is able to transmit torsional shear forces as well as direct shear forces.

Location of center of mass (wall A as reference axis), $\bar{x}_{cg} = 75$ ft

Location of center of stiffness:

Wall	R	Dist. to wall A (ft)	R/∑ R × dist.
A	0.218	0	0
B	0.268	50	13.4
C	0.203	100	20.3
D	0.311	150	46.5
	1.000		80.2

$$\bar{x}_r = \sum R_y x / \sum R_y = \frac{80.2}{1.000} = 80.2 \text{ ft from } A$$

eccentricity $= \bar{x}_r - \bar{x}_{cg} = 80.2 - 75 = 5.2$ ft
min. $e = 0.05 \times 150 = 7.5$ ft (per UBC governs)

Polar moment of inertia:

Wall	R_y	$\bar{x}$ (ft)	$\bar{x}^2$ (ft²)	$R_y\bar{x}^2$ (ft²)
A	0.218	80.2	6432	1409
B	0.268	30.2	912	245
C	0.203	−19.8	392	79
D	0.311	−69.8	4872	1510
				3243

$\bar{x}$ = distance from center of stiffness to wall in question $= \bar{x}_r - x$
J_r = polar moment of inertia
$= \sum (R_y\bar{x}^2 + R_x\bar{y}^2)$
With parallel configuration, $\bar{y} = 0$;
thus, $J_r = 3243$ ft²

TORSIONAL SHEAR FORCE AND TOTAL IN-PLANE SHEAR:

Wall	R_y	$\bar{x}$(ft)	$R_y\bar{x}$	V_T	V_D	$V_D + V_T$
A	0.218	80.2	17.56	1828	20,160	21,988
B	0.268	30.2	8.09	842	20,260	21,102
C	0.203	−19.8	−4.02	−418	16,985	16,985[a]
D	0.311	−69.8	−21.64	−2252	23,795	23,795[a]

[a]Negative torsional shears not considered.

$$V_T = (R_y\bar{x})\frac{P_y e_x}{J_r} = \frac{45{,}000 \times 7.5}{3243}(R_y\bar{x}) = 104.07(R\bar{x})_y$$

Determine force carried by each pier in wall C, assuming a force distribution as listed in the preceding table. Note that these values are to be multiplied by 1.5 when checking in-plane shear capacity; thus, 1.5 × 16,985 = 25,500 lb.

Wall C pier	% Shear to each pier	Total force to each pier
1	16.7	4258
3	36.7 × 61.9 = 22.7	5789
5	63.3 × 61.9 = 39.2	9996
6	21.4	5457
∑	100.0	25,500

Forces would be distributed to each pier in walls A, B, and D in a similar fashion.

EXAMPLE 10-2

Rearrange the same brick walls in the form of a square box, 50 ft on a side.

(*a*) *Assuming that the diaphragm is still reinforced concrete, determine the shear distribution to the walls for ground motion in both the north-south and east-west directions.*

(*b*) *Figure the loads transferred to each of the individual piers within wall* C *of the box-type structure formed in part (a). Determine whether or not any of these are overstressed and specify any shear reinforcing required.*

Solution:

Assume that walls are arranged in box shape as shown:

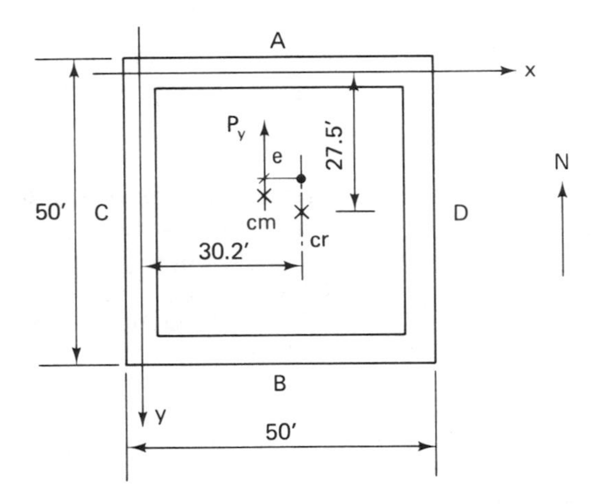

(*a*) *Determine force transferred to* C *and* D *walls for N-S ground motion. Reference axis is along walls* A *and* C.

LOCATION OF CENTER OF MASS:

$$\bar{x}_{cg} = 25'$$

$$\bar{y}_{cg} = 25'$$

Wall	R_y	x (*ft*)	R_x	y (*ft*)	$R_y x$ (*ft*)	$R_x y$ (*ft*)	$\bar{x}$ (*ft*)	$\bar{x}^2$ (*ft*^{2})	$\bar{y}$ (*ft*)	$\bar{y}^2$ (*ft*^{2})	$R_y\bar{x}^2$ (*ft*^{2})	$R_x\bar{y}^2$ (*ft*^{2})
A		—	0.218	0	—	0	—	—	27.5	756.2	—	164.9
B		—	0.268	50	—	13.4	—	—	−22.5	506.2	—	135.7
C	0.203	0	—	0	0	—	30.2	912.0			185.1	—
D	0.311	50	—	0	15.5	—	−19.8	392.0			121.9	—
	0.514		0.486		15.5	13.4					307.0	300.6

$$\bar{x}_r = \frac{\sum R_y x}{\sum R_y} = \frac{15.5}{0.514} = 30.2 \text{ ft from wall } C$$

$$\bar{y}_r = \frac{\sum R_x y}{\sum R_x} = \frac{13.40}{0.486} = 27.5 \text{ ft from wall } A$$

$$J_r = \sum (R_y\bar{x}^2 + R_x\bar{y}^2) = 307.0 + 300.6 = 607.6 \text{ ft}^2$$

Eccentricity:

$$e_x = \bar{x}_r - \bar{x}_{cm} = 30.2 - 25 = 5.2 \text{ ft vs } 0.05 \times 50 = 2.5 \text{ ft} \qquad \therefore \text{use } 5.2 \text{ ft}$$

$$e_y = \bar{y}_r - \bar{y}_{cm} = 27.5 - 25 = 2.5 \text{ ft} \qquad \therefore \text{use } 2.5 \text{ ft}$$

Torsional moment:

N-S is 15,000 × 5.2 ft = 78,000 ft-lb

E-W is 15,000 × 2.5 ft = 37,500 ft-lb

DETERMINATION OF TORSIONAL SHEAR:

Wall	R_y	$\bar{x}$ (*ft*)	$R_y\bar{x}$ (*ft*)	R_x	$\bar{y}$ (*ft*)	$R_x\bar{y}$ (*ft*)	*Torsional shear, V_T* N–S	E–W
A	—	—	—	0.218	27.5	6.00	770	370
B	—	—	—	0.268	−22.5	6.03	774	372
C	0.203	30.2	6.13				787	378
D	0.311	−19.8	6.15				789	379

For North-South direction:

$$\text{Walls } C \text{ and } D\text{: } V_T = R_y\bar{x}\frac{P_y e}{J_r} = R_y\bar{x}\left(\frac{78{,}000}{607.6}\right) = 128.3R_y\bar{x}$$

$$\text{Walls } A \text{ and } B\text{: } V_T = 128.3\ R_x\bar{y}$$

(should be the same numerical value)

For East-West direction:

$$\text{Walls } A \text{ and } B\text{: } V_T = R_x\bar{y}\frac{P_y e}{J} = R_x\bar{y}\left(\frac{37{,}500}{607.6}\right) = 61.7\ R_x\bar{y}$$

$$\text{Walls } C \text{ and } D\text{: } V_T = 61.7\ R_y\bar{x}$$

Total shear N-S Ground motion:

Wall	*In-plane shear* — *Direct, V_D* Roof	+ *Wall*	*Torsional, V_T*	*Total shear*
A	—		774[b]	774[b]
B	—		774	774
C	5900	7850	787	14,537 (design values)
D	9100	9800	787	18,900[a] (design values)

[a]Negative V_T not considered.
[b]Values rounded off to be equal.

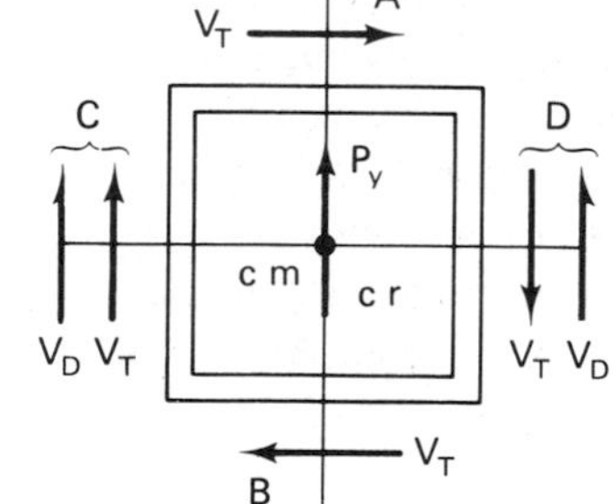

Example: 0.203/0.514 × (300 lb/ft × 50 ft) = 5900 lb to wall *C* from diaphragm.

Inertial wall load in C from Example 10-1(c) = 7850 lb. Direct shear walls A, B, and D determined in similar fashion.

Total shear E-W ground motion:

	In-plane shear			
Wall	Direct, V_D Roof + Wall		Torsional, V_T	Total shear
A	6,750	10,350	372	17,472 } design
B	8,250	8,200	372	16,450[a] } values
C	—	—	379	379
D	—	—	379	379

[a]Negative Torsional Shears not permitted.

(*b*) *Forces taken by individual piers in wall C.*

For in-plane shear capacity $V = 1.5 \times 14{,}537 = 21{,}800$:

Pier	$h/2d$	d (in.)	t (in.)	dt (in.^{2})	% shear	Shear to pier (lb)	$v = V/dt$ (lb/in.^{2})	$v_{\text{allowable}}$ (lb/in.^{2})
1	0.83	72	8.5	612	16.7	3650	6.0	24.0
3	0.50	48	8.5	408	22.7	4950	12.1	28.0
5	0.33	72	8.5	612	39.2	8550	14.0	29.0
6	1.50	96	8.5	816	21.4	4650	5.7	23.0

Note that v_{allow} is based on $M/Vd = h/2d$ for fixed end piers. See Figure 7-8 for this determination. Also, $v_{\text{allowable}}$ can be increased by 1.33 for seismic shears. Since the allowable shear on the masonry exceeds the actual shear stress on each of piers 1, 3, 5, and 6 in wall C, no shear steel is required for strength.

Reinforcing details

When one considers the stresses imposed by the in-plane horizontal force, P_i, it becomes readily apparent that a masonry wall, in order to function properly as a shear wall, must be capable of resisting (1) diagonal tensile stresses caused by the direct and torsional shear forces; plus (2) vertical tensile or compressive stresses at the wall edges or jambs induced by the bending moment due to P_i. Also, the element must be able to overcome any tendency to develop a sliding failure of the base or in the horizontal bed joints. As a final consideration, if the wall carries vertical loads, it must possess the capacity, as a bearing wall, to resist the resulting compressive stresses. In addition, the vertical axial load on the pier may be increased (or decreased) due to the overturning effect of the upper stories. (Refer to the section "Pier Axial and Flexural Stresses" in Chapter 7; also refer to the section "Allowable Design Stresses" in Chapter 7 for shear steel requirements and Example 12-2 in Chapter 12.)

One way of placing the shear steel in a masonry wall may be seen in Figure 10-31. The horizontal bars are anchored with a 90° bend up or down in order to develop their tensile capacity. Where boundary reinforcement may be required to provide the necessary tensile resistance, the details shown on the pier elevation in Figure 10-32 would be typical for that particular condition. Note the position of the jamb steel and shear reinforcement in these two figures. Observe that a limited amount of space, and access to it, is available within the grouted core for positioning this steel. This fact should always be kept in mind when specifying bar sizes and spacings. Too often the reinforcing is detailed in such a way that it becomes impossible to place it within the limited space provided.

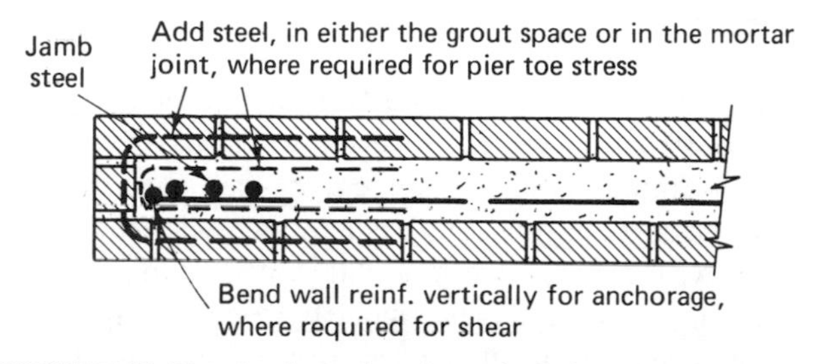

FIGURE 10-31. Jamb steel at the end of piers—cross section.

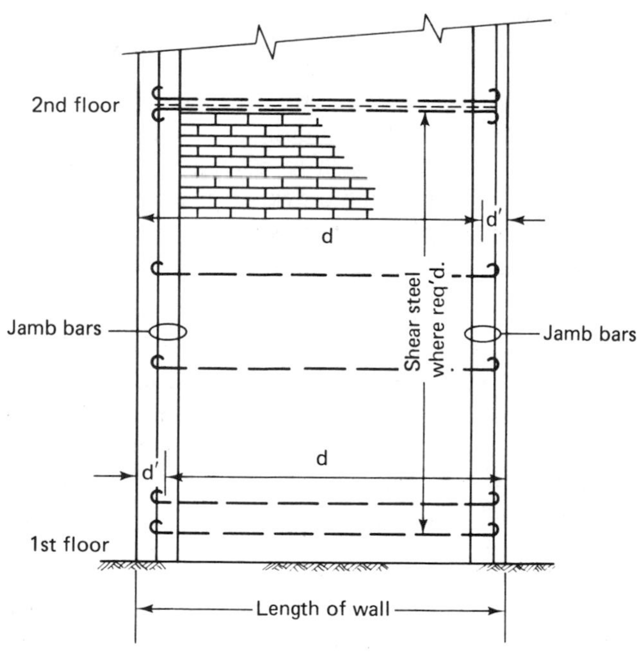

FIGURE 10-32. Boundary (jamb) reinforcement in a shear wall—elevation.

SELECTION OF THE STRUCTURAL SYSTEM

Objective

The objective of any structural system, among other things, of course, is to reduce the project cost to a minimum without compromising its function, quality, or reliability. Final selection of materials and systems has to be made with due consideration of the cost of construction, architectural requirements, resistance to rough usage, fire and other safety hazards, and low maintenance and operating costs over the life cycle of the facility, as well as the availability of funds. Thus, within these limitations, the most efficient structural system possible should be chosen. Further, this system must be compatible with the desired architectural treatment.

Economic aspects

Usually, the major structural and architectural components of a building that most affect the cost of construction are the exterior walls, partitions, floor and roof decks, and the structural framing system. In some instances, the type of foundation may be a major factor in the cost analysis. Simple detailing, with the arrangement of spaces that are compatible with a repetitive modular construction, all contribute greatly to reducing total building costs. This is particularly true of masonry construction. Allowing the contractor some option in the selection of certain materials and details as well as construction procedures will also help to lower costs.

Structural concepts

The participation of all disciplines of the design team in the conceptual planning and selection of basic construction materials at an early stage will provide many opportunities to ensure that the most efficient design at the lowest construction cost will be obtained. Procedures in the approach to develop a concept will vary, depending upon the type of facility, of course, and the experience of the individuals on the design team. In a general way, one approach to a solution might be achieved as described in the next paragraphs.

LAYOUT

A coordinated plan should be developed which provides a layout that offers all the functional requirements that must be incorporated into the particular building structure. Also, the following items, among others, should be examined in detail, so that a sketch would be helpful in demonstrating where improvements might be made. For instance, the structural plan ought to be as symmetrical as possible to avoid, or at least minimize, torsional forces and to eliminate any need for seismic separation joints between adjacent shear walls at wings. If this is not feasible, separations must be provided to allow free deflection to occur. Separating walls and partitions should be spaced so as to be entirely compatible with the modular pattern of the masonry units being utilized. The mechanical layout should be examined with a view toward keeping it simple and located so as to reduce the duct work needed for the air-conditioning and heating system. Also architecturally, the modular layout reduces construction costs and usually conforms better to the esthetics of the exterior appearance.

Basic Materials

The architect's selection of the appropriate materials has a profound effect upon the structural system and its cost. In general, the approved structural systems are broadly classified on the following basis: (1) a box system, (2) a ductile moment-resisting space frame, (3) a moment-resisting frame and shear walls, (4) a vertical load-carrying frame and shear walls and (5) a braced frame system. Any of these structural systems may be used in combination with a wide variety of floor, roof, and partition components comprised of various materials. Walls may be bearing, nonbearing, shear, or simply nonstructural in function, such as partitions or curtain walls. There are many different types of decking components (both floor and roof) available, utilizing various materials in different combinations. Often, through slight adjustments in details, various types of components can be made to serve dual functions. For example, a masonry shear wall may also be used as a bearing wall. In general, however, final selections should conform as closely as possible to the preliminary design (sizing of principal members) and cost studies of the original structural estimate.

Damage control features

No structure can feasibly be designed so that no damage whatsoever will occur during a catastrophic earthquake. The main intent of the seismic provisions is to preclude the possibility of a disastrous failure or collapse. A number of things can be done, without materially increasing construction costs, to limit the damage that otherwise might be quite expensive to repair following a strong earthquake. An important factor to keep in mind is the nature and geometry of the building and its probable response to earthquake motion. It should be assumed that deflections (story drift) may be several times that which would result from the Code-specified lateral forces. A list of features that can aid in avoiding excessive damage would include the following:

1. Provide details which allow structural movement, without damage, to nonstructural elements. Damage to such items as piping, glass, plaster, and partitions due to excessive deflections may constitute a major financial loss even though the structural damage is minor. Special care in detailing is required to minimize this type of damage.
2. Breakage of glass windows can be minimized by providing adequate clearance at edges to allow for frame distortion.
3. Damage to rigid or nonstructural partitions can be largely eliminated by locating them away from columns and by providing a detail at their top or edges which will permit relative motion between the partitions and the floor above.
4. In piping installations, the expansion loops and bellows joints used to accommodate temperature movement can often be adapted to handle the relative seismic deflections between adjacent equipment items attached to floors.

5. Fasten shelving to walls to prevent overturning.

6. If paint and plaster repairs are undertaken without regard to structural rehabilitation after damage from an earthquake, the structure may be left cracked and vulnerable to further damage and possible collapse in the event of a subsequent strong earthquake.

7. Provide structural symmetry as nearly as possible, since past experience has shown that buildings which are unsymmetrical in plan have greater susceptability to earthquake damage than do symmetrical structures. The effect of dissymmetry is to induce torsional oscillations of the structure. Dissymmetry in plan can be eliminated, or improved, by separating L-, T-, E-, and U-shaped buildings into distinct units by the use of seismic joints at the junctions of the individual wings or portions. In regular structures, dissymmetry can also be caused by eccentric structural elements. Such a condition could exist, for example, in a store building with a flexible front resulting from large openings, and an essentially stiff (solid) rear wall. Buildings with this type of structural dissymmetry can usually be improved by modifying the stiffness of the rear wall, or by the judicial insertion of rigid structural partitions so as to make the center of rigidity of the vertical resisting elements more nearly coincide with the center of mass.

8. Very careful attention must be paid to the connection details between the floor and roof diaphragms in shear walls, because such connections are the most vulnerable structural elements in the whole system. Without adequate connections, the box-type structural system will not function to resist lateral forces during strong earthquakes, or heavy wind storms, for that matter. It pays, therefore, to be extremely conservative when detailing these elements. Structural connections are more fully discussed in Chapter 11.

9. With regard to materials, ductility is a highly desirable characteristic in seismic response, for it permits large distortions without total sudden collapse, unlike brittle materials. Brittle failures are explosive in nature and usually occur suddenly and completely, without warning. Ductility can be built into masonry structures provided that they are reinforced in the proper amounts and in the right locations.

10. Since seismic forces or loadings increase with the mass, it pays to minimize weight as much as possible. Therefore, due consideration should be given to the use of lightweight materials (e.g., lightweight concrete masonry units or lightweight concrete floor planks) wherever possible.

11. Differential movements of foundations are a source of structural damage during seismic motions, especially if the structure is too rigid to accommodate such movements. So foundation design must include those features which will minimize relative displacements.

12. Damping characteristics have a major effect on a building's response to ground motion, since even a small amount of damping significantly reduces the maximum deflections due to resonant response. This property is often expressed as a percentage of critical damping. The type of material used has a bearing on this; for instance, reinforced concrete or masonry has a higher degree of damping than does structural steel.

PROBLEMS

10-1. Given the building shown, with 9-in. brick walls. The walls are supported by a plywood diaphragm. Roof DL = 13 lb/ft²; roof LL = 20 lb/ft² (reducible). Determine the shear force (per linear foot) which the diaphragm carries to the side walls (neglect any opening).

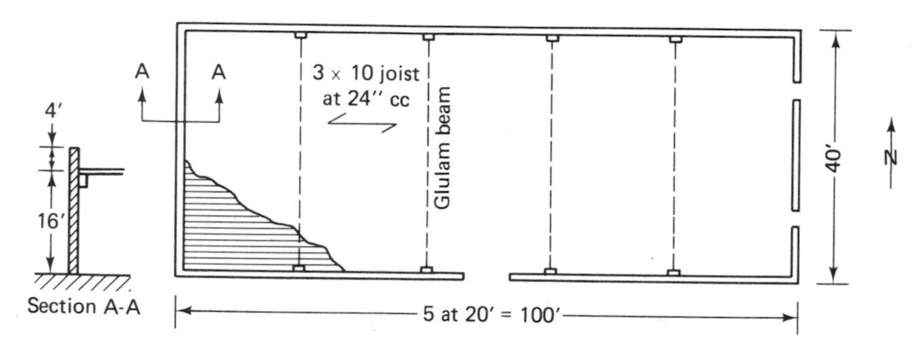

10-2. For the building shown in Problem 10-1, specify the plywood panel layout and nailing schedule using 8d nails.

10-3. In Problem 10-2, where can some of the nailing schedule be reduced and the blocking be eliminated?

10-4. From the results of the previous problems, determine the required diaphragm reinforcing in the masonry bond beam.

10-5. Find the relative rigidities of the walls shown by three different methods described in this chapter. Compare the result in terms of the percent of the base shear each would receive if they were arranged in a parallel configuration.

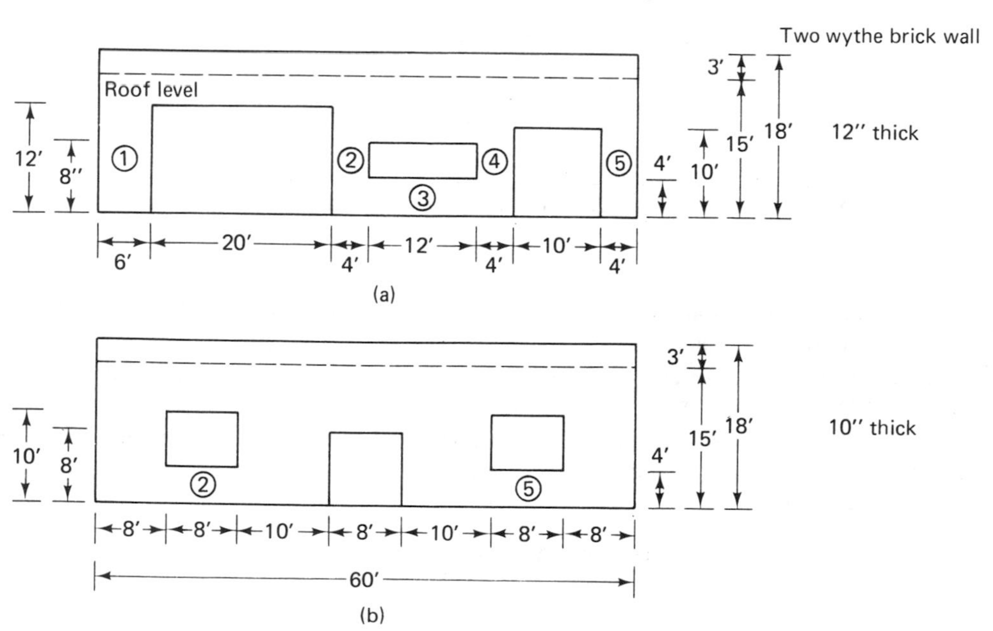

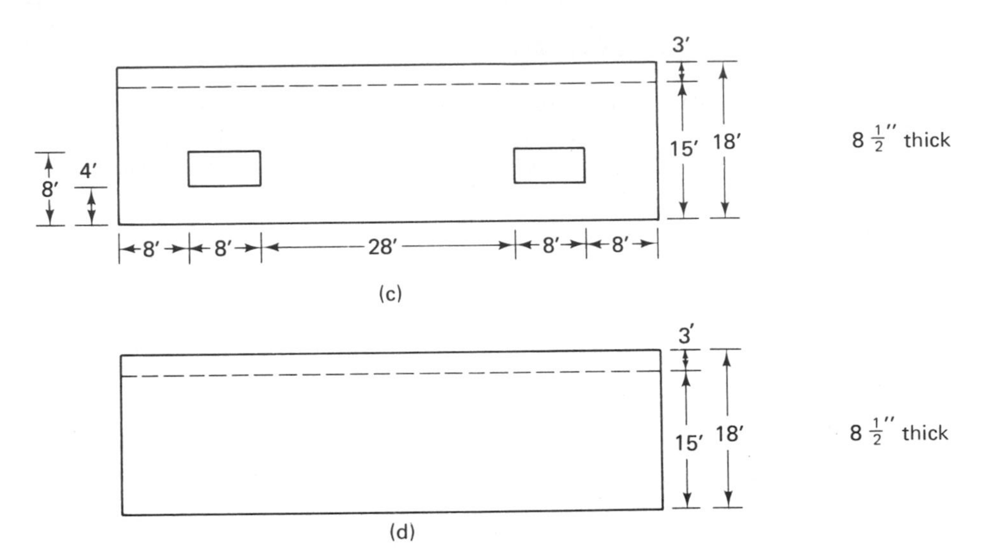

(c)

(d)

10-6. Determine the shear distribution to piers 1, 2, 4, and 5 of wall A. (Use method III.)

10-7. Assume that the four walls of Problem 10-5 are arranged as shown below. The walls are supported by a plywood diaphragm. Given the roof LL, 20 lb/ft² (reducible); roof DL, 16 lb/ft².

$$\text{unit weight of wall } A = 88 \text{ lb/ft}^2$$
$$B = 95 \text{ lb/ft}^2 \quad \text{average values}$$
$$C = 88 \text{ lb/ft}^2 \quad \text{(opening effect is considered)}$$
$$D = 93 \text{ lb/ft}^2$$

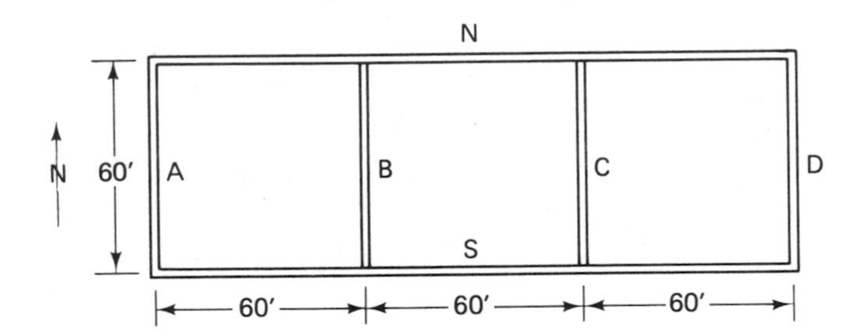

Determine (a) the unit shear force in walls A, B, C, and D at midheight; and (b) the forces on the piers of walls B and C.

10-8. Same as Problem 10-7, but change the plywood diaphragm to a reinforced concrete slab. (roof DL = 70 lb/ft².)

10-9. Rearrange the walls of Problem 10-5 as shown. The walls are supported by a rigid diaphragm. Determine the shear-force distribution in both directions. (Take the torsional effect into account.)

$$\text{Given: roof DL} = 70 \text{ lb/ft}^2$$
$$\text{roof LL} = 20 \text{ lb/ft}^2$$

unit weight of $A = 88\ \text{lb/ft}^2$
$B = 95\ \text{lb/ft}^2$
$C = 88\ \text{lb/ft}^2$
$D = 93\ \text{lb/ft}^2$

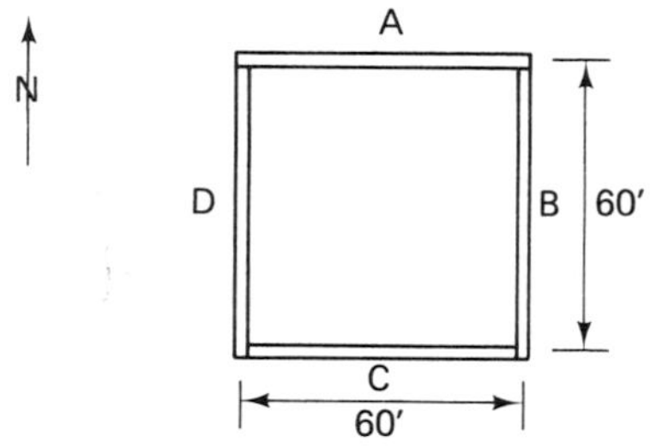

N

A

D

B 60′

C

60′

10-10. Referring to Problem 10-9 and its results, specify the shear reinforcement and jamb steel required for each of the piers of wall *A*. No inspection is provided.

11

Connections and Joints

The proper design and construction of connections needed to transmit direct and torsional shears, axial loads, or bending moments, separately or in combination thereof, from the horizontal resisting elements (floor and roof diaphragms) to the vertical resisting components (shear walls) of the structure is an absolutely essential feature of lateral force design. Unless positive means are provided for transferring shear from the plane of the diaphragm into the shear walls and also for transferring perpendicular or seismic forces from walls into the supporting diaphragms, the box shear-wall system will simply not function. In that event, the walls could fall away from the roof or floors, thereby allowing these elements to collapse, causing the building to fail catastrophically. When designing connections or ties, one should trace the various forces through their potential paths from assumed point of application to the ground, and then provide the appropriate connection detail that will be compatible with these basic assumptions of force distribution along the various loading paths. Since joints and connections directly affect the integrity of the whole structure, their design and fabrication must be structurally adequate for the function intended. When designing and detailing, it is well to keep in mind that the lateral forces are not static, as often assumed for convenience, but rather are dynamic in nature and, to a great extent, rather unpredictable. It is therefore incumbent upon the designer to be extremely conservative when defining what force magnitudes are to be resisted by the connecting elements. The overdesign of key connections generally does not impose any great economic burden; it simply provides good insurance against a malfunction of the structure when subjected to seismic or high wind forces. Even if only 1% of the needed connections are omitted or are in error, possibly the structure will turn out to be 100% in error. It will thus be the primary objective of this chapter to present some of the fundamental aspects involved when designing these important structural components and to portray a few typical details in the process.

Another aspect of reinforced masonry construction demanding attention, but often neglected, involves the provisions that must be made for certain types of separations and joints. Seismic separations are often needed in unsymmetrically shaped building plans. Cracking control joints are provided in masonry walls to accommodate a temperature differential, a change in moisture content, or when the masonry restrains the movement of adjoining elements. Details of these various types of joints are presented in the latter part of this chapter.

CONNECTION RESISTANCE

The forces to be considered when designing connections include gravity loads, temporary erection loads, those caused by differential settlements, horizontal loads normal to the wall (wind or seismic), horizontal forces that act parallel to the wall, creep, shrinkage, and differential thermal forces. These may act separately or in combinations thereof. Bond beams, serving as the flange or chord of a horizontal diaphragm must have continuous reinforcement in order to resist the axial tensile chord stresses induced by the horizontal diaphragm action. To provide that continuity, the reinforcement must therefore be adequately spliced. Or if the diaphragm chord is a timber member, bolted splices must be devised so that the total axial tensile force can be transferred across the joint. On the other edges of the diaphragm, the connection must be strong enough to transfer the total direct and torsional (if any) shear stresses carried from the diaphragm web into the wall. The minimum design shear force to be transferred from diaphragm to shear wall is set at 200 lb/ft by the UBC. Also, the connection must be capable of developing a minimum force of 200 lb/ft normal to the plane of the wall. Shear transfer from wood diaphragms can be accomplished with shear bolts and strap ties, or for concrete diaphragms by extending rebar dowels into the floor slabs. Joints and connections should be located, when possible, at points that may be readily analyzed and detailed and pose no serious construction difficulty.

STRENGTH OF CONNECTIONS

Connections can be formed with weldments, bolts, reinforcement or dowel anchors, and also by such mechanical devices as embedded steel shapes, plates, or studs. The strength of key connections generally should be sufficient to develop the useful strength of the structural elements that it is joining, regardless of the computed stress in the element. As a matter of good practice, design forces for joints and connections between lateral force resisting elements could be taken at about $1\frac{1}{2}$ times the calculated shear when using the Code-prescribed lateral loads, except that the connection need not be designed to develop forces greater than the ultimate capacity of the connected elements.

GENERAL DESIGN CONSIDERATIONS

Connection and joint details should be avoided if they result in stress concentrations that could cause spalling or splitting of the concrete block face shells or brick edges at the contact surfaces. Liberal chamfers, adequate reinforcement, and cushioning materials represent some ways in which stress concentrations may be avoided or at least minimized. Direct bearing of heavy concentrated loads on the face shell of concrete masonry units must be avoided. Welding to any embedded metal items that might cause damage to the adjacent masonry through spalling should also be avoided. This is particularly important where the expansion of the heated metal is restrained by the masonry.

All bolts and dowels that are embedded in the masonry must be solid-grouted in place, with not less than $\frac{1}{4}$ in. of grout between the bolt or the dowel and the masonry unit. To obtain the size and quantity of bolts needed to effect the in-plane shear transfer from diaphragm boundary to shear wall, refer to Table 11-1, which gives the shear permitted on bolts embedded in reinforced masonry.

TABLE 11-1

Allowable Shear on Bolts for All Masonry Except Gypsum and Unburned Clay Units
(*UBC Table 24-G*)

Diameter of bolt (in.)	*Embedment (in.)*	*Solid masonry (shear, lb)*	*Grouted masonry (shear, lb)*
$\frac{1}{2}$	4	350	550
$\frac{5}{8}$	4	500	750
$\frac{3}{4}$	5	750	1100
$\frac{7}{8}$	6	1000	1500
1	7	1250	1850[a]
$1\frac{1}{8}$	8	1500	2250[a]

[a]Permitted only with not less than 2500 lb/in.^{2} units.

At the tops of piers and columns, vertical bolts must be located at least 4 in. in from the interior face of the masonry and contained within the horizontal column ties. Since the Code prohibits cross-grain bending in wood edge members, the connection details must be devised so that such a stress is not imposed upon the wood member, as will be shown later in the chapter.

EMPIRICAL CODE LIMITATIONS

There are numerous specifications in codes that pertain, in particular, to connections of various types. Some of these are rather arbitrary in nature, and would not be apparent to the uninitiated when designing a connection simply by considering the principles of structural mechanics alone. It is the intent of this section to call attention to some of the more significant of these empirical or arbitrary limitations to assist the designer in preparing acceptable connection details.

According to UBC Section 2310, masonry walls must be anchored to all floors and roofs that provide lateral support for the wall. Such anchorage has to provide a positive direct tie capable of transferring the horizontal forces specified by the Code for wind or seismic, with a minimum force of 200 lb/ft foot of wall. Walls must be designed to resist bending between anchors where the anchor spacing exceeds 4 ft. This means, then, that the wall has to be reinforced so that it will literally span horizontally between these connection points. The required anchors in masonry walls composed of hollow units or cavity walls have to be embedded in the reinforced grouted cores of the wall. Section 2310 then refers to Sections 2312(j)2D and 2312(j)3A. The former calls for floor and roof diaphragms to be designed to resist those forces set forth in Table 5-9. Also, it calls for diaphragms supporting masonry walls to have continuous ties between diaphragm chords in order to distribute, into the diaphragm, the anchorage forces specified by the Code. Added chords may be used to form subdiaphragms to transmit the anchorage forces to the main cross ties. Note that Section 2312(j)2D also requires that diaphragm deformations be considered in the design of supported walls.

Section 2312(j)3A deals with special anchorage of wood diaphragms. It calls for wood diaphragms that support masonry walls to have the type of anchorage conforming to Section 2310, while a further stipulation states that any anchorage in zones 2, 3, and 4 cannot be accomplished with toenails or nails subjected to withdrawal. Furthermore, cross-grain bending or cross-grain tension is prohibited by this section in the wood-framing members connected to masonry walls. This prohibition was the direct result of observations of failures in the San Fernando earthquake disaster of 1971. Formerly an allowable cross-grain flexural stress of 60 lb/in.^{2} was permitted in Douglas fir wood ledgers.

CONNECTION TYPES AND DETAILS

Reinforced masonry is an extremely versatile material, and as such it can be used in combinations with many other materials, such as cast-in-place or precast concrete slabs, prestressed concrete planks, concrete block or tile filler slabs, steel bar joists with steel or wood decking, and wood decks. Connection details will vary considerably, depending upon their function as well as the type of material used for the floor or roof diaphragm. However, if the box-type lateral-force-resisting system is to function properly, regardless of the type of material utilized, it must be tied together so that it will act as an integrated unit. For one thing, the walls must be attached to the diaphragm so that they can resist design forces directed both normal to the wall and parallel to it. A very common type of failure occurrence during an earthquake takes place when the walls pull away from the roof or floors, thereby causing a serious collapse. This is why the wall-to-diaphragm connection is so critical. Some examples of this type of connection will be seen in a later section of this chapter.

Where weld plates are used, proper weld techniques are of prime importance (see Figure 11-1 for a typical detail). Welding to reinforcing steel requires quality weld control, in addition to proper anchorage of the reinforcing bars. All such welds should develop at least 125% of the specified yield strength of the bar. On the other hand, cast-in-place dowels need develop only the specified yield strength of the embedded bars.

Provisions for misalignment of precast members must be made at the joint. This is usually done by providing adequate tolerances in the connection details and recognizing in the structural design that such tolerances exist.

FIGURE 11-1. Weld plate-slab connection, concrete diaphragm.

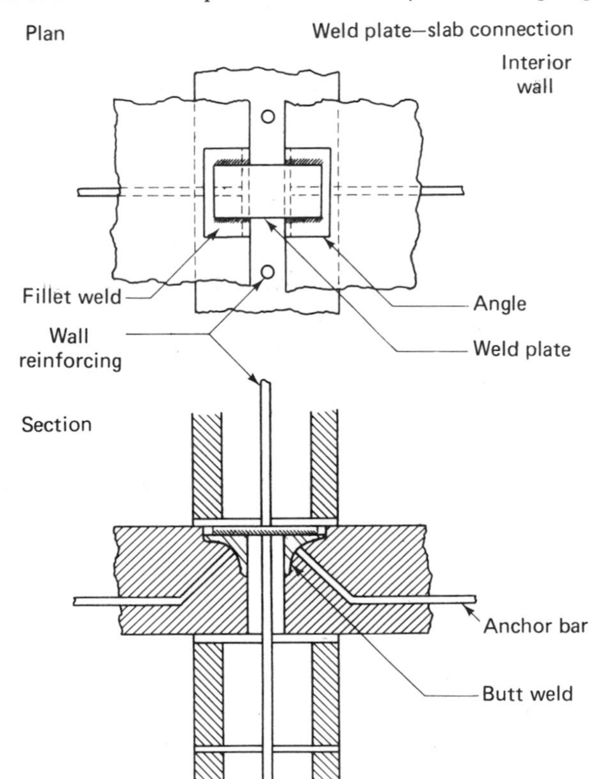

Conditions of design

There are, no doubt, as many different ways to provide good connection details as there are needs for them or engineers to devise them. The form they assume depends upon the type of connection needed (i.e., diaphragm shear anchorage, beam support, foundation tie-down, etc.), the kind of materials to be joined, and the structural function (shear transfer, moment development or combination thereof, etc.) Regardless of these conditions, however, there are certain fundamentals that must be examined and analyzed when designing any type of connection. For instance, if we examine a diaphragm connection (floor or roof to wall), we observe that all the following force patterns must be provided for in some way before the detail will prove satisfactory:

1. Resistance to the diaphragm in-plane shear force at the boundary member that is parallel to the shear wall is generally achieved, at least with timber, through the shear on bolts securing the boundary member to the masonry wall. Allowable values for these were given in Table 11-1. For concrete, this anchorage may be achieved with embedded weld plate anchors.
2. Walls must be firmly anchored to their supporting diaphragm so as to resist the forces acting perpendicular to the wall. Without this tie, as previously mentioned, the wall and roof or floor will simply separate, with disastrous results. This tie may be achieved with joist anchors or tie straps in the case of timber members. Or again, dowels extending from a concrete slab into the wall a distance equal to their development lengths will also provide this anchorage.
3. The vertical roof and floor gravity loads must be carried into the masonry wall. This may be accomplished by direct bearing on the wall, as in the case of a concrete slab, beam, or truss. However, wood joists, purlins, or beams may be hung from a wood ledger by means of metal hangers.
4. For beams connected to the side of a wall, the beam reaction must be transferred through bolts between clip angle and masonry wall. No moment need be developed in this connection, although the wall itself must be reinforced sufficiently to accommodate the eccentrically applied reaction.
5. Walls must be adequately anchored to the foundation to resist overturning and shear due to the effect of the horizontal loads. This may be accomplished with dowel anchors. Or in lighter construction, specially designed holddown anchors at each wall or pier boundary may provide this resistance.

Having established these conditions, structural engineers often find that the final solution for connection details severely taxes his imagination and ingenuity. There is often more than one satisfactory solution for a particular connection detail, so it might be imprudent to label any specific one as "typical." Consequently, those illustrated in the following figures are intended only to account for the various force paths and to demonstrate one way (not necessarily the most efficient) in which they may be accommodated, while conforming to the empirical limitations prescribed by the Code.

Diaphragm connections

CAST-IN-PLACE CONCRETE SLABS

There are several different types of concrete floor or roof diaphragms in use today. These would include precast slabs, cast-in-place slabs, or prestressed concrete planks. Details vary, depending upon the type of slab utilized and the individual preference of the structural engineer or manufacturer.

Where a cast-in-place concrete slab provides the floor surface, typical floor to wall connections are seen in Figure 11-2. Note that the tie shown in Figure 11-2a is obtained through the use of shear dowels extending between slab and wall. These dowels could bend upward or downward.

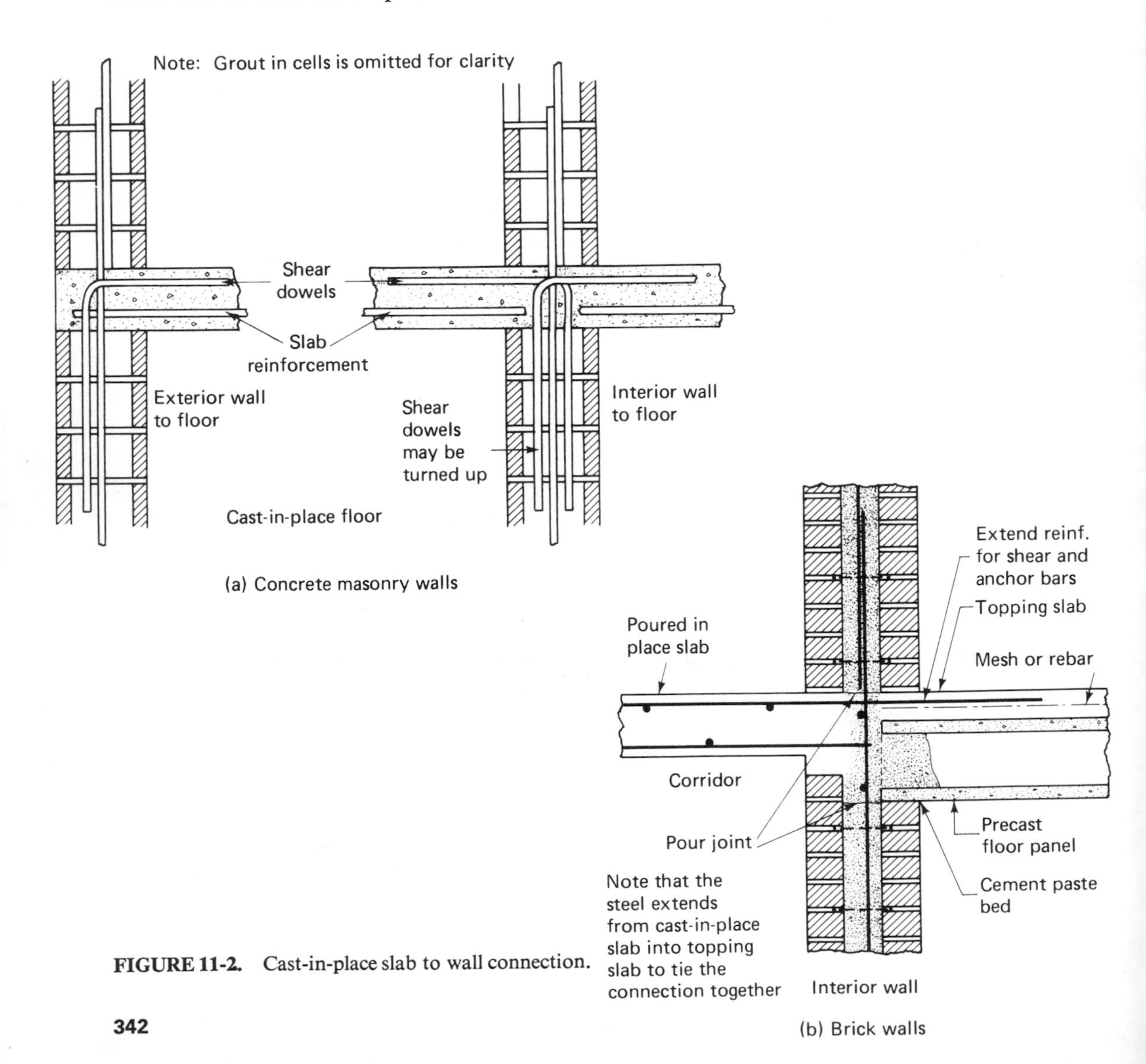

FIGURE 11-2. Cast-in-place slab to wall connection.

Precast Concrete Slabs

Where precast panels are provided, various types of connections are feasible. Figure 11-3 shows various ways to anchor weld plates into a precast slab. The anchor plate, anchored within the grout core by a rebar welded to the plate, (Figure 11-3b–d) secures the slab to an adjacent one or to the wall with a weld to the embedded angle. On the other hand, the reinforcing in a precast slab can be doweled directly into the grout core, as illustrated in Figure 11-4, thereby accomplishing the tie by anchorage bond.

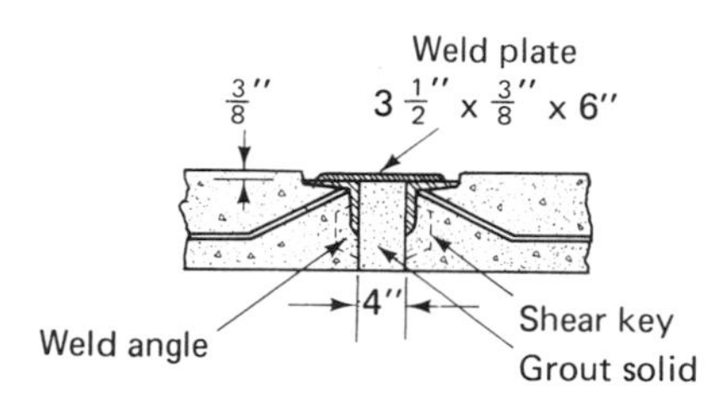

(a) Welded connection—Two precast floor slabs

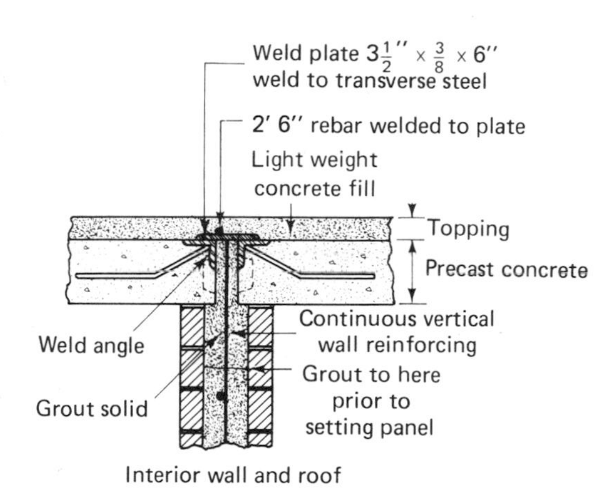

(b) Roof slab with a topping—Interior wall

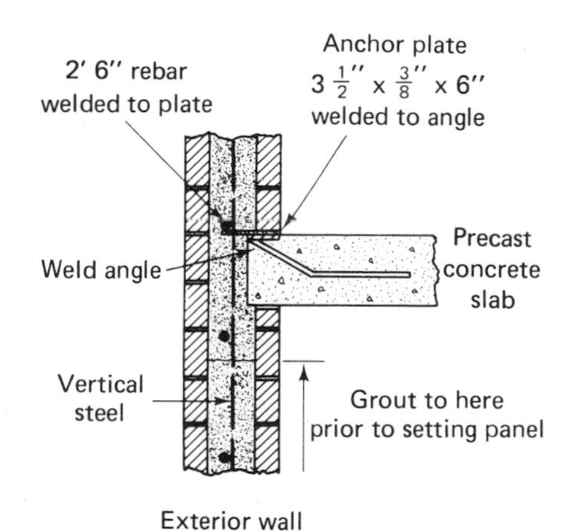

(c) Welded connection of precast slab—Exterior wall

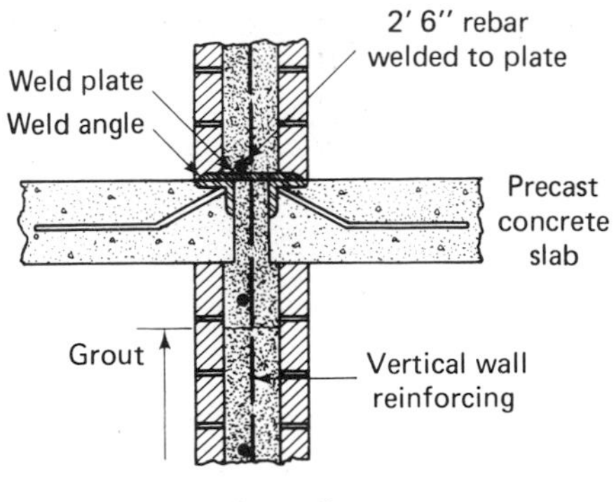

(d) Precast slabs—Interior wall

FIGURE 11-3. Weld plate connection—precast concrete slabs.

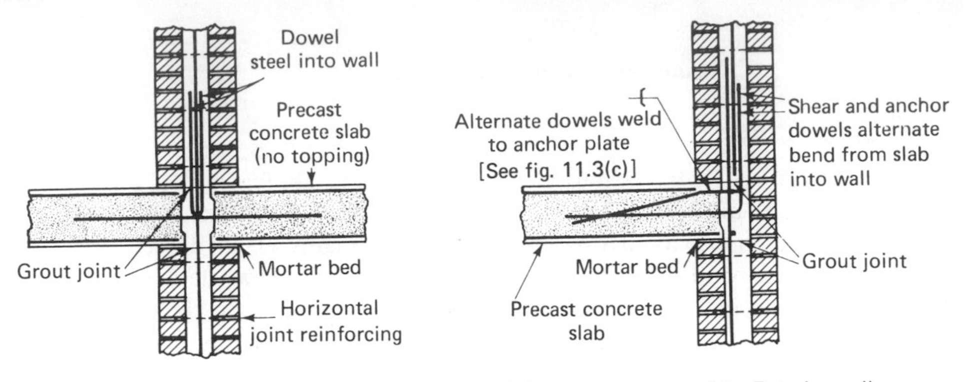

FIGURE 11-4. Precast slabs on brick walls.

Prestressed Concrete Planks

Because they are readily transportable and easy to place on the masonry walls, prestressed planks of various widths (12 to 48+ in.) are often used. To ensure the continuity needed to provide diaphragm action, in some cases a concrete topping slab (say, $2\frac{1}{2}$ in. thick) is poured on top of the in-place planks. Typical connections to the walls may be seen in Figures 11-5 (concrete block wall) and 11-6 (brick wall). The shear dowels located in the topping slab bend into the wall (exterior wall, Figure 11-5a and 11-6a), or into the adjacent topping slab (interior wall, Figure 11-5b and 11-6b) to provide the diaphragm anchorage.

Figures 11-7 and 11-8 show various support conditions for precast, prestressed planks. Figure 11-7 shows different details at a girder support, and Figure 11-8 those for a supporting wall. Where a topping exists, continuity is simply achieved with shear dowels which extend continuously across the joint (Figure 11-7a). Where no topping is present, an adequate tie can be achieved either with (1) dowel bars, extending between adjacent planks and which lie in the joints between planks (Figure 11-7b), or (2) embedded flat bars and weld plates (Figure 11-8).

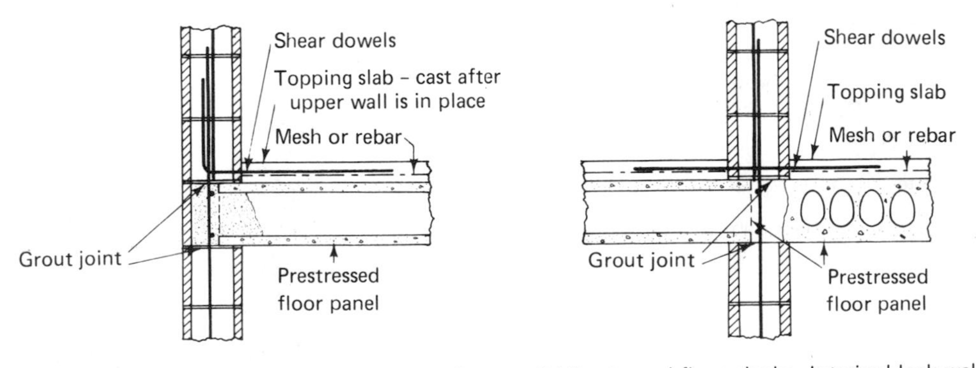

FIGURE 11-5. Prestressed planks on concrete block walls.

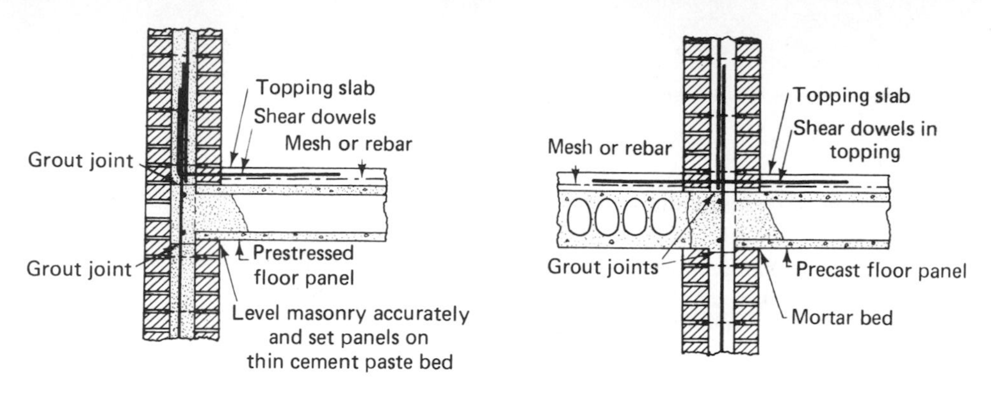

(a) Precast floor planks—Exterior brick wall

(b) Prestressed floor planks—Interior brick wall

FIGURE 11-6. Prestressed planks on brick walls.

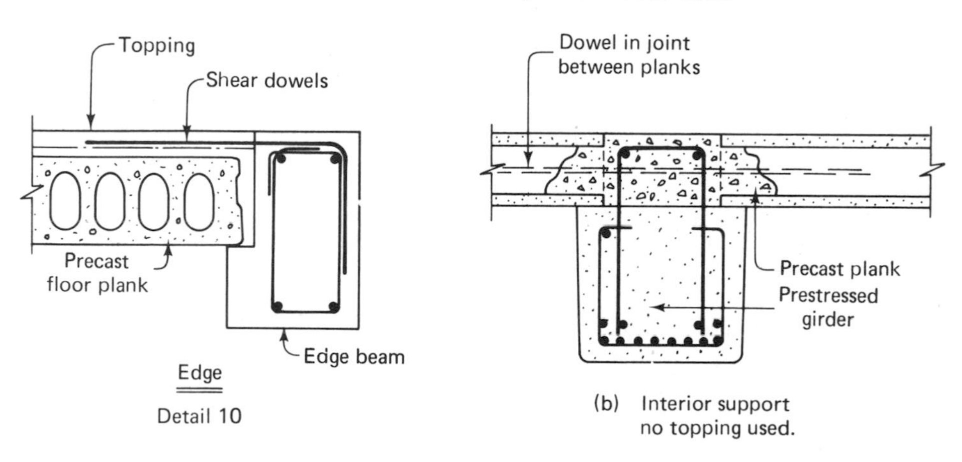

(a) Exterior support
topping used

(b) Interior support
no topping used.

FIGURE 11-7. Prestressed concrete plank to girder connection.

(No concrete topping required)

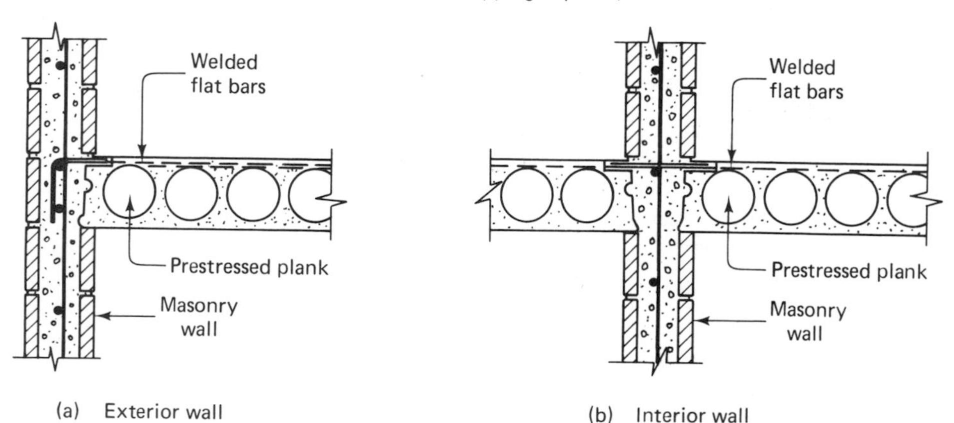

(a) Exterior wall

(b) Interior wall

FIGURE 11-8. Prestressed plank to wall connection.

Block and Beam

An interesting variation of concrete diaphragm systems is known as a *block-and-beam* floor system. This scheme consists of 8 × 8 × 24 in. filler blocks which butt against each other and bear on the flange of a web trussed steel joist or a prestressed concrete beam spaced on about 2 ft 4 in. centers. A reinforced topping, at least 2 in. thick, is then poured over the top of the filler block to provide the continuity needed for effective diaphragm action. Figure 11-9 shows some typical details.

Steel Decking

If a type of steel decking forms the floor or roof diaphragm, some form of connection detail similar to that shown in Figure 11-10 may be utilized successfully. There are a number of different types of steel diaphragms manufactured, most of which involve special connection details, however. So in many cases it would be best to consult the manufacturer of the system being used for their recommendations regarding diaphragm connection details which have been approved by ICBO. The details shown below utilize a concrete fill and wall anchor dowels to achieve diaphragm continuity. Other steel deckings, adequately welded to the supporting members, develop the diaphragm shear without a concrete fill.

Wood

By far the most commonly used material for roof diaphragms, at least in single-story masonry construction, are the plywood systems described in Chapter 10. Where this type of construction is employed, some typical details, such as are shown in

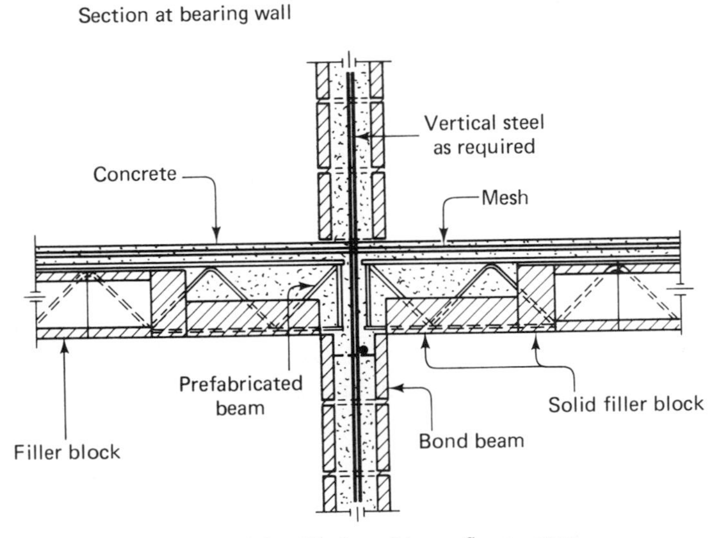

FIGURE 11-9. Block and beam floor system.

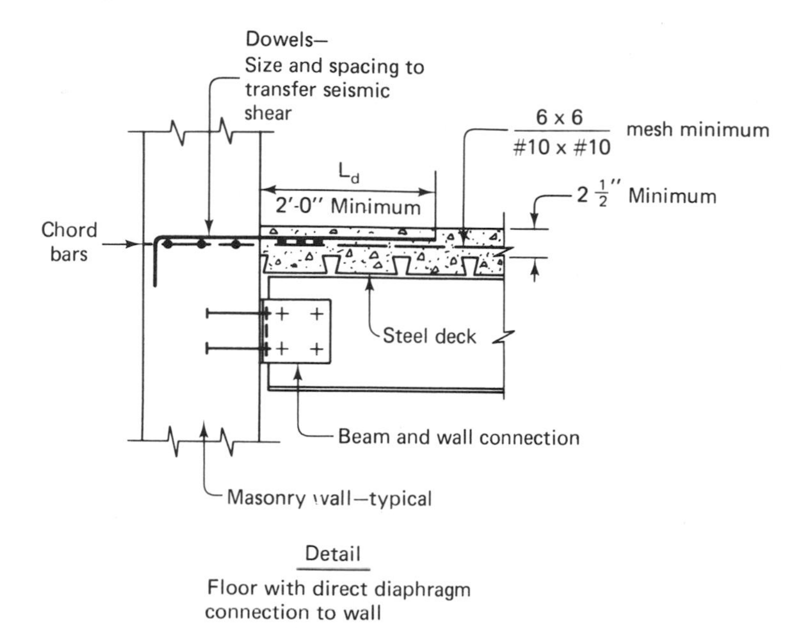

FIGURE 11-10. Steel deck diaphragm connections.

Figures 11-11 and 11-12, have been used to conform to the Code requirements for timber diaphragms.

These details represent only a very few examples of possible ways to tie the lateral-force-resisting system together into an integrated unit. What can be achieved in this regard is limited only by the ingenuity of the structural engineer. Observe that each component performs a very specific function, or it can even serve more than a single purpose. For instance, in Figure 11-11a the joist hanger carries the joist reaction into the wood ledger, which, in turn, transfers it into the wall through the anchor bolts. These fasteners also function as shear transfer bolts for the diaphragm. Bolt spacing thereby is governed by that force which is the vector sum of the vertical load plus the horizontal diaphragm force being transferred to the wall. The joist anchor ties resist lateral forces acting normal to a wall which spans vertically between diaphragms. Consequently, it ties the wall and diaphragm together so that the wall can transfer its inertial seismic effect or wind loading into that diaphragm; thus the three-directional loading aspect is provided for properly. The large malleable iron washers, sometimes used between the bolt head and the connected wood member (Figure 11-11a), by resisting the bending caused by the lateral force acting perpendicular to the member, eliminate the Code-forbidden cross-grain bending in that wood member.

In another type of connection, exemplified in Figure 11-11c or d, a wood or steel ledger supports the vertical load and its bolting need be designed for vertical

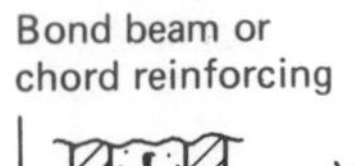

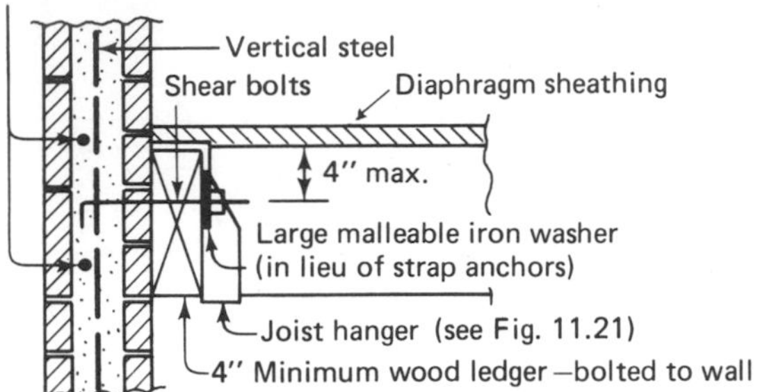

(a) Joists perpendicular to wall
Joist hanger supports

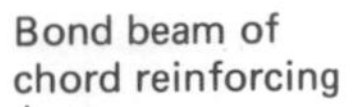

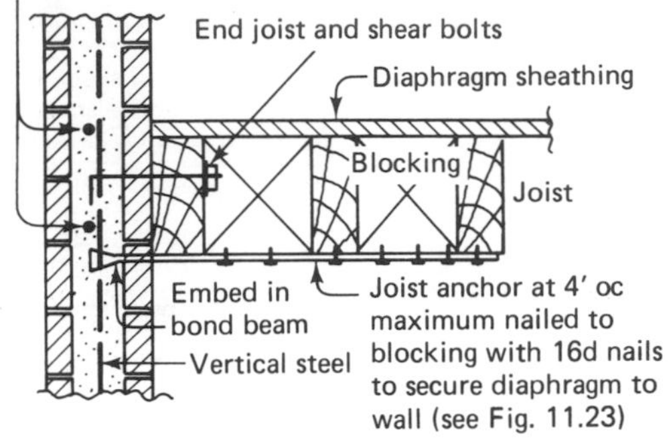

(b) Joists parallel to wall

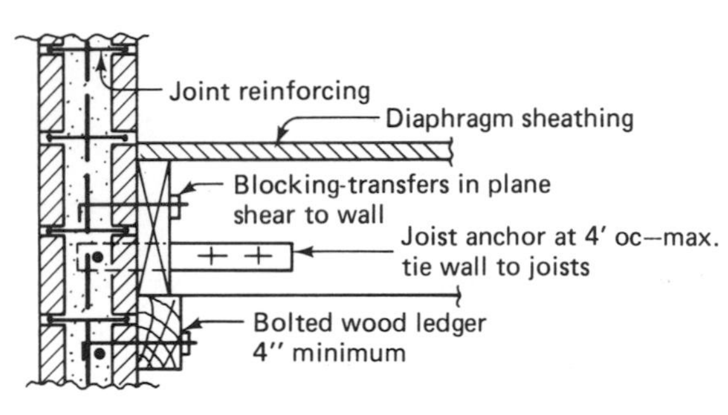

(c) Wood ledger joist support

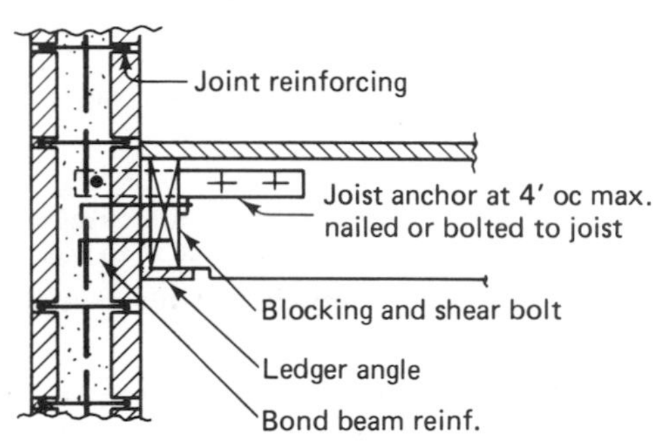

(d) Steel ledger joist support

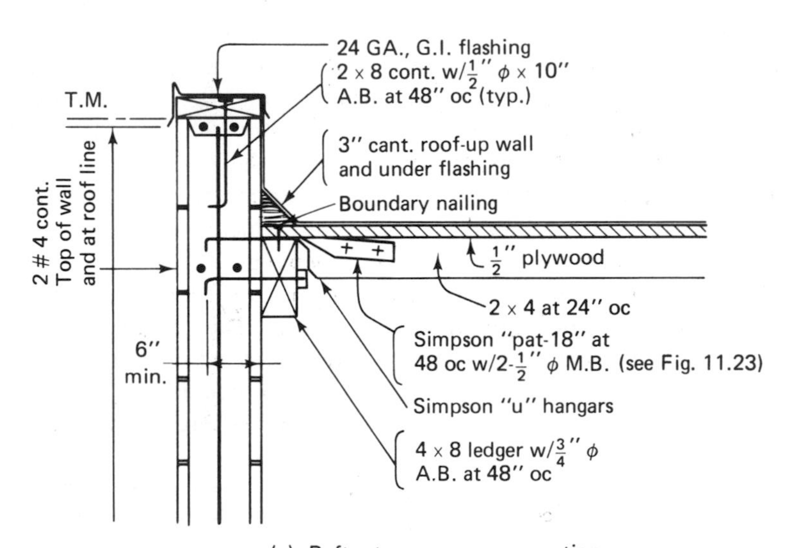

(e) Rafter to masonry—connection

FIGURE 11-11. Bolted connections—timber to masonry.

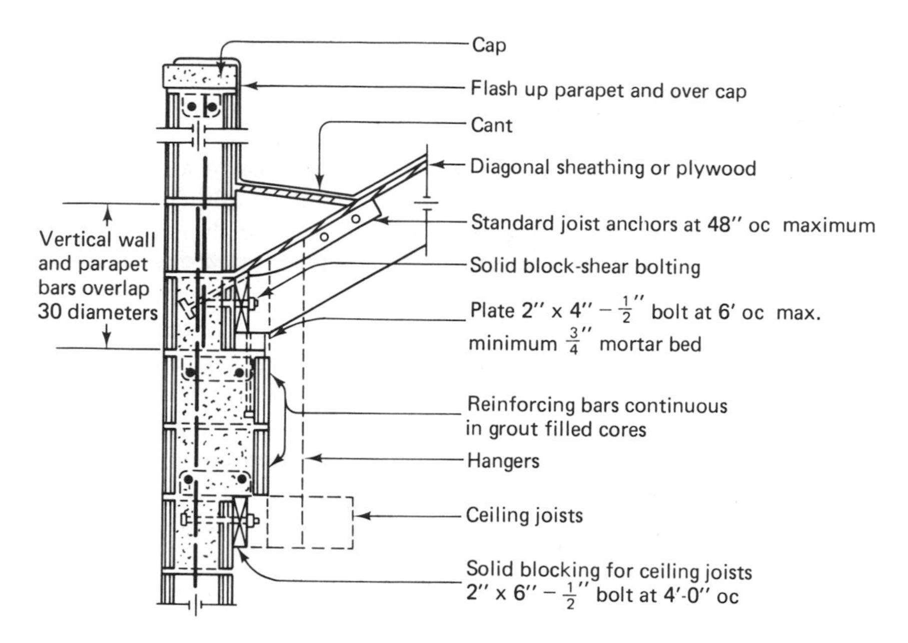

(a) Exterior wall support

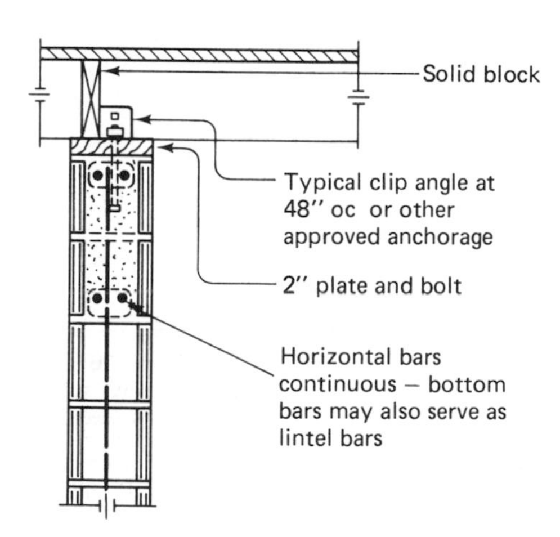

(b) Interior wall support

FIGURE 11-12. Bond-beam supports.

load only. The in-plane shear is developed through the solid blocking, and thus the block bolting needs to be spaced to transfer the horizontal diaphragm shear only. The values specified in Table 11-1 would be used to determine these various bolt spacings. In Figure 11-12a, the vertical reaction from the joist is carried on the projecting bond beam, whereas an interior wall support connection is portrayed in Figure 11-12b.

Beam, girder, and truss connections

Numerous possibilities exist for securing a beam or truss to a masonry wall or column, depending, among other factors, upon the magnitude of the reaction, the type of vertical load-carrying element being connected (i.e., steel or glu-lam beam, wood or steel truss, etc.) and the space available. Figures 11-13, 11-14, and 11-15 present possible means of achieving the connection of beams or trusses to a masonry wall. Eccentricity is minimized in the schemes shown in Figure 11-13. Pilasters offer a way to obtain concentric support where more space is needed for a particular type of connection, as shown in Figure 11-14. Since only a vertical reaction is being transferred, only nominal lateral restraint is needed. However, in other instances, the beam or truss may be subjected to a lateral force as well. This condition could develop where a wall spans horizontally between pilasters, and the pilaster in turn spans from floor to roof truss. Consequently, this connection must resist the lateral thrust of the pilaster reaction occurring at the roof line.

Figure 11-15 represents a way to secure a beam to the side of a masonry wall. Note that in such cases the wall must be reinforced to develop the additional moment, caused by the eccentrically applied beam reaction. Since the maximum wall moment probably occurs at or near the midheight, the design moment to consider here would amount to that shown in Figure 11-16.

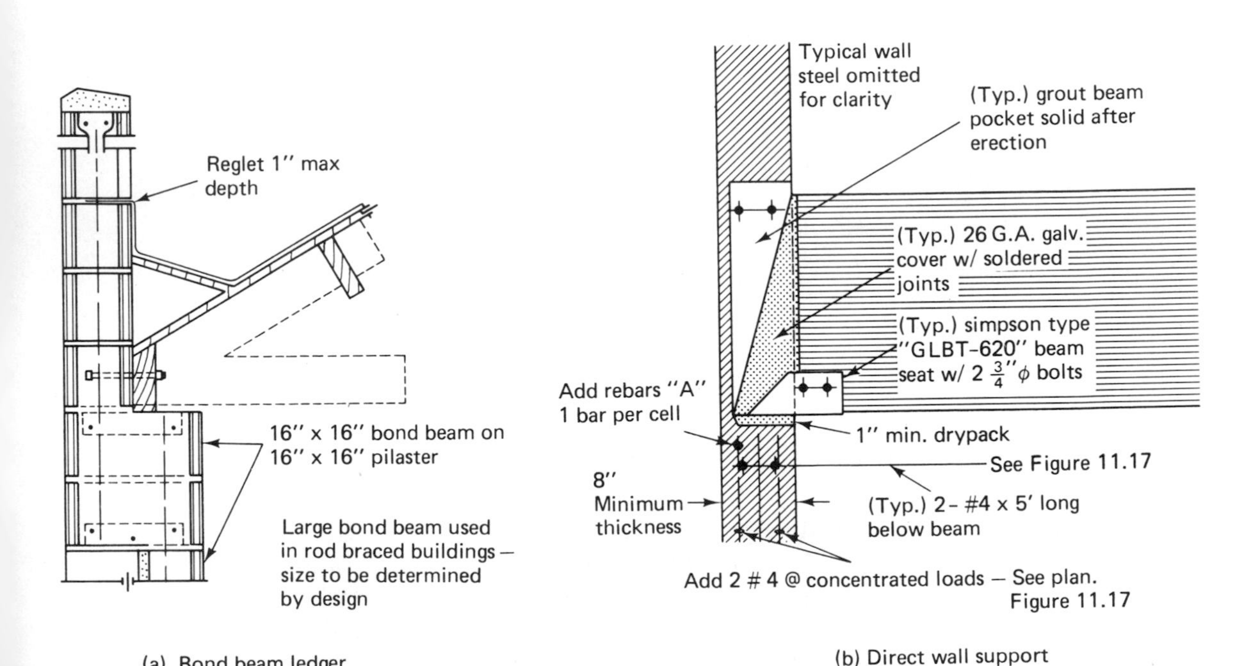

FIGURE 11-13. Beam or truss supports—minimum eccentricity.

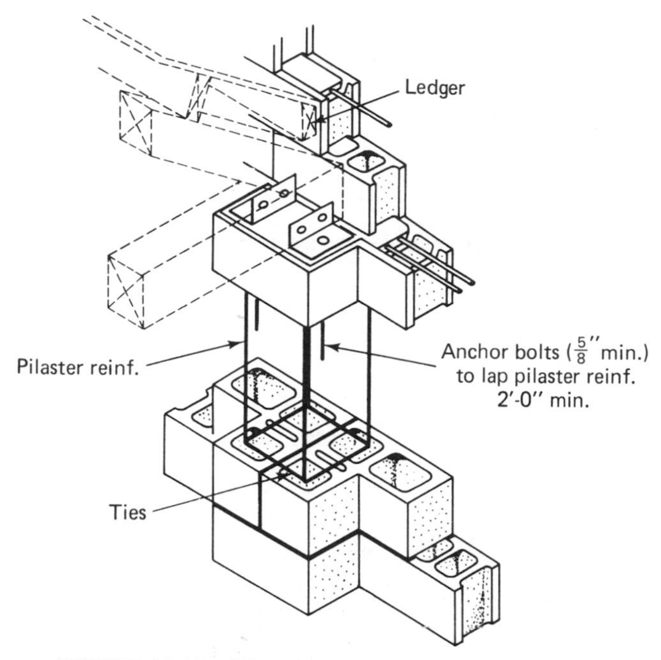

FIGURE 11-14. Pilaster support—minimum eccentricity.

Assuming that the support condition resembles that portrayed in Figures 11-13b or 11-15, the wall could be reinforced for this additional moment, by adding the vertical A bars as shown in Figure 11-17. These can be selected from the chart provided, depending upon the vertical wall reaction and the eccentricity. Note that the minimum wall thickness, in this instance, is 8 in. nominal.

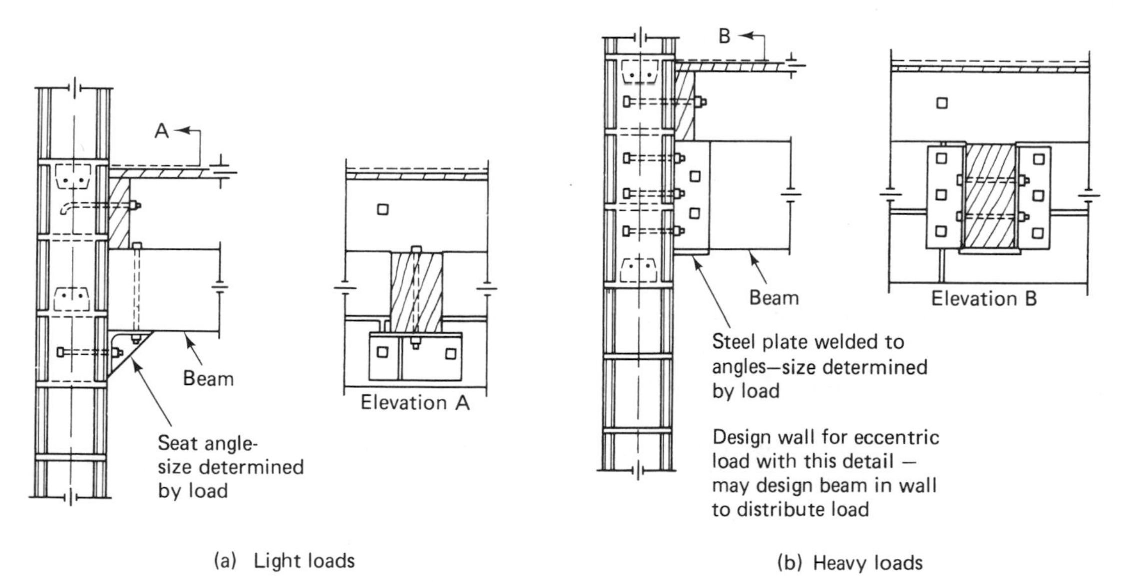

FIGURE 11-15. Eccentric beam to wall connection.

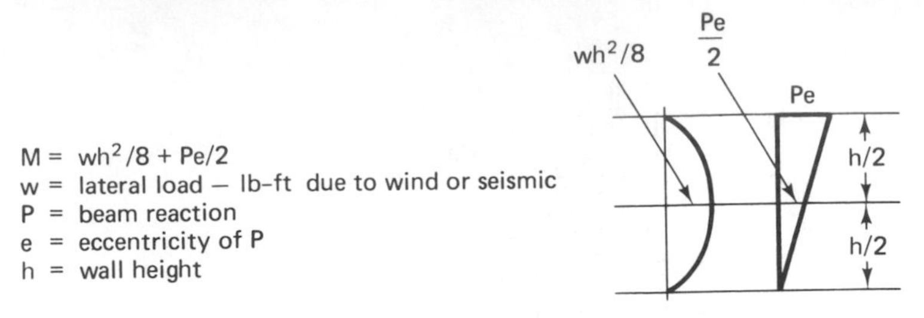

FIGURE 11-16. Moment diagrams for eccentrically loaded wall.

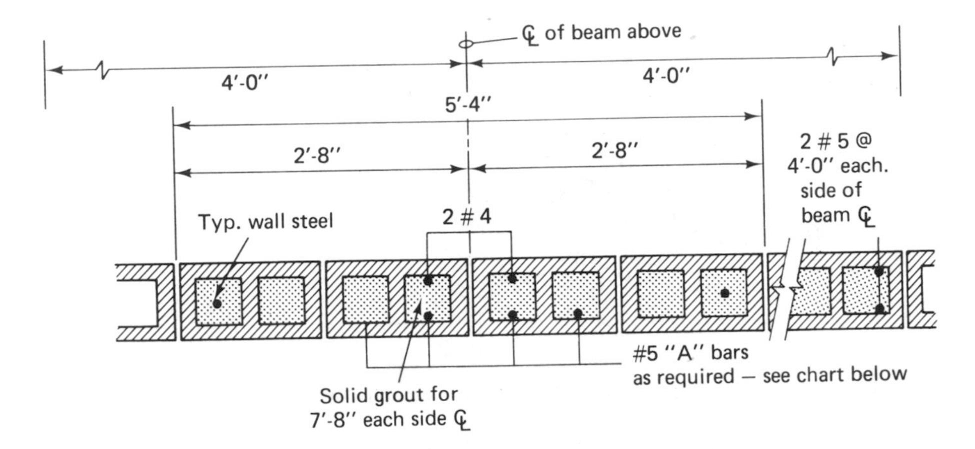

(a) Plan detail

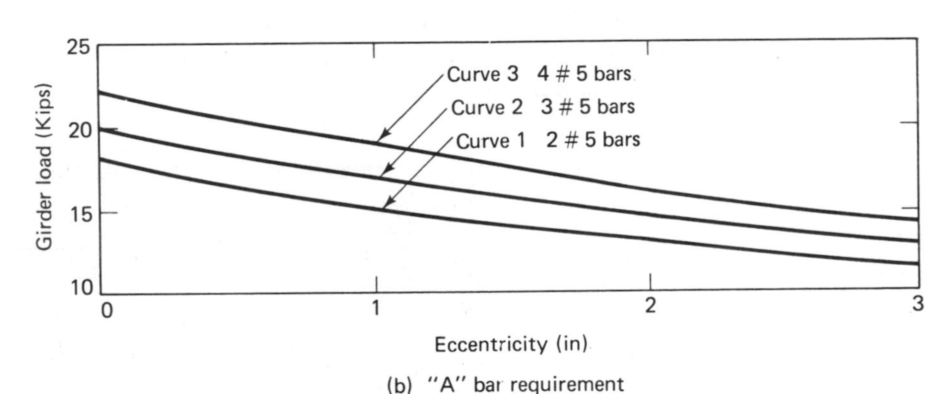

(b) "A" bar requirement

FIGURE 11-17. Additional wall reinforcement at beam or truss support.

Foundation anchor details

As with all of the previously described connections, there are many different ways to support a wall, pilaster, or column on the underlying substructure. Usually the foundation in masonry construction consists of a continuous concrete wall footing or grade beam, although isolated spread footings are sometimes used, especially under interior columns. For lighter construction (one- and two-story), where typically a slab on grade provides the ground floor surface, the details given in Figure 11-18 could be utilized. The reinforcement at the top of the foundation wall

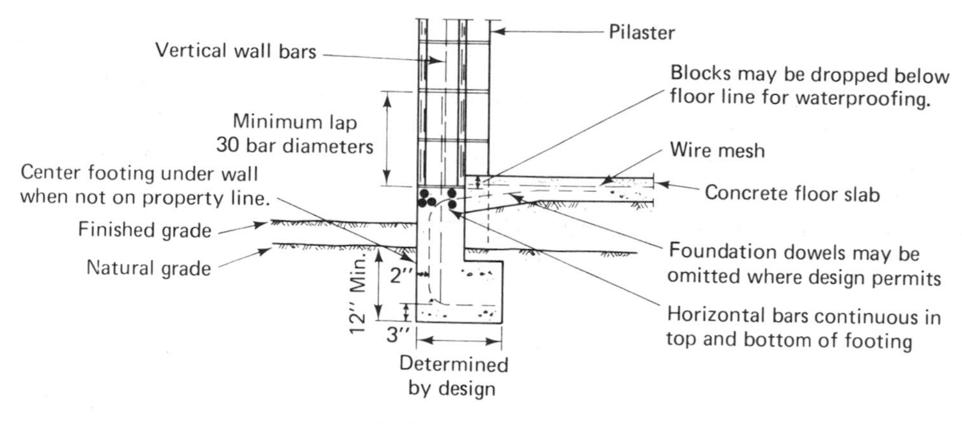

(a) Concrete foundation wall on property line

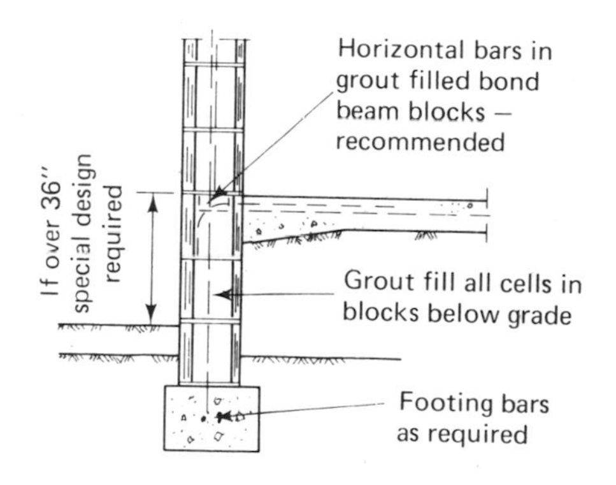

(b) Concrete block foundation wall

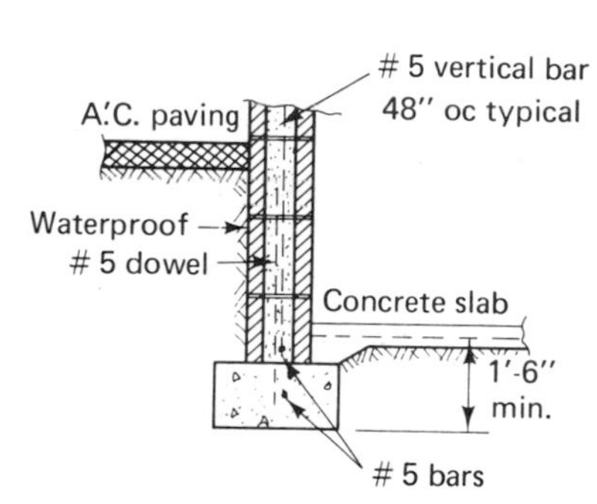

(c) Foundation detail connection of floor to wall with exterior earth fill at a higher level.

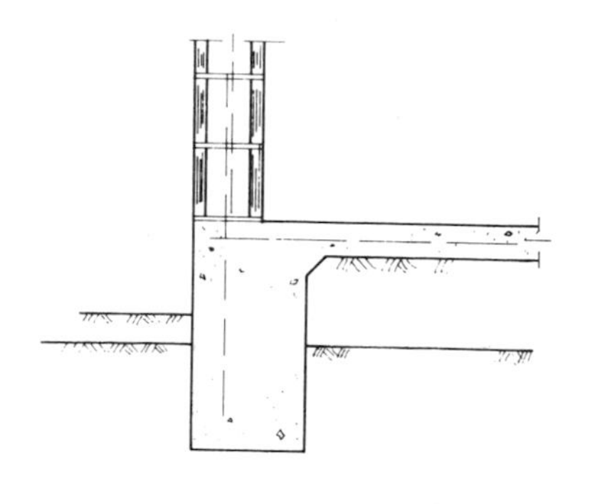

(d) Foundation and slab poured integrally

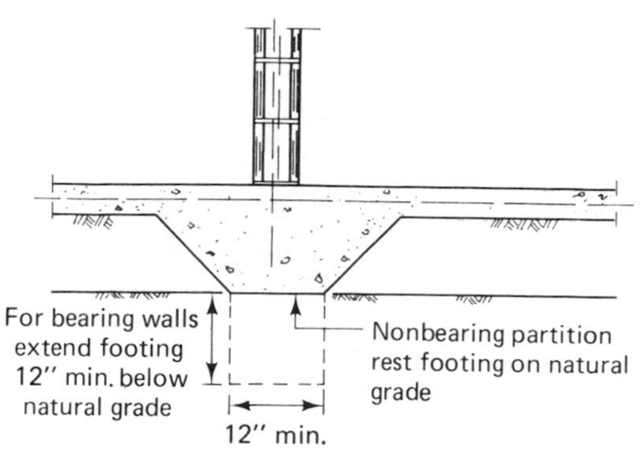

(e) Interior non-bearing-wall-footing

FIGURE 11-18. Foundation details—light commercial construction.

near the floor line serves to tie the building together at this level so that it actually becomes the functional equivalent of a bond beam. Note the location of the longitudinal bars in the bottom of the footing. They are placed there to prevent the footing from cracking structurally in the event of the occurrence of any differential settlement, or to enable the footing to span across any soft spots in the underlying soil. The UBC also requires that walls be designed to resist bending between anchors, where the anchor spacing exceeds 4 ft.

Very careful attention must be paid to the manner in which the walls in the first story of a multistory building are anchored to the substructure. There are several ways to provide such a strucural connection (e.g., by using rebar dowels embedded a sufficient distance into both the foundation and the grouted portion of the wall so as to fully develop the bars). Refer to Figure 11-19 which portrays a foundation anchor detail for a commercial building. Holddown anchors may be used to tie wood stud walls to the foundation, as shown in Figure 11-20.

FIGURE 11-19. Foundation anchor detail. Dowels projecting from footing to lap the tension steel of a multistory shear-wall (bearing-wall) building. (Courtesy Masonry Institute of America.)

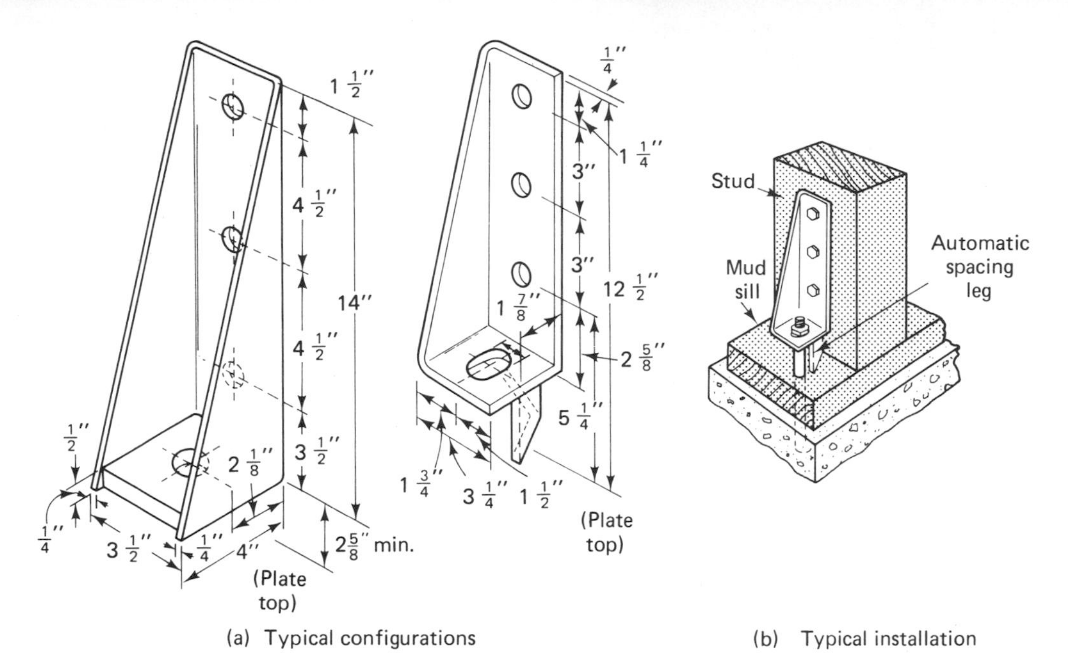

(a) Typical configurations (b) Typical installation

FIGURE 11-20. Holddown anchor.

Connection accessories

To facilitate construction, many different types of prefabricated metal connectors have been devised, most of which are proprietary items. The examples shown carry ICBO Research Committee approval, since they have all been thoroughly tested. Allowable load tables are provided with most of these accessories, based upon these test results. Included in this particular product line are such connectors as various types of holddown anchors (Figure 11-20), joist and beam hangers (Figure 11-21), beam seats (Figure 11-22), purlin anchors (Figure 11-23), ties, straps, and other items. An example of a load table for one of these accessories is shown in Table 11-2.

TABLE 11-2

Load Table for Simpson Purlin Anchors

Model	*Material galv.*	*Length (in.)*	*Connectors to purlins*	*Design loads*[a] *Other*	*Roof*
PA 18	12 ga. × $2\frac{1}{16}$ in.	$18\frac{1}{2}$	12–16d	1.6 kips	2.1 kips
PA 23	12 ga. × $2\frac{1}{16}$ in.	$23\frac{3}{4}$	18–16d	2.4 kips	3.2 kips
PA 28	12 ga. × $2\frac{1}{16}$ in.	29	24–16d	3.1 kips	4.1 kips
PAT 18	12 ga. × $2\frac{1}{16}$ in.	$18\frac{1}{2}$	2-$\frac{1}{2}$ in. MB	1.6 kips	2.0 kips
			7–16d	940 lb	1175 lb
PAT 23	12 ga. × $2\frac{1}{16}$ in.	$23\frac{3}{4}$	3-$\frac{1}{2}$ in. MB	2.4 kips	3.0 kips
			13–16d	1740 lb	2175 lb
PAT 28	12 ga. × $2\frac{1}{16}$ in.	29	4-$\frac{1}{2}$ in. MB	3.1 kips	3.9 kips
			19–16d	2550 lb	3190 lb

[a]Design load increases of one-third allowed for seismic.

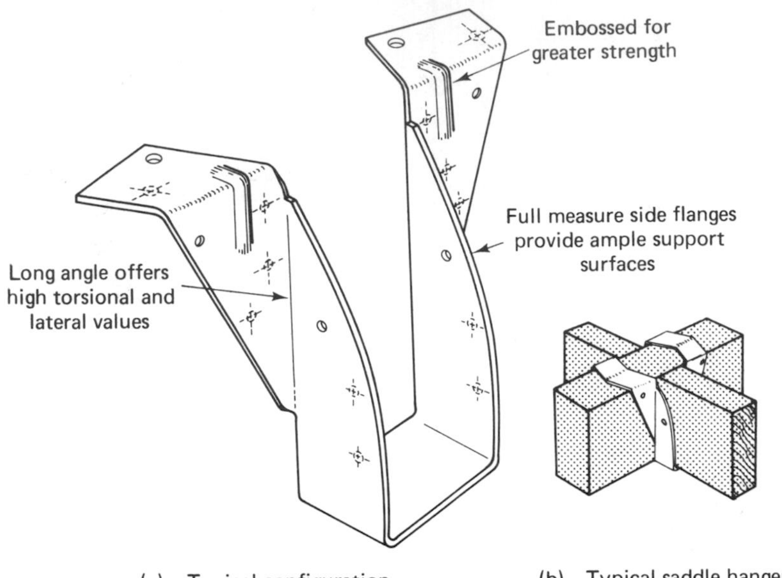

(a) Typical configuration (b) Typical saddle hanger

FIGURE 11-21. Joist hanger.

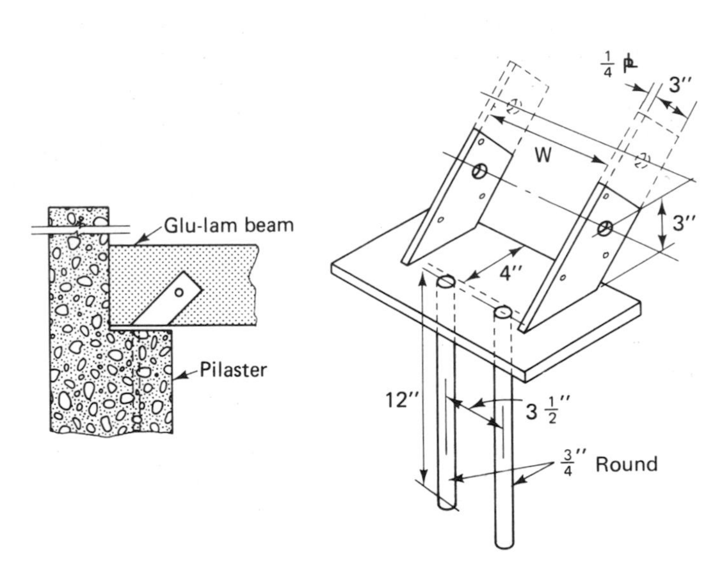

FIGURE 11-22. Beam seat.

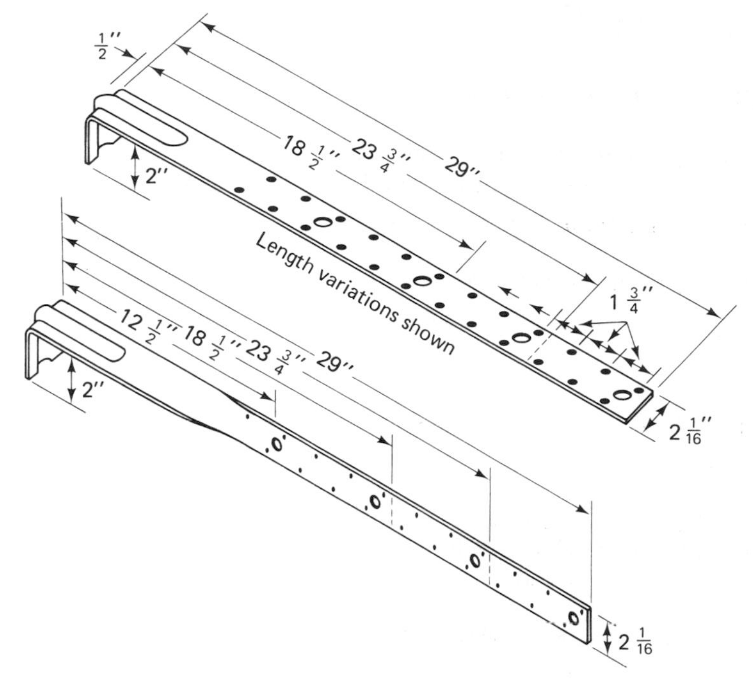

(a) Typical configuration

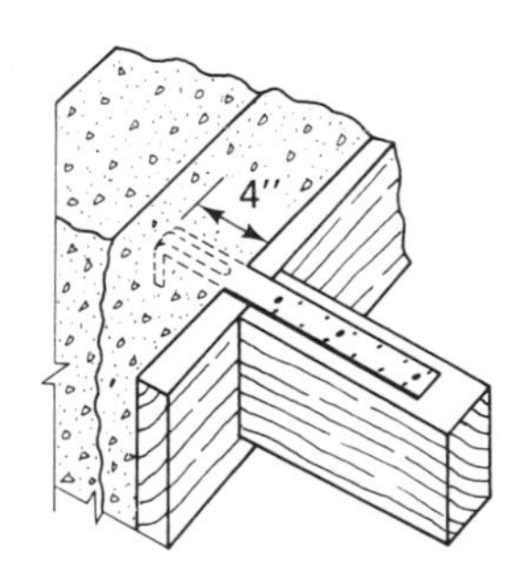

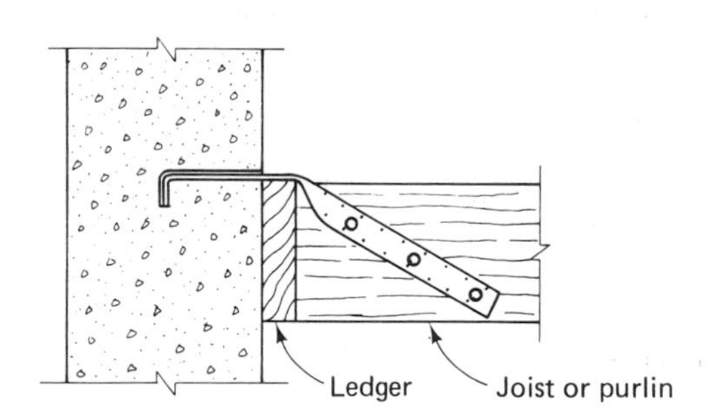

Cross section view of installation

Provides 4″ of embedment

(b) Typical installation

FIGURE 11-23. Purlin anchors.

Example of connection detailing

The following example should demonstrate some of the aspects encountered when designing connections needed to ensure lateral stability under seismic or wind loadings. Other examples appear in Case Study 1 in Chapter 12.

EXAMPLE 11-1: Ledger Beam Connection

Analyze the typical ledger beam connection shown in Figure 11-24.

Solution:

4 × 12 purlin to masonry wall.

LOADS PARALLEL TO WALL:

vertical load/bolt = 2640 lb/4 bolts = 660 lb/bolt
shear load/bolt = (200 lb × 8 ft)/4 bolts = 400 lb/bolt
resultant: 400 lb

660 lb $R = \sqrt{660^2 + 400^2} =$ 772 lb/bolt

CHECK ALLOWABLE:

Wood: parallel to grain: $p = 1470$ lb × 1.33/bolt > 400 lb/bolt ∴ OK
perpendicular to grain: $q = 930$ lb × 1.25/bolt > 660 lb/bolt ∴ OK
Masonry: 1100 lb/bolt > 772 lb/bolt (with 5 in. embedment) ∴ OK

Use purlin hanger (per Figure 11-21); allowable $P = 3155$ lb > 2640 lb (per ICBO Research Report).

LOADS PERPENDICULAR TO WALL:

total perpendicular load = 8 ft × 501 lb/ft = 4008 lb total
perpendicular load/bolt = 4008/4 = 1002 lb/bolt

STRAPS:

Provide 1 strap anchor per purlin with 4 $\frac{1}{2}$-in. ϕ bolts to purlin.

Capacity: 3900 × 1.33 = 5190 lb > 4008 lb ∴ OK (see Figure 11-24)

ANCHOR BOLT PULLOUT:

Approximate analysis: allowable masonry tension = 12 lb/in.^{2}
Conical area $= \pi R\sqrt{R^2 + h^2}$
$= \pi \times 5\sqrt{5^2 + 5^2} = 111$ in.^{2}
$P_\perp$ maximum = 111 in.^{2} × 12 lb/in.^{2} = 1332 lb/bolt
1332 lb/bolt × 1.33 > 1002 lb/bolt ∴ OK

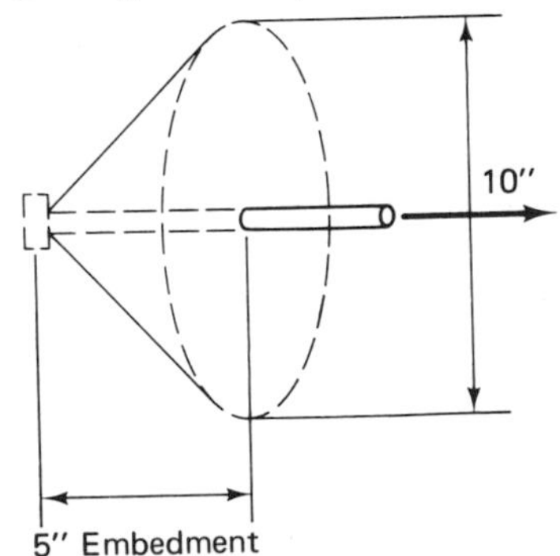

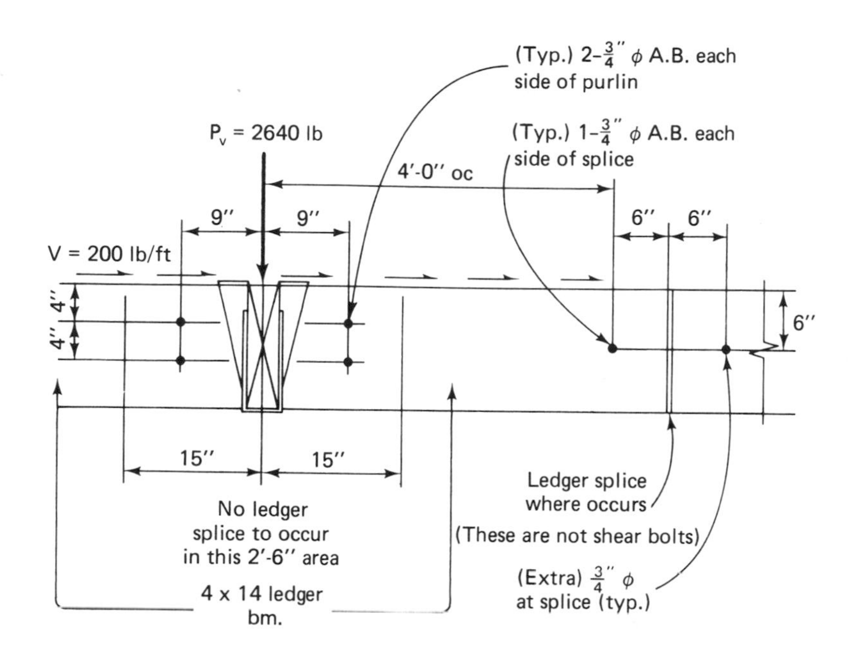

(a) Elevation

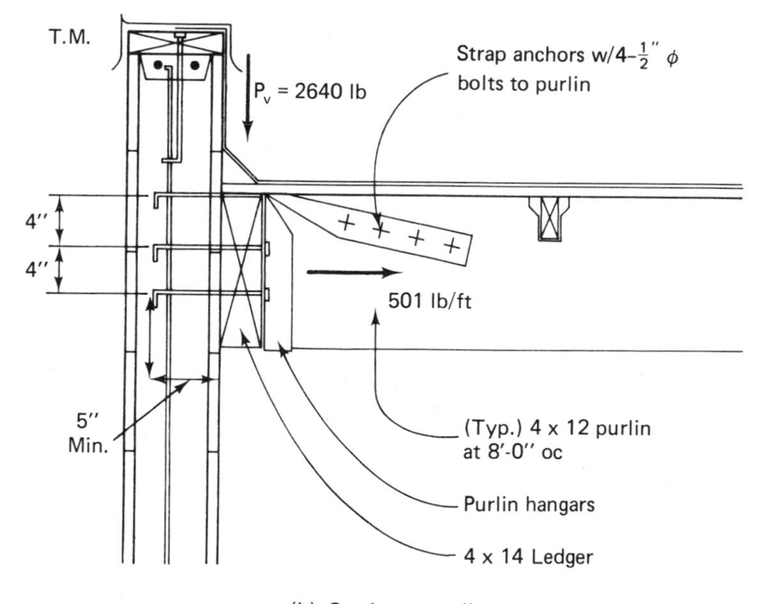

(b) Section at purlin

FIGURE 11-24. Ledger beam connection.

SUMMARY:

1. Use 4 $\frac{3}{4}$-in. ϕ bolts for each purlin at 8 ft centers, spaced as shown, with 5 in. minimum embedment.
2. Purlin hangers support purlins for vertical load and the horizontal in plane shear load. The 4 bolts in the ledger are OK.
3. The strap anchors at each purlin with 4 $\frac{1}{2}$-in. ϕ bolts are OK for the load perpendicular to the wall. Also, alternately, if the detail of the purlin hanger or the nailing of sheathing to ledger can transfer the perpendicular wall load of the ledger, the four bolts per purlin would be adequate for the perpendicular load.

MISCELLANEOUS FEATURES

Seismic separations

An earthquake can cause structures that are not sufficiently separated to pound against each other severely while vibrating during a ground movement. This heavy hammering of one building against the other can cause considerable damage. Thus, adequate clearance must be provided so that each can move laterally without interference or contact. Any rocking or settling of the foundation must also be taken into account as well as the actual deflection of the structure itself. The separation at the top of the shorter structure should at least equal the sum of the total deflections (drift at top), measured from the base of the two buildings, plus any flexural deflections occurring (column lengthening or shortening), plus any movement due to foundation rotation. For a typical masonry shear-wall building, less than 80 ft high, an arbitrary rule of thumb for such a gap is 2 in. for the first 20 ft of height above the ground, plus $\frac{1}{2}$ in. for each additional 10 ft of height.

As observed previously, symmetry is a highly desirable characteristic of efficient seismic design. However, such a configuration is not always possible. In such cases, junctures between distinct parts of a building, such as in L-, T-, or H-shaped plans may be designed with flexible joints which permit relative movement. Thus, each part of the building will vibrate independently of each other, minimizing torsional shear forces. Seismic joints can be covered with flexible and waterproof materials, making them architecturally acceptable. A good example of this lies in the pattern of expansion joints as used at the Veterans Administration hospital complex in Sepulveda, shown in Figure 11-25. This plot shows the complex of 26 reinforced brick buildings, one to seven stories high, which suffered no structural damage in the San Fernando earthquake of 1971, although the five other major hospitals in the Valley, of other types of construction, were badly damaged, four of which were completely demolished. It is noteworthy that the many wings and apparent intersecting portions were structurally separated by joints approximately 5 in. wide. There was some evidence that some adjacent multistory portions bumped together at the top, however, with no apparent distress.

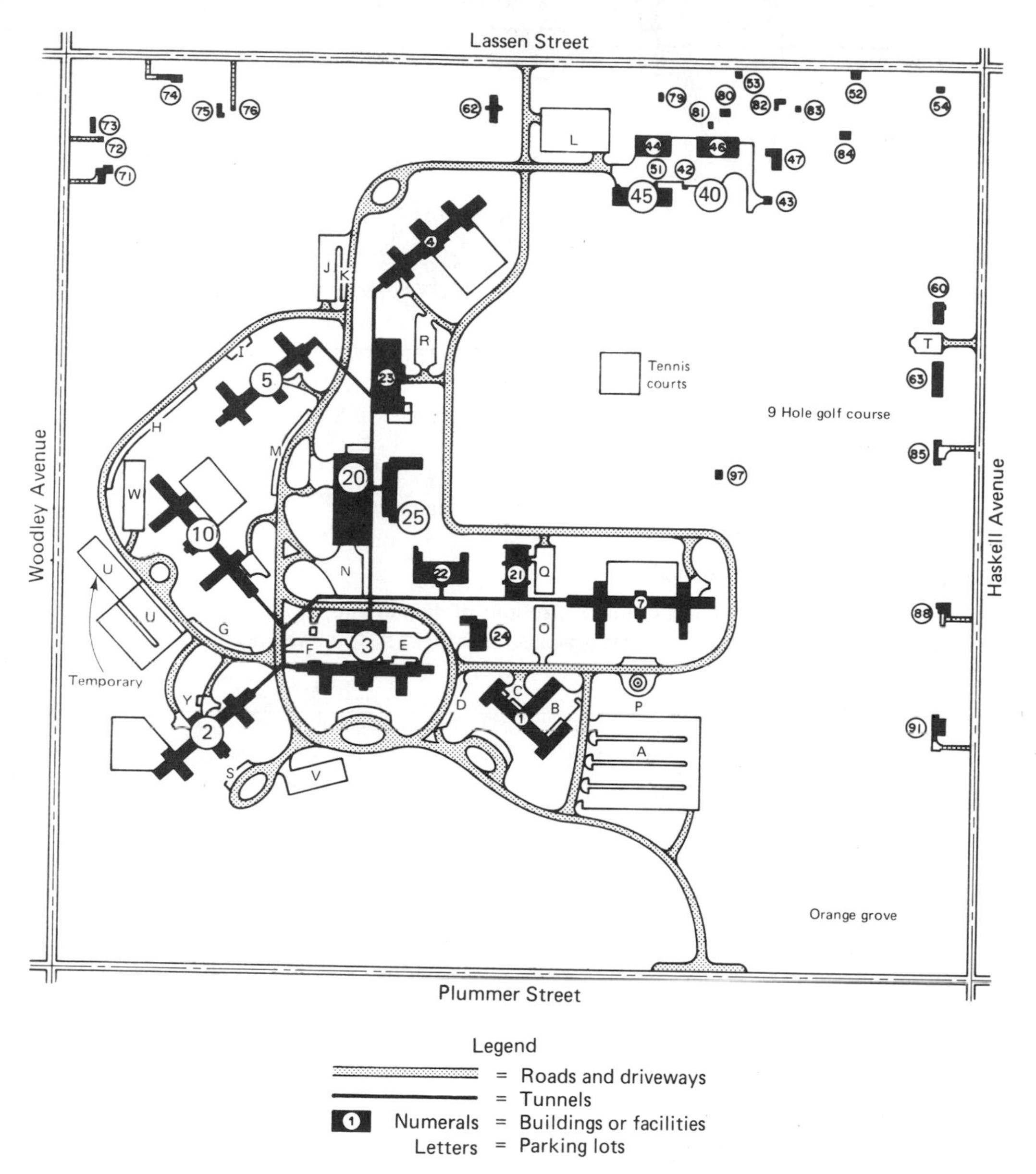

FIGURE 11-25. Veterans Administration Hospital buildings.

For smaller buildings, or where expansion joints may not be feasible, a drag strut may provide a solution for irregular or unsymmetrical roof and floor plans. Such an element ties the diaphragms in the various segments of the L or T together to overcome any deflection incompatibility. One type of connection of this sort may be seen in Figure 11-26, with the design details being described in Example 11-2.

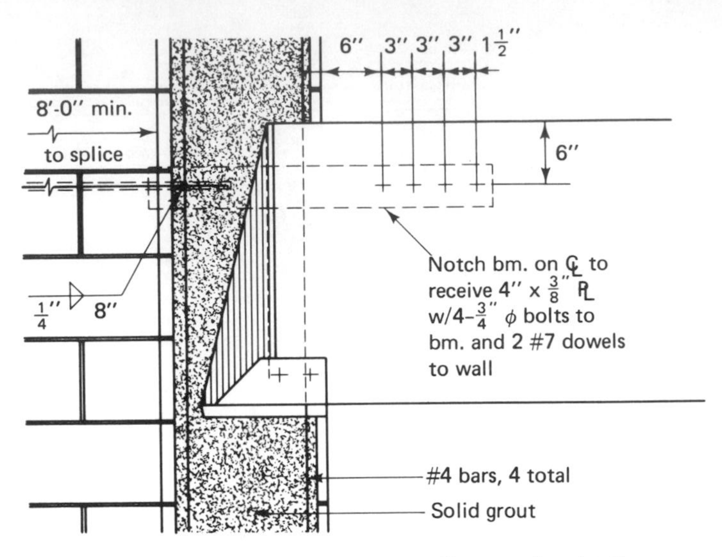

FIGURE 11-26. Drag strut-to-wall connection detail.

EXAMPLE 11-2: Drag Strut to Wall

A drag strut will be provided as shown so that deflections of diaphragms A, B, and C will be compatible (Figure 11-27). Assume a roof dead load of 13 lb/ft². The total wall height is 16 ft (including a 4 ft parapet wall).

SEISMIC LOAD FACTORS:

N-S walls:

$$\begin{aligned} V &= ZIKCSW \\ &= (1)(1)(1.33)(0.14)W = 0.186W \end{aligned}$$

$\quad$ ↖ max. $C \times S$ value

DIAPHRAGMS A AND B–C:

Diaphragm A:

$$13 \text{ lb/ft}^2 \times 60 \text{ ft} \times 0.186 = 145 \text{ lb/ft}$$

Diaphragms B, C:

$$13 \text{ lb/ft}^2 \times 120 \text{ ft} \times 0.186 = 291 \text{ lb/ft}$$

Walls:

$$2[(53 \text{ lb/ft}^2)(16)^2/(12 \times 2)]0.186 = 210 \text{ lb/ft}$$

To N-S walls

$$\text{Section } A\text{: } w = (145 \text{ lb/ft} + 210 \text{ lb/ft}) = 355 \text{ lb/ft}$$
$$B\text{–}C\text{: } w = (291 \text{ lb/ft} + 210 \text{ lb/ft}) = 501 \text{ lb/ft}$$

Similarly, for the east-west seismic direction, the values of 331 lb/ft and 501 lb/ft are obtained.

ROOF SHEARS:

$$\text{Line } A\text{: } V_A = \frac{wL}{2} = \frac{(355 \text{ lb/ft})(70 \text{ ft})}{2} = 12{,}425 \text{ lb}$$

$$v_A = \frac{V_A}{L_A} = \frac{12{,}425 \text{ lb}}{60'} = 207 \text{ lb/ft}$$

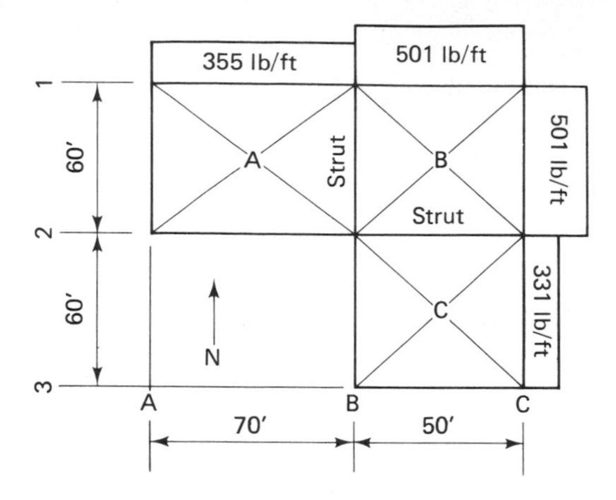

FIGURE 11-27. Drag strut in L-shaped building plan.

$$\text{Line } B\text{: } V_B = \frac{(355\text{ lb/ft})(70\text{ ft})}{2} + \frac{(501\text{ lb/ft})(50\text{ ft})}{2} = 24{,}950\text{ lb}$$

$$v_B = \frac{24{,}950\text{ lb}}{120\text{ ft}} = 208\text{ lb/ft}$$

$$\text{Line } C\text{: } V_C = \frac{(501\text{ lb/ft})(50\text{ ft})}{2} = 12{,}525\text{ lb}$$

$$v_C = \frac{12{,}525\text{ lb}}{120\text{ ft}} = 104\text{ lb/ft}$$

WALL SHEAR AT DIAPHRAGM LEVEL:

Line A: 1.5(207 lb/ft) = 311 lb/ft
Line B: 1.5(208 lb/ft) = 312 lb/ft
Line C: 1.5(200 lb/ft) = 300 lb/ft (minimum v = 200 lb/ft)

STRUT REACTION (LATERAL):

R = 35 ft × 355 lb/ft + 25 ft × 501 lb/ft = 25.0 kips (N-S)
R = 30 ft (331 lb/ft + 501 lb/ft) = 25.0 kips (E-W)

Refer to drag strut wall connection in Figure 11-27.

STRUT PLATES:

Try $\frac{3}{8}$ in. thick × 4 in. wide plate with $\frac{3}{4}$-in. ϕ bolts (A-36: allow $F_t = 0.6F_y$).

gross area = 4 in. × 0.375 = 1.50 in.^{2}
total diameter of bolt hole = 0.75 + 0.125 in. = 0.875 in.
net area = 1.50 − (0.875)(0.375) = 1.17 in.^{2} ⟵ governs
max. allow. net area = 0.85 × 1.50 = 1.28 in.^{2}
tension stress = 25.0/1.17 = 21.4 kips/in.^{2}
allow. stress = 1.33 × (0.60 × 36) = 28.7 kips/in.^{2} ∴ OK

Thus $\frac{3}{8}$ in. × 4 in. plate with $\frac{3}{4}$ in. ϕ bolts is OK.

BOLTS REQUIRED (SHEAR GOVERNS):

area = 1 $\frac{3}{4}$-in. $\phi = \pi(0.75)^2/4 = 0.442$ in.^{2}

For A-325 bolts, allow $F_y = 15$ kips/in.^{2}.

$$\text{load/bolt allowed} = 0.442 \times 15 = 6.6 \text{ kips/bolt}$$
$$\text{min. no. bolts reqd.} = 25.0/6.6 \text{ kips} = 3.8, \text{ say 4, bolts}$$

Cracking control joints

The cracking of masonry walls is caused by the development of tensile stresses within the wall assembly which exceed the low ultimate tensile strength of the masonry materials. These tensile stresses may be caused by wall movements due to a temperature differential or change in moisture content, or when the masonry restrains the movement of adjoining elements. There are several ways to control this unsightly occurrence: (1) material specifications to limit the drying shrinkage of concrete masonry units, (2) reinforcement in strategic locations to cut down on crack propagation, (3) bond beams, and (4) cracking control joint details to accommodate the movement (i.e., to reduce the restraint). Any crack control measure undertaken, however, must be compatible with the provisions made for seismic resistance, since any control joint provides for a weakened plane separation of the masonry.

The spacing and location of these joints depends upon a number of factors, such as (1) the length of the wall, (2) the rigidity of the restraint, (3) the distribution of seismic forces, and (4) the resulting unit stresses. They should be located at junctions of walls (both bearing and nonbearing) or columns or pilasters, in walls weakened by chases and openings, and at changes in wall thickness. In long walls, typically, they are spaced at approximately 20 ft intervals, although this may vary from anywhere between 15 ft and about 50 ft. Also, the spacing depends upon the vertical spacing of the joint reinforcement, since this reinforcement does tend to distribute the local shrinkage stresses as well as the linear shrinkage of the masonry unit itself. In establishing this joint reinforcement spacing, the designer should also consider the presence of large openings and the height/length ratio of the wall. Another rule of thumb calls for a limit on control joint spacing of approximately four times the diaphragm height or 100 ft, whichever is the smaller value. One excellent article describing this phenomenon, written by Copeland, is "Procedures for Controlling Cracking in Masonry."

A control joint detail in a masonry wall must provide an uninterrupted weakened plane for the full height of the wall, including bond beams and foundation walls if constructed of masonry. However, the bond beam, which is the chord tension member of the diaphragm must remain continuous for the entire length of the member. These control joints are generally about $\frac{3}{8}$ in. wide and about $\frac{3}{4}$ in. deep, with the recess being filled with a flexible caulking compound. Finally, it should be observed that bending moment or diagonal tension cannot be transferred across a crack control joint.

Expansion joints

Often clay brick or tile walls do not require joints for cracking control, since very little drying shrinkage occurs. Rather, movements in brick walls are caused by temperature differentials or volume changes resulting from chemical action. The thermal coefficient of expansion for masonry is 0.000004/°F, whereas the volume change due to moisture expansion may reach as high as 0.0002 × wall length.

Brick expansion joints must be continuous and vertical so that they will permit relative movement. To accommodate thermal and moisture expansion, they must provide a complete separation in lieu of a weakened plane. Spacings may vary between 32 and about 100 ft depending upon: (1) the expansion joint width (limited to a maximum of 1 in.), (2) the amount of expansion expected, as well as the maximum anticipated temperature change. The joints should be constructed in such as way that they will at least partially close under any linear expansion of the wall. Consequently, they must be free of extruded mortar or other obstructions. Also, they should be filled with a sealant for weatherproofing. One rule of thumb calls for an accumulated expansion joint width of 1 in. for every 87 ft of wall, for each 100°F anticipated temperature differential. Supposedly the 1-in. width permits a $\frac{1}{2}$-in. expansion of the wall. Brick expansion joints do not transfer any stress; thus, they must be located at points where both the shear and bending moment are expected to remain zero.

PROBLEMS

11-1. Given a building plan as shown, determine (a) the dimensions of the A36 steel plate for strut *B*; and (b) the bolted connection needed assuming double shear (use A-325).

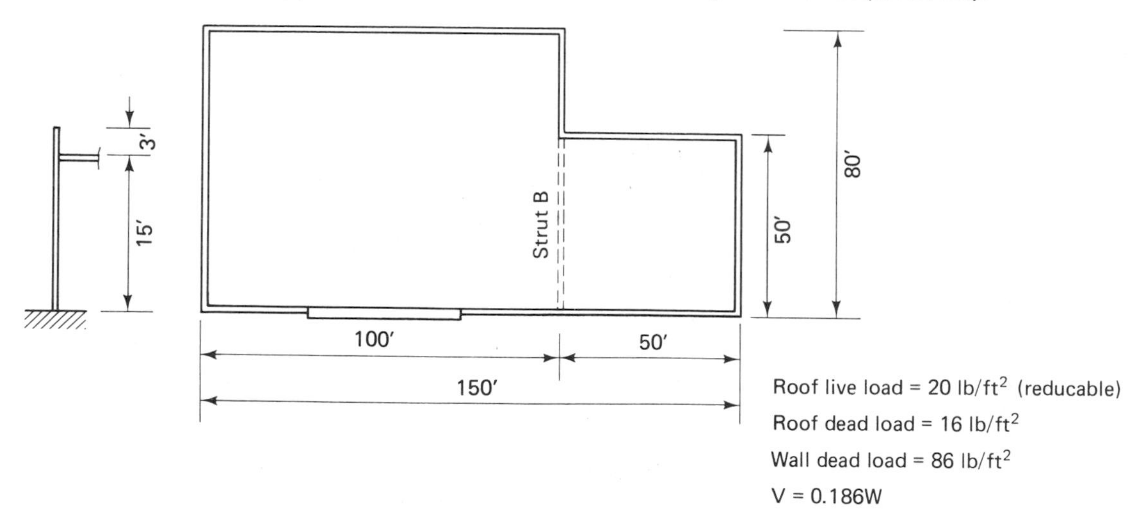

11-2. For the building shown in Problem 11-1, design:

(a) The connection between the diaphragm and wall in the transverse direction.

(b) Do part (a) for the longitudinal direction.

12

Applied Design—One- and Two-Story Buildings

The previous chapters in this text have been devoted to the development of basic design fundamentals of reinforced masonry construction, along with an analysis of its box-type lateral-force-resisting system. This appraisal was based partially upon the masonry materials and assemblage properties, and partly upon empirical limitations imposed by the Code, plus certain rules of thumb dictated by long-time professional practice. This approach has resulted in a form of masonry construction which, in the western United States at least, is both structurally sound and economically viable. These previous discussions were divided essentially along the following concepts:

1. Material and assemblage properties and their resulting structural performance. Covered in this appraisal were the various clay and concrete masonry units, the mortar, grout, and assemblages of grouted brick or concrete block prisms (Chapters 1 through 4).
2. Seismic theory and how it provides the foundation for the current SEAOC and UBC seismic provisions (Chapter 5).
3. Reinforced masonry design theory and how it applies, within the limits of the Code provisions, to the design of basic structural elements, such as

bearing and nonbearing walls, shear walls, columns or pilasters, beams or lintels, and their connection details (Chapters 6 through 11).

4. Aspects of the masonry box-type lateral-force-resisting system with its diaphragm and shear-wall elements, which are designed to maintain the building's stability against wind or seismic pressures (Chapter 10).

We will now put this knowledge to use in the integrated analysis and design of a complete building structure. It will be the intent of this chapter to apply these principles in the examination of certain case studies involving one- and two-story buildings, whereas Chapter 13 takes up the case of the high-rise masonry building, with all its ramifications of lateral stability. In this chapter we will consider a one-story commercial building first (Case Study 1). Although a complete design will not be presented here (e.g., the foundation and certain wood roof details are omitted), all aspects relating to the masonry behavior or the lateral force analysis will be offered. Hopefully, this will be sufficient to provide the student, or the practitioner unfamiliar with these principles, with a logical insight into the intricacies of a properly engineered reinforced masonry structure. The experience gained should enable one, after some practice, to prepare a sound design along with a thoroughly detailed set of plans for a buildable, safe, and economical structure, in light of the current state of the art. There is absolutely no reason whatsoever why some masonry buildings should collapse, in certain parts of the country, simply because they are subjected to severe wind pressures, except for the fact that they often are not properly tied together and reinforced. Although the emphasis throughout this text is on seismic-resistant design, the same principles can be readily applied in those areas where the wind factors predominate. Only the wind dynamics and force magnitudes have to be altered to reflect this form of natural phenomenon.

A partial seismic analysis of a two-story structure appears further on in this chapter (Case Study 2). It is presented to show the effects of overturning, which can become significant in buildings of more than one story. This stems from the fact that overturning moments cause a significant axial compressive load on the first-story piers, in addition to the dead and design live loads present. Also, this example will be used to demonstrate how to handle those torsional stresses which come into play when the mass center and the center of rigidity of the shear-wall elements do not coincide under a rigid roof or floor diaphragm.

These design studies have been prepared in compliance with the empirical limits and specifications of the current UBC. However, since many jurisdictions adopt different editions or revisions to the UBC (or some other code) the designer must verify details of code provisions for specific locale, especially for occupancy, fire, zoning, etc. These differences generally do not vary the principle nor requirements of seismic or masonry design. For example, the masonry revisions for the 1979 edition of the UBC are generally for slight clarification.

The seismic provision revisions in UBC Chapter 23 now delete the S, or site factor, from the equation for seismic force assumption on a building portion, and vary some of the C_p factors so there is a slight revision of magnitude. Actually, the site factor times C_p for most sites had been $1.5 \times 0.20 = 0.30$ for walls. The deletion of S and change of C_p from 0.2 to 0.3 therefore represents no actual change in load magnitude. Note also that C_p for items 4, 6, and 9 in Table 5-9 have all been increased to 0.3 in the 1979 UBC.

The diaphragm response assumption is revised slightly as follows to presumably assign the dynamic load or response a bit more correctly, although not greatly.

> **D. Diaphragms.** Floor and roof diaphragms and collectors shall be designed to resist the forces determined in accordance with the following formula:
>
> $$F_{px} = \frac{\sum_{l=x}^{n} F_l}{\sum_{l=x}^{n} w_l} w_{px} \qquad \text{(UBC 12-9)}$$
>
> Where F_l = the lateral force applied to level l.
> w_l = the portion of W at level l.
> w_{px} = the weight of the diaphragm and the elements tributary thereto at level x, including 25 percent of the floor live load in storage and warehouse occupancies.
>
> The force F_{px} determined from Formula (12-9) need not exceed $0.30ZIw_{px}$. When the diaphragm is required to transfer lateral forces from the vertical resisting elements above the diaphragm to other vertical resisting elements below the diaphragm due to offsets in the placement of the elements or to changes in stiffness in the vertical elements, these forces shall be added to those determined from Formula (12-9).
>
> However, in no case shall lateral force on the diaphragm be less than $0.14ZIw_{px}$.

CASE STUDY 1: ONE-STORY COMMERCIAL BUILDING

Design procedure

To facilitate an understanding of what does constitute proper design procedure for small buildings, this procedure will be divided into several distinctive steps. In actual practice, these various steps may not be so clearly delineated nor distinctly separated, but at this stage, at least, this step-by-step procedure is recommended in order that the student may acquire a solid "feel" for what constitutes soundly engineered masonry design procedures.

1. *Establish the design criteria,* including: live and dead load mag-

nitudes; allowable stresses per the applicable Code, based upon the appropriate f'_m; the governing lateral load pressure (i.e., wind or seismic); the type of diaphragm or lateral bracing to be employed; the location and dimensions of all shear walls, both transversely and longitudinally. Check to ensure that the selected framing system does provide for a *continuous* path of resistance extending from the point of load application to the foundation (e.g., wall to pilaster, pilaster or wall to top and bottom diaphragm support, diaphragm connection to shear wall, shear wall to foundation).

2. *Design the walls to carry the vertical live and dead loads plus the lateral forces acting normal to them.* Consider combined bending and axial load effects, whether the former is generated either by an eccentric vertical load or by the normal lateral load. Check minimum wall steel requirements, per the Code percentage requirements.

3. *Size and reinforce all pilasters and columns.* Consider the bending effect where the loads are eccentrically applied. Also take into account bending effects upon these elements caused by the normal lateral loads where applicable. This can become a critical case, for instance, when pilasters are located adjacent to large wall openings, or when the wall is reinforced to span horizontally between pilasters. Size anchors and bearing plates under the truss or beam reactions carried by the column or pilaster.

4. *Compute flexural and shear (if any) reinforcement for all beams and lintels.* Provide vertical and horizontal bars around all openings. Consider adequate embedment lengths in all cases.

5. *Perform a lateral force analysis based upon either wind or seismic forces, depending upon whichever is the more severe.* Distribute diaphragm shears to the shear walls either on the basis of tributary areas or relative shear-wall rigidities, as the diaphragm stiffness dictates. Distribute shear force to individual piers contained within the shear wall. If a masonry bond beam serves as the diaphragm flange member, check the adequacy of the reinforcement and continuity details for developing its capacity.

6. *Check adequacy of all piers for resistance to both shear and boundary flexural resistance.* Add vertical jamb steel and web reinforcement where needed. Provide proper anchorage for each.

7. *Detail all connections, placing particular emphasis on diaphragm connections to the walls.* Do not overlook the necessity to secure the walls to the floor or roof to prevent the walls from pulling away from this support and destroying lateral stability.

8. *Design the roof framing.* Design the diaphragm elements including deck thickness, edge-member thickness, as well as all plywood or sheathing boundary and field nailing. Consider subdiaphragm requirements where appropriate.

These steps will now be followed in detail in the case study of a one-story commercial building. Note that for reasons of clarity and understanding, more detail is given here than would ordinarily be found in a typical set of design calculations in actual practice.

Analysis and Design

STEP 1: CRITERIA

The plan and elevation of a one-story commercial building, located in seismic zone 4, are shown in Figure 12-1. The walls will be concrete block (8 in. partially grouted), with glu-lam beams supporting rafters at 24-in. cc and covered with $\frac{1}{2}$-in. plywood. The glu-lam beams are supported at the interior building center line on pipe columns. No special inspection is to be provided, so half-stresses for masonry apply.

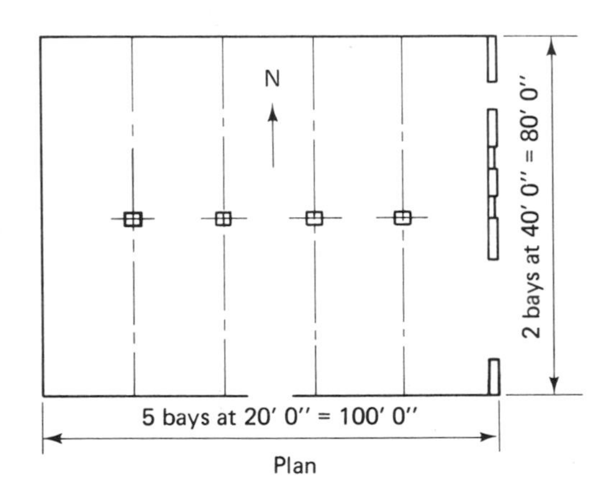

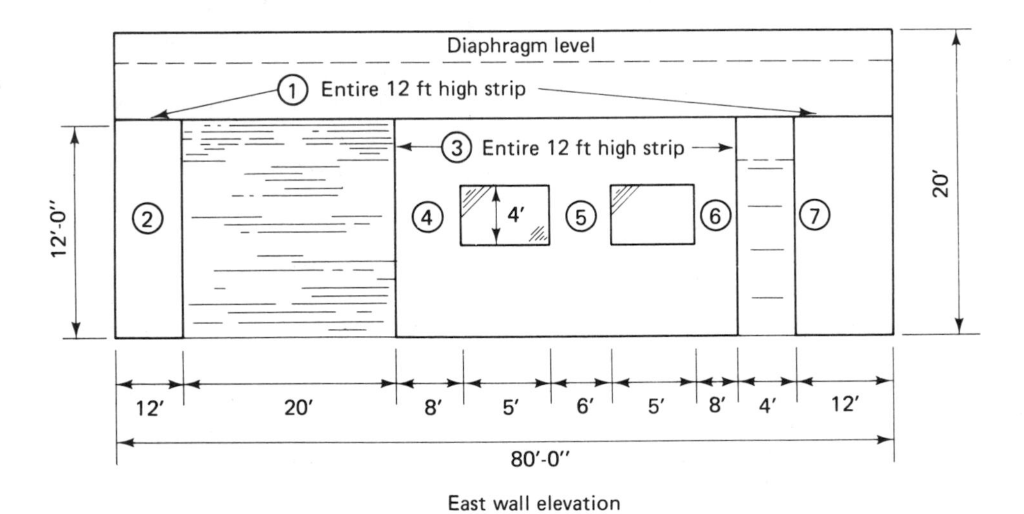

FIGURE 12-1. One-story commercial building—plan and elevation (not to scale).

Allowable Stresses:

$f'_m = \frac{1}{2} \times 1350$ lb/in.^{2} (partially grouted concrete block) $= 675$ lb/in.^{2}
max. f_m or $F_b = 0.33 \times 675 = 225$ lb/in.^{2} also $\times \frac{4}{3} = 300$ lb/in.^{2} for wind or seismic loads

$F_a = 0.20 \times 675 = 135$ lb/in.^{2}

$f_s = 20{,}000$ lb/in.^{2} also $\times \frac{4}{3} = 26{,}667$ lb/in.^{2} for wind or seismic

$$n = \frac{E_s}{E_m} = \frac{30{,}000{,}000}{1000 f'_m} = 44$$

Loads:

1. Dead—roof:

Roofing	2.0 lb/ft^2
2 × 4 at 24 in. cc	0.8
$\frac{1}{2}$-in. plywood	1.5
Purlins	1.5
Misc.	1.7
Total DL	7.5 lb/ft^2

2. Live—roof (per UBC):

20 lb/ft^2 (under 200 ft^2)
16 lb/ft^2 (up to 600 ft^2)
12 lb/ft^2 (over 600 ft^2)

3. Lateral—seismic per UBC [Section 2312(d)]:

Walls: $V = ZIC_pSW_p = (1)(1)(0.20)(1.5)W_p = 0.3W_p$

Building: $V = ZIKCSW = (1)(1)(1.33)(0.14)W = 0.186W$

where $(CS)_{\max} = 0.14$

Z = seismicity factor = 1 for seismic zone 4
C_p = acceleration coefficient for building component = 0.2 for exterior walls (Table 5-9)
I = building importance factor = 1.0 for commercial building
C = acceleration coefficient for total building = $1/(15\sqrt{T})$
S = site structure response factor = 1.5 when T is not properly substantiated
K = ductility factor = 1.33 for box shear-wall building
T = period of building = $0.05\, h_n/\sqrt{D}$, where h_n = building height and D = plan dimension of building parallel to direction of seismic load

4. Lateral—wind: 15 lb/ft^2 at less than 30 ft above the ground for a 20-lb/ft^2 wind pressure area; see Figure 5-3 and Table 5-6.

***Framing Schemes*:**

Vertical load framing scheme:

Plywood spans to rafters.

Rafters span to glu-lam and to end walls.

Glu-lam spans from wall to center post to wall.

Wall rests on foundation (lintels spanning over openings).

Foundation supported on ground.

Lateral load framing scheme:

Wall spans from bottom (pinned at footing) to top connection (window openings spanning to jamb).

Parapet cantilevers above support at roof line.

Top wall connections transmit wall loads to diaphragm.

Diaphragm spans to shear wall connections (with bond beams as chords).

Shear walls transmit in-plane shear to foundations through piers which receive equal in-plane top deflections.

STEP 2: WALL DESIGN

(A) End Walls (Bearing):

Vertical loads:

Roof loads (10 ft tributary width):

$$\begin{aligned} \text{dead load} &= 7.5\ \text{lb/ft}^2(10) = 75\ \text{lb/ft} \\ \text{live load} &= 20.0\ \text{lb/ft}^2(10) = \underline{200\ \text{lb/ft}} \\ \text{total load} &= 275\ \text{lb/ft} \end{aligned}$$

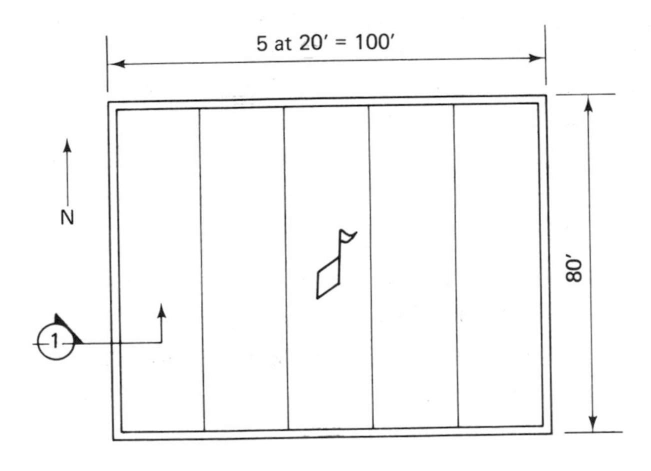

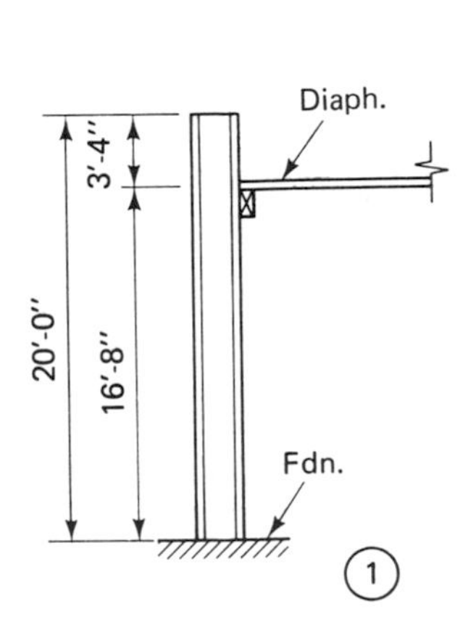

Wall loads:

$$\text{Dead load at midheight}^1 = 75\ \text{lb/ft}^2\ \frac{16.67}{2} = 625\ \text{lb/ft}$$

$$\text{Dead load of parapet} = 3.33\ \text{ft}(75\ \text{lb/ft}^2) = \underline{250\ \text{lb/ft}}$$

$$\text{Total DL} = 875\ \text{lb/ft}$$

Lateral forces on wall:

Wind = 15 lb/ft²

Seismic:

$$F_p = 0.3W_p = 0.3(75\ \text{lb/ft}^2) = 22.5\ \text{lb/ft}^2.$$

Therefore, seismic force governs at 22.5 lb/ft².

Check h/t ratio:

$$h/t = \frac{16.67 \times 12}{8\ \text{in. (use nominal)}} = 25$$

Maximum h/t permitted for bearing walls is 25 ∴ OK

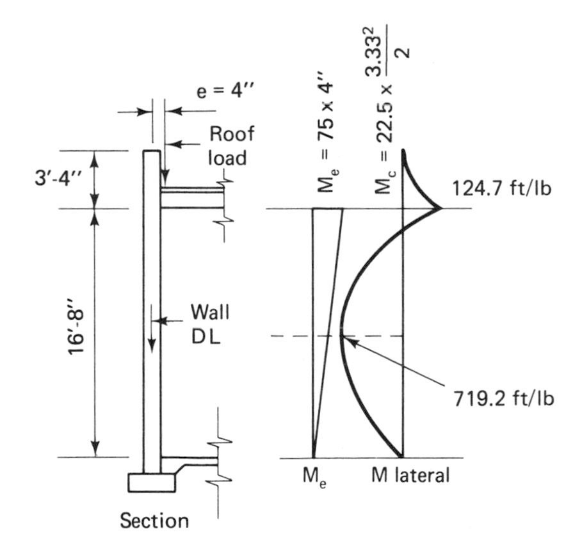

Loading perpendicular to wall:

1. Consider the vertical dead load at eccentricity $e = 4$ in. + lateral-load effects. Thus, moment at mid-height of wall due to lateral load is

$$\frac{Wl^2}{8} - \frac{M_c}{2} = 22.5 \times \frac{(16.67)^2}{8} - \frac{124.7}{2} = 719.2\ \text{ft-lb}$$

[1]Assumes wall steel spacing at 16 in. cc; thus weight = 75 lb/ft². Subsequent calculations show that this could be reduced to about 62 lb/ft². However, in view of the numerous approximations involved, the design will not be revised here.

The total design moment becomes

$$719.2 + \frac{M_e}{2} = 719.2 + 12.5 = 732 \text{ ft-lb}$$

Check combined stress on section (uncracked, unreinforced section):

$$f_a = \frac{P}{A} = \frac{75 + 875}{12 \times 4.0 \text{ in.}} = 20 \text{ lb/in.}^2$$

(note that the equivalent solid thickness for an unreinforced section is 4.0 in., from Table 7-4)

Note that actual Mc/I will be slightly less, since section is hollow rather than solid.

$$f_b = \frac{Mc}{I} = \frac{6M}{bt^2} = \frac{6 \times 732 \times 12}{12 \times (7.6)^2} = 76 \text{ lb/in.}^2$$

$$\frac{P}{A} \pm \frac{Mc}{I} = 20 \text{ lb/in.}^2 - 76 \text{ lb/in.}^2 = 56 \text{ lb/in.}^2$$ tension, which is too high, since this would produce cracked section

Thus consider cracked section. Note that the previous calculation was only performed to demonstrate the point that with a seismic lateral load, a cracked section will result. This need not be carried out in every solution.

$$F_a = 0.2f'_m\left[1 - \left(\frac{h}{40t}\right)^3\right] \times \frac{1}{2} = \frac{0.2(1350)}{2}\left[1 - \left(\frac{16.67 \times 12}{40 \times 8}\right)^3\right]$$

$$= 135(0.756) = 102 \text{ lb/in.}^2$$ for no special inspection

Assume

$$A_s = \frac{732 \times 12}{0.9 \times 3.8 \times 26{,}667} = 0.096 \text{ in.}^2\text{/ft}$$

$$np = \frac{44 \times 0.096}{12 \times 3.8} = 0.093$$ and from Table C-1, $j = 0.88$ and $2/jk = 6.5$

$$f_s = \frac{732 \times 12}{0.88 \times 0.096 \times 3.8} = 27{,}360 \text{ lb/in}^2 > 26{,}667 \text{ lb/in.}^2 \qquad \therefore \text{ N.G.}$$

$$f_m = \frac{732 \times 12}{12 \times (3.8)^2} \times 6.5 = 330 \text{ lb/in.}^2 > 300 \text{ lb/in.}^2 \qquad \therefore \text{ N.G.}$$

Stresses are too high and may be made acceptable by using "inspected masonry" with the higher allowable stresses or by providing additional reinforcement.

If additional steel is provided so as to decrease $2/jk$:

$6.5 \times \frac{300}{330} = 5.9$, so with a $2/jk = 5.8$ in chart, find $np = 0.130$; thus

$$p = \frac{0.130}{44} = 0.0030 \text{ and } j = 0.868$$

$$A_s = 0.0030 \times 12 \times 3.8 = 0.135 \text{ in.}^2$$

No. 6's at 40 in. cc ($A_s = 0.132$ in.^{2})

$$f_s = \frac{732 \times 12}{0.868 \times 3.8 \times 0.132} = 20{,}175 \text{ lb/in.}^2 \qquad \therefore \text{ OK}$$

$$f_m = \frac{732 \times 12}{12 \times (3.8)^2} \times 5.8 = 294 \text{ lb/in.}^2 \qquad \therefore \text{ OK}$$

$$f_a = \frac{75 + 875}{12 \times 4.7} = 16.8 \text{ lb/in.}^2$$

where 4.7 in. is equivalent solid thickness for wall grouted at 40 in. cc.

$$\frac{f_a}{F_a} + \frac{f_b}{F_b} = \frac{16.8}{102 \times 1.33} + \frac{294}{300} = 0.12 + 0.98 = 1.10$$

combined stress too high so increase A_s to 0.237 in.^{2}
(Assume No. 8 at 40 in. cc)

Assume increase

$$np = \frac{0.237}{12 \times 3.8} \times 44 = 0.229 \qquad \text{from Table C-1, } \frac{2}{jk} = 4.9,\ j = 0.838$$

$$f_m = \frac{732 \times 12}{12 \times (3.8)^2} \times 4.9 = 248 \text{ lb/in.}^2$$

$$\frac{f_a}{F_a} + \frac{f_b}{F_b} = \frac{16.8}{102 \times 1.33} + \frac{248}{300} = 0.12 + 0.83 = 0.95$$

The additional steel eliminates the overstress. This is probably preferable to alternative ways of accomplishing this stress reconciliation, such as:

(a) Solid grouting, thereby increasing f'_m from 1350 to 1500 lb/in.^{2}, and also adding area of masonry with the additional grouting.

(b) Continuous inspection, which provides for full allowable stresses rather than half-stresses used here.

(c) Two mats of steel in wall, thereby changing d to 5.3 in.

Minimum reinforcement:

Vertical: No. 8 at 40 in. cc (0.237 in.^{2}/ft) more than adequate.
Horizontal: Minimum = 0.064 × 20 ft = 1.28 in.^{2} (see Table D-1).

1 No. 5 in footing	= 0.31 in.^{2}		
2 No. 5's in bond beam	= 0.62		
1 No. 5 top of wall	= 0.31		
9 × 0.0345 joint reinforcing at 2 ft	= 0.31	(Table D-2)	∴ OK
	1.54 in.^{2}		

2. Another load combination to consider would be that of full live and dead load without a one-third stress increase.

Midheight of wall:

$$M_e = 275 \text{ lb/ft} \times 4 \text{ in.}/2 = 550 \text{ in.-lb}$$

$$f_a = \frac{275 + 875}{12 \times 4.7} = 20.4 \text{ lb/in.}^2$$

$$f_b = \frac{M_e \times 6}{2 \times b \times t^2} = \frac{550 \times 6}{12 \times (7.6)^2} = = 4.76 \text{ lb/in.}^2$$

Top of wall:

$$f_a = \frac{275}{12 \times 4.7} = 4.9 \text{ lb/in.}^2$$

$$f_b = 9.5 \text{ lb/in.}^2 \text{ (since } M_e = 2 \times 550 \text{ in.-lb)}$$

Thus, since no tension occurs on the section with this loading combination, the lateral-loading condition governs.

(B) Nonbearing side walls. The side walls will be subjected to lateral-load moments due to seismic forces acting perpendicular to the wall (732 ft-lb/ft).

From Curve B-2: for

$$\frac{K}{F_b} = 0.75 \times \frac{732 \times 12}{12 \times (3.8)^2} \times \frac{1}{0.225} = 170$$

find $np \sim 0.125$ indicating that $F_b = 300$ lb/in.² governs, rather than $f_s = 26{,}700$ lb/in.²

Thus $p = 0.125/44 = 0.0028$ and $A_s = 0.0028 \times 12 \times 3.8 = 0.129$ in.²/ft.

Or from Curve B-3: for

$$K = \frac{732 \times 12}{12 \times (3.8)^2} = 50.7 \quad \text{and} \quad F_b = 300 \text{ lb/in.}^2$$

find $np \sim 0.12$.

Or from:

$$F_b = \frac{M}{bd^2}\left(\frac{2}{jk}\right), \qquad 300 = 50.7\left(\frac{2}{jk}\right) \qquad \therefore \frac{2}{jk} = 5.9$$

and from Table C-1, for $2/jk = 5.9$, find $np = 0.124$.

Provide No. 6's at 40 in. ($A_s = 0.133$ in.²/ft).

(C) Parapet wall:

Moment:

Seismic again governs:

$$F_p = ZIC_pSW_p = 1.0 \times 1.0 \times 1.0 \times 1.5 \times 62 \text{ lb/in.}^2 = 93 \text{ lb/ft}^2$$

$$M = \frac{wl^2}{2} = 93 \times \frac{(3.33)^2}{2} = 516 \text{ ft-lb/ft}$$

Continue No. 6's at 40 in. cc up through parapet. This reinforcement is more than adequate, since this provides moment resistance in excess of 516 ft-lb/ft.

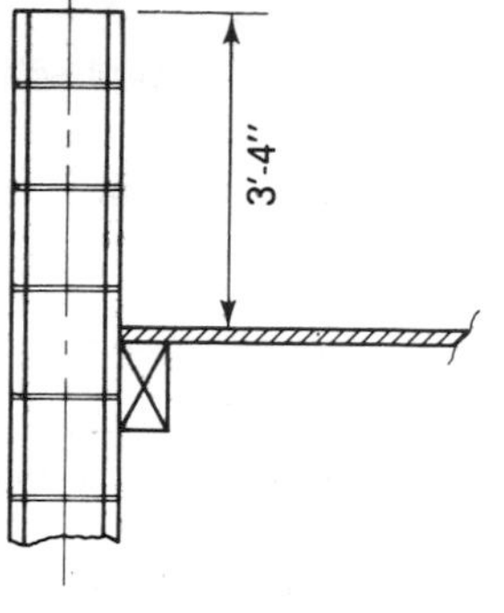

Crosssection of typ. parapet

Shear:

$$v = \frac{V}{bjd} = \frac{93 \times 3.33 \times 40/12}{8.25 \times 0.9 \times 3.8} = 36.6 \text{ lb/in.}^2$$

$25 \times 1\frac{1}{3} = 34 \text{ lb/in.}^2$ close enough $\therefore$ OK without stirrups

(Note flexural shear width = 8.25″ as determined in Chapter 7.)

Bond:

$$u = \frac{V}{\Sigma_0 jd} = \frac{93 \times 3.33 \times 40/12}{2.36 \times 0.9 \times 3.8} = 128 \text{ lb/in.}^2 \text{ versus } 100 \times 1\tfrac{1}{3} = 133 \text{ lb/in.}^2$$

$\therefore$ O.K.

When considering the parapet itself, a C_p value of 1.0 was used in this case. However, this high factor need not extend to any other portion of the structure, such as connections at the diaphragm level. The 1979 UBC revises this minimum figure downward to 0.8.

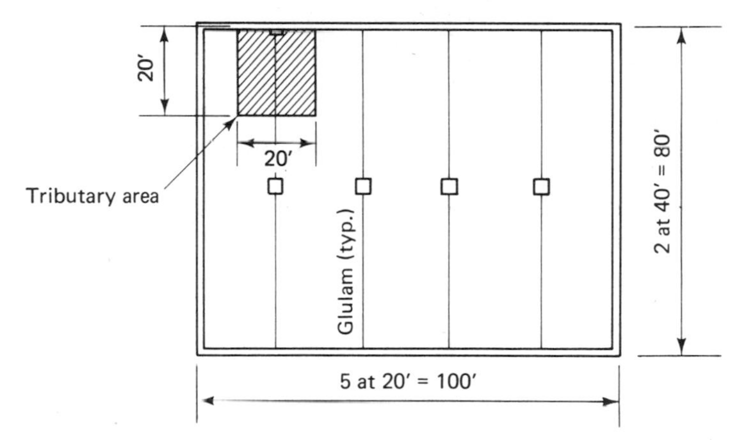

STEP 3: PILASTER SIZE

(A) Projecting Pilaster:

16 × 16 pilaster size.

Loads:

Dead loads:

Roof: 7.5 lb/ft² + 2.5 (Glu-Lam) = 10 lb/ft² × 400 ft² = 4,000 lb

Pilaster: $\frac{16 \text{ in.} \times 16 \text{ in.}}{144} \times 118 \text{ lb/ft}^3 = 210 \text{ lb/ft} \times \frac{16.67}{2} = 1{,}750 \text{ lb}$

Parapet: 62 lb/ft² (3.33 ft)(16/12) = 275 lb

Live loads:

Roof: 16 lb/ft² × 400 = 6,400 lb

Total load 12,425 lb

Lateral loads:

Wind: 15 lb/ft²

Seismic: $F_p = ZIC_pSW_p = 0.3W_p = 0.3(210 \text{ lb/ft-pilaster wt.})$
$= 63 \text{ lb/ft.}$

Loading combinations

1. Total vertical load and moment due to roof load at $e = 8$ in.:
No stress increase:

$$M = 10{,}400 \times \frac{8}{12} = 6933 \text{ ft-lb}$$

$$f'_m = 1500/2 \text{ lb/in.}^2 \text{ (solid grouted cores)}, \ n = 40$$

$$f_s = 16{,}000 \text{ lb/in}^2.$$

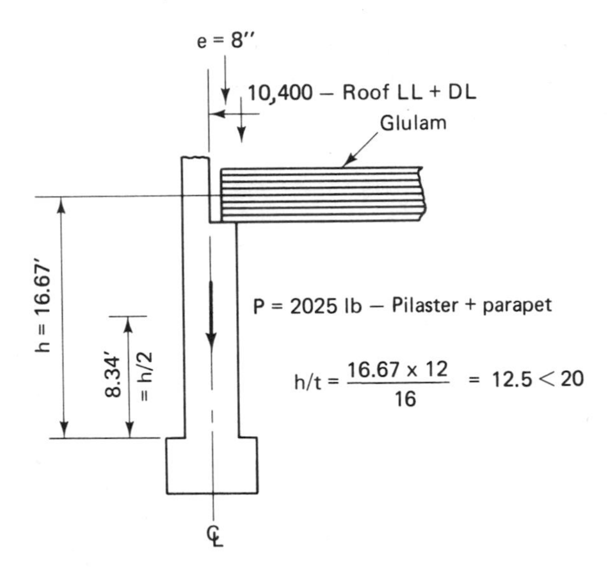

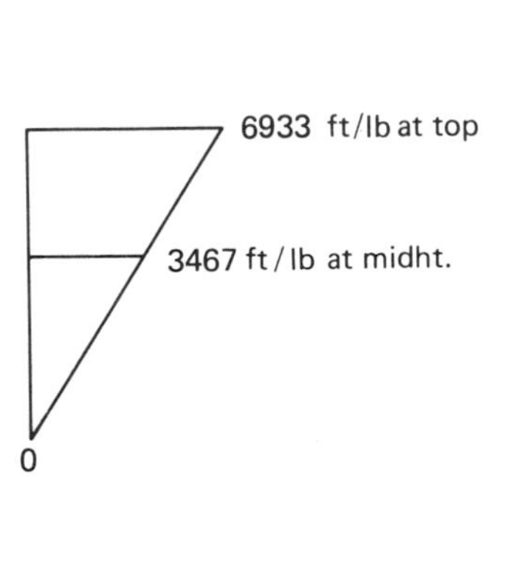

Considering loading conditions at two locations:

a. Top of column

$$P = (12{,}425 - 1750) = 10{,}675 \text{ lb}$$

$$M = 6933 \text{ ft-lb}$$

No column reduction for $P\Delta$ effect

b. Column midheight

$$P = 12{,}425 \text{ lb}$$

$$M = 3467 \text{ ft-lb}$$

Column reduction for $P\Delta$ effect

$$R = \left[1 - \left(\frac{h}{40t}\right)^3\right] = \left[1 - \left(\frac{16.67 \times 12}{40 \times 16}\right)^3\right] = 0.97$$

It would appear that location (a) is more critical and it therefore will be the basis for the pilaster design, unless loading case (2), dead load + lateral load, proves more critical.

Assume four No. 6's for longitudinal reinforcement; thus,

$$p_g = \frac{4 \times 0.44}{15.62 \times 15.62} = 0.0072$$

$$F_a = (0.18 f'_m + 0.65 p_g f_s) = (0.18 \times 750 + 0.65 \times 0.0072 \times 16{,}000) = 210 \text{ lb/in.}^2$$

$$f_a = \frac{10{,}675}{15.62 \times 15.62} = 43.8 \text{ lb/in.}^2 \qquad \therefore\ f_a/F_a = \frac{43.8}{210} = 0.21 \text{ at top}$$

$$F_b = \tfrac{1}{3} \times 1500 \times \tfrac{1}{2} = 250 \text{ lb/in.}^2$$

From interaction equation:

$$f_b \text{ remaining for flexure} = \left(1 - \frac{f_a}{F_a}\right) F_b = (1 - 0.21)250 = 197 \text{ lb/in.}^2$$

2. DL + roof DL moment (at $e = 8$ in.) + moment due to lateral load (allow $\frac{1}{3}$ stress increase)

$$M_e = 4000 \text{ lb} \left(\frac{8}{12}\right) = 2667 \text{ ft-lb at } h$$

$$= 1333 \text{ ft-lb at } h/2$$

$$M_{lat} = \frac{wL^2}{8} = \frac{63(16.67)^2}{8} = 2188 \text{ ft-lb}$$ assuming pilaster carries only its own area (i.e., no wall load) at 63 lb/ft²

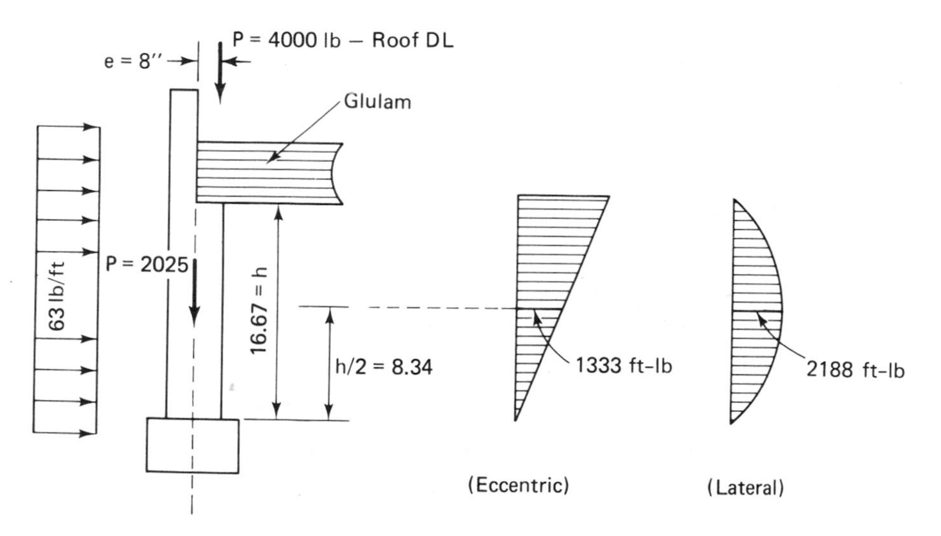

If the wall were reinforced to span horizontally, then the pilaster lateral load would stem from the full tributary width of the wall; i.e., ℄ to ℄ of pilasters.

Therefore,

$$\text{total moment} = 1333 + 2188 = 3521 \text{ ft-lb}$$

$$\text{total load} = 4000 + 1750 + 275 = 6025 \text{ lb}$$

Thus, it can be seen that loading case (1a) is the critical one and it becomes the design criterion:

$$P = 10{,}400 \text{ lb}$$

$$M = 6933 \text{ ft-lb}$$

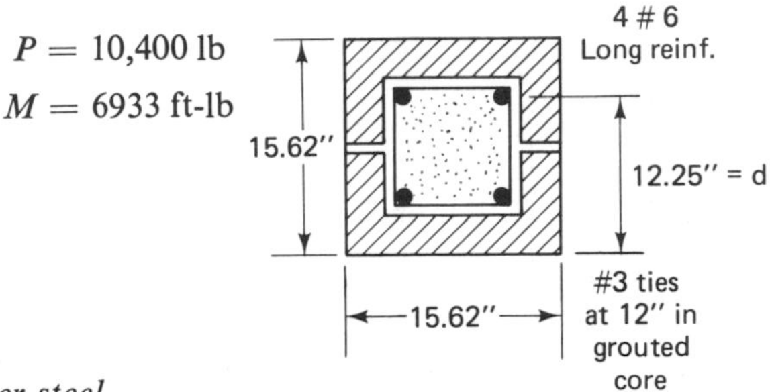

Determination of pilaster steel

(a) Min. $A_s = 0.005 \times 15.62 \times 15.62 = 1.22 \text{ in.}^2 < 1.76 \text{ in.}^2$ assumed $\quad \therefore$ OK

(b) Check for combined stress:

$$p = \frac{2 \times 0.44}{15.62 \times 12.25} = 0.0046 \text{ and } np = 0.184 \text{ for a } d \sim 12.25 \text{ in.}$$

From Table C-1, for $np = 0.184$, $2/jk = 5.22$. Thus,

$$f_m = \frac{M}{bd^2} \times \frac{2}{jk} = \frac{6933 \times 12 \times 5.22}{15.62 \times (12.25)^2} = 185 \text{ lb/in.}^2 < f_b = 197 \text{ lb/in.}^2 \qquad \therefore \text{ OK}$$

$$f_s = \frac{6933 \times 12}{0.88 \times 0.9 \times 12.25} = 8575 \text{ lb/in.}^2 \qquad \therefore \text{ OK}$$

(c) Column ties:

(1) 16 bar diameter $= 16 \times \frac{3}{4} = 12$ in $\leftarrow$ governs. Provide No. 3's at 12 in. cc within the grout space.

(2) 48 tie diameter $= 48 \times \frac{3}{8} = 18$ in.

(3) Minimum column dimension $= 16$ in.

(B) Wall as "Flush Pilaster" (Alternative to Projecting Pilaster) Architecturally speaking, the correct definition of pilaster implies a projection. The flush type is seldom used since it really does not provide any stiffness to the wall.

Bearing plate design for glu-lam:

Loads: LL $= 20$ ft $(20$ ft$)(16$ lb/ft$^2) = 6400$ lb

DL $= 20(20)(10$ lb/ft$^2) = 4000$ lb

total load $= 10{,}400$ lb

bearing on masonry (allow.) $= 187$ lb/in^2. (Table 7-1, solid grout)

$$\text{Req'd area of plate} = \frac{10{,}400 \text{ lb}}{187} = 55.6 \text{ in}^2.$$

Therefore, use bearing plate dimensions: 4 in. × 16 in. = 64.0 in.^{2}

Bearing plate thickness:

$$M_{plate} = \frac{wL^2}{2} = \frac{187(4.5\text{ in})^2}{2} = 1893\text{ in.-lb/in.}$$

$$S = \frac{M}{f_s} = \frac{1893}{20{,}000} = 0.095\text{ in.}^3$$

$$0.095 = \frac{bd^2}{6} \quad (b = 1\text{ in.})$$

$$\therefore\ d = \frac{\sqrt{6S}}{b} = \frac{\sqrt{6(0.095)}}{1} = 0.754\text{ in.} \sim 0.75\text{ in.}$$

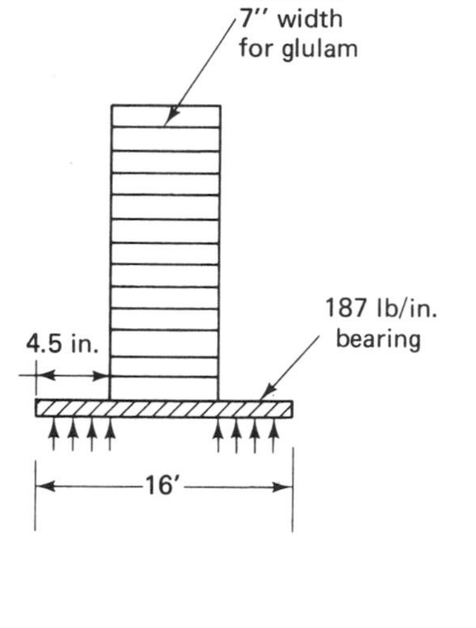

Therefore, use bearing plate: 4 in. × 16 in. × $\frac{3}{4}$ in.

Bearing length (see sketch):

Wall reinforcing:

h/t ratio:

Max $h = t$ (nominal) 20 = 8(20) = 160 in. = 13 ft. 4 in.

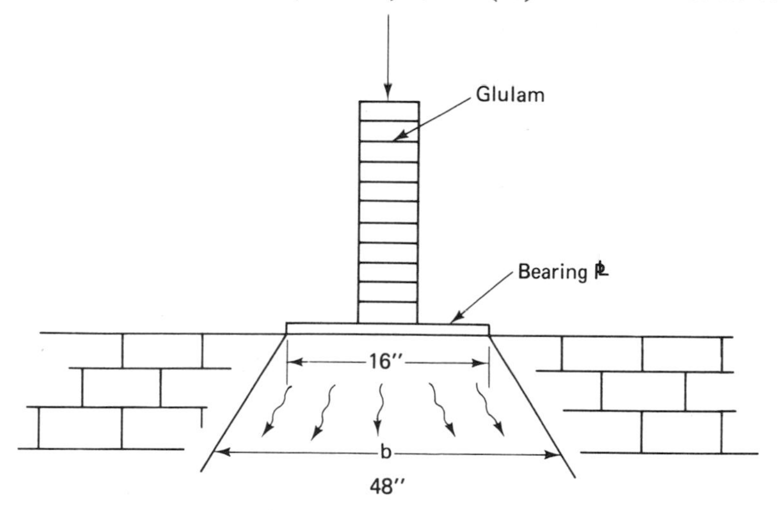

Effective b = 4(8) + 16 = 48″

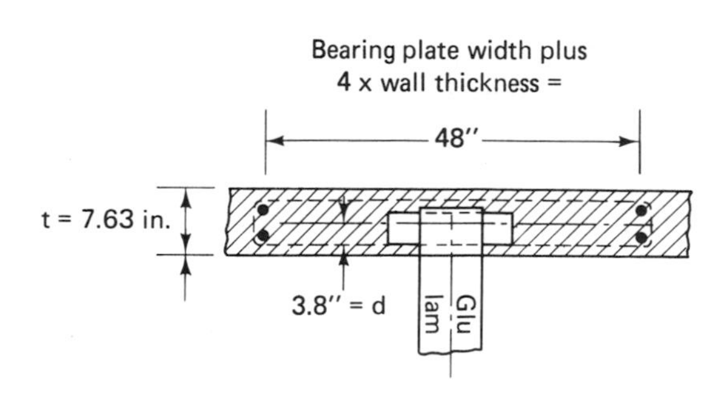

Therefore, provide adequate brace at 13 ft-4 in. height.

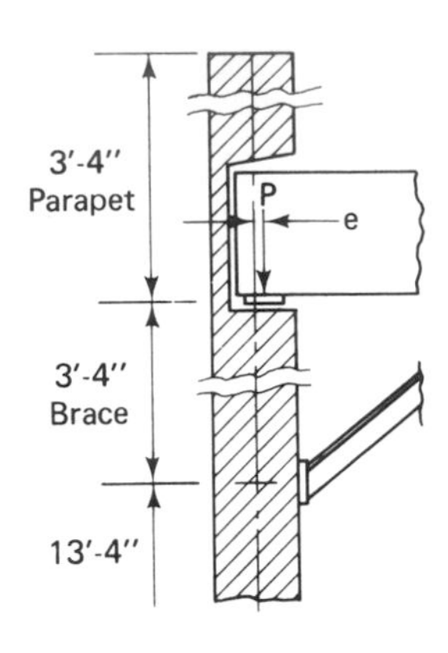

$$R = \left[1 - \left(\frac{20}{40}\right)^3\right] = 0.875$$

Loads:

Roof DL = 4000 lb

Roof LL = 6400 lb

Parapet: 62 lb/ft² (3.33 ft)(4 ft) = 826 lb

Wall DL at midheight between brace and floor:

$$10 \text{ ft} \times 4 \text{ ft} \times 92 \text{ lb/ft}^2 = 3680 \text{ lb}$$

Loading Conditions:

1. Dead load + live load + total load eccentric moment at roof for $e =$ 2 in.: Per detail shown, provide two No. 5 A bars, outside wall face, to resist eccentric load moment. Also provide two No. 4 bars on inside face and two No. 5's as shown at 4 ft-0 in. each side of center line of beam support. The two No. 4 bars on the inside face are provided for the case where no live load from the beam exists, so these must function to resist the lateral load from wind or seismic as the wall spans vertically.

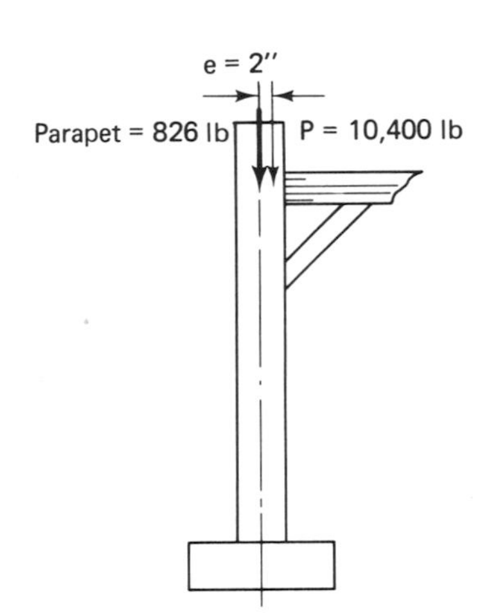

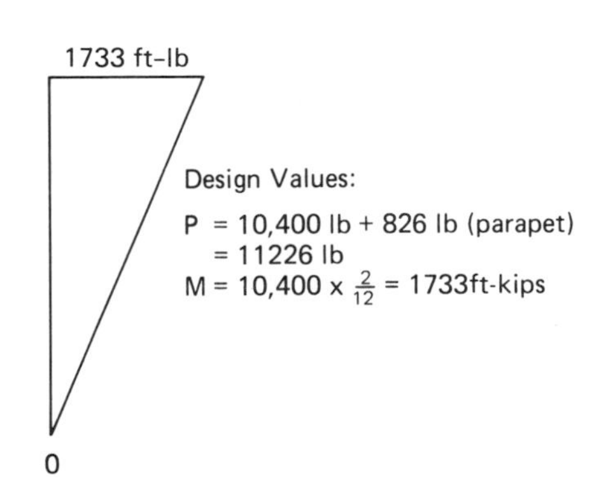

$$p = \frac{0.31 \times 2}{48 \times 5.3} = 0.00244,\ np = 0.097, \text{ thus } \frac{2}{jk} = 6.39 \text{ from Table C-1}$$

$$F_a = 0.2f'_m = 0.2 \times 750 = 150 \text{ lb/in.}^2 \text{ and}$$

$$f_a = \frac{11{,}226}{48 \times 7.6} = 31 \text{ lb/in.}^2$$

$$f_b = \left(1 - \frac{31}{150}\right)250 = 198 \text{ lb/in.}^2$$ allowable flexural stress under combined stress conditions

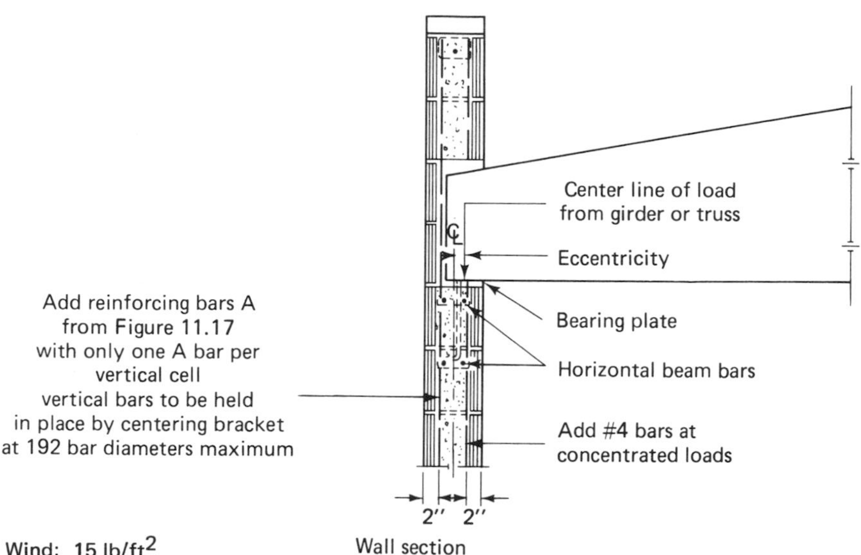

Wall section

Wind: 15 lb/ft^2

Seismic: $F_p = zIC_p sW_p = (1.0)(1.0)(0.2)(1.5)(92 \text{ lb/ft}^2)$

$= 27.6 \text{ lb/ft}^2$ governs

27.6 x 4 ft (tributary width) = 10.4 lb/ft

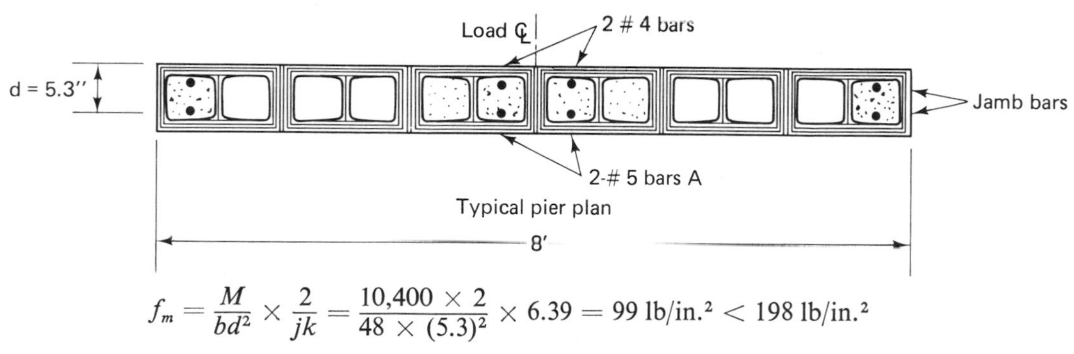

Typical pier plan

$$f_m = \frac{M}{bd^2} \times \frac{2}{jk} = \frac{10{,}400 \times 2}{48 \times (5.3)^2} \times 6.39 = 99 \text{ lb/in.}^2 < 198 \text{ lb/in.}^2$$

Thus the reinforcing, as detailed, is adequate.

2. Dead load + dead-load moment + moment due to lateral load: A conservative estimate for wall moment (masonry design approximations do not warrant a more accurate determination):

$$40 \times \left(\frac{13.33}{2}\right) = \quad 267 \text{ ft-lb}$$

$$110.4 \text{ lb} \times \frac{13.33^2}{10} = \frac{1962}{2229 \text{ ft-lb}}$$

becomes design moment at column midheight
Check the two No. 5's

$$\therefore \; pn = 0.097 \quad \text{and } \frac{2}{jk} = 6.39$$

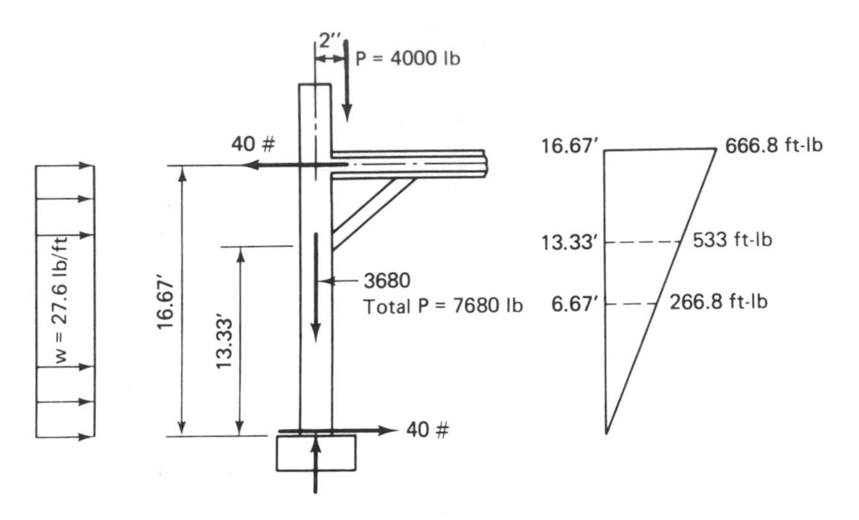

Wing: 15 lb/ft^2

Seismic: $F_p = ZIC_pSW_p = (1.0)(1.0)(0.20)(1.5)(92 \text{ lb/ft}^2) = 27.6 \text{ lb/ft}^2$ governs

(27.6 x 4′ = 110.4)

Tributary width

$$f_a = \frac{7680}{48 \times 7.6} = 21 \text{ lb/in.}^2 \text{ and } F_a = 0.2 \times 750 \times 0.875(=R) = 131 \text{ lb/in.}^2$$

$$\frac{f_a}{F_a} = \frac{21}{131} = 0.16$$

$$f_b = (1.33 - 0.16)(250) = 292 \text{ lb/in.}^2$$

$$f_m = \frac{M}{bd^2} \times \frac{2}{jk} = \frac{2229 \times 12}{48 \times (5.3)^2} \times 6.39 = 127 \text{ lb/in.}^2 < 292 \text{ lb/in.}^2$$

$\therefore$ OK for this loading case

STEP 4: LINTEL DESIGN

Loads:			
	Tributary area	= 10 ft × 20 ft =	200 ft^2
	Roof LL	= 10 ft(20 lb/ft^2) =	200 lb/ft
	Roof DL	= 10 ft(10 lb/ft^2) =	100 lb/ft
	Parapet	= 3.33ft(62 lb/ft^2) =	206 lb/ft
	Wall DL	= 4.67ft(92 lb/ft^2) =	430 lb/ft
			936 lb/ft
		+ misc. =	64 lb/ft
		avg. total =	1000 lb/ft

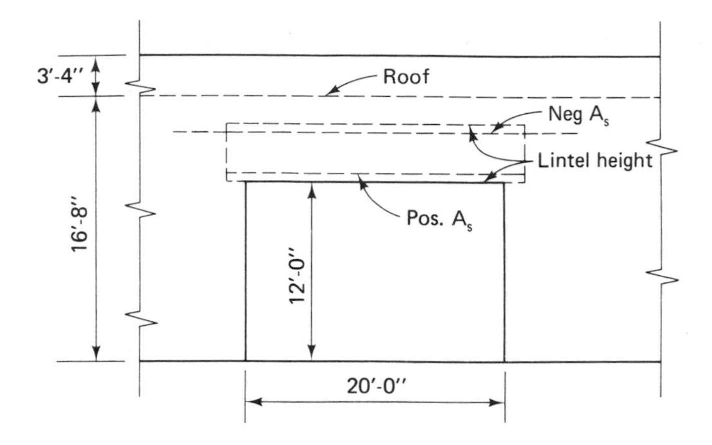

Flexural design (Span = 20 ft):

Note: Since the roof line is only 4 ft 8 in. above the lintel soffit, consider that arching effect not significant, and therefore the lintel carries full uniform load. Also, it is assumed that wall mass on either side of the opening fully restrains the lintel, accounting for the moment coefficients selected. Certainly more conservative values could be used at the discretion of the designer.

$$V = \frac{wL}{2} = \frac{1000\text{ lb/ft}(20\text{ ft})}{2} = 10\text{ kips}$$

$$M^{(+)} = \frac{wL^2}{24} = \frac{1000(20)^2}{24} = 16.7\text{ ft-kips}$$

$$M^{(-)} = \frac{wL^2}{12} = \frac{1000(20)^2}{12} = 33.3\text{ ft-kips}$$

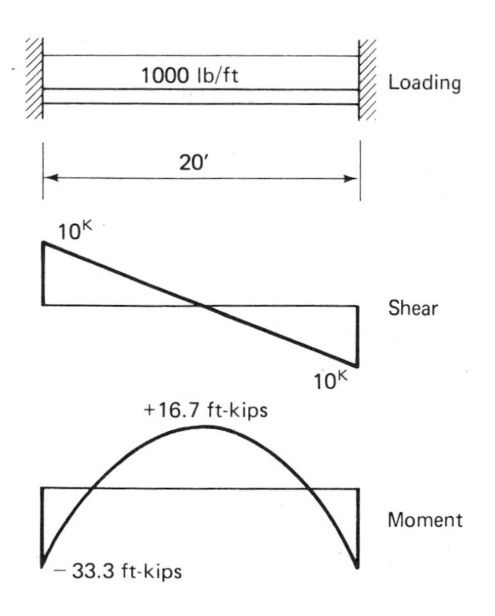

Therefore, from Table 7-1:

$f'_m = 1500$ lb/in.^{2} (solid grout), $F_b = 250$ lb/in.^{2} (no inspection), $n = 40$

$f_s = 20{,}000$ lb/in.^{2}

$v = 25$ lb/in.^{2} (without shear reinf.)

$= 75$ lb/in.^{2} (with shear reinf.)

$u = 100$ lb/in.^{2}

1. *Effective depth:* Solid grout sufficient number of courses so as to preclude need for stirrups, if depth available.

$$d = \frac{V}{bjv} = \frac{10{,}000}{7.63 \times 0.9 \times 25} = 58.2 \text{ in. req.}$$

Distance available from bond beam to opening = 4 ft 8 in. or 56 in.; thus, $d = 52.5$ in. Some question may be raised here, whether, like concrete, V for web reinforcement should be taken at d in. from support face. If this were followed, then:

$$v = \frac{10{,}000 - 52.5/12 \times 1000}{7.63 \times .9 \times 52.5} = 16 \text{ lb/in.}^2$$

$<$ allowable, so no stirrups would be required

With this approach, one could reduce solid grouted portion to 44 in. ($d = 40.5$ in.). Then

$$v = \frac{10{,}000 - 40.5/12 \times 1000}{7.63 \times 0.9 \times 40.5} = 24 \text{ lb/in.}^2$$

< 25 lb/in.^{2}, so no stirrups needed

However, the validity of this assumption has not been tested for grouted masonry, and it would therefore perhaps be preferable to base the determination upon the unit shear stress at the support face. Good practice, it would seem, would dictate a conservative stirrup spacing in masonry beams where they are required. The cost of placing any additional stirrups would be minimal compared with the beneficial effects obtained, including increased ductility.

2. *Stirrup requirements:* If one adopts the more conservative approach and uses V at the support face for the shear, even solid grouting to the top of the bond beam would not preclude the need for stirrups, since available d is 52.5 in. versus required $d = 58.2$ in. for no stirrups.

Use $d = 40.5$ in. (to demonstrate how stirrup spacings may be selected); then $v = 36$ lb/in.^{2} and:

$$s = \frac{A_v f_v}{bv} = \frac{0.11 \times 20{,}000}{7.63 \times 36} = 8.0 \text{ in., where } s_{max} = d/2 \sim 20 \text{ in.}$$

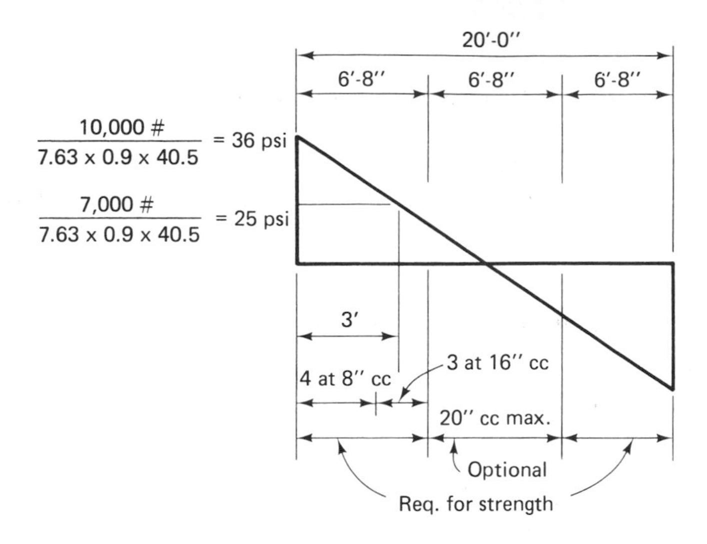

Conservatively, one could specify, measured from the support face:

4 No. 3 at 8 in. cc plus 3 No. 3 at 16 in. cc

This would extend the stirrups a distance of approximately d inches past the point where theoretically they are no longer required for strength. It might be desirable to continue the 16 in. spacing through the intervening 6 ft 8 in. in the center portion of the span to provide for additional strength and ductility, but this would be at the discretion of the designer.

3. *Steel requirements:*

$^{-}A_s$:

$$K = \frac{M^{(-)}}{bd^2} = \frac{33.3 \times 12{,}000}{7.63 \times (40.5)^2} = 32 \qquad \therefore \quad \frac{K}{F_b} = \frac{32}{0.250} = 128$$

Select $np = 0.072$ from Curve B-2.

$p = 0.072/40 = 0.0018$ and $A_s = 0.0018 \times 7.63 \times 40.5 = 0.55$ in.
$\therefore$ 2 No. 5's OK; OK
obtain $f_m = 0.92 \times 250 = 230$ lb/in.^{2} (Curve B-2 for $K/f_b = 128$)
Or from Curve B-4 for $nM/bd^2 = 40 \times 32 = 1280$, $np = 0.073$ (as before).
Check from Table C-1 for $np = 0.072$; find $2/jk = 7.12$ and $j = 0.89$.
Thus, $f_m = M/bd^2(2/jk) = 32 \times 7.12 = 228$ lb/in.^{2} (as before).

$^{+}A_s$: By inspection, two No. 4's will be sufficient at bottom of beam for this $d = 40.5$ in.

Bond:

$$u = \frac{V}{\Sigma_o jd} = \frac{10{,}000}{3.93 \times 0.90 \times 40.5} = 70 \text{ lb/in.}^2 < 100 \text{ lb/in.}^2 \qquad \therefore \text{ OK}$$

Provide "embedment length" for both $^{+}A_s$ and $^{-}A_s$ (Table 9-1).

The spans on the other lintels are short enough so that two No. 4's would suffice in all cases.

STEP 5: SEISMIC ANALYSIS OF BUILDING

(A) Comparison of 1973 and Current UBC Requirements for Seismic Forces:

1973 UBC:

$$V = ZKCW = (1.0)(1.33)(0.10)W = 0.133W$$

Current UBC:

$$V = ZIKCSW = (1.0)(1.0)(1.33)(0.14)W = 0.186W$$

where $(CS)_{\max} = 0.14$

(B) Lateral Force Analysis: Determination of shear forces per foot to each wall (assuming flexible diaphragm):

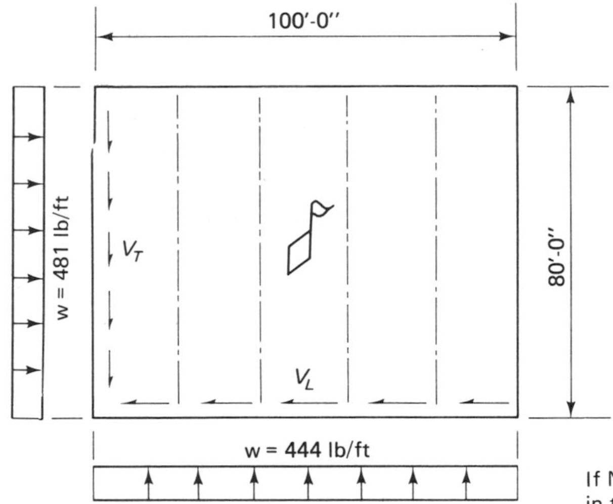

If No. 6 at 24 in.cc were called for in the wall, instead of the No. 8 at 40 in.cc previously specified, then Table 5.2 gives a wall weight of 68 psf. This is done here merely to demonstrate the effect of increasing the number of grouted cores upon the code designated seismic forces.

Wind:

$$w = 15\ \text{lb/ft}^2\left(\frac{16.67}{2} + 3.33\right) = 175\ \text{lb/ft}$$

Seismic:

1. Transverse direction:

			Current Code	'73 Code
Roof:	$(10\ \text{lb/ft}^2)(80\ \text{ft})(0.186 \text{ or } 0.133)$	=	149 lb/ft	106 lb/ft

		Current Code	'73 Code
Walls:	$2(68\ \text{lb/ft}^2)\left(\frac{16.67}{2} + 3.33\right)(0.186 \text{ or } 0.133) =$	295 lb/ft	211 lb/ft
	Total seismic =	444 lb/ft	317 lb/ft

Therefore,

seismic (444 lb/ft) > wind (175 lb/ft) $\quad \therefore$ seismic governs

2. Longitudinal Direction:

		Current Code	'73 Code
Roof:	$(10\ \text{lb/ft}^2)(100\ \text{ft})(0.186) =$	186 lb/ft	133 lb/ft
Walls:	$2(62\ \text{lb/ft}^2)\left(\frac{16.67}{2} + 3.33\right)(0.186 \text{ or } 0.133) =$	295 lb/ft	211 lb/ft
	=	481 lb/ft	349 lb/ft

Therefore,

seismic (481 lb/ft) > wind (175 lb/ft) $\quad \therefore$ seismic governs

Lateral force distribution (tributary area basis):

1. Transverse:

$$V = \frac{(444\ \text{lb/ft})(100\ \text{ft})}{2} = 22{,}200\ \text{lb or } V = \frac{(317\ \text{lb/ft}) \times (100\ \text{ft})}{2} = 15{,}850\ \text{lb}$$

$$v_T = 22{,}200\ \text{lb}/80\ \text{ft} = 277.5\ \text{lb/ft}$$

$$\text{or} \quad \frac{15{,}850\ \text{lb}}{80\ \text{ft}} = 198\ \text{lb/ft (200 lb/ft minimum)}$$

v (design—1973 UBC) = $198 \times 2 = 396$ lb/ft

v (design—Current UBC) = $277.5 \times 1.5 = 416$ lb/ft
or $V = 416 \times 80$ ft $= 33{,}300$ lb for shear load from diaphragm when checking wall shear resistance (Table 24-H UBC)

Design Wall Shear at Midheight:

$$V = 1.5[22{,}200 + 0.186(68\ \text{lb/ft}^2)\left(\frac{16.67}{2} + 3.33\right)(80\ \text{ft}) = 51{,}000\ \text{lb}$$

2. Longitudinal:

$$V = \frac{(481\ \text{lb/ft})(80\ \text{ft})}{2} = 19{,}240\ \text{lb}$$

$$v_L = \frac{19{,}240\ \text{lb}}{100\ \text{ft}} = 192\ \text{lb/ft} < 200\ \text{lb/ft minimum (UBC Sec. 2312)}$$

v (design —Current UBC) = $192 \times 1.5 = 288$ lb/ft
or $V = 288 \times 100$ ft $= 28{,}800$ lb for shear load from diaphragm where checking wall shear resistance

(C) Chord Design:

A typical bond beam detail would call for two No. 4's minimum ($A_s = 0.40$ in.^{2}), which is more than adequate. Provide for adequate splices along the bar lengths (per Table 9-1).

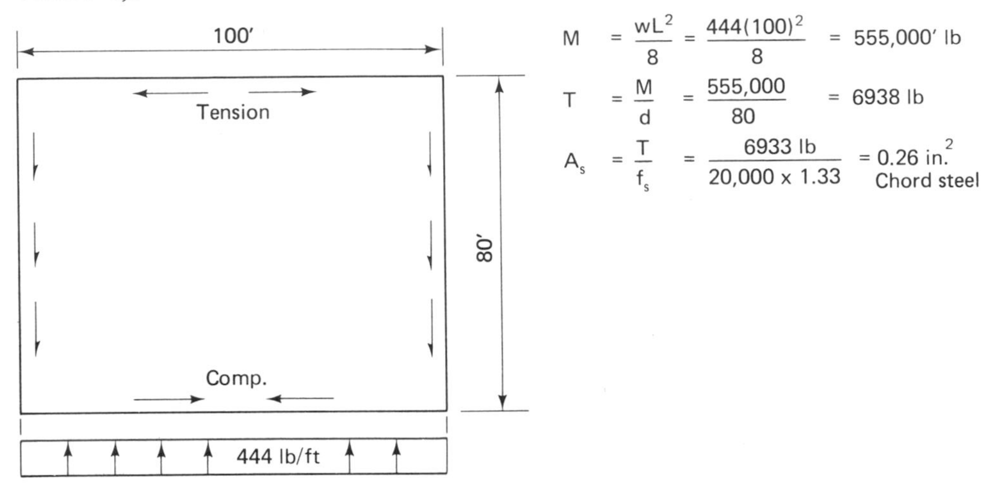

STEP 6: SHEAR AND FLEXURAL STRESS FOR PIERS ALONG EAST WALL

Note: To simplify calculations, the height of the 4 × 8 ft door will be extended to 12 ft 0 in. (i.e., 4 × 12 ft door)

(A) Rigidity of Transverse Shearwall (refer to Table 10-4 or Figure 10-25):

Entire wall (without openings): $\frac{h}{d} = \frac{20.0}{80} = 0.25;\ \Delta_c = 0.0683,\ k_c = 14.64$

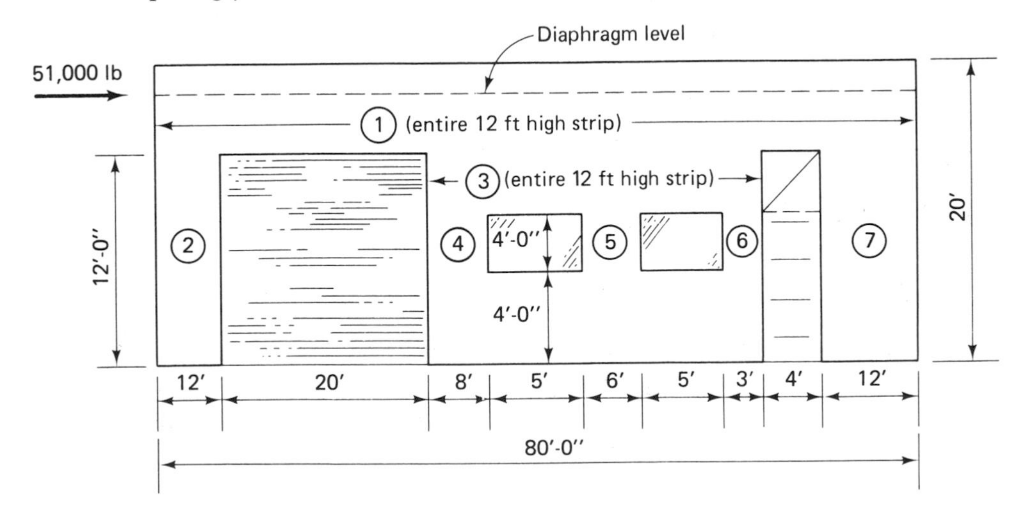

Strip 1
12 ft high: $\left(\frac{h}{d}\right)_1 = \frac{12}{80} = 0.15;\ \Delta_c = 0.0390,\ k_c = 25.64$

Pier 2: $\left(\frac{h}{d}\right)_2 = \frac{12}{12} = 1.0;\ \Delta_F = 0.333,\ k_F = 3.00$

Pier 7: $\left(\frac{h}{d}\right)_7 = \frac{12}{12} = 1.0;\ \Delta_F = 0.333,\ k_F = 3.00$

Strip 3
12 ft high: $\left(\frac{h}{d}\right)_3 = \frac{12}{32} = 0.375;\ \Delta_F = 0.0983,\ k_F = 10.17$

4 ft high: $\left(\frac{h}{d}\right)_{4\text{-}5\text{-}6} = \frac{4}{32} = 0.125;\ \Delta_F = 0.0315,\ k_F = 31.75$

$$\left(\frac{h}{d}\right)_4 = \frac{4}{8} = 0.5;\ \Delta_F = 0.1354,\ k_F = 7.39$$

$$\left(\frac{h}{d}\right)_5 = \frac{4}{6} = 0.67;\ \Delta_F = 0.1918,\ k_F = 5.21$$

$$\left(\frac{h}{d}\right)_6 = 0.50;\ \Delta_F = 0.1354,\ k_F = 7.39$$

$$\therefore\ \Sigma_{4.5.6} = 7.39 + 5.21 + 7.39 = 19.99,\ \Delta_F = 0.0501$$

Strip 3
(with openings): $\therefore\ \Delta_3 = 0.0983 - 0.0315 + 0.0501 = 0.1169$

$$\therefore\ k_3 = \frac{1}{0.1169} = 8.55$$

Strip 1
(with openings): $\Sigma k = k_2 + k_7 + k_3 = 3.00 + 3.00 + 8.55 = 14.55;$

$$\Delta = \frac{1}{14.55} = 0.0687$$

Rigidity of entire wall:

$$\Delta_T = 0.0683 - 0.0390 + 0.0687 = 0.0980 \text{ (entire wall with openings)}$$

$$k_{\text{east wall}} = \frac{1}{0.0980} = 10.20$$

Rigidity of transverse shear walls:

$$k_{\text{east wall}} = 10.20$$

$$k_{\text{west wall}} = 14.64$$

$$\left.\begin{aligned} R \text{ for east wall} &= \frac{10.20}{10.20 + 14.64} = 0.41 \\ R \text{ for west wall} &= \frac{14.64}{10.20 = 14.64} = 0.59 \end{aligned}\right\} \text{If supported by a rigid diaphragm}$$

$$\Sigma = 1.00$$

TABLE 12-1

East-Wall Lateral-Force Distribution
(Lateral force to wall = 51,000 lb)

Pier	Δ_F	k_F	*Lateral force to each pier (%)*	*Lateral force to pier (lb)*
2	0.333	3.00	20.6	10,500
Middle Panel—3	0.1169	8.55	58.8	30,000
7	0.333	3.00	20.6	10,500
		$\Sigma = 14.55$	$\Sigma = 100\%$	$\Sigma = 51{,}000$

TABLE 12-2

Middle-Panel Lateral-Force Distribution
(Lateral force to middle panel 3 = 30,000 lb)

Pier	Δ_F	k_F	*Lateral force to each pier (%)*	*Lateral force to pier (lb)*
4	0.1354	7.39	37	11,100
5	0.1918	5.21	26	7,800
6	0.1354	7.39	37	11,100
		$\Sigma = 19.99$	$\Sigma = 100\%$	$\Sigma = 30{,}000$

% Lateral Force to Middle Panel = 58.8% or 30,000 lb

TABLE 12-3

Shear in Each Pier East Wall

Pier	$\frac{h}{2D}$	*D (ft.)*	*Shear area (in.²)*	*V (lb.)*	*v (lb/in.²)*	*Allow shear without reinf. (× 1.33) (Figure 7-8)*
2	0.50	12	606	10,500	17.3	27.9
4	0.25	8	404	11,100	27.5	30.6
5	0.33	6	303	7,800	25.7	29.3
6	0.25	8	404	11,100	27.5	30.6
7	0.50	12	606	10,500	17.3	27.9

From discussion in Chapter 7 on "Shear Areas," the average in-plane shear area for steel spaced at 24 in. cc = 50.5 in.²/ft of wall length (Table 7-4).

From Table 12-3, the shear stresses in the piers are seen not to exceed the allowable values, and therefore no shear steel is required.

(B) Jamb or Boundary Reinforcement: Moment caused by in-plane shear force on pier equals $Vh/2$.

Piers 2 and 7:

$$V = 10{,}500 \text{ lb}$$

$$M = V\left(\frac{h}{2}\right) = 10{,}500\left(\frac{12}{2}\right) = 63{,}000 \text{ ft-lb}$$

$$d = 144 \text{ in.} - 4 \text{ in.} = 140 \text{ in.}$$

$$\therefore \ A_s = \frac{M}{f_s jd} = \frac{63.0 \times 12}{20 \times 1.33 \times 0.9 \times (140)} = 0.23 \text{ in.}^2$$

Panel 3:

$$V = 30{,}000 \text{ lb}$$

$$M = V\left(\frac{h}{2}\right) = 30{,}000\left(\frac{4}{2}\right) = 60{,}000 \text{ ft-lb}$$

$$\therefore \ A_s = \frac{M}{f_s jd} = \frac{60.0 \times 12}{20 \times 1.33 \times 0.9 \times (380)} = 0.08 \text{ in.}^2$$

Piers 4 and 6:

$$V = 11{,}100 \text{ lb}$$

$$M = V\left(\frac{h}{2}\right) = 11{,}100 \text{ lb}\left(\frac{4}{2}\right) = 22{,}200 \text{ ft-lb}$$

$$\therefore \ A_s = \frac{M}{f_s jd} = \frac{22.2 \text{ ft-kips} \times 12}{20 \times 1.33 \times 0.9 \times (92)} = 0.012 \text{ in.}^2$$

Pier 5:

$$V = 7{,}800 \text{ lb}$$

$$M = V\left(\frac{h}{2}\right) = 7{,}800\left(\frac{4}{2}\right) = 15{,}600 \text{ ft-lb}$$

$$\therefore \ A_s = \frac{M}{f_s jd} = \frac{15.6 \text{ ft-kips} \times 12}{20 \times 1.33 \times 0.9 \times (68)} = 0.12 \text{ in.}^2$$

Therefore, for the jamb steel requirements use one No. 5 bar at each jamb of piers 2 and 7 and 1 No. 4 at each jamb in all other piers.

Step 7: Connection Design

(A) Diaphragm Connections Along East and West Walls:

Load to anchor bolt:

Assume spacing of anchor bolt at 48 in. oc

Load parallel to wall = 277.5 lb/ft

Vertical DL = 10 lb/ft^2(10) = 100 lb/ft

Perpendicular to wall = 481 lb/ft

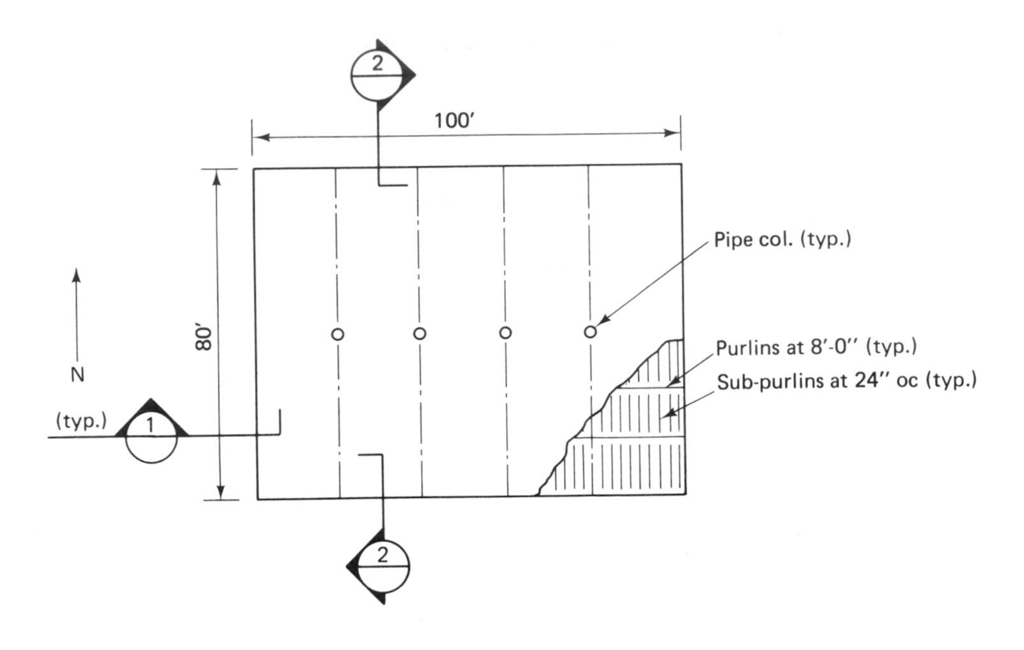

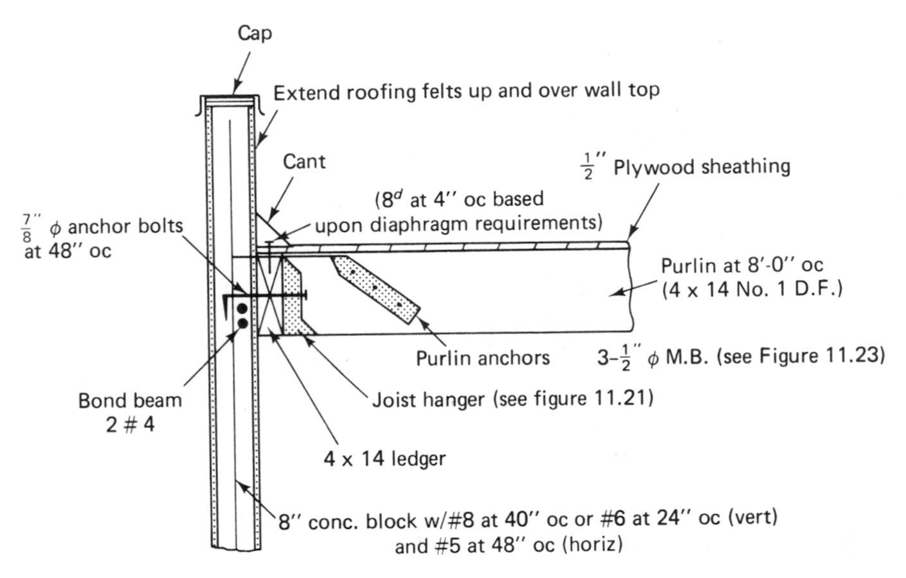

Therefore, for each bolt:

$$277.5 \text{ lb/ft } (4 \text{ ft}) = 1110 \text{ lb}$$

$$100 \text{ lb/ft } (4 \text{ ft}) = 400 \text{ lb}$$

$$\text{resultant} = \sqrt{(1110)^2 + (400)^2} = 1180 \text{ lb}$$

$$\tfrac{7}{8}\ \phi \text{ bolt good for } 1000 \text{ lb} \times 1.33 = 1333 \text{ lb} > 1180 \text{ lb} \qquad \text{OK}$$

Therefore, use $\frac{7}{8}$-in. ϕ anchor bolts at 48 in. oc.

(B) Diaphragm Connections Along North and South Walls:

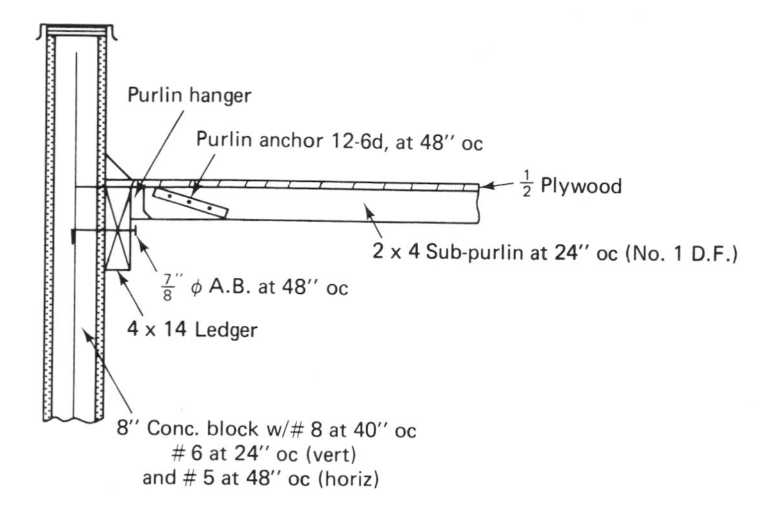

CASE STUDY 2: SEISMIC ANALYSIS OF TWO-STORY BUILDING

Design procedure

Only the seismic analysis (or wind if it governed) of the example two-story building will be offered here. This is presented primarily to show what additional forces are brought to bear upon the lower-story piers through the overturning effect of the upper story. This action imposes an additional axial compressive or tensile stress on each pier within the wall which adds to the already present dead- and live-load compressive stress. Also, the shear-wall elevations in this sample were selected so that the center of rigidity does not occur at the center of mass. Therefore, torsional forces exist, in addition to the in-plane shear forces, and these must be taken into account in the pier shear analysis. As a consequence of all these factors, a number of different loading effects must be accounted for:

1. Vertical dead loads.
2. Vertical live loads,
3. Vertical axial loads due to overturning effect of second story.
4. In-plane shear on pier sections within wall due to lateral load at top of pier.
5. Maximum in-plane moment in pier produced by this lateral load.
6. Out-of-plane bending due to lateral load.
7. Out-of-plane bending due to vertical load eccentricity.

Thus, the pier section must be sufficient (a) to keep the shear stresses below the allowable limits prescribed by the Code with or without web reinforcement, (b) to carry the most severe combination of the vertical loads named in items 1, 2, and 3 above without exceeeding the allowable axial compressive stress, and (c) to resist those tensile forces induced by the in-plane lateral force bending moment, a sufficient amount of properly anchored vertical jamb (boundary) reinforcement must be located within the pier.

Briefly, this seismic analysis entails a determination of:

1. *The total lateral force acting at the roof and second-floor levels.*

2. *The distribution of the diaphragm shears to the parallel shear* walls based upon their relative rigidities, assuming the presence of rigid floor and roof diaphragms. Where the center of mass and the center of rigidity do not coincide, the additional torsional shears must be added to the in-plane shears. Thus, the torsional moment and the polar moment of inertia of the wall system has to be found in order to evaluate the torsional shear stresses.

3. *The internal distribution to each pier within a wall of the total force brought to that shear wall.*

4. *The most severe axial load combination in conjunction with the in-plane bending moment and shear which each pier must be designed to resist.* In addition to the dead and live loads carried to each pier, the axial load imposed upon them due to the upper-story overturning moment must be taken into account.

5. *The most severe combination of axial load with out-of-plane bending.*

The following case study is intended to show how the overturning effects on a building of more than one story can be evaluated. The lateral force on a shear wall will induce an axial force on the pier, P_{ovt} (which will add or subtract from the dead and live loads). In addition, in-plane shear and moment caused by the in-plane seismic shear will be considered in this analysis.

The plan and elevations of the first story of a two-story warehouse building located in seismic zone 4 are shown in Figure 12-2. Assume that the second-story walls have no openings. The reinforced grouted brick masonry building is to be designed to resist the lateral loads designated by the UBC. For an earthquake occurring in the east-west direction, determine:

1. The design lateral forces at each floor level.
2. The distribution of shears at the first-story level.
3. The distribution of the lateral forces carried by the north wall to the individual piers within that wall.
4. The moments and axial loads on each pier in the north wall.
5. The in-plane shear and moment adequacy of the piers in the north wall. Provide any shear steel or boundary steel, if necessary.

Loading:

Roof:

$$\text{Slab DL} = 100 \text{ lb/ft}^2$$
$$\text{LL} = 20 \text{ lb/ft}^2$$

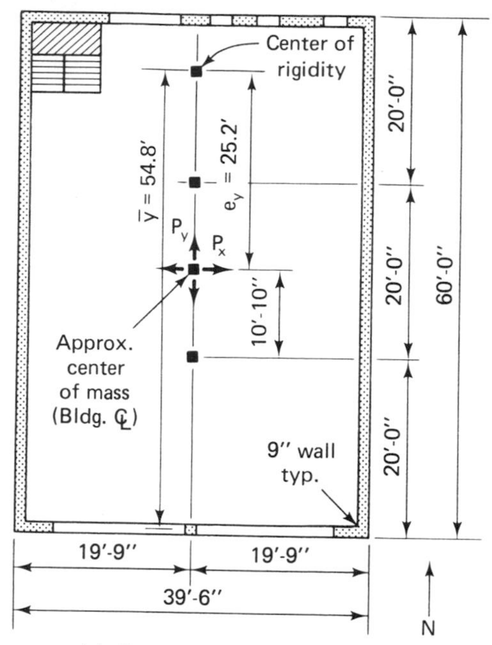

(a) Cross section plan – first story

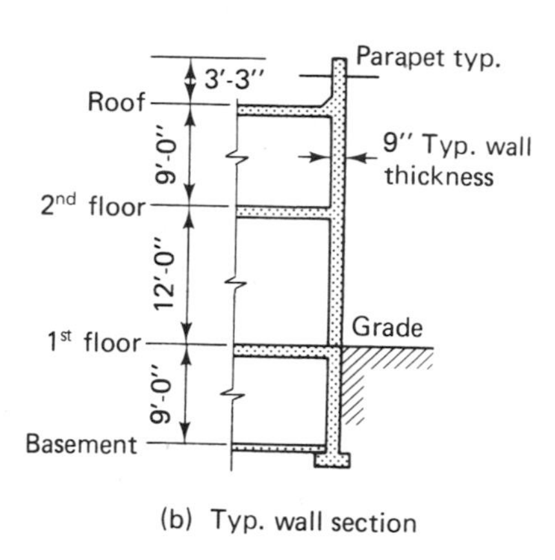

(b) Typ. wall section

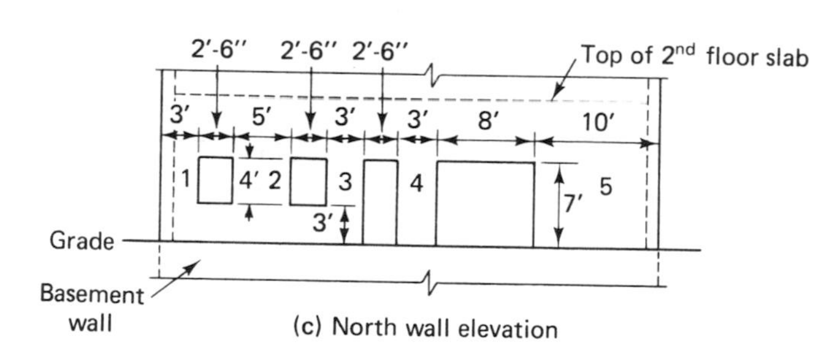

(c) North wall elevation

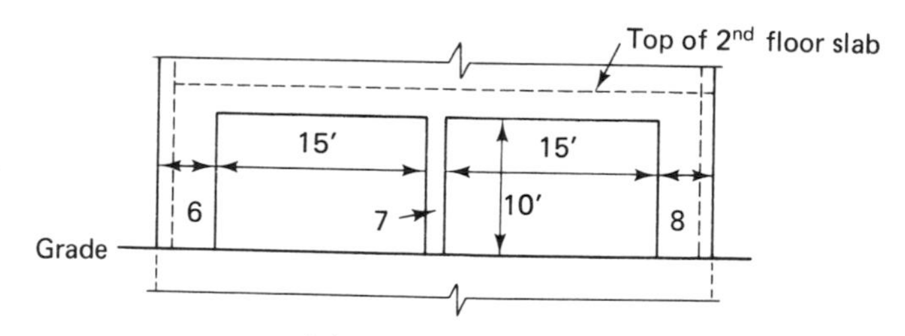

(d) South wall elevation

FIGURE 12-2. Two-story commercial building—plan and elevations.

Floors:

$$\text{Slab DL} = 100 \text{ lb/ft}^2$$

$$\text{Storage LL} = 125 \text{ lb/ft}^2$$

Wall weights:

$$\text{Upper story} = 100 \text{ kips}$$

$$\text{Lower story} = 175 \text{ kips}$$

Design stresses:

$f'_m = 1500$ lb/in.², $n = 20$ (Increase where necessary.)

$f_s = 20{,}000$ lb/in.²

$w = 100$ lb/ft² (9-in. brick walls)

Special inspection provided so full stresses allowed in design.

Design and analysis

STEP 1: LATERAL FORCES ON SHEAR WALLS AT EACH FLOOR LEVEL

(A) Total Loads:

Roof level:

roof DL = (100 lb/ft²)(40 ft)(60 ft) = 240 kips

wall load = 100 kips

2nd-floor level loads:

DL = (100 lb/ft²)(40 ft)(60 ft) = 240 kips

LL = 125 lb/ft²)(40 ft)(60 ft) = 300 kips

wall load = 175 kips

W = total vertical dead load = 240 + 100 + 240 + 0.25(300) + 175 kips = 830 kips total

UBC stipulates that 25% of LL be considered under certain types of storage occupancies when computing W, for determining base shear, V.

(B) Base Shear:

$V = ZIKCSW$

$Z = 1.0$ (seismic zone 4)

$I = 1.0$ (occupancy importance factor)

$K = 1.33$ (box-type structure)

$$T = \frac{0.05h_n}{\sqrt{D}} = \frac{0.05(21 \text{ ft})}{\sqrt{40 \text{ ft}}} = 0.166 \text{ secs}$$

$$C = \frac{1}{15\sqrt{T}} = \frac{1}{15\sqrt{0.166}} = 0.164 > 0.12 \qquad C = 0.12 \text{ governs}$$

$S = 1.5$ (assume max. value) $\qquad CS = 0.12 \times 1.5 = 0.18 > 0.14$ (max.)

$\therefore$ use $CS = 0.14$

$$V = (1.0)(1.0)(1.33)(0.14)(830) = 155 \text{ kips}$$

(C) Design Lateral Forces at Each Floor Level:

$F_t = 0.07TV$ and since $T = 0.166 \text{ s} < 0.7 \text{ s}$ $\quad \therefore\ F_t = 0$, per UBC Section 2312

$$F_x = \frac{(V - F_t)w_x h_x}{\sum_{i=1}^{n} w_i h_i}$$

At the roof level,

$$F_r = \frac{(240 + 100 \text{ kips})(21 \text{ ft})}{(240 + 0.25 \times 300 + 175)(12 \text{ ft}) + (240 + 100 \text{ kips})(21 \text{ ft})} \times (155 - 0)$$
$$= 85 \text{ kips}$$

At the 2nd floor level,

$$F_2 = \frac{(5880)}{5880 + 7140} \times (155 - 0) = 70 \text{ kips}$$

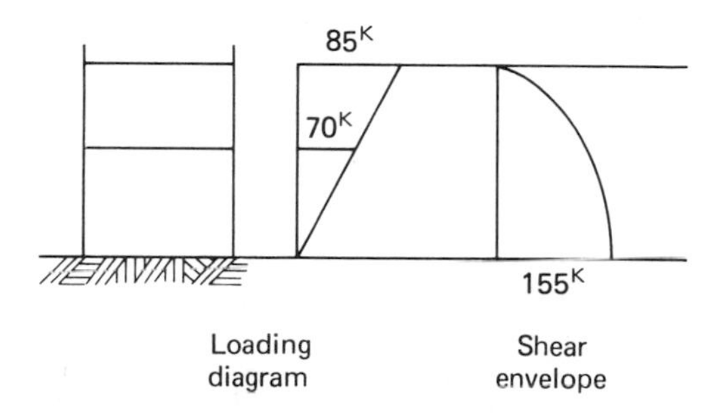

Step 2: Distribution of Shears at the First-Story Level

(A) Locate the Center of Gravity of Lateral Load: Because of its symmetrical layout, the center of gravity will occur close to the center of the building; $\therefore \bar{y}_{cg} = 29.6$ ft (from south wall center line)

(B) Determine the Stiffnesses of the Exterior Walls:

$$\Delta_F = \frac{P}{E_m t}\left[\left(\frac{H}{D}\right)^3 + 3\left(\frac{H}{D}\right)\right] \quad \text{for fixed wall or pier}$$

$$\Delta_c = \frac{P}{E_m t}\left[4\left(\frac{H}{D}\right)^3 + 3\left(\frac{H}{D}\right)\right] \quad \text{for cantilever wall or pier}$$

The equations above represent the total deflection (flexural + shear) of a masonry wall, where $E_v = 0.4E_m$.
For several piers connected along their tops, lateral displacement at the top of the wall was determined as (equation 9, Chapter 10):

$$\Delta = \frac{P}{\sum_{i=1}^{n} k_i} = \frac{P}{k_1 + k_2 + k_{3t} + k_i} = \frac{P}{(1/\delta_1) + (1/\delta_2) + (1/\delta_3) + \ldots + (1/\delta_i)}$$

Note: Pier stiffness is equal to the force required to produce a unit deflection.)
For calculation convenience, use:

$$\left.\begin{aligned} P &= 10^6 \text{ lb}, \ t = 10 \text{ in.} \\ E_m &= 1000f'_m = 1000(1500) = 1{,}500{,}000 \end{aligned}\right\} \text{OK, since only relative rigidities are being sought}$$

For the north and south walls, the piers are fixed.

$$\therefore\ \Delta = \frac{P}{E_m t}\left[\left(\frac{H}{D}\right)^3 + 3\left(\frac{H}{D}\right)\right]$$

$$= \frac{1{,}000{,}000\ \text{lb}}{(1{,}500{,}000\ \text{lb/in.}^2)(10\ \text{in.})}\left[\left(\frac{H}{D}\right)^3 + 3\left(\frac{H}{D}\right)\right] = 0.0667\left[\left(\frac{H}{D}\right)^3 + 3\left(\frac{H}{D}\right)\right]$$

or Table 10-4 or Figure 10-25 could have been used.
Deflection and stiffness of north and south shear walls from Table 12-4:

TABLE 12-4

Pier Deflections and Stiffnesses

Pier	*D* (*in.*)	*H* (*in.*)	$\frac{H}{D}$	$\left(\frac{H}{D}\right)^3$	Δ_F	$k_i = 1/\Delta_F$
1	36	48	1.33	2.35	0.423	2.364
2	60	48	0.80	0.51	0.194	5.155
3	36	48	1.33	2.35	0.423	2.364
4	36	84	2.33	12.65	1.309	0.764
5	120	84	0.70	0.34	0.163	6.144
6	48	120	2.50	15.62	1.540	0.649
7	18	120	6.67	296.74	21.127	0.047
8	48	120	2.50	15.62	1.542	0.649

North:

$$\Delta = \frac{1}{\sum k_i} = \frac{1}{2.364 + 5.155 + 2.364 + 0.764 + 6.144} = 0.060$$

$$\text{and}\quad k = \frac{1}{\Delta} = 16.79$$

South:

$$\Delta = \frac{1}{\sum k_i} = \frac{1}{0.649 + 0.047 + 0.649} = 0.743 \text{ and } k = \frac{1}{\Delta} = 1.34$$

East and west:

$$\text{for } \frac{H}{D} = 0.20,\ \Delta = 0.0406 \text{ for cantilever wall and } k = \frac{1}{\Delta} = 24.66$$

Relative stiffness:

$$R_i = \frac{k_i}{\sum_{i=1}^{n} k_i} = \frac{k_i}{k_1 + k_2 + \ldots + k_i} \text{ or } R_i = \frac{1/\delta_i}{\sum_{i=1}^{n} 1/\delta_i} = \frac{1/\delta_i}{1/\delta_1 + 1/\delta_2 + \ldots + 1/\delta_i}$$

where k_i is stiffness of the individual exterior walls.

To north wall:

$$R_x = \frac{16.79}{16.794 + 1.345} = 0.926$$

To south wall:

$$R_x = \frac{1.345}{16.79 + 1.345} = 0.074$$

(C) *Locate the Center of Rigidity:*

$$\bar{y}_r = \frac{\sum R_x y}{\sum R_x} \text{ (take } y \text{ from center line of south wall)}$$

$$= \frac{(0.926)(59.25) + (0.074)(0)}{0.926 + 0.074} = 54.8 \text{ ft}$$

Therefore, eccentricity,

$$e_y = \bar{y}_r - \bar{y}_{cg} = 54.8 - 29.6 = 25.2 \text{ ft}$$

Check

$$e_{\min} = 5\% = 0.05(59 \text{ ft}) = 2.95 \text{ ft} < 25.2 \text{ ft} \qquad \therefore\ e = 25.2 \text{ ft}$$

The east and west walls are virtually similar; therefore, there is no actual eccentricity in the N-S direction and $\bar{x}_r$ occurs on the center line. However, the Code provides that an accidental eccentricity of 5% must be considered anyway.

(D) *Evaluate the Combined Shear Forces Carried by Each Exterior Wall:* Total shear carried by north and south walls equals 155 kips. However, for shear analysis, $P_x = 1.5V = 1.5(155) = 233$ kips. For overturning and pier moment analysis, $P_x = 155$ kips.

Total shear:

$$(P_x)_i = \frac{R_x}{\sum R_x} P_x + \frac{R_x \bar{y}}{J_r} P_x e_y$$

Polar moment of inertia: $J_r = \sum (R_y \bar{x}^2 + R_x \bar{y}^2)$ (Table 12-5).

TABLE 12-5

Polar Moment of Inertia

Wall	k_i	R_i	$\bar{x}$ (*ft*)	$\bar{x}^2$ (*ft*²)	$\bar{y}$ (*ft*)	$\bar{y}^2$ (*ft*²)	$R_y\bar{x}^2$ (*ft*²)	$R_x\bar{y}^2$ (*ft*²)
N	16.79	0.250	—	—	4.4	19.4	—	4.9
S	1.34	0.020	—	—	54.8	3003	—	60.1
E	24.66	0.365	19.38	375	—	—	136.9	—
W	24.66	0.365	19.38	375	—	—	136.9	—
	$\Sigma = 67.45$	$\Sigma = 1.000$						

$$J_r = (136.9 + 136.9 + 4.9 + 60.1) = 338.8$$

North wall:

$$V = \frac{0.250}{0.250 + 0.020}\begin{Bmatrix}233\\155\end{Bmatrix} + \frac{(.250)(4.4)}{338.8}\begin{Bmatrix}233\\155\end{Bmatrix}(25.2)$$

$$= \begin{Bmatrix}215.7 + 19.1\\143.5 + 12.7\end{Bmatrix} = \begin{Bmatrix}234.8\\156.2\end{Bmatrix} \qquad \therefore \text{ use } V = 234.8 \text{ kips for shear analysis}$$

$$= 156.2 \text{ kips for overturning and flexural effects}$$

South wall:

$$V = \frac{.020}{0.250 + 0.020}\begin{Bmatrix}233\\155\end{Bmatrix} + \frac{(0.020)(54.8)}{338.8}\begin{Bmatrix}233\\155\end{Bmatrix}(25.2)$$

$$= \begin{cases}17.3 + 19.0 = 36.3 \text{ kips for shear}\\11.5 + 12.6 = 24.1 \text{ kips for overturning and flexural effects}\end{cases}$$

For ground motion in the E-W direction: The torsional shear resisted by the east and west walls amounts to:

$$V = 0(\text{no force parallel to walls}) + \frac{0.365(19.38)}{338.8}[233(25.2)] = 122.6 \text{ kips}$$

This would not be critical, since ground motion in the N-S direction would produce higher shears, as seen in the following determination:

$$e_{\min} = 5\% = 0.05(60) = 3 \text{ ft}$$

Therefore, for E or W wall:

$$V = \frac{0.365}{0.365 + 0.365}\begin{Bmatrix}233\\155\end{Bmatrix} + \frac{0.365(19.38)}{338.8}\begin{Bmatrix}233\\155\end{Bmatrix}(3) = \begin{Bmatrix}116.5 + 14.6\\77.5 + 9.7\end{Bmatrix}$$

$$= \begin{Bmatrix}131.1 \text{ kips}\\87.2 \text{ kips}\end{Bmatrix}$$

TABLE 12-6

Combined Shear Forces to Each Wall

Wall	*Shear to wall for for shear stress (kips)*	*Shear to wall for moment analysis*[a] *(kips)*
N	234.8	156.2
S	36.3	24.1
E	131.1	87.2
W	131.1	87.2

[a]Drop 1.5 factor

STEP 3: DISTRIBUTION OF THE LATERAL FORCE CARRIED BY THE NORTH WALL TO THE INDIVIDUAL PIERS WITHIN THE WALL:

TABLE 12-7

Distribution of Shear to Piers in N Wall

Pier	*k*	*R*	*Shear to each pier for shear analysis (kips)*	*Shear to each pier for moment analysis (kips)*
1	2.364	0.141	33.1	22.0
2	5.155	0.307	72.1	48.0
3	2.364	0.141	33.1	22.0
4	0.764	0.045	10.6	7.0
5	6.144	0.366	85.9	57.2
	$\sum = 16.79$	$\sum = 1.000$	234.8	156.2

Example (Pier 1):

$$\frac{2.364}{16.791} = 0.141 \times 234.8 = 33.1 \text{ kips}$$

$$\times\ 156.2 = 22.0 \text{ kips}$$

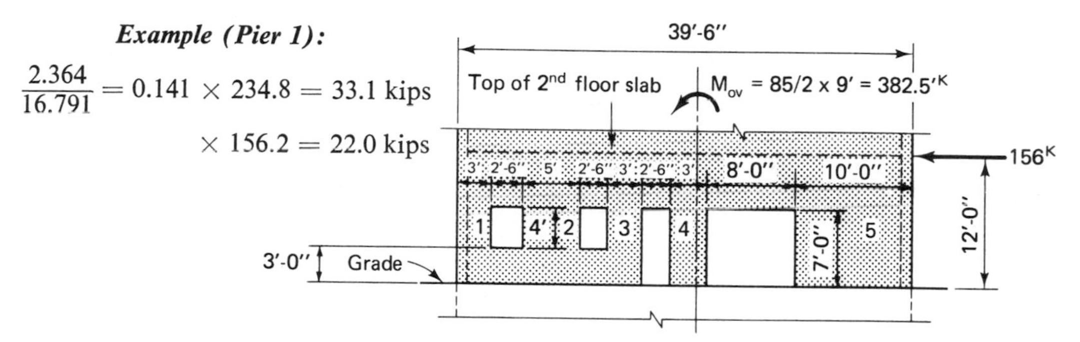

STEP 4: AXIAL LOADS AND MOMENTS CARRIED TO EACH PIER IN NORTH WALL:

(A) Centroid of Net Wall Section (9 in. thick):

TABLE 12-8

Centroid of Wall Section

Pier	*Area* $A = tD$ $= 9D$ (*in.*^{2})	*Distance from left edge of wall to centroid of pier, L* (*in.*)	$A \times L$ (*in.*^{3})
1	324	18	5,832
2	540	96	51,840
3	324	174	56,376
4	324	240	77,760
5	1080	414	447,120
	$\sum A = 2592$		$\sum AL = 638,928$

$$\text{Distance from left edge of wall to centroid} = \frac{\sum AL}{\sum A} = \frac{638{,}928}{2592} = 246.5 \text{ in.}$$

(*B*) *Moment of Inertia of Net Wall Section:*

TABLE 12-9

Moment of Inertia of Wall Section

Pier	Area $A = 9D$ (in.^{2})	Distance from centroid, $\bar{l}$ (in.)	$A\bar{l}^2$ (in.^{4})	$I = \frac{tD^3}{12}$ (in.^{4})	Total $A\bar{l}^2 + I$ (in.^{4})
1	324	228.5	16,916,769	34,992	16,951,761
2	540	150.5	12,231,135	162,000	12,393,135
3	324	72.5	1,703,025	34,992	1,738,017
4	324	6.5	13,689	34,992	48,681
5	1080	167.5	30,300,750	1,296,000	31,596,750
				Total moment of inertia $\sum =$	62,728,344

(*C*) ***Overturning Moment and Resulting Axial Load*** Total M_{ovt} = total V × (vertical distance from V to critical plane where loads being computed) + (applied OTM at second floor level). The critical section in the wall is at sill height, 3 ft above the first-floor level, where the most severe combination of OTM and reduced wall section occurs. Refer to section "Combined Axial and Flexural Stresses" in Chapter 7.

$$P_{ovt} = \frac{M_{ovt}\bar{l}A}{I_n} \text{ on each individual pier}$$

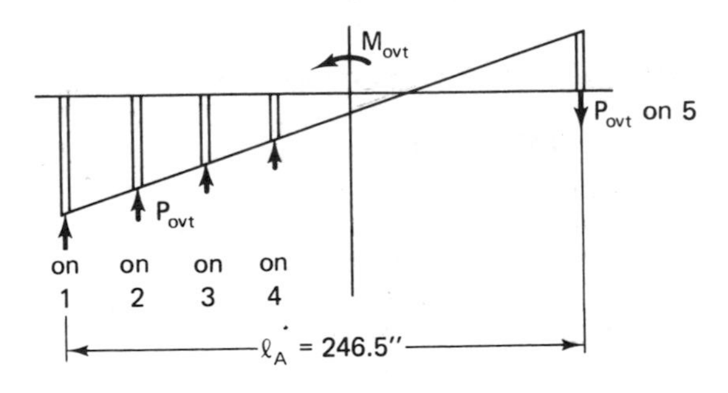

Assume that in the second story, the east and west walls have the same stiffness, then V to each wall $= \frac{85}{2} = 42.5$ kips and M_{ovt} at the second floor level = 42.5 kips × 9 ft = 382.5 ft-kips and M_{ovt} at sill height in the first story would be $M_{ovt} = 156.2 \times 9 + 382.5$ ft-kips = 1788.5 ft-kips.[2]

[2]Refer to Table 12-6.

(D) Pier Loads, Moments, and Shears:

TABLE 12-10

Pier Loads, Moments and Shears

Pier	*Effective load width (ft)*	P_D[1] *(lb)*	P_L[1] *(lb)*	P_{ovt} *(lb)*	V_E *for moment (lb)*	M_E *(in.-lb)*	V_E *for shear analysis (lb) (at 1.5)*
1	4.25	14,025	6,163	25,320	22,000	528,000	33,100
2	7.50	24,750	10,875	27,794	48,000	1,152,000	72,100
3	5.50	18,150	7,975	8,034	22,000	528,000	33,100
4	8.25	27,225	11,963	720	7,000	294,000	10,600
5	14.00	46,200	20,300	61,868	57,200	2,402,400	85,900

[1]The tributary width for vertical live and dead loads, in this instance, is considered to be 10 ft.

Dead load to piers:

First Story

$$\text{Wall DL at sill height} = 750 \text{ lb/lin. ft}$$

Second Story

$$\text{Wall DL} = 550 \text{ lb/lin. ft}$$

$$\text{Floor DL} = 10 \text{ ft} \times 100 = 1000 \text{ lb/lin. ft}$$

Roof

$$\text{Roof DL} = 10 \text{ ft} \times 100 = \underline{1000 \text{ lb/lin. ft}}$$

$$\text{Total} = 3300 \text{ lb/lin. ft}$$

Live load to piers:

Floor

$$10 \text{ ft} \times 125 = 1250 \text{ lb/lin. ft}$$

Roof

$$10 \text{ ft} \times 20 = \underline{200 \text{ lb/lin. ft}}$$

$$\text{Total} = 1450 \text{ lb/lin. ft}$$

$$P_D = \text{effective loading width to pier} \times 3300 \text{ lb/lin. ft}$$

$$P_L = \text{effective loading width to pier} \times 1450 \text{ lb/lin. ft}$$

The effective loading width to the pier equals the width of pier plus one-half of each adjacent opening length. The axial load due to overturning becomes:

$$P_{ovt} = \frac{(1788.5 \text{ ft-kips})(1000 \text{ lb/kips})(12 \text{ in./ft})lA}{62{,}728{,}344} = 0.342\, \dot{l}A$$

$$M_E = \text{for pier fixed at each end} = V_{\text{pier}}\frac{h}{2}$$

Example—Pier 1:

$$P_D = (4.25 \text{ ft})(3300) = 14{,}025 \text{ lb}$$

$$P_L = (4.25 \text{ ft})(1450 \text{ lb/lin. ft}) = 6163 \text{ lb}$$

$$P_{\text{ovt}} = 0.342 \text{ lb/in.}^3(228.5 \text{ in.})(324 \text{ in.}^2) = 25{,}320 \text{ lb}$$

$$M_E = (22.0)(1000 \text{ lb/kips})\left(\frac{48 \text{ in.}}{2}\right) = 528{,}000 \text{ in.-lb}$$

One should consider the effect of certain load combinations, such as:

1. DL + LL only: no allowable stress increase.
2. $\text{DL} + \frac{\text{LL}}{4} + P_{\text{ovt}} + V_E$ on pier $+ V_E\frac{h}{2}$, with a $\frac{1}{3}$ stress increase for seismic loadings. 25% of LL is included when analyzing lateral seismic effects, since this is the portion of storage LL assumed in place (per UBC) when calculating the total base shear.
3. $0.9 \text{ DL} - P_{\text{ovt}} + V_E$ on pier $+ V_E(h/2)$ with $\frac{1}{3}$ stress increase for seismic. This would follow somewhat the pattern established by the UBC for seismic loading effects in strength design of reinforced concrete buildings.

STEP 5: COMPRESSIVE AND SHEAR ADEQUACY OF THE PIERS IN THE NORTH WALL (PROVIDE SHEAR STEEL AND BOUNDARY STEEL AS REQUIRED):

TABLE 12-11

Allowable Pier Shear Stresses for North Wall
(Based upon $f'_m = 3000$ lb/in.² and $n = 10$)

Pier	*M/Vd = H/2D*	*Allowable shear, v, lb/in.²* No reinforcement	Reinforced
1	0.67	50	86
2	0.40	50	96
3	0.67	50	86
4	1.17	34	75
5	0.35	50	97

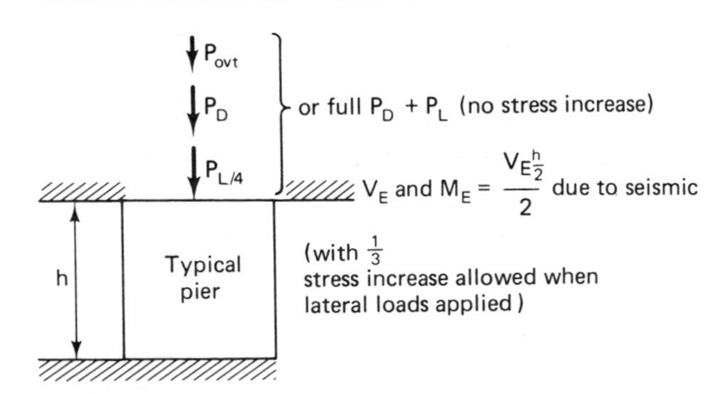

TABLE 12-12

Load Combinations and Resulting Stresses

Pier	$P_D + P_L$ (lbs)	$0.75\left(P_D + \frac{P_L}{4} + P_{ovt}\right)$ (lbs)	$0.75M_E$ (in.-lb)	$0.75V_E$[a] (lbs)	$fa = P/9D$[b] (lb/in.^{2})	$v = \frac{0.75V_E}{9D}$ (lb/in.^{2})	A_s, at jamb (in.^{2})	Combined-stress (lb/in.^{2}) Allow f_b	Actual f_m
1	20,188	30,664	396,000	24,825	94.6	76.6	0.69	758	495
2	35,625	41,447	864,000	54,075	76.8	100.1	0.86	785	382
3	26,125	21,133	396,000	24,825	80.6	76.6	0.68	802	495
4	39,188	23,202	220,500	7,950	120.0	24.5	0.38	793	331
5	66,500	84,857	1,801,800	64,425	78.6	59.7	0.86	782	263

[a] V on each pier was increased by 50% for shear stress check only.

[b] Since axial stresses, f_a, are considerably below the allowable wall stress F_a on each pier, normal wall reinforcement is OK for the vertical loads.

Example (from Table 12-12):

Boundary (jamb) reinforcement:

$$A_s = \frac{M}{f_s jd} = \frac{M}{18{,}000d}; \text{ where pier effective depth, } d = \text{pier length } (D) - 4 \text{ in.}$$

$$j = 0.9$$

$f_s = 20{,}000$ lb/in.^{2} ($\frac{1}{3}$ increase provided for by taking $0.75M_E$ for design value)

Piers 1 and 3: Provide 2 No. 5's at each pier boundary ($A_s = 0.62$ in.^{2})

Piers 2 and 5: Provide 2 No. 6's at each pier boundary ($A_s = 0.88$ in.^{2})

Pier 4: Provide 2 No. 4's at each pier boundary ($A_s = 0.40$ in.^{2})

Combined axial stress and bending:

$$\frac{f_a}{F_a} + \frac{f_b}{F_b} \leq 1.00$$

($\frac{1}{3}$ stress increase taken into account by multiplying P and M by 0.75)

Check on Pier 2.

Stress remaining for flexure:

$$f_b = \left(1.00 - \frac{f_a}{F_a}\right)F_b = \left(1.00 - \frac{76.8}{600}\right)(900)$$

$= 785$ lb/in.^{2} where

$$F_a = 0.2f'_m$$

$= 0.2(3000) = 600$ lb/in.^{2} and

$F_b = \frac{1}{3}f'_m = 1000$ lb/in.^{2} but max permitted by UBC is 900 lb/in.^{2}

Compare with:

$$f_m = \frac{M}{bd^2} \times \frac{2}{jk} \quad \text{where } p = \frac{0.88}{9 \times 56} = 0.00175,\ np = 0.0175$$

$$= \frac{864000}{9 \times (56)^2} \times 12.47 = 382 \text{ lb/in.}^2$$

Web reinforcement:

$$\text{Shear spacing } s = \frac{A_v f_v}{vb} = \frac{0.31 \times 20000}{v \times 9} = \frac{688.9}{v}$$

Piers 1, 2, 3 and 5 require horizontal shear steel as follows:

Piers 1 and 3: $s = 688.9/76.6 = 8.99$ in., say No. 5's at 9 in. cc
Pier 2: $s = 688.9/100.1 = 6.81$ say No. 5's at 6 in. cc
Pier 4: Since $v_{\text{allow}} < 34$ lb/in.², no shear steel is required
Pier 5: $s = 688.9/59.7 = 11.53$ in., say No. 5's at 11 in. cc

Note that when the allowable shear stress on the masonry is exceeded, the entire shear force acting on the pier must be carried by the shear steel alone.

It would be necessary to increase f'_m to 3000 lb/in.² on the north and south walls because of the severe shear requirement placed upon masonry shear walls there, caused by the stipulation that each wall be designed to resist 1.5 × computed lateral shear force on the pier or wall section. This demonstrates the design versatility obtained through the use of reinforced brick masonry. By specifying higher-strength units in certain locations, such as on these two walls, the designer avoids the need to increase the wall thickness to an uneconomical dimension. Also, the higher-strength bricks are usually locally available. Further, undoubtedly the solid east and west walls and the upper-story walls could still be constructed with a design f'_m of 1500 lb/in.².

Note that $(P_D + P_L)$ governs for piers 3 and 4. This becomes immediately evident by comparing these vertical load magnitudes with $0.75(P_D + P_L/4 + P_{\text{ovt}})$. Allowing for the 1/3 stress increase by taking 0.75 times the design loads, shears, or moments facilitates immediate comparisons such as these. All stress comparisons are then made with the basic allowable values. Consequently, if the axial stress magnitudes were critical, then conceivably these two piers would have to be reinforced sufficiently to sustain the design dead and live loads, acting as reinforced walls or as flush pilasters within the walls. The other piers would be designed to sustain the vertical load combination of $(P_D + P_L/4 + P_{\text{ovt}})$.

With regard to the load combination of $(P_D - P_{\text{ovt}})$, valves in Table 12-10 show that no tension will occur on piers 3 and 4 and only a small tensile force occurs on piers 1 and 2; however, on pier 5 a design tensile force of 20,288 lb (= 0.9 × 46,200 − 61,868) would occur. This factor must be considered in the actual design of the pier and footing, although it was not included in this example.

In any event, all piers need be checked for compliance with the code interaction formula; e.g., allow f_b versus actual f_m in Table 12-12. Furthermore, the need for shear steel in all piers must be determined by comparing $v_{\text{allowable}}$ on the masonry versus v_{actual}. Even where shear steel is provided (piers 1, 2, 3, and 5 since $v_{\text{actual}} > v_{\text{allow}}$), the shear stress on the section must not exceed $2\sqrt{f'_m}$ (= 110 lb/in.² to 75 lb/in.², depending upon the $H/2D$ ratio of the pier. See Figure 7-8b). Finally, the jamb steel requirements must be provided for, should this added vertical steel at the pier boundaries (A_s in Table 12-12) be needed. Note, for instance, that on the basis of $A_s = M/(f_s jd)$, the jamb steel (A_s) for pier 2 was found to be about 0.86 in.².

If the combined force system is analyzed as discussed in the section "Pier Axial and Flexural Stresses," Chapter 7, the results indicate that minimum reinforcing is adequate. This result therefore takes into account the influence of the axial compressive force on the tension produced by bending. To demonstrate what happens when significant tension does occur under this combined force system, the moment on pier 2 will be approximately doubled.

$P = 41.5 \text{ kips}, M = 1700 \text{ in. kips}$

$$f_m = f_a + f_b \qquad f_a = \frac{41{,}500}{9 \times 60} = 76.8 \text{ lb/in.}^2$$

$$= 76.8 + 785 \qquad F_a = 0.2(3{,}000) = 600 \text{ lb/in.}^2$$

$$= 861.6 \text{ lb/in.}^2 \qquad F_b = 900 \text{ lb/in.}^2$$

$$f_b = \left(1.00 - \frac{76.8}{600}\right)900$$

$$= 785 \text{ lb/in.}^2$$

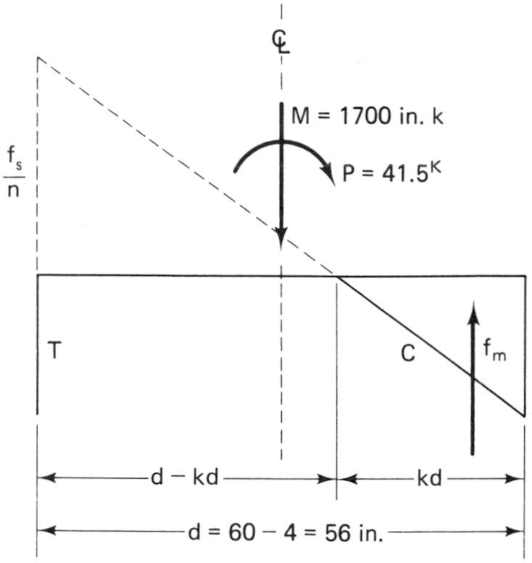

$$C = (f_m/2)(kd)(t) = \frac{861.6}{2}(kd)9 = 3877kd$$

$\sum M$ about $T = 0 \quad \therefore$

$$3877\,kd(56 - kd/3) = 41{,}500(30 - 4) + 1{,}700{,}000$$

$$kd(56 - kd/3) = 716.8$$

$$(kd)^2 - 168(kd) + 2150 = 0$$

$$\therefore\ kd \sim 14 \text{ in. } \& \ k = 0.25$$

$$\sum F_y = 0 \qquad \therefore$$

$$T + 41{,}500 = 3877(14) = 54{,}278$$

$$T = 12{,}778 \text{ lb}$$

$$f_s = \frac{1 - 0.25}{0.25} \times 10(861.6) = 25{,}845 \text{ lb/in.}^2 < 26{,}677 \text{ lb/in.}^2 \qquad \therefore \text{ OK}$$

$$A_s = \frac{12{,}778}{25{,}845} = 0.49 \text{ in.}^2$$

In the same section of Chapter 7, an approximate approach was mentioned wherein k is simply assumed to be about 0.3, hence $j = 0.9$. Near ultimate capacity, the application of C would hence be about $0.1d$ from the compression face. The moment to be resisted by the tension steel would be reduced by the moment of the vertical dead load acting at the centroid, or about $0.4d$. The steel then is assumed to resist the remaining overturning moment at 26,667 lb/in.². This would not be mathematically correct, but would give a consistent factor of safety relation, i.e., ultimate condition to working stress. For example:

$$A_s = \frac{M - 0.4d(P)}{f_{s_{\text{allow}}} \times jd} = \frac{1700 - 0.4 \times 56 \times 41.5}{26.7 \times 0.9 \times 56} = 0.57 \text{ in.}^2$$

13
High-Rise Masonry

One might question why high-rise masonry should be used. For one thing, it would appear evident that a group of high-rise buildings, properly designed and constructed, is more conservative of national resources than a whole series of separate smaller buildings that might provide the same amount of useful floor space. Certainly, reinforced masonry bearing walls are ideally suited for high-rise construction. They have been used throughout the world for all types of buildings, from the smallest and simplest shelters to some of the most significant monuments and public buildings. The development of high-strength concrete block and brick, combined with the improvements in grouting and reinforcing techniques, have made masonry bearing walls practical for such multistory construction. The concept essentially combines reinforced masonry bearing walls with concrete floor and roof slabs. As has been thoroughly described in Chapter 10, the basic concept here involves that of designing every floor to act as a horizontal diaphragm in transferring wind or seismic loads to the transverse shear walls, which in turn carry these forces to the foundation, as shown in Figure 13-1.

The shear walls are reinforced to develop the moment and shear forces brought about through this action, and it is then not necessary to depend upon the massiveness of these walls to provide the stability against overturning as was done in an earlier day (e.g., the 6-ft-thick Monadnock Building bearing walls).

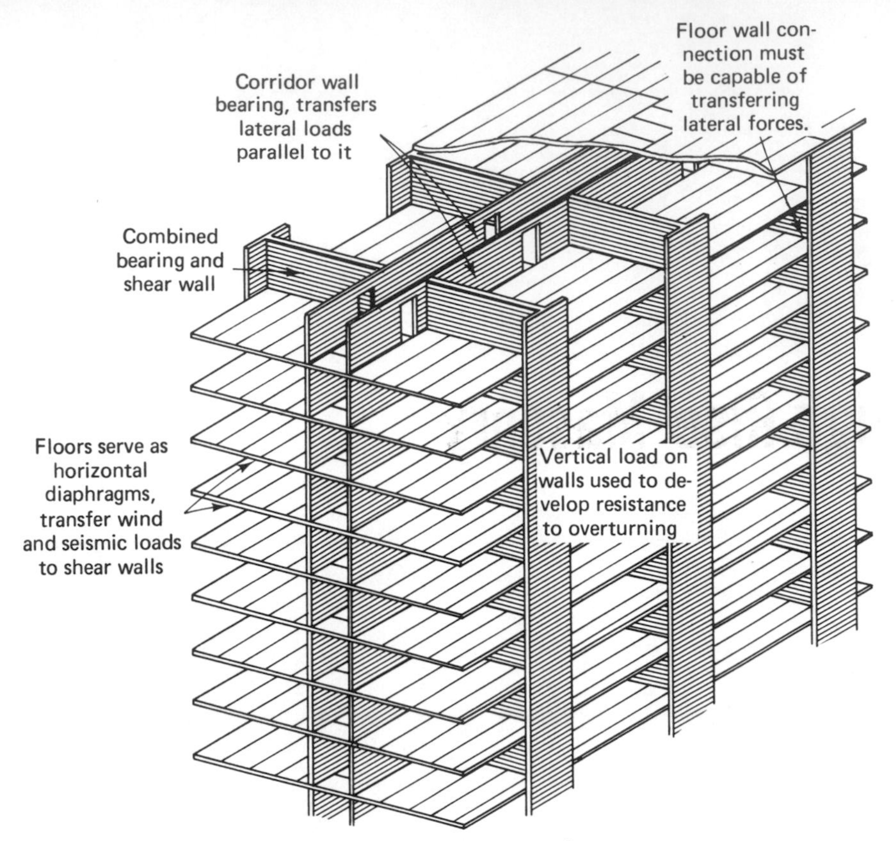

FIGURE 13-1. High-rise concept in masonry.

FEATURES

High-rise masonry construction has several desirable features, among which are (1) simplicity of design, (2) excellent environmental characteristics, (3) speed and ease of construction, and (4) reduced building costs. With its repetitive floor arrangement, masonry bearing-wall buildings offer a great deal of simplicity in architectural detailing and layout. Designers have considerable freedom in planning a building within this concept. Since the interior partitions and corridors may provide structural capability, architects are free to treat the exterior as they deem most appropriate without the inhibition of the structural need for any type of external framing. Buildings may be rectangular, square, or circular in plan; however, it is highly desirable to maintain as much symmetry as possible, to preclude difficulties when providing for seismic resistance. Since all exterior walls need not be load-bearing or shear-resisting, glass or veneers of various types may be used in achieving the desired architectural esthetics. In some cases, as in hotels or apartments, a balcony beyond the exterior curtain wall may be easily provided by cantilevering the floor slab out beyond the bearing-wall support. Further, because of the many different types of masonry units available, a wide variety of architectural styles and plan layouts may be readily achieved. Such versatility provides for the formation of many interesting patterns in which masonry units may be laid. The Holiday Inn Motel, a round bearing-wall multistory structure shown in Figure 13-2, provides an excellent example of this versatility.

The environmental characteristics of reinforced masonry perhaps are not as well known as are their structural capabilities. These will be discussed in more detail in Chapter 16; however, it would be appropriate at this point to mention briefly some of the qualities. For instance, the sound-absorbing qualities of concrete masonry surfaces are rated well above the effective levels recommended by sound engineers. Sound transmitted from room to room through walls, ceilings, and floors can be minimized through the use of masonry bearing walls, concrete floors, and acoustical ceilings. Further, reinforced masonry possesses those inherent fire-resistant characteristics that make fire resistance ratings easily achievable. In the design of heating and air-conditioning systems for a building, the *thermal inertia* of the building materials is a significant phenomenon. This term relates to the material heat-storage capacity, which is high in solid masonry walls. This means that they possess the ability to dampen the effect of the maximum heat gain or loss in cyclic changes. The greater the heat-storage capacity (thermal inertia), the smaller the instantaneous rate of heat flow into the interior (conductivity). This certainly reduces the capacity and size requirement for cooling equipment and is a distinct advantage of reinforced masonry wall construction.

FIGURE 13-2. Holiday Inn Motel—a round bearing-wall multistory structure.

Construction advantages are numerous. For one, construction time is minimized, primarily because of the typically repetitive nature of the layout. When the walls are placed, floors can be placed immediately, thereby permitting a continuous work schedule. These floors then provide a working area for the next story above, eliminating the need for costly exterior scaffolding. The use of masonry bearing wall construction provides many other opportunities for reducing construction costs. There are no columns or projecting beams to form. The elimination of beams permits a reduction in floor to floor heights of 1 ft or more, and this could add up to a reduced building height of 10 ft in an 8- or 10-story building. The use of precast floor slabs permits the achievement of a degree of efficiency of on-site labor that is not obtainable when building frames have to be constructed. By enclosing each floor as the work progresses, craftsmen of other trades can install utilities and finish interiors while masons are adding additional stories above. Furthermore, the structural masonry wall has a surface that can be painted, stained, or left natural. No other surface treatment need be applied except in areas where furring is required for heating and air-conditioning ducts. The soffit of the precast floor deck serves as a finished ceiling for the rooms below, or it may be sprayed with a paint or acoustical plaster to provide an inexpensive and attractive finish. Last, but not least, occupancy of lower stories is possible even before the upper portion of the building is completed. This allows the owner to begin receiving rental income while the building is still under construction.

DESIGN FACTORS

Floor units

A tentative floor slab thickness may be selected on the basis of the moment and shear requirements. Subsequently, the adequacy of this thickness for strength must be measured against the need to satisfy other requirements, which include:

1. An immediate and a long-term floor deflection limit to avoid unsightly cracking or damage to nonstructural elements, such as ceilings or partitions. The Code permits the use of the gross area when calculating the moment of inertia for this determination provided that $pf_y < 500$. Further, the long-term deflection is often approximated as twice the immediate deflection.
2. The specified fire-rated separation between floors (often a governing factor for floor thickness).
3. Conformance with modular masonry course heights.
4. A sufficient space to allow the incorporation of piping and/or conduit within the slab itself.
5. The occurrence of large lifting stresses when precast slabs are placed upon the supporting walls.

Bearing stresses of the floor slab against the wall must be checked. Where concrete masonry is used, only the face shell of the block should enter into the area determination, since all the cells are frequently not grouted, certainly not at the time the floor is placed on the wall. Even though the bearing stresses may be low, it is necessary to achieve good shear transfer through the joint. Bearing stresses at the corridors can be critical and therefore must be carefully examined.

Other slab conditions that should be investigated, where they exist, would include the following:

1. The effect of all actual support conditions and their length/width ratios, upon slab moments and deflections.
2. The actual maximum bearing stresses between the slab and the wall due to the various support restraint conditions.
3. The effect of openings in the slabs.

No live load reduction should normally be taken for the slab design, because the individual slab areas are usually relatively small. However, the reduction, as permitted by the UBC, can be safely applied to the design of walls and foundations, since simultaneous live loadings of maximum intensity at all levels, as specified by code, are not likely to occur. Furthermore, the walls usually possess an excellent ability to distribute any localized live load of very high intensity.

Shear-wall configurations

A key element in any successful high-rise masonry building involves the proper floor plan layout, according to an article written by James Kesler, consulting structural engineer in Los Angeles entitled "A Look at Load-Bearing Masonry Design," published by the Masonry Institute of America. For one thing, layouts should provide sufficient longitudinal shear-wall lengths. The configurations shown in Figure 13-3, for example, achieve this requirement, with no loss in economy by providing a desirable balance in shear-wall lengths between the two principal axes of the building, including some lying within the interior corridors. This type of configuration allows some of the exterior walls to become nonstructural in function, thereby permitting greater flexibility in the development of the exterior architectural treatments. In addition, uniform wall lengths between openings, having returns wherever possible, represent another very effective measure to bolster seismic or wind resistance. This scheme can result in a drastic reduction in the maximum in-plane shear stress on these walls, which in turn could produce decreases in wall thicknesses needed, as well as lowering the amount of wall reinforcement. In many cases where the walls must act as individual vertical cantilevers, the wall returns function as flanges, thereby greatly reducing the masonry flexural compressive stresses. Figure 13-4 provides an example of how effective these wall returns are in reducing the compressive stresses. Wall returns also facilitate placement of the boundary reinforcement by providing added grout space at

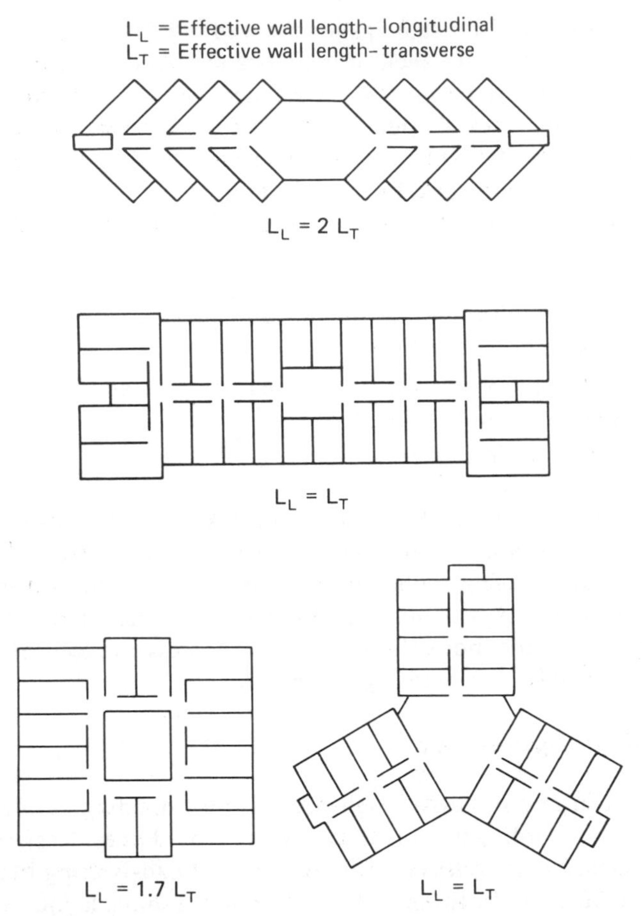

FIGURE 13-3. Typical high-rise floor plans (Kesler).

the wall boundaries. Some increase in the resistance to elastic buckling can be expected as well.

The grouping of openings in shear walls is desirable, since this usually results in longer individual wall panels, again reducing shear and flexural stresses. Also, if these openings can be staggered from floor to floor, the cantilever moments in individual wall panels can be significantly reduced, permitting the entire wall to be analyzed as a perforated plate. This produces some savings in reinforcing as well as in foundation sizes. However, if this stagger cannot be conveniently achieved, the coupling of individual wall panels by means of stiff interconnecting beams or spandrels provides an effective alternative. One method for analyzing this type of coupled shear-wall system was described in Chapter 10, and an example of its use is shown in the next section.

It is exceedingly desirable in seismic design to minimize weight in masonry construction, since the magnitude of the design seismic forces is directly related to the

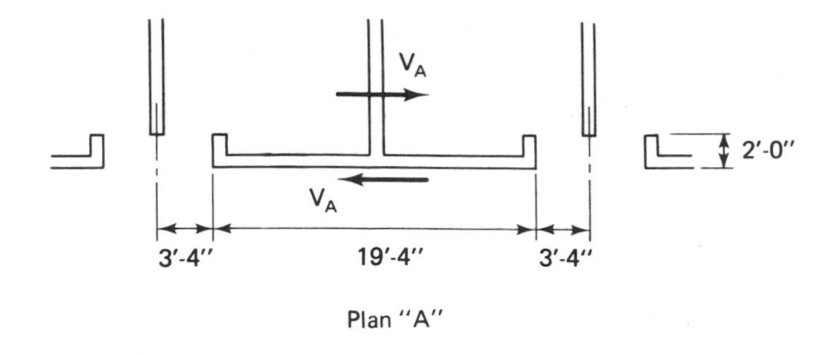

Plan "A"

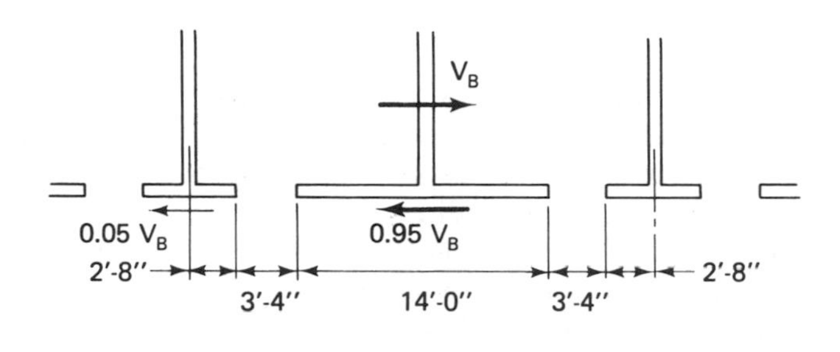

Plan "B"

$V_A = V_B$
and equal ht.

$v_A = 0.76\ v_B$
f_m for A = 0.42 f_m for B
A_s for A = 0.59 A_s for B

FIGURE 13-4. Effectiveness of wall returns (Kesler).

building mass. Furthermore, lateral stability is obtained from the resistance of the bearing shear walls, so the massiveness of the structure is not a desirable characteristic, as it is in the case of unreinforced masonry. The use of lightweight concrete masonry units and lightweight concrete floor slabs helps greatly in this regard. This generally results in no wall cost premium, and certainly reduces foundation sizes.

Shear wall rigidities

In Chapter 10, a procedure for computing wall rigidities was outlined which, if applied to high-rise shear walls, could actually be used at each story. This concept does assume that the walls are fixed at each floor. An approach such as this seems reasonable when some of the following conditions are approached: (1) the shear-wall arrangement is relatively uniform, (2) shear-wall widths are relatively wide and fairly constant from ground to roof, (3) the wall arrangement is reasonably symmetrical, (4) the floors and spandrels are stiff relative to the walls, thereby providing sufficient restraint at each floor level, and (5) the openings are small enough so that they have a negligible effect upon the shear distribution. On the other hand, where the shear walls, story by story, are unequal in length or where the floor provides little rotational restraint to the walls, the relative wall rigidities may have to be considered at each level individually, since this factor can vary considerably at different floor levels. Alternatively, the relative rigidity of the full shear-wall height (ground to roof) may be used, by calculating its value at the roof level only. This presupposes an infinitely stiff diaphragm, an assumption that is often questionable.

In high-rise buildings, horizontal shear distribution is further complicated by the relation between wall and diaphragm deflection. Near the ground the bearing shear walls are stiffer relative to the diaphragm, whereas toward the roof level, the diaphragm is generally the stiffer element. As a result, the diaphragm may be considered as rigid at the roof, progressing to a more flexible element in the lower stories in comparison with the supported shear walls. Consequently, the stiffer walls accept a greater portion of the horizontal shear load in the upper stories, this being pro-rated approximately on the basis of relative wall rigidities. In the lower stories, all walls share the load more or less on the basis of contributing tributary areas. Displacement of walls at the top will be uniform, but nonuniform in the lower stories. The result of all of this is some loss in building stiffness and a call upon the floor diaphragm to shift the load from one shear wall to another. It is therefore imperative in high-rise masonry bearing-wall construction to provide an adequate and continuous diaphragm. In all cases, however, one should exercise careful judgment when distributing the shears on the basis of relative wall rigidities only.

Should one make a determination that it becomes reasonable to assess the relative stiffness of the full height shear wall by measuring the deflection at the roof under an inverted triangular load, the following calculations could be made:

1. For solid shear wall:

Δ total $=$ Δ moment $+$ Δ shear

Δ (for triangular load) at top of wall

$$= \frac{V}{bE_m}\left[2.20\left(\frac{H}{D}\right)^3 + 2\left(\frac{H}{D}\right)\right] \doteq \frac{102.15V}{bE_m},$$

where $\dfrac{H}{D} = 3.51$

Δ (for uniform load) at top of wall

$$= \frac{V}{bE_m}\left[1.5\left(\frac{H}{D}\right)^3 + 1.5\left(\frac{H}{D}\right)\right]$$

$$= \frac{70.2V}{bE_m} \text{ or } 68.5\% \text{ of triangular seismic loading}$$

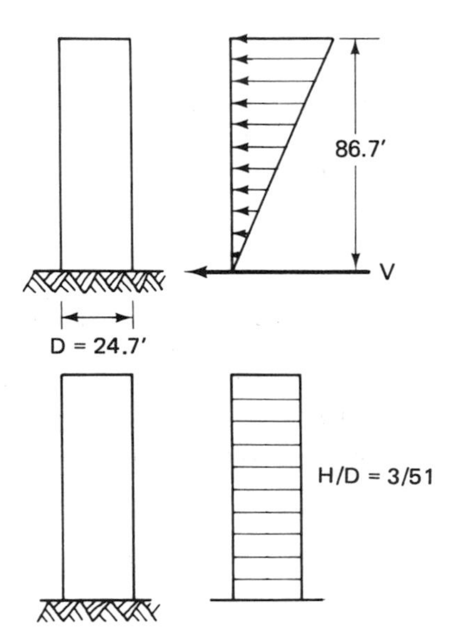

2. For coupled shear wall—uniform loading: Refer to discussion per Coull and Choudbury in Chapter 10. This analysis is based upon a uniform load distribution. Walls are not considered coupled by corridor slab. I_2 and I_1 are moment of inertia of shear walls, where $t = 7.6$ in.

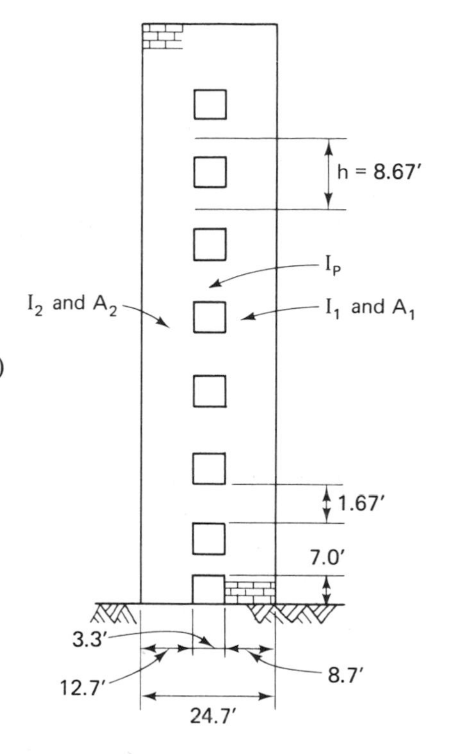

$$I_2 = \frac{7.6}{12} \times \frac{12.7^3}{12} = 108.4 \text{ ft}^4;$$

$$A_2 = 0.635 \times 12.7 = 8.07 \text{ ft}^2$$

$$I_1 = 0.635 \times \frac{8.7^3}{12} = 34.8 \text{ ft}^4;$$

$$A_1 = 0.635 \times 8.7 = 5.53 \text{ ft}^2$$

$$I = I_1 + I_2 = 143.0 \text{ ft}^4 \text{ and}$$

$$A = A_1 + A_2 = 13.6 \text{ ft}^2$$

$$\mu = 1 + \frac{Al}{A_1 A_2 l^2} \quad \text{(where } l = \text{distance between center of gravity of wall sections} = 14.0 \text{ ft)}$$

$$= 1 + \frac{13.6 \times 143.0}{(8.07)(5.33)(14)^2} = 1.23$$

$$\alpha = \left(\frac{12 I_p l^2 \mu}{h b^3 I}\right)^{1/2} = 0.124$$

where $I_p = \dfrac{0.635 \times 1.67^3}{12} = 0.246 \text{ ft}^4$

$b = 3.3$ ft between walls

$h = 8.67$ ft story height

$$\alpha H = 0.124 \times 86.7 = 10.75$$

K_4 from Figure 10-22 $= 0.20$

$$y \text{ at top} = \frac{1}{8}\frac{wH^4}{EI}K_4 = \frac{1}{8}\frac{VH^3}{E_m(bI/b)} \times K_4$$ this is only deflection due to moment

$$\text{Total } y = \frac{73.0V}{bE_m} + \text{shear } \Delta$$

$$= \frac{73V}{bE_m} + \frac{1.5 \times 3.51V}{bE_m} \times \frac{24.7}{21.7} = \frac{79.0V}{bE_m}$$

3. Coupled Shear Wall-Triangular Loading:

$$\Delta = \frac{79.0V}{bE_m} \times \underbrace{\frac{1}{0.685}}_{\text{assuming same relation between loading types exists}} = \frac{115.3V}{bE_m} \quad \text{or} \quad R = 0.0087$$

4. Solid Shear Wall-Triangular Loading:

$$\Delta = \frac{102.15V}{bE_m} \qquad \therefore R = 0.0098$$

Thus, coupled shear wall rigidity as evaluated on this basis, becomes about 89% of that for a solid wall having the same dimensions.

It has already been observed that the inertia effects of a vibrating building, as specified in the UBC, produce an equivalent inverted triangular load condition. Under this loading, the deflection of the top of a rectangular wall section loaded in its plane, in terms of the height/length ratio H/D, considering both bending and shear deflections, was seen in the previous example to be

$$\Delta = \frac{V}{bE_m}\left[2.20\left(\frac{H}{D}\right)^3 + 2.0\left(\frac{H}{D}\right)\right] \qquad \begin{aligned} 2.20\left(\frac{H}{D}\right)^3 &= \text{bending contribution} \\ 2.0\left(\frac{H}{D}\right) &= \text{shear contribution} \end{aligned}$$

The bending deflection portion should be modified to reflect the increased moment of inertia of an *H*- or *L*-shaped wall plan compared with that of a simple rectangular section. Figures 13-5 and 13-6, developed by Kesler, provide a means of converting the depth of an *H* or *L* section to an equivalent rectangular section having the same moment of inertia (uncracked) as that of the actual wall shape.

When flange effects are to be considered, the limit of flange width must be established, one limit being the permissible shear stress between flange and web. For this shear to be less than the average web shear, the following expression must be satisfied:

$$\frac{VQ}{It} \leq \frac{V}{dt} \qquad \text{or} \qquad \frac{Qd}{I} \leq 1$$

This relationship, represented in Figure 13-7, provides shear limits for a specified depth/width ratio. Shear stresses between flange and web will seldom be critical for I-shaped sections, but will be a limiting factor for L-shaped sections, where the flange width exceeds 0.4 times the depth. On the other hand, one rule of thumb, which has been used effectively, suggests that an effective flange width of six times the wall thickness on each side of the wall be used, where flange effects are to be recognized in the calculations. Since shear flow in the flange will also limit the effective flange width, another recommendation calls for the effective flange width to not exceed $H/2$, where H is the height of the cantilever wall. Actually, the effective flange width cannot exceed the center-to-center spacing between walls forming the webs. Regardless of method adopted, some realistic appraisal of effective flange width must be made. If the width used is too low, flexural and shear stresses could be higher than anticipated; if assumed too high, the wall may be overdesigned, to the detriment of other walls. The effectiveness of the flange will depend upon the adequacy of the shear connection between

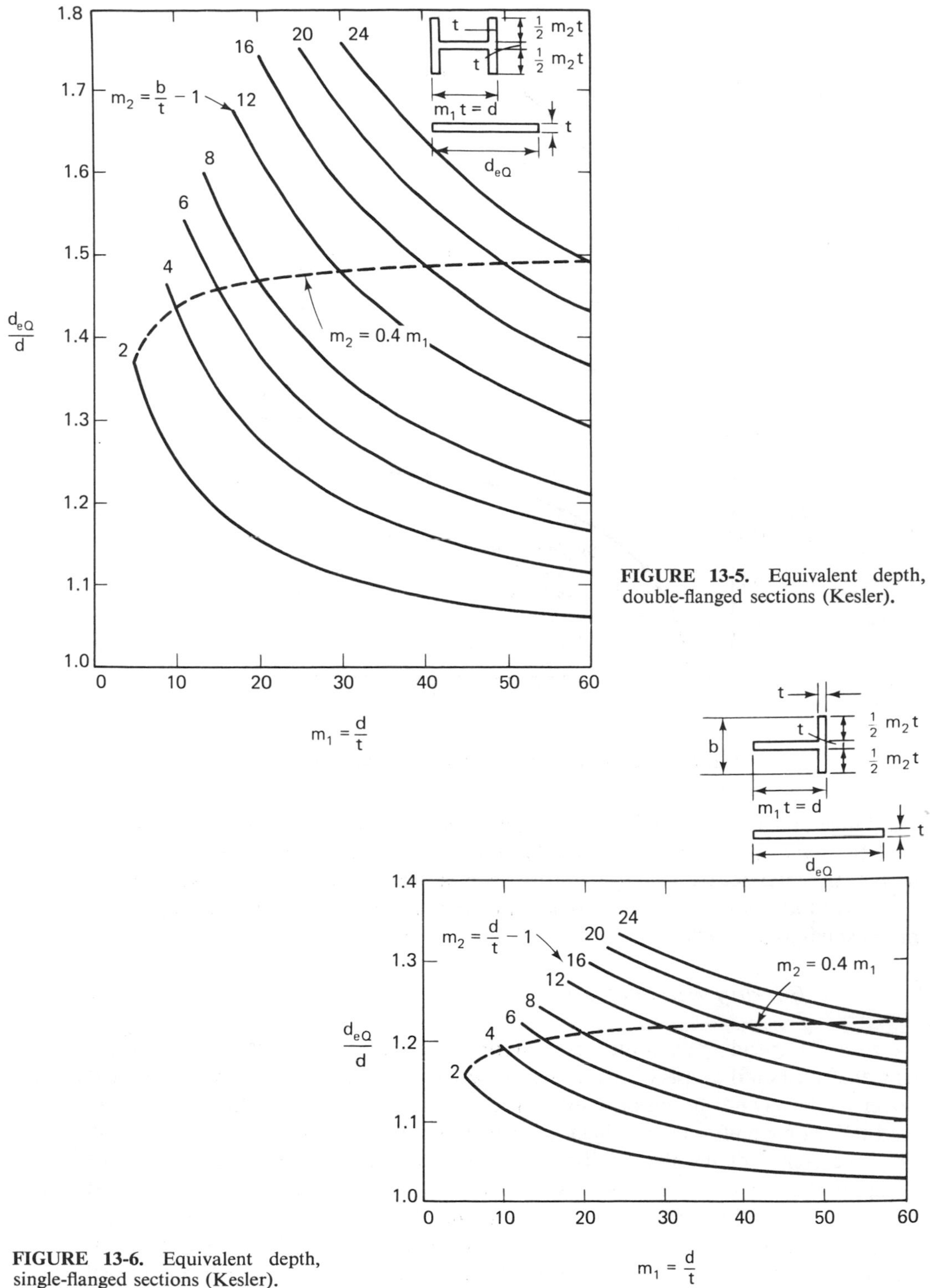

FIGURE 13-5. Equivalent depth, double-flanged sections (Kesler).

FIGURE 13-6. Equivalent depth, single-flanged sections (Kesler).

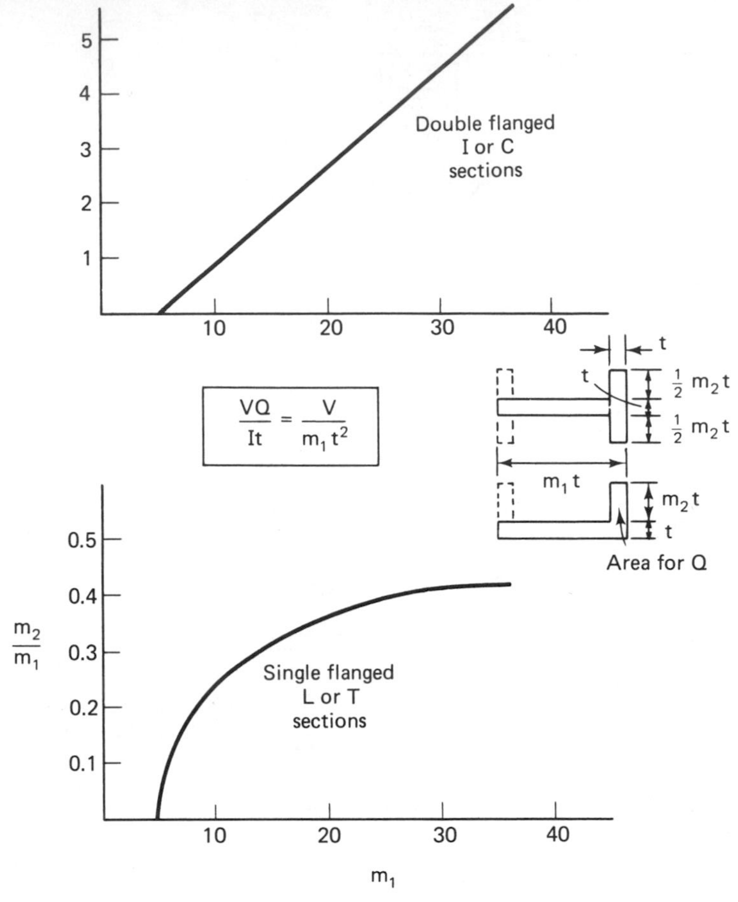

FIGURE 13-7. Shear limits for depth/width ratio (Kesler).

the flange and the web. Shrinkage, temperature change, and incomplete mortar in the head joint will all tend to weaken this joint. Where the flange is considered in the design, the intersection of flange and web can be made adequate for shear by using one of the following: (1) full head joints with joint reinforcing, (2) interlocking blocks, or (3) solid-grout interlocks.

Bearing wall analysis

Depending upon the type of floor system selected, vertical wall loads may be transferred on the basis of either one- or two-way slab action. Loading at different upper levels may not be uniform, but at the critical lower levels the distribution does tend to become uniform where high interconnected walls are present. Also, the probable occurrence of unbalanced live or seismic loads should be recognized in the design of the interior walls as well as in the floor system itself. In the lower stories at least, interior wall stresses may be evaluated on the basis of an uncracked section, since the axial compressive stresses will exceed the flexural tensile stresses by a considerable margin. Thus, one would expect to find that interior wall design is governed more by the axial live and dead loads and the h/t ratio rather than by any flexural requirements. In exterior walls, however, the ratio of moment to axial load is considerably greater than for the interior walls, so a cracked section analysis seems called for over a considerable portion of the upper height of the building. Finally, where vertical loads are eccentrically applied, the induced moment along the height of the wall must be added to that produced by the lateral-load effects, when analyzing the adequacy of this element.

The seismic effect on openings must also be carefully considered. In addition, the possibility of severe cracking occurring in stiff lintels and piers should be recognized and provided for in the analysis. Also, where cracking does occur, the altered properties of the members must be taken into consideration when analyzing seismic response, in order to provide for the transfer of flexural, shear, and axial compressive stresses of the members.

When considering the piers within each wall, one needs to consider the moments both parallel and perpendicular to it caused by the lateral loads. The combined effect of these and the vertical axial loads may increase the f'_m that will be needed in the lower stories. Then a decision has to be made as to whether (1) to specify a higher-strength masonry, (f'_m), to preclude the need to increase the wall thickness; (2) to increase the wall thickness in this region; or (3) to increase the number of grouted cells in hollow unit masonry. An example of the latter might call for grouting at 4 ft cc (maximum spacing) in the upper stories, a somewhat reduced grout spacing at the intermediate levels (say, 32 in. cc), and solid grouted walls in the lower stories, thus accommodating the structural requirements throughout, which attests to the versatility of reinforced masonry design.

Reinforcing details

In those high-rise masonry buildings where the shear walls are subjected to moments and shears in the plane of the wall which are exceedingly high, it becomes somewhat difficult to provide sufficient amounts of reinforcing needed to develop the high flexural tensile stresses. There simply is not enough space in the wall. To overcome some of the difficulty, the following techniques could be considered:

1. Specify a higher-strength steel, say $f_y = 60{,}000$ lb/in.^{2}, to reduce the quantity of steel required, and thereby lessen the congestion within the wall.
2. Use compressive reinforcing in concrete block masonry as an alternative to higher-strength block and grout. Conceivably, the compression steel should be limited to wall corners and intersections, where it can be easily tied, or to relatively thick walls, where close ties can be utilized.
3. Provide horizontal reinforcing in excess of the Code minimum where shear stresses are high. Tests conducted at Cal Poly, Pomona (described in Chapter 17) demonstrate the effectiveness of this shear reinforcing, not only for providing additional strength, but also for improving the ductile performance of the wall subsequent to cracking. It is absolutely essential, however, that the shear steel be properly anchored.

Due consideration should also be given to the placement of the reinforcing if problems arising on the job are to be avoided. This is particularly true of the vertical jamb reinforcing, both as to its location and splicing. Some typical details appear in Figure 13-8. In concrete masonry, placement of vertical reinforcing is affected by the size of the block and whether the units have open or closed ends, as well as by the block module. For the usual 8 × 8 × 16 in. block, the amount of reinforcing that can be placed in one cell is limited. There must be enough room to insert a vibrator

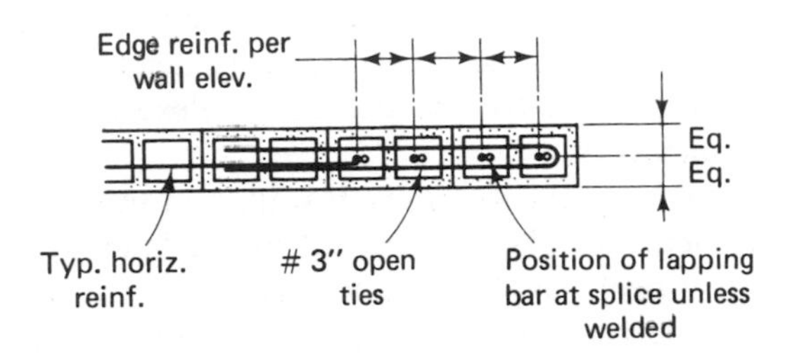

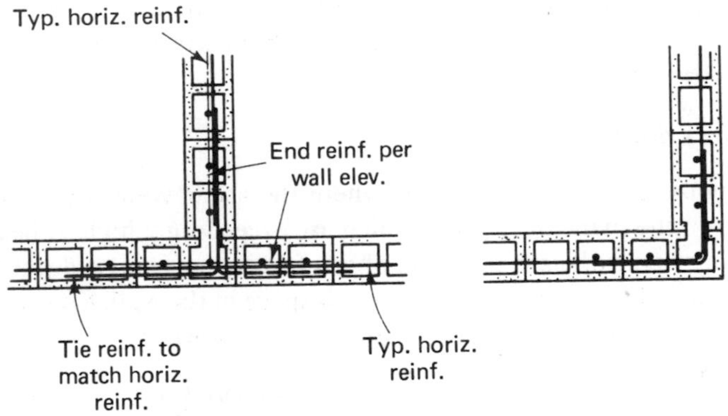

FIGURE 13-8. Typical wall reinforcing details.

during grouting. The presence of lap splices must also be taken into account. These can be accommodated if only one bar is used per cell for No. 8 bars and larger, with the maximum practical bar size being a No. 11. Where No. 10 or No. 11 bars are used, welded splices are advisable, because with these larger bars the length of lap required is quite long, and misalignment may prevent insertion of the vibrator to consolidate the grout. The length of lap for tensile splices, as recommended by ACI Committee 531, where all bars are spliced in the same location, is shown in Table 13-1.

TABLE 13-1

Tensile Lap Splice (in.) in Concrete Masonry
(per ACI Committee 531)

$L = f_s A_s / \sum_0 \mu \times 4/3 \times 1.2^a = 0.4 f_s D_b / \mu$
$\mu = 160$ lb/in.^{2}

	$f_s = 20$ kips/in.^{2} $L = 50D_b$	$f_s = 24$ kips/in.^{2} $L = 60D_b$
No. 3	19	23
No. 4	25	30
No. 5	31	38
No. 6	38	45
No. 7	44	53
No. 8	50	60
No. 9	56	68
No. 10[b]	63	75
No. 11[b]	69	83

[a]The 1.2 factor is included as satisfying the need for special precaution where all bars are spliced at the same location.

[b]For No. 10 and 11 bars, the length of lap required is so long that the cost difference for welding the splice is small and justifies welding for these sizes. Staggering of welds is desirable.

FLOOR SYSTEMS

Types

Floor systems have a multiple function in that they not only support the vertical loads but they also serve as the horizontal diaphragms which distribute the lateral forces to the shear walls. There are several different types of floor systems in common use; these include, but are not limited to, (1) cast-in-place slabs, (2) precast prestressed Tees, (3) prestressed planks, (4) steel bar joists supporting metal decking and concrete topping slabs, (5) lift-on precast slabs, and (6) concrete block or tile filler slabs. These will be described in further detail in the next paragraphs.

Cast-in-place slabs

The simple cast-in-place slab floor system is sometimes used where there are irregular floor plans or odd spans that would make it difficult to form repetitive precast units. Building sites that are steep or difficult to reach also make this type of floor decking desirable. It provides a good structural tie without requiring any special type of connection. This can be seen in Figure 13-9.

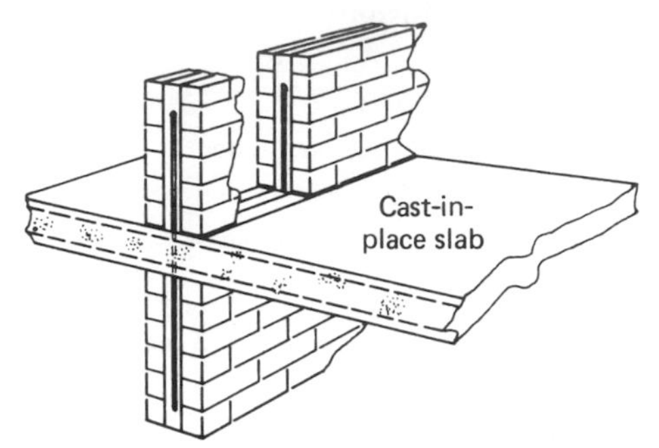

FIGURE 13-9. Cast-in-place slab.

Prestressed elements

Precast prestressed Tees such as those shown in Figure 13-10 represent a different type of system. They provide economical long spans giving unobstructed space below that can be readily subdivided with movable partitions. They also provide adequate space for utilities, which can be located between the stems of the Tees. A topping slab provides both a finish floor surface and a tie, so that the system functions as a diaphragm. It is essential that this poured-in-place topping slab be thick enough so that the reinforcement within it can achieve adequate bond to the concrete, and also that conduit may be placed as required. These are desirable for use in hotels or apartment house complexes. They were used, for instance, in the 7-story, 9-story, and 17-story Mayfair complex built in Denver, Colorado, several years ago.

Prestressed planks, such as Spancrete or Flexcore, and others constitute another very economical type of floor system, particularly for longer spans. They can be used, for instance, on sites that have such a limited working area such that on-site casting is not feasible. They are also adaptable to longer spans, and these prestressed units possess the added feature that they exhibit considerably less deflection than do nonprestressed slabs. As with the precast Tees, a topping slab is poured on top to provide the finish floor surface (Figure 13-11). There are many proprietary units made in this particular configuration. Some of these can be utilized as a structural diaphragm with or without the use of a concrete topping.

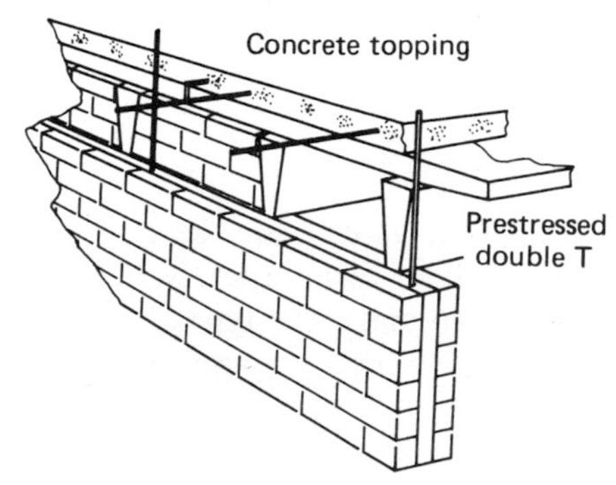

FIGURE 13-10. Prestressed Tees.

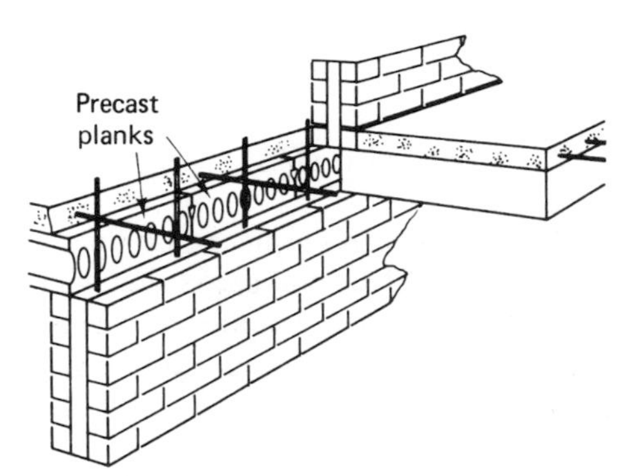

FIGURE 13-11. Prestressed planks.

Steel bar joists

Another system employs steel bar joists which support a concrete floor slab or a metal deck with a concrete topping surface (Figure 13-12). This system is especially useful where it is desirable to provide a plenum between the ceiling and floor deck above for heating and air-conditioning ducts, besides other utilities, such as plumbing and electrical conduits.

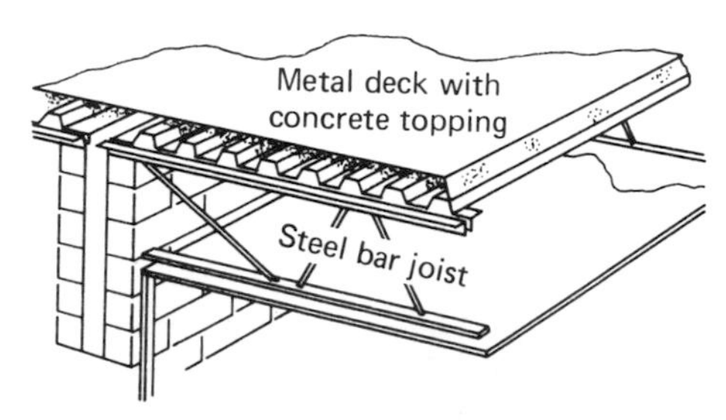

FIGURE 13-12. Bar joist and metal deck.

Lift-on slab

A lift-on slab, usually of full room size, has proven to be economical and efficient on sites where there is access for both casting and crane handling. They do not require a topping or finish since the floor surface and ceiling finish are an integral part of the monolith (Figure 13-13).

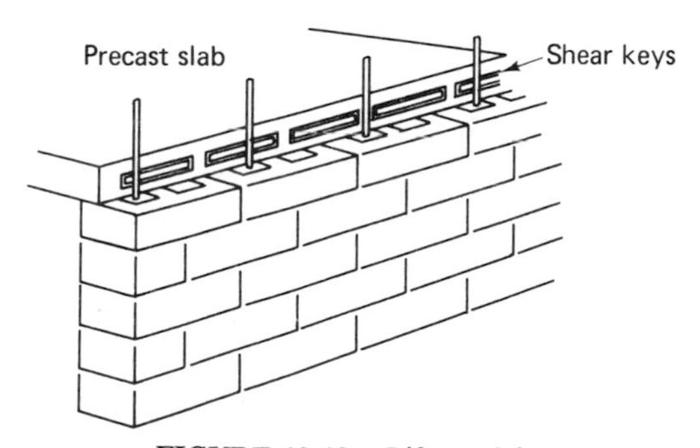

FIGURE 13-13. Lift-on slab.

Block and beam

The block-and-beam system, used extensively in the Pacific Northwest, shown in Figure 13-14, utilizes steel bar joists spaced on 2 ft 4 in. cc spanning between bearing walls. Hollow concrete filler blocks placed against each other are then set on the flange of the tension chord of the beam. A topping slab is poured on top of the filler block to form the finished wearing surface of the T section and the floor diaphragm. This system has an advantage that the block cores provide room for mechan-

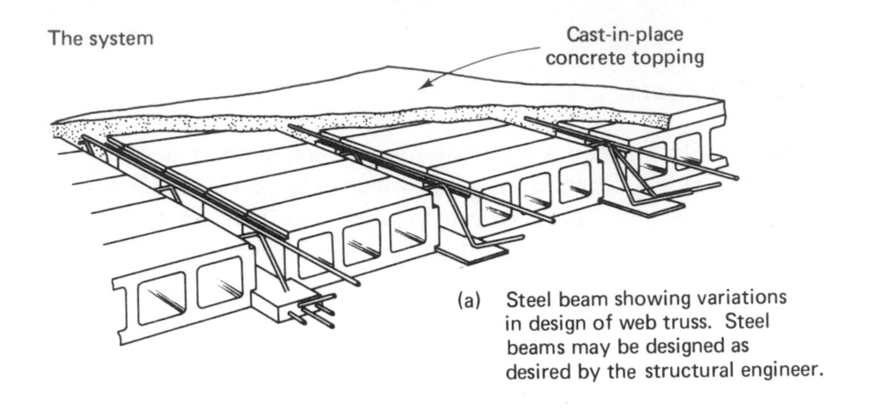

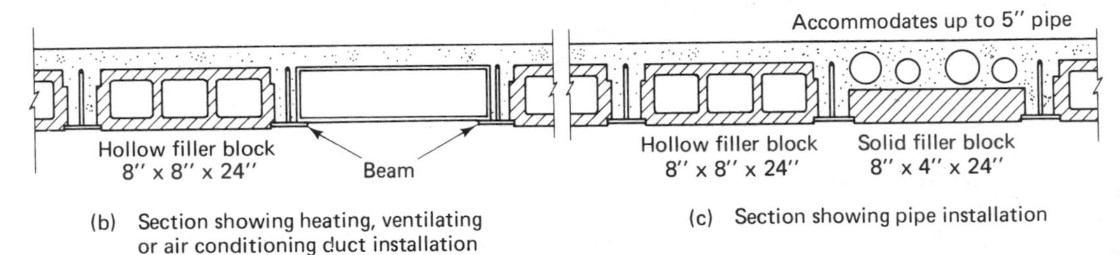

FIGURE 13-14. Block-and-beam system.

ical pipes and electrical conduits as well as a place to locate sheet-metal ducts for heating, ventilating, or air conditioning. Furthermore, this type of construction reduces sound transmission and is quite fireproof.

CONSTRUCTION TECHNIQUES

Setting the floor slabs

As with any relatively new or specialized construction technique, numerous problems are encountered initially and procedures have to be developed which can overcome these in a practical and efficient way. This has been true in the case of high-rise masonry. Often, different methods are adaptable to the same situation, and in many cases it is a matter of personal preference as to which one should be used. In other cases, the type of construction equipment available or job location might influence a decision. At any rate, the contractor should be allowed as much leeway as possible in deciding what particular method should be used.

Several techniques have been developed for placing the reinforcing as well as grouting. One sequence for concrete masonry involves laying walls to story height and placing the slabs on the ungrouted masonry. Then the vertical reinforcing is dropped into the block cells, making certain that it projects above the finish floor slab so as to serve as dowels for the next story walls. This is a very efficient method, but it is susceptible to the fact that the unreinforced wall possesses relatively little stability and therefore can be easily knocked over when the floor slabs were being placed onto it. A safer method involves building the masonry walls to full height with the reinforcement located in the walls as the masonry units (brick or block) are laid up. Grouting is then accomplished in full height lifts. This method requires that the dowels projecting above the slab be rather precisely centered in the wall so that when the slabs are lowered into position, the projecting dowels did not interfere with them.

In another procedure, the grouting is stopped a short distance below the floor line, dowels are placed in the grout space, and this space is solid-grouted after the slabs have been positioned. Thus, the slabs can be placed without interference from prepositioned dowels. A typical section may be seen in Figure 13-15. This method also possesses the advantage that it is easier to place the bond beam steel at the top of the wall. The steel extending along the wall at the floor slab level serves to tie all the slabs together. Another method of tying the slabs together lies through the use of weld plates embedded in the slabs so that the weld plates can be used as splices to adjacent panels, tying all together directly. Also, as seen in Figure 13-16, slab edges usually have shear keys. Grout is inserted in each joint, typically 4 in. wide and 6 in. deep, and vibrated. An expansive admixture is frequently used to prevent shrinkage and improve the bond at joints.

To bring the floor panels to a proper elevation, they must be shimmed so that they are level. This can be accomplished in several ways. One method is to provide a continuous mortar bed around the walls to receive the slab, and provide support posts, cut to the proper length. The slabs are lowered onto them squeezing the mortar out. This procedure provides for a continuous uniform bearing. It also seems to be an easy way to provide a watertight joint in lieu of filling this joint later by tuck pointing. Other schemes involve the use of steel or Neoprene shims of varying thicknesses, located at certain points along the wall to receive the slab at the correct floor elevation, after which grouting is accomplished. This makes for a rather irregular joint, however, requiring tuck pointing afterward. Also, if some of the grout runs down the

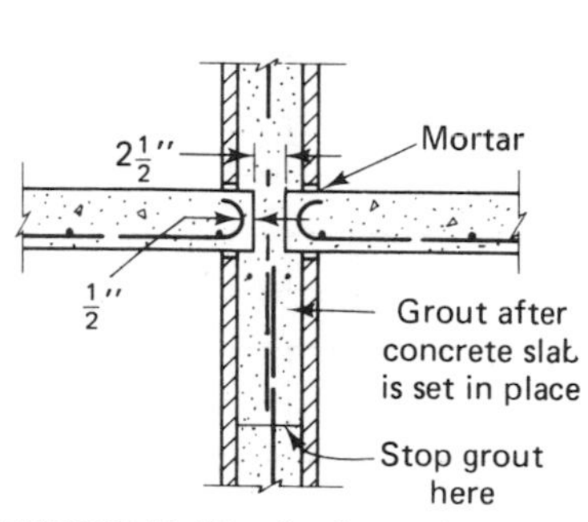

FIGURE 13-15. Section at interior wall.

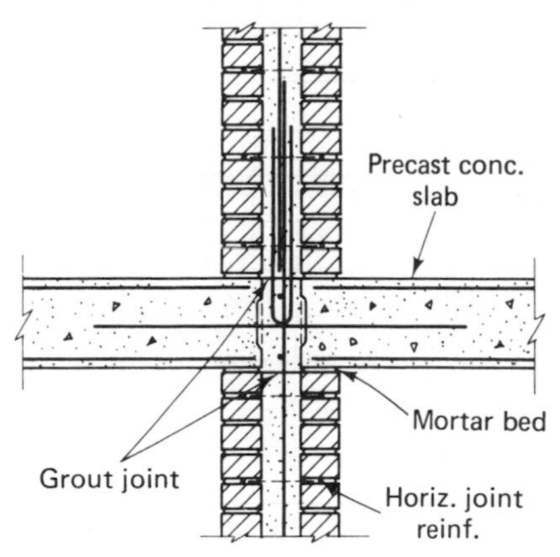

FIGURE 13-16. Shear keys at slab edges.

face of the wall, this does tend to cause problems as far as the finish wall surface is concerned. Where precast planks are used, the use of a small bed or coating of neat cement paste to fill the joint, but thin enough so that it does not change the elevation, seems to achieve adequate provision for bearing and shear along the line of support.

Scheduling

Perhaps the key to the degree of efficiency possible with this system of construction is the ease of scheduling. It is absolutely necessary for general contractors to coordinate their work very closely with all the specialty contractors on a job. A typical bearing-wall job schedule is shown in Figure 13-17.

The LeBaron Hotel in Northern California provides a good example of how quickly a larger reinforced masonry structure can be completed with the proper scheduling. This is a 10-story 350-room hotel which was complete in just 6 months and 4 days. After the foundation was poured, the mason contractor erected the masonry walls to one-story height on about one-half of the building plan. Then the contractor transferred his crew to the other half of the building to raise those walls, while the precast floor slabs were set by crane on the previously constructed walls. The bricklayers could then erect the second-story walls on the deck provided by the

Typical bearing wall job schedule

Item	1st Month				2nd Month				3rd Month				4th Month			
Excavate footings																
Form and cast footings																
Cast ground slab																
Masonry walls					1	2	3	4	5	6	7	8				
Slabs:																
Casting slab																
Form, cast and cure																
Erect and grout					*	1	2	3	4	5	6	7	8	R		
Woodwork and trim																
Plumbing, rough																
Plumbing, finish																
Electrical, rough																
Electrical, finish																
Roofing																
Paint, exterior																
Paint, interior																

*High early strength cement can be used on last slabs cast to allow for earlier erection.

FIGURE 13-17. Typical bearing-wall job schedule.

first-story floor slabs. By using this seesaw method, the bricklayers were able to work steadily and consistently, moving up floor by floor without any work stoppage. Then as the building progressed in height, other trades, such as plumbers, carpenters, electricians, plasterers, painters, and so forth, were able to do their work because no exterior shoring was needed, which always causes slowdowns and disruptions among the crafts. So it can be seen that this floor-to-floor repetition leads to a great deal of efficiency. It was almost like building 10 one-story buildings, one on top of the other. The crane, which is present on these types of jobs for the purpose of setting the floor slabs (see the Catamaran Hotel project, Figure 13-18), is also available for setting lumber, plumbing fixtures, and other items for the work of other crafts.

FIGURE 13-18. Catamaran Hotel, under construction, San Diego. (Courtesy Masonry Institute of America.)

MODERN HIGH RISE MASONRY BUILDINGS

Although the content of this textbook is primarily structural in nature, it might be appropriate at this point to show some of the multivaried types of architectural expressions that can be achieved through the use of reinforced masonry. The combination of shapes, sizes, and textures of masonry units enables the architect to achieve very dramatic or contrasting effects in walls. Reinforced masonry has been used to construct apartments, dormitories, hotels, restaurants, medical buildings, office buildings, and many other different types of high-rise structures. Any layout that involves the use of a repetitive layout, story to story, for the internal room arrangement is ideally suited for high-rise masonry construction. Some interesting and varied examples of these are shown in Figures 13-19 through 13-23. These include the Wilshire Plaza Office Building in Los Angeles, California; the Pasadena Hilton Hotel in Pasadena, California; the Marina Condominiums, Marina del Rey, California; masonry bearing-wall construction in Switzerland; and the Bahia Hotel, San Diego, California. Thus, it can be seen that modern masonry has reached new heights, and there is certainly no shortage of elegant masonry structures that fulfill many different types of functions and provide varied architectural effects.

FIGURE 13-19. Wilshire Plaza (office building), Los Angeles. (Courtesy Julius Shulman.)

FIGURE 13-20. Pasadena Hilton Hotel, Pasadena, California. This concrete masonry load-bearing-wall structure with prestressed hollow core floors is 13 stories high with 264 rooms covering 150,000 ft^2. It also includes 12 banquet and meeting rooms and underground parking. A floor, including all masonry walls, was completed every seven working days. Architect: Curtiss and Rasmussen, AIA, Akron, Ohio; Structural Engineer: Hale and Kullgren Inc., Akron, Ohio; Resident Structural Engineer: Socoloske, Zelner and Assoc., Los Angeles, Calif.; General Contractor: W. J. Moran Co., Alhambra, Calif. (Courtesy W. L. Dickey.)

FIGURE 13-21. Marina Condominiums, Marina Del Rey. (Courtesy W. L. Dickey.)

FIGURE 13-22. Masonry bearing-wall construction in Switzerland. The thrifty and economy-minded Swiss recognized some 10 to 15 years ago the benefits of this bearing-wall construction concept, and have had an increasing program of such bearing-wall structures, approaching the 20-story limit which Mr. Lechner of Switzerland has estimated as an economic limit or maximum (the structural maximum is, of course, higher). Many of these bearing walls shown in the buildings illustrated, consist of 6-in. or $4\frac{3}{4}$-in. thick interior bearing walls with exterior bearing walls of 7 in. or 5 in. thicknesses, with insulation provided. (Gutschick, the Postrasse and Zentrum and Mergelacker.) Although not designed for UBC Zone 3 earthquake requirements, they are designed for quite high wind loads that approach Zone 3 requirements. This program demonstrates the effectiveness of thin bearing walls when braced laterally by intersecting walls.

FIGURE 13-23. Bahia Hotel, San Diego. A typical competitive mid-rise bearing-wall structure in Southern California.

DESIGN OF A 17-STORY BRICK APARTMENT BUILDING

The following design example illustrates the steps involved in the application of UBC requirements to the design of a multistory masonry bearing-wall structure. It is intended to show and clarify the type of design calculations needed for such an analysis and to interpret certain of the UBC seismic requirements that may not be readily understood without some explanation. Instead of being an imaginary problem, it is in the plan, layout, and elevation of a specific building, which was the first of a series of brick load-bearing high-rise buildings in Denver that included several 23-story structures (see Figure 13-24).

The calculation steps are accompanied by clarifying "Notes to The Designer" and appropriate UBC references. Except for the clarifications, these are presented in the manner in which they might appear in a structural design office.

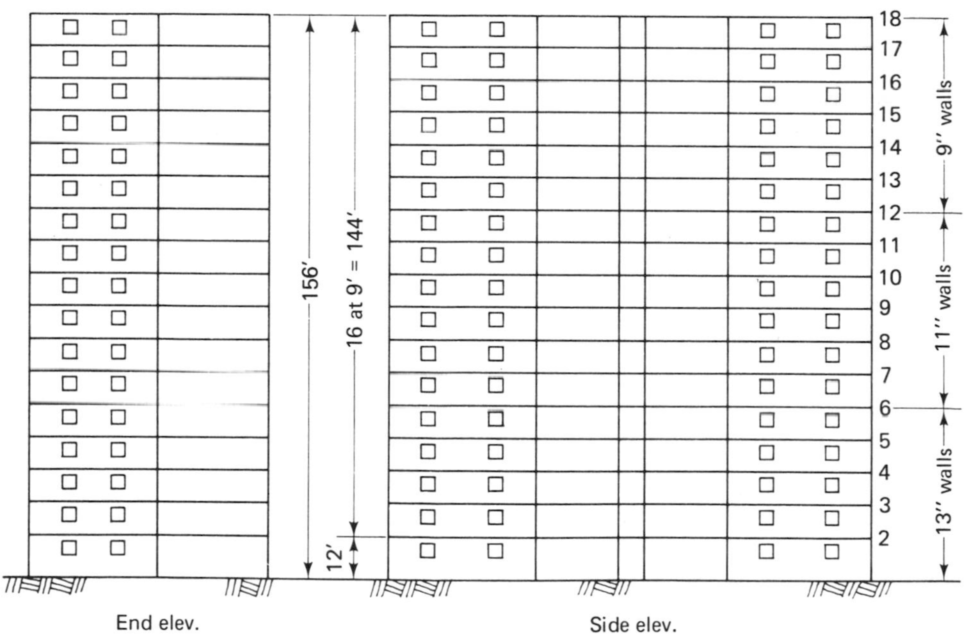

FIGURE 13-24. Seventeen-story brick apartment building.

DESIGN CALCULATIONS

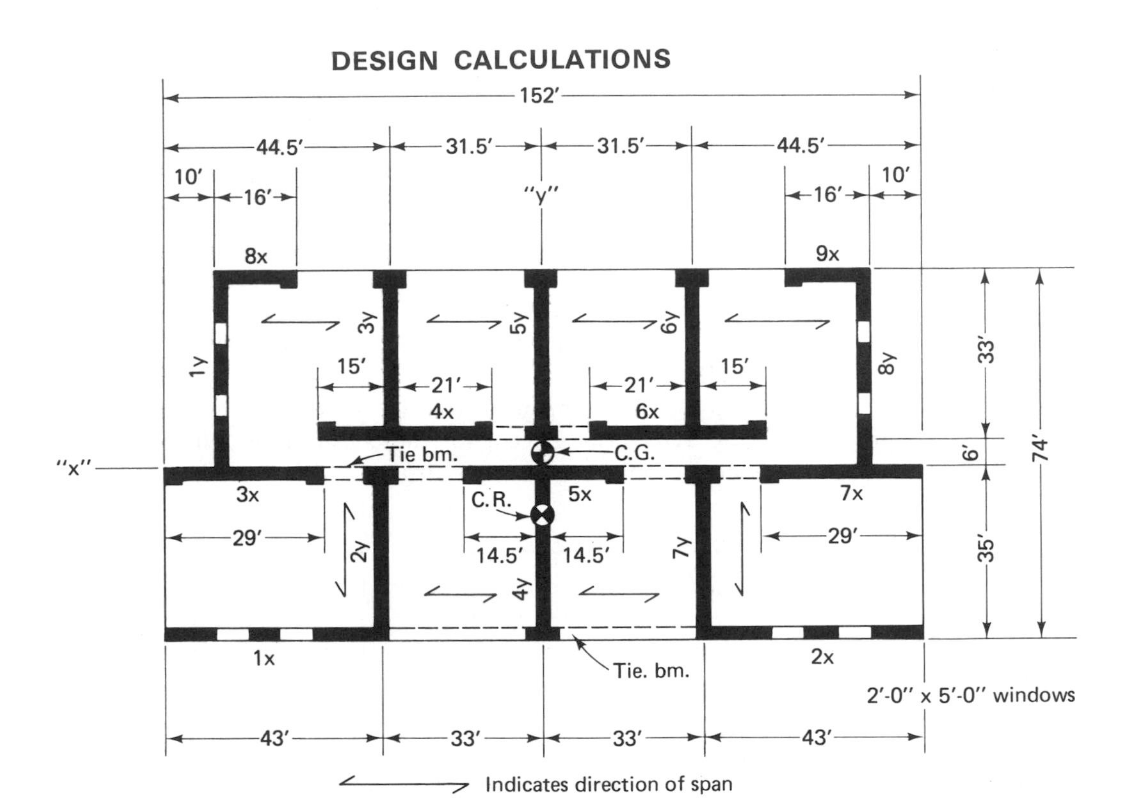

Typical shear wall plan

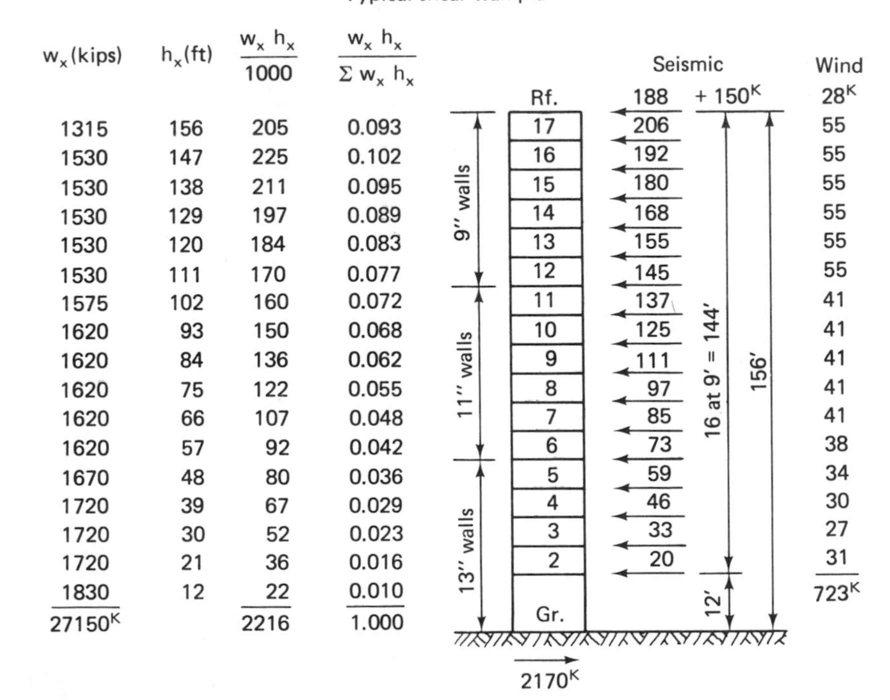

w_x(kips)	h_x(ft)	$\frac{w_x h_x}{1000}$	$\frac{w_x h_x}{\Sigma w_x h_x}$
1315	156	205	0.093
1530	147	225	0.102
1530	138	211	0.095
1530	129	197	0.089
1530	120	184	0.083
1530	111	170	0.077
1575	102	160	0.072
1620	93	150	0.068
1620	84	136	0.062
1620	75	122	0.055
1620	66	107	0.048
1620	57	92	0.042
1670	48	80	0.036
1720	39	67	0.029
1720	30	52	0.023
1720	21	36	0.016
1830	12	22	0.010
27150^K		2216	1.000

NOTES TO DESIGNER

		UBC REFERENCE
Design Code	1976 Uniform Building Code	
Masonry	6000-lb/in.^{2} brick units—ASTM A62	
	$f'_m = 2600$ lb/in.^{2}	Sec. 2404(c) 3
Vertical Loads	Dead Loads	Sec. 2302–2305
	Floor panels + $2\frac{1}{2}$ in. topping = 85 lb/ft^2	
	Partitions (nonbearing) = 15 lb/ft^2	
	Ceiling = 5 lb/ft^2	
	105 lb/ft^2	
	Live Loads	Table 23-A
	For apartments = 40 lb/ft^2	
	Reduction for walls and footings	
	$= 40\% \times 40 = 16$ lb/ft^2	Sec. 2306
Seismic Factors	Zone 3 $\longrightarrow Z = \frac{3}{4}$	Sec. 2312 and Figs. 1, 2, 3 UBC
	$I = 1.0$ for normal occupancy	Sec. 2312(k) and Table 23-K
	$K = 1.33$ for bearing wall system	Table 23-I
Wind Factors	25-lb/ft^2 basic	Sec. 2311 and Fig. 4 UBC

Load distribution to the floors: Sec. 2312(e) 1

$$V = F_t + \sum_{i=1}^{n} F_i \qquad \text{Formula (12-5)}$$

$$F_x = \frac{(V - F_t)w_x h_x}{\sum_{i=1}^{n} w_i h_i} \qquad \text{Formula (12-7)}$$

Note that the wind loads are less than one-half of the seismic forces and thus may be ignored. See notes on drift.

DESIGN CALCULATIONS

APPROXIMATE METHOD OF RELATIVE WALL RIGIDITIES

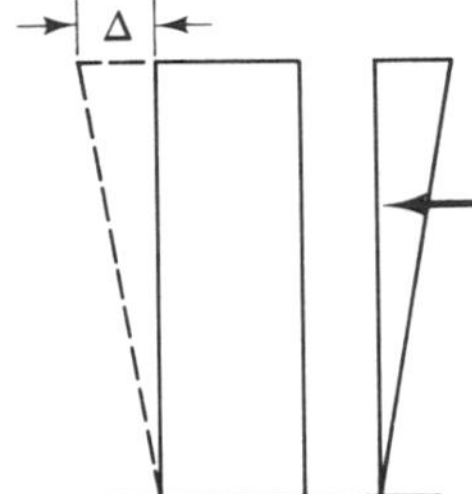

$$\text{Approx. } \Delta = \Delta_v + \Delta_m = \frac{3WH}{AE} + \frac{WH^3}{6EI}$$

Let

$$W = 500 \text{ kips}$$
$$H = 156 \text{ ft} = 1872 \text{ in.}$$
$$E_m = 2600 \text{ kips/in.}^2$$
$$b = 11\text{-in. brick wall (average)}$$

Wall	*Length* L (*in.*)	$A = bL$ (*in.*²)	$I = \frac{bL^3}{12}$ (*in.*⁴ × *10*⁸)	Δ_v + (*in.*)	Δ_m = (*in.*)	Δ (*in.*)	$R = \frac{1}{\Delta}$
1*x*, 2*x*	516	5676	1.259 × 10⁸	0.19	1.67	1.86	0.538
3*x*, 5*x*, 7*x*	348	3828	0.386	0.28	5.44	5.72	0.175
4*x*, 6*x*	432	4752	0.739	0.23	2.85	3.08	0.325
8*x*, 9*x*	192	2112	0.065	0.51	32.41	32.92	0.030
1*y*, 8*y*	477	5247	0.995	0.21	2.11	2.32	0.430
3*y*, 5*y*, 6*y*	396	4356	0.569	0.25	3.69	3.94	0.254
2*y*, 4*y*, 7*y*	420	4620	0.679	0.23	3.10	3.33	0.301

[a]Note that R in this example refers to the individual wall rigidity itself, symbolized by k in previous discussions.

$$\text{Typical floor area} = 74 \times 152 \text{ ft} - 2 \times 10 \times 39 \text{ ft} = 10{,}470 \text{ ft}^2$$

$$\begin{aligned}\text{Length of walls} &= 2(16 + 36 + 43) + 3 \times 29 = 277 \text{ ft} \\ \text{(neglect openings)} &\quad + 2 \times 39.75 + 3(33 + 35) = \underline{283} \\ &\qquad 560 \text{ ft}\end{aligned}$$

Floor dead loads:

$$\begin{aligned}\text{Floor deck} + 2\tfrac{1}{2}\text{ in. topping} &= 85 \text{ lb/ft}^2 \\ \text{Partitions (nonbearing)} &= 15 \\ \text{Ceiling} &= \underline{5} \\ &\quad 105 \text{ lb/ft}^2\end{aligned}$$

Tributary loads for lateral:

$$\begin{aligned}\text{Floor} &= 105 \text{ lb} \times 10{,}470 \text{ ft}^2 = 1100 \text{ kips} \\ \text{9-in. wall} &= 90 \text{ lb} \times 560 \text{ ft} \times 8.5 \text{ ft} = 430 \text{ kips} \\ \text{11-in. wall} &= 110 \times 560 \times 8.5 = 520 \text{ kips} \\ \text{13-in. wall} &= 130 \times 560 \times 8.5 = 620 \text{ kips}\end{aligned}$$

NOTES TO DESIGNER

UBC REFERENCE

Deflection of a fixed cantilever member with a triangular loading. Deflection includes both shear and flexure for determination of relative rigidities.

The value of W can be any arbitrary number since relative rigidities of walls in the structure are desired.

The window openings were ignored for ease of computation. It will be noted that the shear distortion is small when compared to the overall distortion. In a similar manner, little effect will be contributed to the moment of inertia.

The effect of large openings may have an adverse change on the above assumptions and should be evaluated for each project.

For rotational effect

From symmetry $\bar{x}$ is at center of building;
for $\bar{y}$:

$$\text{Center mass—floor} = \frac{1}{10{,}470}\left(-152 \times \frac{35^2}{2} + 132 \times \frac{39^2}{2}\right) = +0.70 \text{ ft.}$$

$$\text{walls} = \frac{1}{560 \text{ ft}}\left(-2 \times 43 \times 34 \text{ ft} + 2 \times 36 \times 7 \text{ ft} + 2 \times 16 \times 39 \text{ ft} - 3 \times \frac{34^2}{2} + 2 \times \frac{39^2}{2} + 3 \times 33 \times 23\right) = 1.59 \text{ ft}$$

$$\bar{y} = \frac{1100 \text{ kips} \times 0.70 \text{ ft} + 520 \text{ kips} \times 1.59 \text{ ft}}{1100 + 520} = 0.99 \text{ ft}$$

Center of rigidity from *x-x:*

$$\bar{y} = \frac{Rd}{\sum R} = \frac{-0.538 \times 2 \times 34 \text{ ft} + 0.325 \times 2 \times 7 \text{ ft} + 0.030 \times 2 \times 39 \text{ ft}}{2 \times 0.538 + 3 \times 0.175 + 2 \times 0.325 + 2 \times 0.030} = -12.85 \text{ ft}$$

$$\text{Eccentricity} = -12.85 - (+0.99) = -13.84 \text{ ft}$$

$$\text{Min. 5\% Code Ecc.} = 0.05 \times 152 \text{ ft} = 7.6 \text{ ft}$$

$$M_x = V \times 13.84 \text{ ft}$$

$$M_y = V \times 7.6 \text{ ft}$$

Load distribution to walls:

Wall	*R*	*d (ft)*	*Rd*	Rd^2	$\frac{R}{\sum R}$	+ $\frac{MRd}{\sum Rd^2}$	= *V*
1*x*	0.538	−21.15	−11.38	240	0.233	0	0.233
2*x*	0.538	−21.15	−11.38	240	0.233	0	0.233
3*x*	0.175	12.85	2.25	29	0.076	0.005	0.081
5*x*	0.175	12.85	2.25	29	0.076	0.005	0.081
7*x*	0.175	12.85	2.25	29	0.076	0.005	0.081
4*x*	0.325	19.85	6.45	128	0.141	0.015	0.156
6*x*	0.325	19.85	6.45	128	0.141	0.015	0.156
8*x*	0.030	51.85	1.56	81	0.013	0.004	0.017
9*x*	0.030	51.85	1.56	81	0.013	0.004	0.017
	2.311			985	1.002		
1*y*	0.430	−65.5	−28.17	1845	0.170	0.037	0.207
8*y*	0.430	+65.5	28.17	1845	0.170	0.037	0.207
3*y*	0.254	−31.5	−8.00	252	0.101	0.010	0.111
5*y*	0.254	0	0	0	0.101	0	0.101
6*y*	0.254	+31.5	8.00	252	0.101	0.010	0.111
2*y*	0.301	−33.5	−10.08	338	0.119	0.013	0.132
4*y*	0.301	0	0	0	0.119	0	0.119
7*y*	0.301	+33.5	10.08	338	0.119	0.013	0.132
	2.525			4870	1.000		
				5855			

NOTES TO DESIGNER

UBC REFERENCE

Horizontal torsional moments to determine center of mass. The center is determined by the summations of areas or wall lengths times the distances to centroids, all divided by the total areas or lengths.

Sec. 2312(e) 5

Load distribution to walls is in proportion to rigidities. Thus, total shear is that from direct shear plus that from the eccentricity.

Sec. 2312(e) 4

$$V = \frac{R}{\sum R} + M\frac{Rd}{\sum Rd^2}$$

Note that the values indicated are relative and are based on the characteristics of the structure only. Thus, this analysis may be performed before actual design loads are computed.

The rotational effects that are mathematically negative are not subtracted from the direct shears due to relative rigidities. The values of $MRd/\sum R^2$ shown as 0 are actually negative values which, per UBC, cannot be considered.

DESIGN CALCULATIONS

Seismic loads

$$\text{Code } T = \frac{0.05 \times 156 \text{ ft}}{\sqrt{74}} = 0.91\text{s—transverse}$$

$$T = \frac{0.05 \times 156 \text{ ft}}{\sqrt{152}} = 0.63\text{s—longitudinal}$$

For an approximation of the true period, use wall $2y$ and the Rayleigh formula using tributary mass and flexural deflection at the top.

$$W = 0.119 \times 27{,}150 = 3230 \text{ kips}$$

$$\Delta_{\text{Top}} = 3230 \times \frac{3.10 \text{ in.}}{500 \text{ kips}} = 20.0 \text{ in.}$$

$$T = 0.25\sqrt{20.0} = 1.12 \text{ sec}$$

Use: $T = 0.9 \times 1.12 = 1.01$ s.

$$T_s = 0.5 \text{ s (bedrock)}$$

$$S = 1.2 + 0.6\left(\frac{1.01}{0.5}\right) - 0.3\left(\frac{1.01}{0.5}\right)^2 \sim 1.20$$

$$C = \frac{1}{15\sqrt{1.01}} = 0.067$$

$$V = \tfrac{3}{4} \times 1.0 \times 1.33 \times 0.067 \times 1.20 \times 27{,}150 \text{ kips} = 2170 \text{ kips}$$

$$F_t = 0.07 \times 1.0 \times 2170 = \underline{150}$$

$$2020 \text{ kips}$$

NOTES TO DESIGNER

UBC REFERENCE

$$T = \frac{0.05\, h_n}{\sqrt{D}}$$

Formula (12-3A)

The period may be calculated by the Code formula

$$T = 2\pi \sqrt{\left(\sum_{i=1}^{n} w_i \delta_i^2\right) \Big/ g\left[\sum_{i=1}^{n-1} F_i \delta_i + (F_t + F_n)\delta_n\right]}$$

Formula (12-3)

or approximated by other methods such as $T = C\sqrt{\Delta}$, where Δ is the deflection from a lateral load equal to the tributary dead load to a structure and C is a constant varying between 0.25 for a member in bending to 0.32 for a single-mass oscillator. In this case the dead load is proportioned to the wall rigidity (neglecting accidental torsion) and the calculated deflection of the wall for an applied load of 500 kips. It should be noted that the period for seismic loads should be somewhat less than the calculated to assure a higher design load.

Note: The period determined is applicable in both directions, since the summation of the rigidities is nearly the same in each direction.

For bedrock T_s cannot be taken as less than 0.5 s.

Sec. 2312(d)

$$S = 1.2 + 0.6\frac{T}{T_s} - 0.3\left(\frac{T}{T_s}\right)^2$$

Formula (12-4A)

$$C = \frac{1}{15\sqrt{T}}$$

Formula (12-2)

$$V = ZIKCSW$$

Formula (12-1)

$$F_t = 0.07TV$$

Formula (12-6)

DESIGN CALCULATIONS

Wind loading = 25 lb/ft² basic

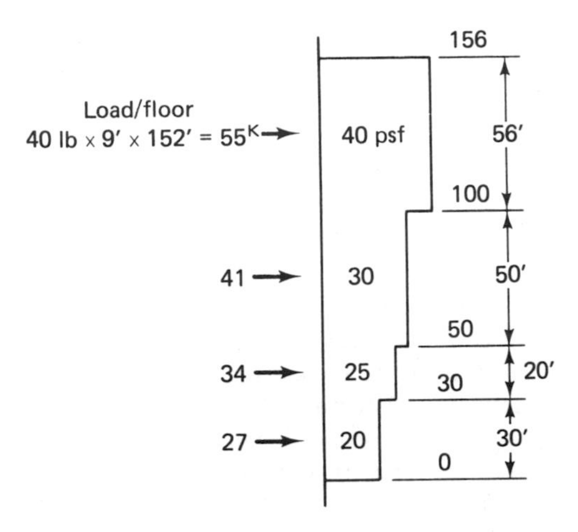

Check drift of building:

Seismic dynamic load analysis; top story critical because of cantilever effect.

$$\text{Dyn. drift} = \frac{5.083 - 4.678}{108 \text{ in.}} = 0.0038$$

$$\text{Code drift} = 0.0038 \times \frac{0.132 \times 2170}{782} = 0.0014 \ll 0.005$$

Wind drift:

Wind load $\sim \frac{1}{3}$ seismic

$$\text{Code drift} = \tfrac{1}{3} \times 0.0014 = 0.0005 \ll 0.0025$$

NOTES TO DESIGNER

UBC REFERENCE

Wind load varies with height and wind zone. Some recommendations consider the terrain and dynamic wind effects to a structure.

Sec. 2311 and Fig. 4 (UBC)

The story-to-story drift for seismic loads is limited to 0.005. This example uses a proportion of the Code to dynamic loads given by the computer output, which follows. The deflections are for the combined-mode in the top stories, which is usually the worst condition.

There is no Code drift limit for wind loads.

Sec. 2312(h)

DESIGN CALCULATIONS

PROGRAM FOR LATERAL LOADING ANALYSES
VERSION 18 JULY 1976

PROJECT: HI-RISE MASONRY BLDG.
(JOB NUMBER: 00000)

NUMBER OF STORIES = 17

OPTION 2: RESPONSE SPECTRUM ANALYSIS—ENTER A(MAX), %DAMPING

STORY (T—B)		HEIGHT (INCHES)	WEIGHT (KIPS)
1	ROOF	108.0	164.0
2	17TH	108.0	191.0
3	16TH	108.0	191.0
4	15TH	108.0	191.0
5	14TH	108.0	191.0
6	13TH	108.0	191.0
7	12TH	108.0	197.0
8	11TH	108.0	202.0
9	10TH	108.0	202.0
10	09TH	108.0	202.0
11	08TH	108.0	202.0
12	07TH	108.0	202.0
13	06TH	108.0	209.0
14	05TH	108.0	215.0
15	04TH	108.0	215.0
16	03RD	108.0	215.0
17	2ND	144.0	229.0

TOTAL WEIGHT = 3403.00 KIPS
TOTAL HEIGHT = 1872.00 INCHES

BUILDING STIFFNESS INFORMATION
YOUNGS MODULUS = 2600
SHEAR MODULUS = 1040

STORY	MOMENT OF INERTIA	AREA
1	55600000.0	3780.000
2	55600000.0	3780.000
3	55600000.0	3780.000
4	55600000.0	3780.000
5	55600000.0	3780.000
6	55600000.0	3780.000
7	67900000.0	4620.000
8	67900000.0	4620.000
9	67900000.0	4620.000
10	67900000.0	4620.000
11	67900000.0	4620.000
12	67900000.0	4620.000
13	80300000.0	5460.000
14	80300000.0	5460.000
15	80300000.0	5460.000
16	80300000.0	5460.000
17	80300000.0	5460.000

A(MAX) = 0.3 GRAVITY
CRITICAL DAMPING = 10

MODE NO. 1
FREQUENCY= 0.985 PERIOD= 1.016

STORY	EIGENVECTOR	MODE SHAPE
1	0.0261281	1.0000
2	0.0259516	0.9204
3	0.0236978	0.8404
4	0.0214437	0.7605
5	0.0192039	0.6811
6	0.0169949	0.6027
7	0.0148365	0.5262
8	0.0131195	0.4524
9	0.0110639	0.0815
10	0.0091141	0.8148
11	0.0072947	0.2515
12	0.0056314	0.1942
13	0.0042225	0.1432
14	0.0029689	0.0992
15	0.0018632	0.0623
16	0.0009904	0.0331
17	0.0003875	0.0126

(*Note:* Refer to top of p. 448 for the rest of Mode No. 1 data.)

NOTES TO DESIGNER

UBC REFERENCE

The dynamic analysis is allowed as an alternative means of design. The design ground motion may be developed for a particular site or may be a general spectrum based on various standard earthquake records.

Sec. 2312(i)

The basis for dynamic design is to determine the ultimate loads to a structure for expected ground motions. The building period is first determined based on tributary mass to an element (shear wall on frame) or the overall structure. The accidental torsion should not be considered as an additional force or mass in this analysis.

Computer programs are available for dynamic calculation and are different in many different offices. A sample from a typical one is shown as an indication.

The computer provides a convenient means of developing forces and overturning effects to walls in a structure. In this case the accidental torsion is added to the direct shear loads, or for wall $2y$, the weight acting is 0.132 times the weight at each floor (see Table, page 440).

PSUEDO-VELOCITY= 20.40 MODE NO. 1 (CONT.)
MASS RATIO= 0.628500872
BASE SHEAR= 698.5645629

STORY (T—B)		FORCE (KIPS)	SHEAR (KIPS)	OVT.MOMENT (IN-KIPS)	DEFLECTION (INCHES)
1	ROOF	82.55	82.55	0.0	5.0815
2	17TH	88.48	171.02	8914.9	4.6765
3	16TH	80.79	251.82	27385.5	4.2706
4	15TH	73.11	324.93	54581.0	3.8645
5	14TH	65.47	390.40	89674.0	3.4608
13	06TH	15.06	676.06	563172.4	0.7275
14	05TH	10.74	686.80	636186.5	0.5043
15	04TH	6.74	693.54	710360.5	0.3165
16	03RD	3.58	697.12	785262.3	0.1682
17	2ND	1.45	698.56	860551.0	0.0638
				961144.3	

MODE NO. 2
FREQUENDY= 5.122 PERIOD= 0.195

STORY	EIGENVECTOR	MODE SHAPE
1	0.0009654	1.0000
2	0.0007492	0.7191
3	0.0004505	0.4324
4	0.0001572	0.1509
5	-0.0001175	-0.1128
6	-0.0003600	-0.3455
7	-0.0005574	-0.5350
8	-0.0007206	-0.6726
9	-0.0008145	-0.7602
10	-0.0008521	-0.7953
11	-0.0008344	-0.7787
12	-0.0007661	-0.7150
13	-0.0006672	-0.6128
14	-0.0005403	-0.4888
15	-0.0003907	-0.3535
16	-0.0002430	-0.2198
17	-0.0001175	-0.1030

```
          .         *
          .      *
          .   *
          . *
         *.
       *  .
     *    .
   *      .
  *       .
  *       .
  *       .
   *      .
    *     .
     *    .
      *   .
        * .
         *.
..........*..........
```

PSUEDO-VELOCITY= 5.58
MASS RATIO= 0.217860130
BASE SHEAR= 344.7864079

STORY (T—B)		FORCE (KIPS)	SHEAR (KIPS)	OVT.MOMENT (IN-KIPS)	DEFLECTION (INCHES)
1	ROOF	-62.23	-62.23	0.0	-0.1415
2	17TH	-52.11	-114.34	-6720.3	-0.1018
3	16TH	-31.34	-145.68	-19068.8	-0.0612
4	15TH	-10.94	-156.61	-34801.9	-0.0214
11	08TH	59.69	145.88	-81760.0	0.1102
12	07TH	54.80	200.64	-66009.9	0.1012
13	06TH	48.55	249.19	-44341.3	0.0867
14	05TH	39.88	289.06	-17429.1	0.0692
15	04TH	28.84	317.90	13789.9	0.0500
16	03RD	17.94	335.84	48123.1	0.0311
17	2ND	8.95	344.79	84393.5	0.0146
				134042.7	

MODE NO. 3
FREQUENCY= 12.145 PERIOD= 0.082

STORY	EIGENVECTOR	MODE SHAPE
1	0.0001717	1.0000
2	0.0000994	0.5364
3	0.0000100	0.0542
4	−0.0000696	−0.3756
5	−0.0001268	−0.6843
6	−0.0001523	−0.8220
7	−0.0001424	−0.7605
8	−0.0001078	−0.5658
9	−0.0000484	−0.2539
10	0.0000210	0.1100
11	0.0000877	0.4600
12	0.0001397	0.7331
13	0.0001709	0.8817
14	0.0001749	0.8896
15	0.0001520	0.7732
16	0.0001104	0.5614
17	0.0000617	0.3041

```
          .         *
          .     *
          .*
      *   .
   *      .
 *        .
 *        .
    *     .
       *  .
          .*
          .    *
          .      *
          .        *
          .        *
          .       *
          .     *
          .  *
..........*..........
```

PSUEDO−VELOCITY= 1.38
MASS RATIO= 0.077920387
BASE SHEAR= 72.2848459

STORY (T—B)		FORCE (KIPS)	SHEAR (KIPS)	OVT.MOMENT (IN-KIPS)	DEFLECTION (INCHES)
1	ROOF	19.75	19.75	0.0	0.0080
2	17TH	12.34	32.08	2132.6	0.0043
3	16TH	1.25	33.33	5597.5	0.0004
4	15TH	−8.64	24.69	9197.0	−0.0030
5	14TH	−15.74	8.96	11863.7	−0.0055
14	05TH	23.03	29.35	−10348.1	0.0071
15	04TH	20.01	49.37	−7177.9	0.0062
16	03RD	14.53	63.90	−1846.2	0.0045
17	2ND	8.38	72.28	5055.1	0.0024
				15464.2	

COMBINED FORCES & SHEARS FOR ALL MODES

STORY (T—B)		FORCE (RSS) (KIPS)	SHEAR (RSS) (KIPS)	DEFL (RSS) (INCHES)	DRIFT (RSS) (INCHES)
1	ROOF	105.241	105.241	5.08249	22.16739
2	17TH	102.970	208.211	4.67796	13.83269
3	16TH	84.611	292.822	4.27104	19.60316
4	15TH	68.724	361.546	3.86453	27.06525
5	14TH	56.219	417.764	3.46087	36.41478
6	13TH	47.357	465.121	3.06312	48.61662
7	12TH	41.685	506.806	2.67483	76.64433
8	11TH	40.690	547.496	2.30102	39.40999
9	10TH	38.699	586.195	1.94182	120.31304
10	09TH	37.096	623.291	1.00111	160 54579
11	08TH	35.197	658.488	1.28806	210.93327
12	07TH	32.470	690.358	0.89204	380.84388
13	06TH	29.586	720.546	0.73268	425.51538
14	05TH	25.181	745.727	0.50907	519.45379
15	04TH	18.792	764.519	0.32047	897.50207
16	03RD	11.911	776.430	0.17114	930.66651
17	2ND	5.925	782.365	0.06647	1051.41235

DESIGN CALCULATIONS

Check diaphragm. Loading in transverse direction. Proportion dead load of floor and walls to total of 1000 kips. Use rigidities of walls as resisting elements without torsion.

$$w = \text{tributary wt. fl.} + \text{wall}$$

$$w_1 = 105 \text{ lb} \times 74 \text{ ft} + 130 \text{ lb} \times 8.5 \text{ ft} \times 1 = 8.88 \text{ kips/ft} \times \frac{63}{2} = 280 \text{ kips}$$

$$w_2 = 105 \times 74 + 130 \times 8.5 \times 2 = 9.98 \text{ kips/ft} \times 15.5 \text{ ft} = 155 \text{ kips}$$

$$w_3 = 105 \times 74 + 130 \times 8.5 \times 3 = 11.09 \text{ kips/ft} \times 18.5 \text{ ft} = 205 \text{ kips}$$

$$w_4 = 105 \times 35 + 130 \times 8.5 \times 2 = 5.88 \text{ kips/ft } 10.5 \text{ ft} = 62 \text{ kips}$$

$$1404 \text{ kips} = 2 \times 702 \text{ kips}$$

Normalize loads to 1000 kips; i.e., 1000/1404 = 0.712

$$w_1 = 400 \text{ kips}$$
$$w_2 = 110 \text{ kips}$$
$$w_3 = 146 \text{ kips}$$
$$w_4 = 44 \text{ kips}$$

Diaphragm stresses:

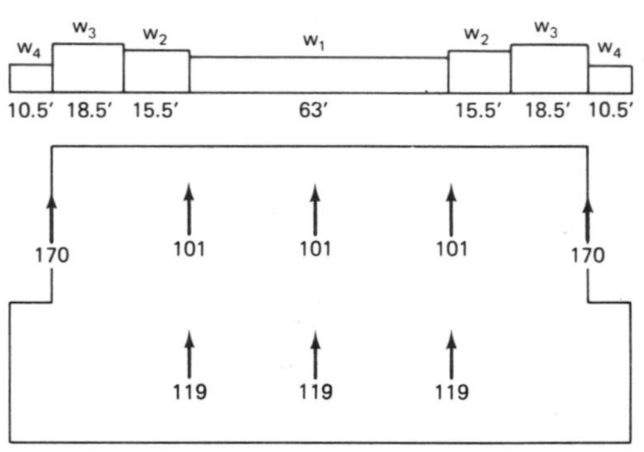

$$ZC_pS = 0.75 \times 0.12 \times 1.20 = 0.108$$

$$V_{\max} = 0.108 \times \frac{130}{1000} \times 2170 \text{ kips} = 30.5 \text{ kips}$$

For shear stress in concrete:

$$v_u = \frac{1.4 \times 30{,}500}{74 \text{ ft}}$$
$$= 577 \text{ lb/ft}$$

with $2\frac{1}{2}$-in. topping

$$v_u = \frac{577}{0.85 \times 2.5 \times 12} = 22.6 \text{ lb/in.}^2 < 2\sqrt{3000}$$

For chord stress:

$$M = 0.108 \times \frac{1178}{1000} \times 2170 = 276 \text{ ft-kips}$$

$$T = C = \frac{M}{h} = \frac{276}{73 \text{ ft}} = 3.78 \text{ kips}$$

$$A_s = \frac{U \times T}{\varphi f_y} = \frac{1.4 \times 3.78}{0.9 \times 40.0} = 0.15 \text{ in.}^2$$

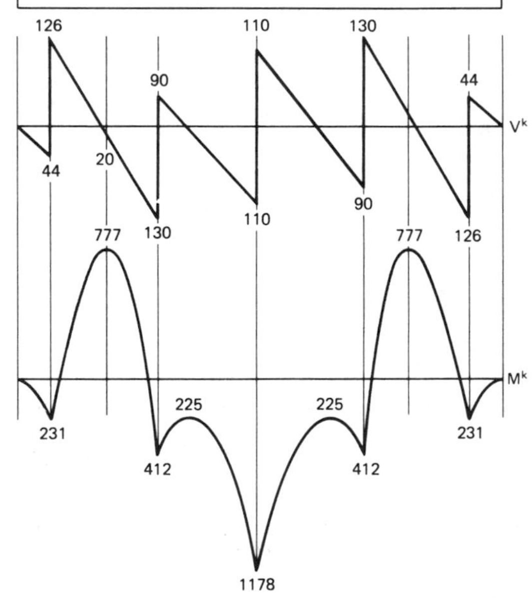

NOTES TO DESIGNER

UBC REFERENCE

The diaphragm stresses must be checked for induced lateral loads.

Sec. 2312(j) 2D

The loads may be proportioned to a unit loading equivalent to the sum of the rigidities of the walls. Note that the extra loads from *accidental* torsion may be neglected. The moment and shear diagrams are force-balanced to these unit loads.

Diaphragm stresses are designed to other code values for elements of structures.

Sec. 2312(g)

$$F_p = ZIC_pSW_p$$

Formula (12-8)

Note that V_{max} is proportioned to the shear diagram and the weight at the second floor, this being the heaviest loading of the floors. This same criterion is used for the moment for determining chord stresses.

Allowable stresses are from the chapter on concrete.

Chapter 26 or ACI 318

The $2\frac{1}{2}$-in. topping concrete and nominal reinforcing at the boundary elements are found to be adequate.

DESIGN CALCULATIONS

Shear-friction transfer:

$$V_u = A_s f_y \varphi \mu,$$

Use $\mu = 0.7$ for concrete, to concrete, and smooth interface at construction joint and for cyclic loading

$$A_s = \frac{30{,}500 \text{ lb} \times 1.4}{40{,}000 \text{ psi} \times 0.85 \times 0.7} = 1.79 \text{ in}^2. \quad \text{or} \quad \frac{1.79}{68 \text{ ft}} = 0.026 \text{ in.}^2/\text{ft} < 0.029$$

Use min. 6 × 6 − 10/10 EWWM.

Required transfer of wall to floor or floor to wall in a direction perpendicular to the wall.

$$F_p = \tfrac{3}{4} \times 1.0 \times 0.20 \times 1.20 \times 130 \text{ lb} \times \frac{8.51 \text{ ft}}{2} = 50 \text{ lb/ft} \quad \text{for the wall to floor}$$

$$\text{or} \quad \tfrac{3}{4} \times 1.0 \times 0.20 \times 1.20 \times 105 \text{ lb} \times \frac{33 \text{ ft}}{2} = 310 \text{ lb/ft} \quad \text{for the floor to wall}$$

$$\text{or Code minimum} = 200 \text{ lb/ft}$$

$$A_s = \frac{310 \times 1.4}{40{,}000 \times 0.85} = 0.013 \text{ in.}^2/\text{ft}^2 < 0.029$$

Check induced moment to floor slab between walls $2y$ and $3y$.

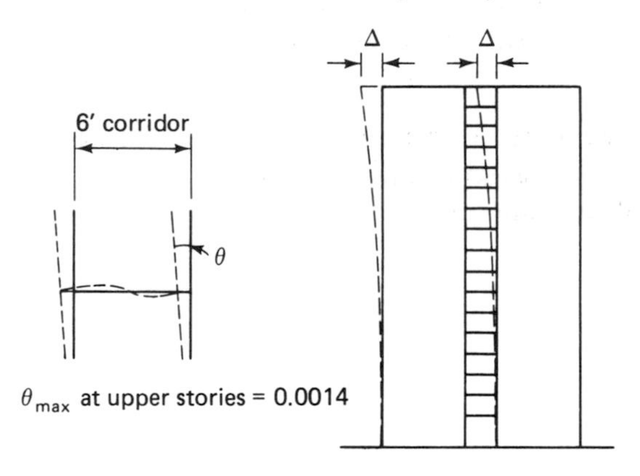

Induced moment to corridor slab by moment area method:

$$M = \frac{6EI\theta}{L}$$

For an 8-in. slab: $I = \dfrac{12 \times 8^3}{12} = 512 \text{ in.}^2$

$$M = \frac{6 \times 3160 \times 512 \times 0.0014}{72 \text{ in.}} = 188 \text{ in.-kips}$$

$$M_{D+L} = 190 \text{ lb} \times 6^2 \times 1.5 = \underline{10}$$

$$198 \text{ in.-kips}$$

NOTES TO DESIGNER

UBC REFERENCE

The shear-friction theory is applied for the transfer of the slab to the wall and also to provide some negative moment transfer of the slab superimposed loads.

Sec. 2611(p)

See also UBC for minimum recommended strut forces of 16,000 lb or $2\frac{1}{2}\%$ of the axial load to a wall, but not less than 1500 lb/ft of wall (based on yield stresses).

$$F_p = ZIC_pSW_p$$

Formula (12-8)

Code minimum transfer of walls to floor or roof is 200 lb/ft.

Sec. 2310

As the walls deflect, induced moments form in the floor slab link between the wall elements. The critical area is usually the top floor.

The wall rotation value is from the previously determined drift.

DESIGN CALCULATIONS

$$V = \frac{2M}{L} = \frac{2 \times 198}{72} = 5.5 \text{ kips}$$

$$V_{D+L} = 190 \text{ lb} \times 3 \text{ ft} \times 3 \text{ ft} = \underline{1.7}$$

$$7.2 \text{ kips}$$

$$V_u = 7.2 \times 1.4 = 10.1 \text{ kips}$$

$$A_s = \frac{M_u}{a_u d} = \frac{198 \text{ in.-kips}/12 \times 1.4}{2.89 \times 7 \text{ in.}} = 1.15 \text{ in.}^2 < 4 \times 0.31 = 1.24 \text{ in.}^2$$

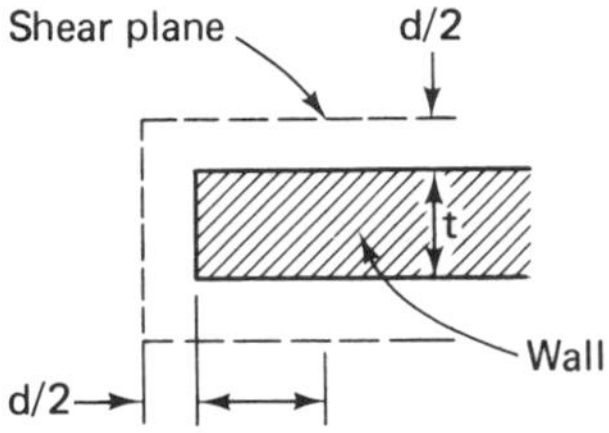

Shear plane

d/2

t

Wall

d/2

Check for ultimate shear to slab:

$$a_u = \frac{1.25 A_s f_y}{0.85 f'_c b} = \frac{1.25 \times 0.31 \times 4 \times 60.0}{0.85 \times 3.0 \times 12} = 3.04 \text{ in.}$$

$$M_u = A_s f_y\left(d - \frac{a_u}{2}\right) = 1.25 \times 0.31 \times 4 \times 60\left(7 - \frac{3.04}{2}\right)$$

$$= 510 \text{ in.-kips}$$

use 8-in. slab with four No. 5's top and bottom of slab at wall

$$V_u = \frac{2 \times 510}{72 \text{ in.}} = 14.16 \text{ kips}$$

$$V_{D+L} = 1.7 \times 1.4 = \underline{2.40}$$

$$16.56 \text{ kips}$$

$$v_u = \frac{16{,}560}{0.85 \times 7 \text{ in.} \times 3(9 + 7) \text{ in.}} = 58 \text{ lb/in.}^2 < 4\sqrt{f'_c}$$

Note: Actual seismic loads could raise the distortions of the walls and adjacent slabs by two to three times.

Check wall $2y$ for shear stress.

$$V_{gr} = 287.6 \text{ kips at ground story (ground to 2nd floor)}$$

$$v = \frac{1.5 \times 287{,}600}{13 \times 12 \times 35 \text{ ft}} = 79.0 \text{ lb/in.}^2 < 100 \text{ lb/in.}^2$$

Allowable stress:

$$\frac{M}{Vd} = \frac{374{,}814}{287.6 \times 35 \times 12} = 3.1 > 1.0$$

$$v = 1.5\sqrt{2600} = 76.5 \text{ lb/in.}^2 > 75 \text{ lb/in.}^2 \text{ max.}$$

$$v_{\text{allow}} = 75 \times 1.33 = 100 \text{ lb/in.}^2$$

NOTES TO DESIGNER

UBC REFERENCE

The maximum shear that can be developed is the maximum moment at each end divided by the span.

The effective width of slab reacting with a wall is suggested as one-third of the corridor span.

The critical shear plane around a pier has been studied for a number of years.

Sec. 2626(e) 5

The allowable shear in slabs with reverse loadings has been suggested as $2\sqrt{f_c'}$ versus $4\sqrt{f_c'}$.

Note: Shear stresses for seismic loads must be increased by a factor of 1.5.

Footnote 3
Table 24-H

$$\text{Spacing No. 5 bars} = \frac{0.31 \times 24{,}000 \times 1.33}{79 \times 13} = 9.66 \text{ in.}$$

Use No. 5 horizontal bars at $9\frac{1}{2}$ in. oc.

$$V_6 = 266.3 \text{ kips} \quad \text{at 6th story}$$

$$v = \frac{1.5 \times 266{,}300}{11 \times 12 \times 35} = 86.5 \text{ lb/in.}^2 < 100 \text{ lb/in.}^2$$

$$V_{12} = 183.3 \text{ kips} \quad \text{at 12th story}$$

$$v = \frac{1.5 \times 183{,}300}{9 \times 12 \times 35} = 72.7 \text{ lb/in.}^2 < 100 \text{ lb/in.}^2$$

Check wall 2y for combined loads:

Wt. wall	$= 90 \text{ lb} \times 35 \times 6 \times 8.5 \text{ ft}$	$= 160$ kips
	$110 \times 35 \times 6 \times 8.5$	$= 195$
	$130 \times 35(4 \times 8.5 + 12)$	$= 210$
DL floors	$= 105 \text{ lb} \times 35 \times 17(\frac{33}{2} + 4) \text{ ft}$	$= 1280$
		1845 kips
Reduced LL floors	$= 16 \text{ lb} \times 35 \times 17 \times 20.5 \text{ ft}$	$= 195$
		2040 kips

Seismic overturning at base = 374,800 in.-kips

Combined vertical loads:

$$\frac{P}{A} = \frac{1{,}845{,}000}{13 \text{ in.}(420 + 2 \text{ ft} \times 64)} = -259 \text{ lb/in.}^2$$

$$\frac{Mc}{I} = \frac{37.48 \times 10^7 \times 210}{14.92 \times 10^7} = \frac{\pm 528}{-787} \text{ lb/in.}^2 \text{ or } +269 \text{ lb/in.}^2$$

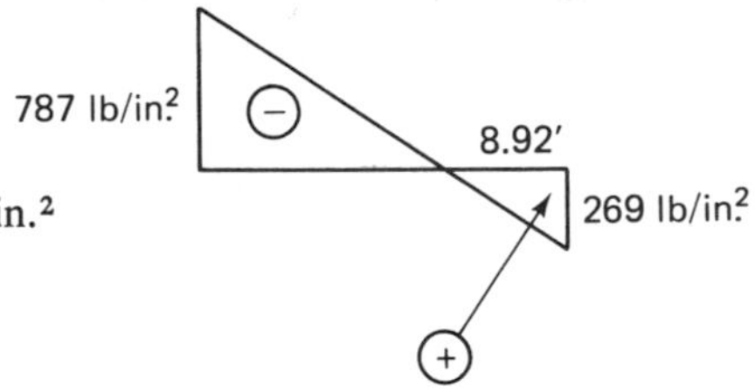

787 lb/in.²
⊖
8.92′
269 lb/in.²
⊕

$$\text{Allow } F_b = 0.33 \times 2600 \times 1.33 = 1141 \text{ lb/in.}^2$$

$$\text{or } 900 \times 1.33 = 1200 \text{ lb/in.}^2 \text{ max.}$$

For axial load:

$$\frac{P}{A} = \frac{2{,}040{,}000}{13 \times 420 \text{ in.}} = 374 \text{ lb/in.}^2$$

$$\text{Allow } F_a = 0.20 \times 4000\left[1 - \left(\frac{8.5 \times 12}{40 \times 13}\right)^3\right]$$

$$= 794 \text{ lb/in.}^2 > 374 \text{ lb/in.}^2 \qquad \text{OK}$$

Tension steel for uplift:

$$\text{Average tension on end ft. of wall} = 254 \text{ lb/in.}^2 = \frac{8.92 - 0.5}{8.92} \times 269$$

$$A_s = \frac{254 \times 13(2 \times 77)}{24{,}000 \times 1.33} = 15.88 \text{ in.}^2 < 16.0 \text{ in.}^2$$

NOTES TO DESIGNER

UBC REFERENCE

Spacing of horizontal steel for full shear stress per Table 24-H:

$$v = \frac{V}{bjd} \quad \text{or} \quad vb = \frac{V}{jd}$$ Formula (18-1)

$$A_v = \frac{V \cdot s}{f_v jd} = \frac{vbs}{f_v} \quad \text{or} \quad s = \frac{A_v f_v}{vb}$$ Formula (18-2)

Note: Check for minimum reinforcing. Sec. 2418(j) 3

Note: Walls $1y$, $8y$, $1x$, and $2x$ may be overstressed in shear with the window openings, and thicker walls may be required. The additional bending stresses through the window piers must be checked for combined loads.

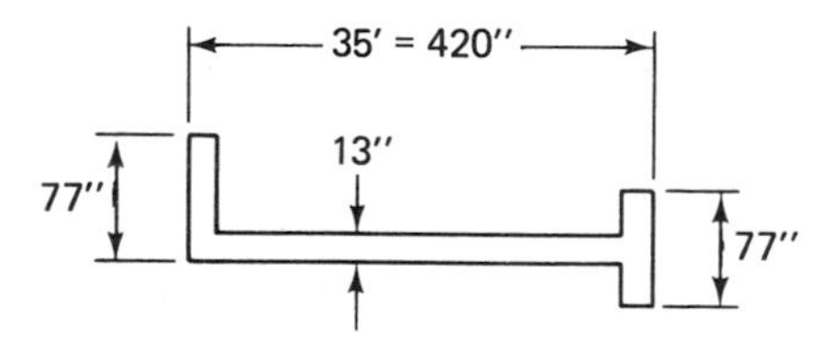

$$I = I_0 + Ad^2$$

$$= \frac{13 \times 420^3}{12} + 2 \times 13(77 - 13)(210 - 6.5)^2$$

$$= 8.03 \times 10^7 + 6.89 \times 10^7 = 14.92 \times 10^7 \text{ in.}^4$$

Allowable $F_b = 0.33 f'_m$, but < 900 lb/in.^{2} (in flexure). 6000-lb/in.^{2} units are adequate. Table 24-H

Allowable $F_a = 0.20 f'_m \left[1 - \left(\frac{h}{40t}\right)^3\right]$ (for axial loading) Formula (18-6) in Sec. 2418(j) 2

For next ft. of wall, $f = 224$ lb/in.^{2}

$$A_s = \frac{224 \times 13 \times 12}{24{,}000 \times 1.33} = 1.09 \text{ in.}^2 < 1.20 \text{ in.}^2$$

Check walls $2y$ and $3y$ for combined loads to footing. Try 12 ft wide $\times$ 86 ft long ftg.
Wall $2y$:

$$\text{Seismic O.T.M.} = 374{,}800 \text{ in.-kips} = 31{,}235 \text{ ft-kips}$$

$$\text{DL walls + floor} = 1845 \text{ kips}$$

Wall $3y$:

$$\text{Seismic O.T.M.} = \frac{0.111}{0.132} \times 31{,}235 = 26{,}265 \text{ ft-kips}$$

$$\text{D.L. walls + floor} = 2470 \text{ kips}$$

$$\text{wall } 4x\text{, DL wall} = 190 \text{ kips}$$

$$\text{wall } 1x\text{, DL wall} = 110 \text{ kips}$$

$$\text{Wt. ftg. (approx.)} = 470 \text{ kips}$$

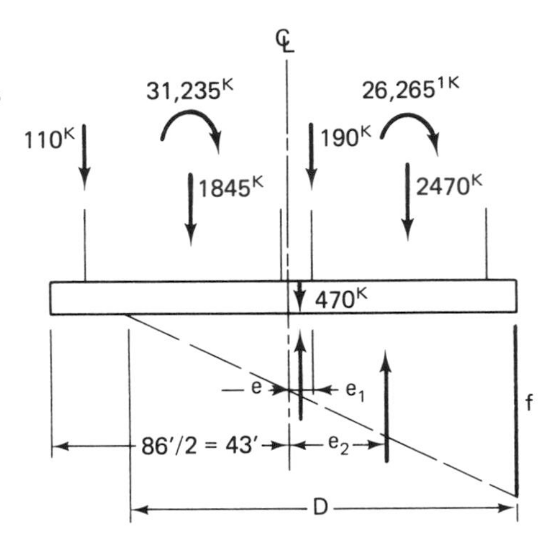

Check eccentricity for D.L. about wall $4x$:

$$e_1 = \frac{110(-41) + 1845(-23.5) + 2470(+16)}{110 + 1845 + 470 + 190 + 2470} = -1.65 \text{ ft}$$

$$e \text{ from center line} = 43 \text{ ft} - (32.5 + 6) - 1.65 = 2.85 \text{ ft}$$

Check eccentricity for O.T.:

$$\sum \text{O.T.M.} = 57{,}500 \text{ ft-kips}$$

$$\sum P_{D.L.} = 5085 \text{ kips}$$

$$e_2 = \frac{57{,}500}{5085} = 11.30 \text{ ft} - \text{since seismic can act either way}$$

$$e = 11.30 + 2.85 = 14.15 \text{ ft}$$

$$D = \left(\frac{L}{2} - e\right)3 = \left(\frac{86}{2} - 14.15\right)3 = 86.55 \text{ ft} \sim 86$$

$$f = \frac{2P}{WD} = \frac{2 \times 5085}{12 \times 86 \text{ ft}} = 9.85 \text{ kips/ft}^2 < 8.0 \times 1.33 = 10.67 \text{ kips/ft}^2$$

Use 12-ft-wide $\times$ 86-ft-long cont. footing

NOTES TO DESIGNER

UBC REFERENCE

For uplift provide 16 No. 9 bars (60 gr) at each end of wall into cross wall designed as a pilaster. For remainder of wall space, No. 7 bars at 6 in.oc for first foot from end, and increase spacing for stress needs. — Sec. 2418(k)

Use an assumed soil bearing value of 8000 lb/ft^2 with a 33% increase for seismic loadings. — Sec. 2303(d)

The footing should be designed for shear and flexure. — Chapter 26

Note: Walls 1y, 8y, 1x, and 2x will have high uplift forces from overturning. To reduce the uplift to the foundation only, it may be necessary to perform a yielding soil period determination. This operation will increase the overall period and reduce the seismic forces from overturning.

PRECAST BUILDING CONSTRUCTION

According to UBC Section 2314(b), a vertical load-carrying space frame is designed to accommodate all vertical loads. To complete the system, shear walls may be provided. A ductility factor K of 1.00 applies to such a structure per Table 5-8. One way to create a building such as this entails the use of prestressed precast roof and floor planks, with concrete columns and girders to carry the vertical loads. The framing elements are tied together by a poured topping into a homogenous structure with drag struts or load collector ties. The masonry shear walls, which provide the cantilever lateral load resistance, may be constructed on site or they may consist of prefabricated reinforced masonry wall panels transported from the manufacturing plant to the job site. Some details of these prefabricated high-rise wall panels are pictured in Figure 13-25. The precast columns may also be designed and constructed

FIGURE 13-25. Prefabricated high-rise wall panels: (a) prefabricated column enclosures; (b) beveled ends of the spandrel enclosures that will rotate 90° and fit against the bevel of the column, as shown in (c).

(a)

(b)

(c)

to act as edge members for the shear walls. Such plant prefabrication offers the advantages of lower initial cost and shorter construction time on a building that possesses versatility in space function as well as architectural expression. Figure 13-26 shows a recent example of this type of construction.

In the Tomax type of automated wall construction, the machine manufactures panels in 6-, 8-, or 12-in. thicknesses up to 12 × 24 ft in size. In this process, blocks are arranged in the proper sequence on a feed belt according to the design plan, and are fed onto a layout conveyor base. High-strength mortar is then fed into the head joint by vibrating fingers to assure a solid fill. At the same time, mortar is vibrated on to form a complete and uniform bed joint, so all face shells and cross webs are completely covered with mortar, thereby achieving excellent bond between masonry units. Thus, much higher wall strengths than those set by hand are obtained. This cycle takes about 5 s per block, and the procedure is continued until the desired length of panel is on the base. A carriage clamp then raises the entire course and places it on the panel pallet, a wheeled metal beam mounted on rails. Subsequent courses are placed on top of this course, with vibration to control the height, and this procedure is repeated until the desired panel height is achieved. The panels are then moved from the panel pallet 24 h later and placed in a storage area until they are delivered to the job site. Up to 3000 ft^2 of panels can be produced in an 8-h shift with a seven-worker crew, and up to eight or nine panels an hour can be set on the job site. This type of prefabricated wall panel readily lends itself to the deep-wall-beam design concept described in Chapter 7.

FIGURE 13-26. Precast concrete and reinforced masonry construction. This is an example of a commercial building consisting of a complete vertical load-carrying frame of precast concrete elements with the lateral resistance provided by masonry shear walls. (Courtesy W. L. Dickey.)

14
Retaining Walls and Fences

In Chapter 13 we saw how reinforced masonry is ideally suited to high-rise building construction, where, if properly reinforced, the bearing walls serve a dual function by acting as shear walls as well. Thus, the structure no longer must depend upon its massiveness for stability. Rather, the relatively thin walls can be reinforced in such a way so that they provide resistance to the overturning effects of lateral forces. In light of this, then, we will procede to examine its potential for use in retaining-wall structures, whereby one may utilize the flexural capability of a properly reinforced masonry stem wall to resist the overturning effect of the retained earth behind it, rather than by achieving this resistance through the mass of masonry alone. As we shall see, reinforced masonry retaining walls are quite sound structurally. Besides, because of the many different masonry unit shapes and colors available, the exterior faces of such structures show very pleasing architectural appearances.

TYPES OF MASONRY RETAINING WALLS

There are about four basic shapes or forms that are readily adaptable to masonry retaining-wall construction. These would include the following types.

Gravity

This type depends upon the massiveness of the masonry itself to achieve lateral stability against earth pressure through dead weight alone. Typically, it is unreinforced so that it must be sized and shaped in such a way that tensile stresses are low or nonexistent. Usually, this calls for a trapezoidal cross-sectional form. Since the advent of modern reinforced masonry, however, this type is not often used.

Cantilever

The cross section of this type consists of an L or an inverted T shape. The vertical stem wall acts structurally as a cantilever beam, and as such must be reinforced to withstand the rather considerable tensile stresses present due to the pressure exerted by the retained earth. This stem wall can be readily constructed with masonry units (brick or concrete block) and properly reinforced. The base slab (usually reinforced concrete) essentially maintains lateral stability by restraining the wall against both sliding and overturning. Normally, the L-shaped retaining wall is used when it is located on a property line where the base slab cannot extend beyond the stem face. Otherwise, it would be preferable to use the inverted T shape, to minimize the slab flexural stresses.

Counterfort or buttress

The third type is somewhat similar in shape to the cantilever; however, structurally it behaves quite differently. Located at intervals along the length of the wall are vertical supports which receive a reaction from the stem wall, since it spans horizontally between these supports. Thus, the wall stem must be reinforced horizontally so that it safely can carry the earth pressure in this direction. If the supports are located behind the wall (within the retained earth) they are called *counterforts*, and thus serve as tension ties, whereas if they occur on the exposed face, they are called *buttresses*, in which case they act as compression struts. The wall stem may be assumed to act as a continuous beam spanning across these several supports, and reinforced accordingly for the positive moments in the center of the span, and negative bending moments at the counterfort or buttress supports. Actually, this type would more than likely be employed in the higher walls, where the size of a cantilever stem and amount of vertical reinforcing required would become prohibitive. Refer to Figure 14-1 for sketches of these three types.

Supported walls

Actually, a basement or subterranean garage wall forms a type of retaining wall, although it is not always thought of as such. In some cases, the wall spans vertically between floors, acting as a continuous beam, where several levels exist, as portrayed in Figure 14-2, or from floor to footing for the single-level basement. In others, it may be reinforced to span horizontally between intersecting walls.

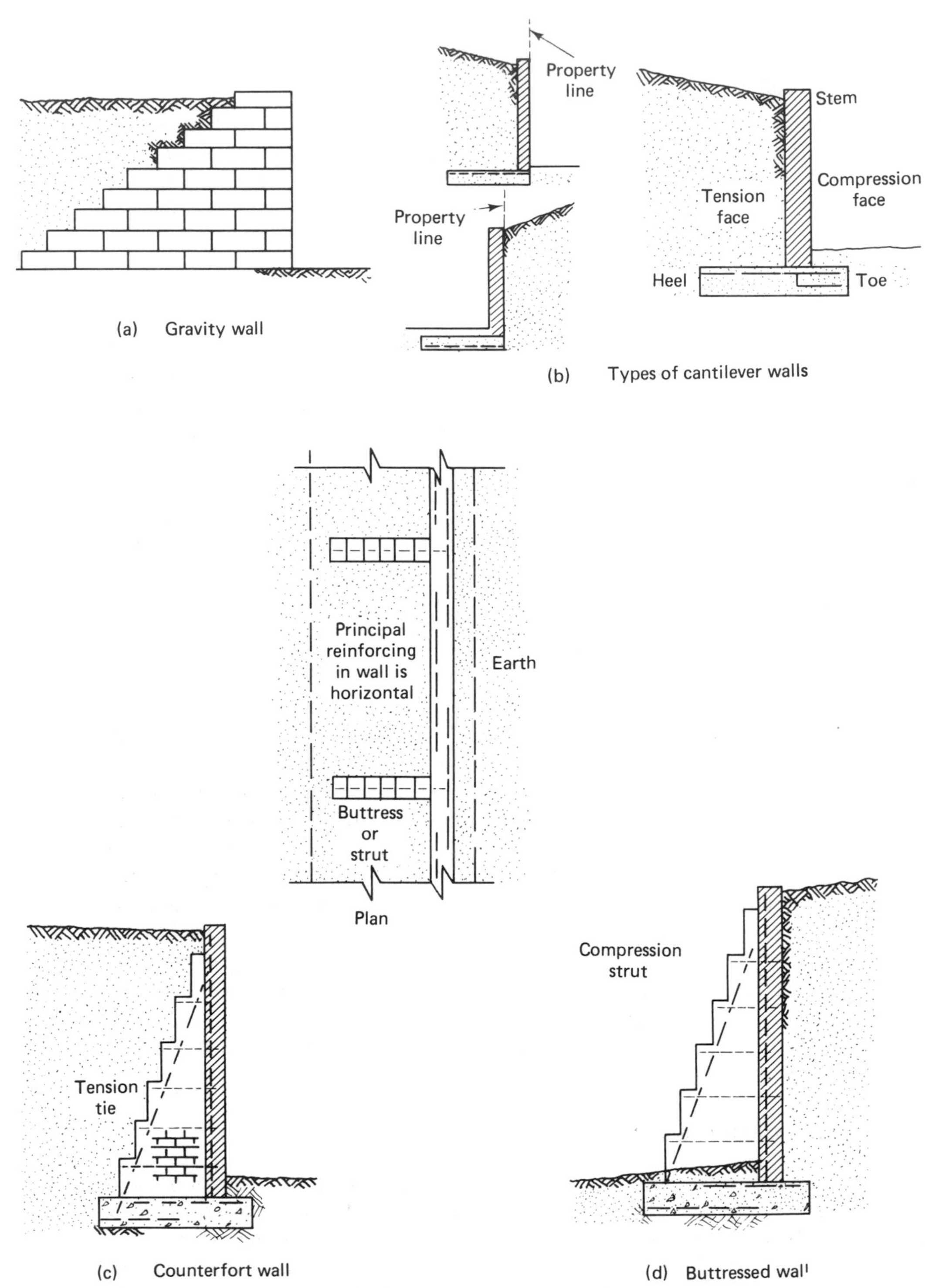

FIGURE 14-1. Basic retaining wall types.

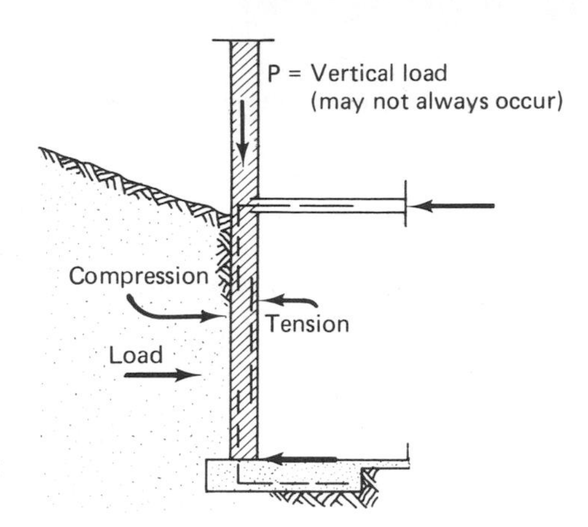

FIGURE 14-2. Supported retaining wall.

A more refined analysis would recognize that the wall actually acts as a flat plate supported on four sides, under various edge restraint conditions. Note that under these circumstances, the wall generally will be carrying at least a vertical dead, and possibly a live, load as well. When the wall spans vertically, the combined effects of vertical axial compressive stress and the flexural stress should be analyzed.

SELECTION OF WALL TYPE

The selection of the type of masonry retaining wall depends upon several factors, such as (1) height of wall; (2) magnitude of earth pressure, including any moisture effects; (3) the presence of any surcharge; and (4) soil bearing characteristics, among others. But, typically, one encounters the cantilever shape simply because reinforced masonry is often economically adaptable to this form. It should be noted that in this type of wall, the cantilever stem will tend to deflect outward slightly as a result of the earth pressure. As this is a bit unsightly and it appears to the layman as a potentially dangerous distortion, it is good practice to batter the front face to offset this movement, much the same as one builds a camber into the bottom chord of a roof truss. This means that a built-in tilt in the reverse direction should be provided, say at least $\frac{1}{2}$ in. in 12 in.

Another factor to consider is the strengthing effect of corners on the capacity of a retaining wall. Refer to Figure 14-3, which shows schematically how walls

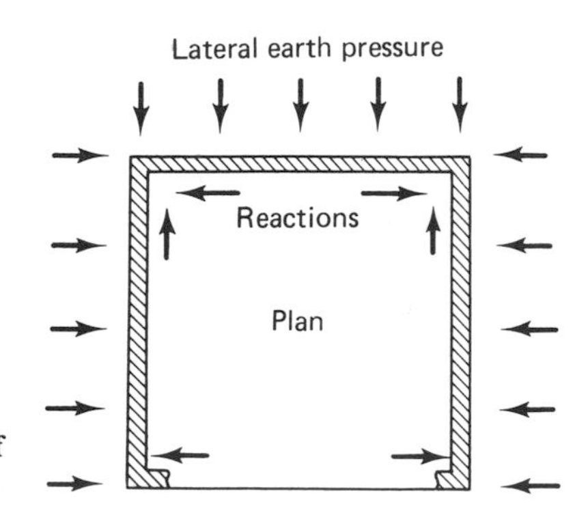

FIGURE 14-3. Intersecting effect of retaining walls.

extending in different directions tend to support one another to some extent. Unless properly reinforced, those intersecting points of stiffness or support may crack. Should the length of the wall be not more than three times the height, the wall moments may be obtained from a PCA publication titled "Rectangular Concrete Tanks—IS 003." This treatise describes four conditions of support, and observes that the wall will be subjected to triangular earth pressure plus any surcharge present.

LATERAL EARTH PRESSURE

Cantilever walls

EQUIVALENT FLUID PRESSURE

It is not the intent of this text to present a detailed dissertation on the various earth-pressure theories as they have evolved from the science of soil mechanics. Rather this must be left to texts specifically written for this purpose, to which readers are referred for any expansion beyond the following brief discussion.

Many minor retaining-wall structures, whose heights are not great (15 ft), can be safely designed by using Code-specified earth-pressure values, without having to obtain the soil characteristics through a geotechnical exploration and then resorting to an acceptable theoretical analysis to obtain the lateral earth-pressure effect. The Code factors presume very static and stable soil conditions, where the water pressure is negligible. Further, it is assumed that the wall tilts forward slightly, thereby relieving the soil pressure somewhat to produce what is known as active earth pressure on the stem. The Code approach for estimating backfill pressure is empirical in nature and is based upon an equivalent fluid pressure expression:

$$P = KW_e \frac{H^2}{2} \tag{1}$$

where P = total lateral force
K = lateral pressure coefficient
W_e = weight of backfill, lb/ft^3
H = height of wall above base, ft

ARBITRARY DESIGN METHOD (LOS ANGELES BUILDING CODE)

KW_e becomes the equivalent fluid pressure term. The "Arbitrary Design Method," spelled out in the 1972 Los Angeles City Building Code, contains Table 14-1, which lists these values for various backfill slopes. Note that the basic value, under a level backfill condition, is 30 lb/ft^3.

The expressions in the discussion accompanying Table 14-1 give the magnitude and location of the lateral force transmitted to the wall by a surcharge load imposed by isolated or continuous wall footings. Also spelled out are the conditions for bearing pressure and overturning, plus that of friction and lateral soil pressure. These are listed under Section 91.2309 which follows Table 14-1 in this discussion.

The "Arbitrary Design Specifications" permit one to select allowable foundation pressures for various soil types when no geotechnical investigation is made at the site and certain buildings of Type V construction (see Chapter 16) are involved. These are listed in Table 14-2. These specifications also provide allowable values for frictional resistance and lateral bearing values for various soil types. These are itemized in Table 14-3.

Restrained walls

The lateral soil pressure on a rather unyielding restrained wall is larger than the active soil pressure on a cantilever wall. These would include, among other cases, basement walls supported by concrete slabs. Under these circumstances, the lateral earth pressure approaches a soil-at-rest condition. Soil mechanics tells us that the lateral force coefficient for soil-at-rest is $K_r = (1 - \sin \phi)$, where ϕ is the angle of repose for the soil in this condition. For a coarse-grained soil, $\phi \sim 30°$, thus $K_r \sim 0.50$; whereas $\phi \sim 15°$ for a plastic clay, yielding a value of $K_r \sim 0.75$. Translating these figures into an equivalent fluid pressure, one obtains a valve of 55 lb/ft² for a coarse-grained soil (level backfill) and 75 lb/ft² for a plastic clay soil.

EXCERPTS FROM CITY OF LOS ANGELES BUILDING CODE

Sections 91.2309, 91.2802, and 91.2803 taken from the 1972 City of Los Angeles Building Code, are reproduced on the following pages to aid in preparing a masonry retaining-wall design whose height does not exceed about 15 ft.

EXCERPTS FROM CITY OF LOS ANGELES BUILDING CODE (LABC) 1972 EDITION

SEC. 91.2309—RETAINING WALLS

(a) Design. Retaining walls shall be designed to resist the lateral pressure of the retained material determined in accordance with accepted engineering principles.

The soil characteristics and design criteria necessary for such a determination shall be obtained from a special foundation investigation performed by an agency acceptable to the Department. The Department shall approve such characteristics and criteria only after receiving a written opinion from the investigation agency together with substantiating evidence.

EXCEPTION: Freestanding walls which are not over 15′ in height or basement walls which have spans of 15′ or less between supports may be designed in accordance with Subsection (b) of this Section.

(b) Arbitrary Design Method. Walls which retain drained earth and come within the limits of the exception to Subsection (a) of this section may be designed for an assumed earth pressure equivalent to that exerted by a fluid weighing not less than shown in Table 23-E. A vertical component equal to one-third of the horizontal force so obtained may be assumed at the plane of application of the force.

The depth of the retained earth shall be the vertical distance below the ground surface measured at the wall face for stem design or measured at the heel of the footing for overturning and sliding.

TABLE 14-1

(*LABC Table 23-E*)

Surface slope of retained material, horiz. to vert.*	*Equivalent fluid weight (lb/ft³)*
Level	30
5 to 1	32
4 to 1	35
3 to 1	38
2 to 1	43
1½ to 1	55
1 to 1	80

*Where the surface slope of the retained earth varies, the design slope shall be obtained by connecting a line from the top of the wall to the highest point on the slope, whose limits are within the horizontal distance from the stem equal to the stem height of the wall.

(c) Surcharge. Any superimposed loading, except retained earth, shall be considered as surcharge and provided for in the design. Uniformly distributed loads may be considered as equivalent added depth of retained earth. Surcharge loading due to continuous or isolated footings shall be determined by the following formulas or by an equivalent method approved by the Superintendent of Building.

Resultant lateral force:

$$R = \frac{0.3Ph^2}{x^2 + h^2}$$

Location of lateral resultant:

$$d = x\left[\left(\frac{x^2}{h^2} + 1\right)\left(\tan^{-1}\frac{h}{x}\right) - \left(\frac{x}{h}\right)\right]$$

where R = resultant lateral force measured in pounds per foot of wall width

P = resultant surcharge load of continuous or isolated footings measured in pounds per foot of length parallel to the wall

x = distance of resultant load from back face of wall measured in feet

h = depth below point of application of surcharge loading to top of wall footing measured in feet

d = depth of lateral resultant below point of application of surcharge loading measured in feet

$\left(\tan^{-1}\frac{h}{x}\right)$ = angle in radians whose tangent is equal to $\left(\frac{h}{x}\right)$

Loads applied within a horizontal distance equal to the wall stem height, measured from the back face of the wall, shall be considered as surcharge.

For isolated footings having a width parallel to the wall less than three feet. *R* may be reduced to $\frac{1}{6}$ the calculated value.

The resultant lateral force *R* shall be assumed to be uniform for the length of footing parallel to the wall, and to diminish uniformly to zero at the distance *x* beyond the ends of the footing.

Vertical pressure due to surcharge applied to the top of the wall footing may be considered to spread uniformly within the limits of the stem and planes making an angle of 45° with the vertical.

(d) Bearing Pressure and Overturning. The maximum vertical bearing pressure under any retaining wall shall not exceed that allowed in Division 28 of this Article except as provided for by a special foundation investigation. The resultant of vertical loads and lateral pressures shall pass through the middle one-third of the base.

(e) Friction and Lateral Soil Pressures. Retaining walls shall be restrained against sliding by friction of the base against the earth, by lateral resistance of the soil, or by a combination of the two. Allowable friction and lateral soil values shall not exceed those allowed in Division 28 of this Article except as provided by a special foundation investigation.

When used, keys shall be assumed to lower the plane of frictional resistance and the depth of lateral bearing to the level of the bottom of the key. Lateral bearing pressures shall be assumed to act on a vertical plane located at the toe of the footing.

(f) Construction. No retaining wall shall be constructed of wood.

SEC. 91.2802—FOUNDATION ANALYSIS

(a) General. The classification of the foundation material under every building shall be based upon the examination of test borings or excavations made at the site. The extent and number of the test borings or excavations shall be sufficient to provide the data necessary to classify the foundation materials under the entire building. The location of the test borings or excavations and the nature of the subsurface materials shall be indicated on the plans.

EXCEPTION: The requirements of this Subsection shall not apply to any building constructed in accordance with the arbitrary requirements of Division 48 (Wood Frame Dwellings).

(b) Foundation Materials. The foundation of every structure shall be a uniform natural deposit of rock, gravel, sand, clay, silt or combination thereof which does not contain and which does not overlie strata containing more than 10% by dry weight of organic matter.

EXCEPTION: Foundations may be artificial fill or nonuniform areas of dis-similar materials, provided due allowance is made for the effect of differential settlement.

(c) Effect of Change in Moisture Content. Due allowance shall be made in determining the capacity of foundations for the effect of possible change in moisture content.

(d) Effect of Pressure on Foundations. Where footings are to be placed at varying elevations or at different elevations from existing footings, the effect of adjacent loads shall be included in the foundation analysis.

(e) Load Distribution. A load upon a foundation stratum shall be assumed as distributed uniformly over an area subtended by planes extending downward from the edges of the footing and making an angle of 60 degrees with the horizontal.

(f) Arbitrary Design Specification. Certain buildings of Type V construction may have footings and foundations designed in accordance with the provisions of Section 91.1708 (Type V Buildings) and Section 91.4807 (Wood Frame Dwellings).

TABLE 14-2

Allowable Foundation Pressure
(*LABC Table 28-A*)
(*Kips per Square Foot—1 Kip = 1,000 pounds*)

Class of material			
Rock—depth of embedment shall be to a fresh unweathered surface except as noted	*Value at min. depth*	*Increase for depth*	*Maximum value*
*Massive crystalline bedrock; basalt, granite and diorite in sound condition	20		20
*Foliated rocks; schist and slate, in sound condition	8		8
*Sedimentary rocks; hard shales, dense siltstones and sandstones, thoroughly cemented conglomerates	6		6
Soft, or broken bedrocks; soft shales, shattered slates, distomaceous shales; other badly jointed (fractured) or weathered rock. 12 in. minimum embedment	2		2

**Note:* The above values apply only where the strata are level or nearly so, and/or where the area has ample lateral support. Tilted strata, and the relationship to nearby slopes should receive special consideration. These values may be increased one-third to a maximum of two times the assigned value, for each foot of penetration below fresh, unweathered surface.

Class of material						
*Soils—minimum depth of embedment shall be one foot below the adjacent undisturbed ground surface**	*Loose*	*Compact*	*Soft*	*Stiff*	*Increase for depth*	*Maximum value*
Gravel, well graded. Well graded gravels or gravel-sand mixtures, little or no fines	1.33	2.0			20	8
Gravel, poorly graded. Poorly graded gravels or gravel-sand mixtures, little or no fines	1.33	2.0			20	8
Gravel, silty. Silty gravels or poorly graded gravel sand silt mixtures	1.0	2.0			20	8
Gravel, clayey. Clayey gravels or gravel-sand clay mixtures	1.0	2.0			20	8
Sand, well graded. Well graded sands or gravelly sands, little or no fines	1.0	2.0			20	6

TABLE 14-2 (Continued)

Class of material						
*Soils—minimum depth of embedment shall be one foot below the adjacent undisturbed pressed surface**	*Loose*	*Compact*	*Soft*	*Stiff*	*Increase for depth*	*Maximum value*
Sand, poorly graded. Poorly graded sand or gravelly sands, little or no fines	1.0	2.0			20	6
Sand, silty. Silty sand, or poorly graded sand-silt mixtures	0.5	1.5			20	4
Sand, clayey. Clayey sands or sand-clay mixtures	1.0	2.0			20	4
Silt. Inorganic silts and very fine sands, rock flour, silty or clayey fine sands with slight plasticity	0.5	1.0			20	3
Silt, organic. Organic silts and organic silt-clays of low plasticity	0.5	1.0	0.5	1.0	10	2
Silt, elastic. Very compressible silts, micaceous or diatomaceous fine sandy or silty soils	0.5	1.0			10	1.5
Clay, lean. Inorganic clays of low to medium plasticity, silty clays, lean clays	1.0	2.0	1.0	2.0	20	3
Clay, fat. Very compressible clays, Inorganic clays of high plasticity			0.5	1.0	10	1.5
Clay, organic. Organic clays of medium to high plasticity, very compressible			0.5			0.5
Peat. Peat and other highly organic swamp soils			0			0

Notes:

1. Value for gravels and sand given are for footings one foot in width and may be increased in direct proportion to footing width to maximum of three times the maximum value, or to the designated maximum value, whichever is the least.
2. Where the bearing values in the above table are used, it should be noted that increased width or unit load will cause increase in settlement.
3. Special attention should be given to the effect of increase in moisture in establishing soil classifications.
4. Minimum depth for highly expansive soils to be one and one-half feet.
5. Increases for depth are given in percentage of minimum value for each additional foot below the minimum required depth.

SEC. 91.2803—FOUNDATION CLASSIFICATION

(a) Foundation Classification. Foundation materials shall be grouped in classes having the designations set forth in Table No. 28-A.

(b) Variation in Soil Strata. The classification of foundation material shall be that of the weakest stratum within a depth below the footing equal to twice the least width of the footing.

TABLE 14-3

Allowable Frictional and Bearing Values for Rock[1]
(*LABC Table 28-B*)

Type	*Friction coefficient*	*Allowable lateral bearing lb/ft²*	*Max. value lb/ft²*
Massive crystalline bedrock	1.0	4,000	20,000
Folliated rocks	.8	1,600	8,000
Sedimentary rocks	.6	1,200	6,000
Soft or broken bedrocks	.4	400	2,000

Allowable Frictional and Lateral Bearing Values for Soils

Frictional resistance—gravels and sands[1]	
Soil type	*Friction coefficient*
Gravel, well graded	0.6
Gravel, poorly graded	0.6
Gravel, silty	0.5
Gravel, clayey	0.5
Sand, well graded	0.4
Sand, poorly graded	0.4
Sand, silty	0.4
Sand, clayey	0.4

[1]Coefficient to be multiplied by the dead load.

Allowable Frictional Resistance
(lb/ft^2)—clay and silt[2]

Soil type	*Loose or soft*	*Compact or stiff*
Silt, inorganic	250	500
Silt, organic	250	500
Silt, elastic	200	400
Clay, lean	500	1000
Clay, fat	200	400
Clay, organic	150	300
Peat	0	0

[2]Frictional values to be multiplied by the width of footing subjected to positive soil pressure. In no case shall the frictional resistance exceed $\frac{1}{2}$ the dead load on the area under consideration.

TABLE 14-3 (Continued)

Allowable Lateral Bearing per Foot of Depth Below Natural Ground Surface (lb/ft²) (Natural Soils or approved compacted fill)

Soil type	*Loose or soft*	*Compact or stiff*	*Max. values*
Gravel, well graded	200	400	8000
Gravel, poorly graded	200	400	8000
Gravel, silty	167	333	8000
Gravel, clayey	167	333	8000
Sand, well graded	183	367	6000
Sand, poorly graded	77	200	6000
Sand, silty	100	233	4000
Sand, clayey	133	300	4000
Silt, inorganic	67	133	3000
Silt, organic	33	67	2000
Silt, elastic	33	67	1500
Clay, lean	267	667	3000
Clay, fat	33	167	1500
Clay, organic	33	—	500
Peat	0	0	0

General Conditions of Use

1. Frictional and lateral resistance of soils may be combined, provided the lateral bearing resistance does not exceed $\frac{2}{3}$ of allowable lateral bearing.
2. A $\frac{1}{3}$ increase in frictional and lateral bearing values will be permitted to resist loads caused by wind pressure or earthquake forces.
3. Isolated poles such as flag poles or signs may be designed using lateral bearing values equal to two times the tabulated values.
4. Lateral bearing values are permitted only when concrete is deposited against natural ground or compacted fill, approved by the Superintendent of Building.

(c) Allowable Foundation Pressures. The design unit pressure upon every foundation shall not exceed the arbitrary values exhibited in Table No. 28-A.

EXCEPTION: The tabulated values may be modified as prescribed in Section 91.2804.

(d) Friction and Lateral Soil Pressures. The design unit values for friction and lateral soil pressures shall not exceed the arbitrary values exhibited in Table No. 28-B.

EXCEPTION: The tabulated values may be modified as prescribed in Section 91.2804.

BRICK RETAINING WALLS

A typical two-wythe brick retaining-wall section appears in Figure 14-4. This type of design configuration readily lends itself to a computer solution, which would yield the dimensions *A* through *F*, and both the bar size and spacing within the

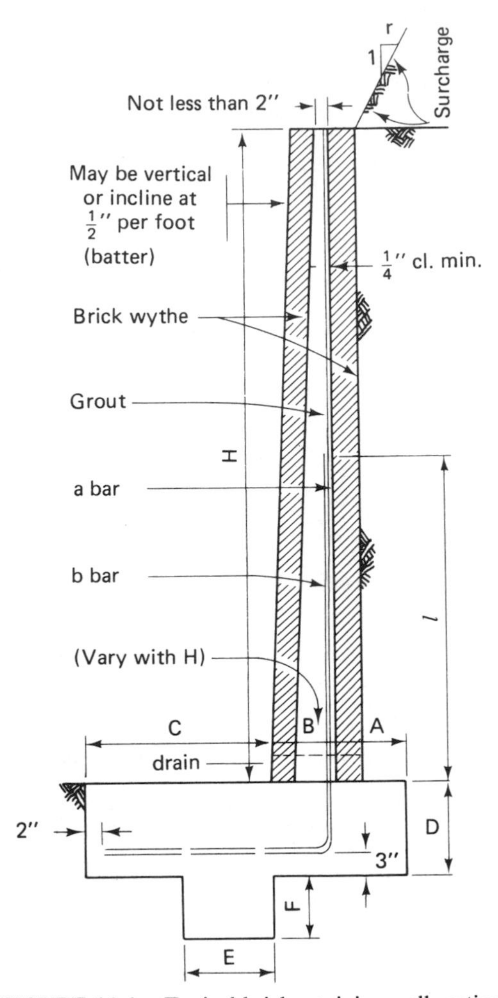

FIGURE 14-4. Typical brick retaining-wall section.

wall (bars *a* and *b*) as well as in the foundation slab, for various wall heights (H), soil characteristics, backfill slopes, and surcharge magnitudes and their locations with respect to the back of the wall. On the other hand, the occasional wall may be designed by hand without too much difficulty, as is shown in the following example.

Certain other factors should be observed, among which are the following:

Reinforcement Vertical reinforcement must be placed within about $\frac{1}{4}$ in. of the masonry face adjacent to the retained earth (tension side), leaving enough space so that the bar will be adequately bonded to the grout.

Brick The masonry units must have full head and bed joints and should be laid up in running bond.

Grout Grouting must be done in accordance with the grouting procedures specified by the UBC.

Drainage Adequate drainage of the backfill is absolutely necessary. This can be accomplished with weep holes through the wall or by a drain on top of the footing extending the full length of the wall.

Shear Key In order to resist any slippage between the masonry stem and the concrete footing, a shear key is often provided. It may simply consist of an extension of the grout core into the top of the footing, or it could be located directly in front of the stem wall. These details are shown in Figure 14-5.

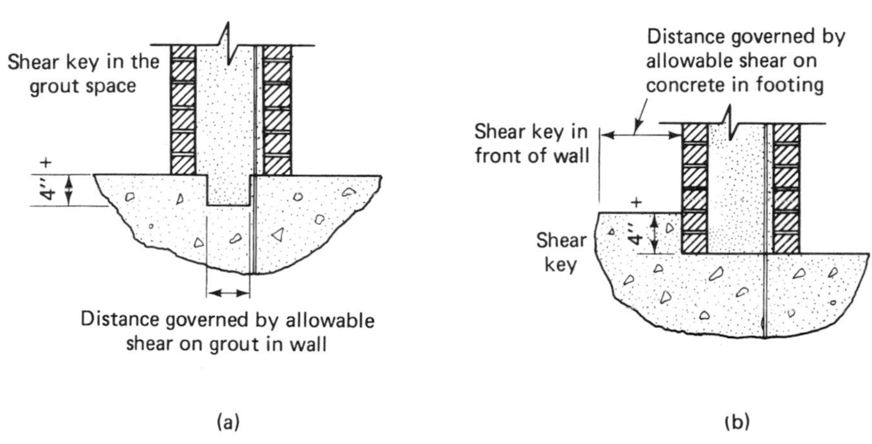

FIGURE 14-5. Shear key details.

EXAMPLE 14-1

A cantilever retaining wall, having a height of 8 ft 0 in., is to be constructed of reinforced grouted brick masonry (Figures 14-4 and 14-5). The backfill has a 2:1 slope. The arbitrary design provisions of the L.A. City Building Code will be followed since $H < 15$ ft. The passive earth pressure is found to be $P_h = 300$ lb/ft².

DESIGN CRITERIA:

Solid-grouted brick
$f'_m = 1500$ lb/in.² Use $\frac{1}{2}$ stresses, since no continuous inspection required, $n = 40$
$f_s = 20{,}000$ lb/in.²
Concrete footing $f'_c = 2000$ lb/in.²
Grout $f'_c = 2000$ lb/in.²
Lateral earth pressure $K_h = 43$ lb/ft² (Table 14-1)
Vertical earth pressure $K_v = \frac{1}{3} K_h = 14.3$ lb/ft²
Passive earth pressure $P_h = 300$ lb/ft²
Allowable soil bearing $= 2000$ lb/ft²
Wt. of soil $= 100$ lb/ft³

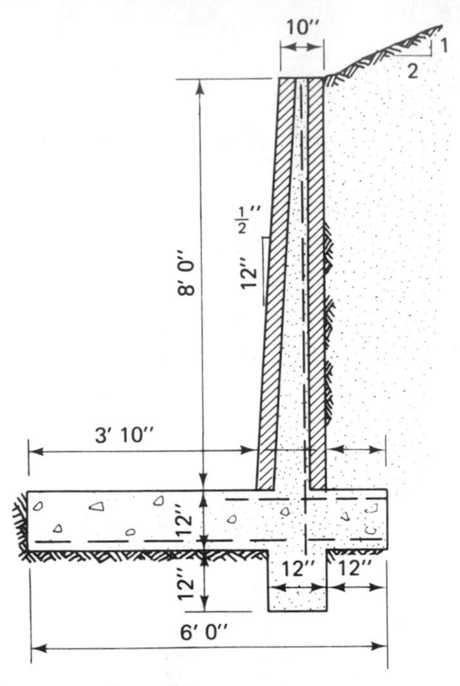

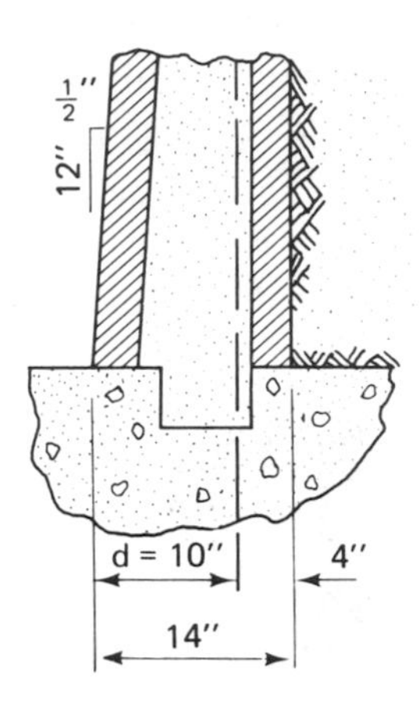

1. Design of Stem Wall

Lateral force on stem:

$$V = \frac{wh^2}{2} = \frac{43 \times 8^2}{2} = 1376 \text{ lb}$$

Moment on stem:

$$M = \frac{wh^3}{6} = \frac{43 \times 8^3}{6} = 3670 \text{ ft-lb}$$

Reinforcing steel in stem: At the top of the wall, a 3-in. grout core, combined with the width of the brick units, provides for a total width of 10 in. Assume (perhaps by comparison with a previous design) a width at the base of the stem of 14 in., leaving an effective depth of about 10 in.

$$K = \frac{M}{bd^2} = \frac{3670 \times 12}{12 \times 10^2} = 36.7; \quad \frac{K}{F_b} = \frac{36.7}{0.250} = 147$$

From Chart B-2 $\qquad pn = 0.080 \qquad \therefore p = 0.002$

$$A_s = pbd = 0.002 \times 10 \times 12 = 0.24 \text{ in.}^2\text{/ft}$$

Use No. 5's at 12 in. oc ($A_s = 0.31$ in./ft).

To save on reinforcing, try cutting alternate bars at 6 ft from top of wall. Check the adequacy of such an arrangement.

$$M = \frac{wh^3}{6} = \frac{43 \times 6^3}{6} = 1548 \text{ ft-lb, and Say } d = 9''$$

$$p = \frac{A_s}{bd} = \frac{.31}{24 \times 9} = .0014$$

$$K = \frac{M}{bd^2} = \frac{1548 \times 12}{12 \times 9^2} = 19.1, \quad \frac{K}{F_b} = \frac{19.1}{0.250} = 76$$

From Chart B-2 for $K/F_b = 76$, $f_m = 165$ lb/in.², $f_s = 20{,}000$ lb/in.² OK. Extend bars 12 diam. or 12 in. past theoretical cutoff thus cut alternate bars 5 ft from top.

Bond stress in bars: At connection with footing,

$$u = \frac{V}{\sum_0 jd} = \frac{1376}{1.96 \text{ in} \times 0.9 \times 10} = 78 \text{ lb/in.}^2 < 100 \text{ lb/in.}^2 \times 1.33 \text{ allowable}$$

Shear at connection of stem to footing: At footing,

$$v = \frac{V}{bjd} = \frac{1376}{12 \times 0.9 \times 10} = 12.7 \text{ lb/in.}^2 < 25 \text{ lb/in.}^2 \qquad \therefore \text{ OK}$$

No special key required between the stem wall and the footing. If a shear key were needed, it might resemble one of the details shown in Figure 14-5.

2. STABILITY OF WALL—OVERTURNING MOMENT AND RESISTING MOMENT

Total lateral load on wall:

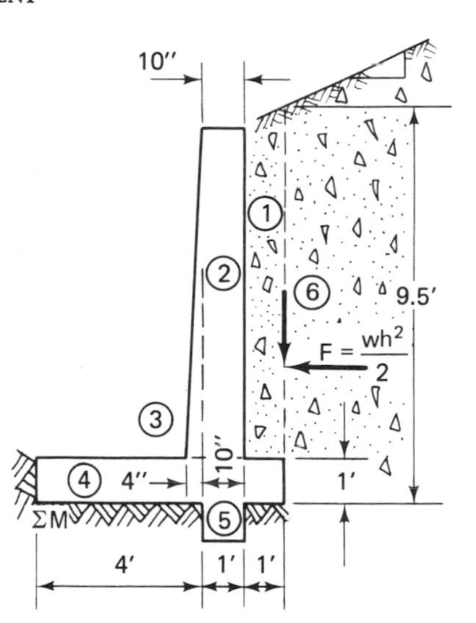

$$K_h = \frac{wh^2}{2} = \frac{43 \times 9.5^2}{2} = 1940 \text{ lb}$$

Vertical component:

$$K_v = \frac{1}{3} K_H = \frac{1940}{3} = 647 \text{ lb}$$

Forces on the wall:

		Wt.	Arm	Moment
(1) Soil	$1 \times 8.25 \text{ ft} \times 100$	= 825	× 5.5	= 4538
(2) Stem wall	$10/12 \times 8 \times 120$	= 800	× 4.6	= 3680
(3) Stem wall	$\frac{1}{2} \times \frac{4}{12} \times 8 \times 120$	= 160	× 4.05	= 648
(4) Footing	$1 \times 6 \times 150$	= 900	× 3	= 2700
(5) Key	$1 \times 1 \times 150$	= 150	× 4.5	= 675
(6) Vertical force		= 647	× 6	= 3882
		W = 3482 lb		M_R = 16123 ft-lb resisting moment

Overturning moment (OTM):

$$M_{\text{ovt}} = \frac{wh^3}{6} = \frac{43 \times 9.5^3}{6} = 6144 \text{ ft-lb}$$

$$= \frac{M_R}{M_{\text{ovt}}} = \frac{16{,}123}{6144} = 2.62 > 1.5 \qquad \therefore \text{ OK}$$

Resisting moment must be at least $1\frac{1}{2}$ times overturning moment per UBC.

Location of resultant on footing:

$$\text{distance from } O = \frac{16{,}123 - 6144}{3482} = 2.86 \text{ ft (within middle third)}$$

The resultant need not fall within the middle third, if the maximum soil pressure does not exceed the allowable. Refer to discussion at the conclusion of this problem.

3. Soil Bearing

$$\text{soil bearing} = \frac{W}{A} \pm \frac{We}{S} = \frac{W}{l} \pm \frac{6We}{l^2}$$

where $A = bl = 1 \text{ ft.} \times l = 1 \times 6 = 6 \text{ ft}^2$

$$S = \frac{bh^2}{6} = \frac{1 \times l^2}{6} = \frac{1 \times 6^2}{6} = 6 \text{ ft}^3$$

$$e = \frac{6}{2} - 2.86 = 0.14 \text{ ft}$$

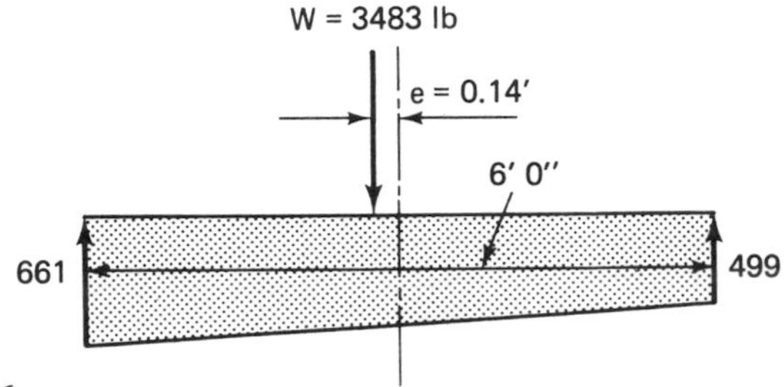

$$\text{soil bearing} = \frac{3482}{6} \pm \frac{3482 \times 0.14 \times 6}{6^2}$$

$$= 580 \pm 81 = 661 \text{ lb/ft}^2 \quad \text{or} \quad 499 \text{ lb/ft}^2$$

allowable soil bearing $= 2000 \text{ lb/ft}^2$ $\quad \therefore$ OK

4. Design of Foundation

a. Toe Design

$$\text{soil pressure at } A = 499 + (661 - 499)\frac{2.16}{6.0}$$

$$= 499 + 58 = 557 \text{ lb/ft}^2$$

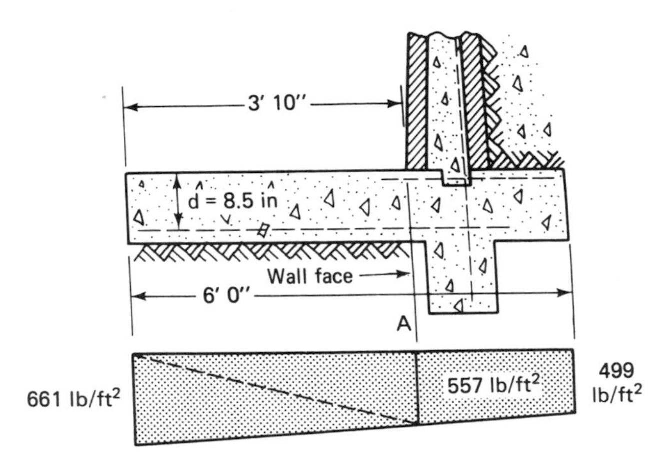

$$\text{moment on toe} = \tfrac{1}{2} \times 661 \times 3.83 \times \tfrac{2}{3} \times 3.83 + \tfrac{1}{2} \times 557 \times 3.83 \times \tfrac{1}{3} \times 3.83 - 150 \times 3.83 \times \tfrac{1}{2} \times 3.83 = 3494 \text{ ft-lb}$$

$$A_s = \frac{M}{f_s jd} = \frac{3494 \times 12}{20{,}000 \times 0.9 \times 8.5} = 0.27 \text{ in}^2/\text{ft}$$

b. Heel Design

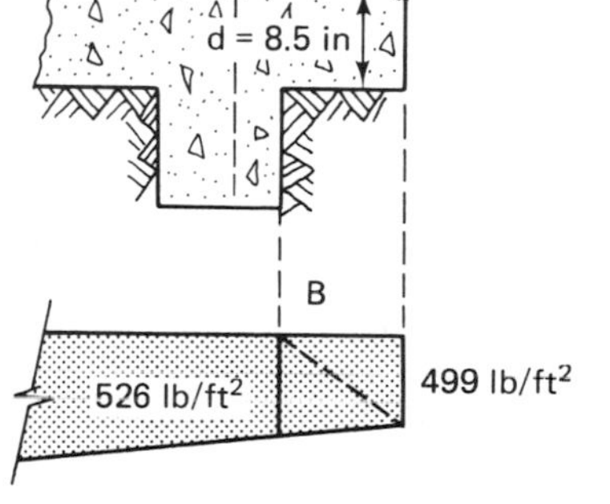

$$\text{soil pressure at } B = 499 + (661 - 499)\tfrac{1}{6} = 526 \text{ lb/ft}^2$$

$$\text{moment on heel} = 150 \times \tfrac{1}{2} + 825 \times \tfrac{1}{2} - \tfrac{1}{2} \times 526 \times \tfrac{1}{3} - \tfrac{1}{2} \times 499 \times \tfrac{2}{3} = 234 \text{ ft-lb}$$

$$A_s = \frac{M}{f_s jd} = \frac{234 \times 12}{20{,}000 \times 0.9 \times 8.5} = 0.018 \text{ in.}^2/\text{ft}$$

$$\text{min. steel} = 0.002bt = 0.002 \times 12 \times 12 = 0.288 \text{ in./ft}$$

Use No. 5's at 12 in. oc ($A_s = 0.31$ in.^{2}/ft)

5. Sliding of Wall

$$\text{lateral force on wall} = \frac{wh^2}{2} = \frac{43 \times 9.5^2}{2} = 1940 \text{ lb}$$

coef. of friction of concrete on earth, $\mu = 0.5$

$$\text{sliding resistance } \mu W = 0.5 \times 3482 = 1741 \text{ lb} < 1940 \text{ lb provide key}$$

Sliding key design—assume a *12 × 12 in. key.*

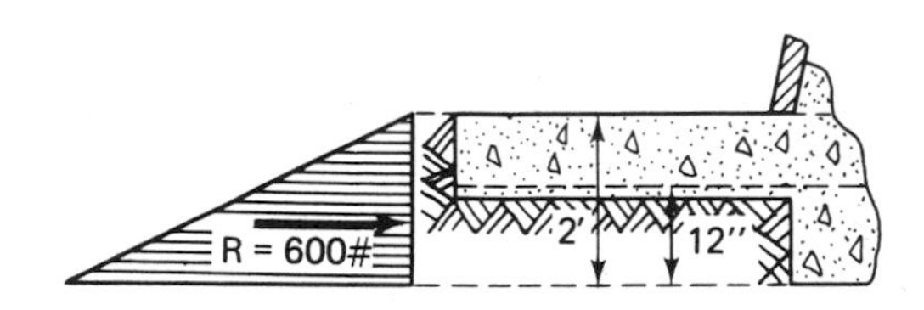

passive resistance of soil = 300 lb/ft²

$$R = \frac{wh^2}{2} = \frac{300 \times 2^2}{2} = 600 \text{ lb}$$

total sliding resistance = 1741 + 600 = 2341 lb > 1940 lb ∴ OK

As noted previously, the resultant of the overturning forces may not always fall within the middle third. If that be the case, the soil pressure diagram would be triangular, as shown: Then

$$\text{eccentricity } e = \frac{l}{2} - \frac{M_R - M_{ovt}}{\sum W}$$

$$\text{length of compression area, } l' = 3(l/2 - e)$$

$$\text{maximum soil pressure} = \frac{2 \sum W}{l'}$$

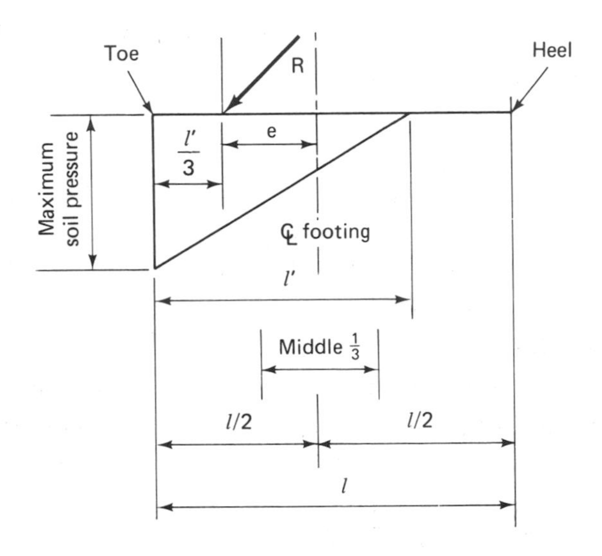

This maximum value must not exceed the allowable soil pressure. If it does, then either the toe or the heel must be extended. This dimensional increase will have the effect of decreasing e and will also increase the length of the compression area, l'.

CONCRETE MASONRY RETAINING WALLS

The design procedure for a retaining wall constructed of reinforced concrete masonry units is somewhat similar to that of reinforced grouted brick. The dimensions and design stresses would simply have to reflect the material characteristics of concrete block. For this reason, it does not appear necessary to also present design calculations exemplifying a concrete block retaining wall. However, since certain of the construction details are quite different, some typical cases are shown in Figure 14-6. These are reproduced from the *Masonry Design Manual* published by the Masonry Institute of America. What is essentially presented therein are the block thickness and the reinforcement requirements (X bars and Y bars) for wall heights

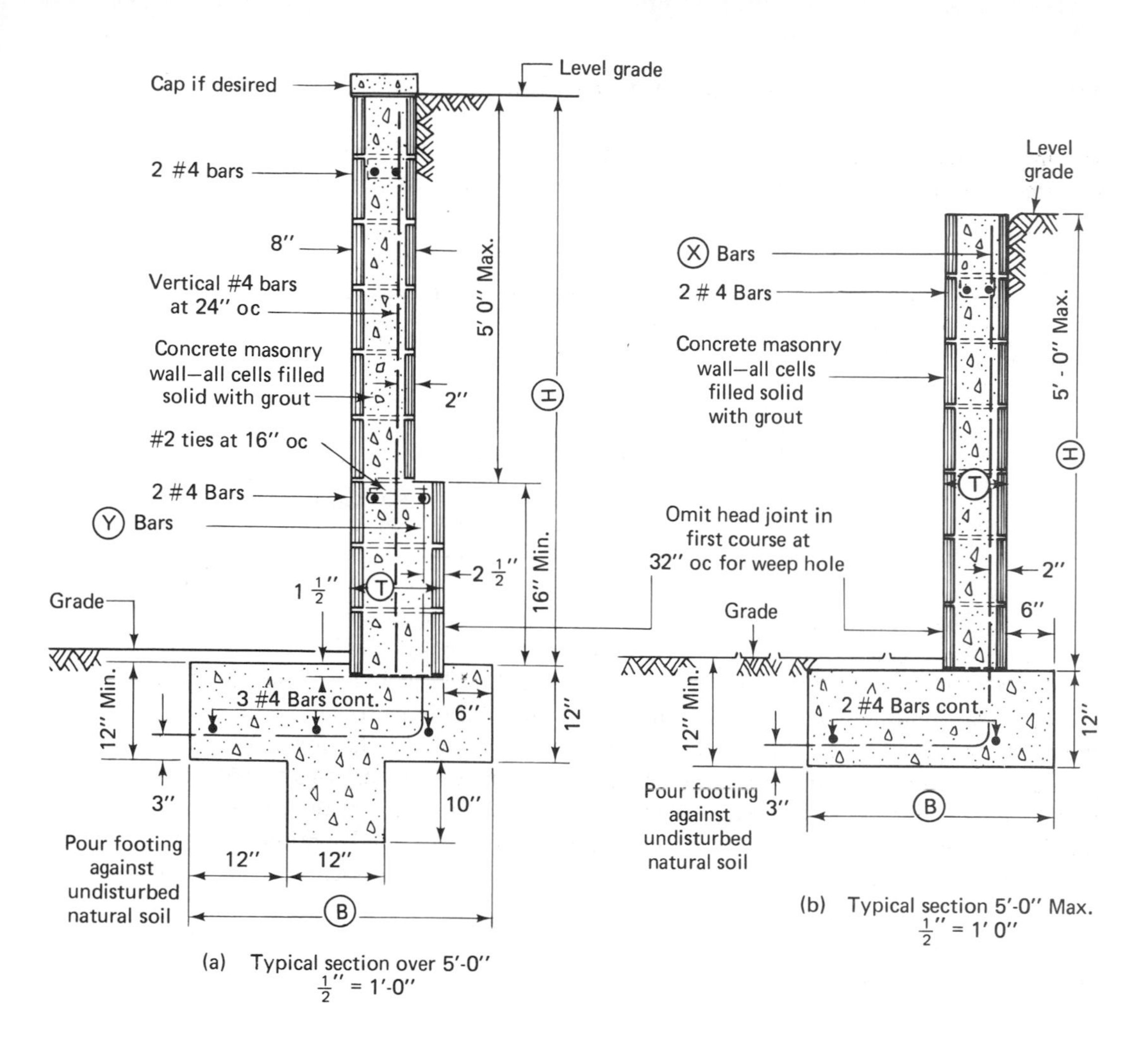

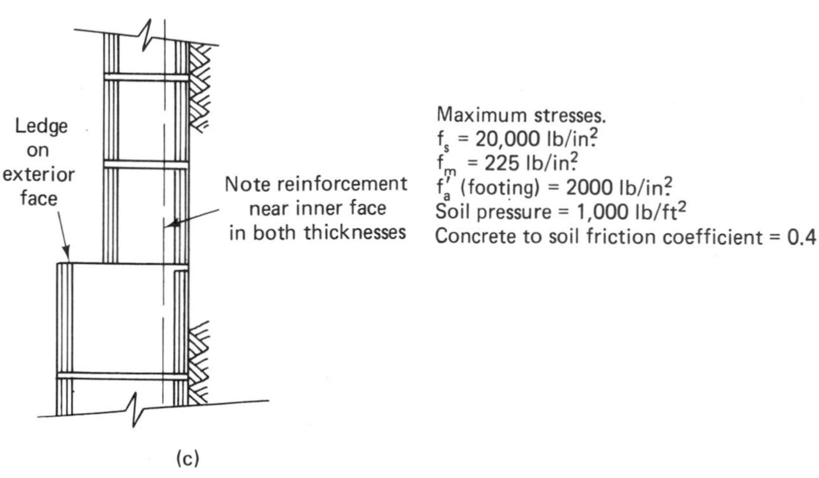

Maximum stresses.
f_s = 20,000 lb/in.²
f_m = 225 lb/in.²
f'_a (footing) = 2000 lb/in.²
Soil pressure = 1,000 lb/ft²
Concrete to soil friction coefficient = 0.4

FIGURE 14-6. Concrete masonry retaining wall details.

varying between 3 and 8 ft maximum, with a level backfill (Table 14-4). Also the required footing length is given. Both L- and inverted T-shaped sections are included.

TABLE 14-4

Concrete Masonry Retaining Wall—Dimensions and Reinforcing

Ⓗ(*ft*)	Ⓣ(*in.*)	Ⓑ	Ⓧ*bars*	Ⓨ*bars*
3	6	1 ft 9 in.	No. 3 at 32 in. oc	—
4	8	2 ft 2 in.	No. 4 at 48 in. oc	—
5	8	2 ft 9 in.	No. 4 at 24 in. oc	—
6	12	3 ft 3 in.	—	No. 4 at 24 in. oc
7	12	3 ft 10 in.	—	No. 4 at 16 in. oc
8	12	4 ft 6 in.	—	No. 5 at 16 in. oc

Note that when *H* exceeds 5 ft, the wall thickness must be increased from 8 to 12 in. for a height of at least 16 in. (two courses of block). The juncture of the 8- and 12-in. thickness is very critical. Unless those ties (No. 2's at 16 in. oc) are carefully grouted in around the pair of continuous No. 4 horizontal bars, it is quite likely that the earth pressure will crack the wall at this joint or even force the upper 8-in. thick portion off lower 12-in. thick wall. Another way to protect against this would be to move the 8-in. portion to the rear so that the inner wall face against the earth is even for the full height of the wall. This would permit the steel to extend continuously near the interior wall face in both thicknesses, as shown in Figure 14-6c, whereas in the condition shown in Figure 14-6a, the steel falls well into the interior of the grout space in the 12-in. portion. Moving the upper part back does leave a ledge on the exterior face, of course, to which some may object.

PROBLEMS

14-1. Given a solid-grouted concrete block retaining wall as shown. No inspection is provided.

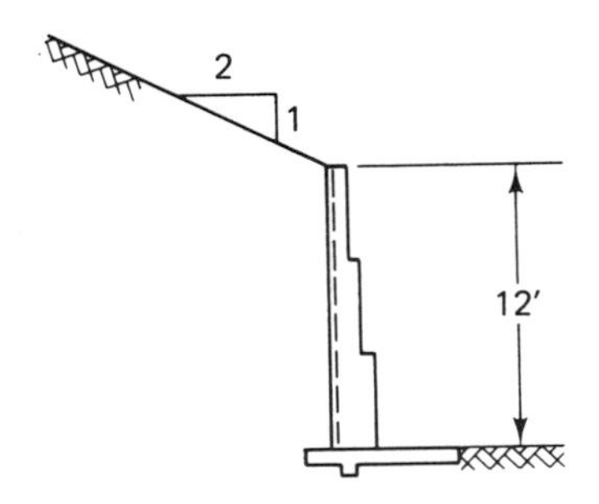

(a) Determine the thickness and reinforcement needed at the base of the wall, assuming balanced condition.
(b) Locate positions where the thickness of the wall can be reduced. Determine the steel required in each part and the steel cutoff points. Check the bond and shear requirement.

14-2. Repeat Problem 14-1 for the wall shown. Determine the area of steel ties needed at X.

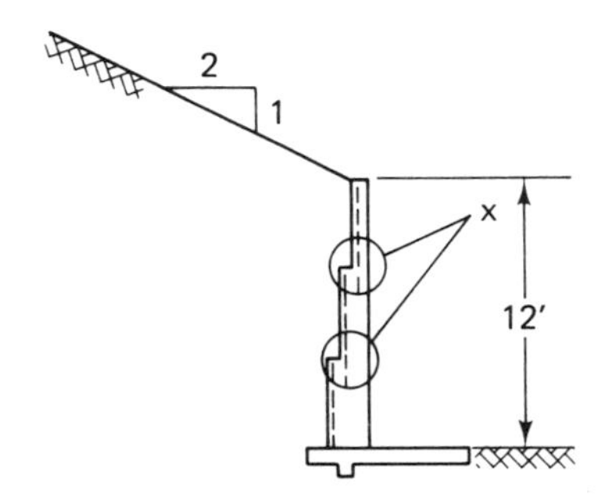

15

Masonry Veneer

Veneer is one of man's oldest techniques for beautifying structures. Many beautiful examples survive today from the golden age of the early Greek, Roman, and Byzantine, and even earlier Mesopotamia civilizations in which veneer was widely used. When man ceased to be a nomad, solely dependent on hunting for survival, discovering agriculture and the luxury of settled habitation, architecture was born. Its early development was slow but man's ambitions soon turned from building a mere shelter to creating a tomb, a monument by which to be remembered, a shrine to the gods, a palace from which to rule or to live in power and grandeur, all built to impress. They were beautified by line and proportion, but also by the mosaicists.

STRUCTURAL VERSUS NONSTRUCTURAL

Veneer is a surface beautification that is not intended nor designed to add to the structural capacity. However, there are certain structural considerations. It is supported by the structure, and local portions of it must therefore support the veneer weight itself. Structural separations that permit differential movements must be provided for in the veneer surface. The deflections of the structure or the expansion and contraction of the veneer due to moisture or temperature volume change must be provided for in the detail provisions of veneer attachment.

The influence of the structure deflections must be recognized, as is demonstrated in the following actual case. A marble exterior veneer was attached to a multistory concrete structure. The facing consisted of 5-ft-wide column area facings and 5-ft-deep spandrel facings. The floor slab consisted of a relatively thin, long-span prestressed flat slab. Owing to creep and load, the flat slab could deflect almost 1 in. in time. However, the spandrel could deflect only about $\frac{1}{8}$ in. without overstressing. The attachments then pulled loose, because they could not accommodate the large differential deflections and some large marble slabs fell to the street below. This caused a rather traumatic reaction in some pedestrians!

Another failure may occur when a stiff veneer facing is applied to a wood structure. The shrinkage and deflection of green or side grain wood can be considerable, whereas the veneer facing will not deflect very much, since it is rather stiff. Hence the attachments may pull loose under those circumstances of differential movement.

DESIGN AND USE

The modern use of veneer was summarized and codified in a revision of the veneer chapter in the 1967 UBC. These are provisions that are rather logical in nature and therefore will be used as a pattern for the following recommended procedures. It must be emphasized that there may be different code requirements in different locations as a result of local experience, materials, and practice that may have developed over the ages of veneer use. The designer must verify those code requirements in force at the building site pertinent to the local materials and methods.

Summary

The requirements of the UBC are divided into the following sections:

1. General requirements and limitations.
2. Definition of the two types, adhered and anchored.
3. Requirements for design, and listing of design criteria for each type.
4. Provision for use of certain Standards that may be used in lieu of design.

General limitations

Some of the arbitrary or general limitations imposed, in the interest of avoiding distress, are as follows:

1. A maximum height of 25 ft above ground for wood support.

2. Anchored veneer supports to be not more than 20 ft maximum for the first story, and not more that 12 ft, or story height, above that.
3. Supports to be incombustible and corrosion-resistant.
4. Adhered units to be not more that 36 in. in greatest dimension, nor more than 720 in.^{2} in total area.
5. Adhered units to weigh not more than 15 lb/ft^2.

General design requirements

Some of the general design type of items that must be considered are:

Differential movements between veneer and support.
Shrinkage of concrete areas or concrete supports.
Deflections of supporting structures.
Creep, as it occurs in prestressed or heavily loaded concrete.
Deflection and shrinkage of wood structures.
Deflections of limber steel frame structures.
Differential temperature/volume change.

Control joints have been effective in permitting the small movements necessary to release stress that such movements or deflections may induce. Also, they must be located in the veneer over any control joints of the underlying supporting structure.

Proper consideration of differential movements is a subject involving many factors, and its precise solution is impossible with the present inadequate material information. Common sense, experience with material performance, and observations of performance of previous installations provide the best guide. For example, exterior veneers should have expansion joints to relieve stress accumulation. How far apart? Somewhere between 12 and 20 ft seems to be a reasonable maximum.

CODE REQUIREMENTS

The following pages provide a reproduction of the UBC provisions concerned with masonry veneer, contained in Chapter 30, along with a commentary to clarify the various provisions. It is followed by the UBC Standard 30-1, which describes the specification type of time-proven methods that may be used in lieu of design. Some examples of these are illustrated in Figures 15-1 through 15-5.

CHAPTER 30

VENEER

Scope

Sec. 3001. (a) **General.** All veneer and its application shall conform to the requirements of this Code. Wainscots not exceeding 4 feet in height measured above the adjacent ground elevation for exterior veneer or the finish floor elevation for interior veneer may be exempted from the provisions of this Chapter if approved by the Building Official.

COMMENTARY

(a) The 4 foot height is exempted from requirements because there is no appreciable hazard to life in the failure of such items.

(b) **Limitations.** Exterior veneer shall not be attached to wood frame construction at a point more than 25 feet in height above the adjacent ground elevation except when approved by the Building Official considering special construction designed to provide for differential movement.

(b) The limitation of 25 feet above adjacent ground is because of the possibility of shrinkage and deflection of the wood construction compared to the negligible vertical deflections that will occur in the stiff masonry veneer. However, if special provisions are made to minimize this type of distress, the building official may approve such details for greater heights

Definitions

Sec. 3002. For the purpose of this Chapter, certain terms are defined as follows:

BACKING as used in this Chapter is the surface or assembly to which veneer is attached.

VENEER is nonstructural facing of brick, concrete, stone, tile, metal, plastic or other similar approved material attached to a backing for the purpose of ornamentation, protection, or insulation.

Adhered Veneer is veneer secured and supported through adhesion to an approved bonding material applied over an approved backing.

Adhered Veneer is sometimes known as adhesive veneer.

Anchored Veneer is veneer secured to and supported by approved mechanical fasteners attached to an approved backing.

Exterior Veneer is veneer applied to weather-exposed surfaces as defined in Section 424.

Interior Veneer is veneer applied to surfaces other than weather-exposed surfaces as defined in Section 424.

Materials

Sec. 3003. Materials used in the application of veneer shall conform to the applicable requirements for such materials as set forth elsewhere in this Code.

For masonry units and mortar see Chapter 24.

For precast concrete units see Chapter 26.

For portland cement plaster see Chapter 47.

Anchors, supports and ties shall be noncombustible and corrosion-resistant.

The materials used in masonry veneer are of the same general character as outlined for masonry units, mortar, grout and so forth as specifically called out in Chapter 24 in more detail.

Design

Sec. 3004. (a) **General.** The design of all veneer shall comply with the requirements of Chapter 23 and this Section.

Veneer shall support no load other than its own weight and the vertical dead load of veneer above.

Surfaces to which veneer is attached shall be designed to support the additional vertical and lateral loads imposed by the veneer.

Consideration shall be given for differential movement of supports including that caused by temperature changes, shrinkage, creep and deflection.

(a) The basic requirements for structural design consideration.

The consideration of differential movement requires careful consideration of the expansion and contraction due to temperature changes, the shrinkage that may occur in concrete or creep that may occur in prestressed concrete structures on which the veneer is mounted. Also, the deflection of the structure under load must be considered because the masonry veneer will be stiff with relatively negligible deflection.

(b) **Adhered Veneer.** Adhered veneer and its backing shall be designed to have a bond to the supporting element sufficient to withstand a shearing stress of 50 pounds per square inch.

(b) The design requirement of adhered veneer.

(c) **Anchored Veneer.** Anchored veneer and its attachments shall be designed to resist a horizontal force equal to twice the weight of the veneer.

(c) This is to provide for security and safety. Some Codes require more than double this value but that is not really necessary.

Adhered veneer

Sec. 3005. (a) **Permitted Backing.** Backing shall be continuous and may be of any material permitted by this Code. It shall have surfaces prepared to secure and support the imposed loads of veneer.

Exterior veneer, including its backing, shall provide a weatherproof covering.

For additional backing requirements, see Sections 1707 (a), 1711 (a), 1711 (b) and 2517 (g).

(b) **Area Limitations.** The height and length of veneered areas shall be unlimited except as required to control expansion and contraction and as limited by Section 3001 (b).

(c) **Unit Size Limitations.** Veneer units shall not exceed 36 inches in the greatest dimension, nor more than 720 square inches in total area and shall weigh not more than 15 pounds per square foot unless approved by the Building Official.

EXCEPTION: Veneer units weighing less than 3 pounds per square foot shall not be limited in dimension or area.

(c) The unit size limitations of 36 inches and 720 square inches in total area as well as 15 pounds per square foot are because of the practical difficulties in installing larger elements, unless one makes special provision in the design. As noted in the Exception, material weighing less than 3 pounds per square foot is not necessarily limited in dimension or area, because this lightweight material can be installed easily in relatively large units.

(d) **Application.** In lieu of the design required by Section 3004 (a) adhered veneer may be applied by one of the methods specified in U.B.C. Standard No. 30-1.

(d) This is the paragraph which permits the use of the standard specification type of detail in lieu of design.

(e) **Plastic Veneer.** Plastics used as veneer shall conform to the provisions of Chapter 52. When used within a building, plastic veneer shall comply with the interior finish requirements of Chapter 42. All plastic veneer shall be installed in an approved manner.

Anchored veneer

Sec. 3006. (a) **Permitted Backing.** Backing may be of any material permitted by this Code. Exterior veneer including its backing shall provide a weatherproof covering.

(a) The weatherproof requirements are generally spelled out in Chapter 17.

(b) **Height and Support Limitations.** Anchored veneer shall be supported on footings, foundations, or other noncombustible supports.

Where anchored veneer is applied more than 25 feet above the adjacent ground elevation, it shall be supported by noncombustible, corrosion-resistant, structural framing having horizontal supports spaced not over 12 feet vertically above the 25-foot height.

Noncombustible, noncorrosive lintels and noncombustible supports shall be provided over all openings where the veneer unit is not self-spanning. The deflections of all structural lintels and horizontal supports required by this Subsection shall not exceed 1/500 of the span under full load of the veneer.

(b) These requirements are primarily to provide that the stiff, unyielding masonry veneer not be subjected to the distress caused by flexible deflecting supports which, by differential movement, may impose excessive stress on the anchorages. Incombustible and corrosion resistant lintels can be provided by appropriate metals or by providing reinforcing within the veneer which provides an internal or self-supporting lintel or support.

(c) **Area Limitations.** The area and length of anchored veneer walls shall be unlimited, except as required to control expansion and contraction and by Section 3001 (b).

(d) **Application.** In lieu of the design required by Section 3004 anchored veneer may be applied by one of the methods specified in U.B.C. Standard No. 30-1.

(d) This provides for the time-proven specification type of installations listed in the Standards volume of the Uniform Building Code.

UNIFORM BUILDING CODE STANDARD NO. 30-1
(1970 Edition, Standard No. 30-1)

VENEER APPLICATION

Based on recommended standards of the International Conference of Building Officials

(See Sections 3005 and 3006, Uniform Building Code, Volume 1)

SCOPE

Sec. 30.101. This Standard provides methods for application of adhered and anchored veneer where designs are not provided.

LIMITATIONS

Sec. 30.102. Veneer applied in accordance with this Standard shall be limited as to size, area, height, length and location in accordance with Chapter 30 of the Uniform Building Code, Volume I.

ADHERED VENEER

Sec. 30.103. One of the following application methods may be used:

1. A paste of neat portland cement shall be brushed on the backing and the back of the veneer unit. Type S mortar then shall be applied to the backing and the veneer unit. Sufficient mortar shall be used to create a slight excess to be forced out the edges of the units. The units shall be tapped into place so as to completely fill the space between the units and the backing. The resulting thickness of mortar in back of the units shall be not less than one-half inch ($\frac{1}{2}''$) nor more than one and one-fourth inches ($1\frac{1}{4}''$). See Figure 15-1.

2. Units of tile, masonry, stone or terra cotta, not over one inch ($1''$) in thickness shall be restricted to eighty-one square inches (81 sq. in.) in area unless the back side of each unit is ground or box screeded to true up any deviations from plane. These units and glass mosaic units of tile not over two inches by two inches by three-eighths inch ($2'' \times 2'' \times \frac{3}{8}''$) in size may be adhered by means of portland cement. Backing may be of masonry, concrete or portland cement plaster on metal lath. Metal lath shall be fastened to the supports in accordance with the requirements of Chapter 47 of the Uniform Building Code, Volume I. Type S mortar shall be applied to the backing as a setting bed. The setting bed shall be a minimum of three-eighths-inch ($\frac{3}{8}''$) thick and a maximum of three-fourths-inch ($\frac{3}{4}''$)

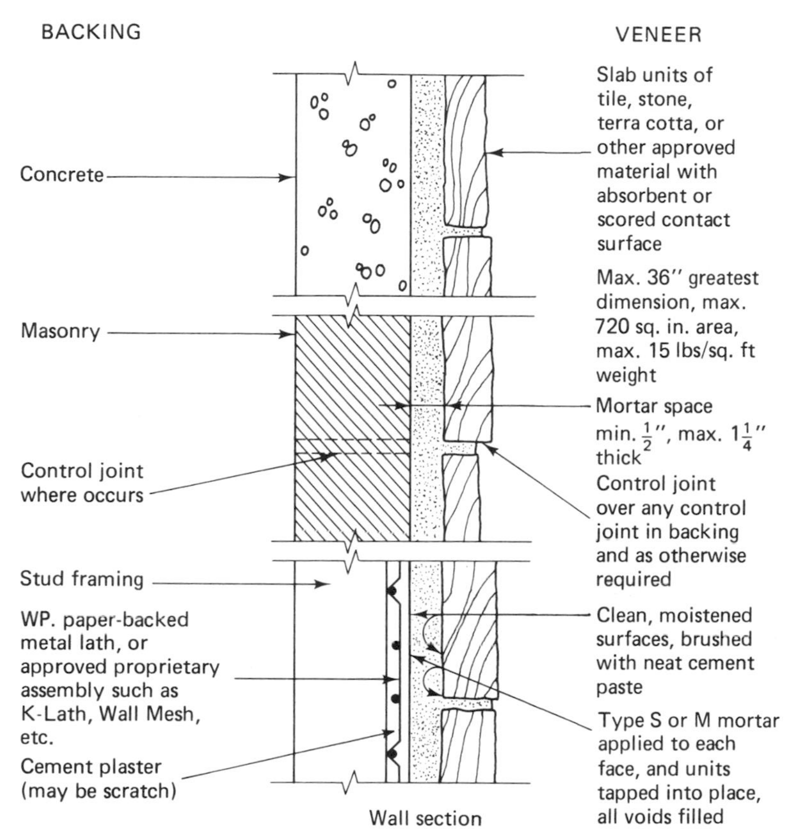

FIGURE 15-1. Masonry veneer units adhered to concrete, masonry, or plaster backing (Reference: Uniform Building Code Standard 30-1-70, Section 30.103, 1).

thick. A paste of neat portland cement or half portland cement and half graded sand shall be applied to the back of the exterior veneer units and to the setting bed and the veneer pressed and tapped into place to provide complete coverage between the mortar bed and veneer unit. A portland cement grout shall be used to point the veneer.

ANCHORED VENEER

Sec. 30.104. (a) **Masonry and Stone Units [Five Inches (5″) Maximum in Thickness].** Masonry and stone veneer not exceeding five inches (5″) in thickness may be anchored directly to structural masonry, concrete, or studs in one of the following manners:

1. Anchor ties shall be corrosion-resistant and if made of sheet metal, shall have a minimum size of No. 22 gauge by one inch (1″) or if of wire, shall be a minimum of No. 9 gauge. Anchor ties shall be spaced so as to support not more than two square feet (2 sq. ft.) of wall area but not more than twenty-four inches (24″) on center horizontally. In Seismic Zone No. 3, anchor ties shall be provided to horizontal joint reinforcement wire of No. 9 gauge or equivalent. The joint reinforcement shall be continuous with butt splices between ties permitted. See Figure 15-2.

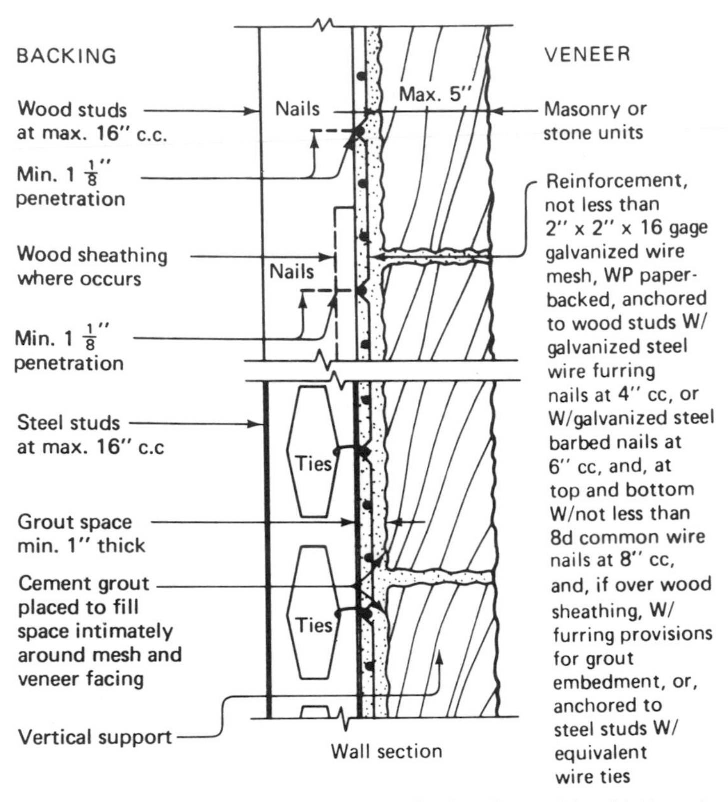

FIGURE 15-2. Masonry or stone veneer units (maximum 5 in. thick) anchored to stud backing (paper-backed mesh) [Reference: Uniform Building Code Standard 30-1-70, Section 3.0.104(a)2].

When applied over stud construction, the studs shall be spaced a maximum of sixteen inches (16″) on centers and approved paper shall first be applied over the sheathing or wires between studs except as otherwise provided in Section 1707, Uniform Building Code, Volume I, and mortar shall be slushed into the one-inch (1″) space between facing and paper.

As an alternate, an air space of at least one inch (1″) may be maintained between the backing and the veneer in which case temporary spot bedding may be used away from the ties to align the veneer. Spot bedding at the ties shall be of cement mortar entirely surrounding the ties. See Figure 15-3.

2. Veneer may be applied with a one-inch (1″) minimum grouted backing space which is reinforced by not less than two-inch by two-inch (2″ × 2″) No. 16 gauge galvanized wire mesh placed over waterproof paper backing and anchored directly to stud construction.

The stud spacing shall not exceed sixteen inches (16″) on center. The galvanized wire mesh shall be anchored to wood studs by galvanized steel wire furring nails at four inches (4″) on center or by barbed galvanized nails at six inches (6″) on center with a one and one-eighth-inch ($1\frac{1}{8}$″) minimum penetration. The galvanized wire mesh may be attached to

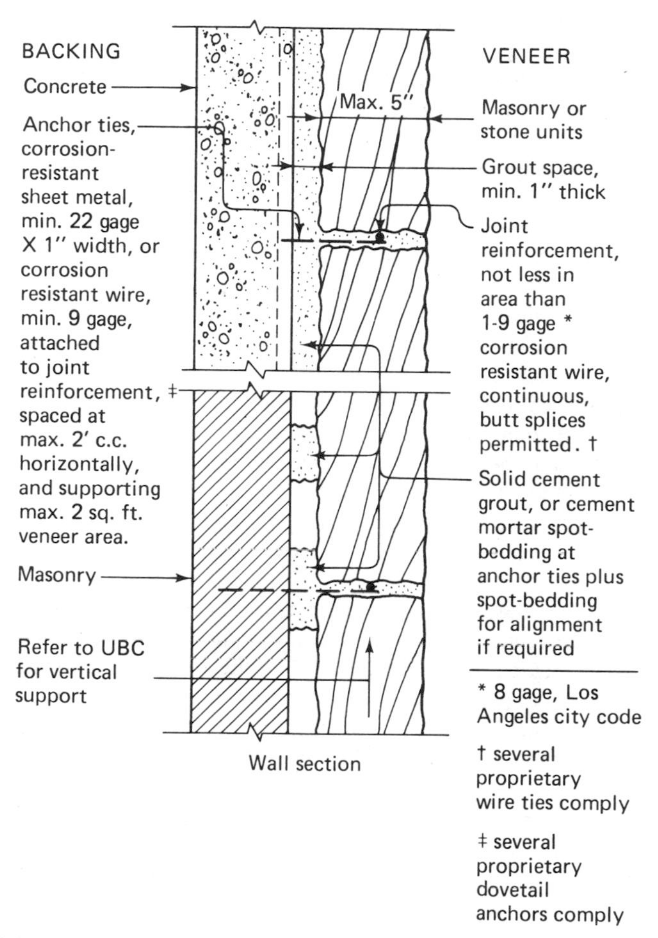

FIGURE 15-3. Masonry or stone veneer units (maximum 5 in. thick) anchored to concrete or masonry backing [Reference: Uniform Building Code Standard 30-1-70, Section 30.104(a)1].

steel studs by equivalent wire ties. If this method is applied over solid sheathing, the mesh must be furred for embedment in grout. The wire mesh must be attached at the top and bottom with not less than 8-penny common wire nails. The grout fill shall be placed to fill the space intimately around the mesh and veneer facing.

(b) **Stone Units [Ten Inches (10″) Maximum in Thickness].** Stone veneer units not exceeding ten inches (10″) in thickness may be anchored directly to structural masonry, concrete or to studs:

1. **With concrete or masonry backing.** Anchor ties shall be not less than No. 12 gauge galvanized wire or approved equal formed as an exposed eye and extending not less than one-half inch ($\frac{1}{2}$″) beyond the face of the backing. The legs of the loops shall be not less than six inches (6″) in length bent at right angles and laid in the masonry mortar joint and

spaced so that the eyes or loops are twelve-inches (12″) maximum on center in both directions. There shall be provided not less than a No. 12 gauge galvanized wire tie or approved equal threaded through the exposed loops for every two square feet (2 sq. ft.) of stone veneer. This tie shall be a loop having legs not less than fifteen inches (15″) in length so bent that it will lie in the stone veneer mortar joint. The last two inches (2″) of each wire leg shall have a right angle bend. One inch (1″) of cement grout shall be placed between the backing and the stone veneer.

2. **With stud backing.** A two-inch by two-inch (2″ × 2″) No. 16 gauge galvanized wire mesh with two layers of waterproof paper backing shall be applied directly to wood studs spaced a maximum of sixteen inches (16″) on center. On studs the mesh shall be attached with two-inch (2″) long galvanized steel wire furring nails at four inches (4″) on center providing a minimum one and one-eighth-inch ($1\frac{1}{8}$″) penetration into each stud and with 8-penny common nails at eight inches (8″) on center into top and bottom plates. The galvanized wire mesh may be attached to steel studs with equivalent wire ties. There shall be not less than a No. 12 gauge galvanized wire or approved equal looped through the mesh for every two square feet (2 sq. ft.) of stone veneer. This tie shall be a loop having legs not less than fifteen inches (15″) in length, so bent that it will lie in the stone veneer mortar joint.

The last two inches (2″) of each wire leg shall have a right angle bend. One-inch (1″) minimum thickness of cement grout shall be placed between the backing and the stone veneer.

(c) **Slab Type Units [Two Inches (2″) Maximum in Thickness].** For veneer units of marble, travertine, granite or other stone units of slab form, ties shall engage drilled eyes of corrosion-resistant metal dowels located in the middle third of the edge of the units spaced a maximum of eighteen inches (18″) apart around the periphery of each unit with not less than four ties per veneer unit. Units shall not exceed twenty square feet (20 sq. ft.) in area.

If the dowels are not tight fitting, the holes may be drilled not more than one-sixteenth inch ($\frac{1}{16}$″) larger in diameter than the dowel with the hole countersunk to a diameter and depth equal to twice the diameter of the dowel in order to provide a tight-fitting key of cement mortar at the dowel locations when the mortar in the joint has set.

All veneer ties shall be corrosion-resistant metal capable of resisting in tension or compression a force equal to two times the weight of the attached veneer.

If made of sheet metal, veneer ties shall be not smaller in area than one-sixteenth inch by one inch ($\frac{1}{16}$″ × 1″) or if made of wire, not smaller in diameter than No. 9 gauge wire.

(d) **Terra Cotta or Ceramic Units.** Tied terra cotta or ceramic veneer units shall be not less than one and one-fourth inches ($1\frac{1}{4}$″) in thickness with projecting dovetail webs on the back surface spaced approximately eight inches (8″) on centers. The facing shall be tied to the backing wall with noncorrosive metal anchors not less than No. 8 gauge wire installed at the top of each piece in horizontal bed joints not less than twelve inches (12″) or more than eighteen inches (18″) on centers; these anchors shall be secured to one-fourth-inch ($\frac{1}{4}$″) galvanized pencil rods which pass through the vertical aligned loop anchors in the backing wall. The veneer ties shall have sufficient strength to support the full weight of the veneer in tension. The facing shall be set with not less than a two-inch (2″) space from the backing wall and the space shall be filled solidly with portland cement grout and pea gravel. Immediately prior to setting, the backing wall and the facing shall be drenched with clean water and shall be distinctly damp when the grout is poured. See Figure 15-4.

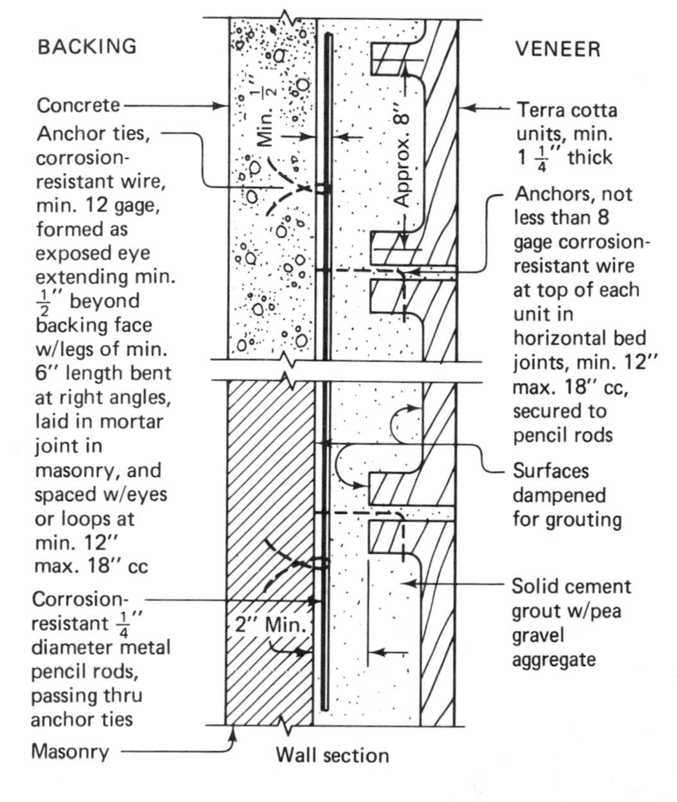

FIGURE 15-4. Terra-cotta veneer units anchored to concrete or masonry backing [Reference: Uniform Building Code Standard 30-1-70, Section 30.104(d)].

SUMMARY

Veneer is a surface application that does not contribute to structural capacity. However, it may be designed by masonry structural design methods described in previous chapters. The design criteria prescribed in the UBC are:

1. Develop 50 lb/in.^{2} in shear and tension between the veneer and support material for adhered veneer.
2. Develop two times the veneer weight in structural support for anchored veneer.

However, in lieu of design, one may use the specification type of installations listed in the UBC Standards (see Figures 15-1, 15-2, 15-3, and 15-4). These have been used successfully for many years after a period of development by trial and error. Certain modifications may be made, such as use of newer, specially developed cements or adhesives. A series of special cementing adhesives is covered in American National Standards Institute specifications, for example, the ANSI A108 and A118 series.

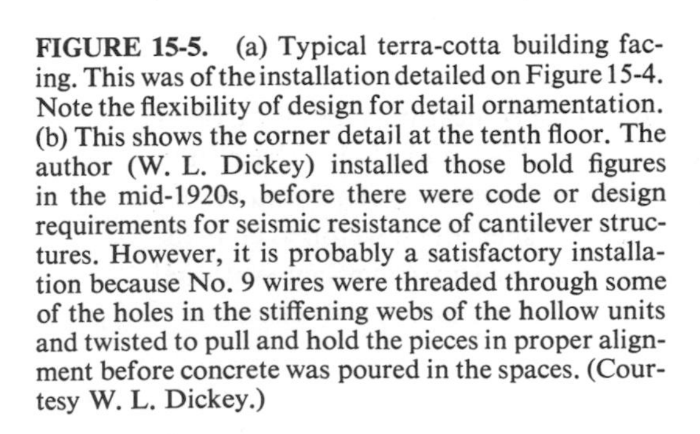

FIGURE 15-5. (a) Typical terra-cotta building facing. This was of the installation detailed on Figure 15-4. Note the flexibility of design for detail ornamentation. (b) This shows the corner detail at the tenth floor. The author (W. L. Dickey) installed those bold figures in the mid-1920s, before there were code or design requirements for seismic resistance of cantilever structures. However, it is probably a satisfactory installation because No. 9 wires were threaded through some of the holes in the stiffening webs of the hollow units and twisted to pull and hold the pieces in proper alignment before concrete was poured in the spaces. (Courtesy W. L. Dickey.)

16

Environmental Features of Masonry Construction

There is need for consideration of items other than the structural resistance to load. The purpose of a building is to provide shelter and a controlled environment. Fire resistance is certainly one factor that must be included to meet these needs properly. The intrinsic importance of fire resistance varies in different types of occupancies, and the codes for design practice have developed tables for its evaluation. Another environmentally important function is to provide relief from undesirable sound. Still a third factor, due to the awareness of energy conservation, involves heat conductivity, with its consideration of comfort and energy savings. Also, the basic function of weatherproofing must be assured. These items may dictate the selection of the type of material and size rather than the load requirements. They should be checked prior to any stress analysis (e.g., the stress analysis of a 6-in. wall would be wasted if a later check showed that 8 in. is necessary for the required fire rating).

FIRE RESISTANCE

Requirements

The hazards of various occupancies or uses are recognized by the Uniform Building Code in its classification of such occupancies. Some of the factors that are interrelated in the consideration of hazards are (1) combustibility of contents, (2) the type of personnel occupancy, (3) the size of the hazardous area, (4) the fire resistance of the structure, (5) the location and accessibility for fire fighters, (6) fire zones or hazards, and (7) similar factors. These are tabulated for clarity and for interrelationship in various tables of the UBC. The goal of the tabulation is to classify and grade or rate the hazards and then meet them with consistent resistance or protection.

Table 16-1 shows typical fire-resistive requirements for exterior walls, and for openings in the exterior walls, for various types of occupancy classification. These uses, or occupancies are described briefly in the table.

In addition to the "occupancy" or classification of building use, the table lists the fire rating required. The rating is the period of time that the wall will resist fire and the transmission of flame and temperature when exposed to a "standard fire." In this fire test, the temperature of the fire space on one side of a test wall is raised in a carefully controlled manner while thermocouples check the temperature rise on the opposite face. The rating or endurance of the wall is the time until failure by (1) passage of flame through wall, (2) collapse, or (3) temperature rise on the exterior face, whichever occurs first; or (4) failure under hose stream test.

Table 16-2 lists the basic allowable one-story floor area permitted for various types of construction for the different occupancies: for the specific hazard imposed or, conversely, the type of construction required for sizes of buildings. The larger areas require more fire-resistant type of construction, and these requirements are shown in Table 16-3. UBC Chapters 18 through 21 contain detailed requirements for these different types of construction. These basic areas may be increased for various factors that would influence or increase the safety, for example:

1. Buildings over one story in height may contain twice the maximum area noted.
2. Public yards or streets on two sides of the area will permit an increase up to 50%.
3. Public yards or streets on three or more sides will permit the area to be increased up to 100%.
4. Sprinklers provide for greater safety, so their use permits increasing the area greatly.

Table 16-4 lists the maximum height of buildings permitted for various types of construction for the different occupancies. These heights may be increased if the building is sprinklered.

The body of local codes and ordinances will contain more details of requirements of occupancies and exceptions, and hence, in practice, the codes having jurisdiction over the building site must be studied carefully for a specific application.

The preceding items determine the fire rating of certain elements and the type of construction that is necessary for the proper function of the building. The requirements for the type of construction in Table 16-3 is a listing of the fire-resistive

TABLE 16-1

Wall and Opening Protection of Occupancies Based on Location on Property
(*Partial UBC Table 5-A. Obtain current revisions for specific locale*)

TYPES IV AND V Construction: For exterior wall and opening protection of Types IV and V buildings, see this table. Type V construction is not permitted within fire zone 1. Exceptions to limitation for Types IV and V construction, as provided in Sections 1109, 2103, and 2203 apply. For Types I, II, and III construction, see Sections 1803, 1903, and 2003.

Group	*Description of occupancy*	*Fire zone*	*Fire resistance of exterior walls*	*Openings in exterior walls*
A	Any assembly building with a stage and an occupant load of 1000 or more in the building		Not applicable [See Section 602(a)]	
B (see also Section 702)	1—Any assembly building with a stage and an occupant load of less than 1000 in the building	1	2 h less than 20 ft 1 h elsewhere	Not permitted less than 5 ft Protected less than 20 ft
	2—Any assembly building without a stage and having an occupant load of 300 or more in the building, including such buildings used for educational purposes not classed as a Group C or Group F, Division 2 Occupancy	2 and 3	2 h less than 10 ft 1 h elsewhere	Not permitted less than 5 ft Protected less than 10 ft
	3—Any assembly building without a stage and having an occupant load of less than 300 in the building, including such buildings used for educational purposes and not classed as a Group C or Group F, Division 2 Occupancy	1	2 h less than 20 ft 1 h elsewhere	Not permitted less than 5 ft Protected less than 20 ft
		2	2 h less than 5 ft 1 h elsewhere	Not permitted less than 5 ft
		3	2 h less than 5 ft 1 h less than 10 ft	Protected less than 10 ft
	4—Stadiums, reviewing stands, and amusement park structures not included within Group A or Divisions 1, 2 and 3, Group B, Occupancies	1	2 h less than 20 ft 1 h elsewhere	Protected less than 20 ft
		2	1 h	Protected less than 10 ft
		3	1 h less than 10 ft	
C (see also Section 802)	1—Any building used for educational purposes through the 12th grade by 50 or more persons for more than 12 h/week or 4 h in any one day	1	2 h less than 20 ft 1 h elsewhere	Not permitted less than 5 ft Protected less than 20 ft
	2—Any building used for educational purposes through the 12th grade by less than 50 persons for more than 12 h/week or 4 h in any one day	2	2 h less than 10 ft 1 h elsewhere	Not permitted less than 5 ft
	3—Any building used for day care purposes for more than six children	3	2 h less than 5 ft 1 h less than 10 ft	Protected less than 10 ft

Notes: (1) See Section 504 for type of walls affected and requirements covering percentage of openings permitted in exterior walls.

(2) For additional restrictions, see Chapters under Occupancy, Fire Zones, and Types of Construction.

(3) For walls facing streets, yards, and public ways, see Part V.

(4) Openings shall be protected by a fire assembly having a $\frac{3}{4}$-h fire-protection rating.

TABLE 16-1 (Continued)

Group	Description of occupancy	Fire zone	Fire resistance of exterior walls	Openings in exterior walls
D (see also Section 902)	1—Mental hospitals, mental sanitariums, jails, prisons, reformatories, houses of correction, and buildings where personal liberties of inmates are similarly restrained		Permitted in Types I and II buildings only [See Section 902(b)]	
	2—Nurseries for full-time care of children under kindergarten age, hospitals, sanitariums, nursing homes with nonambulatory patients, and similar buildings (each accommodating more than five persons)	1	2 h less than 20 ft 1 h elsewhere	Not permitted less than 5 ft Protected less than 20 ft
		2 and 3	2 h less than 5 ft 1 h elsewhere	Not permitted less than 5 ft Protected less than 10 ft
	3—Nursing homes for ambulatory patients, homes for children of kindergarten age or over (each accommodating more than five persons)	1	2 h less than 20 ft 1 h elsewhere	Not permitted less than 3 ft Protected less than 20 ft
		2 and 3	1 h	Not permitted less than 3 ft Protected less than 10 ft
E (see also Section 1002)	1—Storage and handling of hazardous and highly inflammable or explosive materials other than flammable liquids		Not permitted in Fire Zones Nos. 1 and 2	
		3	4 h less than 5 ft 2 h less than 10 ft 1 h less than 20 ft	
	2—Storage and handling of Class I, II and III flammable liquids as specified in UBC Standard 10-1, dry cleaning plants using flammable liquids, paint stores with bulk handling, paint shops and spray painting rooms and shops 3—Woodworking establishments, planning mills, box factories, buffing rooms for tire rebuilding plants and picking rooms, shops, factories or warehouses where loose combustible fibers or dust are manufactured, processed, generated or stored, and pin refinishing rooms. 4—Repair garages	1	4 h less than 20 ft 1 h elsewhere	Not permitted less than 5 ft Protected less than 20 ft
		2	4 h less than 5 ft 2 h less than 10 ft 1 h elsewhere	
		3	4 h less than 5 ft 2 h less than 10 ft 1 h less than 20 ft	
	5—Aircraft repair hangars		Not permitted in Fire Zones Nos. 1 and 2 except as set forth in Sections 1602(c) and 1603(c).	
		3	1 h less than 60 ft	Protected less than 60 ft

TABLE 16-2

Basic Allowable Floor Area for Buildings One Story in Height in Fire Zones 1 and 2. For Buildings Located in Fire Zone 3 the Basic Area May Be Increased $33\frac{1}{3}$ % (square feet)
(*UBC Table 5-C*)

	Types of construction								
	I	*II*			*III*		*IV*	*V*	
Occupancy	*F.R.*[a]	*F.R.*[a]	*1-h*	*N*[a]	*1-h*	*N*[a]	*H.T.*[a]	*1-h*	*N*[a]
A-1	Unlimited	22,500	Not permitted						
(A) 2-2.1	Unlimited	22,500	10,100	Not permitted	10,100	Not permitted	10,100	7,900	Not permitted
(A) 3-4	Unlimited	22,500	10,100	6,800	10,100	6,800	10,100	7,900	4500
E	Unlimited	34,000	15,200	10,100	15,200	10,100	15,200	11,800	6800
(I) 1-2	Unlimited	11,300	5,100	Not permitted	5,100	Not permitted		3,900	Not permitted
(I) 3	Unlimited	11,300	Not permitted[b]				5,100		
(H) 1-2[c]	11,250	9,300	4,200	2,800	4,200	2,800	4,200	3,300	1900
(H) 3-4-5[c]	Unlimited	18,600	8,400	5,600	8,400	5,600	8,400	6,600	3800
(B) 1-2-3[d]	Unlimited	30,000	13,500	9,000	13,500	9,000	13,500	10,500	6000
B-4	Unlimited	45,000	20,300	13,500	20,300	13,500	20,300	15,800	9000
R-1	Unlimited	22,500	10,100	6,800[e]	10,100	6,800[e]	10,100	7,900	4500[e]
R-3	Unlimited								
M[f]	See Chapter 15								

[a]N, no requirement for fire resistance; F. R., fire resistive; H. T., heavy timber.
[b]See Section 902(b).
[c]For additional limitations in Fire Zones No. 1 and No. 2 see Sections 1602 and 1603.
[d]For open parking garages see Section 1109.
[e]For limitation and exceptions see Section 1302(b).
[f]For agricultural buildings also see Appendix, Chapter 15.

TABLE 16-3

Types of Construction—Fire-Resistive Requirements[a,b] (In hours)
(*UBC Table 17-A*)

	Type I	*Type II*			*Type III*		*Type IV*	*Type V*	
	Noncombustible				*Combustible*				
Building element	*Fire resistive*	*Fire resistive*	*1-Hr.*	*N*	*1-Hr.*	*N*	*H.T.*	*1-Hr.*	*N*
Exterior bearing walls	4 Sec. 1803(a)	4 1903(a)	1	N	4 2103(a)	4 2103(a)	4 2103(a)	1	N
Interior bearing walls	3	2	1	N	1	N	1	1	N
Exterior nonbearing walls	4 Sec. 1803(a)	4 1903(a)	1	N	4 2103(a)	4 2103(a)	4 2103(a)	1	N
Structural frame[c]	3	2	1	N	1	N	1 or H.T.	1	N
Partitions, permanent	1	1	1	N	1	N	1 or H.T.	1	N
Shaft enclosures	2	2	1	1	1	1	1	1706	1706
Floors	2	2	1	N	1	N	H.T.	1	N
Roofs	2 Sec. 1806	1 1906	1 2006	N	1	N	H.T.	1	N
Exterior doors and windows	Sec. 1803(b)	1903(b)	2003	2003	2103(b)	2103(b)	2103(b)	2203	2203

[a]For details, see chapters under Occupancy and Types of Construction; for exceptions, see Section 1705.

[b]See Table 16-2 notes.

[c]Structural frame elements in the exterior wall shall be protected against external fire exposure as required for exterior bearing walls or the structural frame, whichever is greater.

TABLE 16-4

Maximum Height of Buildings[a]
(*UBC Table 5-D*)

	Types of construction								
	I	*II*			*III*		*IV*	*V*	
	F.R.	*F.R.*	*1-h*	*N*	*1-h*	*N*	*H.T.*	*1-h*	*N*
	Maximum height in feet								
	Unlimited	160	65	55	65	55	65	50	40
Occupancy	*Maximum height in stories*								
A-1	Unlimited	4				Not permitted			
(A) 2, 2.1	Unlimited	4	2	Not permitted	2		2	2	Not permitted
(A) 3-4	Unlimited	12	2	1	2	1	2	2	1
E	Unlimited	4	2	1	2	1	2	2	1
(I) 1	Unlimited	3	1	Not permitted	1	Not permitted	1	1	Not permitted
(I) 2	Unlimited	3	2	Not permitted	2	Not permitted	2	2	Not permitted
(I) 3	Unlimited	2				Not permitted			
(H) 1	Unlimited	2	1	1	1	1	1	1	1
(H) 2-3-4-5	Unlimited	5	2	1	2	1	2	2	1
(B) 1-2-3	Unlimited	12	4	2	4	2	4	3	2
(B) 4	Unlimited	12	4	2	4	2	4	3	2
(R) 1	Unlimited	12	4	2	4	2	4	3	2
(R) 3	Unlimited	3	3	3	3	3	3	3	3

[a]See Table 16-2 notes.

requirements for the various elements in the different types. Obviously, many of these requirements (in hours) will determine the details or type of masonry that is to be used in the building (i.e., exterior bearing walls, interior bearing walls, exterior nonbearing walls, partitions, floors, roofs, and exterior doors and windows).

The various hour ratings of protection for structural parts, with various noncombustible insulating materials, are listed in Table 16-5, and for walls and partitions in Table 16-6. These are ones that have been verified by exposure to the ASTM standard fire test.

TABLE 16-5

Minimum Protection of Structural Parts Based on Time Periods for Various Noncombustible Insulating Materials
(*Excerpts from UBC Table 43-A*)

Structural parts to be protected	*Item number*	*Insulating material used*	*Minimum thickness of insulating material for following fire-resistive periods (in.)*			
			4 h	*3 h*	*3 h*	*1 h*
Steel columns and all members of primary trusses	1	Grade A concrete, members 6 × 6 in. or greater (not including sandstone, granite and siliceous gravel)	$2\frac{1}{2}$	2	$1\frac{1}{2}$	1
	2	Grade A concrete, members 8 × 8 in. or greater (not including sandstone, granite and siliceous gravel)	2	$1\frac{1}{2}$	1	1
	3	Grade A concrete, members 12 × 12 in. or greater (not including sandstone, granite and siliceous gravel)	$1\frac{1}{2}$	1	1	1
	4	Grade B concrete and Grade A concrete excluded above, members 6 × 6 in. or greater	3	2	$1\frac{1}{2}$	1
	5	Grade B concrete and Grade A concrete excluded above, members 8 × 8 in. or greater	$2\frac{1}{2}$	2	1	1
	6	Grade B concrete and Grade A concrete excluded above, members 12 × 12 in. or greater	2	1	1	1
	7	Clay or shale brick with brick and mortar fill	$3\frac{3}{4}$			$2\frac{1}{4}$
	8	4-in. Hollow clay tile in two 2-in. layers; $\frac{1}{2}$-in. mortar between tile and column; $\frac{3}{8}$-in. metal mesh (wire diameter = 0.046 in.) in horizontal joints; tile fill	4			
	9	2-in. Hollow clay tile; $\frac{3}{4}$-in. mortar between tile and column; $\frac{3}{8}$-in. metal mesh (0.046-in. wire diameter) in horizontal joints; Grade A concrete fill; plastered with $\frac{3}{4}$ in. gypsum plaster	3			
	10	2-in. Hollow clay tile with outside wire ties (0.08 in. diameter) at each course of tile or $\frac{3}{8}$-in. metal mesh (0.046-in. diameter wire) in horizontal joints; Grade A concrete fill extending 1 in. outside column on all sides			3	

TABLE 16-6

Rated Fire-Resistive Periods for Various Wall and Partitions[a]
(*Excerpts from UBC Table 43-B*)

Material	*Item number*	*Construction*[b]	*Minimum finished thickness face-to-face (in.)*[c]			
			4 h	*3 h*	*2 h*	*1 h*
Brick of clay or shale	1	Solid units (at least 75% solid)	8		6[d]	4
	2	Solid units plastered each side with $\frac{5}{8}$-in. gypsum or portland cement plaster. Portland cement plaster mixed 1 : $2\frac{1}{2}$ by weight, cement to sand			$4\frac{3}{4}$[e]	
	3	Hollow brick units[f] at least 71% solid		8		
	4	Hollow brick units[f] at least 71% solid, plastered each side with $\frac{5}{8}$-in. gypsum plaster	$8\frac{3}{4}$			
	5	Hollow (rowlock[g])	12		8	
	6	Hollow (rowlock[g]) plastered each side with $\frac{5}{8}$-in. gypsum or portland cement plaster. Portland cement plaster mixed 1 : $2\frac{1}{2}$ by weight, cement to sand	9			
	7	Hollow cavity wall consisting of two 4-in. nominal clay brick units with air space between	10			
	8	Hollow brick units at least 60% solid, cells filled with perlite loose fill insulation	8			
	9	4-in. nominal thick units at least 75% solid backed with a hat shaped metal furring channel $\frac{3}{4}$ in. thick formed from 0.021-in. sheet metal attached to the brick wall on 24-in. centers with approved fasteners; and $\frac{1}{2}$-in. Type X gypsum wallboard attached to the metal furring strips with 1-in.-long Type S screws spaced 8 in. on center.			5[e]	
	10	Cavity wall consisting of two 3-in. nominal thick solid clay units with air space		8		
Hollow clay Tile, non-load-bearing (end or side construction)	11	One cell in wall thickness, units at least 50% solid, plastered each side with $\frac{5}{8}$-in. gypsum plaster				$4\frac{1}{4}$
	12	Two cells in wall thickness, units at least 45% solid				6
	13	Two cells in wall thickness, units at least 45% solid. Plastered each side with $\frac{5}{8}$-in. gypsum plaster			7	
	14	Two cells in wall thickness, units at least 60% solid. Plastered each side with $\frac{5}{8}$-in. gypsum plaster			5	
Hollow clay tile, load-bearing (end or side construction)	15	Two cells in wall thickness, units at least 40% solid				8
	16	Two cells in wall thickness, units at least 40% solid. Plastered one side with $\frac{5}{8}$-in. gypsum plaster			$8\frac{1}{2}$	
	17	Two cells in wall thickness, units at least 49% solid			8	
	18	Three cells in wall thickness, units at least 40% solid			12	
	19	Two units and three cells in wall thickness, units at least 40% solid		12		
	20	Two units and four cells in wall thickness, units at least 45% solid	12			
	21	Two units and three cells in wall thickness, units at least 40% solid. Plastered one side with $\frac{5}{8}$-in. gypsum plaster	$12\frac{1}{2}$			
	22	Three cells in wall thickness, units at least 43% solid. Plastered one side with $\frac{5}{8}$-in. gypsum plaster		$8\frac{1}{2}$		

TABLE 16-6 (Continued)

Material	Item number	Construction[b]	Minimum finished thickness face-to-face (in.)[c]			
			4 h	3 h	2 h	1 h
	23	Two cells in wall thickness, units at least 40% solid. Plastered each side with $\frac{5}{8}$-in. gypsum plaster		9		
	24	Three cells in wall thickness, units at least 43% solid. Plastered each side with $\frac{5}{8}$-in. gypsum plaster	9			
	25	Three cells in wall thickness, units at least 40% solid. Plastered each side with $\frac{5}{8}$-in. gypsum plaster	13			
	26	Hollow cavity wall consisting of two 4-in. nominal clay tile units (at least 40% solid) with air space between. Plastered one side (exterior) with $\frac{3}{4}$-in. portland cement plaster and other side with $\frac{5}{8}$-in. gypsum plaster. Portland cement plaster mixed 1 : 3 by volume, cement to sand	10			
Combination of clay brick and load-bearing hollow clay tile	27	4-in. brick and 8-in. tile	12			
	28	4-in. brick and 4-in. tile		8		
	29	4-in. brick and 4-in. tile plastered on the tile side with $\frac{5}{8}$-in. gypsum plaster	$8\frac{1}{2}$			
Concrete masonry units[h]	30	Expanded slag or pumice	4.7	4.0	3.2	2.1
	31	Expanded clay or shale	5.7	4.8	3.8	2.6
	32	Limestone, cinders or air cooled slag	5.9	5.0	4.0	2.7
	33	Calcareous or siliceous gravel	6.2	5.3	4.2	2.8
Solid concrete	34	Horizontal reinforcement not less than 0.25% and vertical reinforcement not less than 0.15% (three-fourths as much for welded wire fabric) — Grade A concrete	$6\frac{1}{2}$	6	5	$3\frac{1}{2}$
		Grade B concrete	$7\frac{1}{2}$	$6\frac{1}{2}$	$5\frac{1}{2}$	4[e]
Hollow gypsum tile	35	3-in. tile not less than 70% solid				3[e]
	36	3-in. tile plastered one side with $\frac{5}{8}$-in. gypsum plaster			$3\frac{5}{8}$[e]	
	37	4-in. tile plastered one side with $\frac{1}{2}$-in. gypsum plaster		$4\frac{1}{2}$[e]		
	38	3-in. tile plastered both sides with $\frac{1}{2}$-in. gypsum plaster		4[e]		
	39	4-in. tile plastered both sides with $\frac{1}{2}$-in. gypsum plaster	5[e]			
Glazed or unglazed facing tile, non-load-bearing	40	One 2-in. unit cored 15% maximum and one 4-in. unit cored 25% maximum with $\frac{3}{4}$-in. mortar filled collar joint. Unit positions reversed in alternate courses		$6\frac{3}{8}$		
	41	One 2-in. unit cored 15% maximum and one 4-in. unit cored 40% maximum with $\frac{3}{8}$-in. mortar filled collar joint. Plastered one side with $\frac{3}{4}$-in. gypsum plaster. Two wythes tied together every fourth course with No. 22 gauge corrugated metal ties		$6\frac{3}{4}$		
	42	One unit with three cells in wall thickness, cored 29% maximum			6	
	43	One 2-in. unit cored 22% maximum and one 4-in. unit cored 41% maximum with $\frac{1}{4}$-in. mortar filled collar joint. Two			6	

TABLE 16-6 (Continued)

Material	Item number	Construction[b]	Minimum finished thickness face-face (in.)[c] 4 h	3 h	2 h	1 h
		wythes tied together every third course with No. 22 gauge corrugated metal ties				
	44	One 4-in. unit cored 25% maximum with $\frac{3}{4}$-in. gypsum plaster on one side			$4\frac{3}{4}$	
	45	One 4-in. unit with two cells in wall thickness, cored 22% maximum				4
	46	One 4-in. unit cored 30% maximum with $\frac{3}{4}$-in. vermiculite gypsum plaster on one side			$4\frac{1}{2}$	
	47	One 4-in. unit cored 39% maximum with $\frac{3}{4}$-in. gypsum plaster on one side				$4\frac{1}{2}$
Exterior or interior walls	84	$2\frac{1}{4}$- × $3\frac{3}{4}$-in. clay face brick with cored holes over $\frac{1}{2}$-in. gypsum sheathing on exterior surface of 2- × 4-in. wood studs at 16-in. on center and two layers $\frac{5}{8}$-in. Type "X" gypsum wallboard on interior surface. Sheathing placed horizontally or vertically with vertical joints over studs nailed 6-in. on center with $1\frac{3}{4}$-in. by No. 11 gauge by $\frac{7}{16}$-in. head galvanized nails. Inner layer of wallboard placed horizontally or vertically and nailed 8-in. on center with 6d cooler nails. Outer layer of wallboard placed horizontally or vertically and nailed 8-in. on center with 8d cooler nails. All joints staggered with vertical joints over studs. Outer layer joints taped and finished with compound. Nailheads covered with joint compound. No. 20 gauge corrugated galvanized steel wall ties $\frac{3}{4}$ × $6\frac{5}{8}$-in. attached to each stud with two 8d cooler nails, every sixth course of bricks			$10\frac{1}{8}$	

[a]Generic fire resistance ratings (those not designated by company code letter) as listed in the Design Manual—Fire Resistance 1975–76 Edition as published by the Gypsum Association—may be accepted as if herein listed.

[b]Staples with equivalent holding power and penetration may be used as alternate fasteners to nails for attachment to wood framing.

[c]Thicknesses shown for brick and clay tile are nominal thicknesses unless plastered, in which case thicknesses are net. Thicknesses shown for solid or hollow concrete masonry units are "equivalent thicknesses" as defined in UBC Standard 24-4. Thickness includes plaster, lath and gypsum wallboard where mentioned and grout when all cells are solidly grouted.

[d]Single-wythe brick.

[e]Shall be used for nonbearing purposes only.

[f]Hollow brick units 4- by 8- by 12-in. nominal with two interior cells having a $1\frac{1}{2}$-in. web thickness between cells and $1\frac{3}{4}$in. thick face shells.

[g]Rowlock design employs clay brick with all or part of bricks laid on edge with the bond broken vertically.

[h]See also footnote No. 2. The equivalent thickness may include the thickness of portland cement plaster or 1.5 times the thickness of gypsum plaster applied in accordance with the requirements of Chapter 47 of the Code.

The ratings of the elements stem from their resistances to exposure to a standard fire. This fire temperature is raised at a specified rate according to ASTM E119.

Failure to that exposure is measured by:

1. Passage of flame.
2. Thermal rise of 250°F above ambient.
3. Collapse under load.
4. Failure to resist the standard hose stream test.

Table 16-6 lists items 30 through 33, which contain the term "equivalent thicknesses." Most of the previous items are specific descriptions of types of detail wall construction as they were actually tested, with the actual thickness listed under the appropriate endurance. However, the hollow concrete masonry thicknesses shown represent equivalent thicknesses, on a more engineered basis, as noted below.

The relation between equivalent thickness, material type, and endurance rating is based upon the plotting of results of many fire tests and noting that the variation may be expressed consistently by the term

$$R = CKT^n$$

where R = endurance rating
C = shape factor
K = constant, dependent on the material
T = equivalent thickness
n = coefficient to provide a curve shape to fit the test results

Equivalent thickness

Equivalent thickness is the solid thickness that would be obtained if the same amount of material contained in a hollow unit were recast as a solid unit without core holes. The percent solid can be calculated by the ratio of net area to the gross area, or the net volume values, as determined by ASTM C140 "Methods of Testing Concrete Masonry Units," to the gross volume. For example, if we assume that an 8-in. hollow masonry wall is constructed of expanded slag units which are 55% solid, the estimated fire resistance of the wall would be determined by using the equivalent thickness of 0.55×7.625 in., or 4.19 in. Referring to item 30 for expanded slag aggregate, we find that 4.0 in. would provide 3 h and 4.7 would be required for a 4-h rating. Therefore, the 4.19 in. of equivalent thickness would be adequate for a 3-h fire resistance, but not for a 4-h rating requirement.

If it were desired to provide 4-h resistance, one could solid-grout the wall, which would provide 7.6 in. of actual thickness, which would be more than adequate (4.7 in. required). Alternatively, one could apply $\frac{5}{8}$ in. of cement plaster or $\frac{3}{8}$ in. of gypsum to the hollow masonry wall in order to increase the rating to 4 h. The footnotes clarify that the equivalent thickness for brick and clay tile includes the actual thicknesses of lath, gypsum wall boards, grout if all cells are solidly grouted and portland cement plaster, or 1.5 times the thickness of gypsum plaster if used instead. This additional credit for gypsum plaster thickness is justified because gypsum is more effective in resisting heat flow than are masonry materials. It has good thermal resistance, and in addition contains water of crystallization, which absorbs the heat in latent heat of vaporization, slowing heat transmission. Apparently, it is actually more than twice as good as masonry of equal thickness.

Grout contributes to fire resistance only when it is continuous (solid-grouted), because if there were large areas ungrouted, they would permit "hot spots" to occur, places where the heat would flow through the wall more rapidly and raise the local temperature above that allowed for the fire rating required.

The Western States Clay Products Association obtained approval for a hollow clay unit (Table 16-7) which is similar in shape to a hollow concrete block unit. This approval contained a table of fire ratings or the fire-resistive periods of walls of various thicknesses and applications of plaster.

TABLE 16-7

Fire-Resistive Periods—Clay Walls
(In hours)

		$\frac{5}{8}$-in. plaster	
Wall of hollow brick	*No plaster (h)*	*1 side (h)*	*2 sides (h)*
10-in. not continuously grouted	3	4	4
8-in. solid-grouted	4	4	4
8-in. grouted at reinforcing, and perlite-filled	4	4	4
8-in. not continuously grouted	3	4	4
6-in. solid-grouted	3	4	4
6-in. not continuously grouted	1	2	3
4-in. solid-grouted	1	2	3

Influence of fire ratings

An example of the influence of these "nonstructural" items upon the structural design and selection of material follows. Assume that a property owner desires to build a hotel apartment building 50 × 100 ft on a 100 × 100 ft corner lot. Financing and economics, indicate that the building should be about 10 stories high. The scheme proposed consisted of a load-bearing concrete block system previously used satisfactorily by the owner. A check is to be made to determine if the building may be Type I; Type II fire resistive or 1 h; or Type III 1 h.

Table 16-1 indicates the occupancy to be R-1 (requires not less than a 2-h exterior wall, or 1 h if clearance is 20 ft or more). Table 16-1 also indicates that the area for R-1 occupancy may be:

Unlimited for Type I construction:
22,500 ft^2 for Type II fire-resistive
10,100 ft^2 for Type II, 1 h
10,100 ft^2 for Type III, 1 h

These areas are for one-story buildings. They may be doubled for multistory structures. They may be increased 100% due to side-yard clearance on three sides, which are more than 40 ft wide. These increases permit increase of the areas to:

Unlimited for Type I construction:
90,000 ft^2 for Type II fire-resistive
40,400 ft^2 for Type II, 1h or III, 1 h

The maximum height limits as shown in Table 16-4 are:

Unlimited for Type I construction:
12 stories for Type II, fire-resistive
4 stories for Type II or III, 1 h

The proposed building is to be 10 floors 50×100 ft = 50,000 ft^2 total area. Type II, 1 h or III, 1 h construction is hence ruled out because of the four stories and the 40,000-ft^2 limit.

The building may be Type I or Type II fire resistive construction and it is noted that interior bearing walls must have 3-h fire rating for Type I, and 2-h rating for Type II. It is intended to use local lightweight block of expanded clay aggregate. Table 16-6 indicates that these blocks will require equivalent thicknesses of 4.8 and 3.8 in., respectively. The 8-in. block will provide the 3.8 in. for 2-h rating but the wall must be solid-grouted to provide the 3-h rating. The use of Type II construction, with the 2-h ungrouted block, will hence reduce the weight of the structure, and the consequent lateral earthquake loads. This will represent a reduction in the cost of walls, reinforcing, and foundations which the engineer may take advantage of in providing a professional service to the client.

These factors of occupancy, area, height, and type of construction must be considered before final layout and design is begun.

ACOUSTICAL CHARACTERISTICS

Need

In addition to their structural properties, materials of construction must also be considered for acoustical properties, which may influence the material selection, the size, type and location of wall required, and consequently the structural design. This discussion will simply describe some of the acoustical considerations pertinent to the selection of a masonry element. The design of acoustics of halls, auditoriums, or rooms; the problem of reverberation; and the interrelation of the many complex factors of complete acoustical design will not be covered in this brief presentation of material characteristics. There are certain acoustical requirements established by code agencies. These are relatively new and consequently are apt to change rapidly as use dictates improvements or deletions. Therefore, the specific requirements for a particular project in a specific jurisdiction must be determined. That is, what are the current requirements for the particular project?

The problem

The noise problem encountered in multifamily housing is considerably greater than that in single-family residences. Apartment dwellers are not only exposed to the outside noise of automobiles, planes, gasoline-powered tools, and so on, but also to noise that is generated within the apartment building itself. So they can be subjected to noise from occupants of adjoining units or other neighbors, and this may frequently be most annoying. People like to converse comfortably in the room they occupy and they do not want to hear conversation coming from adjacent rooms. Above all, they do not ordinarily want people in adjacent rooms to hear their conversations. Studies and surveys show conclusively that people want apartments and townhouses to be soundproofed between living units, and such desires will frequently determine the selection of materials of construction.

Three methods are commonly used to minimize unwanted sound or noise. One method is to eliminate the source. However, this would hardly be possible, or impractical at least, in multifamily housing. The second method of reducing noise is to lower the sound level within a room by the use of materials that will absorb the sound energy instead of reflecting it back into the room. The third means of eliminating noise is to prevent the sound waves from being transmitted from one area into an adjoining area by using sound barriers that have sufficient mass to lower the sound before it penetrates the barrier. Masonry walls provide excellent barriers with less cost than do many other materials.

Sound absorption and noise reduction coefficient (NRC)

Sound absorption is the property of controlling sound within a room so that it does not reflect back into the room but, rather, will be absorbed by the wall and ceiling surfaces. This absorption of sound not only reduces the level of sound within a room, but also, as a primary function, prevents the sound from bouncing around within a space. This makes it easier for people to hear distinctly. However, sound-absorption materials such as acoustical tile or acoustical plaster cannot resist the transmission of sound through a wall to any great extent. To absorb sound usefully a material must create a frictional drag on the sound energy. This is done by materials with a porous texture, which will not reflect, or bounce back, the sound.

This type of absorption is also accomplished with carpets, furniture, human bodies, and similar material (i.e., anything that will impede the flow of sound and prevent its "bouncing back").

Sound is absorbed by any surface that dissipates the energy and converts it to heat primarily by the use of acoustically porous materials. If the surfaces of a room were capable of absorbing all sound generated within the room, they would have a sound absorption coefficient of 1. If only 50% of the sound were absorbed, the coefficient would be 0.50. A commonly used measurement of sound absorption is the *noise reduction coefficient* (NRC). It is established by measuring the sound absorption coefficient at certain frequencies of sound (i.e., at 250, 500, 1000, and 1000 hertz, or cycles per second). The NRC is the average of those four measured coefficients. Noise reduction coefficients for some materials are presented in Table 16-8.

TABLE 16-8

Noise Reduction Coefficient (NRC) of Building Materials and Furnishings

Material	*NRC*	*Material*	*NRC*
Brick			
Unglazed	0.04	Plaster on brick or concrete blocks	0.03
Unglazed, painted	0.02	Plywood paneling on furring strips	0.13
Carpet, heavy on concrete	0.30	Drapes	
on 40-oz pad or foam rubber	0.55	Light fabric, 10 oz/yd^2	0.14
Concrete block		Medium fabric, 14 oz/yd^2	0.40
Coarse	0.40	Heavy fabric, 18 oz/yd^2	0.55
Medium	0.35		
Fine	0.30		
Painted	0.10	Furnishings (values in absorption per square foot of floor area)	
Concrete floor	0.01		
Asphalt tile floor on concrete	0.03	bed	0.80
		sofa	0.85
Wood floor	0.08	wood table, chairs, etc.	0.20
		leather-covered upholstered chair	0.50
Marble or glazed tile	0.01	cloth-covered upholstered chair	0.70
Glass			
Single-strength window	0.12		
Heavy plate, large panes	0.04		
Gypsum wall board on 2 × 4 in. studs	0.07		
Gypsum wall board on concrete	0.03		

To estimate the average sound absorption (average NRC) in a room, one can multiply the NRC values of each material in the room by its surface area, sum these products, and then divide that sum by the total surface area in the room. The noise reduction provided by the sound absorption characteristics of furnishings within a room can be estimated by evaluating the average NRC of the unfurnished room and that of the carpeted and furnished room. These two NRC values are then averaged to obtain the final noise reduction coefficient for that room as furnished.

Sound transmission

Sound transmission or sound transmission resistance deals with the problem of sound traveling through barriers from one space to another. An example in our modern world concerns the problem of preventing outside sound from traveling into homes, churches, schools, offices, and so on. Effective barriers must be dense so that they can effectively stop the passage of sound or noise. Actually, sound waves do not pass through a wall like water through a sieve. Rather, they cause the wall to vibrate like a diaphragm. This vibration, in turn, causes the air on the other side of the wall to vibrate, thereby creating sound waves. Sound waves are made up of waves of energy, and to reduce them we must transform that energy to heat. This is done effectively by a heavy mass which will stop the particles of energy from further vibration. So the greater the inertia or resistance to vibration of a wall, the greater its ability to prevent the transfer of sound. Observe, however, that this transmission loss is not directly proportional to the weight per unit area but is proportional to the logarithm of the weight. The initial doubling of the weight produces the greatest increase in transmission loss. Subsequent increases in density result in proportionally fewer increases in transmission resistance.

Another effective method of producing high transmission loss, besides increasing material density, lies in a method of discontinuous construction. Separating a wall into two or more distinct layers divided by an air space results in high sound transmission losses. Up to a width of about 24 in., the greater the air space, the more transmission loss there will be.

Sound transmission class (STC)

A measure of the effectiveness of wall construction to resist sound transmission is the *sound transmission class*. This single-number coefficient simplifies the use of the resistance capacity, and is simply the result of a measurement of sound levels made for the reduction of each of the 16 frequencies which vary over a range 125 to 4000 Hz. The reduction of sound in decibels for each of the frequencies is plotted and then a series of sound transmission class contours are drawn, being adjusted to provide the best fit. The single number which is used as the coefficient (STC) is the value of the contour average at the frequency of 500 Hz (Figure 16-1).

Frequencies and decibels

A single-number coefficient is used although it is recognized that sound at different frequencies of vibration and at different decibels (dB) will be reduced to different levels. One *decibel* is defined as the minimum noise level audible to the human ear. It is a measure of objective sound, not loudness or subjective response to sound. The decibel scale has proven suitable for measuring ratios of sound intensity, since the response of the human ear is approximately proportional to the logarithm of sound energy. It remains a good measure of the loudness or the strength of sound

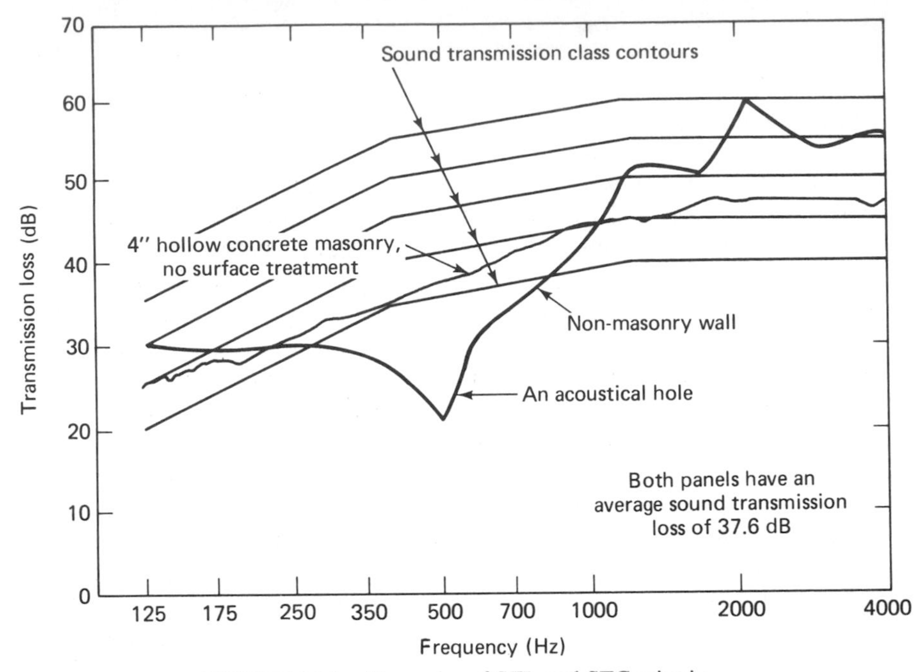

FIGURE 16-1. Examples of STL and STC criteria.

vibration. The scale is set up so that when the sound increases from 10 to 20 dB, the intensity of sound is increased by 10 times. Figure 16-2 gives a descriptive value to the different vibration ranges. Table 16-9 lists the sound transmission class limitations that are imposed by FHA for providing proper sound isolation for apartments. Finally Table 16-10 shows the STC ratings for various of walls.

FIGURE 16-2. Description of decibels.

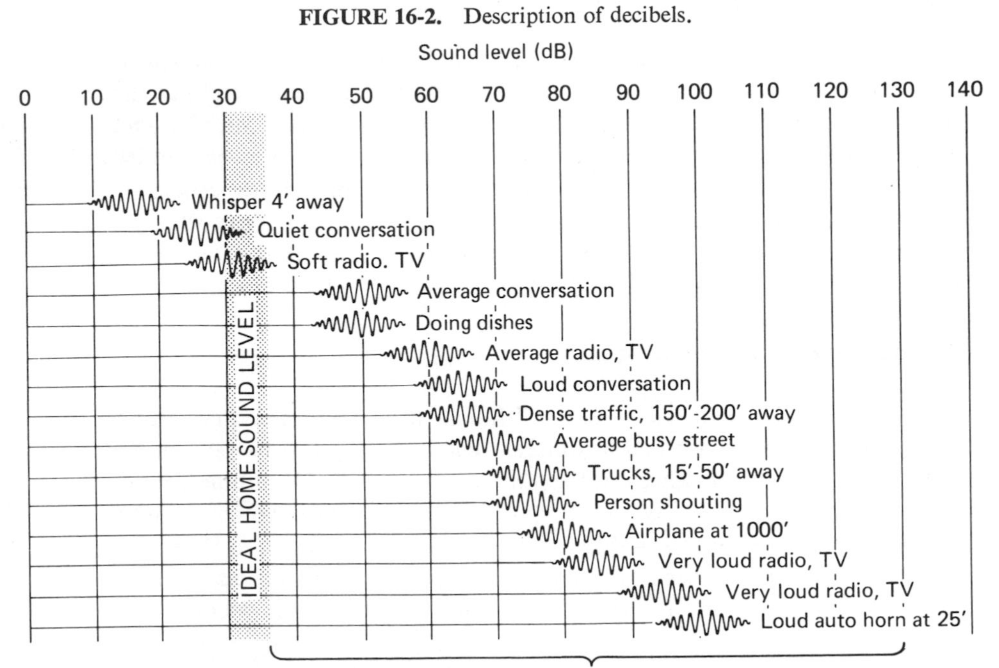

TABLE 16-9

Sound Transmission Class (STC) Limitations[a]

Location of partition	Low background noise		High background noise	
	Bedroom adjacent to partition	*Other rooms adjacent to partition*	*Bedroom adjacent to partition*	*Other rooms adjacent to partition*
Living unit to living unit	50	45	45	40
Living unit to corridor	45	40	40	40
Living unit to public space (average noise)	50	50	45	45
Living unit to public space and service areas (high noise)	55	55	50	50
Bedrooms to other rooms within same living unit	45	NA	40	NA

[a]FHA requirements for apartment quieting, from FHA publication 2600.

TABLE 16-10

STC Values of Masonry Walls

Wall thickness (in.)	*STC*	*Wall description*	*Weight of wall (lb/ft^2)*	*Test*
4		Hollow concrete block, unplastered, ungrouted		
	40	Lightweight units, unpainted	18.0	KAL 359-1-66
	41	Normal-weight units, painted both sides (two coats)	26.5	TL 67-99
4		Hollow concrete block, painted both sides (two coats)		
	43	Lightweight units, ungrouted	22.0	KAL 1379-5-72
	44	Normal-weight units, ungrouted	29.0	KAL 1379-3-72
4		Hollow concrete block, plastered both sides, ungrouted		
	44	Normal-weight units	34.8	TL 67-102
4	45	Face brick wall: $3\frac{3}{4}$ in. thick	38.7	TL 67-70
4		Hollow concrete block, $\frac{1}{2}$-in. gypsum board on resilient channels each side		
	47	Lightweight blocks, ungrouted	26.0	KAL 1379-4-72
	48	Normal-weight blocks, ungrouted	32.0	KAL 1379-2-72
4	48	Hollow concrete block, plastered both sides, ungrouted	30.0	KAL 359-7-66
4	50	Brick wall, brick $3\frac{5}{8}$ in. thick $\frac{1}{2}$-in. sand plaster, on one face	42.4	TL 69-283
6	44	Hollow concrete block, unpainted, unplastered, ungrouted	21.0	KAL 359-4-66
6		Hollow concrete block, painted both sides (two coats)		
	46	Lightweight blocks, ungrouted	28.0	KAL 933-2-70
	48	Normal-weight blocks, ungrouted	39.0	KAL 1379-1-72
6	53	Hollow concrete block, two coats, paint one side $\frac{1}{2}$-in. gypsum board on resilient channel other side	27.0	KAL 359-6-66
8	45	Hollow concrete block, unpainted, unplastered, ungrouted	36.0	KAL 359-3-66
8	48	Hollow concrete block, painted both sides, two coats, ungrouted	33.5	TL 67-61
8	49	Hollow concrete block, unpainted, unplastered, ungrouted,	42.8	KAL 1144-2-71
8	50	Hollow concrete block, exterior wall painted outside (two coats)	45.6	TL 67-93
		Gypsum board on furring strips inside, ungrouted		

TABLE 16-10 (Continued)

Wall thickness (in.)	*STC*	*Wall description*	*Weight of wall (lb/ft²)*	*Test*
8	50	Face brick and structural clay tile composite brick dimensions: 3¾-in. thick, tile dimensions: 4-in. thick	63.8	TL 67-65
8	51	Hollow concrete block, cells filled with zonolite, unpainted	39.6	KAL 1144-4-71
8	52	Solid face brick Brick dimensions: 2¼ × 3¾ × 8¼ in.	83.3	TL 67-68
8	52	Hollow concrete block, normal weight block, ungrouted, unpainted	53.0	KAL 1144-3-71
8	53	Solid brick wall; wall 9¼ in. thick ½-in. gypsum board on furring strips one side	86.7	TL 69-287
8	55	Hollow concrete block, lightweight block, solid grouted, painted both sides (2 coats), reinforced	73.0	KAL 1023-3-71
8	56	Hollow concrete lightweight, ungrouted ⅝-in. gypsum board on resilient channels one side Lightweight block	40.4	KAL 933-1-70
8		Hollow concrete lightweight block, solid grouted, reinforced		
	56	½-in. gypsum plaster both sides	79.0	KAL 1023-9-71
	60	½-in. gypsum board on resilient channels, both sides	77.0	KAL 1023-3-71
10	59	Reinforced brick masonry, solid grouted, wall 9½ in. thick brick dimensions: 2½ × 3⅝ × 7⅝ in.	94.2	TL 70-6
12	55	Solid concrete block, unpainted, unplastered	121.0	NGC 3002
12	58	Solid concrete block, ⅝-in. gypsum board on resilient channels, one side	124.0	NGC 3003
		Sources of data: KAL: Kodaras Acoustical Lab., Elmhurst, New York TL: Riverbank Acoustical Lab., Geneva, Illinois NGC: National Gypsum Company, Buffalo, New York		

Flanking path control

The transmission of sound from one room to another may occur not only through the barrier partitions, but also through windows, doors, convenience outlets, ventilators, or plumbing systems. These "sound leaks" are commonly called *flanking paths*. When a field test is performed to demonstrate compliance with the standards, it is stated that all sound transmission from the room containing the source to the receiver room shall be considered to be transmitted through the barrier test partition. The contradictions are immediately apparent. A sound-rated separating wall could be selected that meets the requirements of the standard. However, flanking paths may exist which would leak additional sound into the receiving room and thus the completed building would not be in compliance. This is recognized in the requirement that a 50-dB STC rated wall may be permitted to show 45 under field test.

Impact noise control

In addition to a minimum STC of separating walls and floors, standards require a minimum *impact insulation class* (IIC) for floor and ceiling assemblies. This is measured by placing a standard tapping machine on the floor being evaluated.

The noise produced is measured in the room below. From the reduction of sound, a one-number rating, the impact insulation class is determined similar to the STC single-number coefficient. This impact resistance is really a measure of the influence of carpets or coverings on the floors, especially those with a cushion-backed material.

Reduction of community noise equivalent level (CNEL)

Some areas require that exterior community sound levels not exceed a certain decibel level in any habitable room. This requires that the barrier walls have certain resistance to sound transmission, and such factors should be considered in the design.

HEAT TRANSMISSION—ENERGY CONSERVATION

The nomenclature frequently used in heat transmission calculations include the following:

> *k*, the conductivity, is the Btu transmitted per hour per square foot per degree difference in temperature for a 1-inch thickness of a homogeneous material between the surface of the warmer side and the surface on the cooler side.
>
> *C*, the conductance, is the Btu transmitted per hour per square foot per degree difference in temperature for a stated thickness of an obstruction (wall) between the surface of the warmer side and the surface of the cooler side.
>
> *f*, the film or surface conductance: f_i, designates the inside surface and f_o the outside surface in Btu per hour per square foot per degree difference in temperature.
>
> *a*, the conductance of an air space expressed in Btu per hour per square foot per degree difference in temperature.
>
> *U*, the thermal transmittance or overall coefficient of heat transmission, is the amount of heat transmitted in Btu per hour per square foot per degree difference in temperature between the air on the warmer side of an obstruction and the air on the cooler side.
>
> *R*, the resistance or resistivity, is the reciprocal of transmission, conductance or conductivity; that is, $1/k$, $1/C$, $1/f$, $1/a$, $1/U$, etc.

In the past, the heat flow or transmission calculations have been based on classic uses of *R*, resistance, or *C*, conductance, and differentials of temperature, etc. The conductance factors were based upon tests of conditions of steady state heat flow through materials. The calculations were based on consideration of wind effect, surface factors, resistance through various materials, and the differentials of temperature.

There have been many hysterical attempts to use those simple transmission calculations as measure of *Energy Conservation*. However, these are only a small part of the true performance and total considerations. One factor is that the heat flow through a solid wall is only a small portion of the total heat "loss" of a building—doubling that small part of the resistance does not double the total "conservation."

Another fallacy is that real life is not static, it is dynamic. The daily fluctuation of temperature through midnight and noon is extremely wide, e.g., in areas of alternating hot days, cool nights. Thermal inertia plays a big part in the performance and is influenced by the climatic conditions at sites. Some considerations of these interrelations have attempted to use "degree-days" as a classification of climate, but this omits proper consideration of the cyclic variations of temperature, sun, color, exposure, heat gain, etc.

The subject of design for Energy Conservation is at present so colored with political and fluctuating inadequate code provisions that it will not be covered at this time in this text, although it is, admittedly, a potent factor in the overall function and suitability of walls.

17

Reinforced Masonry Research–Past and Present

EARLY RESEARCH ATTEMPTS

Until reinforced masonry design techniques and construction practices are based upon far more realistic appraisals of the material's performance under various loading conditions and occupancy uses, the use of masonry will continue to be severely limited and heavily penalized by the use of high factors of safety in empirical codified rules, many of which have no basis in fact whatsoever. Much information and data are still needed if safe, yet economically competitive, masonry structures are to be built in the future. This can only evolve, at this stage at least, from research that has a definite "applied" emphasis. This means that to really advance the state of the art, these programs must be designed to produce the kind of results that can be readily codified and incorporated into practical office design procedures and sound construction practices.

However, a considerable amount of study has already been expended on the basic masonry materials (i.e., mortar, grout, and masonry units and assemblages). Early contributions to the advancement of the state of the art by the industry itself were made both by the National Brick Manufacturers Research Foundation and the Brick Manufacturers Association of America. Formed in 1932, the former set forth immediately to enlist the advice of a consulting committee comprised of leading authorities on structural materials. This committee was known as the Reinforced Brick Masonry Research Board. In addition to its active participation in the initiation, correlation, and supervision of early masonry research, this board developed drafts of recommended design practice and standard specifications for reinforced brick masonry which have had considerable influence on subsequent building-code requirements, specifications, and construction methods for this material. One of their reports described some very early attempts at masonry research by Marc Brunel, who is credited with the discovery of reinforced brick masonry in the early 1800s. Sometime during that period he built a reinforced brick masonry structure in an effort to determine the strength imparted to the masonry by the reinforcement. This was followed about a year later by Colonel Pasley of the Corps of Royal Engineers, who conducted a series of tests in an attempt to analyze the behavior of reinforced brick beams and how the reinforcement strengthened them. His reinforced beam (18 in. wide × 12 in. deep), on a 10-ft span, carried about 10 times the load on the unreinforced beam of similar size. These results, which apparently paralleled those of Brunel, settled any dispute which had existed prior to that time regarding the effectiveness of the reinforcement. Although the manner in which the masonry and steel resisted the forces was not made apparent from these tests, they did stimulate an increased interest in this seemingly new type of construction.

Later, in 1851, reinforced brick beam tests were conducted at the Great Exposition in London. These were unique in that the construction involved the first known use of a new type of cement in masonry, known commercially as "portland cement." The tests were apparently successful if for no other reason than the fact that they resulted in the immediate widespread use of portland cement in Europe, and to a lesser degree in the United States. N. B. Corson later reviewed the data from these tests, as well as those of Brunel. Armed with this information, plus test results on other unreinforced beams and arches, he computed the tensile stress of unreinforced masonry flexural members and then recommended an allowable tensile stress for use in the design of masonry lintels. This appears to be the first recorded technical discussion regarding the relation between masonry tensile strength and mortar strength. But Corson apparently failed to recognize the effect of the reinforcement on tensile resistance.

In the United States, Hugo Fillippi, a Chicago consulting engineer at the time, built and tested a group of reinforced brick masonry beams in 1913. Later, in 1919, L. J. Mensch, another consulting engineer from Chicago, also tested some reinforced brick beams, unique because the reinforcement was located in a bed of mortar below the brick masonry. Unfortunately, none of the data from these tests seems to have been reported on, so little information was received by others interested in this type of construction.

In 1923, a report by A. Brebner, of the Public Works Department of the Government of India, describing an extensive series of tests on reinforced masonry elements aroused considerable interest in reinforced brick masonry. Some 282 specimens were built and tested over a 2-year period. These included reinforced brick masonry slabs of varying thicknesses, as well as reinforced brick beams, columns,

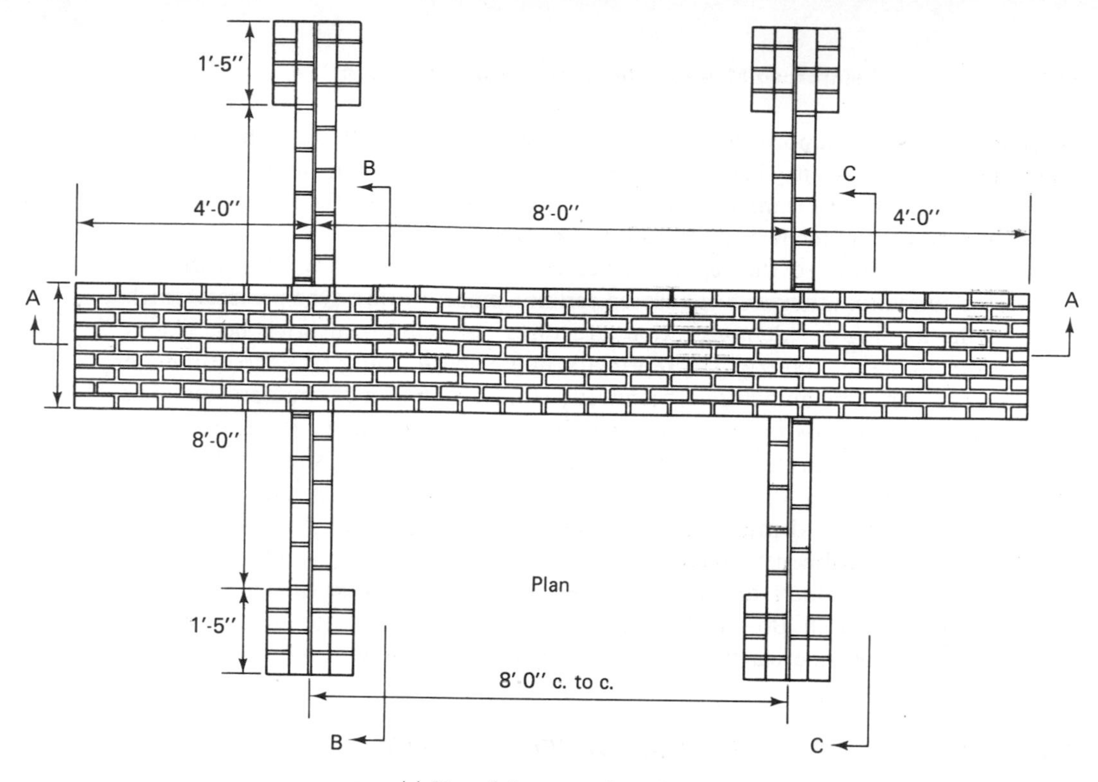

(a) Plan of demonstration structure

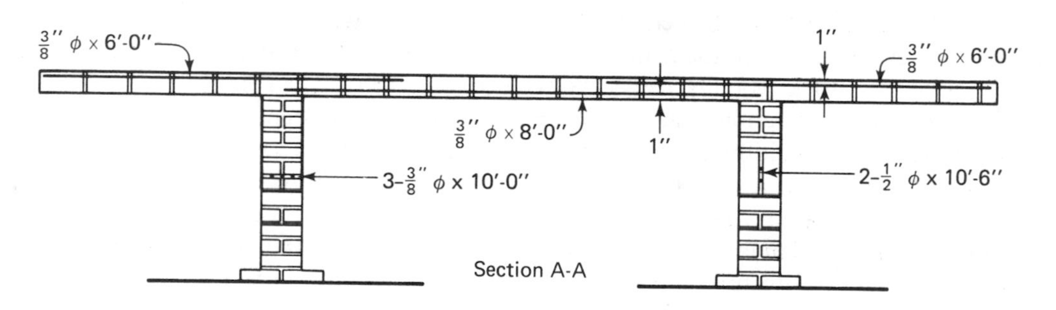

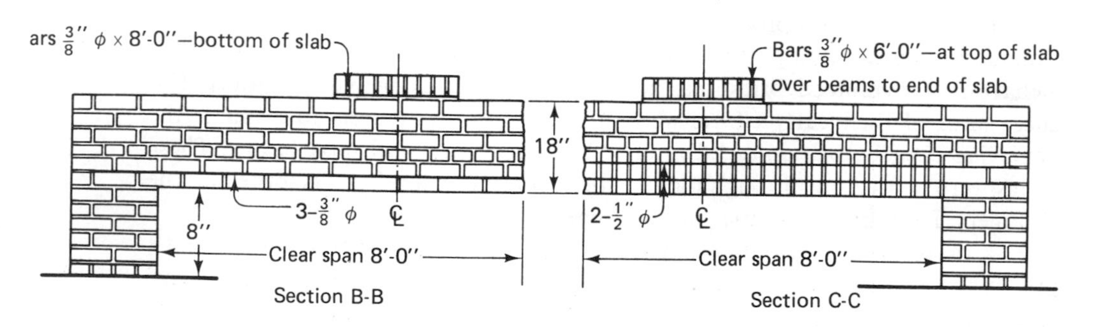

(b) Section of demonstration structure

FIGURE 17-1. Early masonry demonstration structure. Note that brick headers such as those used here should not be used in recommended grouted construction.

and arches. These tests appear to constitute the first attempt to both study the behavior of reinforced brick masonry in an organized manner and to formally publish an analysis of the test results. So it might be said that this research marks the beginning stage in the modern development of reinforced brick masonry.

In addition to the reports issued by the National Brick Manufacturers Research Foundation, they also sponsored certain demonstration-type projects which involved the loading of "demonstration structures" of reinforced brick beams and slabs throughout the country. Their primary intent in so doing was "to stimulate interest among brick manufacturers, architects, engineers, mason contractors, building inspectors, etc. in improving brick masonry design." One of these structures, a slab and beam and column element is shown in Figure 17-1. These demonstrations provided dramatic illustrations of the capacity of reinforced brick masonry, and no doubt succeeded in creating and maintaining a significant interest in this type of construction. In 1941 the Board widened its scope of activities to embrace research on all types of structural clay products and in the process changed its name to the Structural Clay Products Research Board. This group was an early ancestor of the present Masonry Institute of America, whose function was described previously.

MODERN MASONRY RESEARCH DEVELOPMENTS

Initial industry contributions

Around 1950, the Structural Clay Products Research Foundation was organized by several brick and tile manufacturers in the United States and Canada to continue what the Brick Manufacturers Association of America had begun in an earlier day. One of its functions was to develop and carry out research programs on both reinforced and unreinforced masonry. Other industry agencies sponsoring or carrying out early test programs include the Clay Brick and Tile Association, the National Concrete Masonry Association, the Structural Clay Products Institute, and the Associated Brick Manufacturers of Southern California. The latter organization had a dual function: (1) to promote the use of reinforced masonry in California by providing a consulting service to architects and engineers regarding the proper use of masonry both as to design procedure and construction practice, and (2) to sponsor and conduct research relating to the material properties of brick masonry as well as the behavior of reinforced brick masonry elements such as beams and walls. For instance, some of their early efforts were directed toward attempts to evaluate the load-carrying capacity of reinforced grouted brick masonry walls and beams. These programs were carried out through the facilities of the Raymond G. Osborne and the Smith-Emery Testing Laboratories in Los Angeles.

In addition to the industry-sponsored efforts, the federal government's interest was early manifested by the activities of the National Bureau of Standards' work in this field, particularly in their studies on mortar workability and retentivity, as well as in their tests on the physical and mechanical properties of brick. In addition, this agency carried out early programs involving the compressive strength of clay brick walls (1929) and shear tests on reinforced brick masonry beams (1932) under the supervision of J. W. McBurney, A. H. Stang, and D. E. Parsons, all pioneers in the field of masonry research. Beginning in 1933 in California, the then State Division of Architecture (now Office of State Architect) was extremely instrumental in developing improved masonry design and construction techniques, particularly in the area of reinforced masonry walls. Most of their work was conducted in various university laboratories, and these will be referred to in more detail in subsequent paragraphs.

Then, of course, the various ASTM committees were active over the years in establishing standards and specifications for masonry material ingredients, such as aggregates, mortar, grout, and masonry units of all types. The UBC further expanded these specifications in their standards, which embrace all types of reinforced masonry materials as well as reinforced masonry assemblages.

Initial academic programs

BEAM, WALL, AND COLUMN TESTS (1930 TO 1950)

Early Studies Masonry research began to flourish on a fairly sustained basis in the early 1930s in many of the nation's colleges and universities. In 1928, M. Vaugh, at the University of Missouri, prepared an analysis of several brick masonry beams. One of the most extensive of the early investigations involved the factors affecting the flexural strength of unreinforced masonry, carried out under the direction of Raymond E. Davis at the University of California at Berkeley in 1929 and 1930. This program dealt with the flexural, shear, and bond strengths of brick and hollow tile masonry. Another significant program of that era was conducted by John W. Wittemore and Paul Deer at Virginia Polytechnic Institute. In 1932, they looked into the performance characteristics of reinforced brick masonry slabs of various thicknesses. The behavior of these masonry slabs was compared to reinforced concrete slabs having the same size, type of reinforcement, percentage of steel, and effective depth. Among other things these early investigators concluded was that reinforced brick masonry slabs perform in a manner very similar to that of reinforced concrete slabs, and therefore can be subjected to similar analysis. Furthermore, they decided that the reinforced concrete design formulas were adaptable to the design of reinforced brick masonry slabs, provided that the masonry material properties are injected into them.

In 1936, a full-size reinforced brick masonry wall was tested to determine the effect of the spacing of vertical reinforcement in it. This was an extremely significant test, because at that time no one had a valid idea of precisely how a masonry wall should be reinforced. The test itself was conducted by Smith-Emery Testing Laboratories, with Charles A. Fork as the supervising engineer. The Clay Products Institute of Southern California sponsored the program. The test demonstrated that the masonry is quite capable of spanning between vertical reinforcing bars, 24 in. cc in this case. See Figure 17-2 for an example of this early forerunner of brick masonry wall tests. Further along, in 1939, F. B. Thomas and L. B. Simms reported on the strength of some reinforced brick masonry elements in bending and shear.

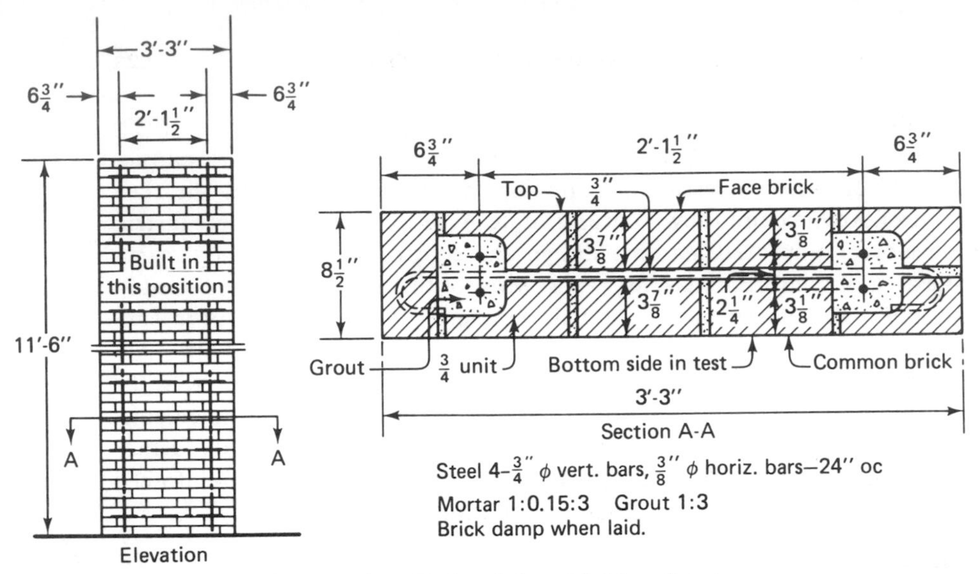

FIGURE 17-2. Early reinforced brick wall test.

Lehigh University An extensive series of column tests was conducted by Inge Lyse at Lehigh University in 1933. Some 33 brick masonry columns were loaded, some of which were unreinforced. Others had untied longitudinal reinforcement, while another group contained longitudinal reinforcement which was laterally stayed with column ties. The conclusions reached, highly significant at the time, were that (1) with sufficient lateral ties, small longitudinal reinforcing bars attain their yield strength at the ultimate capacity of the column; (2) on the other hand, the larger bars did not develop their full strength when the column capacity was reached; (3) the lateral ties were quite effective in enabling the smaller longitudinal reinforcement to reach its yield strength; (4) lateral reinforcement had little, if any, direct effect upon the actual ultimate capacity of the column, but it did determine the failure mode; (5) columns with no lateral reinforcement collapsed immediately and completely upon reaching the maximum load; and (6) both brick strength and mason workmanship displayed a marked effect upon column strength.

University of Wisconsin In 1934, M. O. Withey at the University of Wisconsin supplemented the Lehigh program by subjecting a similar group of both unreinforced and reinforced brick columns to a series of axial loads. A total of 32 brick columns in all were tested, with percent of longitudinal reinforcing varying between 0 and 4% and lateral reinforcing ranging between 0 and $1\frac{1}{2}$%. See Figure 17-3 for a typical column elevation and sections. Withey concluded from these tests that the strength of a reinforced masonry column derives from three components: the strengths of the masonry and the longitudinal reinforcement plus the lateral restraint provided by the hoop ties located in the mortar joints. In his paper, he proposed a formula involving these parameters to define the capacity of the column.

A significant series of beam tests was devised in 1949 at the University of Wisconsin, under the direction of K. F. Wendt and G. W. Washa. Their objective was to examine the plastic flow characteristics of three different types of masonry beams,

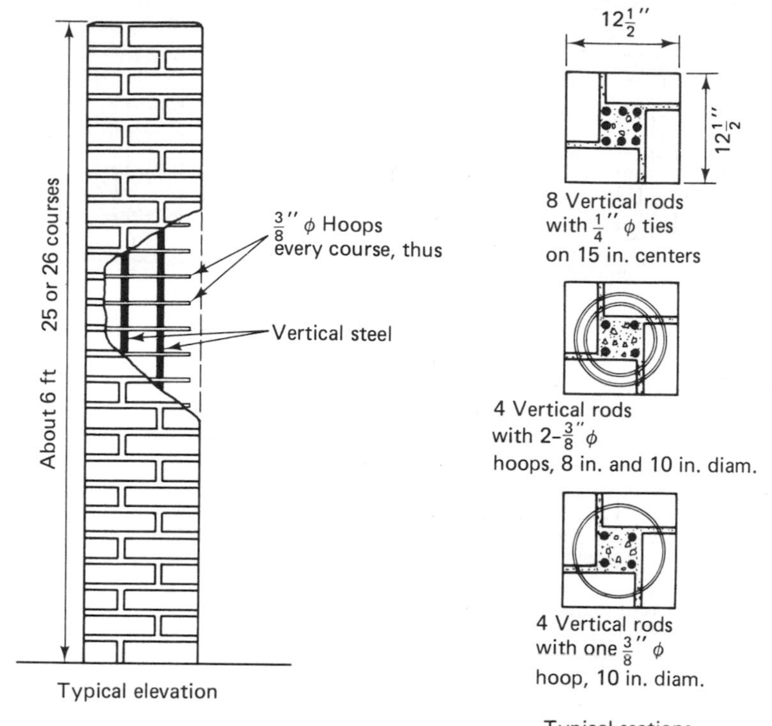

FIGURE 17-3. University of Wisconsin—details of test columns.

having high span/depth ratios, which were subjected to sustained loads. Deflection-time curves show that, for about the first 180 days, the curves approximated straight lines. When compared to reinforced concrete of similar construction, the results of these studies indicated that the plastic flow deflections of reinforced brick beams were considerably less, being on the ratio of approximately 1:6.

University of Southern California In 1951, an extensive investigation involving the behavior of reinforced grouted brick masonry elements under flexural, in-plane shear, and axial compressive loadings was instituted at the University of Southern California under the supervision of Robert R. Schneider. The program was sponsored by the Associated Brick Manufacturers of Southern California, whose director, Norman W. Kelch, provided the basic conception of the overall project. Tests were conducted on three basic groups of assemblies: (1) reinforced grouted brick beams, (2) shear on bolts in a reinforced brick wall, and (3) compression and shear-wall panels of reinforced brick. A major part of the beam investigation was devoted to a determination of the resistance offered by reinforced brick beams to diagonal tension stresses both with and without web reinforcement present. Several variables were introduced into these specimens, including such factors as proper and improper construction workmanship procedures, types of mortar mixes, different beam widths, and beams consisting of a combination of common and face brick. In addition to a determination of shearing strength as a measure of resistance to diagonal tension, other factors were evaluated, such as bond stress, modulus of rupture, and deflection characteristics of the flexural members. Figures 17-4a and b

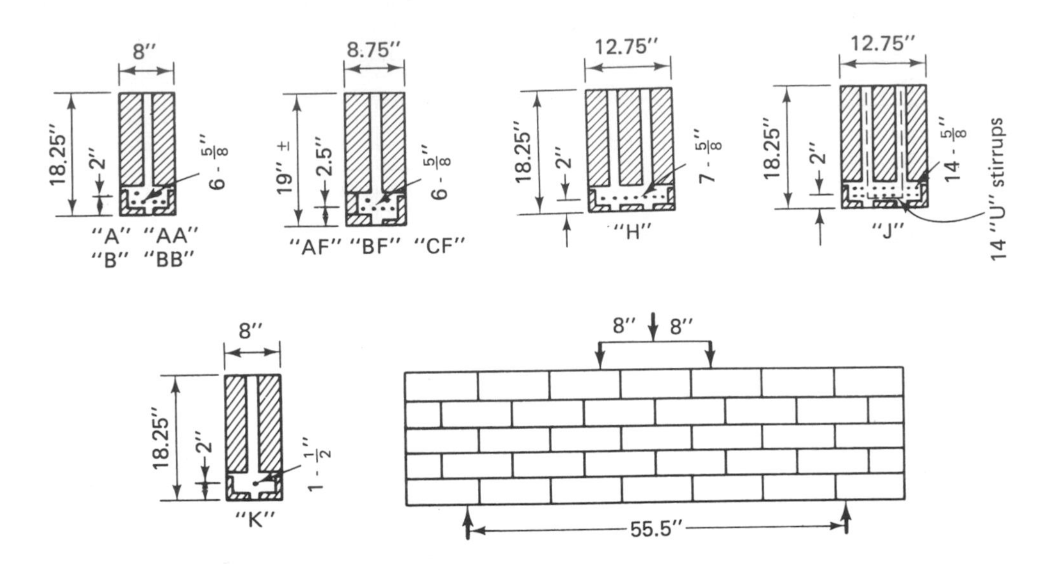

Grout joints in above beams all 1.5" wide

(a)

(b)

FIGURE 17-4. University of Southern California beam test specimens, 1951: (a) short beam details; (b) short beam under test load; (c) long-span beam details; (d) long beam in test rig. (Courtesy R. Schneider.)

FIGURE 17-4 (continued)

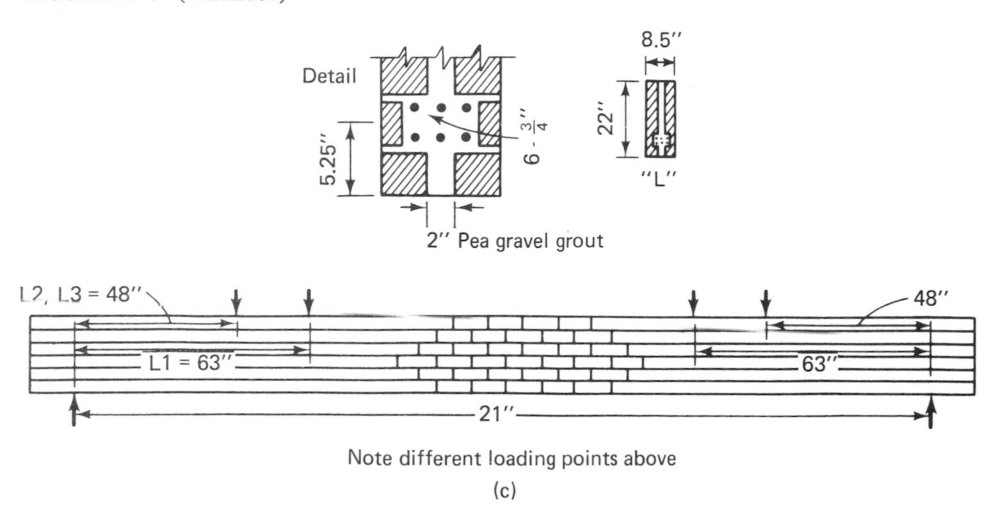

(c)

(d)

show some details of these beam test specimens. In essence, it was shown that beams having adequate longitudinal steel and no web reinforcement all failed in diagonal tension. Depending upon the particular variable injected into the test specimen, the unit shearing stress, taken as a measure of diagonal tension, varied between 118 lb/in.^{2} for the poor work specimens to 225 lb/in.^{2} for the proper workmanship group. The latter also employed the use of a mortar mix similar to that specified for what was then current practice (1 part cement, 0.3 part lime, and 4.5 parts sand). For three-wythe beams, the average unit shearing stress reached 224 lb/in.^{2} where web reinforcement was not present and 408 lb/in.^{2} when it was provided. No conclusions were drawn on any of these tests regarding the manner in which the stirrups resist diagonal tension.

It was simply observed that the shear steel contributed significantly to the ultimate capacity of the flexural member. Figures 17-4c and d show the details for long-span beams loaded to failure to observe their flexural behavior.

The objective of the second group of tests was to determine the resistance to shearing forces of bolts embedded in both the grout and mortar within a reinforced wall panel. Also, the deflection of these bolts was observed when they were subjected to large shear forces. Bolt sizes varied from $\frac{1}{2}$ in. to $\frac{3}{4}$ in. in diameter. It was found that the Code-specified embedment for each bolt size was sufficient, when embedded in grout, to prevent excess crushing of the brick directly below the bolt.

The purpose of the third group of tests was to evaluate the behavior of wall panels of various configurations when subjected to different types of loadings. The piers portrayed in Figure 17-5a simulated both a vertical and a horizontal wall span subjected to a lateral load. The wall panels shown in Figures 17-5b and c were designed to measure the effectiveness of a nonreinforced wall panel in resisting in-plane diagonal tension introduced by diagonal forces. These racking tests produced average shearing stresses of between 116 lb/in.^{2} for the poor work specimens to 120 lb/in.^{2} for the good work specimens again, with a modern type of mortar. An increase in the grout space from $1\frac{1}{2}$ to $2\frac{1}{2}$ in., whereby a pea gravel grout could be used, resulted in a shearing stress of about 190 lb/in.^{2}.

Bearing- and Shear-Wall Investigations

Virtually all the programs previously described involved (1) tests to determine the basic physical and mechanical properties of mortar, grout and the masonry units themselves, or (2) a study of the behavior of reinforced flexural members, with a few column tests thrown in for good measure. So as data was developed through such basic "applied" research, the apparent erratic performance of some of the early reinforced brick beams and other structural elements was explained to a certain extent, and gradually several of the variables affecting their strength were identified and to some degree evaluated. However, a surge of reinforced masonry construction really began in the early 1940s and has continued unabated ever since. The advent of the shear-wall system, accompanying these developments, brought with it a need for more sophisticated knowledge regarding its behavior, particularly as it bears upon its stability under the lateral forces of wind or earthquakes. Thus, many subsequent programs were designed in an attempt to isolate certain parameters involved, to evaluate them, and then to codify the results into more rational expressions and limits. For all the early emerging building codes were filled with many empirical limitations, which were based upon little more than a "feel" for the performance, or an observance of how the earlier structures behaved in the past under earthquake or high-wind conditions.

Stanford University The forerunner of the early investigations involving the behavior of "full-size" reinforced masonry walls was that conducted by J. R. Benjamin and H. A. Williams, beginning around 1955, at Stanford University. The entire program was funded by the federal government and the extensive findings were

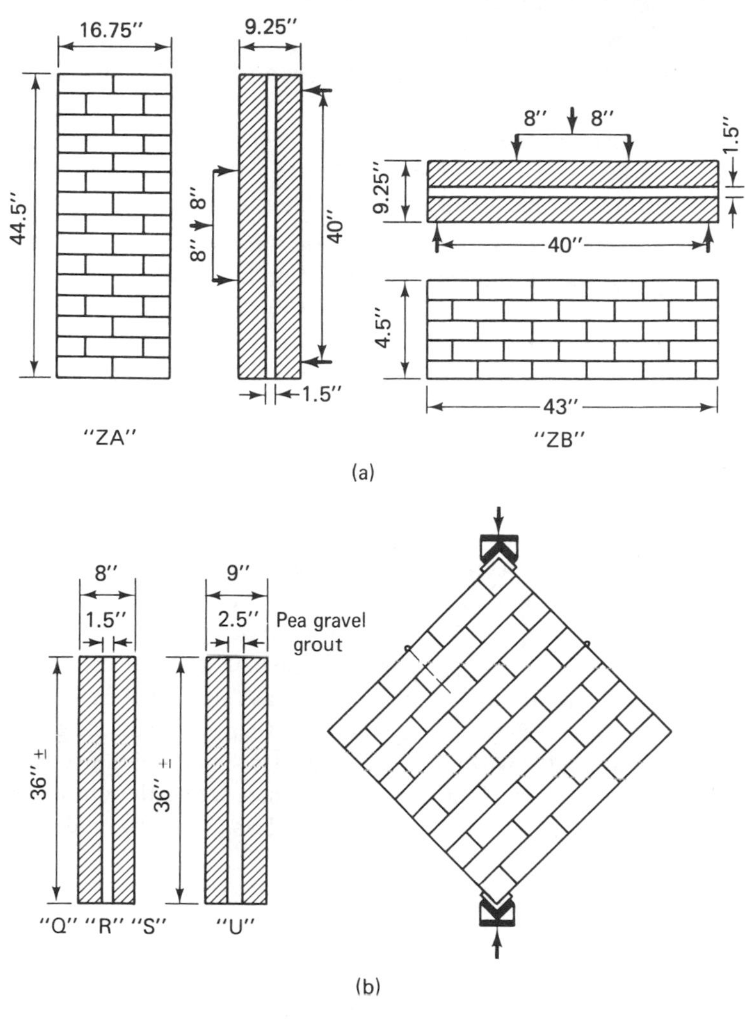

(c)

FIGURE 17-5. University of Southern California wall test specimen, 1951. Photo shows wall panel after overload in test rig. (Courtesy R. Schneider.)

reported in the *ASCE Structural Journal* in 1958. There, reinforced brick shear walls were all encased within a reinforced concrete frame which actually resisted the overturning moment caused by the externally applied horizontal in-plane load.

University of Southern California Another early investigation, sponsored by the then California State Division of Architecture, was carried out under the supervision of Robert R. Schneider at the University of Southern California in 1956. The twofold objective of this investigation was (1) to evaluate the behavior of full-size piers subjected to lateral loads, and (2) develop suitable testing techniques whereby the results would constitute a suitable measure of the lateral in-plane load-carrying capacity of reinforced piers. See Figure 17-6 for the shear jig that was devised for this purpose. In an attempt to identify and evaluate some of the variables affecting the structural behavior of such elements, two groups of shear panels, each consisting of five specimens, were constructed. They were 3 ft 2 in. wide × 7 ft 9 in. high and consisted of two wythes of brick separated by a $2\frac{1}{2}$-in. grout space, forming a 9-in.-thick wall. The boundary reinformement at each vertical wall edge for one group consisted of one No. 8, and for the other, two No. 9's. A third group of five specimens was made similar to the latter, except that the height was decreased to 2 ft 11 in., the jamb steel consisting of two No. 9 bars. It was found that in these shorter panels, which were relatively stiff, the average ultimate shear stress on the gross area was about $105\ \text{lb/in.}^2$, a diagonal tensile failure being evident in all cases. For the two groups of higher piers, however, flexural stresses predominated, and therefore the principal tendency toward failure was initiated by general yielding in the boundary reinforcement. The average

FIGURE 17-6. Shear-wall test jig. (Courtesy R. Schneider.)

shear stress at failure on the first group of five averaged around 37 lb/in.^{2}, whereas the second group attained a failure stress of 73 lb/in.^{2}. It should have been apparent from these results, although seemingly it was not at that time, that shear-wall elements should not be assigned a single allowable shear value, regardless of their height/depth ratio (h/d), as was the then-current practice. It was not until 1969 that recognition was made of the phenomenon and incorporated into the UBC. The basic change in philosophy that this concept represents was initially suggested by results obtained in reinforced concrete beam tests, and it was finally formalized in the code through information gathered in a subsequent masonry shear-wall investigation, as described in a later section of this chapter.

These results also demonstrated the need to recognize the flexural stresses occurring in relatively high piers due to overturning and to accommodate them by placing adequate vertical reinforcement at the boundaries of the pier.

Following this pilot program, an extensive investigation into the behavior of masonry shear walls subjected to lateral in-plane shear loads was conducted by the same investigator in 1959. It was jointly sponsored by the California State Division of Architecture, the Associated Brick Manufacturers of California, the Concrete Masonry Association, and a private concrete block manufacturer. It had as its objectives the following: (1) to obtain shear strength values as a measure of the diagonal tensile resistance of masonry shear walls; (2) to evaluate the effects of cement and lime on mortar strength as it, in turn, affects the shear resistance of the composite wall; (3) to determine the relative percentages of total shear sustained individually by the mortar and grout, and (4) to study the behavior of walls with different-size opening configurations. Also, some attention was paid to the different Code requirements for minimum amounts of reinforcing [i.e., $0.002bt$ (UBC) versus $0.003bt$ (OSA).]. Specimens were constructed with clay brick (ASTM C62-MW Grade), hollow concrete block (ASTM C90-Grade A), and a proprietary concrete masonry unit (Shel-brik). Five groups of walls, each consisting of three similar specimens, all 8 × 8 ft × 9 in. thick, were built. This was one of the earliest attempts to circumvent scale effect by testing what were referred to at that time as "full-size" walls. See Figure 17-7 for wall details. Both solid walls and walls with openings were constructed. Different mortar mix proportions were utilized in the solid brick walls. All these specimens were unique in another sense, in that the resistance to overturning by the lateral load was developed internally by means of a holddown anchor welded to the boundary steel grouted within the wall (see loading diagram, Figure 17-8a), rather than by using the typical external holddown bracing which imposes a very heavy compressive force at the point of application of the lateral force. The influence of this heavy load concentration, Figure 17-8b, upon the actual shear stress is a matter of conjecture, so the values obtained with this type of external device should be viewed more in the context of an index rather than representing a true value. The jamb bars (in this case four No. 8's) had to have enough area to enable the wall to resist the tensile forces caused by the overturning moment.

Five series of concrete block walls, each consisting of three similar specimens, were also built. The variables in these specimens consisted of the bond pattern

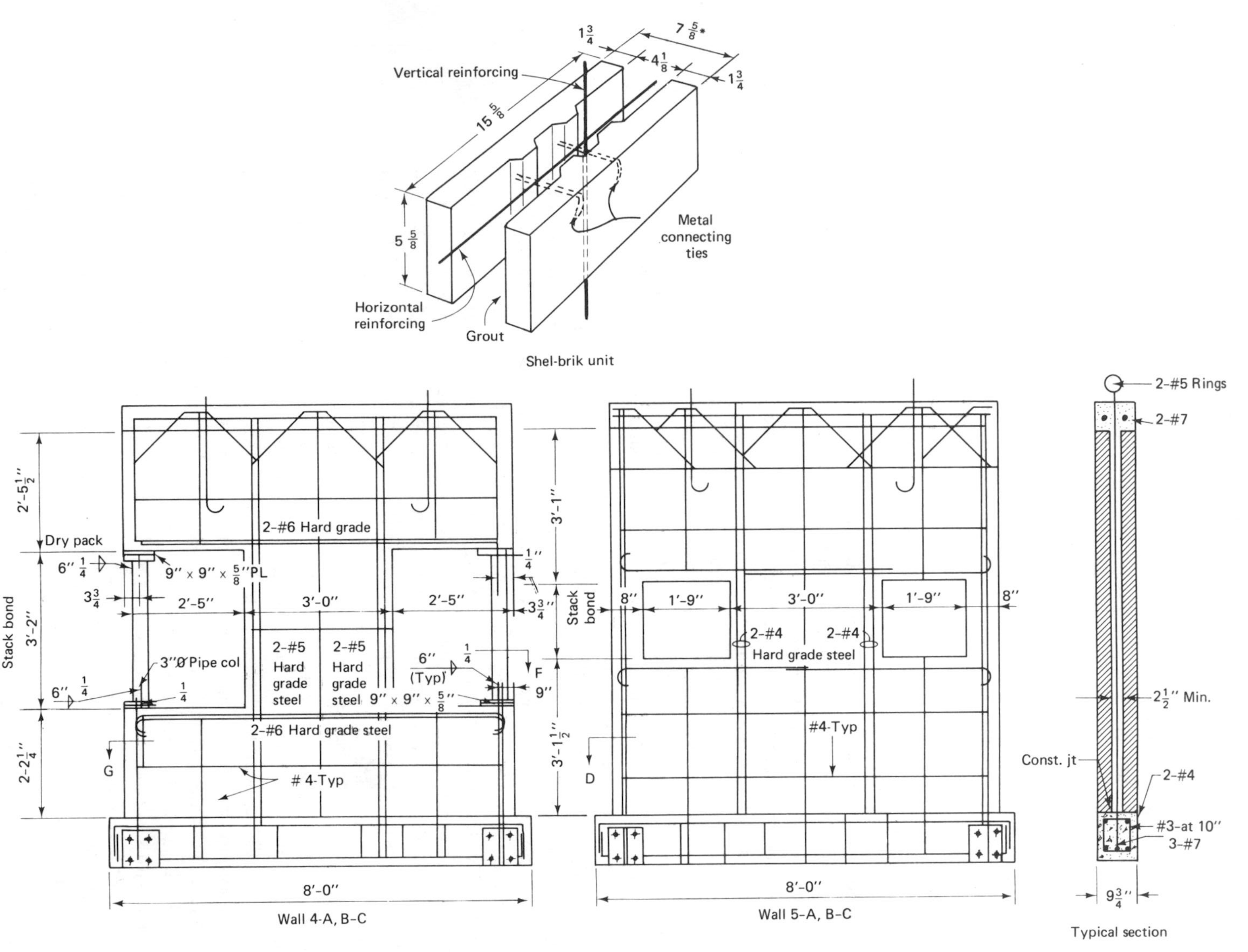

FIGURE 17-7. Shear-wall test specimens, University of Southern California, 1959.

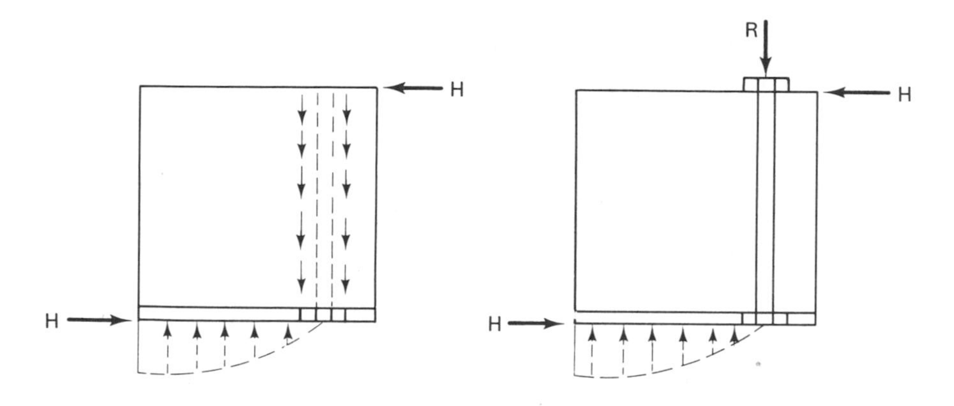

FIGURE 17-8. Holddown loading diagrams.

(running versus stack bond) and the amount of reinforcement (0.003*bt* versus 0.002*bt*). Refer to Figure 17-9, which shows the manner in which the walls were tested.

In most cases the nature of the crack patterns indicated that the mode of failure was essentially that of diagonal tension. However, in some instances, the opening up of the bed joints on the tension side of the masonry simultaneously with the elongation of the tension reinforcing, caused a concentration of high crushing forces at the toe which seemed to be the dominant factor in the failure. The average shear stress on the gross wall section for both the solid brick and concrete block walls was about 143 lb/in.[2], demonstrating conclusively that the solid grouted masonry walls possess adequate ability to sustain horizontal in-plane shear loads. Furthermore, since

FIGURE 17-9. Shear-wall test procedure, University of Southern California, 1959. (Courtesy R. Schneider.)

the three groups of brick walls were each constructed with different mortar proportions, the test results indicated that the mortar-mix proportion, under these test conditions, had little effect on the shear resistance of the wall. In one case a straight lime mortar was used, which possesses virtually no shear strength whatsoever.

The presence of the openings in the brick wall group created piers within the wall itself having different *h*/*d* ratios from that of the solid walls, which possessed a 1:1 ratio. The individual piers between the openings averaged about 268 lb/in.^{2} on the gross wall section for an *h*/*d* of about 1:1 and 284 lb/in.^{2} with the *h*/*d* of 0.53:1. It was observed that the net area reduction at the opening level, in cases where the walls had opening discontinuities, was considerably more than the corresponding reduction in load-carrying capacity as compared to a solid brick wall. This again points out the fallacy, as suggested in the pilot program, in assigning a constant allowable shear stress, independent of the *h*/*d* ratio of the pier. The shear resistance is a variable, one that is significantly affected by the *h*/*d* ratio of the pier. This highly significant concept is readily emphasized in these and other tests, and is now reflected in the allowable code stresses. Actually, the *h*/*d* factor appears in the Code through the parameter, *M*/*Vd*, which will be discussed in more detail later. This constitutes an excellent example of the immediate use of applied research in developing a code parameter.

In the case of the concrete block walls, the amount of field reinforcing did not appear to have any appreciable effect upon the ultimate capacity, as measured in terms of the unit shear stress on the gross area. However, the added reinforcement did seem to increase the load required to produce an initial crack in the wall. It might also be mentioned that some of the test panels were subjected to static load reversals; however, these walls demonstrated no particular change in horizontal shear resistance or wall stiffness characteristics compared to those that were loaded monotonically.

Structural Engineers Association of California (SEAOC) A rather interesting pier test program appeared sometime in 1957 when the Los Angeles Board of Education proposed that the Structural Engineers Association of California devise a series of tests to prove the adequency of rehabilitating unreinforced brick shear walls in existing pre-1933 school buildings through a special method developed by the Board. It was essential that the then California Division of Architecture approve this procedure before it would certify that these unreinforced masonry buildings possessed adequate seismic resistance. The Division refused to do this until the method had been thoroughly tested. Construction essentially involved the application of at least 1 in. of plaster, reinforced with a mat of galvanized welded wire fabric, on each side of the old lime mortar walls. The effectiveness of this plaster diaphragm depends, to a large extent, upon the bond of the plaster to the brick.

Test parameters in the 20 wall specimens constructed were: (1) the gauge of the mesh (2 in. × 2 in. × 12/12 and 2 in. × 2 in. 16/16 WWF) plus the existence of a plaster return at the ends of walls. Sixteen specimens were built, approximately 4 ft × 4 ft × 13 in. in size (four of which comprised a control group; i.e., no plaster or mesh). Another four were approximately 8 ft high × 4 ft × 13 in. in size (to observe the influence of the shape factor). Construction simulated the old pre-1933 type of lime mortar masonry as closely as possible. See Figure 17-10 for wall details. An external yoke provided the means of holding the panel so as to resist the overturning moment developed by a horizontal load applied at the top of the wall. Refer to Figure 17-8 for the loading diagram for these tests, supervised by R. Schneider.

The average ultimate shear stress registered by the unplastered control group amounted to about 27 lb/in.^{2}, whereas the average value of the plastered piers, of similar size, ranged between 58 and 71 lb/in.^{2}, depending upon the test variable involved. This increase in load capacity dramatically attested to the effectiveness of this method of externally reinforcing the old masonry walls. Thus, these results demonstrated that the shear resistance of the specimens depended to a large degree upon the capacity of the reinforced plaster to withstand diagonal tension. It also appeared that the gauge of the mesh had no great effect on the shear resistance. Load-deflection characteristics also substantiated an adequate performance by showing that

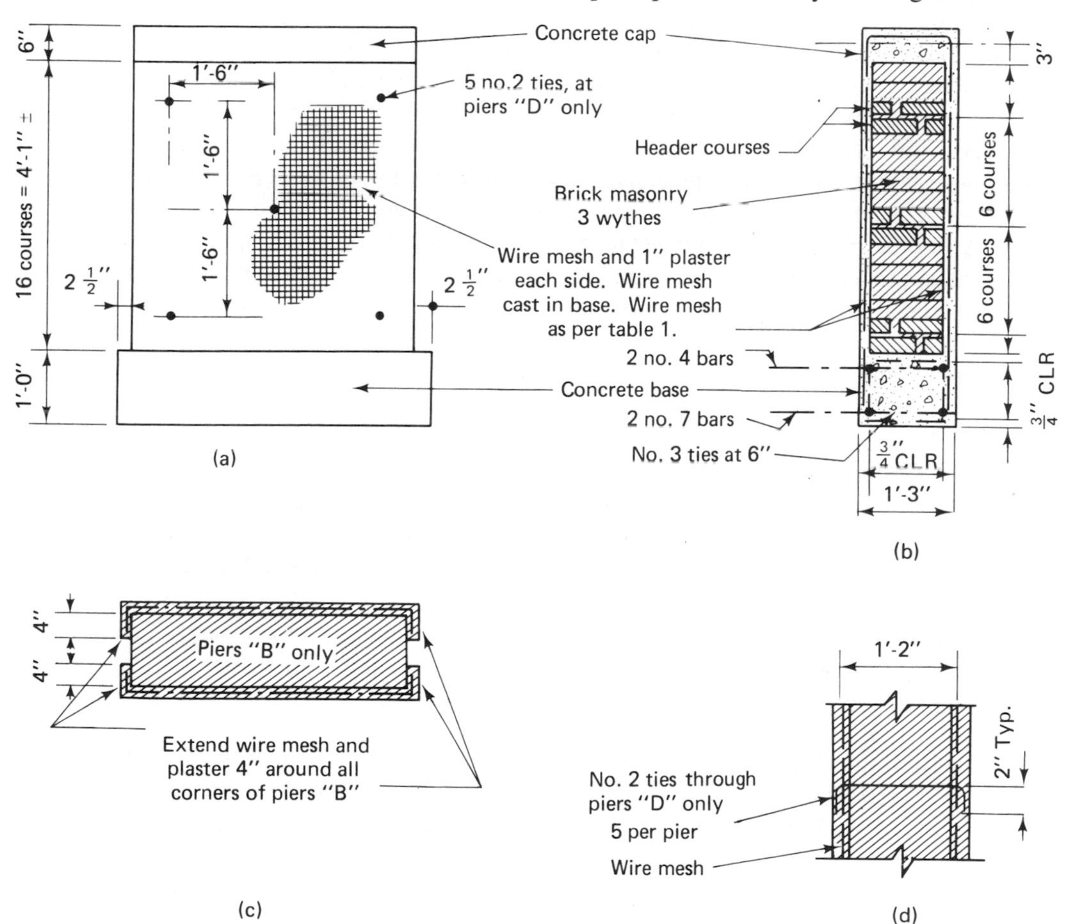

FIGURE 17-10. Plastered pier test specimens, SEAOC,1957. (Courtesy R. Schneider)

the panels exhibited sufficient stiffness. Finally, it was noted that the reinforced plaster returns around the ends of the piers did not appreciably alter the load-carrying capacity. For reasons not entirely clear at this later time, the Division of Architecture refused to extend its approval, so the Board of Education did not promote the method any further. However, interest is being revived once again, this time by the City of Los Angeles in its attempt to find a structurally sound, but economically feasible method of revitalizing numerous pre-1933 masonry buildings so that they may offer at least some seismic resistance.

Research in the 1960s and early 1970s

The advent of the computer ushered in an entire new era in the field of material research. In the past, most of the applied loadings were of necessity static in nature and the volume of material that could be digested or analyzed by hand was highly limited. However, with the computer, great volumes of data could be reduced, reported on, and sound conclusions reached in relatively short periods of time. Furthermore, very sophisticated methods of loading were developed; consequently, very large test specimens of various shapes and sizes could be utilized for the first time. Computer programs could control, not only the magnitude of the loads, but also the velocity with which they were applied as well as the frequency of the loading cycles. Thus, it is not surprising to see that masonry test programs grew in number and complexity. A number of the more significant ones will be briefly reported upon herein. Because of space limitations, a description of other later programs conducted by various principal investigators must be omitted here. However, details could be obtained from the report on any specific investigation of interest. Most of the significant research reports are summarized in a literature survey prepared by R. Mayes and R. Clough at the University of California, Berkeley, in 1975.

MONOTONIC SHEAR-WALL LOADINGS

Scrivener J. C. Scrivener, starting sometime in 1966 at the University of Canterbury, Christchurch, New Zealand, began a series of racking tests on concrete masonry wall panels. There were three sets of tests altogether, using 8 ft 8 in.-high by 8 ft-long panels. In this series, he sought to (1) measure the factors influencing flexural failure, and the influence of different reinforcing distributions on the shear failure mode; and (2) compare the performance of a partially grouted wall versus that of a solid grouted one. Among other things, he concluded that:

1. Both the stiffness and failure load increase with an increase in the amount of vertical steel, the latter being most effective when located at the periphery of the wall. Also, the inclusion of horizontal bond beams did not significantly alter the stiffness or failure load.
2. Higher shear failure loads were evidenced by those specimens having higher percentages of reinforcing, up to 0.3% of the gross area. Above this percentage, additional reinforcing had little effect on the failure load. Horizontal shear stress values of up to 170 lb/in.^{2} were obtained on the

wall reinforced with 0.3% or more. Also, with this much reinforcement present, the ultimate load was much higher than that causing the first crack.

3. A solid-grouted masonry wall will withstand higher loads at first crack and ultimate, and also is much stiffer than a partially grouted wall. A strength increase of between 20 and 50% was observed. However, based on net area strength, the partially grouted masonry performed well.

Schneider In 1969, Robert R. Schneider, Department of Civil Engineering, California State Polytechnic University, Pomona, carried out an extensive test program developed by Walter L. Dickey, Consulting Structural Engineer for the Masonry Institute of America. The program dealt essentially with a thorough study of the shear behavior of concrete masonry piers functioning within the confines of a shear-wall system. The overall objective of this program was to conduct an evaluation of the ability of reinforced concrete masonry piers to resist lateral forces by measuring the influence exerted by certain design or construction parameters upon this ability and to observe the resulting failure mechanism. Those parameters deemed most relevant to pier behavior were (1) a/D ratio ($a = h/2$ for restrained piers and $a = h$ for cantilever piers), (2) presence of web reinforcement (horizontal and vertical), (3) amount of jamb reinforcement, (4) effect of axial compressive load, and (5) behavior of partially grouted wall sections. Several full-size panels, approximately 10 ft 8 in. square, were constructed within each test group, each group incorporating a single parameter for observation. See Figure 17-11 for an illustration of a typical wall panel.

FIGURE 17-11. Typical wall panel specimen, Schneider and Dickey, 1969.

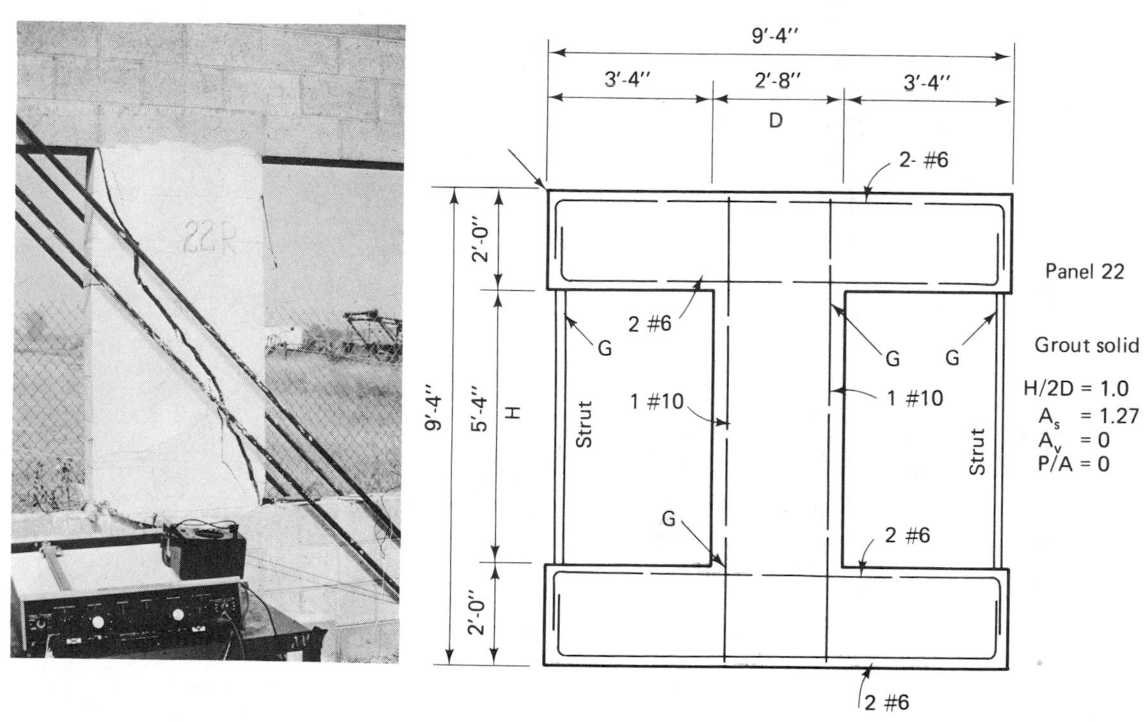

A diagonal loading frame, complete with hydraulic jacking devices, was used to apply the load along a diagonal of the wall similar to earlier ASTM racking tests. See Figure 17-12 for a diagrammatic representation of this loading method. Shearing resistance, as related to the a/D ratio, was measured in terms of the unit shear stress observed over the cross-sectional area of the pier. The stiffness was viewed from the load deflection characteristics exhibited by each individual pier. See Figure 17-13, which shows a typical wall panel located in the loading device.

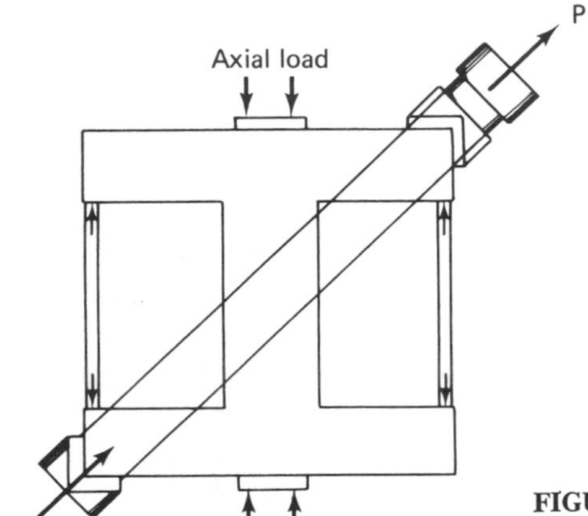

FIGURE 17-12. Diagonal frame-loading diagram.

FIGURE 17-13. Loading device, Schneider, 1969.

A great deal of information and data were obtained from these tests. For complete details, refer to the report submitted by the investigator, titled "Shear in Concrete Masonry Piers," issued by the Masonry Institute of America. Some of the more significant findings, noted in that report, may be summarized in the following statements:

1. Shear strength definitely increases with a decrease in the a/D ratio, and this rate of increase jumped sharply below an a/D ratio of about 0.5:1. Above this value, the rate of stress decrease becomes considerably less, and the a/D-ratio curves appear to approach a constant stress level of about 100 lb/in.^{2}. Within the range considered, shear values varied between 103 lb/in.^{2} for an a/D of 1.5 and 290 lb/in.^{2} for a/D of 0.17. On the other hand, the cantilever piers did not exhibit much of a stress change, varying from 90.4 lb/in.^{2} for an a/D of 1.0 to 60.1 lb/in.^{2} for an a/D of 3.0. Refer to Figure 7-9, which shows this stress versus a/D ratio.
2. The presence of a heavy amount of horizontal web reinforcement (No. 5's at 8 in. cc) did improve the behavior of the pier as compared to those without any web reinforcement. Shear values ranged from 243 lb/in.^{2} for an a/D of 0.5 to 127 lb/in.^{2} for an a/D of 1.5. This differential becomes more pronounced as the a/D ratio decreases as was seen in Figure 7-10. The upper ratio for which any appreciable difference in shear strength existed appeared to be somewhere between 1.75 and 2. For the lightly reinforced piers (No. 4's at 24 in. cc), a slight increase in shear strength over those without any web reinforcement was shown for a/D ratios below about 1.0. Above that ratio, there appears to be virtually no change.
3. The presence of vertical shear steel in the pier did not significantly increase pier resistance, as measured by the unit shear stress at the ultimate load. Furthermore, the failure mechanism and the cracking patterns exhibited by the piers in this group suggest that the vertical reinforcement will not restrain the propagation of the diagonal cracks during the loading sequence as effectively as does the horizontal shear steel. Finally, the load-deflection curves indicated that the vertical reinforcement does not impart as much stiffness to a wall as the horizontal shear steel does. In the case of the cantilever piers, web reinforcement, either vertical or horizontal, contributed little toward either pier shear strength or stiffness.
4. When the jamb reinforcement was doubled over that provided in the control group, the shear resistance in the restrained piers increased only slightly, if at all.
5. An external axial compressive load of about 400 lb/in.^{2} was imposed upon the top of some of the walls, to simulate vertical floor or roof loads carried by the pier. This load was applied before the application of the diagonal load. Results seem to indicate that some increase in shear strength and pier stiffness may be expected when an external axial compressive load is applied.

6. The partially grouted walls sustained considerably less load (about 25%) than that carried by the grouted walls, as would be expected. The most distinct aspect of the difference in behavior between these two walls, however, stems from the manner in which the ungrouted piers failed. They deflected drastically and failed suddenly, denoting a brittle fracture. No load was sustained after the ultimate had been reached, as was the case with the solid-grouted walls.
7. The energy-absorbing ability of an adequately reinforced masonry pier was well demonstrated in these tests, as the shape of the load deflection curves indicated. It would appear that these piers were able to absorb a considerable amount of inelastic strain energy without collapsing or even spalling seriously. Thus, concrete masonry, if properly reinforced, exhibits a tendency toward a ductile-like behavior throughout a loading sequence. It can sustain significant proportions of the ultimate load well into the inelastic region (beyond the first major crack), while undergoing rather large lateral deflections. It also exhibits effective damping characteristics, especially after cracking.
8. The ultimate diagonal tensile strength of a concrete masonry assemblage, regardless of the amount of reinforcing present, appears to be somewhere between 260 and 280 lb/in.^{2}. These results were obtained by subjecting 4×4 ft assemblages to a racking test (i.e., compressive loading along a diagonal).
9. Although the amount and orientation of the shear reinforcing influenced shear capacity, as was indicated proviously, the bar spacing within the field of the wall did not prove to be a significant factor, at least within the spacings utilized in these tests, this being approximately 4 ft 0 in. cc.

Meli The aim of an extensive investigation carried out by R. Meli, research professor at the Institute of Engineering, National University of Mexico, beginning in 1969, was to observe the behavior of masonry walls, insofar as their shear resistance, ductility, and energy-absorption characteristics are concerned, when subjected to in-plane lateral forces. A total of 56 walls, 2 by 2 m in size, were constructed using both concrete block and hollow and solid clay brick masonry units. Each was built on a stiff concrete beam. Most were tested as cantilevers (Figure 17-14a)

FIGURE 17-14. Loading diagrams, Meli, 1969.

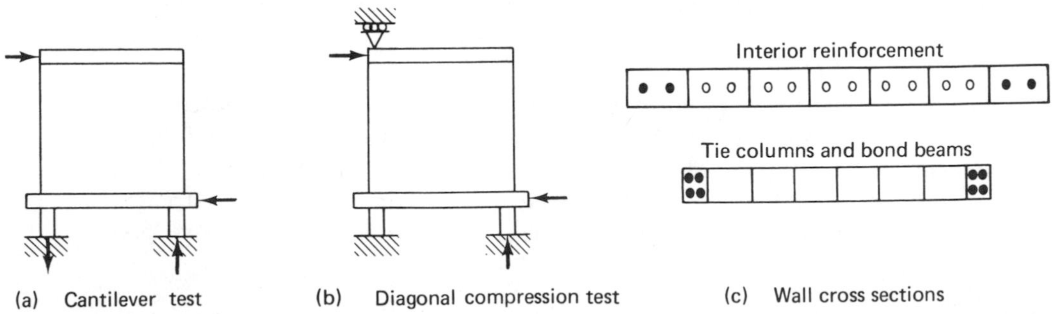

Interior reinforcement

Tie columns and bond beams

(a) Cantilever test (b) Diagonal compression test (c) Wall cross sections

although a few walls were subjected to a diagonal compressive load as shown in Figure 17-14b. Two types of reinforcement were provided: (1) interior rebar grouted within the holes or cells of the units, or (2) tie columns and bond beams of the same thickness as the wall, comprising an enclosing boundary frame as shown in Figure 17-14c. Brick walls were all single-wythe in thickness. A vertical load was imposed on several of the walls prior to the application of the lateral load.

Walls encased within the frame behaved as a monolithic unit until the load was large enough to cause an initial separation on the lower tensile corner and later in the opposite corner. Subsequent progressive flexural cracking in both the frame and the wall resulted in a serious reduction in stiffness. Table 17-1 lists the failure modes

TABLE 17-1

Failure Modes (Meli, 1969)

Type of unit	*Other test characteristics*	*Precompression (kg/cm²)*
Interior reinforcement concrete block	Bending failure	0
		3.5
		10
	Shear failure: only	0
	extreme holes	3.5
	reinforced	10
	Intermediate reinforcement:	0
	fully grouted	0
hollow clay brick	Shear failure: only	0
	extreme holes	6
	reinforced	12.5
	Intermediate reinforcement	0
Tie columns and bond beams hollow clay brick		0
	Intermediate reinforcement	0
solid brick	Shear failure	0
		3
perforated brick	Shear failure	0
		3
hollow clay brick	Shear failure	0
		3
sand lime brick	Shear failure	0
		3

for the various monotonically loaded test specimens subjected to different precompression load conditions, and Figure 17-15 portrays typical cracking patterns at failure for the various types of wall panels.

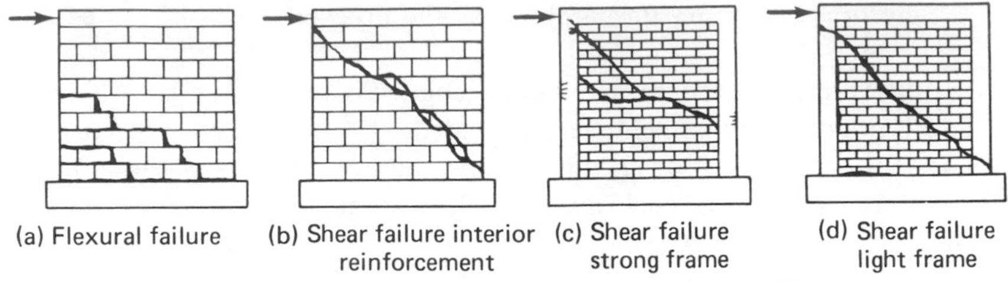

FIGURE 17-15. Cracking patterns at failure, Meli.

In "cantilever-type" specimens, the test variables, besides the type of masonry unit, consisted of the reinforcement ratio and the imposition of a vertical precompression load of both low and high intensities. In some instances the vertical load was left off entirely, however. Cracking patterns at failure are shown in Figures 17-15c and d. Influence of the means of reinforcing of the masonry walls on their behavior is summarized in Figure 17-16.

Meli states that the strength and general behavior of the cantilever walls failing in flexure (low reinforcement ratio and little or no precompression load) could be predicted from the hypotheses used for reinforced concrete analysis. Their capacity was limited by yielding of the jamb steel, which was followed by a considerable amount of inelastic distortion. Failure was due to crushing of the compression corner or to rupture of the extreme bars. Precompression on the wall increased the strength and, at the higher vertical stresses at least, caused a change in failure mode to that of brittle shear. High reinforcement ratios also produced a failure of this sort. For the walls with interior reinforcement and no precompression load, behavior was rather ductile in nature, due no doubt, to the friction exerted between the diagonal cracks and also the dowel action of the vertical steel.

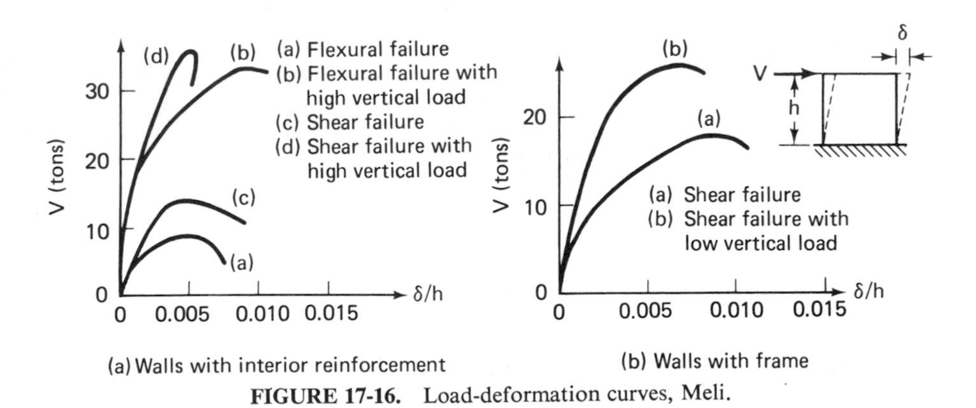

FIGURE 17-16. Load-deformation curves, Meli.

Meli found that the property that could best be related to wall strength was the average shear stress obtained on his diagonal compression tests of small square panels, one and one-half masonry units long and as high as they were long (Figure 17-17). For those walls, tested in diagonal compression, the average shear stress at diagonal cracking was found to be about 85% of that obtained on the small

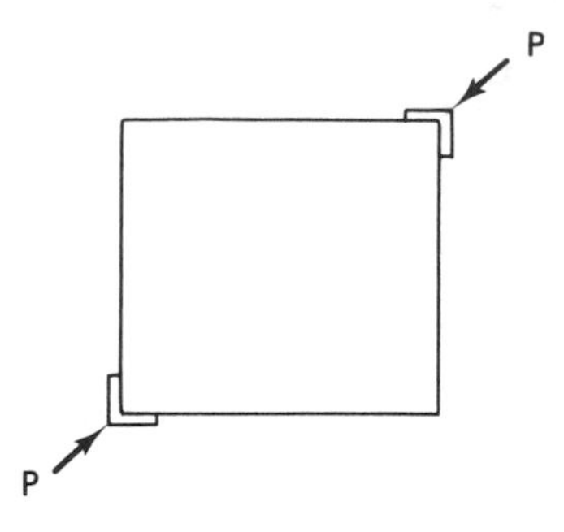

FIGURE 17-17. Racking load diagram.

assemblages. On the other hand, for walls tested as cantilevers, flexural stress reduced the shear at diagonal cracking to about 50% of that of the small wall panels.

The postcracking behavior of the frame-contained walls depends primarily upon the strength, stiffness, and ductility of the exterior columns. This behavior may be summarized as follows:

1. A shear failure at a load just above the cracking load was evidenced in the case of a light frame. However, the frame provided sufficient confinement to permit high angular deformations to take place.
2. A strong column has sufficient shear strength to prevent the propagation of a diagonal crack into the column, causing a deviation of the diagonal crack to a more horizontal position (refer to Figure 17-15a).
3. In the case of a strong frame, the ultimate capacity is much higher than the cracking load and is limited by bending failure of the short columns.
4. Ductile behavior was evidenced, but it was reduced where low-strength masonry units were used.
5. Where a precompression axial load was imposed, both cracking and maximum load increased, but at the expense of ductility.

Blume Significant work was done by John Blume and Associates for the Western States Clay Products Association in 1968. The investigation dealt with an evaluation of the shear strength of grouted brick masonry walls. It was also desired to observe the influence of various amounts, as well as orientations, of the reinforcing upon the strength and ductility of the panels. Shear was determined (as the measure of diagonal tension) on some 84 brick masonry panels each 4 × 4 ft in size. They consisted of 67 solid building brick panels, 14 hollow brick panels, and 3 hollow brick panels with all cells filled with grout. A typical panel is shown in Figure 17-18. The panels were all loaded on the diagonally opposite corners, thereby producing a tensile failure along the loading diagonal, as was shown in Figure 17-17. Shear stresses were then computed on the basis of the gross wall section. These are utilized to index the diagonal tension resistance of the panel. Among the more significant findings stemming from results on the grouted solid brick panels were the following:

1. The mode of failure for each panel test was a crack along the loaded diagonal caused by diagonal tension stresses.

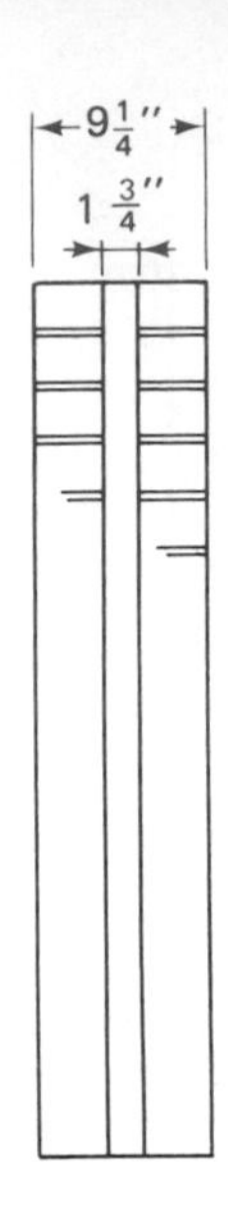

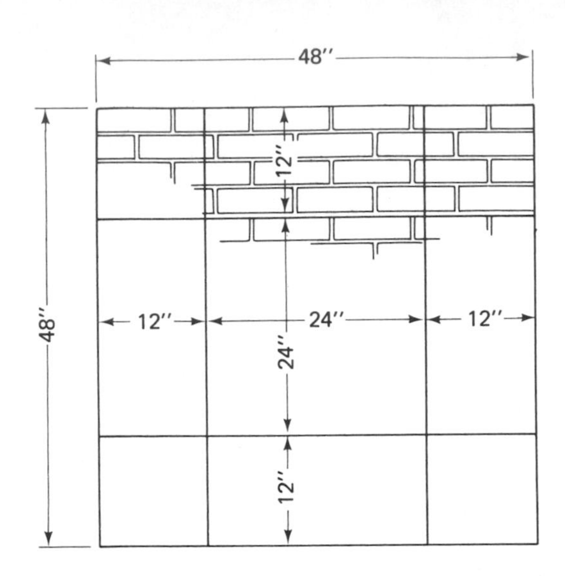

Reinforcement: #4 bars

Ratio of $\frac{\text{steel area}}{\text{wall cross section}}$ (%):

Horizontal: 0.09
Vertical: 0.09
Total: 0.18

Brick properties:
Size: $3\frac{3}{4}'' \times 2\frac{5}{8}'' \times 8\frac{1}{4}''$
Compression: 7730 lb/in.^{2}
Rupture: 825 lb/in^2
Suction: 54 gr/min./30 in.^{2}

FIGURE 17-18. Racking test panels, Blume.

2. The average maximum shear stress for the unreinforced grouted panels varied between 314 and 426 lb/in.^{2}.
3. As the percentage of steel increased above 0.2%, the maximum shear strength and energy capacity increased.
4. With a $1\frac{3}{4}$-in. grout space and only small percentages of reinforcing (less than a total of 0.2%), the reinforced panels failed at values slightly less than those of unreinforced panels, but the wall shear strength still averaged about 300 lb/in.^{2}.
5. A steel ratio as low as 0.045% was sufficient to prevent the brick panels from separating after cracking.
6. As the brick shear strength increased, the shear strength and energy capacity of the panels increased. A brick compressive strength of between 6,000 and 10,000 lb/in.^{2} seemed to produce the optimum energy capacity.
7. With mortar proportioned according to the then-current UBC, the shear capacity and the energy capacity of the panels were considerably greater than for panels utilizing a weaker mortar.
8. An increase in grout strength produced an increase in shear strength and energy capacity. The intentionally weakened mortar specimens (300 lb/in.^{2})

developed diagonal tension cracks along the mortar joint as opposed to those through the brick in the case where mortar conforming to UBC standards was utilized.

9. Increasing the grout core thickness from $1\frac{3}{4}$ to $2\frac{1}{2}$ in., resulted in an increase in the shear strength and energy capacity of the panel.
10. The anchorage of reinforcing bars to simulate normal wall continuity had little effect on the results over nonanchored bars terminating with plain ends in the masonry.
11. Pouring the grout 24 h after the wall panels were constructed produced higher strength values than were obtained where grouting was done immediately following panel construction.

The purpose of the hollow brick panel tests was to investigate physical characteristics of such panels, that is, their maximum shear stress, strain at maximum stress, ultimate energy, and ductility. These test results produced the following conclusions:

1. With wall panels having grout in the end cell only, the average maximum shear stress, based upon net area, was 123.4 to 390.8 lb/in.^{2}, depending upon the type of masonry units utilized.
2. For solid-grouted wall panels, the maximum shearing stress average ranged between 265.9 to 387.0 lb/in.^{2}, again depending upon the type of masonry unit used. The measured energy capacity in those panels was between 2.2 and 4.2 times greater than those panels having grout on the end cells only.

The tests to evaluate the damping characteristics were performed on three 4 ft square grouted brick masonry panels vibrating freely in the transverse mode. Damping was expressed as a percentage of critical damping. The conclusions reached were:

1. The maximum value of damping, just prior to any cracking or failure in the case of the unreinforced grout brick panels, varied between 8.7 and 9.4% of critical damping.
2. Mortar and grout compressive strengths have no apparent effect on the value of damping.
3. As the brick compressive strength decreased from 6470 to 3280 lb/in.^{2}, the value of damping increased from 8.7 to 9.4% of critical damping.

On the basis of the test results and conclusions reached, Blume made certain comments and recommendations, among which were the following:

1. The most important property of walls or wall elements that resist lateral forces in the plane of the walls is that of diagonal tension resistance, as measured by the in-plane shear stress.

2. The minimum recommended grout core thickness is 2 in. for high-lift grouting (4-ft lifts). It was further recommended that at least $\frac{3}{8}$ in. and preferably $\frac{1}{2}$ in. of grout cover be provided between the near face of the reinforcing steel and the interface of both wythes.
3. Wall steel in excess of the minimum specified by the Code is not justified on the basis of the slight increase in ductility and energy absorption capacity that is obtained, unless, of course, it is required for flexural stresses due to overturning or lateral forces normal to the wall. The development of ductility and increased energy absorption capacity in flexure can be obtained by providing the amount of tension steel necessary to ensure that it will yield before the failure of the masonry unit itself takes place (i.e., by underreinforcing).
4. The arbitrary use of large percentages of steel area or closely spaced bars in grouted masonry should not be encouraged. Excessive steel may detract from the attainment of a complete bond between the brick wythes and the grout core.
5. The relatively high damping values obtained on these panels constitute a very desirable characteristic, in that it indicates a reduction in the response to earthquakes.

Piper and Trautsch This pair of investigators performed a study in 1970 relating to the shear strength of long walls, with the test apparatus illustrated in Figure 17-19. The variables included were the applied compressive stress, length and

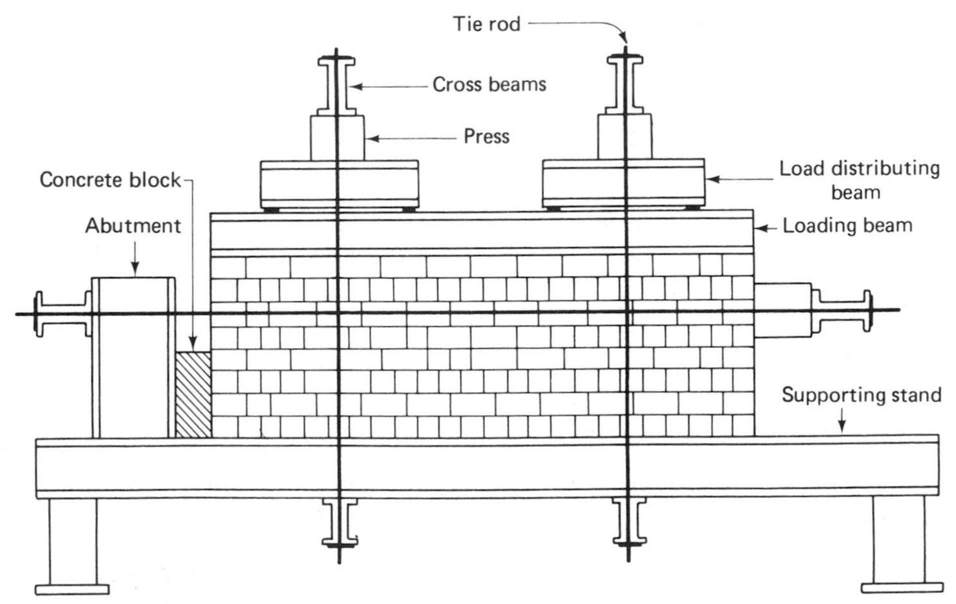

FIGURE 17-19. Shear test apparatus, Piper and Trautsch, 1970.

thickness of the wall specimens, and mortar strength. Because of the manner in which the load was applied, the results obtained in this study can be considered only as a comparative behavior index. At any rate, the authors concluded that, in essence, the shear strength of the beams increased as the mortar strength or axial compressive stress increased. This test actually serves more as a measure of the horizontal bed joint or sliding mode of failure, rather than that of a diagonal tension or toe-crushing mode of failure.

Cyclic Shear Load Tests

It is obviously closer to seismic reality to subject a shear wall to in-plane dynamic cyclic reversals of load. This procedure, of course, requires very sophisticated and expensive equipment to accomplish the task, and relatively few laboratories are equipped to handle this type of experimental method. Thus far at least, four major investigations have been carried out and reported in the literature. They include the work of Williams, Meli, Mayes and Clough, and Priestly and Bridgeman. Details on some of these are described in the next sections.

Williams A major work was conducted by D. Williams in 1971 at the University of Canterbury, Christchurch, New Zealand, on the seismic behavior of reinforced masonry shear walls. The overall objective of this important and very extensive investigation was to make an evaluation of the post-elastic performance of reinforced masonry shear walls under cyclic loading conditions. More specifically, certain behavior patterns exhibited by these reinforced masonry walls were sought, such as:

1. Their ductility capabilities.
2. Their stiffness degradation and load deterioration characteristics under cyclic loading.
3. Their failure mechanism and ultimate load capacities.
4. The effect of certain test parameters upon the above, such as wall geometry, amount and disposition of reinforcing, and magnitude of bearing load.

Static cyclic load tests: Table 17-2 summarizes the details of the various wall specimens subjected to static cyclic loadings. Horizontal shear reinforcing was placed in two of the walls (A-1 and A-2) to measure its effect. The brick specimens were made with a cored brick unit having a compressive strength of 7500 lb/in.^{2}. Concrete masonry walls consisted of $3\frac{5}{8} \times 3\frac{5}{8} \times 11\frac{5}{8}$ in. concrete block, having a compressive strength of 5400 lb/in.^{2}. The lime mortar proportion consisted of 1 part cement : 0.5 lime : 5 sand. All cores in every wall, except for the unreinforced cores in brick wall No. 4, were filled with the grout, having a mix proportion consisting of 1 part cement : 3 sand, which developed a compressive strength of 2800 lb/in.^{2}.

The lateral load was applied to each wall in cyclic sequences which were carried to some constant deformation that was a multiple of the initial yield or maxi-

TABLE 17-2

Static Test Wall Details

Material	*Designation*	*Height*	*Length*	*Nominal aspect ratio*	*Vertical reinforcing*[a]	*Reinforcing (%)*[b]	*Bearing stress (lb/in.²)*[b]
Brick	1	3 ft 9 in.	3 ft 8 in.	1	4 $\frac{3}{8}$-in. bars uniformly distributed	0.24	0
Brick	2	3 ft 9 in.	3 ft 8 in.	1	4 $\frac{3}{8}$-in. bars uniformly distributed	0.24	125
Brick	3	3 ft 9 in.	3 ft 8 in.	1	4 $\frac{3}{8}$-in. bars uniformly distributed	0.24	250
Brick	4[c]	3 ft 9 in.	3 ft 8 in.	1	4 $\frac{3}{8}$-in. bars uniformly distributed	0.24	500
Brick	5	3 ft 9 in.	3 ft 8 in.	1	4 $\frac{3}{8}$-in. bars uniformly distributed	0.24	500
Concrete block	CB 1	4 ft 0 in.	4 ft 0 in.	1	4 $\frac{3}{8}$-in. bars uniformly distributed	0.26	0
Concrete block	CB 2	4 ft 0 in.	4 ft 0 in.	1	4 $\frac{3}{8}$-in. bars uniformly distributed	0.26	125
Concrete block	CB 3	4 ft 0 in.	4 ft 0 in.	1	4 $\frac{3}{8}$-in. bars uniformly distributed	0.26	250
Concrete block	CB 4	4 ft 0 in.	4 ft 0 in.	1	4 $\frac{3}{8}$-in. bars uniformly distributed	0.26	500
Brick	A1	3 ft 9 in.	3 ft 8 in.	1	2 $\frac{7}{8}$-in. bars on periphery	0.67	250
Brick	A2	3 ft 9 in.	3 ft 8 in.	1	2 $\frac{7}{8}$-in. bars on periphery & 2$\frac{5}{8}$ -in. bars horizontally	0.67 & 0.33	250
Brick	B1	3 ft 11 in.	2 ft 2 in.	2	2 $\frac{3}{8}$-in. bars on periphery	0.20	250
Brick	B2	3 ft 11 in.	2 ft 2 in.	2	2 $\frac{3}{8}$-in. bars on periphery	0.20	500
Brick	B3	3 ft 11 in.	2 ft 2 in.	2	2 $\frac{3}{8}$-in. bars on periphery	0.20	125
Brick	B4	3 ft 11 in.	2 ft 2 in.	2	4 $\frac{3}{4}$-in. bars on periphery	1.63	250
Brick	D1	3 ft 2 in.	6 ft 1 in.	0.5	6 $\frac{3}{8}$-in. bars uniformly distributed	0.22	0
Brick	D2	3 ft 2 in.	6 ft 1 in.	0.5	6 $\frac{3}{8}$-in. bars uniformly distributed	0.22	250

[a]All reinforcing bars deformed mild steel butt-welded to base; $\frac{3}{8}$-in. bars anchored into top beam with standard 180° hook (1 in. radius, 3 in. turn down); other vertical bars and horizontal bars anchored with 90° bend and 8-in. extension.

[b]Based on gross horizontal section.

[c]Only reinforced cores grouted.

mum load deformation. After several cycles had been applied, each wall was then monotonically loaded to failure. The ultimate strength capacities are summarized in Table 17-3.

TABLE 17-3

Static Test Results

Wall	*Area (in.²)*	*Theoretical yield load (kips)*	*Experimental maximum load (kips)*	*Shear strength (lb/in.²)*	*Predicted behavior*
CB1	172	10	11.2	65	Flexural
CB2	172	20	20.5	119	Flexural
CB3	172	30	29.5	174	Transitional
CB4	172	40	44.7	260	Shear
1	186	10	12.5	67	Flexural
2	186	20.6	20.5	110	Flexural
3	186	31.3	30.6	165	Transitional
4	186	52.6	32.8	176	Shear
5	186	52.6	39.5	212	Shear
A1	178	48	37.5	210	Shear
A2	178	48	40.0	225	Shear
B1	108	9.7 (10.8)[a]	10.6	98	Flexural
B2	108	16.7	16.6	154	Transitional
B3	108	6.2 (7.3)[a]	7.4	68	Flexural
B4	108	25.2	16.0	148	Shear
D1	300	30	30.0	100	Flexural
D2	300	98	70.5	235	Shear

[a]Theoretical ultimate load.

With respect to the failure mechanism, it was observed that when the walls exhibited a flexural type of behavior, the initial cracking, occurring mainly in the horizontal mortar joints near the base of the wall, was produced by vertical movements occurring in the brickwork as it accommodated the deformations of the elongated steel. This pattern is exemplified in Figure 17-20a. After yielding, the load remained steady at the yield level while deformations increased, until failure was brought about by crushing, usually accompanied by diagonal cracking at the toe of the wall. The more pronounced the flexural behavior, the greater was the ductility exhibited by the wall. In contrast, shear behavior was characterized by initial diagonal cracking (Figure 17-20b), producing a less stiff wall and one that had virtually no constant load plateau. Rather, the load tended to drop sharply from the maximum as the deformation increased. Failure was suddden and extensive, principally due to the disintegration of the masonry at the toe of the wall. These observations corroborate those of Meli. For all cases, the toe failure appeared to coincide with a tensile splitting of the units, following the crushing in adjacent mortar joints, as shown in Figure 17-20c.

Some of the walls exhibited what is described in Table 17-3 as a "transitional" behavior. In these instances, the initial behavior was flexural in nature;

(a)

(b)

(c)

FIGURE 17-20. Wall failure patterns.

however, at yield there was cracking along the compression diagonal. Subsequently, at higher loads, this produced a more shearlike behavior, except that the panels did demonstrate a constant load plateau similar to flexural behavior, accompanied by a load deterioration when the wall was cycled at a constant amplitude. This load degradation was presumed to have been caused by progressive deterioration of the masonry along the shear cracks.

The load-deflection cycles produced some interesting results. A stiffness degradation with load repetition was convincingly apparent. Stiffness degradation, as used in this context, must be construed as a dual-phase phenomenon: (1) degradation that occurs before ultimate or yield load is attained (usually before any substantial cracking occurs); and (2) that which takes place during post-elastic deformations after substantial cracking has occurred. This is a very important parameter in measuring the overall inelastic dynamic behavior of a masonry structure. For should a sufficient number of shear resisting elements suffer significant stiffness degradation, the frequency of vibration of the overall structure decreases, and this could definitely affect the shear force that the structure ultimately resists. In the flexural walls (Figure 17-21), the major stiffness loss seemed to occur between the first and second cycles of each deformation amplitude, and additional load cycles at the same deformation indicated that a relatively stable condition had been achieved, a highly significant aspect from a seismic point of view. The more flexural the behavior, the less pronounced the stiffness

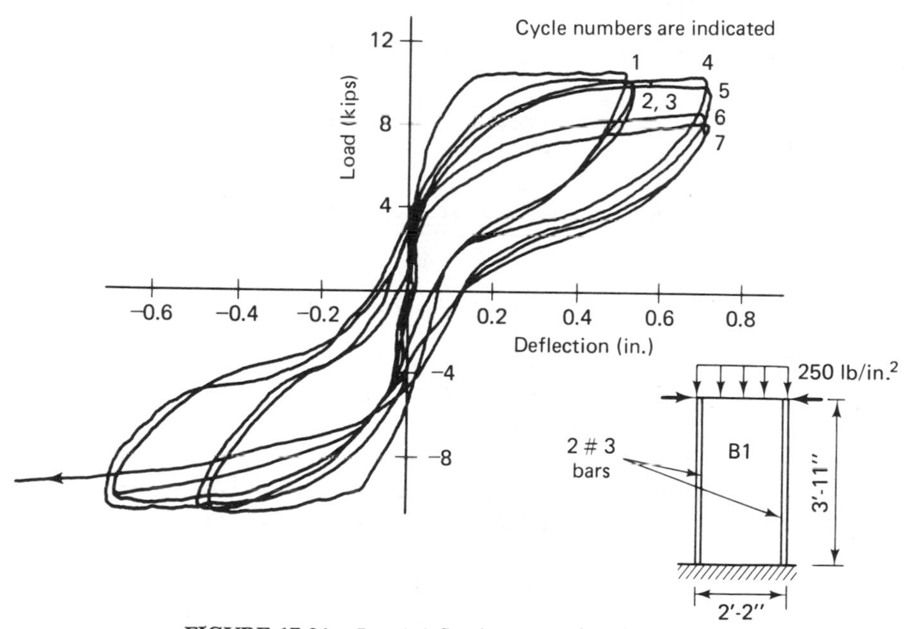

FIGURE 17-21. Load-deflection curve for yield failure.

degradation. In contrast, the initial stiffness degradation exhibited by the walls failing predominantly in shear was very large. Severe load reduction and further stiffness degradation occurred during each subsequent cycle, as seen in Figure 17-22. Deterioration of the reaction corner caused this reduction in shear resistance; whereas in flexural behavior, post-elastic deflections were due primarily to the steel yielding. Therefore, the load capacity of the reaction corner, for the latter condition, was not impaired until very large displacements were reached. Williams attributes the stiffness loss on load reversal and subsequent cycling to three factors: (1) the opening and closing of cracks, (2) a general deterioration of the load-resisting mechanism, and (3) a softening of the steel as a result of post-elastic cyclic loading.

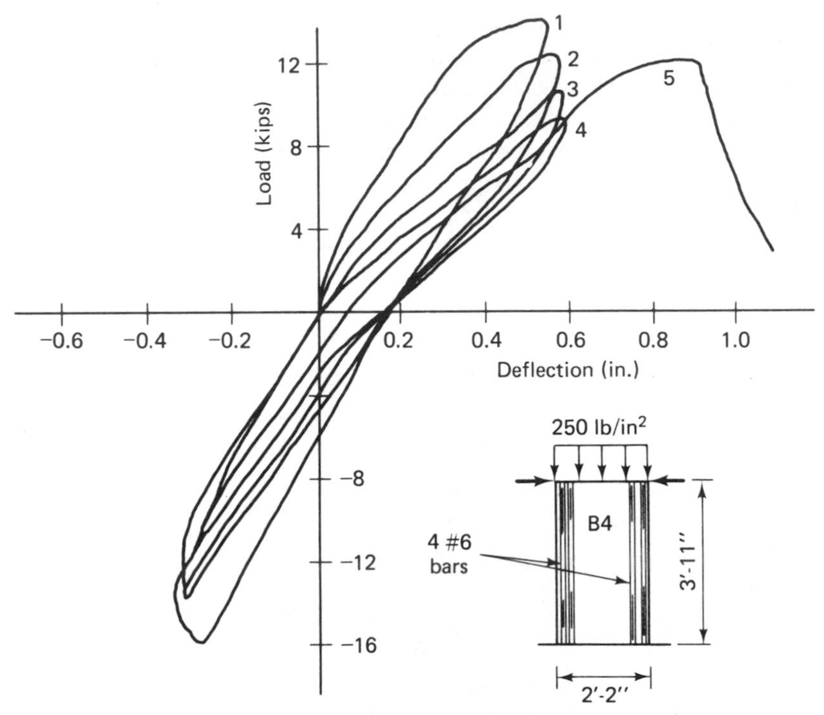

FIGURE 17-22. Load-deflection curve for shear failure.

Every wall was found to be capable of supporting the bearing load applied to it until severe damage had drastically reduced its lateral strength. The interaction of horizontal and vertical loads, within the range of bearing loads utilized in these tests (0 to 500 lb/in.^{2}), indicated that lateral strength, as well as the ultimate capacity, would increase with increasing bearing load, as shown by the test results in Table 17-3. Of greater interest, however, was the post-elastic cyclic behavior, whereby increasing bearing loads tended to cause a more shearlike behavior.

Regarding the effect of wall geometry, it would appear that as the aspect ratio (height/width between 0.5 and 2) increased, the wall exhibited a more flexural-type behavior. Thus, walls with a very high aspect ratio could be regarded as long and shallow beams, having a characteristic flexural behavior. It was observed that for walls having this type of behavior pattern, the ratio of lateral deflection to bending strain increases with increasing aspect ratio, thereby evidencing a large ductility capability. On the other hand, low-aspect-ratio walls show an essentially shear-type deformation and thus may be considered as pure shear-resisting elements demonstrating a nonductile-type behavior. Since these walls have a very high shear strength because of their large cross-sectional areas, the shear strength generally does not limit the capacity of the wall. More than likely, structural instability of the foundation or overturning becomes the primary strength resisting criterion. At any rate, the upper limit on the shear capacity after post-elastic cyclic deformation is governed by a shear transfer in the reaction corner and dowel action, provided that adequate web reinforcement is present.

It was found that an increase in the amount of vertical reinforcing increased the horizontal load required to produce yielding in the steel, thus raising the flexural strength of the wall without altering the shear strength appreciably. The other factors remaining the same, this has the ultimate effect of increasing the tendency for a shear failure, which should be avoided because of its brittle nature. It would appear also that, from the inelastic performance of some of these walls, a concentration of reinforcing at the jambs or wall boundaries will not necessarily produce the most suitable type of earthquake-resistant structure, even though this is certainly the most effective vertical steel for flexural resistance. The inclusion of horizontal shear steel appeared to increase the shear strength of the wall by about 7% and to cause a displacement plateau at the maximum load during the first load cycle. However, subsequent load cycles indicated that the shear reinforcing was apparently not too effective.

In conclusion, Williams stated that the static test results indicate that it should be possible to design reinforced masonry shear walls so that a ductile and generally satisfactory post-elastic behavior is achieved. With walls having an aspect ratio of 2 or more, provided that the shear strength exceeds the ultimate flexural capacity, such behavior should prevail. Furthermore, low bearing loads, low flexural strength associated with light reinforcing, and high aspect ratios all enhance the prospect of this flexural-type behavior, which in turn is characterized by a minor stiffness degradation and negligible load deterioration with load repetition. However, he points out that the energy-dissipating capacity of shearlike wall behavior, for those walls where a flexural-type behavior cannot be obtained, should not be overlooked. Then, such factors as steel distribution, anchorage details, and grouting effectiveness would appear to highly influence the energy dissipation in this shear mode.

Dynamic cyclic load tests: Williams also subjected four brick walls to a dynamic cyclic loading to observe how the dynamic application affected the stiffness degradation and load deterioration properties of reinforced brick walls. Of the four walls tested, the three that showed a high degree of structural deterioration in the static tests behaved in a similar fashion when subjected to dynamic load conditions. Thus, for walls where shear effects predominated, the static and dynamic behaviors were similar. However, the flexural-type brick wall results indicated, in contrast to its satisfactory ductile behavior when tested statically, a severe and unexpected loss of structural capability when subjected to dynamic load repetitions.

In an overall review of his findings, Williams suggests that the prime object in designing a shear wall is to ensure that the comparatively low shear strengths of the masonry are not exceeded. He states that this aim could be fulfilled by limiting the amount of flexural reinforcing steel in the structural element, so that the forces associated with the element's flexural strength do not exceed its shearing strength. He points out that excess flexural steel may prove disastrous, as a potentially ductile situation could become a potentially dangerous brittle one. This may be likened to the restriction of the steel ratio in the ultimate strength design of reinforced concrete flexural members, such that the capacity of the beam is governed by yielding of the reinforcing steel (ductile behavior). Furthermore, he says that for ductile behavior, the wall geometries must be of such proportion that sufficient post-elastic deformation can occur without shear distortion becoming excessive enough to initiate a shear failure. It is noted that in these tests, true flexural behavior, accompanied by large ductility performance, was unobtainable in walls having an aspect ratio of 1 or less. However, it is conceivable that walls in this geometric range which demonstrate apparent ductile characteristics (shear displacement at constant load) may still be capable of dissipating energy by inelastic deformation. Perhaps the high in-plane lateral strength of such walls relieves the need for a high degree of ductility anyway.

Williams recommends that, since a sufficiently ductile behavior in reinforced masonry cannot presently be satisfactorily predicted, a working stress design method must be utilized. This method consists of designing a structure so that the stresses sufficient to cause material damage will not be exceeded when the structure is subjected to the prescribed Code loadings. Williams states that for stiff masonry structures, this method is irrational. Since no ductility can be assumed, the structure must be designed to withstand the real forces associated with the design earthquake. These may be several times the Code-specified values, which have proved satisfactory for ductile frame structures. Unfortunately, the lack of appreciation of this fact has led to the mistaken belief that there is an inherent weakness in masonry under seismic conditions; whereas, in fact, the earthquake forces that must be resisted by this type of structure have been grossly underestimated. The same argument applies to stiff reinforced concrete shear walls. Factors of safety and the low Code-specified allowable stresses have often compensated for the underestimated dynamic loads, but the merit of a more rational design approach cannot be overemphasized, Williams observes. So he concludes that although static tests have shown encouraging results with regard to the attainment of ductility, until the structural deterioration revealed in the dynamic test is prevented, it is recommended that for seismic design of reinforced masonry, the working stress approach must be retained.

Meli In addition to the static racking tests, previously described, Meli and his coworkers conducted some cyclic load tests on reinforced masonry walls. Both "cantilever" (Figure 17-14a) and "diagonal compression" (Figure 17-14b) loadings were applied. The 9 ft 0 in. × 9 ft 0 in. test specimens consisted of 26 concrete block walls with interior reinforcement and four brick walls encased within a concrete frame, as shown in Figure 17-15. The load deformation curves (Figure 17-23) changed significantly between the first and second cycles, but after about the sixth cycle, the curve tended to exhibit a more stable pattern. The amount of deterioration depended mainly upon the type of reinforcement and the failure mode, as well as on the type of masonry unit and the presence of any vertical load.

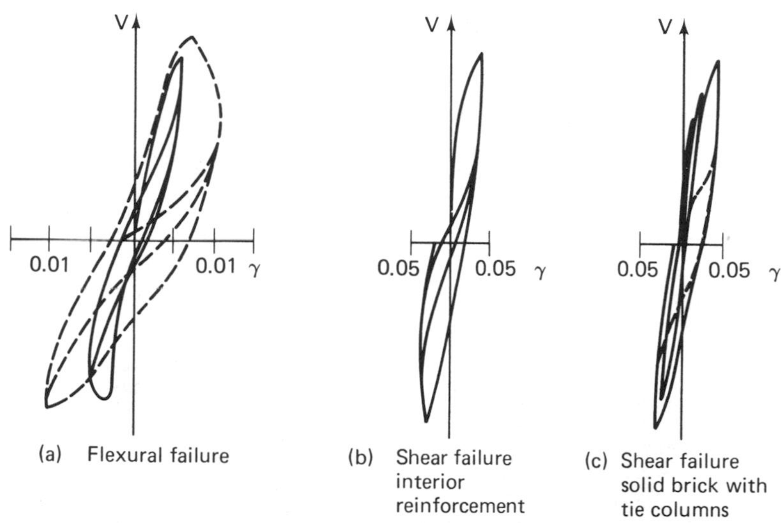

FIGURE 17-23. Typical load-distortion curves under alternating loads.

Meli decided that his observations of wall behavior under the cyclic dynamic loadings justified certain conclusions, as follows:

1. Behavior of walls with interior reinforcement, where failure is governed by flexure, is nearly elastoplastic with remarkable ductility and small deterioration being exhibited under alternating load, except at very high deformations, where significant deterioration is caused by progressive crushing and shearing at the compression corner.
2. After flexural yielding, even though stiffness is significantly reduced, the ultimate strength was unaffected. Where failure is governed by shear, a lowered ductility is evidenced. Extreme deterioration occurs after the appearance of diagonal cracking. Increasing the amount of interior wall reinforcement evidently does little to improve the situation. This performance actually deteriorates into a brittle one where high vertical loads are applied.
3. Because of the desire to build into a wall as much ductility as possible, the wall geometry and reinforcement must be chosen, where possible, to fail in bending.
4. In the frame-encased walls of solid units, because of the confinement provided, considerable ductility was achieved, much more so than with the interior-reinforced specimens. Deterioration was still significant in these elements, although it was considerably less severe than in the diagonally cracked walls, again probably because of the confinement provided by the

frame. At any rate, they exhibited improved behavior under cyclic dynamic loads over that of unreinforced or interior reinforced walls.

5. When a low-level precompression (40 to 60 lb/in.^{2}) was applied, deterioration decreased for all types of reinforcement and failure modes.

Mayes and Clough In 1975 these investigators reported on the results gathered in their analysis made with 6-in. concrete block units. The eight specimens (Figure 17-24) consisted of four pair of identical panels, the test variables incorporated being tabulated in Table 17-4, along with some of the final strength results. One of each pair was tested at an input displacement frequency of 0.02 Hz, the other at 3 Hz.

FIGURE 17-24. Typical double-pier test specimen and apparatus, Mayes and Clough, 1975.

TABLE 17-4

Material and Pier Properties

Test	Frequency[a] (Hz)	Bearing stress[b] (lb/in.²)	Vertical reinforcement[c]	Horizontal reinforcement[d]	Prism strength (lb/in.²)	Mortar strength (lb/in.²)	Grout strength (lb/in.²)	Reinforcement stresses[e] Horizontal (kips/in.²)	Reinforcement stresses[e] Vertical (kips/in.²)	Average ultimate shear strength (lb/in.²)[f]
1	0.02	250	2 No. 6	—	2280	4955	4510	—	43.1 (64.9)	125
2	3	250	2 No. 6	—	2280	4955	4510	—	43.1 (64.9)	161
3	0.02	125	2 No. 4	—	2115	3985	3420	—	54.1 (83.4)	135
4	3	125	2 No. 4	—	2115	3985	3420	—	54.1 (83.4)	119
5	0.02	0	2 No. 6	—	2430	5260	6150	—	78.1 (108.8)	99
6	3	0	2 No. 6	—	2430	5260	6150	—	78.1 (108.8)	113
7	0.02	250	2 No. 6	1 No. 5	2630	5610	4330	67.8 (94.6)	78.1 (108.8)	203
8	3	250	2 No. 6	1 No. 5	2630	5610	4330	67.8 (94.6)	78.1 (108.8)	229

[a]Frequency of the sinusoidally applied actuator displacement.
[b]Bearing stress based on the gross area (192 in.²).
[c]Vertical reinforcement in each jamb of the piers.
[d]Horizontal reinforcement at the one-fourth points of each pier.
[e]The top value is the yield stress; the value in parentheses is the ultimate strength.
[f]Average ultimate shear stess = $(P_1 + P_2)/(2 \times \text{gross area})$.

The test apparatus is shown in Figure 17-25. To provide an indication of stiffness degradation occurring between different sequences of loading, the authors established what they termed a stiffness coefficient, K_1. It was spelled out as follows:

$$K_1 = \frac{|P_1| + |P_2|}{|d_1| + |d_2|}$$

where P_1 and P_2, one in either direction, are approximately 90% of the mean of the peak ultimate loads, maintained for more than one cycle of input displacement; and d_1 and d_2 are the corresponding average lateral displacements. K_1 for the right-side pier was plotted against average lateral displacement and shear force as shown in Figure 17-26.

FIGURE 17-25. Single-pier test apparatus developed by Mayes and Clough and used after test comparisons verified consistent relation to double-pier specimen results. The single-pier specimens could be built, handled, and tested much more rapidly. (Courtesy R. Mayes.)

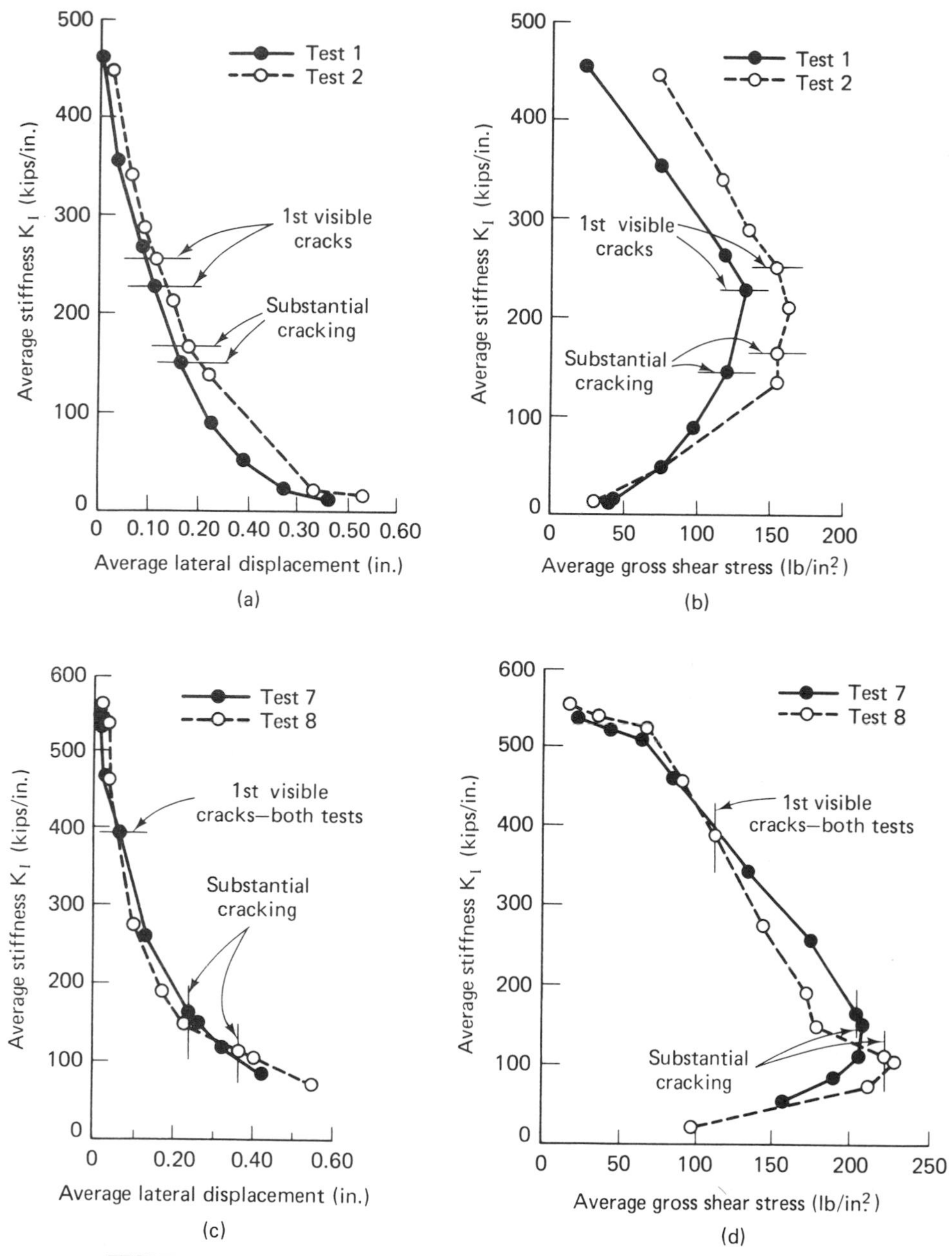

FIGURE 17-26. Typical stiffness coefficient, K_1, Mayes and Clough.

Based upon their findings, the authors made the following observations:

1. Pier strength was affected by the rate of loading, the precompression stress, and the amounts of vertical and horizontal reinforcement.
2. Piers exhibiting a shear failure had a peak ultimate strength of between 13 and 23% less in the "pseudo-static" test (0.02 Hz) than under the

corresponding dynamically applied load (3 Hz). On the other hand, piers that failed in a combination of flexure and shear reached a peak ultimate strength of 16% greater under the static test. Thus, static tests do not always produce conservative indications of ultimate strengths.

3. At a shear stress of 20 to 25 lb/in.^{2}, the piers tested with horizontal reinforcement were 16% stiffer than piers without this shear steel.
4. As the shear stress increased from 25 to 50 lb/in.^{2}, piers with no horizontal steel suffered a 14 to 20% decreased in stiffness, whereas piers with horizontal reinforcement had a corresponding decrease of only 4%.
5. The piers tested did exhibit a ductile-type behavior under gradually increasing lateral displacements.

Priestly and Bridgeman These investigators in 1974 devised a rather thorough analysis of the effect of both vertical and horizontal reinforcement on the shear strength of cantilever piers. They also examined the prospect of providing a confining plate in the mortar joints at the compression toes of the piers and found that this did considerably improve the inelastic behavior of the piers.

All in all, they tested 14 reinforced grouted brick walls (F2-12 Series—70.5 × 70.9 × 8.7 in. in size and F13-15 Series—64.2 × 59.4 × 8.7 in. in size), two wythes thick, separated by a $2\frac{1}{2}$-in. grout space. In addition, four walls were made with solid-grouted concrete block (CBU 1-2 Series—40.9 × 41.5 × 5.5 in. in size and CBU 3-4 Series—59.8 × 61.4 × 5.5 in. in size). Walls F13-15 and CBU 3-4 had confining plates located at the compression toe. The principal test variable was the amount of reinforcing, although two walls (F9 and F10) had a precompression vertical load applied to them. Vertical and horizontal reinforcing was evenly distributed within the field of the wall, except in the case of walls F2 and CBU 1, wherein the vertical steel was concentrated at the panel boundaries. The cantilever panels were tested cyclically, as shown in Figure 17-27, at very low rates of loading.

Table 17-5 gives the failure loads and compares them with the theoretical capacities, in flexure or shear, of walls without the confining plates in the critical mortar

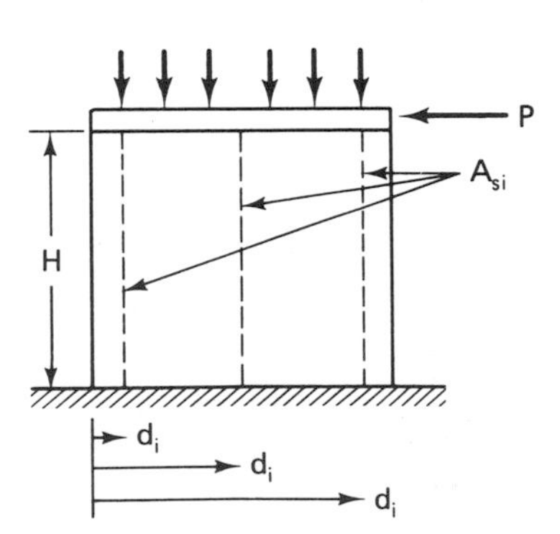

FIGURE 17-27. Cantilever load test, Priestley and Bridgeman, 1974.

TABLE 17-5

Failure Loads for Walls Without Confining Plates[a]

	Theoretical loads (kips)				Experimental loads (kips)						
Wall	*Yield,* P_y[a]	*Ultimate flexural* P_u	*Shear capacity,* P_v[b]	*Failure mode*	*First loading,* P_1	*First load reversal,* P_2	*Second load reversal,* P_3	P_1/P_u	P_2/P_u	P_3/P_u	*Shear stress at max. load* ($lb/in.^2$)
F2	29.0	29.0	—	Flexure	30.6	28.1	19.1	1.05	0.97	0.66	49
F3	20.7	27.2	—	Flexure	32.1	15.7	22.0	1.18	0.58	0.81	52
F4	31.0	40.0	—	Flexure	47.2	—	—	1.18	—	—	77
F5	20.7	27.2	34.1	Flexure	31.9	15.1	7.9	1.17	0.55	0.29	52
F6	31.0	40.0	39.6	Flexure	44.3	—	—	1.11	—	—	73
F7	49.0	74.2	—	Shear	62.1	40.0	9.9	0.84	0.54	0.13	100
F8	47.7	71.7	34.2	Shear	65.0	22.0	11.7	0.91	0.31	0.16	106
F9	69.0	103.9	—	Shear	100.0	100.0	28.1	0.96	0.96	0.27	163
F10	92.9	132.0	—	Shear	119.8[c]	119.8[c]	100.0	0.91	0.91	0.76	196
F11	40.9	65.2	111.7	Flexure	67.4	46.5	—	1.03	0.71	—	115
F12	60.7	94.6	148.8	Flexure	96.0	78.9	—	1.01	0.83	—	163
CBU 1	51.3	65.6	139.4	Bond	55.3	39.3	—	0.84	0.60	—	247
CBU 2	38.4	65.6	139.4	Bond	56.2	—	—	0.86	—	—	255

[a] P_y, theoretical load at which tensile jamb steel first yields; P_u, flexural capacity based upon ultimate strength concepts (per reinforced concrete) using values of $e_u = 0.003$ and $f_m' = 3000$ lb/in.^{2}.
[b] Based on horizontal steel only.
[c] Maximum capacity of load system.

joints. Flexural failures occurred on walls containing lower to moderate amounts of reinforcement (F2 to F6). Where the ultimate shear strength governed (F7 to F10), diagonal tension cracks formed and failure was sudden, involving large horizontal displacements. The walls in this group without any bearing load (F7 and F8) failed soon after the formation of the first diagonal crack at the rather low shear stress of 103 lb/in.^{2}, so apparently the precompression increased the shear resistance.

It would appear that the authors have evolved a very significant hypothesis here when considering the effectiveness of vertical versus horizontal steel in resisting shear. They determined that the principal stresses near the center of the panel are inclined at about 45° to the vertical, resulting in the formation of the diagonal cracks at 45°, whereupon the imposed load is no longer a true shear load. This is due to the fact that the imposed load then tends to move the top half of the wall horizontally past the bottom, thus widening the crack horizontally rather than in a direction perpendicular to the crack. As shown in Figure 17-28, any properly anchored horizontal steel crossing

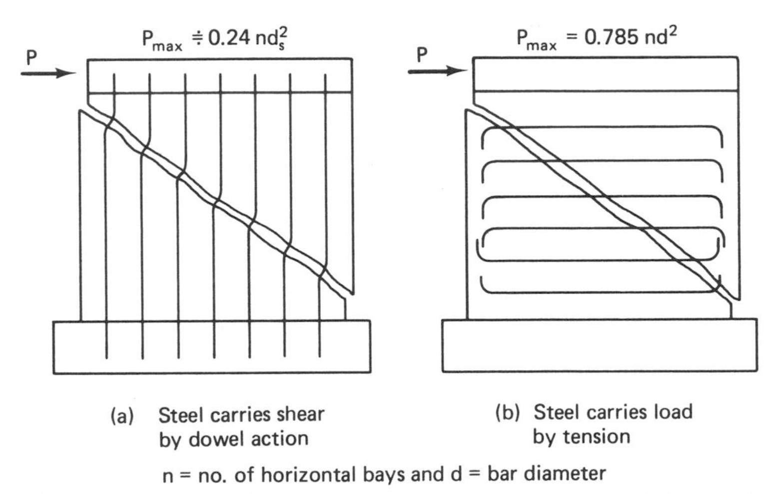

FIGURE 17-28. Relative effectiveness of vertical and horizontal shear steel.

the crack resists the lateral movement in direct tension, whereas the vertical steel must carry the load strictly by a dowel action. The authors assign a shear capacity to n horizontal bars of diameter d_s crossing the crack equal to $0.785nd_s^2f_y$ (where f_y = yield stress of the bars). The vertical steel appears to be nowhere near as effective in resisting this lateral movement, and the resistance they assign to the vertical steel amounts to $0.240nd_s^2f_y$. Theoretically, this makes the horizontal steel roughly three times as efficient in this function as is the vertical.

On the basis of their results, the investigators had the following comments:

1. Shear steel is effective in improving the ultimate shear capacity of masonry walls, provided that sufficient shear steel is available to develop the full ultimate flexural load.

2. Horizontal steel is over three times as efficient as vertical steel in transferring the shear force across a diagonal crack.
3. Load degradation was sufficiently eliminated at ductility factors of up to 5 by the inclusion of stainless steel confining plates in the bottom few mortar courses at each end of the wall. They served to confine the crushing zone, thereby eliminating vertical tension cracks in the masonry. Further, they provided needed restraint against buckling of the compression steel. For this reason the plate should probably extend from the edge of the wall to at least past the second vertical bar to provide this buckling support.
4. For walls with adequately confined crushing zones, it is suggested that a conservative design approach would involve the use of the first-yield load rather than ultimate flexural capacity.
5. Reinforced grouted brick masonry walls appeared to possess a somewhat better capability in resisting the horizontal in-plane shear loads than did the walls composed of concrete masonry units. This is possibly due to the poor adhesion between the concrete block face shells and the grout, and also perhaps because of the greater confinement attainable in grouted brick construction.

Miscellaneous Tests

Shear-Wall Models (Sinha and Hendry) The most comprehensive research performed on the shear strength of model structures was done by Hendry and his coworkers at the University of Edinburgh. Models of portions of a building were utilized because of the fact that obviously large sections are very difficult to handle, besides which the cost often becomes prohibitive. Most of this work was actually done on unreinforced brick masonry using $\frac{1}{3}$- and $\frac{1}{6}$-scale models.

In 1969, B. P. Sinha and A. W. Hendry reported on a program whose primary objective was to investigate the rigidity and shear strength of a single-story brick shear wall structure ($\frac{1}{6}$-scale model) consisting of three shear panels with a floor slab above, stiffened by cross walls and having door openings between two of the panels (Figure 17-29). Vertical bearing stresses imposed upon each of the test specimens ranged between 55 and 147.5 lb/in.^{2}. The racking load produced ultimate shear stresses in the range between 77.7 and 119.3 lb/in.^{2}. Sinha and Hendry observed that the failure of story height shear walls with openings, when subjected to a racking load, is generally due to a breakdown of the bond at the interface, leading to the propagation of diagonal cracks down through the vertical and horizontal mortar joints, although they occasionally pass through the masonry units. They also discovered that precompression (vertical bearing load) increases the shear strength of the brick panels up to a certain point, depending upon the compressive strength of the panel itself. They further discovered that the effective rigidity and shearing modulus of the brick panels decreased in a nonlinear fashion with an increase in the racking load and a decrease in compressive load. Several other model analyses, too numerous to

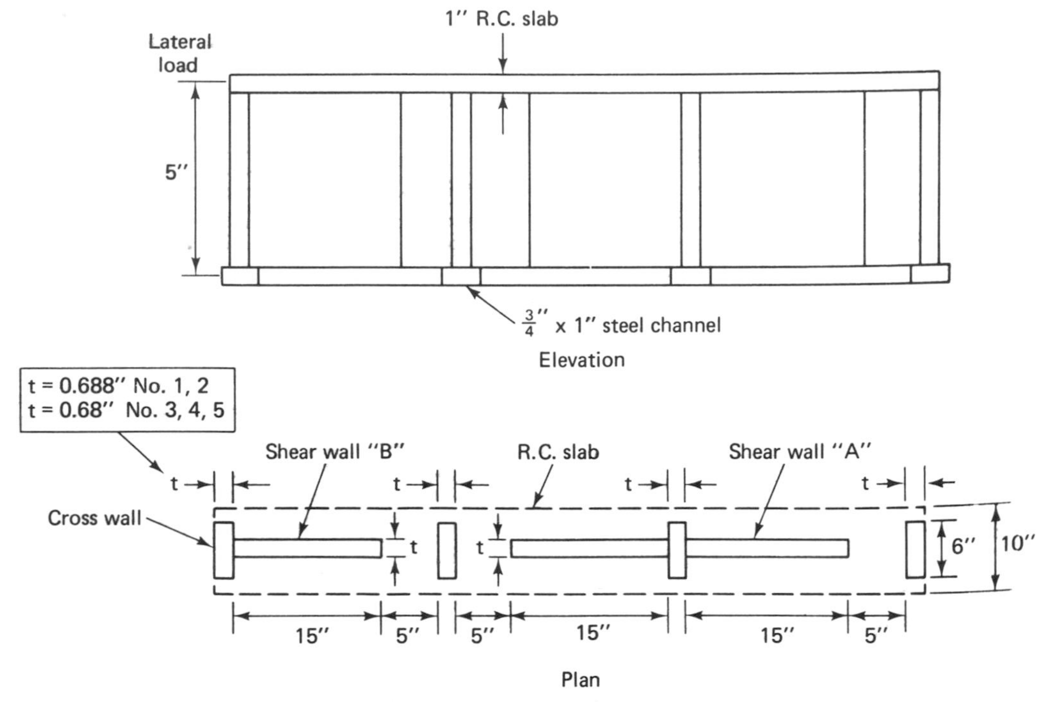

FIGURE 17-29. Model shear wall test structure.

identify here, have been made by these investigators. Details may be found in Mayes and Clough's literature survey.

Shear on Bolts (Kelly, Pittelko, Fritz, and Forssen) Embedment of bolts in a reinforced masonry wall often provides the means for transferring the floor or roof level shear from the diaphragm boundary members into the shear wall itself. If this connection is not adequate, then actually or functionally there is no shear wall, and the system is unable to resist the lateral deformations produced by wind or seismic action. The only way that allowable values for this shear on bolts transfer (expressed in total pounds per a given bolt size) can be established for the Code is by actual field observations.

For this reason, KPFF performed a test program in 1975 utilizing $\frac{1}{2}$-in.-diameter bolts embedded in a 4-in.-thick hollow clay block wall. In the initial portion, loads were applied separately in both the horizontal (for seismic or wind) and the vertical (for dead and live loads) directions through a wood ledger. Load-deflection curves were obtained along with the failure loads, which led to the values recommended to ICBO. Failure was evidenced through either yielding of the bolt due to bending or crushing of the wood ledger or a combination of these occurrences. Thus, where a wood ledger was used, the capacity of the masonry was not fully utilized because of the earlier failure of other materials. At any rate, a value of 550 lb for

$\frac{1}{2}$-in.-diameter bolts embedded $2\frac{1}{2}$ in. was recommended in the test report; however a 420-lb value was actually allowed when the load acts in the vertical direction and 320 lb for the horizontal direction. Along with this, it was also requested that 4-in. high-strength brick masonry be allowed in bearing walls and that the *h*/*t* ratio for bearing walls be increased to 27 from 25.

In the second part of the program, a steel ledger replaced the wood. Thus, bending of the bolt was minimized, since this placed the point of load application much closer to the wall. Both $\frac{1}{2}$-in. and $\frac{3}{4}$-in.-diameter bolts were utilized. The former was located in the hollow clay block face shell and the latter in both the face shell and in the mortar. See Figure 17-30 for the installation details. As a result of these tests, ICBO recommended 640 lb for a $\frac{1}{2}$-in.-diameter J bolt, and 1275 lb for the $\frac{3}{4}$-in. size in the 4-in. masonry units, with a $2\frac{1}{2}$-in. embedment.

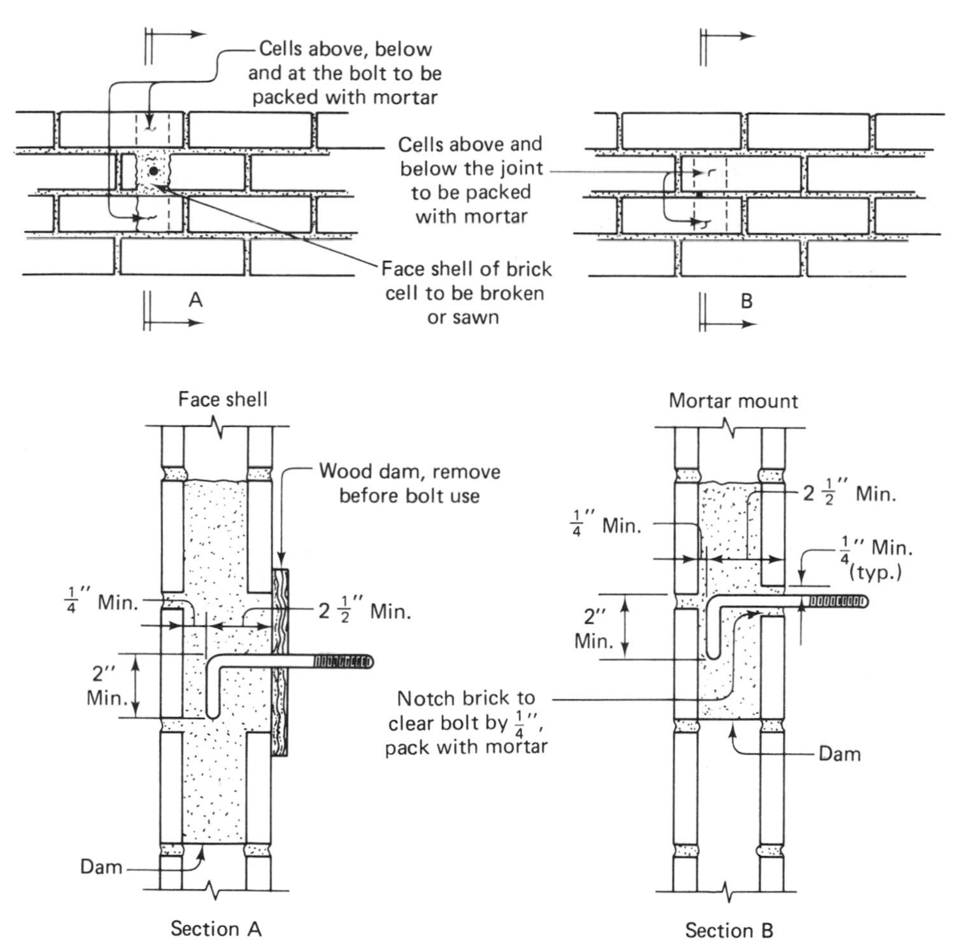

FIGURE 17-30. One-half or three-quarter inch diameter bolt installation.

CURRENT RESEARCH

Multistory building piers

This ongoing program being carried out by Mayes and Clough at the Earthquake Engineering Research Center (EERC) is primarily oriented toward multistory buildings. To reduce costs, a single pier test (versus the previously described double piers shown in Figure 17-25) was devised to be tested as shown in Figure 17-31. The primary objective being sought here is to determine what constitutes the amount of pier reinforcement necessary to ensure a ductile performance and how it should be distributed. In addition, an attempt is being made to develop an ultimate strength design concept for reinforced masonry that can be codified where seismic design

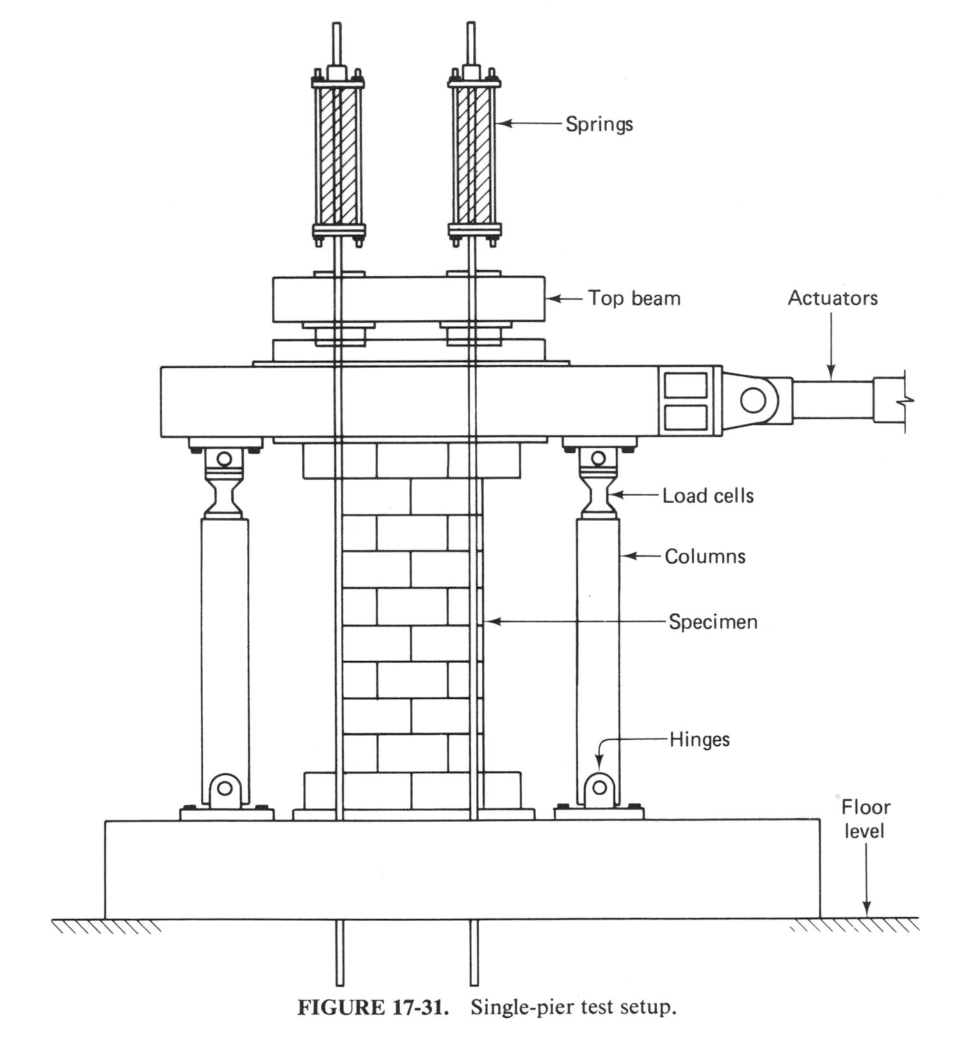

FIGURE 17-31. Single-pier test setup.

codes apply. At this point, a total of 80 full-size pier tests are planned for one series, of which 57 belong in the first phase. This group is composed of hollow clay brick, two-wythe grouted solid brick, and concrete block specimens.

An attempt will be made to correlate results obtained on the single pier results and those performed on a group of 4-ft-square panels resembling those of the previously described Blume tests. If successful correlations are obtained, it would be the goal then to revive the results of Blume's program in an attempt to utilize them in certain future code developments.

Single-story dwellings

This program, funded by HUD, is exclusively oriented toward single-story masonry dwellings. The overall objective is to establish appropriate design criteria and construction procedures for such structures. It has three major thrusts:

1. To determine reinforcement requirements, if any, for the adequate in-plane resistance to seismic action of typical masonry walls in houses; primarily for zones 1 and 2.
2. To determine the reinforcement requirements, if any, for the out-of-plane resistance of these same elements.
3. To evaluate the adequacy of typical connection details for masonry housing construction (i.e., roof diaphragm to walls and walls to footings, etc.).

Shaking-table tests are also planned to measure the recommendations for structural adequacy during earthquakes.

18

Quality Control, Specifications, and Inspection

WHY INSPECTION?

There are many reasons for inspection. Inspection bridges the gap between the dream of the designer and the completion of the final structure. Because a structure is designed in a certain manner does not ensure that the final physical structure will necessarily resemble that of the original design, and if it does not, it may not function properly. This is especially significant with masonry with its many variations. For instance, it consists of units that may vary enough in their properties and methods to influence the end result. Furthermore, since it is a hand-placed assemblage, it must be carefully observed during placement to see that the construction is suitable.

There is another important factor regarding the need for inspection in the area of legal responsibility. There are many lawsuits for "errors and omissions" just as there have been many suits for the so-called "malpractice" of the medical profession, so the cost of insurance has soared. There is the growing implication of legal responsibility by engineers for work in which they have participated as designers. Therefore, it is vital that they be assured that those materials which they have specified actually have been used, and their prescribed methods and details carefully followed.

Another important reason for inspection of work that is constructed under the UBC stems from the provision that the design stresses for work not continuously inspected are reduced by 50%. This, of course, is a very important factor in the design considerations (i.e., whether or not the savings effected in a more efficiently designed structure will more than offset the inspection costs). Also, more sophisticated design procedures have evolved which use masonry more effectively. Higher strengths and improved reinforcing placement may be used. However, these improved techniques require that certain items be assured which may be different from some local customary methods. Consequently inspection is necessary to ensure that the actual design provisions are followed.

WHAT IS INSPECTION?

Inspection is the observing of the construction process to confirm that it complies with the plans and specifications. Since there are so many aspects that may affect the final construction, a good record must be kept; hence many inspectors develop a checklist that can serve as a reminder of the items that must be checked and observed or confirmed and recorded. The inspection authority does not include direction of the work as a supervisor nor any authority to make revisions. Apparent deviations must be reported promptly to the responsible design professional for decision as to what should be done. Furthermore, those items that would be pertinent to the scheduling of the job and the quality of the masonry must be noted in the record. For example, weather can have a potent effect on the quality of masonry. Extremely hot dry weather would have one effect, and cold, rainy, or freezing weather, or heavy winds, would produce other effects. Such diaries are extremely valuable in the event lawsuits occur. They are, of course, of very little later importance otherwise. Good records should be kept, with the hope they would not be needed for such lawsuits or other distress that might occur.

The attitude toward inspection has been changing over the years. In the early days inspectors were considered more as policemen or spies. A greater respect for the activity has been developing. It has become a profession, one that is worthy of registration. Furthermore, good inspectors will continue to further their education by taking courses that will update their knowledge. Design considerations will change.

SPECIFICATIONS

The contract documents contain plans with details and the specifications. The purpose of the plans is to show dimensions, locations, sizes, and other pertinent information needed for the completion of the work. Also, some brief specifications involving certain methods and materials often appear in the general notes shown on drawings. However, the detailed material property requirements, along with the sampling and testing procedures, as well as any special construction methods called for to complete the work will be outlined in detail in the specifications. However, on small jobs, all the specifications may appear on the drawings in the general notes. Generally, the job specifications will refer to the UBC or ASTM Standards for materials and for testing. Furthermore, the specifications are intended to assure that the project, as it has been designed, drawn, and detailed, meets all the governing building code provisions. It is assumed that inspectors would be familiar with those code requirements and they can quickly review the specifications to see if any additional requirements or deviations exist. A designer may have made some specific exception or used some method not contained in the body of a code, but included in some special approval.

MATERIAL STANDARDS

The materials and test methods for evaluating them are described in detail in the UBC or ASTM Standards. The practical problem in the field is to assure that the quality of the material is in compliance, and that the testing is properly done. These items are discussed in more detail in portions of other chapters. There are compilations of the masonry standards available that will expedite the verification of use for masonry.

BRICK

A brick of fired clay includes building brick (C-62), face brick (C-216), hollow brick (C-652), reinforceable brick per WSCPA Standard, structural tile (C-34), and glazed tile (C-216), all of which will have certain specified strengths and physical properties, generally tested in accordance with ASTM C-67. Inspectors should assure that the materials have been tested and certified previously for compliance, or that job-site samples are made and tested for the specific job to assure compliance. Inspectors should familiarize themselves with these specifications, methods of sampling, and allowable or acceptable tolerances.

Some of the statements in the standards are indefinite and must be carefully interpreted. For example, the tolerance for warpage of face brick states that "tolerances for distortion or warpage of face or edges of individual brick from a plain surface and from a straight line respectively shall not exceed the maximum for the type specified as prescribed in Table 7" (Table 18-1). This was intended to mean the measured deviation was from a surface such as a table top. Some people have tried to interpret that statement to mean a deviation from a straight line, averaged between the maximum plus and minus deviations.

Another item of confusion or misinterpretation has been the waiver of CB ratio for various areas. The waiver in the ASTM Standard refers to a map. The southern California area is shown to have a weathering index of less than 50. So in these areas, MW brick may be used for SW brick, if it meets the other physical

TABLE 18-1

Tolerances on Distortion (Brick)

Maximum face dimension [in. (mm)]	*Maximum permissible distortion [in. (mm)]* Type FBX	Type FBS
8 (203) and under	$\frac{1}{16}$ (1.6)	$\frac{3}{32}$ (2.4)
Over 8 to 12 (203 to 305), incl.	$\frac{3}{32}$ (2.4)	$\frac{1}{8}$ (3.2)
Over 12 to 16 (305 to 406), incl.	$\frac{1}{8}$ (3.2)	$\frac{5}{32}$ (4.0)

properties of the SW grade. However, within that general area there are certain localities, such as mountain areas, that would show an extremely high weathering index. Within those localized areas the durability factor, or the CB ratio, could not be waived because there is the hazard of deterioration due to freezing and thawing.

Dimensional tolerances

Another item that may cause inspectors difficulty is the dimension tolerances that are permitted in ASTM. Those for brick are shown in Table 18-2. Inspec-

TABLE 18-2

Tolerances on Dimensions (Brick)

Specified dimension [in. (mm)]	*Maximum permissible variation from specified dimension, plus or minus [in. (mm)]* Type FBX	Type FBS
3 (76) and under	$\frac{1}{16}$ (1.6)	$\frac{3}{32}$ (2.4)
Over 3 to 4 (76 to 102), incl.	$\frac{3}{32}$ (2.4)	$\frac{2}{16}$ (3.2)
Over 4 to 6 (102 to 152), incl.	$\frac{2}{16}$ (3.2)	$\frac{3}{16}$ (4.7)
Over 6 to 8 (152 to 203), incl.	$\frac{5}{32}$ (4.0)	$\frac{4}{16}$ (6.4)
Over 8 to 12 (203 to 305), incl.	$\frac{7}{32}$ (5.6)	$\frac{5}{16}$ (7.9)
Over 12 to 16 (305 to 406), incl.	$\frac{9}{32}$ (7.1)	$\frac{3}{8}$ (9.5)

tors may have the problem of enforcing aesthetic suitability if the units are provided at those maximum limits.

One example would occur in stack bond wall patterns with units being placed one above the other. A dimensional tolerance in the brick length of $\frac{5}{8}$ in., with a joint width of $\frac{1}{2}$ in. would indeed produce undesirable irregularities in appearance. Fortunately, this extreme problem does not occur too frequently. Normally, most of the brick in only one lot, for example, will be all longer or all shorter than the specified length dimension, with the difference between individual units being small. It is well to clarify suitability of these tolerances and the acceptability of units prior to the start of construction to minimize disruption of job progress later.

Sample panel

One of the best methods of considering size-tolerance discrepancy is by the use of a sample panel. This is built prior to starting the job to show the variations in color that might occur, as well as any variations in size that might occur which are acceptable for the job. The sample panel may be built separately and maintained during the job or may be built in final position. If it is built in its final position in the wall and its extent noted, then there will be an easy comparison between the rest of the wall and the approved portion.

MORTAR

The aggregates of mortar are rather easily checked, and the proportioning aspects are mentioned briefly here. Specific details on mortar requirements were presented in Chapter 4. There is a wide variation in the grading of sands, and unfortunately most projects do not have much control over the grading of sands used for mortar. Therefore, sand proportions must be adjusted within the Code specified range (i.e., from $2\frac{1}{4}$ to 3 times the amount of cementitious material) so that a good workable mix is obtained. Workability is one of the most important properties of the mortar, and that is adjusted by revising the amount of aggregates present, depending upon the type, coarseness, and size of the sand particles. Some agencies (OSA, for instance) insist upon calibrated boxes for measuring the mortar sand. However, the designer, who has never seen the sand aggregate that is delivered for a particular job, would not necessarily know what the proportion ratio should be that should be carefully measured. Box calibrated measurements might be used at the start of a job to check the size and count of shovel measurement, but shovel measurements would be adequate during the progress of the job. Table 18-3 (repeated from Chapter 4) shows how these shovel measurements are used.

TABLE 18-3

Mortar Proportions (Parts by Volume)

Mortar type	Minimum compressive strength at 28 days (lb/in.²)	Portland cement	Hydrated limes or lime putty[a]		Masonry cements	Shovel count at 7–8/ft³		Parts (ft³)		Damp loose aggregate
			Min.	Max.		2¼ to 3		2¼ to 3		
M	2500	1	—	¼	—	21	28	2.81	3.75	Not less than 2¼ and not more than 3 times the sum of the volumes of the cement and lime used
		1	—	—	1	34	45	4.5	6	
S	1800	1	¼	½	—	21–25	28–34	2.8–3.4	3.7–4.5	
		½	—	—	1	25	34	3.4	4.5	
N	750	1	½	1¼	—	25–38	34–51	3.4–5.1	4.5–6.7	
		—	—	—	1	17	22	2.25	3	
O	350	1	1¼	2½	—	38–59	51–79	5.1–7.9	6.7–10.5	

[a]When plastic or waterproof cement is used as specified in Section 2403(p), hydrated lime or putty may be added but not in excess of one-tenth the volume of cement.

METHODS

Inspectors must be familiar with the many different methods or techniques of installation in the various forms of construction. For example, consider the "shoved" joint. This provides that the head joint mortar is pressed against both head faces of the adjacent units, and simultaneously the units are pressed and squeezed onto the bed joint. This pressure assures filling of the joint space, and increases the bond of the mortar to the units. The units immediately remove some of the water that has provided plasticity. If any of the units are subsequently moved, the bond that had been developed will be broken, resulting in a weaker and a leaky joint.

The moisture content of masonry units is also important and must be checked. Concrete units should be dry, not wetted with additional water, because that will tend to cause expansion and subsequent shrinking and cracking. Additional water is necessary only in a hot, dry, wind exposure. On the other hand, the clay product surfaces should be moistened slightly so that there will be better wetting of the surface and bonding. Also, the clay unit may have to be wetted to reduce excessive absorption that may occur. The excessive absorption would reduce the bond and strength of the mortar joint. The adequacy of wetness of the units may be checked by the circle-and-drop test which will serve as an indicator, as described in Chapter 2.

The shoving of units is particularly important for items such as toothed joints, shown in Figure 18-1. The difficulty is that one unit in place cannot be pressed down upon the one under it. This type of joint is not permitted in masonry unless

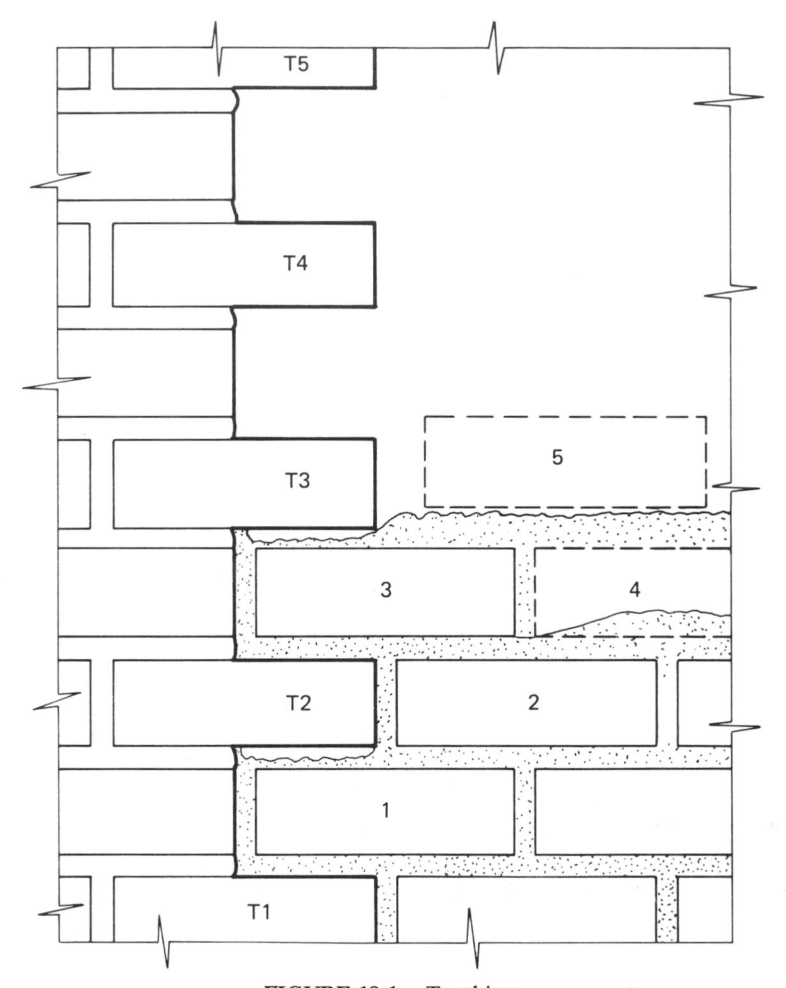

FIGURE 18-1. Toothing.

special care is taken. If the surfaces to receive mortar are troweled with mortar, (buttered), and then if mortar is pressed into the joints tightly so that it is pushed in, it will squeeze out sideways against the masonry unit surfaces and develop a bond.

PLACING THE REINFORCEMENT

It is, of course, essential that reinforcing details be closely followed as shown in the drawings. One important item is the placement of reinforcing steel in the masonry wall. A good example of that importance is shown by considering the location of the rebar in a retaining wall, noted in Figure 18-2. In one actual case when the tension steel was located properly, the effective depth was 10 in., as designed. However, the contractor located the bars in the center of the wall, reducing the

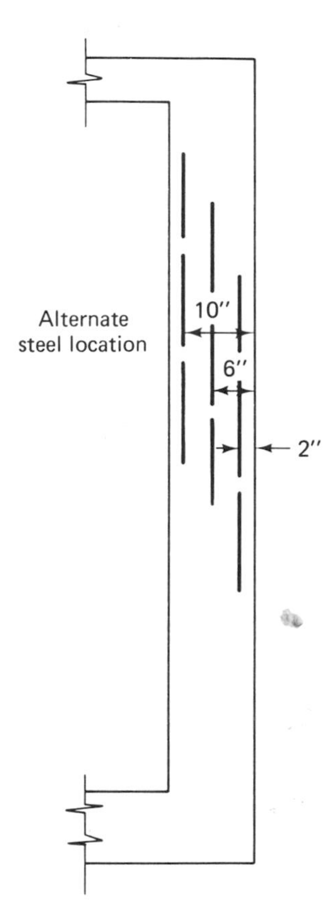

FIGURE 18-2. Reinforcing steel in retaining wall.

effective depth to about 6 in. This latter condition occurred because the draftsman had omitted a dimension to the steel in one of his details. The strength ratio at this location equals $6^2/10^2$, or 0.36 of the design strength, which resulted in a subsequent wall failure. Here is a good argument for proper inspection. Any qualified inspector would immediately question the fact that there was no dimension to the steel and would have called this to the attention of the designer. Further, the inspector should know something about where the steel should be placed under various conditions.

An even more serious discrepancy in the strength occurs if the steel is placed on the wrong face, as has been done in some basement retaining walls. In that case the strength ratio would drop to $2^2/10^2$, or 0.04 of the intended design strength! No question about the end result should this error go unnoticed.

Another matter of steel installation involves supporting of the steel in proper position until the grout has set. This can be done by wiring the steel in place or by using chairs or positioners, as shown in Figure 18-3. Another method, which is not generally viewed favorably, although it is used frequently, involves merely placing the bar in position while the grout is stiffening, thereby depending only on the stiffness of the grout to hold the steel in position.

The placement of anchor bolts is another extremely important field detail.

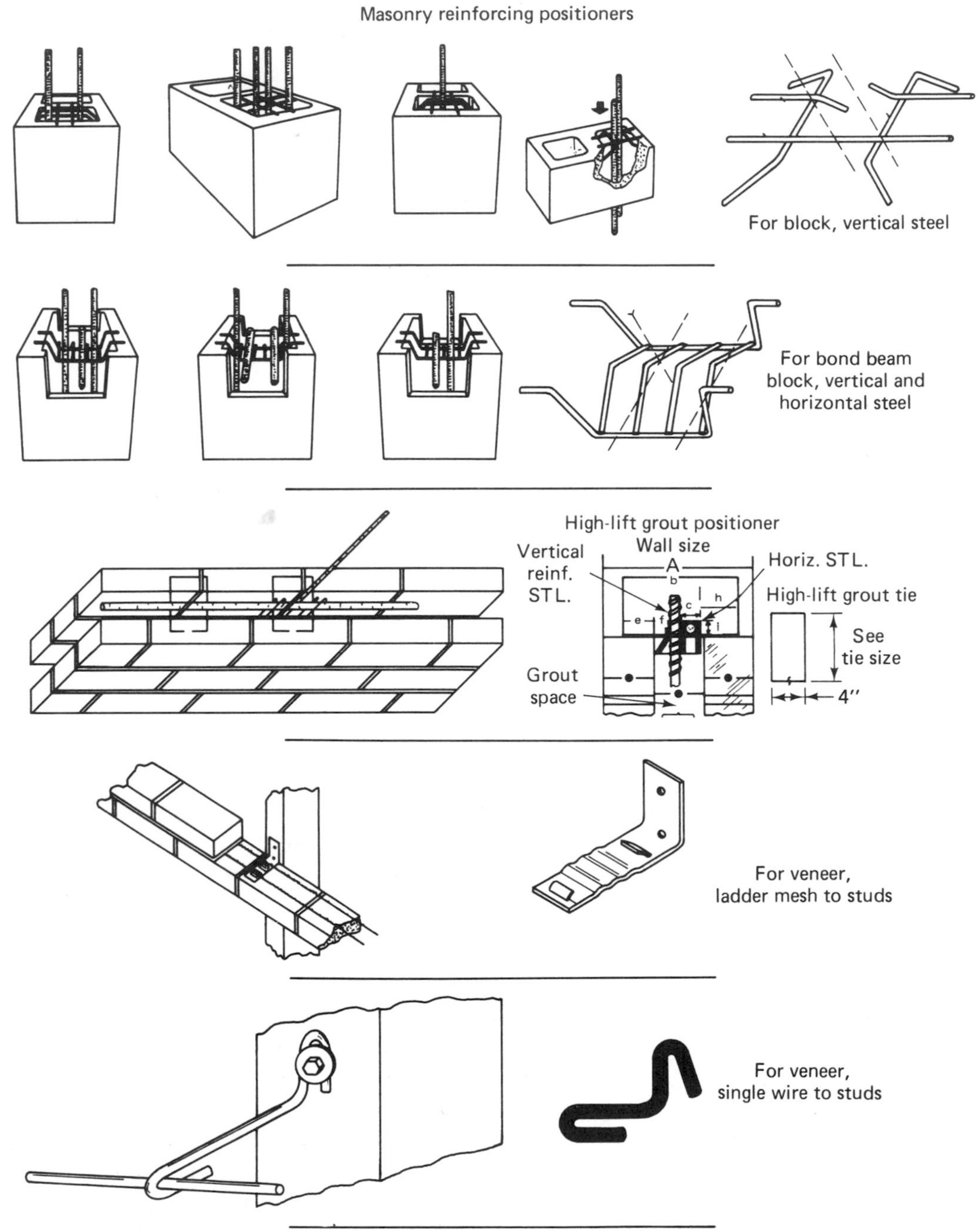

Masonry reinforcing positioners

For block, vertical steel

For bond beam block, vertical and horizontal steel

High-lift grout positioner

Wall size

A

Vertical reinf. STL.

Horiz. STL.

High-lift grout tie

See tie size

Grout space

4"

For veneer, ladder mesh to studs

For veneer, single wire to studs

FIGURE 18-3. Steel placement accessories. There are many mechanical aids for placement provided in various local areas. The above is an example of some of the many available in the Western United States.

The mortar and/or grout must embed the bolt firmly to develop it fully. Bolts properly placed are far stronger than the design values permitted in the UBC. However, placement is often subject to inadequate bedding or seating. Here again, it is up to the inspector to prevent such deficiencies.

There is also the implied general responsibility of the inspector to assure that the construction complies with the local code jurisdiction, even though some of these provisions might not be spelled out explicitly in the drawings and specifications, unless they are specifically exempted.

Grouting methods are frequently subject to confusion and misapplication. These procedures, fully described in the UBC, will be summarized here for clarification, and to emphasize key items of each.

1. Two-wythe, low lift
   Lay up about one course; grout and puddle; repeat; repeat.
2. Two-wythe, high lift
   Lay up full height, with ties and cleanouts; clean out; grout full height in lifts; consolidate and reconsolidate each lift.
3. Hollow unit, low lift
   Lay up 4 ft; grout and puddle; repeat; repeat.
4. Hollow unit, high lift
   Lay up full height; clean out; grout in lifts, with consolidation and reconsolidation.

The first method developed for widespread use in reinforced masonry was *two-wythe low-lift grouting*, Section 2413(c). The essential feature is that one wythe is carried up about 18 in., then the other is laid up and grouted in lifts not to exceed 8 in. or six times the width of the grout space.

The vertical steel must be in place and braced before the start of grouting, but the horizontal steel is best laid in the fresh grout as the work proceeds.

Two-wythe high-lift grout, Section 2413(d), was developed as an improvement, in speed and in quality. The wythes are laid up about 16 in. high at a time and wire ties are placed from wythe to wythe in repeated operations to the top of the wall. These have been laid up and grouted as high as 30 ft successfully on some jobs. Cleanouts are provided so the debris at the bottom may be removed. These are closed and the grout poured in 4-ft lifts, consolidated for good flow and placement, then reconsolidated after water volume loss while still plastic. These 4-ft lifts are continued to the top of the wall. Special inspection must be provided during the pouring. If full stresses are desired in lieu of one-half stresses, the entire masonry installation must be fully inspected as well. Vertical barriers are provided at 25 ft or less to control the lateral flow of grout and to facilitate the control of the height of the lift.

Hollow unit low-lift grouting has been the long-time method, specified in Sec. 2415(a), (b) for "block" construction. The units are laid up to about 4-ft heights and then grouted. No cleanouts are used because the droppings do not accumulate in the hollow unit walls as much as in the two wythe. The mortar bed joints are "buttered" on so that they do not fall off into the grout cores as much as when a bed is "strung" or "thrown." Also, the cross webs tend to prevent squeezings from falling so freely, and the hollow unit bed joints contain much less mortar to drop than do the two wythe beds.

Hollow unit high-lift grouting was developed as an improvement and was incorporated into Section 2415(b) 3 and 5 more by implication, or by permission, than by being specified step by step. Essentially the practice, as interpreted, is to lay up the block full height, with cleanouts. Then clean out the hollow cells, close the holes, and grout in 4-ft lifts, all the while consolidating for good placement and then reconsolidating to offset the water volume loss, continuing to the top of the wall. There have been many attempts to clarify the rather loose wording in the codes describing this procedure, so it is anticipated that clarifications will be forthcoming in future codes.

The bed joints of grouted construction described above should be held back from the grout space to minimize droppings into the space. In fact, the bed joint may be partially filled from the grout space with fluid grout for depths of up to about 1 in. Certainly it would be better to err by having mortar gaps rather than having joints that are too full.

Appendix

LIST OF TABLES AND DESIGN CURVES

LIST OF COMMONLY USED SYMBOLS AND THEIR MEANINGS

A_g = gross area of masonry section
A_{st} = total area of longitudinal reinforcement in a column
A_s = total area longitudinal tensile reinforcement
A'_s = area of compressive reinforcement in flexural members
A_v = area of web reinforcement
b = width of rectangular beam or width of flange of T-beam or total width of reinforced masonry column
b' = width of web in T-beams (used in flexural computations)
C = resultant of compressive masonry forces
C and C_p = force factor in seismic response equation
d = effective depth of flexural members; i.e., distance from extreme compression fiber to centroid of longitudinal reinforcement
d' = distance from extreme masonry compression fiber to centroid of compressive reinforcement
D or d = overall length of pier section in a shear wall, in.
e = eccentricity of the resultant load on a column or wall measured from gravity axis, $= M/P$
ϵ_m = unit longitudinal strain in masonry, in./in.
ϵ_s or ϵ'_s = unit longitudinal strain in tensile or compressive steel, in./in.
E_m = modulus of elasticity of masonry, lb/in.^2
E_s = modulus of elasticity of steel, lb/in.^2
f_a = computed axial stress in masonry, lb/in.^2
f_b = allowable flexural stress in masonry when axial compressive stress exists, lb/in.^2
f_m = flexural compressive stress in extreme fiber of masonry, lb/in.^2
f'_m = specified ultimate axial compressive stress of masonry, lb/in.^2
f_s = flexural stress in tensile reinforcement or axial compressive stress in column reinforcement, lb/in.^2
f'_s = stress in compressive reinforcement in flexural members, lb/in.^2
f_t = flexural tensile stress in masonry (uncracked section), lb/in.^2
f_v = stress in web reinforcement, lb/in.^2
F_a = allowable axial stress in masonry, if member were carrying axial load only, lb/in.^2
F_b = allowable flexural stress in masonry, if member were carrying bending only, lb/in.^2
F_t = seismic force assumed acting at top of structure, kips
g = ratio of distance (gt) between bars at opposite faces of columns to overall column dimension (t)
h = actual unsupported height of wall or column; horizontal clear distance in wall between supporting vertical elements, ft
h' = effective length of wall or column, considering end restraints, ft
I = moment of inertia, in.^4
I = importance factor in seismic calculations
j = ratio of distance between centroid of compression and centroid of tension forces in a beam to the effective depth, d
J = torsional moment of inertia
k = ratio of distance between extreme compression fiber and neutral axis to effective depth of flexural section, d

j_b, k_b = same ratios at the balanced state
$K = \frac{1}{2} f_m jk$; used in flexural computations as a measure of masonry stress
K = framing response factor for seismic calculations
$K_b = \frac{1}{2} F_b j_b k_b$
M = external bending moment, in.-kips
M_m = allowable moment as governed by masonry, in.-kips
M_s = allowable moment as governed by reinforcing steel, in.-kips
n = ratio of modulus of elasticity of steel (E_s) to that of masonry (E_m)
p = ratio of area of tensile reinforcement in beams, columns, or walls (A_s) to the effective area of masonry (bd)
p_g = ratio of area of total vertical reinforcement in columns or walls to that of the gross masonry area (A_g)
p_b = balanced steel ratio in working stress design
P = total axial load on column or wall, lb or kips
P' = basic allowable axial load on a concentrically loaded long column, lb or kips
r = radius of gyration of effective column area
R = reduction factor for compressive capacity in walls or columns
$\sum_0$ = sum of perimeters of bars, in.
s = spacing of stirrups, in.
S = elastic section modulus of flexural member, in.^{3}
S = site response factor for seismic calculations
t = overall dimension of columns; also thickness of wall, in.
T = resultant of tensile steel forces, lb
T = period of building, seconds
u = bond stress per unit of surface area of bar, lb/in.^{2}
v = actual unit shearing stress in masonry beam, lb/in.^{2}
$v_{allowable}$ = allowable shearing stress in masonry, lb/in.^{2}
V = total shear on flexural member, lb or kips
V = base shear on building for seismic calculations, kips
w = uniformly distributed load, lb/ft
W = total uniformly distributed load, lb
Z = ratio of distance between extreme compression fiber and resultant of compressive forces to distance kd, where compressive reinforcement is present
Z = zone factor for seismic intensity

A-1, A-2, AND A-3

TABLE A-1

Allowable Working Stresses in Unreinforced Unit Masonry[a]

	Type M	*Type S*	*Type M or type S mortar*				*Type N*		
Material	*Compression*[b]	*Compression*[b]	*Shear or tension in flexure*[c,d]		*Tension in flexure*[e]		*Compression*[b]	*Shear or tension in flexure*[c,d]	
Special inspection required	No	No	Yes	No	Yes	No	No	Yes	No
Solid brick masonry									
4500 plus lb/in.^{2}	250	225	20	10	40	20	200	15	7.5
2500–4500 lb/in.^{2}	175	160	20	10	40	20	140	15	7.5
1500–2500 lb/in.^{2}	125	115	20	10	40	20	100	15	7.5
Solid concrete unit masonry									
Grade A	175	160	12	6	24	12	140	12	6
Grade B	125	115	12	6	24	12	100	12	6
Grouted masonry									
4500 lb/in.^{2}	350	275	25	12.5	50	25			
2500–4500 lb/in.^{2}	275	215	25	12.5	50	25			
1500–2500 lb/in.^{2}	225	175	25	12.5	50	25			
Hollow unit masonry[f]	170	150	12	6	24	12	140	10	5
Cavity wall masonry solid units[f]									
Grade A or 2500 lb/in.^{2} plus	140	130	12	6	30	15	110	10	5
Grade B or 1500–2500 lb/in.^{2}	100	90	12	6	30	15	80	10	5
Hollow units[f]	70	60	12	6	30	15	50	10	5
Stone-masonry									
cast stone	400	360	8	4	—	—	320	8	4
natural stone	140	120	8	4	—	—	100	8	4
Gypsum masonry	20	20	—	—	—	—	20		
Unburned clay masonry	30	30	8	4	—	—			

[a]UBC Table 24-B.

[b]Allowable axial or flexural compressive stresses in pounds per square inch gross cross-sectional area (except as noted). The allowable working stresses in bearing directly under concentrated loads may be 50% greater than these values.

[c]This value of tension is based on tension across a bed joint (i.e., vertically in normal masonry work).

[d]No tension allowed in stack bond across head joints.

[e]The values shown here are for tension in masonry in the direction of running bond (i.e., horizontally between supports).

[f]Net area in contact with mortar or net cross-sectional area.

TABLE A-2

Allowable Shear on Bolts for All Masonry Except Gypsum and Unburned Clay Units[a]

Diameter of bolt (in.)	*Embedment*[b] *(in.)*	*Solid masonry (shear, lb)*	*Grouted masonry (shear, lb)*
$\frac{1}{2}$[c]	4	350	550
$\frac{5}{8}$	4	500	750
$\frac{3}{4}$[c]	5	750	1100
$\frac{7}{8}$	6	1000	1500
1	7	1250	1850[d]
$1\frac{1}{8}$	8	1500	2250[d]

[a]UBC Table 24-G.

[b]An additional 2 inches of embedment shall be provided for anchor bolts located in the top of columns or pilasters for buildings located in seismic zones II or III.

[c]Special installation in approved hollow units permits higher values and $2\frac{1}{2}$-in. embedment.

[d]Permitted only with not less than 2500-lb/in.^{2}

TABLE A-3

Allowable Steel Stresses, f_s

Tensile stress:	*lb/in.*^{2}
For billet-steel or axle steel reinforcing bars of structural grade	20,000
For deformed bars with a yield strength of 60,000 lb/in.^{2} or more and in sizes No. 11 and smaller	24,000
Joint reinforcement, 50% of the minimum yield point specified in UBC Standards for the particular kind and grade of steel used, but in no case to exceed	30,000
For all other reinforcement	20,000
Compressive stress in column verticals:	
40% of the minimum yield strength, but not to exceed	24,000
Compressive stress in flexural members:	
For compression reinforcement in flexural members, the allowable stress shall not be taken as greater than the allowable tensile stress shown above—See Chapters 6 and 9.	
The modulus of elasticity of steel reinforcement may be taken as 30,000,000 lb/in.^{2} so $n = 30{,}000{,}000 \div 1000f'_m$.[a]	

[a]This figure is based upon an ancient value for the concrete modulus (E_c) which has been subsequently drastically refined. However, since no accurate values based upon valid test results have been established for E_m, the next figure being proposed for inclusion in the Code will probably be a value $\sim 600f'_m$, another fictitious number.

DESIGN AIDS

Charts and tables are included to serve as design aids and suggested short cuts in the design calculations of reinforced masonry elements. Different designers, under various design circumstances, may prefer one particular table or chart over another, so several different kinds are shown herein.

Method A

Use np as in Curve B-5 or Table C-1.

Given: $M = 8367$ in.-lb, $b = 12$ in., $d = 3.8$ in., $f_s = 20{,}000 \times \frac{4}{3} = 26{,}667$ lb/in.^{2}, $f'_m = 1500/2 = (0.333 \times 1500/2 = 250) \times \frac{4}{3} = 333$ lb/in.^{2}, $n = 30{,}000{,}000/(1000 \times 750) = 40$.

Estimate steel: Min. $A_s = 0.0007 \times 12 \times 7.6 = 0.064$ in.^{2}/ft or $A_s = 8367/0.9 \times 3.8 \times 26{,}667 = 0.092$ in.^{2}. Use larger amount of A_s.

Calculate: $np = nA_s/bd = 40 \times 0.092/12 \times 3.8 = 0.080$

Select from C-1 or *B-5:* $j = 0.891$, $2/kj = 6.85$

Calculate: $M_m = 333 \times 12 \times 3.8^2/(6.85) = 8400$ in.-lb OK

$M_s = 0.092 \times 0.891 \times 3.8 \times 26.667 = 8306$ in.-lb Low but OK

Use No. 5's at 40 in. cc.

Method B

Use Curve B-1.

Given: Same as in Method A.

Estimate steel: See Method A and calculate np.

Enter Curve B-1: At $d = 3.8$ in. go to intersection with $np = 0.080$. Drop down to $M = 8367/12 = 700$ ft-lb. Read $f_m = 330$ lb/in.^{2}

$$\text{From method A, } f_m = \frac{8367}{12(3.8)^2} \times 6.85 = 331 \text{ lb/in.}^2$$

Method C

Use Curve B-2. See item "Nondimensionalized Design Curve" discussion and example C following.

Method D

Use Curves B-3 and B-4.

Given: Same as Method A.

Calculate: $K = M/bd^2 = 8367/(12 \times 3.8^2) = 48$ (for Curve B-3)

$nK = nM/bd^2 = 40 \times 48 = 1920$ (for Curve B-4)

Enter Curve B-3: At $K = 48$, go to intersection with $f_m = 333$, read $np = 0.070$

Calculate: $p = 0.070/40 = 0.0018$

$A_s = 0.0018 \times 12 \times 3.8 = 0.08$ in.^{2}

Try No. 5's at 48 in. cc. ($A_s = 0.31/4 = 0.077$ in.^{2})

Check: $f_s = M/A_sjd = 8367(0.077 \times 0.9$ (approx.) $\times 3.8) = 31{,}800$ lb/in.^{2} Too high. Increase steel by $0.077 \times 31{,}800/26{,}667 = 0.092$ in.^{2} Use No. 5's at 40 in. cc. ($0.31/3.33 = 0.09$ in.^{2})

Or, enter B-4: At $nK = 1920$ and $f_s = 26{,}667$ lb/in.^{2} to read $np = 0.08$

Calculate: $p = 0.08/40 = 0.0020$

$A_s = 0.0020 \times 12 \times 3.8 = 0.091$ in.^{2}

Use No. 5's at 40 in. cc.

NONDIMENSIONALIZED DESIGN CURVES

A nondimensionalized K/F_b versus np curve proves to be an extremely simple, yet very versatile design tool, as exemplified by Curve B-2. Note that the ordinate has been nondimensionalized by dividing K by the allowable masonry stress, F_b. Thus, the curve can be utilized for any value of F_b as long as it is defined as $\frac{1}{3}f'_m$, and the modulus of elasticity of the masonry, E_m, remains as $1000f_m$. Further, curve A is based upon an allowable steel stress of 20,000 lb/in.^{2}, whereas curve B provide values for an allowable steel stress of 24,000 lb/in.^{2}. Note that the swing point on the curves is the np value for balanced design; that is, that fictitious condition which occurs upon the simultaneous attainment of the allowable F_b in the masonry and the allowable f_s in the steel. Below that point, the allowable steel stress is reached first and it therefore governs. Thus f_m, which is some ratio of F_b, can be taken from the appropriate curve, A or B. Above this swing point, the allowable masonry stress governs and then the corresponding steel stress can also be taken from this curve. Note that a single steel stress curve provides the answer for f_m regardless of whether or not the allowable design stress was specified at 20,000 or 24,000 lb/in.^{2}. Thus it is seen that a single set of nondimensionalized curves will replace a large number of curves or charts which are based upon various values of f'_m, F_b, n, and f_s.

EXAMPLE A

Consider an 8-in. partially grouted concrete block wall, subjected to a lateral force only, having an h/t of 16. Design stresses are to be based upon a no inspection condition. Determine the steel requirement for lateral forces on the wall.

Seismic $F_p = ZIC_pSW_p = 1.0 \times 1.0 \times 0.2 \times 1.5\ W_p = 0.3(75$ lb/in.$^2) = 22.5$ lb/in.^{2}

Wind at 15 lb/in.^{2}, so seismic governs.
For $h/t = 16$, then $h = 10$ ft 8 in.

$$M = Wl^2/8 = 22.5 \times (10.67)^2/8 = 320 \text{ ft-lb/ft and } K = M/bd^2 = \frac{320 \times 12}{12 \times (3.8)^2} = 22.2$$

Since the allowable stress can be increased by $\frac{1}{3}$, take $\frac{3}{4}K = 16.65$ and then $K/F_b = 16.65/0.225 = 74$. With this as the ordinate, find 0.041 on the pn abcissa (B-2 curve—B); thus, $p = 0.00093$. Since f_m stress lies below balanced state, f_s governs at $20{,}000 \times \frac{4}{3} =$

26,667 lb/in.^{2}. For $K/F_b = 74$, find f_m on abcissa $(0.66 \times 225) = 149$. Therefore, f_m actual $= 149 \times \frac{4}{3} = 198$ lb/in.^{2}.

EXAMPLE B

Design a 9-in. brick beam to carry a moment of 50 ft-kips. Use a balanced design condition, with $f'_m = 1500$ lb/in.^{2} (special inspection): On Curve B-2 (B) at the balanced design point, $pn = 0.083$ and $K/F_b = 148.2$. Since $F_b = \frac{1}{3} \times 1500 = 500$ lb/in.^{2}, then $K = 148.2 \times 0.500 = 74.1$; thus, $bd^2 = M/K = (50 \times 12{,}000)/74.1 = 8097$. For $b = 9$ in., required $d = 30$ in. and $A_s = 0.0041 \times 9 \times 30 = 1.11$ in.^{2}.

EXAMPLE C

A solid-grouted concrete masonry beam has $d = 28.5$ in. and two No. 4's as tensile reinforcement. Determine the allowable moment capacity and the material stresses. No inspection is provided.

Find $np = 0.40 \times 40/(7.63 \times 28.5) = 0.073$. From the curve, note that 0.073 intersects curve B-2 (B) below $K/F_b = 148$; thus $f_s = 20{,}000$ lb/in.^{2} governs the moment capacity.

For this value of np, find $K/F_b = 131$ and K then is $131 \times 0.250 = 32.75$, with allowable moment equal to $32.75 \times 7.63 \times 28.5^2 = 203.0$ in.-kips. Masonry stress is 0.92×250 (abcissa of curve B for ordinate of 131) $= 230$ lb/in.^{2}.

EXAMPLE D

Determine the capacity of a 9-in. reinforced brick wall spanning vertically between foundation and roof. Say $f'_m = 1800$ lb/in.^{2} with special inspection provided. Total wall steel will conform to UBC minimum requirements (i.e., $0.002bt$), so place $0.0013bt$ vertically. What is the permitted height under these circumstances?

$0.0013bt$ in 9-in. wall gives $A_s = 0.140$ in.^{2}/ft. $np = 0.140/(12 \times 4.5) \times 16.7 = 0.043$. Find $K/F_b = 77.5$. Thus, $K = 77.5 \times 0.600 \times \frac{4}{3} = 61.84$ and $M = 61.84 \times (12/12) \times 4.5^2 = 1252.3$ ft-kips/ft of wall, which weighs approximately 100 lb/ft.^{2}. Then $w = 0.30 \times 100$ (see Example A) $= 30$ lb/ft^2. $M = wh^2/8$ or $h^2 = 8 \times (1252.4/30)$; thus, $h = 18.3$ ft. Maximum permitted height is $25 \times 9/12 = 18.75$ ft. Masonry stress f_m (from abcissa curve B, for $K/F_b = 77.5) = 0.67 \times 600 \times \frac{4}{3} = 535$ lb/in.^{2}.

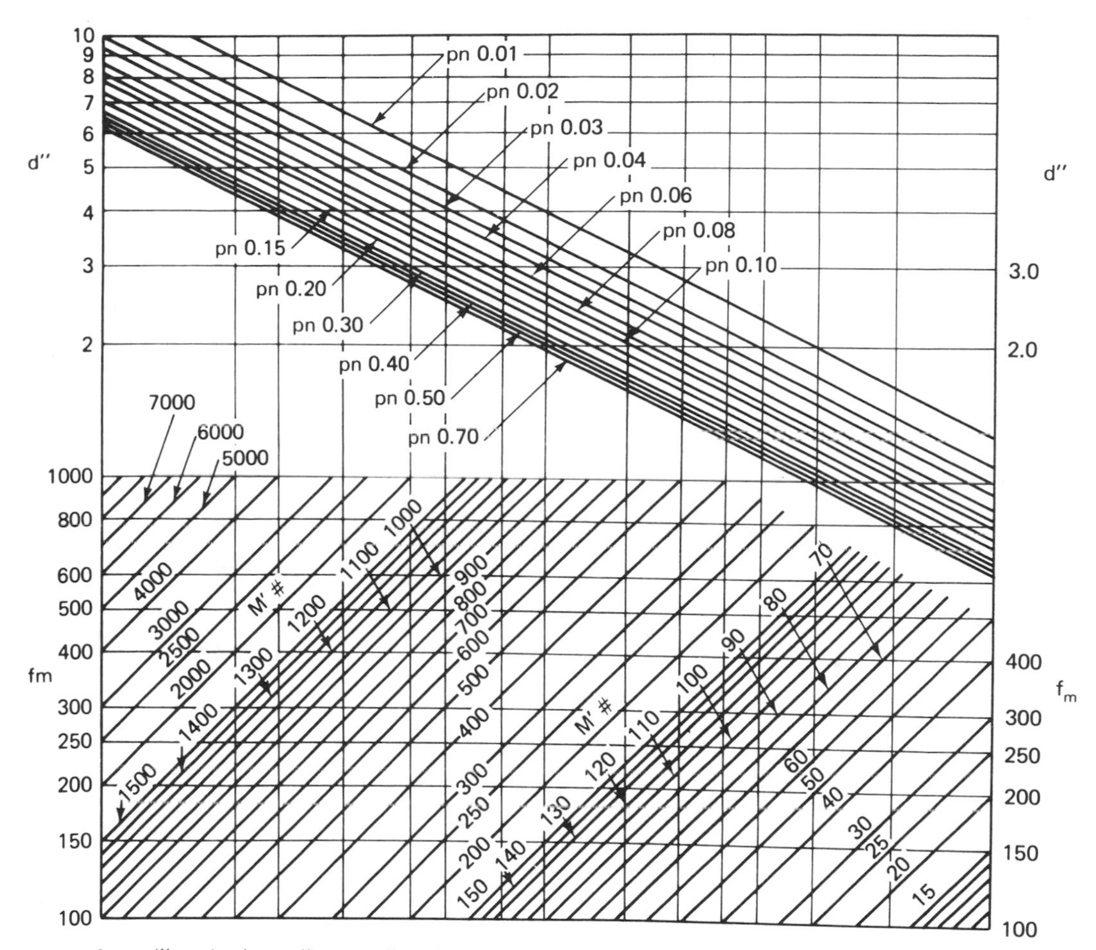

Starting from d'', go horizontally to pn line, then go vertically down to intersect M' #, then proceed horizontally to f_m lb/in.^{2} For d'' larger than 10'' use d''/10 and M' #/100 then read f_m lb/in.^{2} direct.

CURVE B-1. Flexural chart, *np* and *M*.

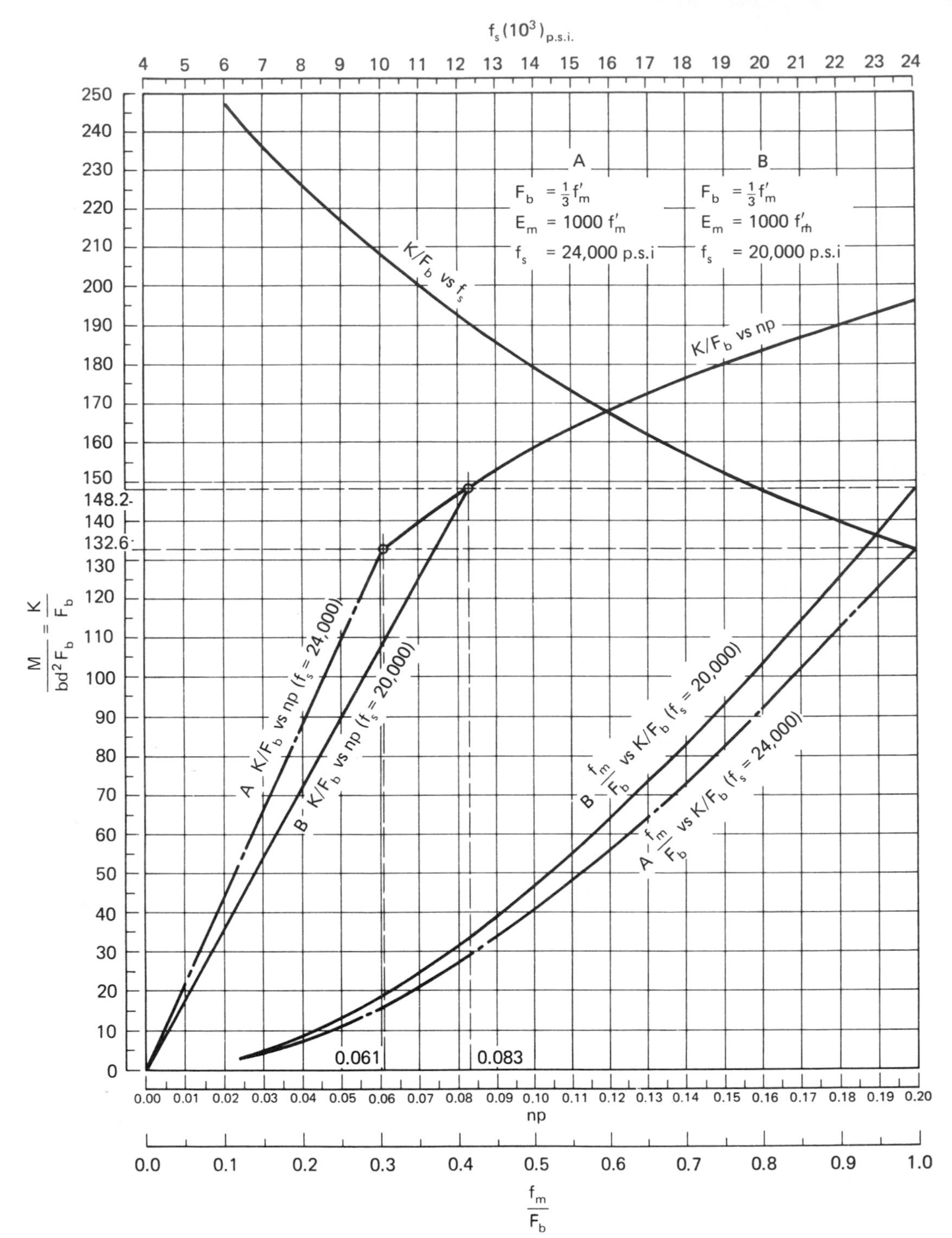

CURVE B-2. K/F_b versus np and f_m and f_s.

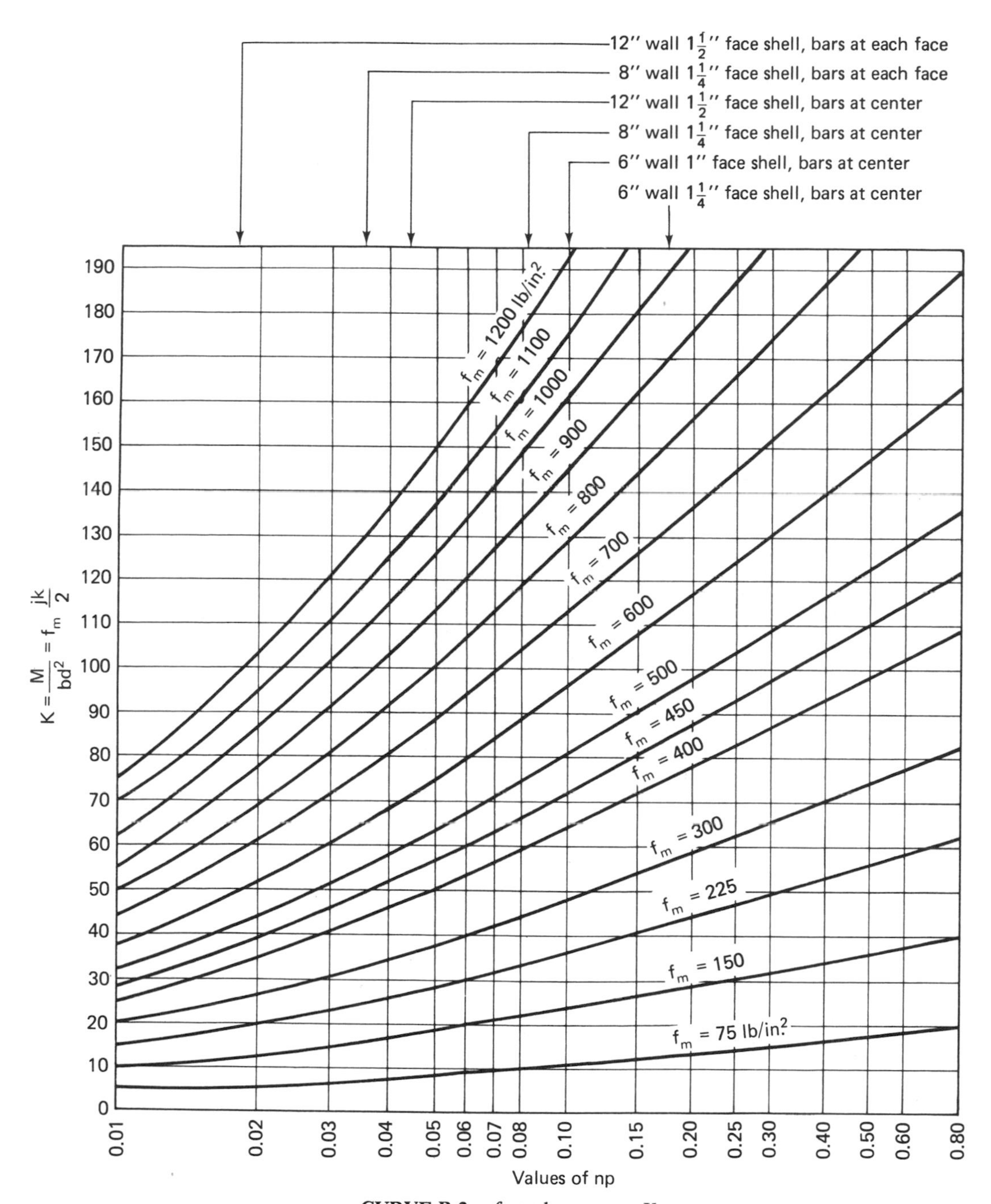

CURVE B-3. f_m and np versus K.

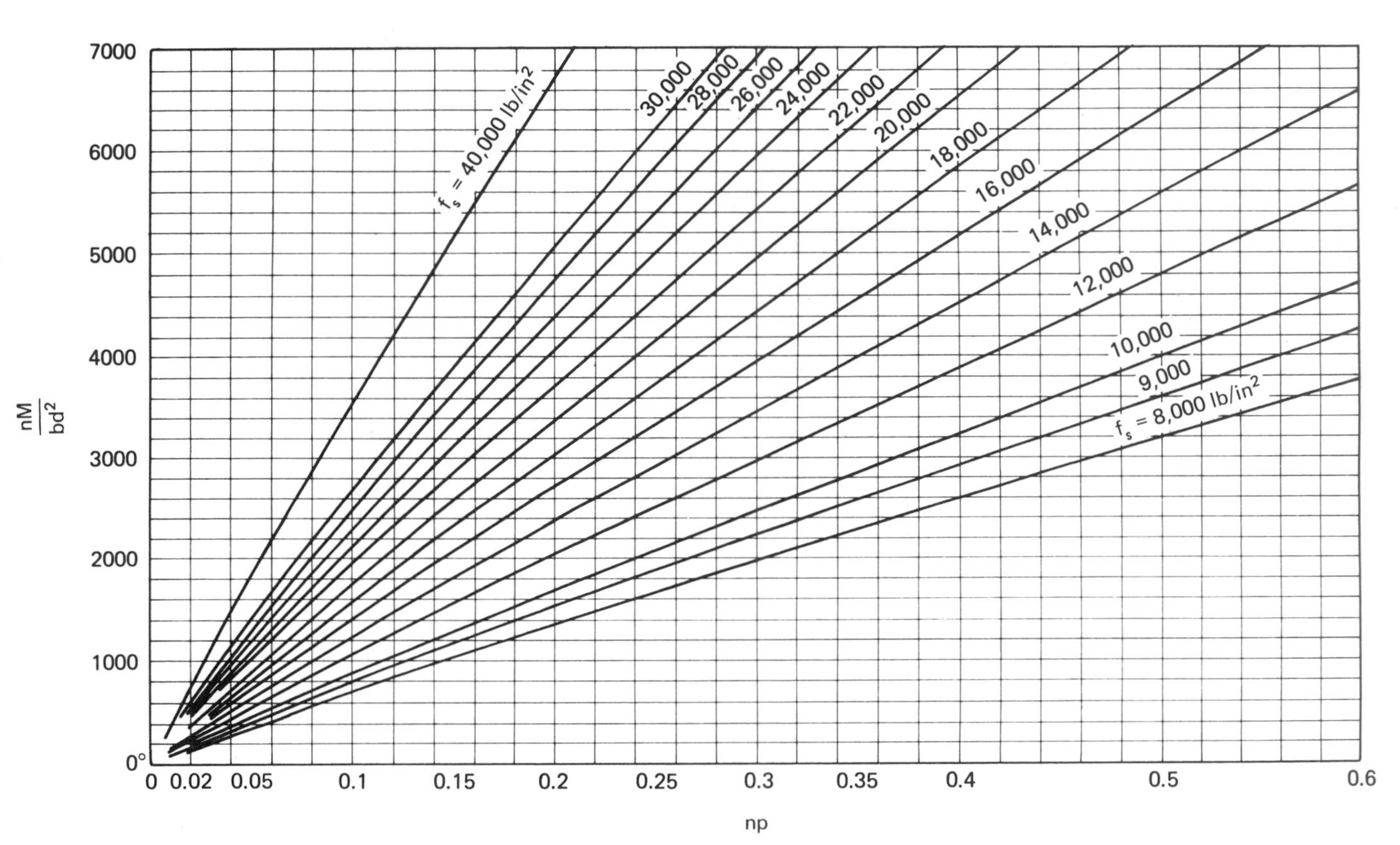

CURVE B-4. nM/bd^2 versus np and f_s.

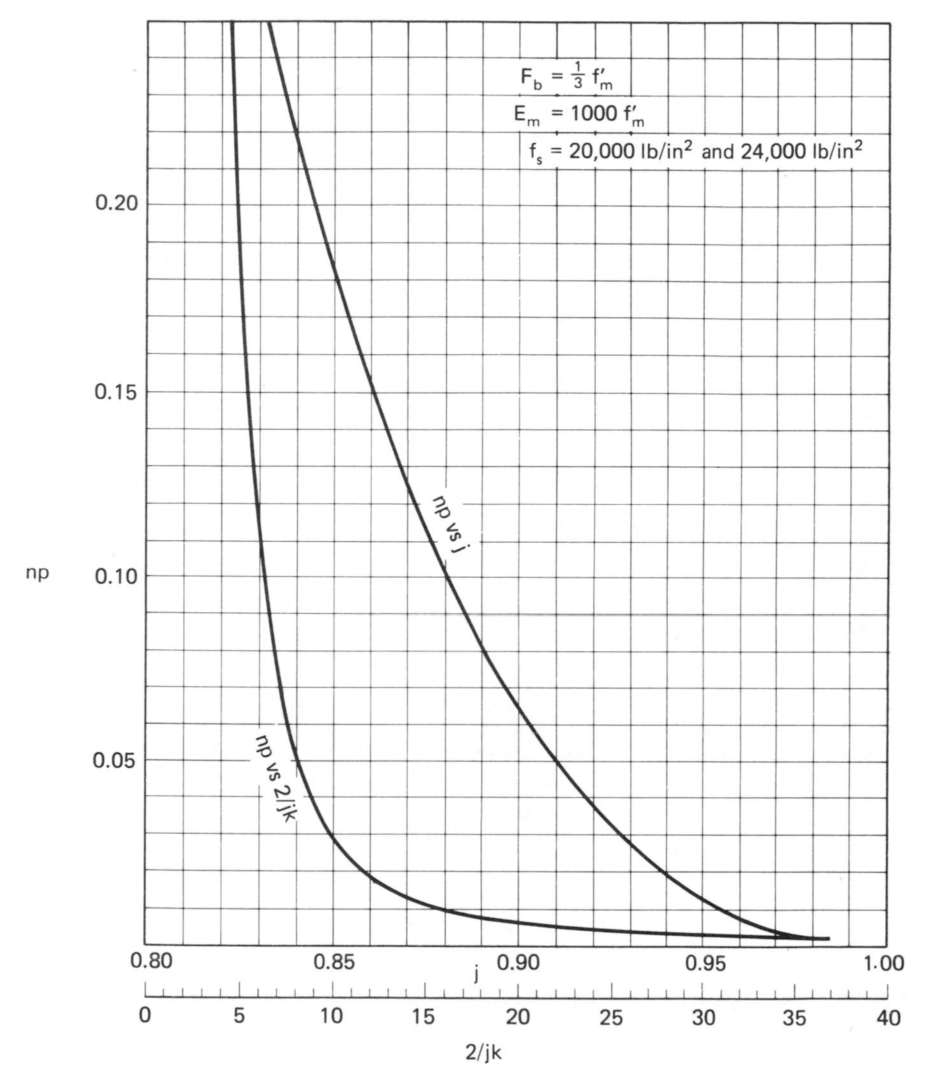

CURVE B-5. *np* versus *j* and 2/*jk*.

C-1

TABLE C-1

Table of Flexure Coefficients

$$p = \frac{A_s}{bd} \quad n = \frac{E_s}{E_m} \quad k = \sqrt{2np + (np)^2} - np$$

$$j = 1 - \frac{k}{3} \quad f_m = \frac{M}{bd^2}\left(\frac{2}{kj}\right) \quad f_s = \frac{M}{A_s jd}$$

np	k	j	$2/kj$	np	k	j	$2/kj$	np	k	j	$2/kj$
0.010	0.132	0.956	15.93	0.115	0.378	0.874	6.04	0.190	0.455	0.848	5.18
0.015	0.159	0.947	13.30	0.120	0.384	0.872	5.97	0.195	0.459	0.847	5.14
0.020	0.181	0.940	11.76	0.125	0.390	0.870	5.89	0.200	0.463	0.846	5.11
0.025	0.200	0.933	10.73	0.130	0.396	0.868	5.82	0.220	0.479	0.840	4.96
0.030	0.217	0.928	9.95	0.135	0.402	0.866	5.75	0.240	0.493	0.835	4.85
0.035	0.232	0.923	9.36	0.140	0.407	0.864	5.68	0.260	0.507	0.831	4.75
0.040	0.246	0.918	8.89	0.145	0.413	0.862	5.62	0.280	0.519	0.827	4.66
0.050	0.270	0.910	8.14	0.150	0.418	0.861	5.56	0.300	0.531	0.823	4.58
0.060	0.291	0.903	7.60	0.155	0.423	0.859	5.51	0.350	0.557	0.814	4.41
0.070	0.311	0.897	7.19	0.160	0.428	0.857	5.46	0.400	0.580	0.807	4.27
0.080	0.328	0.891	6.85	0.165	0.433	0.856	5.41	0.450	0.600	0.800	4.17
0.090	0.344	0.885	6.58	0.170	0.437	0.854	5.36	0.500	0.618	0.794	4.07
0.100	0.358	0.880	6.34	0.175	0.442	0.853	5.31	0.550	0.634	0.788	4.00
0.105	0.365	0.878	6.24	0.180	0.446	0.851	5.26	0.600	0.649	0.784	3.94
0.110	0.372	0.876	6.14	0.185	0.451	0.850	5.22	0.700	0.675	0.775	3.82

D-1 AND D-2

STEEL AREA AND SPACING TABLES

This table is intended to aid in determining the areas of steel required for various thicknesses of masonry and for various jurisdictions. The Uniform Building Code requires the area of steel in masonry to be not less than *0.002* times the wall area, with not less than one-third of the steel in either direction.

Title 21 requires *0.003* times the wall area for total steel.

The values in the upper table are for the conditions of $\frac{1}{3}$, $\frac{2}{3}$, and total area required above.

Having determined the desirable area of steel for one direction, by use of the lower spacing table, then subtract this amount from the total area required, giving the amount of steel required in the other direction. The most desirable spacing and size can then be selected.

These two quantities can be revised and adjusted easily by using the lower table for selecting the most effective bar combinations for both directions.

The double wires listed are "ladder bar" type of reinforcement, which is available in several proprietory types, and the area given is the area of both wires (i.e, the contribution of the double wire to the total steel in that direction). The spacings are set up in 4-in. modules up to 4 ft and 8-in. modules up to 8 ft. The wall thickness shown is actual, not nominal, so it is on the safe side.

EXAMPLE:

Given: Nominal 8-in. block wall. Select ladder bar reinforcing and vertical bar reinforcing to be suitable for walls under the Uniform Building Code.

Solution:

Enter the upper table to determine the total steel required. Use actual thickness of $7\frac{1}{2}$-in. for nominal 8-in. wall. On the line for factor of 1.0×0.002, find values of 0.168 for 7 in. and 0.012 for $\frac{1}{2}$ in., which equal a total value of 0.180 in.^{2} (0.168 plus 0.012). This is the minimum total steel required, and there may not be less than 0.060 in.^{2} in either vertical or horizontal directions.

Now enter the lower table to determine spacing. If one provides Heavy Duty Wall Mesh ($2\frac{3}{16}$) at 16 in., follow the line for $2\frac{3}{16}$ and find a value of 0.041 under the 16-in. heading. If No. 5 bars at 8 ft 0 in. are used for horizontal reinforcing, follow the line for No. 5 and find a value of 0.038 under the 96-in. heading. The sum of these first two values is 0.079, which exceeds the minimum allowable value. The vertical reinforcing, therefore, should exceed a value of 0.101 (0.180 less 0.079).

Using the most desirable spacing of 48 in. on center, follow the 48-in. column down to find the minimum standard bar size suitable. The No. 6 bar has a value of 0.110, which exceeds the arbitrary steel area requirements, and thus would be the recommended reinforcing. The structural stress requirements should be checked for the specific loading applied also.

TABLE D-1

Area of Steel Required by UBC or Title 21 (Arbitrary Percentages)

Factor:	*Wall thickness*												
UBC at 0.002 Title 21 at 0.003	$\frac{1}{2}$ in.	1 in.	2 in.	3 in.	4 in.	5 in.	6 in.	7 in.	8 in.	9 in.	10 in.	11 in.	12 in.
$\frac{1}{3} \times 0.002$	0.004	0.008	0.016	0.024	0.032	0.040	0.048	0.056	0.064	0.072	0.080	0.088	0.096
$\frac{1}{3} \times 0.003$	0.006	0.012	0.024	0.036	0.048	0.060	0.072	0.084	0.096	0.108	0.120	0.132	0.144
$\frac{2}{3} \times 0.002$	0.008	0.016	0.032	0.048	0.064	0.080	0.096	0.112	0.128	0.144	0.160	0.176	0.192
1.0×0.002 $\frac{2}{3} \times 0.003$	0.012	0.024	0.048	0.072	0.096	0.120	0.144	0.168	0.192	0.216	0.240	0.264	0.288
1.0×0.003	0.018	0.036	0.072	0.108	0.144	0.180	0.216	0.252	0.288	0.324	0.360	0.396	0.432

TABLE D-2

Area of Steel per Foot

Size steel	Diameter	Area	Steel spacing: 8 in. (0 ft 8 in.)	12 in. (1 ft 0 in.)	16 in. (1 ft 4 in.)	20 in. (1 ft 8 in.)	24 in. (2 ft 0 in.)	28 in. (2 ft 4 in.)	32 in. (2 ft 8 in.)	36 in. (3 ft 0 in.)	40 in. (3 ft 4 in.)	44 in. (3 ft 8 in.)	48 in.[a] (4 ft 0 in.)	56 in. (4 ft 8 in.)	64 in. (5 ft 4 in.)	72 in. (6 ft 0 in.)	80 in. (6 ft 8 in.)	88 in. (7 ft 4 in.)	96 in. (8 ft 0 in.)
2 No. 9	0.148	0.0345	0.052	0.034	0.026	0.021	0.017	0.015	0.013	0.012	0.010	0.009	0.0086	—	—	—	—	—	—
2 No. 8	0.162	0.0412	0.062	0.041	0.031	0.025	0.021	0.018	0.015	0.014	0.012	0.011	0.010	—	—	—	—	—	—
$2\frac{3}{16}$	0.1875	0.0552	0.083	0.055	0.041	0.033	0.028	0.024	0.021	0.018	0.017	0.015	0.014	—	—	—	—	—	—
$2\frac{1}{4}$	0.250	0.098	0.147	0.098	0.073	0.059	0.049	0.042	0.037	0.033	0.029	0.027	0.024	—	—	—	—	—	—
$2\frac{5}{16}$	0.312	0.152	0.229	0.15	0.114	0.092	0.076	0.065	0.057	0.051	0.046	0.042	0.038	—	—	—	—	—	—
No. 2	$\frac{1}{4}$	0.049	0.073	0.05	0.036	0.029	0.024	0.021	0.018	0.016	0.015	0.013	0.012	—	—	—	—	—	—
No. 3	$\frac{3}{8}$	0.110	0.165	0.11	0.083	0.066	0.055	0.047	0.041	0.037	0.033	0.030	0.027	0.024	0.021	0.018	0.016	0.015	0.014
No. 4	$\frac{1}{2}$	0.196	0.293	0.20	0.147	0.118	0.098	0.084	0.073	0.065	0.059	0.054	0.049	0.042	0.037	0.033	0.029	0.027	0.024
No. 5	$\frac{5}{8}$	0.307	0.460	0.31	0.230	0.184	0.154	0.132	0.115	0.102	0.092	0.084	0.077	0.066	0.057	0.051	0.046	0.042	0.038
No. 6	$\frac{3}{4}$	0.442	0.663	0.44	0.332	0.265	0.221	0.189	0.166	0.147	0.133	0.120	0.110	0.095	0.083	0.074	0.066	0.060	0.055
No. 7	$\frac{7}{8}$	0.601	0.900	0.60	0.450	0.361	0.300	0.258	0.226	0.200	0.180	0.164	0.150	0.129	0.112	0.100	0.090	0.082	0.075
No. 8	1.0	0.786	1.180	0.79	0.590	0.471	0.392	0.337	0.295	0.261	0.236	0.214	0.196	0.168	0.147	0.131	0.118	0.107	0.098
No. 9	1.128	1.000	1.50	1.00	0.750	0.600	0.500	0.428	0.375	0.333	0.300	0.273	0.250	0.214	0.187	0.167	0.150	0.136	0.125

[a]Recommended spacing

E-1

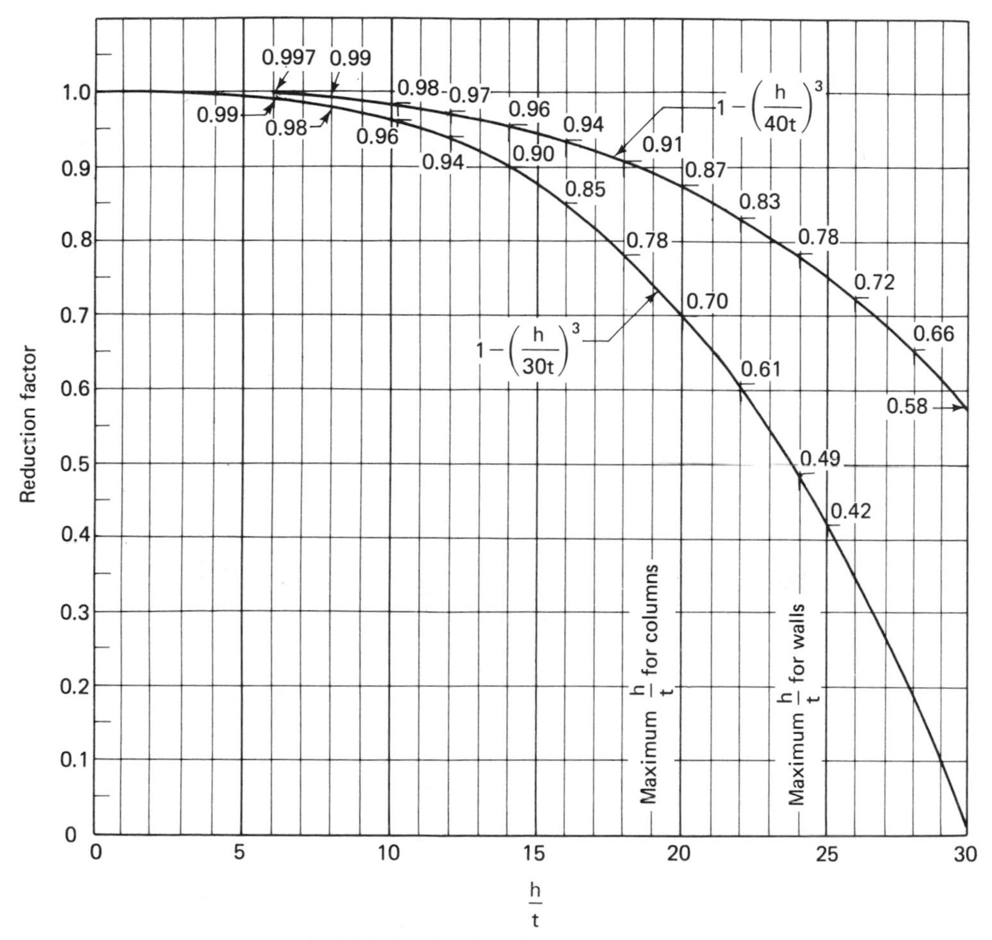

DIAGRAM E-1. Axial load reduction factor curve.

F-1, F-2, AND F-3

TABLE F-1

Load carried by reinforcing (at 0.65 × 20,000): Max., at 4%		Min., at ½%		Nominal sizes (in.) within ½ in. of actual sizes	A_g net area (in.)	Load carried by masonry at various f'_m values (at $0.18f'_m$ kips)												
						675	750	1350	1500	1800	2000	2500	3000	3500	4000	5000	6000	f'_m
A_s at 4%	P (kips)	A_s at ½%	P (kips)			121	135	243	270	374	360	450	540	630	720	900	1080	$0.18f'_m$
3.2	42	0.40	5.2	8 × 12	80	9.7	10	19	21	26	29	36	43	50	58	72	86	
4.0	52	0.50	6.5		100[a]	12.1	13.5	24.3	27.0	32.4	36.0	45.0	54.0	63.0	72.0	90.0	108	
4.6	60	0.57	7.7	8 × 16	115	13	15	27	31	37	41	52	62	72	83	103	124	
4.8	62	0.60	7.8		120	14	16	29	32	39	43	54	65	76	86	108	130	
5.2	67	0.65	8.4	12 × 12	130	15	17	32	35	42	47	58	70	82	94	117	140	
6.2	78	0.75	9.7		150	18	20	36	40	48	54	67	81	94	108	135	162	
7.2	93	0.90	11.7	12 × 16 or 8 × 24	180	22	24	43	48	58	65	81	97	113	129	162	194	
8.0	104	1.0	13.0		200	24	27	48	54	64	72	90	108	126	144	180	216	
9.6	124	1.2	15.6	16 × 16	240	29	32	58	64	77	86	108	129	151	172	216	259	
10.8	140	1.3	17.5	12 × 24	270	32	36	65	73	87	97	121	145	170	194	243	291	
12.0	156	1.5	19.5	16 × 20	300	36	40	73	81	97	108	135	162	189	216	270	324	
14.4	187	1.8	23.4	16 × 24	360	43	48	87	97	116	129	162	194	226	259	324	388	
15.2	197	1.9	24.7	20 × 20	380	46	51	92	102	123	136	171	205	239	273	342	410	
16.0	208	2.0	26.0		400	48	54	97	108	129	144	180	216	252	288	360	432	
18.4	239	2.3	30.0	20 × 24	460	55	62	112	124	149	165	207	248	290	331	414	496	
20.0	260	2.5	32.5		500	60	67	121	135	162	180	225	270	315	360	450	540	
22.0	286	2.7	35.7	24 × 24	550	66	74	133	148	178	198	247	297	346	396	495	594	

[a]The values for 100 in.^{2} are carried past the decimal point, but such precision is not valid when considering the design values compared to true column capacity.

TABLE F-2

Load on Longitudinal Bars (kips)—Tied Masonry Columns[a]

Bar size	*Number of bars* 2	4	6	8	10	12
	$f_y = 40{,}000$ lb/in.^{2}; $0.4f_y = f_s = 16{,}000$ lb/in.^{2}					
No. 3	2.3	4.6	6.9	9.2	11.4	13.7
No. 4	4.1	8.3	12.5	16.6	20.8	25.0
No. 5	6.4	12.9	19.3	25.8	32.2	38.7
No. 6	9.1	18.3	27.5	36.6	45.8	54.9
No. 7	12.5	25.0	37.4	49.9	62.4	74.9
No. 8	16.4	32.9	49.3	65.7	82.2	98.6
No. 9	20.8	41.6	62.4	83.2	104.0	124.8
No. 10	26.4	52.5	79.2	105.7	132.1	158.5
No. 11	32.4	64.9	97.3	129.8	162.2	194.7

Bar size	*Number of bars* 2	4	6	8	10	12
	$f_y = 60{,}000$ lb/in.^{2}; $0.4f_y = f_s = 24{,}000$ lb/in.^{2}					
No. 3	3.4	6.9	10.3	13.7	17.2	20.6
No. 4	6.2	12.5	18.7	25.0	31.2	37.4
No. 5	9.6	19.3	29.0	38.7	48.4	58.0
No. 6	13.7	27.5	41.2	54.9	68.6	82.4
No. 7	18.7	37.4	56.2	74.9	93.6	112.3
No. 8	24.6	49.3	73.9	98.6	123.2	147.9
No. 9	31.2	62.4	93.6	124.8	156.0	187.2
No. 10	39.6	79.2	118.9	158.5	198.1	237.7
No. 11	48.6	97.3	146.0	194.7	243.4	292.0

[a]Load taken by longitudinal bars (kips) $= 0.65 f_s A_s \div 1000$

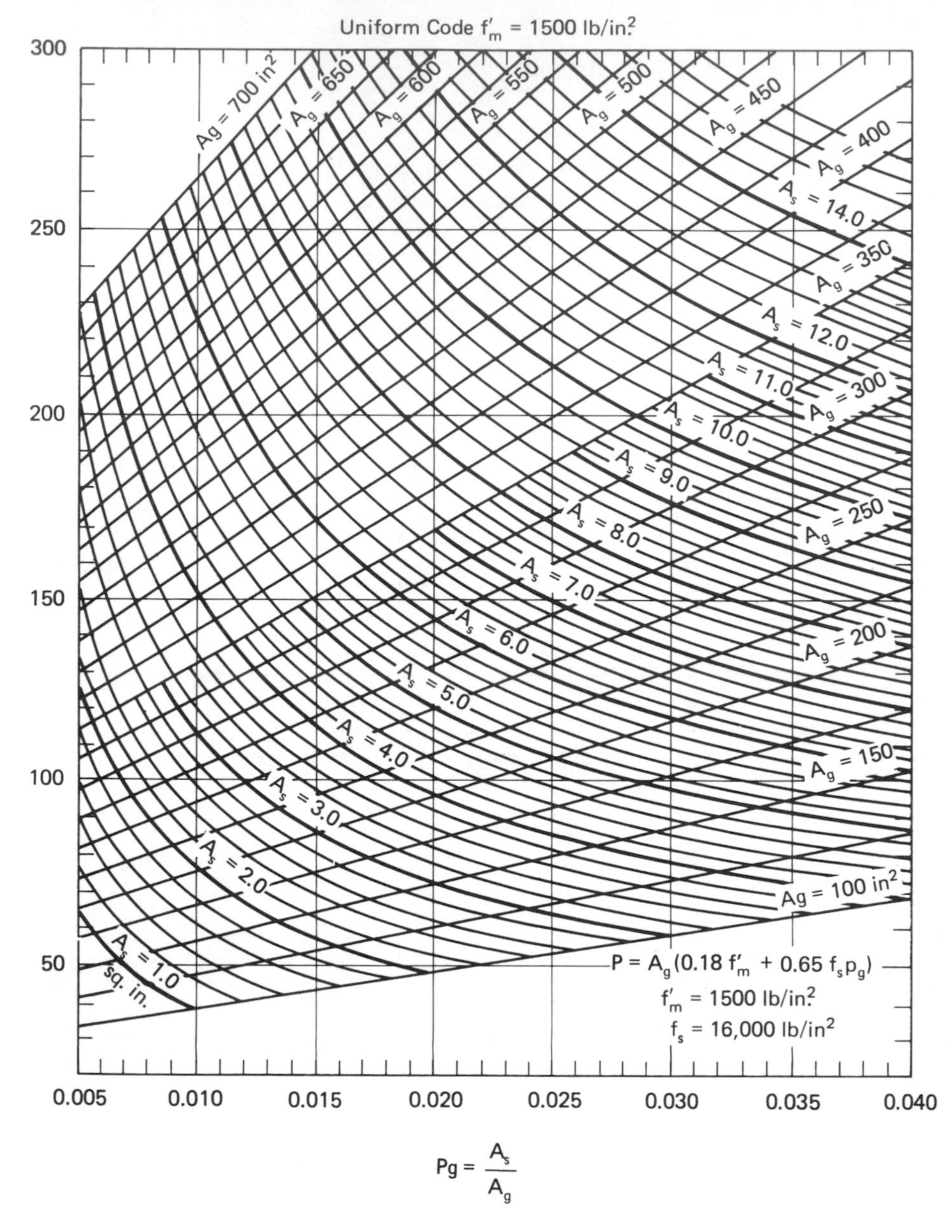

Note: With hard-grade steel vertical reinforcement and f_s = 20,000 lb/in.^{2}, the basic allowable load may be obtained by applying the factors given below to the basic allowable loads for f_s = 16,000 lb/in.^{2} Direct interpolation between values to obtain factors for intermediate values of p_g will be sufficiently accurate for practical design purposes.

p_g	0.005	0.010	0.015	0.020	0.025	0.030	0.035	0.040
Load factor	1.040	1.065	1.090	1.105	1.120	1.130	1.140	1.150

CURVE F-3. Basic allowable axial loads on reinforced concrete masonry columns and pilasters. (NCMA based upon UBC f'_m = 1500 lb/in.^{2}.)

G-1

TABLE G-1

Système Internationale unites (Metric)

English measurement to S.I. (metric) measurement

Unit	*Exact conversion*		*Approximate conversion*
Length			
1 mile	1.609 344	kilometers	1.6 km or $1\frac{1}{2}$ km
1 yard	0.914 4	meter	0.9 m or 1 m
1 foot	0.304 8	meter	0.3 m or $\frac{1}{3}$ m
1 inch	25.40	millimeters	25 mm or $\frac{1}{10}$ m
Speed			
1 mile per hour	1.609 344	kilometers per hour	1.6 km/h or $1\frac{1}{2}$ km/h
1 foot per second	0.304 8	meter per second	0.3 m/s or $\frac{1}{3}$ m/s
Area			
1 acre	4 046.856	square meters	4 000 m^2
1 square foot	0.092 9	square meter	$\frac{1}{10}$ m^2 or 1 000 cm^2
1 square inch	6.452	square millimeters	6 cm^2 or 650 mm^2
Weight or Mass			
1 ounce (avdp)	28.35	grams	30 g
1 pound	0.453 59	kilogram or 435.59 grams	$\frac{1}{2}$ kg or 500 g
1 kip	453.59	kilograms	500 kg or 0.5 Mg
1 Ton (short)	907.18	kilograms or 907 megagrams	1 Mg
Volume			
1 cubic yard	0.764 6	cubic meter or 764.56 liters	$\frac{3}{4}$ m^3 or 750 liters
1 cubic foot	0.028 3	cubic meter or 28.217 liters	$\frac{1}{35}$ m^3 or 30 liters
1 cubic inch	16.387	cubic centimeters	16 cm^3 or 16 000 mm^3
1 gallon	3 785.4	cubic centimeters or 3.785 liters	4 000 cm^3 or 4 liters
1 quart	946.35	cubic centimeters or 0.94 635 liter	1 000 cm^3 or 1 liter
Density			
1 pound/cubic foot	0.016 02	gram/cubic meter	16 kg/m^3 or 16 g/liter
	16.018	kilograms/cubic meter or (grams/liter)	
1 pound/gallon	0.119 8	gram/cubic meter	120 kg/m^3 or $\frac{1}{10}$ g/cm^3
	119.83	kilograms/cubic meter or (grams/liter)	
Force			
1 pound force	4.448	newtons	$4\frac{1}{2}$ N
1 kip force	4.448	kilonewtons	4500 N
Pressure			
1 pound/square inch	6 894.8	pascals	7 000 Pa
1 kip/square inch	6.895	megapascals	7 M Pa
Energy			
1 Btu	1 054.35	joules or 1.054 kilo joule	1 kj
Temperature			
°Fahrenheit	[(°F − 32)5/9] °Celsius		

TABLE G-1 (Continued)

Unit	Exact conversion		Approximate conversion
	S.I. (metric) measurement to English measurement		
Length			
1 kilometer	0.621 4	mile	$\frac{5}{8}$ mi or 0.6 mi
1 meter	3.280 8	feet or 39$\frac{3}{8}$ inches	3 ft 3 in. or 3 ft +
1 centimeter	0.393 7	inch	0.4 in. or $\frac{3}{8}$ in.
1 millimeter	0.039 4	inch	$\frac{1}{32}$ in.
Speed			
1 kilometer per hour	0.621 4	mile per hour	$\frac{5}{8}$ mph or 0.6 mph
1 meter per second	3.280 8	feet per second or 39.375 inches per second	3 ft/s or 1 yd/s
Area			
1 square kilometer	0.386 1	square mile or 247.1 acres	$\frac{1}{3}$ mile^2 or 250 acres
1 square meter	1.196	square yards or 10.764 square feet	1.2 yd^2 or 10 ft^2
1 square centimeter	0.155	square inch	$\frac{1}{6}$ in.^{2}
Weight or Mass			
1 gram	0.035 27	ounce (avdp)	$\frac{1}{30}$ oz
1 kilogram	2.205	pounds	2$\frac{1}{4}$ lb or 2 lb
1 megagram	2.205	kips or 2 205 pounds	2 kips or 2 000 lb
1 gigagram	1 102	ton or 2 205 000 pounds	1 000 tons or 2 million lb
Volume			
1 cubic meter	35.315	cubic feet or 264.17 gallons	35 ft^3 or 265 gal
1 litre	0.035 3	cubic foot or 0.264 2 gallon or 61.024 cubic inches	$\frac{1}{4}$ gal or 1 qt or 60 in.^{2}
1 cubic centimeter	0.061	cubic inch	$\frac{1}{16}$ in.^{3}
Density			
1 gram/cubic centimeter	8.345	lb/gal or 62.428 lb/cubic foot	8$\frac{1}{2}$ lb/gal or 62 lb/ft^3
1 kg/cubic meter	0.008 354	lb/gal or 0.062 428 lb/cubic foot	$\frac{1}{8}$ oz/gal or $\frac{1}{16}$ lb/ft^3
Force			
1 newton	0.224 8	pound force	$\frac{1}{4}$ pound force
1 kilonewton	224.8	pound force	225 pound force
Pressure			
1 pascal	0.000 145	pound/square inch	
1 kilopascal	0.145	pound/square inch	$\frac{1}{7}$ lb/in.^{2}
1 megapascal	145	pounds/square inch	150 lb/in.^{2}
Energy			
1 joule	0.000 948 45	Btu	1/1 000 Btu
1 000 joule	0.948 45	Btu	1 Btu
Temperature			
°Celsius	[(1.8°C) + 32] °Fahrenheit		

H-1, H-2, AND H-3

TABLE H-1

UBC Standards and ASTM Specifications Applicable to Clay, Brick, and Tile (Chapter 2)

Reference	*UBC Stand.*	*ASTM Spec.*
Testing:		
Sampling & Testing Brick & Structural Clay Tile	24-24	C-67
Masonry Units:		
Building brick tile	24-1	C-62
Facing brick	24-1	C-216
Structural clay—load-bearing wall tile	24-8	C-34
Facing tile		C-212
Ceramic glazed facing tile		C-126
Nonload bearing	24-9	C-56
Paving brick		C-62 or C-216
Hollow brick	24-1*	C-652
Sand lime building brick	24-22	
Gypsum tile	24-11	C-52

*Also Western States Clay Products Association Specification for Hollow Clay Units Specifically Designed for Reinforced Masonry.

TABLE H-2

UBC Standards and ASTM Specifications Applicable to Concrete Masonry (Chapter 3)

Reference	*UBC Stand.*	*ASTM Spec.*
Testing:		
Sampling & Testing Concrete Masonry Units	24-7	C-140
Raw Materials:		
Portland cement		C-150 or C-175
Hydrated lime	24-18	C-207
Pozzolons		C-68
Aggregates		C-33 or C-331
Masonry Units:		
Hollow load bearing concrete masonry units	24-4	C-90
Solid load bearing concrete masonry units	24-5	C-145
Concrete building brick	24-3	C-55
Nonload bearing concrete masonry units	24-6	C-129

TABLE H-3

UBC Standards and ASTM Specifications Applicable to Mortar, Grout, and Reinforcement (Chapter 4)

Reference	*UBC Stand.*	*ASTM Spec.*
Mortar:		
Raw Materials:		
Similar to those for concrete masonry, except add masonry cement	24-16	C-91
Aggregate for masonry mortar	24-21	C-144
Mortar Mixes	24-20	C-270 or C-476
Field Tests	24-22	
Grout:		
Raw Materials:		
Similar to those for concrete masonry except add portland blast furnace slag cement		C-595
Aggregates for masonry grout	24-23	C-404
Field Tests:	24-22	
Low Lift Placement	*UBC Section 2413 (c)	
High-Lift Placement	OSA Specs. and UBC Section 2413 (d)	
Reinforcement:		
Rebar for concrete	26-4	A-615
Cold drawn steel wire for concrete reinforcement	24-15	A-82

*Uniform Building Code and Office of State Architect Specifications.

Bibliography

ACI Committee 531. "Concrete Masonry Structures—Design and Construction." *American Concrete Institute Journal, Proceedings*, Volume 67, 1979.

American Society for Testing and Materials. *Annual Book of ASTM Standards*, Philadelphia, Pa., 1975.

Amrhein, James E. *Reinforced Masonry Engineering Handbook*, Masonry Institute of America, Los Angeles, Calif., 1978 (3rd Edition).

Army, Navy, and Air Force. *Seismic Design for Buildings*, Tri-Services Manual TM 5–809–10, April 1973.

Benjamin, J. R., and H. A. Williams. "Behavior of One-Story Brick Shearwalls," *Proceedings of ASCE, Journal of Structural Division*, Volume 84, No. ST4, 1954.

Blume, John A., and Associates. *Shear on Grouted Masonry Wall Elements*, Western States Clay Products Association, San Francisco, Calif., 1968.

Blume, John A., Nathan M. Newmark, and Leo H. Corning. *Design of Multi-Story Reinforced Concrete Buildings for Earthquake Motions*, Portland Cement Association, Chicago, Ill., 1961.

Carney, J. M. *Plywood Diaphragm Construction*, American Plywood Association, Tacoma, Wash., 1966.

Copeland, R. E. "Procedures for Controlling Cracking in Concrete Masonry," *Concrete Products*, September 1964.

COULL, ALEXANDER, AND J. R. CHOUDBURY. "Stresses and Deflections in Coupled Shearwalls," *American Concrete Institute Journal, Proceedings*, Volume 64, 1967, pp. 65–72.

DERCHO, ARNALDO T., DONALD M. SCHULTZ, AND MARK FINTEL. *Analysis and Design of Small Reinforced Concrete Buildings for Earthquake Forces*, Portland Cement Association, Chicago, Ill., 1974.

DICKEY, WALTER L. "Concrete Masonry Construction," *Handbook of Concrete Engineering*, edited by Mark Fintel, Van Nostrand Reinhold Company, New York, 1974.

DICKEY, WALTER L. *Inspectors Masonry Handbook*, Masonry Institute of America, Los Angeles, Calif., 1973.

DICKEY, WALTER L. Masonry Chapter, *Structural Engineering Handbook*, edited by E. H. Gaylord and E. N. Gaylord, McGraw-Hill Book Company, New York.

DICKEY, WALTER L. *Masonry Veneer*, Masonry Institute of America, Los Angeles, Calif., 1974.

DICKEY, WALTER L. *Multi-Story Load Bearing Brick Walls*, Masonry Institute of America, Brick Institute of California, Los Angeles, Calif., 1968.

DICKEY, WALTER L. *1974 Masonry Codes and Specifications*, Masonry Industry Advancement Committee, Los Angeles, Calif.

DICKEY, WALTER L. "Reinforced Masonry Revisited," *Proceedings of International Conference On Masonry Structures*, 1967.

DICKEY, WALTER L., AND JAMES E. AMRHEIN. *Masonry Design Manual*, 2nd ed., Masonry Industry Advancement Committee, Los Angeles, Calif., 1972.

EARTHQUAKE ENGINEERING RESEARCH CENTER, *A Literature Survey—Transverse Strength of Masonry Walls*, Report No. UCB/EERC-77/07, March 1977.

GROVES, JAMES G., AND HARRY C. PLUMMER. *Principles of Clay Masonry Construction*, Structural Clay Products Institute, McLean, Va., 1970.

INTERNATIONAL CONFERENCE OF BUILDING OFFICIALS. *Uniform Building Code*, Whittier, Calif., 1976, 1979.

INTERNATIONAL CONFERENCE OF BUILDING OFFICIALS. *Uniform Building Code Standards*, Whittier, Calif., 1976.

KARIOTIS, JOHN, AND JAMES KESLER. *Reinforced Load Bearing Block Walls for Multi-Story Construction*, Concrete Masonry Association of California, Los Angeles, Calif., 1967.

KELLY, PITTELKO, FRITZ, AND FORSSEN. *Allowable Bolt Values-Test Report*, Western States Clay Products Association, San Francisco, Calif., 1975.

KESLER, JAMES J. *A Look at Load Bearing Masonry Design*, Masonry Institute of America, Los Angeles, Calif., 1971.

LOS ANGELES DEPARTMENT OF BUILDING AND SAFETY. *Official Building Code, City of Los Angeles*, Building News, Inc., Volume C, 1972.

MACKINTOSH, ALBYN. *Design Manual*, National Concrete Masonry Association, McLean, Va., 1968.

MACLEOD, IAN A. *Shear Wall-Frame Interaction*, Portland Cement Association, Skokie, Ill., 1970.

MAYES, RONALD L., AND RAY W. CLOUGH. *A Literature Survey—Compressive, Tensile, Bond and Shear Strength of Masonry*, University of California, Berkeley, Calif., 1975.

MAYES, RONALD L., YUTARO OMOTE, AND RAY W. CLOUGH. *Cyclic Shear Tests of Masonry Pier Test Results*, University of California, Berkeley, 1976.

MELI, R. "Behavior of Masonry Walls Under Lateral Loads," *Proceedings of Fifth World Conference on Earthquake Engineering*, Rome, 1972.

PLUMMER, HARRY C. *Brick and Tile Engineering*, Structural Clay Products Institute, Washington, D.C., 1950.

PLUMMER, HARRY C., AND JOHN A. BLUME. *Reinforced Brick Masonry and Lateral Force Design*, Structural Clay Products Institute, Washington, D.C., 1953.

PRIESTLY, M. J., AND D. O. BRIDGMAN. "Seismic Resistance of Brick Masonry Walls," *Bulletin of the New Zealand National Society for Earthquake Engineering*, Volume 7, No. 4, December 1974.

RANDALL, FRANK A., AND WILLIAM C. PANARESE. *Concrete Masonry Handbook*, Portland Cement Association, Skokie, Ill., 1976.

SAHLIN, SVEN. *Structural Masonry*, Prentice-Hall, Inc., Englewood Cliffs, N.J., 1971.

SCHNEIDER, ROBERT R. *Behavior of Reinforced Plastered Brick Piers under Lateral Loads*, Los Angeles Board of Education, and Structural Engineers Association of Southern California, Los Angeles, Calif., 1957.

SCHNEIDER, ROBERT R. *Investigation of Reinforced Brick Masonry Under Lateral Loads*, State of California, Division of Architecture, Sacramento, Calif., 1956.

SCHNEIDER, ROBERT R. *Lateral Load Tests on Reinforced Grouted Masonry Shear Walls*, State of California, Division of Architecture, Sacramento, Calif., 1959.

SCHNEIDER, ROBERT R. *Shear on Concrete Masonry Piers*, California State Polytechnic University, Pomona, 1969.

SCHNEIDER, ROBERT R. *Shear Tests on Surewall Bonded Walls*, California State Polytechnic University, Pomona, 1973.

SCRIVNER, J. C. "Concrete Masonry Wall Panels—Static Racking Test with Predominent Flexural Effect," *New Zealand Country Construction*, July 1966.

SCRIVNER, J. C. "Static Racking Tests on Masonry Walls," *Designing, Engineering and Construction of Masonry Products*, Gulf Publishing Co., Houston, Tex., 1969.

SINHA, B. P., AND A. W. HENDRY. "Further Tests on Model Brick Walls and Piers," *Proceedings of British Ceramic Society*, No. 17, February 1970.

STANG, A. H., D. E. PARSONS, AND J. W. MCBURNEY. "Compressive Strength of Clay Brick Walls," *Journal of Research*, Bureau of Standards, Department of Commerce, Volume No. 4, Research Paper 108, October 1929.

STATE OF CALIFORNIA ADMINISTRATIVE CODE. *Clay Brick Masonry—High Lift Grouting Method*, Office of State Architect, Sacramento, Calif., 1975.

STATE OF CALIFORNIA ADMINISTRATIVE CODE. *Filled Cell Concrete Masonry—High Lift Grouting Method*, Office of State Architect, Sacramento, Calif., 1976.

STRUCTURAL CLAY PRODUCTS INSTITUTE. *Recommended Practice for Engineered Brick Masonry*, McLean, Va., 1969.

STRUCTURAL ENGINEERS ASSOCIATION OF CALIFORNIA, SEISMOLOGY COMMITTEE. *Recommended Lateral Force Requirements and Commentary*, Structural Engineers Association of California, Los Angeles, Calif., 1969, 1970, 1971, 1973, 1975.

TEAL, EDWARD J. "Seismic Design Practice for Steel Buildings," *American Institute of Steel Construction, Engineering Journal*, Fourth Quarter, 1975.

WESTERN STATES CLAY PRODUCTS ASSOCIATION. *Tall, Thin Brick Walls*. San Francisco, Calif., 1976.

WILLIAMS, D. *Seismic Behavior of Reinforced Masonry Shearwalls*, University of Canterbury, Christchurch, New Zealand, 1971.

WINTER, GEORGE, AND ARTHUR NILSON. *Design of Reinforced Concrete Structures*, McGraw-Hill Book Company, New York, 1964.

Index

A

D

E

N

O

P

Q

R

S

V

W

Y